A2-Level Mathematics

A2 Maths is seriously tricky — no question about that.
To do well, you're going to need to revise properly and practise hard.

This book has thorough notes on everything in modules C3, C4, S2 and M2.
It'll help you learn the stuff you need and take you step-by-step through loads of examples.

It's got practice questions... lots of them. For every topic
there are warm-up and exam-style questions.

And of course, we've done our best to make the whole thing vaguely entertaining for you.

Complete Revision and Practice

Covers modules C3, C4, S2 and M2.

Contents

Core Mathematics C3 & C4

Statistics S2

Contents

Contributors:
Andy Ballard, Mary Falkner, Paul Jordin, Dave Harding, Sharon Keeley-Holden,
Claire Jackson, Simon Little, Ali Palin, Andy Park, David Ryan, Lyn Setchell,
Caley Simpson, Jane Towle, Chris Worth, Jonathan Wray, Dawn Wright

Proofreaders:
Paul Garrett, Dan Heller, Glenn Rogers and James Yates

Published by CGP.

ISBN: 978 1 84762 588 5

Groovy Website: www.cgpbooks.co.uk

Printed by Elanders Ltd, Newcastle upon Tyne.

Based on the classic CGP style created by Richard Parsons.

Functions and Mappings

This page is for AQA C3, Edexcel C3, OCR C3 and OCR MEI C3

A2 Maths is for all those people who liked AS Maths so much they came back for more. Good for you, I like you already.

Values in the Domain are Mapped to values in the Range

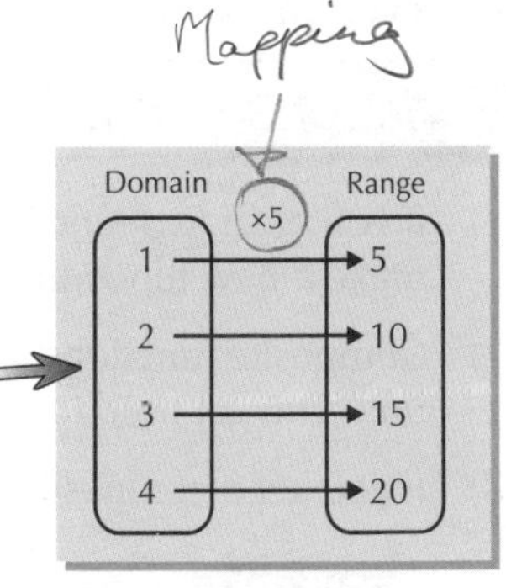

1) A mapping is an operation that takes one number and transforms it into another. E.g. 'multiply by 5', 'square root' and 'divide by 7' are all mappings.
2) The set of numbers you can start with is called the domain, and the set of numbers they can become is called the range. Mappings can be drawn as diagrams like this: You can also draw mappings as graphs (see below).
3) The domain and / or range will often be the set of real numbers, $\mathbb{R}$ (a real number is any positive or negative number (or 0) — fractions, decimals, integers, surds). If x can take any real value, it's usually written as $x \in \mathbb{R}$.
4) You might have to work out the range of a mapping from the domain you're given. For example, $y = x^2$, $x \in \mathbb{R}$ has the range $f(x) \geq 0$, as all square real numbers are positive (or zero).

Other sets of numbers include $\mathbb{Z}$, the set of integers, $\mathbb{N}$, the set of natural numbers (positive integers, not including 0) and $\mathbb{C}$, the complex numbers (made up of 'imaginary' numbers — you don't meet these in C3 or C4).

A Function is a type of Mapping

Functions (e.g. x^2) are usually written as $f(x) = x^2$ or $f : x \to x^2$.

1) Some mappings take every number in the domain to only one number in the range. These mappings are called functions. If a mapping takes a number from the domain to more than one number in the range (or if it isn't mapped to any number in the range), it's not a function.

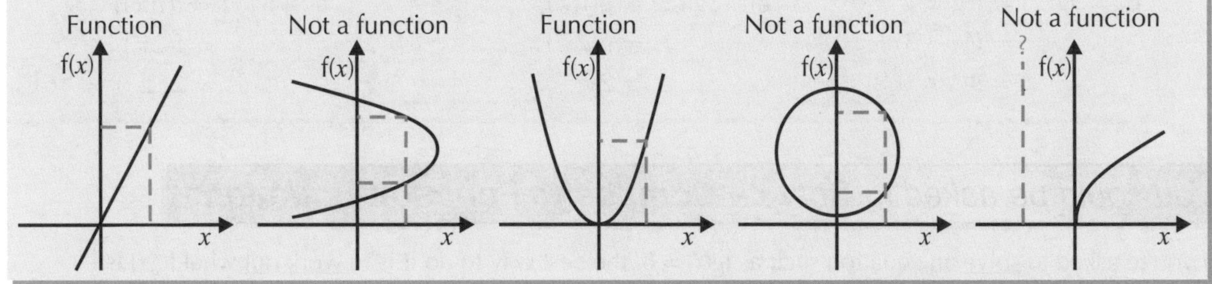

2) For the graphs above, the first and third are functions because each value of x is mapped to a single value of $f(x)$. The second and fourth aren't functions because the values of x are mapped to two different values of $f(x)$. The fifth also isn't a function, this time because $f(x)$ is not defined for $x < 0$.
3) Some mappings that aren't functions can be turned into functions by restricting their domain. For example, the mapping $y = \frac{1}{x-1}$ for $x \in \mathbb{R}$ is not a function, because it's not defined at $x = 1$ (draw the graph if you're not convinced). But if you change the domain to $x > 1$, the mapping is now a function.
4) Some functions can be described as even or odd. Even functions are ones where $f(x) = f(-x)$ — the function is symmetrical about the y-axis, e.g. $\cos x$. For odd functions, $f(x) = -f(-x)$ — the graph for $x > 0$ is reflected in the y-axis then reflected again in the x-axis to give the graph for $x < 0$, e.g. $\sin x$. Both $\sin x$ and $\cos x$ are also periodic — the pattern repeats at regular intervals (for both of these, the interval is 2π or 360°).

Functions can be One-to-One or Many-to-One

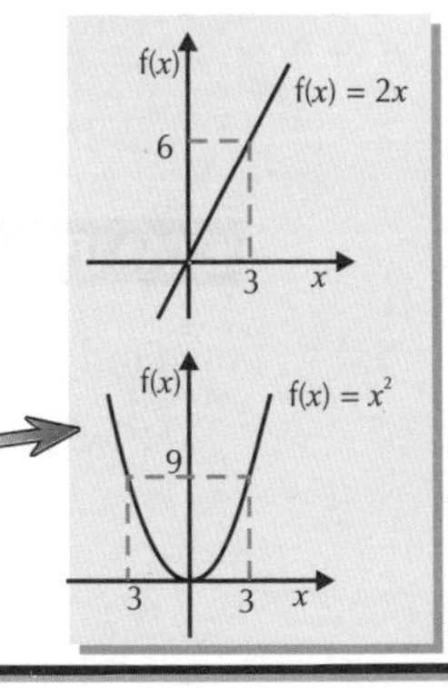

1) A one-to-one function maps one element in the domain to one element in the range — e.g. $f : x \to 2x$, $x \in \mathbb{R}$ is one-to-one, as every x is mapped to a unique value in the range (the range is also $\mathbb{R}$). So only 3 in the domain is mapped to 6 in the range.
2) A many-to-one function maps more than one element in the domain to one element in the range (remember that no element in the domain can map to more than one element in the range, otherwise it wouldn't be a function). $f(x) = x^2$, $x \in \mathbb{R}$ is a many-to-one function, as two elements in the domain map to the same element in the range — e.g. both 3 and −3 map to 9.

Welcome to my domain...

When you're drawing a function or a mapping, you should draw a mapping diagram if you're given a discrete set of numbers (e.g. $x \in \{0,1,2,3\}$), but you should draw a graph if the domain is continuous (e.g. $x \in \mathbb{R}$).

Composite Functions

This page is for AQA C3, Edexcel C3, OCR C3 and OCR MEI C3

You're not done with functions yet. Oh no. You need to know what happens if you put two (or more) functions together. There's only one way to find out...

Functions can be Combined to make a Composite Function

1) If you have two functions f and g, you can combine them (do one followed by the other) to make a new function. This is called a composite function.
2) Composite functions are written $fg(x)$ — this means do g first, then f. If it helps, put brackets in until you get used to it, so $fg(x) = f(g(x))$. The order is really important — usually $fg(x) \neq gf(x)$.
3) If you get a composite function that's written $f^2(x)$, it means $ff(x)$ — you do f twice.

Composite functions made up of three or more functions work in exactly the same way.

EXAMPLE For the functions $f : x \to 2x^3 \{x \in \mathbb{R}\}$ and $g: x \to x - 3 \{x \in \mathbb{R}\}$, find:

a) $fg(4)$ b) $fg(0)$ c) $gf(0)$ d) $fg(x)$ e) $gf(x)$ f) $f^2(x)$.

a) $fg(4) = f(g(4)) = f(4 - 3) = f(1) = 2 \times 1^3 = 2$

b) $fg(0) = f(g(0)) = f(0 - 3) = f(-3) = 2 \times (-3)^3 = 2 \times -27 = -54$

c) $gf(0) = g(f(0)) = g(2 \times 0^3) = g(0) = 0 - 3 = -3$

From parts b) and c) you can see that $fg(x) \neq gf(x)$.

d) $fg(x) = f(g(x)) = f(x - 3) = 2(x - 3)^3$

e) $gf(x) = g(f(x)) = g(2x^3) = 2x^3 - 3$

f) $f^2(x) = f(f(x)) = f(2x^3) = 2(2x^3)^3 = 16x^9$

You could be asked to Solve a Composite Function Equation

If you're asked to solve an equation such as $fg(x) = 8$, the best way to do it is to work out what $fg(x)$ is, then rearrange $fg(x) = 8$ to make x the subject.

EXAMPLE For the functions $f : x \to \sqrt{x}$ with domain $\{x \geq 0\}$ and $g : x \to \frac{1}{x - 1}$ with domain $\{x > 1\}$, solve the equation $fg(x) = ½$. Also, state the range of $fg(x)$.

First, find $fg(x)$: $fg(x) = f\left(\frac{1}{x-1}\right) = \sqrt{\frac{1}{x-1}} = \frac{1}{\sqrt{x-1}}$

So $\frac{1}{\sqrt{x-1}} = \frac{1}{2}$

Rearrange this equation to find x:

$$\frac{1}{\sqrt{x-1}} = \frac{1}{2} \Rightarrow \sqrt{x-1} = 2 \Rightarrow x - 1 = 4 \Rightarrow x = 5$$

To find the range, it's often helpful to draw the graph of $fg(x)$:

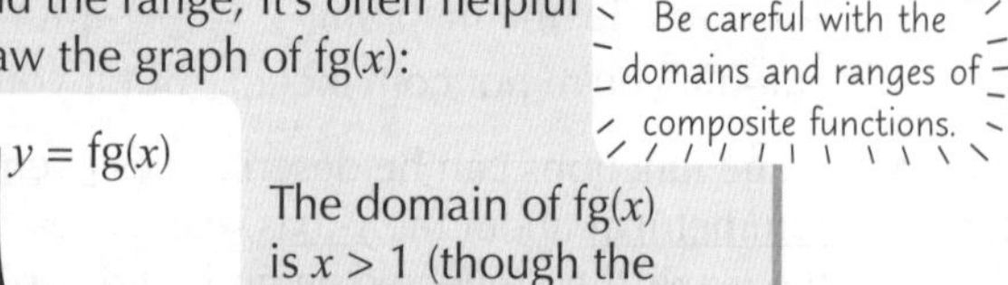

Be careful with the domains and ranges of composite functions.

The domain of $fg(x)$ is $x > 1$ (though the question doesn't ask for this) and the range is $fg(x) > 0$.

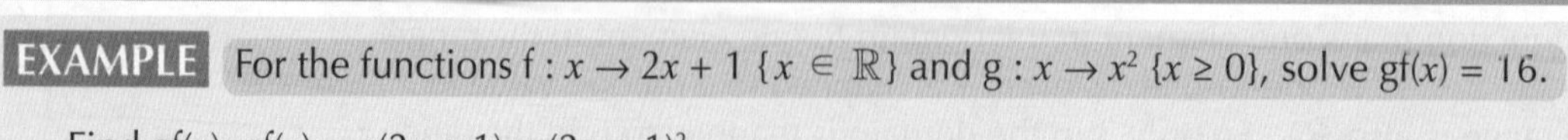

EXAMPLE For the functions $f : x \to 2x + 1 \{x \in \mathbb{R}\}$ and $g : x \to x^2 \{x \geq 0\}$, solve $gf(x) = 16$.

Find $gf(x)$: $gf(x) = g(2x + 1) = (2x + 1)^2$.

Now solve $gf(x) = 16$: $(2x + 1)^2 = 16 \Rightarrow 4x^2 + 4x + 1 = 16$
$\Rightarrow 4x^2 + 4x - 15 = 0$
$\Rightarrow (2x - 3)(2x + 5) = 0$ so $x = \frac{3}{2}$ or $x = -\frac{5}{2}$

Compose a concerto for f(x) and orchestra...

The most important thing to remember on this page is the order you do the functions in — for $fg(x)$ you always do g first as g is next to x. It's like getting dressed — you wouldn't put your shoes on before your socks, as your socks go next to your feet.

Inverse Functions

This page is for AQA C3, Edexcel C3, OCR C3 and OCR MEI C3

Just when you'd got your head around functions, ranges, domains and composite functions, they go and turn it all back to front by introducing inverses.

Only One-to-One Functions have Inverses

1) An inverse function does the opposite to the function. So if the function was '+ 1', the inverse would be '– 1', if the function was '× 2', the inverse would be '÷ 2' etc. The inverse for a function f(x) is written $f^{-1}(x)$.
2) An inverse function maps an element in the range to an element in the domain — the opposite of a function. This means that only one-to-one functions have inverses, as the inverse of a many-to-one function would be one-to-many, which isn't a function (see p.1).
3) For any inverse $f^{-1}(x)$,

Doing the function and then the inverse...

$$f^{-1}f(x) = x = ff^{-1}(x)$$

...is the same as doing the inverse then doing the function — both just give you x.

4) The domain of the inverse is the range of the function, and the range of the inverse is the domain of the function.

EXAMPLE The function $f(x) = x + 7$ with domain $x \geq 0$ and range $f(x) \geq 7$ is one-to-one, so it has an inverse.

The inverse of + 7 is – 7, so $f^{-1}(x) = x - 7$. $f^{-1}(x)$ has domain $x \geq 7$ and range $f^{-1}(x) \geq 0$.

Work out the Inverse using Algebra

For simple functions (like the one in the example above), it's easy to work out what the inverse is just by looking at it. But for more complex functions, you need to rearrange the original function to change the subject.

Finding the Inverse

1) **Replace f(x) with y to get an equation for y in terms of x.**
2) **Rearrange the equation to make x the subject.**
3) **Replace x with $f^{-1}(x)$ and y with x — this is the inverse function.**
4) **Swap round the domain and range of the function.**

EXAMPLE Find the inverse of the function $f(x) = 3x^2 + 2$ with domain $x \geq 0$, and state its domain and range.

1) First, replace f(x) with y: $y = 3x^2 + 2$.

It's easier to work with y than f(x).

2) Rearrange the equation to make x the subject:

$$y - 2 = 3x^2 \Rightarrow \frac{y-2}{3} = x^2 \Rightarrow \sqrt{\frac{y-2}{3}} = x$$

$x \geq 0$ so you don't need the negative square root.

3) Replace x with $f^{-1}(x)$ and y with x:

$$f^{-1}(x) = \sqrt{\frac{x-2}{3}}$$

4) Swap the domain and range: the range of f(x) is $f(x) \geq 2$, so $f^{-1}(x)$ has domain $x \geq 2$ and range $f^{-1}(x) \geq 0$.

You might have to Draw the Graph of the Inverse

The inverse of a function is its reflection in the line $y = x$.

EXAMPLE Sketch the graph of the inverse of the function $f(x) = x^2 - 8$ with domain $x \geq 0$.

It's easy to see what the domains and ranges are from the graph — f(x) has domain $x \geq 0$ and range $f(x) \geq -8$, and $f^{-1}(x)$ has domain $x \geq -8$ and range $f^{-1}(x) \geq 0$.

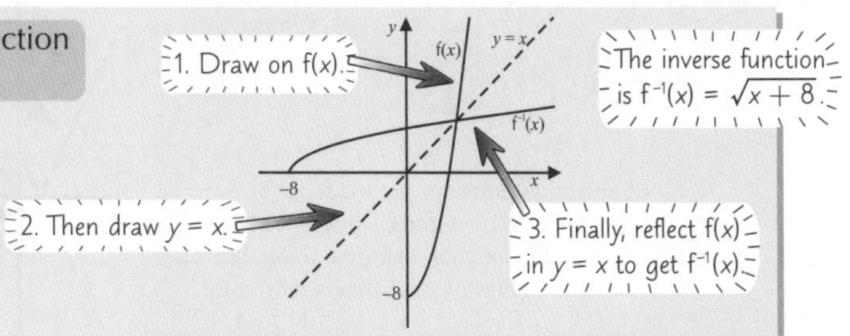

Line y = x on the wall — who is the fairest of them all...

I think I've got the hang of this inverse stuff now — so the inverse of walking to the shops and buying some milk would be taking the money out the till, putting the milk back on the shelf, leaving the shop and walking home backwards. Sorted.

Modulus

This page is for AQA C3, Edexcel C3, OCR C3 and OCR MEI C3

The modulus of a number is really useful if you don't care whether something's positive or negative — like if you were finding the difference between two numbers (e.g. 7 and 12). It doesn't matter which way round you do the subtraction (i.e. 12 – 7 or 7 – 12) — the difference between them is still 5.

Modulus is the *Size* of a number

1) The modulus of a number is its size — it doesn't matter if it's positive or negative. So for a positive number, the modulus is just the same as the number itself, but for a negative number, the modulus is its positive value. For example, the modulus of 8 is 8, and the modulus of –8 is also 8.
2) The modulus of a number, x, is written $|x|$. So the example above would be written $|8| = |-8| = 8$.
3) In general terms, for $x \geq 0$, $|x| = x$ and for $x < 0$, $|x| = -x$.
4) Functions can have a modulus too — the modulus of a function f(x) is its positive value. Suppose f(x) = –6, then $|f(x)| = 6$. In general terms,
5) If the modulus is inside the brackets in the form $f(|x|)$, then you make the x-value positive before applying the function. So $f(|-2|) = f(2)$.

The modulus is sometimes called the absolute value.

$|f(x)| = f(x)$ when $f(x) \geq 0$ and $|f(x)| = -f(x)$ when $f(x) < 0$.

The *Graphs* of *|f(x)|* and *f(|x|)* are *Different*

You'll probably have to draw the graph of a modulus function — and there are two different types.

1) For the graph of $y = |f(x)|$, any negative values of f(x) are made positive by reflecting them in the x-axis. This restricts the range of the modulus function to $|f(x)| \geq 0$ (or some subset within $|f(x)| \geq 0$, e.g. $|f(x)| \geq 1$).
2) For the graph of $y = f(|x|)$, the negative x-values produce the same result as the corresponding positive x-values. So the graph of f(x) for $x \geq 0$ is reflected in the y-axis for the negative x-values.
3) The easiest way to draw these graphs is to draw f(x) (ignoring the modulus for now), then reflect it in the appropriate axis. This will probably make more sense when you've had a look at a couple of examples:

EXAMPLE Draw the graphs of $y = |f(x)|$ and $y = f(|x|)$ for the functions $f(x) = 4x - 5$ and $f(x) = x^2 - 4x$.

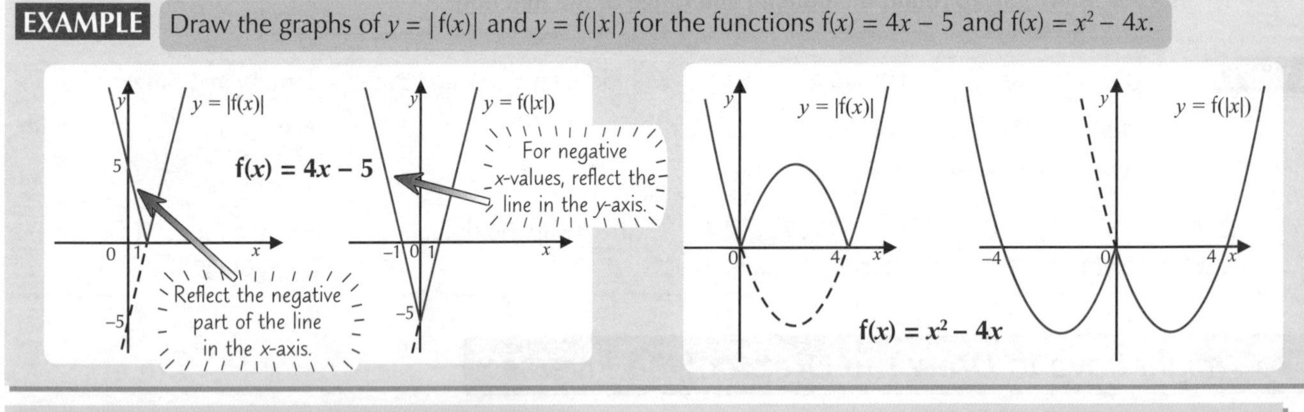

EXAMPLE Draw the graph of the function $f(x) = \begin{cases} |2x+1| & x < 0 \\ \sqrt{x} & x \geq 0 \end{cases}$.

Sometimes functions are made up of two or more parts — for x between certain values, the function does one thing, but for other values of x it behaves differently.

Draw on each part of the graph separately.

$y = |2x + 1|$

$y = \sqrt{x}$

At $x = 0$, $y = |2(0) + 1| = 1$ for the first part of the function and $y = \sqrt{0} = 0$ for the second part of the function — so there'll be a gap in the graph.

Modulus built the city of Mode...

You might have to draw modulus graphs for functions like f(x) = ax + b from scratch. You could be asked for trig graphs and exponentials too. For harder graphs, you'll often be given a graph which you can use as a starting point for the modulus.

Modulus

This page is for AQA C3, Edexcel C3, OCR C3 and OCR MEI C3

An exam question might ask you to solve an equation like '$|f(x)| = n$' (for a constant n) or '$|f(x)| = g(x)$' for a function g. I admit, it would be more exciting to solve a crime, but I'm afraid modulus functions must come first...

Solving modulus functions usually produces **More Than One** solution

Here comes the method for solving '$|f(x)| = n$'. Solving '$|f(x)| = g(x)$' is exactly the same — just replace n with $g(x)$.

Solving Modulus Equations of the form $|f(x)| = n$

1) **First, sketch the functions $y = |f(x)|$ and $y = n$, on the same axes.** — The solutions you're trying to find are where they intersect.
2) **From the graph, work out the ranges of x for which $f(x) \geq 0$ and $f(x) < 0$:**
   E.g. $f(x) \geq 0$ for $x \leq a$ or $x \geq b$ and $f(x) < 0$ for $a < x < b$ — These ranges should 'fit together' to cover all possible x values.
3) **Use this to write two new equations, one true for each range of x...**
   (1) $f(x) = n$ for $x \leq a$ or $x \geq b$ — The original equation '$|f(x)| = n$' becomes '$f(x) = n$' in the range where $f(x) \geq 0$...
   (2) $-f(x) = n$ for $a < x < b$ — ...and it becomes '$-f(x) = n$' in the range where $f(x) < 0$.
4) **Now just solve each equation and check that any solutions are valid — get rid of any solutions outside the range of x you've got for that equation.**
5) **Look at the graph and check that your solutions look right.**

Sketch the Graph to see **How Many Solutions** there are

EXAMPLE Solve $|x^2 - 9| = 7$.

1) First off, sketch the graphs of $y = |x^2 - 9|$ and $y = 7$.
   They cross at 4 different points, so there should be 4 solutions.
2) Now find out where $f(x) \geq 0$ and $f(x) < 0$:
   $x^2 - 9 \geq 0$ for $x \leq -3$ or $x \geq 3$, and $x^2 - 9 < 0$ for $-3 < x < 3$

   $x^2 - 9 = (x + 3)(x - 3)$, so curve crosses x-axis at 3 and -3.
3) Form two equations for the different ranges of x:
   (1) $x^2 - 9 = 7$ for $x \leq -3$ or $x \geq 3$
   (2) $-(x^2 - 9) = 7$ for $-3 < x < 3$
4) Solving (1) gives: $x^2 = 16 \Rightarrow x = 4, x = -4$
   Check they're valid: $x = -4$ is in '$x \leq -3$' and $x = 4$ is in '$x \geq 3$' — so they're both valid.

   Solving (2) gives: $x^2 - 2 = 0 \Rightarrow x^2 = 2$ so $x = \sqrt{2}, x = -\sqrt{2}$.
   Check they're valid: $x = \sqrt{2}$ and $x = -\sqrt{2}$ are both within $-3 < x < 3$ — so they're also both valid.
5) Check back against the graphs — we've found four solutions and they're in the right places. Nice.

EXAMPLE Solve $|x^2 - 2x - 3| = 1 - x$.

1) Sketch $y = |x^2 - 2x - 3|$ and $y = 1 - x$. The graphs cross twice.
2) Looking at where $f(x) \geq 0$ and where $f(x) < 0$ gives...

   $x^2 - 2x - 3 = (x + 1)(x - 3)$, so it crosses axis at -1 and 3.
3) (1) $x^2 - 2x - 3 = 1 - x$ for $x \leq -1$ or $x \geq 3$
   (2) $-(x^2 - 2x - 3) = 1 - x$ for $-1 < x < 3$.
4) Solving (1) using the quadratic formula gives $x = 2.562$, $x = -1.562$.
   $x \leq -1$ or $x \geq 3$, so this solution is not valid... ...but this one is.
   Solving (2) using the quadratic formula gives $x = 3.562$, $x = -0.562$.
   $-1 < x < 3$, so this solution is not valid... ...but this one is.

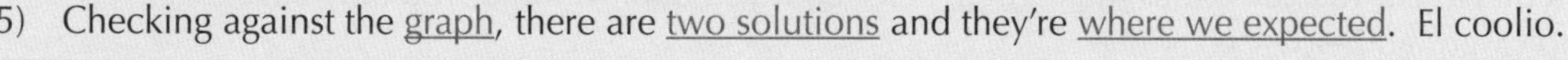

5) Checking against the graph, there are two solutions and they're where we expected. El coolio.

How very interesting...

So if the effect of the modulus is to make a negative positive, I guess that means that |exam followed by detention followed by getting splashed by a car on the way home| = sleep-in followed by picnic followed by date with Hugh Jackman. I wish.

Modulus

This page is for AQA C3, Edexcel C3, OCR C3 and OCR MEI C3

Three whole pages on modulus might seem a bit excessive, but it's a tricky little topic that can easily trip you up if you're not careful. It's better to be safe than sorry, as my Auntie Marjorie would say — and believe me, she would know.

You might come across a modulus in an **Equation** or an **Inequality**

You saw how to solve modulus equations on the previous page, but there are a few more useful relations you can use.

1) If you have $|a| = |b|$, this means that $a^2 = b^2$ (as $-a$ and a are the same when squared). This comes in really handy when you have to solve equations of the form $|f(x)| = |g(x)|$ (see below).
2) Inequalities that have a modulus in them can be really nasty — unfortunately you can't just leave the modulus in there. $|x| < 5$ means that $-5 < x < 5$.

   This is because $|x| < 5$ means that $x < 5$ and $-x < 5$, and $-x < 5$ is the same as $x > -5$. You can then put the two inequalities together to get $-5 < x < 5$.
3) Using this, you can rearrange more complicated inequalities like $|x - a| \leq b$. From the method above, this means that $-b \leq x - a \leq b$, so adding a to each bit of the inequality gives $a - b \leq x \leq a + b$.

EXAMPLE Solve $|x - 4| < 7$.

As $|x - 4| < 7$, this means that $-7 < x - 4 < 7$. Adding 4 to each bit gives $-3 < x < 11$.

Solve a modulus equation by **Squaring Both Sides**

If you have an equation of the form $|f(x)| = |g(x)|$, you can solve it using the method on the previous page, but it can get a bit messy with all the modulus signs flying all over the place. Instead, you can use the fact that if $|a| = |b|$ then $a^2 = b^2$ and square both sides of the equation. You'll end up with a quadratic to solve, but that should be a doddle.

EXAMPLE By squaring both sides, solve $|x + 3| = |2x + 5|$.

First, square both sides: $x^2 + 6x + 9 = 4x^2 + 20x + 25$

Then rearrange and solve: $0 = 3x^2 + 14x + 16$

$= (3x + 8)(x + 2)$

So $x = -\frac{8}{3}$ or $x = -2$.

You could also have solved it by sketching the graphs like you did on the previous page:

EXAMPLE
a) By sketching the graphs, solve $|x + 3| = |2x + 5|$.
b) Hence solve the inequality $|x + 3| < |2x + 5|$.

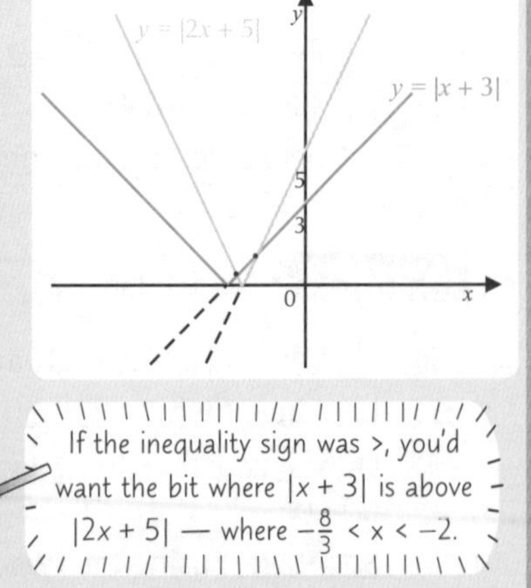

a) Sketch $y = |x + 3|$ and $y = |2x + 5|$. The graphs cross twice.

From the graph, you can see that there's one solution where $x + 3 = 2x + 5$, and another where $x + 3 = -(2x + 5) = -2x - 5$.

Solving $x + 3 = 2x + 5$ gives $x = -2$ and solving $x + 3 = -2x - 5$ gives $x = -\frac{8}{3}$ (these are the solutions you found above — and you can see from the graph that they're in the right places).

b) To solve the inequality $|x + 3| < |2x + 5|$, you have to look at the graphs and work out where the graph of $|x + 3|$ is underneath the graph of $|2x + 5|$: you can see from the sketch that this is true when $x < -\frac{8}{3}$ and when $x > -2$.

If the inequality sign was >, you'd want the bit where $|x + 3|$ is above $|2x + 5|$ — where $-\frac{8}{3} < x < -2$.

If you'd tried to solve $-(x + 3) = -(2x + 5)$ and $-(x + 3) = 2x + 5$, you would have just got the same pair of solutions. This is true for any equation of the form $|f(x)| = |g(x)|$ — it might look more complicated than $|f(x)| = n$ or $|f(x)| = g(x)$, but it's actually a bit easier. You can either square both sides or, if you don't want to, you only have to solve two equations: $f(x) = g(x)$ and $-f(x) = g(x)$.

Be there or b²...

Personally, I think squaring both sides is a bit easier (but then again, I'm a sucker for a quadratic equation). You'll get the same answer whichever way you do it, so it's up to you. By the way, please don't mention penguins to Auntie Marjorie.

Transformations of Graphs

This page is for AQA C3, Edexcel C3, OCR C3 and OCR MEI C3

Back in C1, you came across transformations of graphs — vertical and horizontal translations, stretches and reflections. In C2, you saw the same transformations on trig graphs. As if that wasn't enough for you, you now need to be able to do combinations of transformations — more than one applied to the same graph.

There are Four main Transformations

The transformations you met in C1 and C2 are translations (adding things — a vertical or horizontal shift), stretches or squeezes (either vertical or horizontal) and reflections in the x- or y- axis. Here's a quick reminder of what each one does:

$y = f(x + c)$

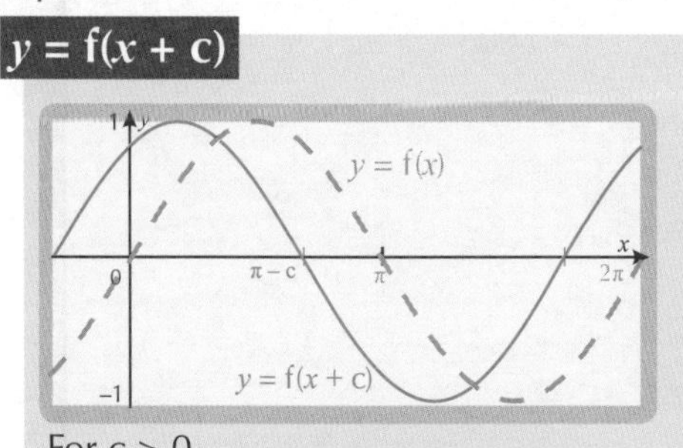

For $c > 0$,
$f(x + c)$ is $f(x)$ shifted c to the left,
and $f(x - c)$ is $f(x)$ shifted c to the right.

$y = f(x) + c$

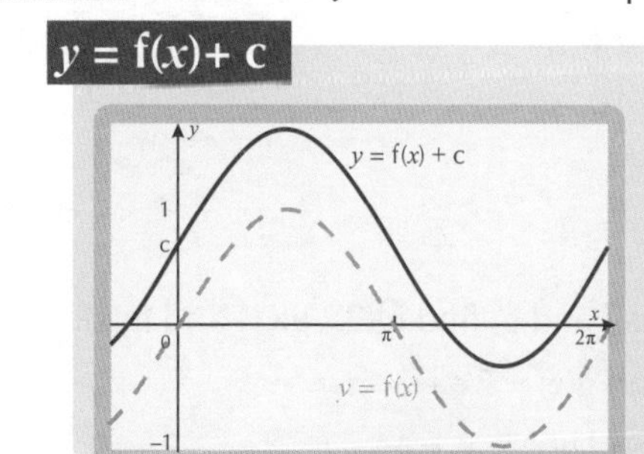

For $c > 0$,
$f(x) + c$ is $f(x)$ shifted c upwards,
and $f(x) - c$ is $f(x)$ shifted c downwards.

All these graphs use $f(x) = \sin x$.

$y = af(x)$

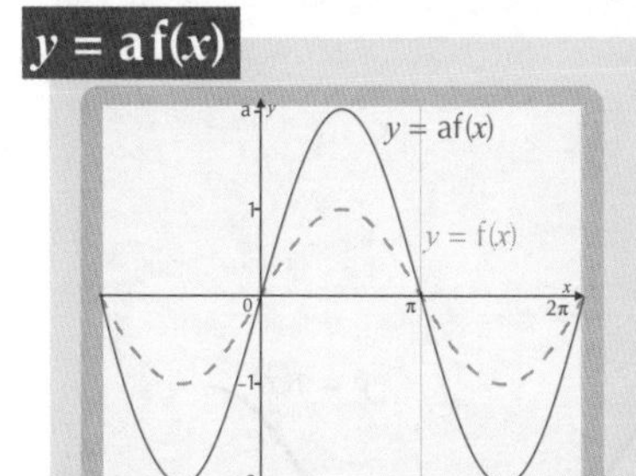

If $a > 1$, the graph of $af(x)$ is $f(x)$ stretched vertically by a factor of a.

If $0 < a < 1$, the graph is squashed.

And if $a < 0$, the graph is also reflected in the x-axis.

Remember that a squash by a factor of a is really a stretch by a factor of $1/a$.

$y = f(ax)$

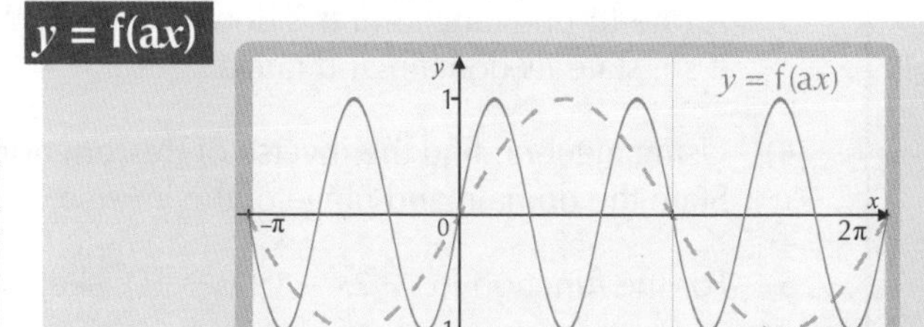

If $a > 1$, the graph of $f(ax)$ is $f(x)$ squashed horizontally by a factor of a.

If $0 < a < 1$, the graph is stretched horizontally.

And if $a < 0$, the graph is also reflected in the y-axis.

Do Combinations of Transformations One at a Time

Combinations of transformations can look a bit tricky, but if you take them one step at a time they're not too bad. Don't try and do all the transformations at once — break it up into separate bits (as above) and draw a graph for each stage.

EXAMPLE The graph shows the function $y = f(x)$. Draw the graph of $y = 3f(x + 2)$, showing the coordinates of the turning points.

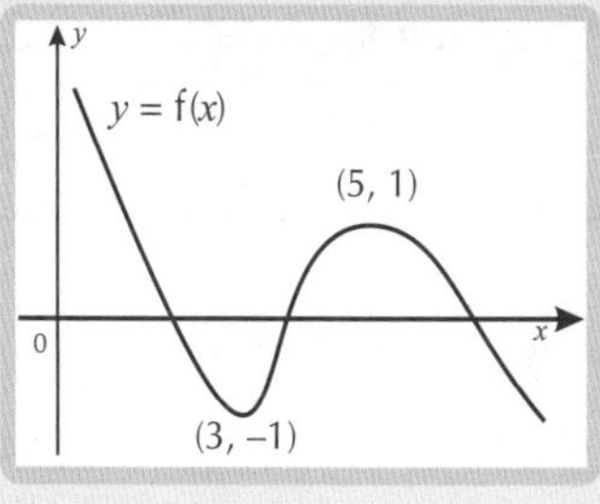

Make sure you do the transformations the right way round — you should do the bit in the brackets first.

Don't try to do everything at once. First draw the graph of $y = f(x + 2)$ and work out the coordinates of the turning points.

The graph is shifted left by 2 units, so subtract 2 from the x-coordinates.

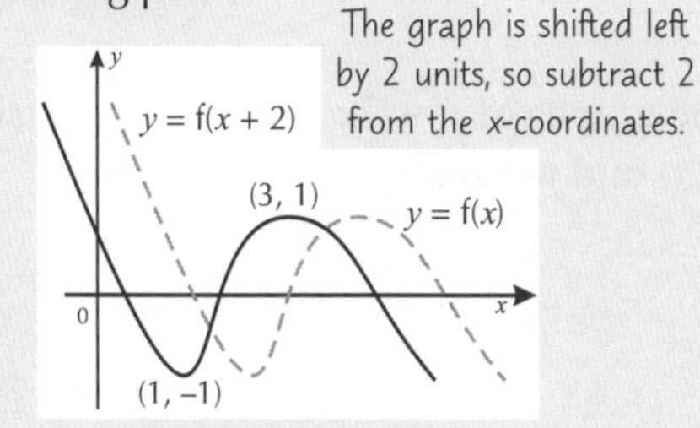

Now use your graph of $y = f(x + 2)$ to draw the graph of $y = 3f(x + 2)$.

This is a stretch in the direction of the y-axis with scale factor 3, so multiply the y-coordinates by 3.

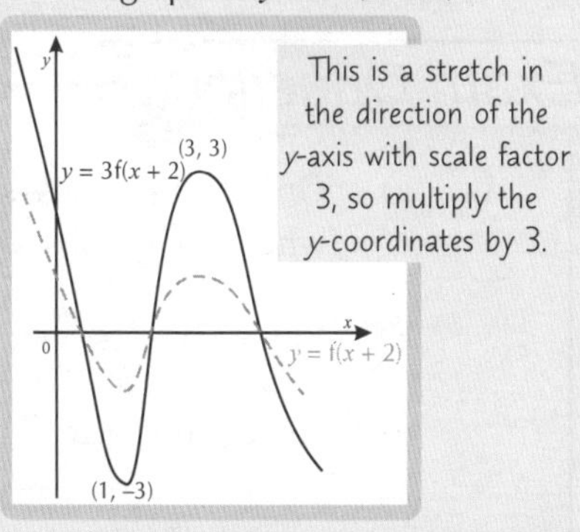

Tea and cake — the perfect combination...

Working out coordinates can be a bit tricky. The easiest way to do it is to work out what you're doing to the graph, then think about what that does to each point. Have a look at your transformed graph and check that the new coordinates make sense.

Core Section 1 — Practice Questions

These questions are for AQA C3, Edexcel C3, OCR C3 and OCR MEI C3

Well, that's the first section over and done with, and what better way to round it off than with some lovely questions. Have a go at these warm-up questions to get you in the mood.

Warm-up Questions

1) For the following mappings, state the range and say whether or not the mapping is a function. If not, explain why, and if so, say whether the function is one-to-one or many-to-one.
   a) $f(x) = x^2 - 16,\ x \geq 0$
   b) $f : x \to x^2 - 7x + 10,\ x \in \mathbb{R}$
   c) $f(x) = \sqrt{x},\ x \in \mathbb{R}$
   d) $f : x \to \frac{1}{x-2},\ x \in \mathbb{R}$

2) For each pair of functions f and g, find fg(2), gf(1) and fg(x).
   a) $f(x) = \frac{3}{x},\ x > 0$ and $g(x) = 2x + 3,\ x \in \mathbb{R}$
   b) $f(x) = 3x^2,\ x \geq 0$ and $g(x) = x + 4,\ x \in \mathbb{R}$

3) A one-to-one function f has domain $x \in \mathbb{R}$ and range $f(x) \geq 3$. Does this function have an inverse? If so, state its domain and range.

4) Using algebra, find the inverse of the function $f(x) = \sqrt{2x - 4},\ x \geq 2$. State the domain and range of the inverse.

5) For the function $f(x) = 2x - 1\ \{x \in \mathbb{R}\}$, sketch the graphs of:
   a) $|f(x)|$
   b) $f(|x|)$

6) Use your graph from part 5) a) to help you solve the equation $|2x - 1| = 5$.

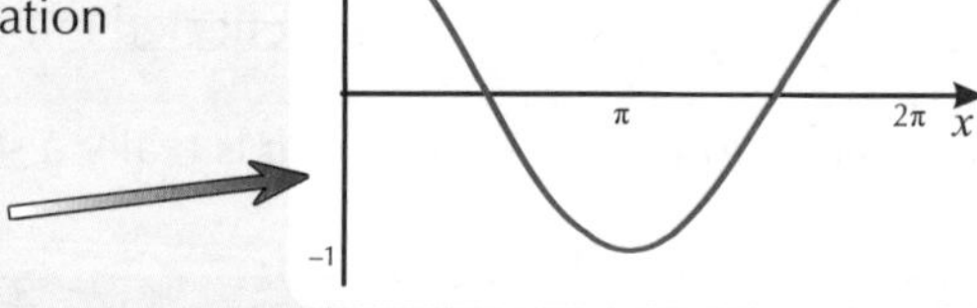

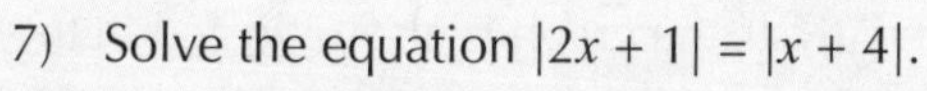

7) Solve the equation $|2x + 1| = |x + 4|$.

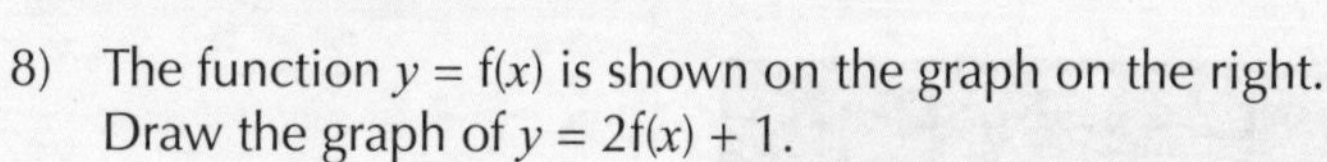

8) The function $y = f(x)$ is shown on the graph on the right. Draw the graph of $y = 2f(x) + 1$.

9) For the following functions, say whether they are odd or even and also decide if they're periodic:
   a) $y = x^2$
   b) $y = \tan x$
   c) $y = x^3$

Now that you're in the functions zone (not to be confused with the twilight zone or the phantom zone), I think you're ready to have a go at some exam-style questions.

Exam Questions

1 a) On the same axes, sketch the graphs of $y = |2x|$ and $y = |x - 1|$, showing clearly the points where the graphs touch the x- and y- axes.
(3 marks)

b) Solve $|2x| = |x - 1|$.
(3 marks)

c) Hence solve $|2x| \leq |x - 1|$.
(2 marks)

2 In words, describe what happens to the curve $y = x^3$ to transform it into the curve $y = 2(x - 1)^3 + 4$.
(6 marks)

Core Section 1 — Practice Questions

These questions are for AQA C3, Edexcel C3, OCR C3 and OCR MEI C3

They were nice questions to ease you in gently. I have to warn you, they get a bit harder on this page. It's nothing you can't handle though. Just arm yourself with a mosquito net, an invisibility cloak and some A2 Maths knowledge and you'll be fine.

3 The functions f and g are given by: $f(x) = x^2 - 3,\ x \in \mathbb{R}$ and $g(x) = \frac{1}{x},\ x \in \mathbb{R}, x \neq 0$.

a) Find an expression for $gf(x)$. *(2 marks)*

b) Solve $gf(x) = \frac{1}{6}$. *(3 marks)*

4 For the functions f and g, where

$$f(x) = 2^x,\ x \in \mathbb{R} \qquad \text{and} \qquad g(x) = \sqrt{3x - 2},\ x \geq \tfrac{2}{3},$$

find:

a) $fg(6)$ *(2 marks)*

b) $gf(2)$ *(2 marks)*

c) (i) $g^{-1}(x)$ *(2 marks)*

(ii) $fg^{-1}(x)$ *(2 marks)*

5 The function f(x) is defined as follows: $f : x \rightarrow \frac{1}{x+5}$, domain $x > -5$.

a) State the range of $f(x)$. *(1 mark)*

b) (i) Find the inverse function, $f^{-1}(x)$. *(3 marks)*

(ii) State the domain and range of $f^{-1}(x)$. *(2 marks)*

c) On the same axes, sketch the graphs of $y = f(x)$ and $y = f^{-1}(x)$. *(2 marks)*

6 The graph below shows the curve $y = f(x)$, and the intercepts of the curve with the x- and y-axes.

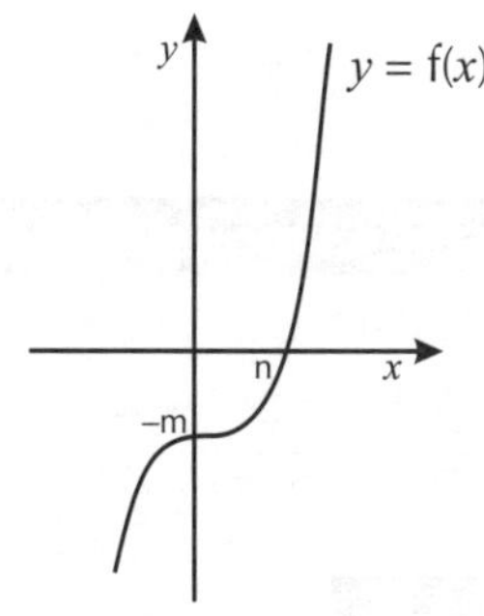

Sketch the graphs of the following transformations on separate axes, clearly labelling the points of intersection with the x- and y-axes in terms of m and n.

a) $y = |f(x)|$ *(2 marks)*

b) $y = -3f(x)$ *(2 marks)*

c) $y = f(|x|)$ *(2 marks)*

Simplifying Expressions

This page is for AQA C4, Edexcel C3, OCR C4 and OCR MEI C4

Oo, my favourite — algebraic fractions. Still, at least they're over with early on, so if they pop up later in A2 Maths you'll know what to do. No, not run away and cower in a corner — use the things you learnt on this page.

Simplify algebraic fractions by *Factorising* and *Cancelling Factors*

Algebraic fractions are a lot like normal fractions — and you can treat them in the same way, whether you're multiplying, dividing, adding or subtracting them. All fractions are much easier to deal with when they're in their simplest form, so the first thing to do with algebraic fractions is to simplify them as much as possible.

A function you can write as a fraction where the top and bottom are both polynomials is called a rational function.

1) Look for common factors in the numerator and denominator — factorise top and bottom and see if there's anything you can cancel.
2) If there's a fraction in the numerator or denominator (e.g. $\frac{1}{x}$), multiply the whole thing (i.e. top and bottom) by the same factor to get rid of it (for $\frac{1}{x}$, you'd multiply through by x).

EXAMPLES Simplify the following:

a) $\frac{3x+6}{x^2-4} = \frac{3(x+2)}{(x+2)(x-2)} = \frac{3}{x-2}$

Watch out for the difference of two squares — see C1.

b) $\frac{2+\frac{1}{2x}}{4x^2+x} = \frac{\left(2+\frac{1}{2x}\right)\times 2x}{x(4x+1)\times 2x} = \frac{4x+1}{2x^2(4x+1)} = \frac{1}{2x^2}$

3) You multiply algebraic fractions in exactly the same way as normal fractions — multiply the numerators together, then multiply the denominators. It's a good idea to cancel any common factors before you multiply.
4) To divide by an algebraic fraction, you just multiply by its reciprocal (the reciprocal is 1 ÷ the original thing — for fractions you just turn the fraction upside down).

EXAMPLES Simplify the following:

a) $\frac{x^2-2x-15}{2x+8} \times \frac{x^2-16}{x^2+3x} = \frac{(x+3)(x-5)}{2(x+4)} \times \frac{(x+4)(x-4)}{x(x+3)} = \frac{(x-5)(x-4)}{2x} \left(= \frac{x^2-9x+20}{2x}\right)$

Factorise both fractions.

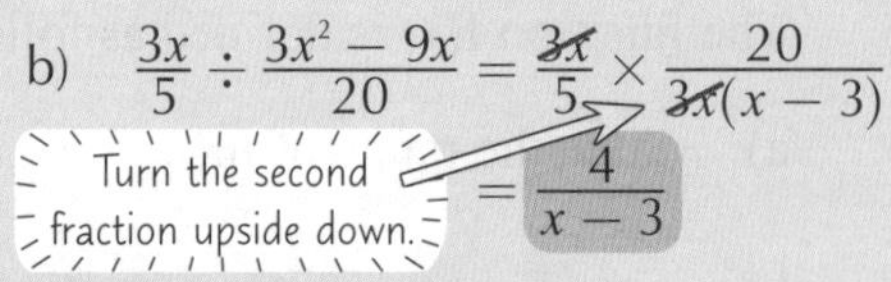

b) $\frac{3x}{5} \div \frac{3x^2-9x}{20} = \frac{3x}{5} \times \frac{20}{3x(x-3)} = \frac{4}{x-3}$

Turn the second fraction upside down.

Add and *Subtract* fractions by finding a *Common Denominator*

You'll have come across adding and subtracting fractions before in C1, so here's a little reminder of how to do it:

EXAMPLE Simplify:

$$\frac{2y}{x(x+3)} + \frac{1}{y^2(x+3)} - \frac{x}{y}$$

The common denominator is the lowest common multiple (LCM) of all the denominators.

① Find the Common Denominator

Take all the individual 'bits' from the bottom lines and multiply them together. Only use each bit once unless something on the bottom line is raised to a power.

The individual 'bits' here are x, $(x + 3)$ and y...

$xy^2(x+3)$

...but you need to use y^2 because there's a y^2 in the second fraction's denominator.

② Put Each Fraction over the Common Denominator

Make the denominator of each fraction into the common denominator.

$$\frac{y^2\times 2y}{y^2x(x+3)} + \frac{x\times 1}{xy^2(x+3)} - \frac{xy(x+3)\times x}{xy(x+3)y}$$

Multiply the top and bottom lines of each fraction by whatever makes the bottom line the same as the common denominator.

③ Combine into One Fraction

Once everything's over the common denominator, just add the top lines together.

All the bottom lines are the same — so you can just add the top lines.

$$= \frac{2y^3 + x - x^2y(x+3)}{xy^2(x+3)} = \frac{2y^3 + x - x^3y - 3x^2y}{xy^2(x+3)}$$

Who are you calling common...

Nothing on this page should be a big shock to you — it's all stuff you've done before. You've been using normal fractions for years, and algebraic fractions work in just the same way. They look a bit scary, but they're all warm and fuzzy inside.

Algebraic Division

This page is for AQA C4, Edexcel C3, OCR C4 and OCR MEI C4

I'll be honest with you, algebraic division is a bit tricky. But as long as you take it slowly and don't rush, it'll all fall into place. And it's really quick and easy to check your answer if you're not sure. What more could you want?

There are some **Terms** you need to **Know**

There are a few words that keep popping up in algebraic division, so make sure you know what they all mean.

1) DEGREE — the highest power of x in the polynomial (e.g. the degree of $4x^5 + 6x^2 - 3x - 1$ is 5).
2) DIVISOR — this is the thing you're dividing by (e.g. if you divide $x^2 + 4x - 3$ by $x + 2$, the divisor is $x + 2$).
3) QUOTIENT — the bit that you get when you divide by the divisor (not including the remainder — see p.12).

Method 1 — **Divide** by **Subtracting Multiples** of the **Divisor**

Back in AS Maths, you learnt how to do algebraic division by subtracting chunks of the divisor. Here's a quick reminder of how to divide a polynomial by $x - k$:

Algebraic Division

1) **Subtract a multiple of $(x - k)$ to get rid of the highest power of x.**
2) **Repeat step 1 until you've got rid of all the powers of x.**
3) **Work out how many lumps of $(x - k)$, you've subtracted, and read off the remainder.**

Have a look back at your AS notes if you can't remember how to do this.

EXAMPLE Divide $2x^3 - 3x^2 - 3x + 7$ by $x - 2$.

① Start with $2x^3 - 3x^2 - 3x + 7$, and subtract $2x^2$ lots of $(x - 2)$ to get rid of the x^3 term. $\Rightarrow (2x^3 - 3x^2 - 3x + 7) - 2x^2(x - 2) = x^2 - 3x + 7$

② Now start again with $x^2 - 3x + 7$. The highest power of x is the x^2 term, so subtract x lots of $(x - 2)$ to get rid of that. $\Rightarrow (x^2 - 3x + 7) - x(x - 2) = -x + 7$

③ All that's left now is $-x + 7$. Get rid of $-x$ by subtracting $-1 \times (x - 2)$. $\Rightarrow (-x + 7) - (-1(x - 2)) = 5$

So $(2x^3 - 3x^2 - 3x + 7) \div (x - 2) = 2x^2 + x - 1$ remainder 5.

Method 2 — use **Algebraic Long Division**

To divide two algebraic expressions, you can use long division (using the same method you'd use for numbers).

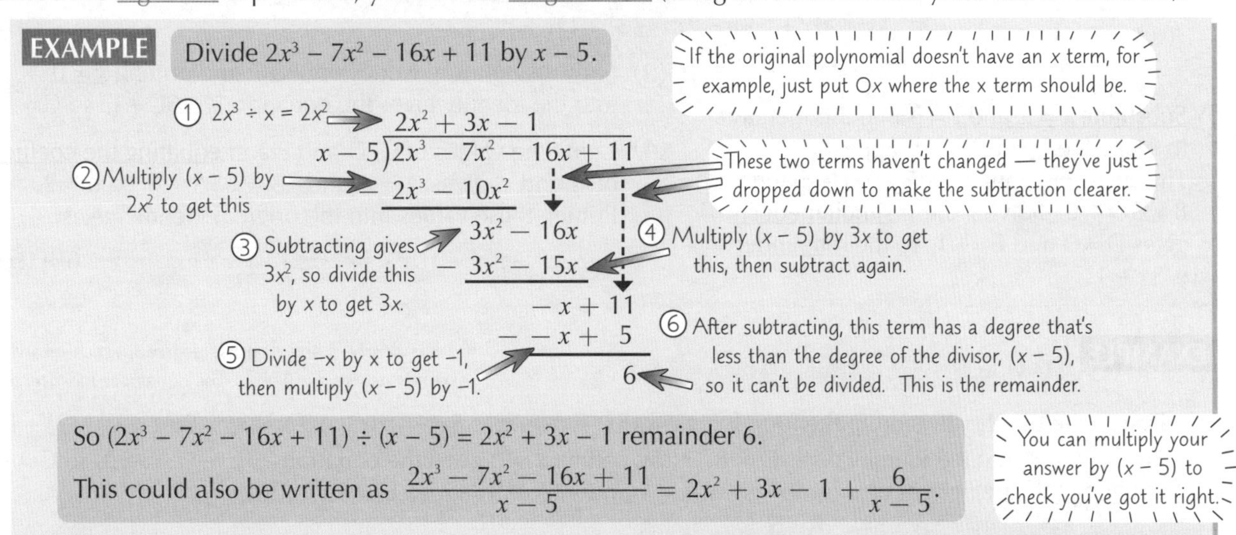

Just keep repeating — divide and conquer, divide and conquer...

For algebraic division to work, the degree of the divisor has to be less than (or equal to) the degree of the original polynomial (for example, you couldn't divide $x^2 + 2x + 3$ by $x^3 + 4$ as $3 > 2$, but you could do it the other way around). If you don't like either of these methods, you'll be pleased to know there's another way to divide coming up on the next page.

Algebraic Division

This page is for AQA C4, Edexcel C3, OCR C4 and OCR MEI C4

I really spoil you — as if two different methods for doing algebraic division weren't enough, I'm going to give you a third. If you're not sure about any of the terms, look back at the definitions on p.11.

Method 3 — use the Formula f(x) = q(x)d(x) + r(x)

There's a handy formula you can use to do algebraic division. It looks like this:

> A polynomial $f(x)$ can be written in the form $f(x) \equiv q(x)d(x) + r(x)$, where $q(x)$ is the quotient, $d(x)$ is the divisor and $r(x)$ is the remainder.

This comes from the Remainder Theorem that you met in AS. It's a good method for when you're dividing by a quadratic — long division can get a bit tricky when the divisor has 3 terms.

You'll be given $f(x)$ and $d(x)$ in the question, and it's down to you to work out $q(x)$ and $r(x)$. Here's how you do it:

Using the Formula

1) **First, you have to work out the degrees of the quotient and remainder, which depend on the degrees of the polynomial and the divisor. The degree of the quotient is $\deg f(x) - \deg d(x)$, and the degree of the remainder has to be less than the degree of the divisor.**
2) **Write out the division using the formula above, but replace $q(x)$ and $r(x)$ with general polynomials (i.e. a general polynomial of degree 2 is $Ax^2 + Bx + C$, and a general polynomial of degree 1 is $Ax + B$, where A, B, C, etc. are constants to be found).**
3) **The next step is to work out the values of the constants — you do this by substituting in values for x to make bits disappear, and by equating coefficients.**
4) **It's best to start with the constant term and work backwards from there.**
5) **Finally, write out the division again, replacing A, B, C, etc. with the values you've found.**

Equating coefficients means comparing the coefficients of each power of x on the LHS and the RHS.

The method looks a bit intense, but follow through the examples below to see how it works.

Start with the Remainder and Work Backwards

When you're using this method, you might have to use simultaneous equations to work out some of the coefficients. Have a look back at your C1 notes for a reminder of how to do this if you need to.

EXAMPLE Divide $x^4 - 3x^3 - 3x^2 + 10x + 5$ by $x^2 - 5x + 6$.

① First, write out the division in the form $f(x) \equiv q(x)d(x) + r(x)$:

$$x^4 - 3x^3 - 3x^2 + 10x + 5 \equiv (Ax^2 + Bx + C)(x^2 - 5x + 6) + Dx + E$$
$$\equiv (Ax^2 + Bx + C)(x - 2)(x - 3) + Dx + E.$$

f(x) has degree 4 and d(x) has degree 2, which means that q(x) has degree 4 − 2 = 2. The remainder has degree 1 or 0 — put in Dx + E, as D can always be 0.

d(x) factorises to give (x − 2)(x − 3).

② Substitute $x = 2$ and $x = 3$ into the identity to make the $q(x)d(x)$ bit disappear. This gives the equations $5 = 2D + E$ and $8 = 3D + E$. Solving these simultaneously gives $D = 3$ and $E = -1$, so the remainder is $3x - 1$.

③ Now, using these values of D & E and putting $x = 0$ into the identity gives the equation $5 = 6C + E$, so $C = 1$.

④ Using the values of C, D and E and equating the coefficients of x^4 and x^3 gives: $1 = A$ and $-3 = -5A + B$, so $B = 2$. Putting these values into the original identity gives:

$$x^4 - 3x^3 - 3x^2 + 10x + 5 \equiv (x^2 + 2x + 1)(x^2 - 5x + 6) + 3x - 1.$$

EXAMPLE Divide $x^3 + 5x^2 - 18x - 10$ by $x - 3$.

f(x) has degree 3 and d(x) has degree 1, which means that q(x) has degree 3 − 1 = 2. The remainder has degree 0.

First, write out the division in the form $f(x) \equiv q(x)d(x) + r(x)$: $x^3 + 5x^2 - 18x - 10 \equiv (Ax^2 + Bx + C)(x - 3) + D$.
Putting $x = 3$ into the identity gives $D = 8$. Now, setting $x = 0$ gives the equation $-3C + D = -10$, so $C = 6$.
Equating the coefficients of x^3 and x^2 gives $A = 1$ and $-3A + B = 5$, so $B = 8$.
So $x^3 + 5x^2 - 18x - 10 \equiv (x^2 + 8x + 6)(x - 3) + 8$.

A reminder about remainders...

The degree of the remainder has to be less than the degree of the divisor, otherwise it would be included in the quotient. E.g. if $r(x) = (x + 1)$ and $d(x) = (x - 3)$, then $r(x)$ can be divided by $d(x)$, giving a remainder of 4 (so $(x + 1)$ wasn't the remainder).

Partial Fractions

This page is for AQA C4, Edexcel C4, OCR C4 and OCR MEI C4

In a similar way to using a hammer to crack a walnut, you can use partial fractions to split up certain algebraic fractions.

'Expressing in Partial Fractions' is the Opposite of Adding Fractions (sort of)

1) You can split a fraction with more than one linear factor in the denominator into partial fractions.

$\frac{7x-7}{(2x+1)(x-3)}$ can be written as partial fractions of the form $\frac{A}{(2x+1)}+\frac{B}{(x-3)}$.

$\frac{9x^2+x+16}{(x+2)(2x-1)(x-3)}$ can be written as partial fractions of the form $\frac{A}{(x+2)}+\frac{B}{(2x-1)}+\frac{C}{(x-3)}$.

$\frac{x^2+17x+16}{(x+2)^2(3x-1)}$ can be written as partial fractions of the form $\frac{A}{(x+2)^2}+\frac{B}{(x+2)}+\frac{C}{(3x-1)}$.

Watch out here — this one doesn't quite follow the pattern.

2) The tricky bit is figuring out what A, B and C are.
You can use the substitution method or the equating coefficients method:

EXAMPLE Express $\frac{9x^2+x+16}{(x+2)(2x-1)(x-3)}$ in partial fractions.

You know that $\frac{9x^2+x+16}{(x+2)(2x-1)(x-3)} \equiv \frac{A}{(x+2)}+\frac{B}{(2x-1)}+\frac{C}{(x-3)}$. Now to work out A, B and C.

1 Add the partial fractions and cancel the denominators from both sides

$$\frac{A}{(x+2)}+\frac{B}{(2x-1)}+\frac{C}{(x-3)} \equiv \frac{A(2x-1)(x-3)+B(x+2)(x-3)+C(2x-1)(x+2)}{(x+2)(2x-1)(x-3)}$$

So the numerators are equal: $9x^2+x+16 \equiv A(2x-1)(x-3)+B(x+2)(x-3)+C(2x-1)(x+2)$

2 Substitute x for values which get rid of all but one of A, B and C...

Substituting $x = 3$ gets rid of A and B: $(9\times 3^2)+3+16 = 0+0+C((2\times 3)-1)(3+2)$
$100 = 25C \Rightarrow C = 4$

Substituting $x = -2$ gets rid of B and C: $(9\times(-2)^2)+(-2)+16 = A((2\times -2)-1)(-2-3)+0+0$
$50 = 25A \Rightarrow A = 2$

Substituting $x = 0.5$ gets rid of A and C: $(9\times(0.5^2))+0.5+16 = 0+B(0.5+2)(0.5-3)+0$
$18.75 = -6.25B \Rightarrow B = -3$

...OR compare coefficients in the numerators

$9x^2+x+16 \equiv A(2x-1)(x-3)+B(x+2)(x-3)+C(2x-1)(x+2)$

x^2 coefficients: $9 = 2A + B + 2C$
x coefficients: $1 = -7A - B + 3C$
constant terms: $16 = 3A - 6B - 2C$

Solving these equations simultaneously gives $A = 2$, $B = -3$ and $C = 4$ — the same as the substitution method.

3 Write out the solution $\frac{9x^2+x+16}{(x+2)(2x-1)(x-3)} \equiv \frac{2}{(x+2)}-\frac{3}{(2x-1)}+\frac{4}{(x-3)}$

Watch out for Difference of Two Squares Denominators

Just for added meanness, they might give you an expression like $\frac{4}{x^2-1}$ and tell you to express it as partial fractions.

You have to recognise that the denominator is a difference of two squares, write it as two linear factors, and then carry on as normal. E.g. $\frac{21x-2}{9x^2-4} \equiv \frac{21x-2}{(3x-2)(3x+2)} \equiv \frac{A}{(3x-2)}+\frac{B}{(3x+2)}$

Not all coefficients are created equal — but some are...

It's worth getting to grips with both methods for step 2. Sometimes one's easier to use than the other, and sometimes you might want to mix and match. It's just another crucial step on the path to going down in history as a mathematical great.

Partial Fractions

This page is for AQA C4, Edexcel C4, OCR C4 and OCR MEI C4

Now things are hotting up in the partial fractions department — here's an example involving a repeated factor.

Sometimes it's best to use Substitution AND Equate Coefficients

EXAMPLE Express $\frac{x^2 + 17x + 16}{(x+2)^2(3x-1)}$ in partial fractions.

You know that $\frac{x^2 + 17x + 16}{(x+2)^2(3x-1)} \equiv \frac{A}{(x+2)^2} + \frac{B}{(x+2)} + \frac{C}{(3x-1)}$. Now to work out A, B and C.

1 Add the partial fractions

You end up with an extra $(x + 2)$ factor in each term that can be cancelled.

$$\frac{A}{(x+2)^2} + \frac{B}{(x+2)} + \frac{C}{(3x-1)} \equiv \frac{A(x+2)(3x-1) + B(x+2)^2(3x-1) + C(x+2)(x+2)^2}{(x+2)^2(x+2)(3x-1)}$$

Cancel the denominators from both sides $x^2 + 17x + 16 \equiv A(3x-1) + B(x+2)(3x-1) + C(x+2)^2$

2 Substitute x for values which get rid of all but one of A, B and C

Substituting $x = -2$ gets rid of B and C: $(-2)^2 + (17 \times -2) + 16 = A((3 \times -2) - 1) + 0 + 0$

$-14 = -7A \quad \Rightarrow \underline{A = 2}$

Substituting $x = \frac{1}{3}$ gets rid of A and B: $\left(\frac{1}{3}\right)^2 + \left(17 \times \frac{1}{3}\right) + 16 = 0 + 0 + C\left(\frac{1}{3} + 2\right)^2$

$\frac{196}{9} = \frac{49}{9}C \quad \Rightarrow \underline{C = 4}$

The trouble is, there's no value of x you can substitute to get rid of A and C to just leave B.

So: **Equate coefficients of x^2**

From $x^2 + 17x + 16 \equiv A(3x-1) + B(x+2)(3x-1) + C(x+2)^2$

Coefficients of x^2 are: $1 = 3B + C$

You know $C = 4$, so: $1 = 3B + 4 \quad \Rightarrow \underline{B = -1}$

3 Write out the solution You now know A, B and C, so: $\frac{x^2 + 17x + 16}{(x+2)^2(3x-1)} \equiv \frac{2}{(x+2)^2} - \frac{1}{(x+2)} + \frac{4}{(3x-1)}$

Divide Before Expressing Improper Fractions as Partial Fractions

The numerator of an improper algebraic fraction has a degree equal to or greater than the degree of the denominator.

E.g. $\frac{x^2 + 4}{(x+3)(x+2)}$ ← degree 2 / degree 2 $\qquad \frac{x^4 + 2x}{(x-1)^2(x+2)}$ ← degree 4 / degree 3

The degree of a polynomial is the highest power of x.

There's an extra step involved in expressing an improper fraction as partial fractions:

1) Divide the numerator by the denominator to get the quotient (q(x)) + a proper fraction (r(x) / d(x))
2) Express the proper fraction as partial fractions.

See pages 11-12 for algebraic division methods.

EXAMPLE Express $\frac{x^4 - 3x^3 - 3x^2 + 10x + 5}{(x-3)(x-2)}$ as partial fractions.

This is $(x - 3)(x - 2)$ multiplied out.

1) First work out $(x^4 - 3x^3 - 3x^2 + 10x + 5) \div (x^2 - 5x + 6)$:
   - Write out the result in the form f(x) ≡ q(x)d(x) + r(x):

     $x^4 - 3x^3 - 3x^2 + 10x + 5 \equiv (x^2 + 2x + 1)(x^2 - 5x + 6) + 3x - 1.$

     Exactly as on page 12 — q(x) = quotient, d(x) = divisor and r(x) = remainder.
   - Divide through by d(x): $\frac{x^4 - 3x^3 - 3x^2 + 10x + 5}{(x-3)(x-2)} \equiv (x^2 + 2x + 1) + \frac{3x-1}{(x-3)(x-2)}$ ← $q(x) + \frac{r(x)}{d(x)}$
2) Now just express the proper fraction as partial fractions: $\frac{x^4 - 3x^3 - 3x^2 + 10x + 5}{(x-3)(x-2)} \equiv (x^2 + 2x + 1) + \frac{A}{(x-3)} + \frac{B}{(x-2)}$

Rid the partial fraction world of improperness — it's only proper...

After you've found the partial fractions, don't forget to go back to the original fraction and write out the full solution...

The Binomial Expansion

This page is for AQA C4, Edexcel C4, OCR C4 and OCR MEI C4

Yeah, I know the binomial expansion. We spent some time together back in AS Maths. Thought I'd never see it again. And then, of all the sections in all the maths books in all the world, the binomial expansion walks into mine...

The Binomial Expansion Formula is pretty useful

The binomial expansion is a way to raise a given expression to any power.

For simpler cases it's basically a fancy way of multiplying out brackets.
You can also use it to approximate more complicated expressions (see p.18).

This is the general formula for binomial expansions:

$$(1+x)^n = 1 + nx + \frac{n(n-1)}{1\times 2}x^2 + \ldots + \frac{n(n-1)\ldots(n-r+1)}{1\times 2\times \ldots \times r}x^r + \ldots$$

The Binomial Expansion sometimes gives a Finite Expression

From the general formula, it looks like the expansion always goes on forever.
But if n is a positive integer, the binomial expansion is finite.

EXAMPLE Give the binomial expansion of $(1 + x)^5$.

You can use the general formula and plug in $n = 5$:

$n(n-1)$

$n = 5$

$$(1+x)^5 = 1 + 5x + \frac{5(5-1)}{1\times 2}x^2 + \frac{5(5-1)(5-2)}{1\times 2\times 3}x^3 + \frac{5(5-1)(5-2)(5-3)}{1\times 2\times 3\times 4}x^4$$
$$+ \frac{5(5-1)(5-2)(5-3)(5-4)}{1\times 2\times 3\times 4\times 5}x^5 + \frac{5(5-1)(5-2)(5-3)(5-4)(5-5)}{1\times 2\times 3\times 4\times 5\times 6}x^6 + \ldots$$
$$= 1 + 5x + \frac{5\times 4}{1\times 2}x^2 + \frac{5\times 4\times 3}{1\times 2\times 3}x^3 + \frac{5\times 4\times 3\times 2}{1\times 2\times 3\times 4}x^4$$
$$+ \frac{5\times 4\times 3\times 2\times 1}{1\times 2\times 3\times 4\times 5}x^5 + \frac{5\times 4\times 3\times 2\times 1\times 0}{1\times 2\times 3\times 4\times 5\times 6}x^6 + \ldots$$
$$= 1 + 5x + \frac{20}{2}x^2 + \frac{60}{6}x^3 + \frac{120}{24}x^4 + \frac{120}{120}x^5 + \frac{0}{720}x^6 + \ldots$$
$$= 1 + 5x + 10x^2 + 10x^3 + 5x^4 + x^5$$

You can stop here — all the terms after this one are zero

The formula still works if the coefficient of x isn't 1.

EXAMPLE Give the binomial expansion of $(1 - 3x)^4$.

This time $n = 4$, but you also have to replace every x in the formula with $-3x$:

$(1-3x)^4$ ← Think of this as $(1 + (-3x))^4$ — you need to put the minus into the formula as well as the $3x$.

$n = 4$ | $n(n-1)$ | Don't forget to square the -3 as well. | Stop here

$$= 1 + 4(-3x) + \frac{4\times 3}{1\times 2}(-3x)^2 + \frac{4\times \cancel{3}\times \cancel{2}}{1\times \cancel{2}\times \cancel{3}}(-3x)^3 + \frac{\cancel{4}\times \cancel{3}\times \cancel{2}\times \cancel{1}}{\cancel{1}\times \cancel{2}\times \cancel{3}\times \cancel{4}}(-3x)^4 + \frac{\cancel{4}\times \cancel{3}\times \cancel{2}\times \cancel{1}\times 0}{\cancel{1}\times \cancel{2}\times \cancel{3}\times \cancel{4}\times 5}(-3x)^5 + \ldots$$
$$= 1 + 4(-3x) + \frac{12}{2}(9x^2) + \frac{4}{1}(-27x^3) + (81x^4) + \frac{0}{5}(-243x^5) + \ldots$$
$$= 1 - 12x + 54x^2 - 108x^3 + 81x^4$$

Make life easier for yourself by cancelling down the fractions before you multiply.

The Binomial Expansion

This page is for AQA C4, Edexcel C4, OCR C4 and OCR MEI C4

Unfortunately, you only get a nice, neat, finite expansion when you've got a positive integer n.
But that pesky n sometimes likes to be a negative number or a fraction. n for nuisance. n for naughty.

If n is Negative the expansion gets more complicated...

EXAMPLE Find the binomial expansion of $\frac{1}{(1+x)^2}$ up to and including the term in x^3.

This is where things start to get a bit more interesting.
First, rewrite the expression: $\frac{1}{(1+x)^2} = (1+x)^{-2}$.

You can still use the general formula. This time $n = -2$:

$n = -2$ ... $n(n-1)$

$$(1+x)^{-2} = 1 + (-2)x + \frac{(-2)\times(-2-1)}{1\times 2}x^2 + \frac{(-2)\times(-2-1)\times(-2-2)}{1\times 2\times 3}x^3 + \ldots$$
$$= 1 + (-2)x + \frac{(-\cancel{2})\times(-3)}{1\times \cancel{2}}x^2 + \frac{(-\cancel{2})\times(-\cancel{3})\times(-4)}{1\times \cancel{2}\times \cancel{3}}x^3 + \ldots$$
$$= 1 + (-2)x + \frac{3}{1}x^2 + \frac{-4}{1}x^3 + \ldots$$
$$= 1 - 2x + 3x^2 - 4x^3 + \ldots$$

Again, you can cancel down before you multiply — but be careful with those minus signs.

With a negative n, you'll never get zero as a coefficient. If the question hadn't told you to stop, the expansion could go on forever.

We've left out all the terms after $-4x^3$, so the cubic equation you've ended up with is an approximation to the original expression. You could also write the answer like this:

$$\frac{1}{(1+x)^2} \approx 1 - 2x + 3x^2 - 4x^3$$

... and if n is a Fraction things can be tricky too

The binomial expansion formula doesn't just work for integer values of n.

EXAMPLE Find the binomial expansion of $\sqrt[3]{1+2x}$ up to and including the term in x^3.

This time we've got a fractional power: $\sqrt[3]{1+2x} = (1+2x)^{\frac{1}{3}}$

So this time $n = \frac{1}{3}$, and you also need to replace x with $2x$:

$n = \frac{1}{3}$ ... $n(n-1)$

$$(1+2x)^{\frac{1}{3}} = 1 + \tfrac{1}{3}(2x) + \frac{\frac{1}{3}\times(\frac{1}{3}-1)}{1\times 2}(2x)^2 + \frac{\frac{1}{3}\times(\frac{1}{3}-1)\times(\frac{1}{3}-2)}{1\times 2\times 3}(2x)^3 + \ldots$$
$$= 1 + \tfrac{2}{3}x + \frac{\frac{1}{3}\times(-\frac{2}{3})}{1\times 2}4x^2 + \frac{\frac{1}{3}\times(-\frac{2}{3})\times(-\frac{5}{3})}{1\times 2\times 3}8x^3 + \ldots$$
$$= 1 + \tfrac{2}{3}x + \frac{(-\frac{2}{9})}{2}4x^2 + \frac{(\frac{10}{27})}{6}8x^3 + \ldots$$
$$= 1 + \tfrac{2}{3}x + (-\tfrac{2}{9}\times\tfrac{1}{2})4x^2 + (\tfrac{10}{27}\times\tfrac{1}{6})8x^3 + \ldots$$
$$= 1 + \tfrac{2}{3}x - \tfrac{4}{9}x^2 + \tfrac{40}{81}x^3 + \ldots$$

Cancelling down is much trickier with this type of expansion — it's usually safer to multiply everything out fully.

Exam questions often ask for the coefficients as simplified fractions.

The Binomial Expansion

This page is for AQA C4, Edexcel C4, OCR C4 and OCR MEI C4

More binomial goodness... this page is so jam-packed with the stuff, there's only room for a one-line intro...

If the *Constant* in the brackets isn't *1*, you have to *Factorise* first

So the general binomial expansion of $(1 + x)^n$ works fine for any n, and you can replace the x with other x-terms, but that 1 has to be a 1 before you can expand. That means you sometimes need to start with a sneaky bit of factorisation.

EXAMPLE Give the binomial expansion of $(3 - x)^4$.

To use the general formula, you need the constant term in the brackets to be 1.
You can take the 3 outside the brackets by factorising:

The aim here is to get an expression in the form $c(1 + dx)^n$, where c and d are constants.

$$3 - x = 3(1 - \tfrac{1}{3}x)$$
$$\text{so} \quad (3 - x)^4 = [3(1 - \tfrac{1}{3}x)]^4 = 3^4(1 - \tfrac{1}{3}x)^4 = 81(1 - \tfrac{1}{3}x)^4$$

Now we can use the general formula, with $n = 4$, and $-\frac{1}{3}x$ instead of x:

$$\left(1 - \tfrac{1}{3}x\right)^4 = 1 + 4\left(-\tfrac{1}{3}x\right) + \frac{4 \times 3}{1 \times 2}\left(-\tfrac{1}{3}x\right)^2 + \frac{4 \times 3 \times 2}{1 \times 2 \times 3}\left(-\tfrac{1}{3}x\right)^3 + \frac{4 \times 3 \times 2 \times 1}{1 \times 2 \times 3 \times 4}\left(-\tfrac{1}{3}x\right)^4$$
$$= 1 - \tfrac{4}{3}x + 6\left(\tfrac{1}{9}x^2\right) + 4\left(-\tfrac{1}{27}x^3\right) + \tfrac{1}{81}x^4$$
$$= 1 - \frac{4x}{3} + \frac{2x^2}{3} - \frac{4x^3}{27} + \frac{x^4}{81}$$

So now we can expand the original expression:

$$(3 - x)^4 = 81\left(1 - \tfrac{1}{3}x\right)^4 = 81\left(1 - \frac{4x}{3} + \frac{2x^2}{3} - \frac{4x^3}{27} + \frac{x^4}{81}\right) = 81 - 108x + 54x^2 - 12x^3 + x^4$$

Some *Binomial Expansions* are only *Valid* for *Certain Values* of x

When you find a binomial expansion, you usually have to state which values of x the expansion is valid for.

If n is a positive integer, the binomial expansion of $(p + qx)^n$ is valid for all values of x.

If n is not a positive integer, the expansion would be infinite. Because you only write out a few terms, the binomial expansion you get is just an approximation. But the approximation is only valid if the sequence converges — this only happens if x is small enough (for larger values of x, the sequence will diverge)...

If n is a negative integer or a fraction, the binomial expansion of $(p + qx)^n$ is valid when $\left|\frac{qx}{p}\right| < 1$.

You can rewrite this as $|x| < \left|\frac{p}{q}\right|$ — just use the version you find easiest to remember.

This means there's a little bit more to do for the two examples on the previous page:

$(1 + x)^{-2} = 1 - 2x + 3x^2 - 4x^3 + \ldots$ This expansion is valid for $|x| < 1$.

$(1 + 2x)^{\frac{1}{3}} = 1 + \frac{2}{3}x - \frac{4}{9}x^2 + \frac{40}{81}x^3 + \ldots$

This expansion is valid if $|2x| < 1 \Rightarrow 2|x| < 1 \Rightarrow |x| < \frac{1}{2}$.

You might already know the rules $|ab| = |a||b|$ and $\left|\frac{a}{b}\right| = \frac{|a|}{|b|}$. If you don't, then get to know them — they're handy for rearranging these limits.

Lose weight and save money — buy no meals...

Two facts: 1) You can pretty much guarantee that there'll be a binomial expansion question on your C4 exam, and 2) any binomial expansion question they can throw at you will feature some combination of these adaptations of the general formula.

Approximating with Binomial Expansions

This page is for AQA C4, Edexcel C4, OCR C4 and OCR MEI C4

Binomial expansions can give you a handy way to estimate various roots.
OK, so it's not that handy... just go with it for now...

To find ***Approximations***, substitute the right value of *x*

When you've done an expansion, you can use it to estimate the value of the original expression for given values of x.

EXAMPLE

The binomial expansion of $(1+3x)^{\frac{1}{3}}$ up to the term in x^3 is $(1+3x)^{\frac{1}{3}} \approx 1 + x - x^2 + \frac{5}{3}x^3$.
The expansion is valid for $|x| < \frac{1}{3}$.
Use this expansion to approximate $\sqrt[3]{1.3}$. Give your answer to 4 d.p.

For this type of question, you need to find the right value of x to make the expression you're expanding equal to the thing you're looking for.

In this case it's pretty straightforward: $\sqrt[3]{1.3} = (1+3x)^{\frac{1}{3}}$ when $x = 0.1$.

$$\begin{aligned}\sqrt[3]{1.3} &= (1+3(0.1))^{\frac{1}{3}} \\ &\approx 1 + 0.1 - (0.1)^2 + \tfrac{5}{3}(0.1)^3 \\ &= 1 + 0.1 - 0.01 + \frac{0.005}{3} \\ &= 1.0917 \text{ (to 4 d.p.)}\end{aligned}$$

Don't forget to use a "≈" here — the answer's an approximation because you're only using the expansion up to the x^3 term.

This is the expansion given in the question, with $x = 0.1$.

In trickier cases you have to do a spot of rearranging to get to the answer.

EXAMPLE

The binomial expansion of $(1-5x)^{\frac{1}{2}}$ up to the term in x^2 is $(1-5x)^{\frac{1}{2}} \approx 1 - \frac{5x}{2} - \frac{25}{8}x^2$.

The expansion is valid for $|x| < \frac{1}{5}$.

Use $x = \frac{1}{50}$ in this expansion to find an approximate value for $\sqrt{10}$.
Find the percentage error in your approximation, to 2 s.f.

First, sub $x = \frac{1}{50}$ into both sides of the expansion:

$$\begin{aligned}\sqrt{\left(1 - 5\left(\tfrac{1}{50}\right)\right)} &\approx 1 - \tfrac{5}{2}\left(\tfrac{1}{50}\right) - \tfrac{25}{8}\left(\tfrac{1}{50}\right)^2 \\ \sqrt{\left(1 - \tfrac{1}{10}\right)} &\approx 1 - \frac{1}{20} - \frac{1}{800} \\ \sqrt{\frac{9}{10}} &\approx \frac{759}{800}\end{aligned}$$

Now simplify the square root and rearrange to find an estimate for $\sqrt{10}$:

$$\sqrt{\frac{9}{10}} = \frac{\sqrt{9}}{\sqrt{10}} = \frac{3}{\sqrt{10}} \approx \frac{759}{800} \quad \Rightarrow \quad \sqrt{10} \approx 3 \div \frac{759}{800} = \frac{800}{253}$$

The percentage error is $\left|\frac{\text{real value} - \text{estimate}}{\text{real value}}\right| \times 100 = \left|\frac{\sqrt{10} - \frac{800}{253}}{\sqrt{10}}\right| \times 100 = 0.0070\%$ (to 2 s.f.)

You never know when you might need to estimate the cube root of 1.3...

That percentage error bit in the second example is one of those ways they might sneak a seemingly unrelated topic into an exam question. The examiners are allowed to stick a bit from any of the earlier Core modules into C4, so don't freak out if they ask you something slightly unexpected — remember, you will have seen it before and you do know how to do it.

Binomial Expansions and Partial Fractions

This page is for AQA C4, Edexcel C4 and OCR MEI C4

Binomial expansions on their own are pretty nifty, but when you combine them with partial fractions (see p.13-14) they become all-powerful. I'm sure there's some sort of message about friendship or something in there...

Split functions into Partial Fractions, then add the Expansions

You can find the binomial expansion of even more complicated functions by splitting them into partial fractions first.

EXAMPLE

$$f(x) = \frac{x-1}{(3+x)(1-5x)}$$

a) $f(x)$ can be expressed in the form $\frac{A}{(3+x)} + \frac{B}{(1-5x)}$. Find the values of A and B.

b) Use your answer to part a) to find the binomial expansion of $f(x)$ up to and including the term in x^2.

c) Find the range of values of x for which your answer to part b) is valid.

a) Convert $f(x)$ into partial fractions:

$$\frac{x-1}{(3+x)(1-5x)} \equiv \frac{A}{(3+x)} + \frac{B}{(1-5x)} \Rightarrow x - 1 \equiv A(1-5x) + B(3+x)$$

Let $x = -3$, then $-3 - 1 = A(1 - (-15)) \Rightarrow -4 = 16A \Rightarrow A = -\frac{1}{4}$

Let $x = \frac{1}{5}$, then $\frac{1}{5} - 1 = B\left(3 + \frac{1}{5}\right) \Rightarrow -\frac{4}{5} = \frac{16}{5}B \Rightarrow B = -\frac{1}{4}$

b) Start by rewriting the partial fractions in $(a + bx)^n$ form:

$$f(x) = -\frac{1}{4}(3+x)^{-1} - \frac{1}{4}(1-5x)^{-1}$$

Now do the two binomial expansions:

$$\begin{aligned}(3+x)^{-1} &= \left(3\left(1+\tfrac{1}{3}x\right)\right)^{-1} \\ &= \tfrac{1}{3}\left(1+\tfrac{1}{3}x\right)^{-1} \\ &= \tfrac{1}{3}\left(1 + (-1)\left(\tfrac{1}{3}x\right) + \frac{(-1)(-2)}{2}\left(\tfrac{1}{3}x\right)^2 + \ldots\right) \\ &= \tfrac{1}{3}\left(1 - \tfrac{1}{3}x + \tfrac{1}{9}x^2 + \ldots\right) \\ &= \tfrac{1}{3} - \tfrac{1}{9}x + \tfrac{1}{27}x^2 + \ldots\end{aligned}$$

$$\begin{aligned}(1-5x)^{-1} &= 1 + (-1)(-5x) + \frac{(-1)(-2)}{2}(-5x)^2 + \ldots \\ &= 1 + 5x + 25x^2 + \ldots\end{aligned}$$

And put everything together:

$$\begin{aligned}f(x) = -\tfrac{1}{4}(3+x)^{-1} - \tfrac{1}{4}(1-5x)^{-1} &\approx -\tfrac{1}{4}\left(\tfrac{1}{3} - \tfrac{1}{9}x + \tfrac{1}{27}x^2\right) - \tfrac{1}{4}\left(1 + 5x + 25x^2\right) \\ &= -\tfrac{1}{12} + \tfrac{1}{36}x - \tfrac{1}{108}x^2 - \tfrac{1}{4} - \tfrac{5}{4}x - \tfrac{25}{4}x^2 \\ &= -\tfrac{1}{3} - \tfrac{11}{9}x - \tfrac{169}{27}x^2\end{aligned}$$

c) Each of the two expansions from part b) is valid for different values of x.
The combined expansion of $f(x)$ is valid where these two ranges overlap, i.e. over the narrower of the two ranges.

The expansion of $(3 + x)^{-1}$ is valid when $\left|\frac{x}{3}\right| < 1 \Rightarrow \frac{|x|}{|3|} < 1 \Rightarrow |x| < 3$.

The expansion of $(1 - 5x)^{-1}$ is valid when $|-5x| < 1 \Rightarrow |-5||x| < 1 \Rightarrow |x| < \frac{1}{5}$.

The expansion of $f(x)$ is valid for values of x in both ranges, so the expansion of $f(x)$ is valid for $|x| < \frac{1}{5}$.

Remember — the expansion of $(p + qx)^n$ is valid when $\left|\frac{qx}{p}\right| < 1$.

Don't mess with me — I'm a partial arts expert...

Here's where it all comes together. This example looks pretty impressive, but if you know your stuff you'll sail through questions like this. I think that's all I've got to say for this page... hmm, looks like I've still got another line to fill...
So, going anywhere nice on your holidays this year? Read any good books lately? (Answer: Yes, this one.)

Proof

This page is for AQA C3 & C4, Edexcel C3, OCR C3 & C4 and OCR MEI C3

Like an annoying child who keeps asking 'But whyyyyyyyy?' sometimes the examiners expect you to prove something is true. The next two pages feature three classic maths ways of proving things, plus a bonus way to disprove stuff.

Direct Proof

You probably won't be tested directly on this for AQA, Edexcel and OCR — it'll just come up within another topic.

A direct proof (or 'proof by direct argument') is when you use known facts to build up your argument and show a statement must be true.

EXAMPLE A definition of a rational number is 'a number that can be written as a quotient of two integers, where the denominator is non-zero'.

Use this definition to prove that the following statement is true:

"The product of two rational numbers is always a rational number."

Take any two rational numbers, call them a and b.

By the definition of rational numbers you can write them in the form $a = \frac{p}{q}$ and $b = \frac{r}{s}$, where p, q, r and s are all integers, and q and s are non-zero.

The product of a and b is $ab = \frac{p}{q} \times \frac{r}{s} = \frac{pr}{qs}$

pr and qs are the products of integers, so they must also be integers, and because q and s are non-zero, qs must also be non-zero.

We've shown that ab is a quotient of two integers and has a non-zero denominator, so by definition, ab is rational.

Hence the original statement is true.

Proof by Contradiction

To prove a statement by contradiction, you say 'Suppose the statement isn't true...', then prove that something impossible would have to be true for that to be the case.

EXAMPLE Prove the following statement: "If x^2 is even, then x must be even."

We can prove the statement by contradiction.

Suppose the statement is not true. Then there must be an odd number x for which x^2 is even.

If x is odd, then you can write x as $2k + 1$, where k is an integer. (This is the definition of an odd number.)

Now, $x^2 = (2k + 1)^2 = 4k^2 + 4k + 1$

$4k^2 + 4k = 2(2k^2 + 2k)$ is even because it is 2× an integer (this is the definition of an even number),
$\Rightarrow 4k^2 + 4k + 1$ is odd (since even + odd = odd).

But this isn't possible if the statement that x^2 is even is true.
We've contradicted the statement that there is an odd number x for which x^2 is even.

So if x^2 is even, then x must be even, hence the original statement is true.

That was proof by contradiction... Oh no it wasn't... Oh yes it was... etc...

One crucial point to remember with proofs is that you have to justify every step of your working. Make sure that you've got a mathematical rule or principle to back up each bit of the proof — you can't take anything for granted.

Proof

This page is for AQA C3 & C4, Edexcel C3, OCR C3 & C4 and OCR MEI C3

And now, part two of our bumper-pack double-bill super-sized prooforama...

Proof by Exhaustion

In proof by exhaustion you break things down into two or more cases. You have to make sure that your cases cover all possible situations, then prove separately that the statement is true for each case.

EXAMPLE Prove the following statement: "For any integer x, the value of $f(x) = x^3 + x + 1$ is an odd integer."

To prove the statement, split the situation into two cases:

(i) x is an even number, and (ii) x is an odd number

(i) If x is an even integer, then it can be written as $x = 2n$, for some integer n (this is the definition of an even number).

Substitute $x = 2n$ into the function: $f(2n) = (2n)^3 + 2n + 1 = 8n^3 + 2n + 1 = 2(4n^3 + n) + 1$

n is an integer $\Rightarrow (4n^3 + n)$ is an integer (as the sum or product of any integers are also integers)
$\Rightarrow 2(4n^3 + n)$ is an even integer (because 2× an integer is the definition of an even number)
$\Rightarrow 2(4n^3 + n) + 1$ is an odd integer (as even + odd = odd)

So $f(x)$ is odd when x is even.

You can use the binomial expansion formula on p.15 to help you find these coefficients.

(ii) If x is an odd integer, then it can be written as $x = 2m + 1$, for some integer m.

Substitute $x = 2m + 1$ into the function: $f(2m + 1) = (2m + 1)^3 + 2m + 1 + 1 = (8m^3 + 12m^2 + 6m + 1) + 2m + 1 + 1$
$= 8m^3 + 12m^2 + 8m + 3 = 2(4m^3 + 6m^2 + 4m) + 3$

m is an integer $\Rightarrow (4m^3 + 6m^2 + 4m)$ is an integer
$\Rightarrow 2(4m^3 + 6m^2 + 4m)$ is an even integer
$\Rightarrow 2(4m^3 + 6m^2 + 4m) + 3$ is an odd integer

So $f(x)$ is odd when x is odd.

We have shown that $f(x)$ is odd when x is even and when x is odd. As any integer x must be either odd or even, we have therefore shown that $f(x)$ is odd for any integer x.

Disproof by Counter-example

Disproof by counter-example is the easiest way to show a mathematical statement is false.
All you have to do is find one case where the statement doesn't hold.

EXAMPLE Disprove the following statement:
"For any pair of real numbers x and y, if $x > y$, then $x^2 + x > y^2 + y$."

To disprove the statement, it's enough to find just one example of x and y where $x > y$, but $x^2 + x \leq y^2 + y$.

Let $x = 2$ and $y = -4$.

Then $2 > -4 \Rightarrow x > y$

but $x^2 + x = 2^2 + 2 = 6$ and $y^2 + y = (-4)^2 + (-4) = 12$, so $x^2 + x < y^2 + y$

So when $x = 2$ and $y = -4$, the first part of the statement holds, but the second part of the statement doesn't.

So the statement is not true.

And that's the proof, the whole proof, and nothing but the proof...

When you're trying to disprove something, don't be put off if you can't find a counter-example straight away. Sometimes you have to just try a few different cases until you find one that doesn't work.

Core Section 2 — Practice Questions

Ah, here we are on another of these soothing green pages. Relax... this is your happy place... nothing to worry about here... enjoy this tranquil blue pool of shimmering warm-up questions.

Warm-up Questions

Q1-3 are for AQA C4, Edexcel C3, OCR C4 and OCR MEI C4

1) Simplify the following:

a) $\frac{4x^2 - 25}{6x - 15}$ b) $\frac{2x + 3}{x - 2} \times \frac{4x - 8}{2x^2 - 3x - 9}$ c) $\frac{x^2 - 3x}{x + 1} \div \frac{x}{2}$

2) Write the following as a single fraction:

a) $\frac{x}{2x + 1} + \frac{3}{x^2} + \frac{1}{x}$ b) $\frac{2}{x^2 - 1} - \frac{3x}{x - 1} + \frac{x}{x + 1}$

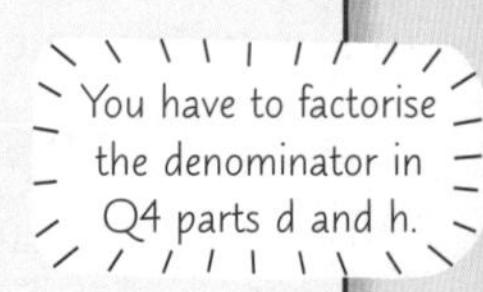

3) Use algebraic long division to divide $x^3 + 2x^2 - x + 19$ by $x + 4$.

Q4-9 are for AQA C4, Edexcel C4, OCR C4 and OCR MEI C4

4) Express the following as partial fractions.

a) $\frac{4x + 5}{(x + 4)(2x - 3)}$ b) $\frac{-7x - 7}{(3x + 1)(x - 2)}$ c) $\frac{x - 18}{(x + 4)(3x - 4)}$ d) $\frac{6 + 4y}{9 - y^2}$

e) $\frac{-11x^2 + 6x + 11}{(2x + 1)(3 - x)(x + 2)}$ f) $\frac{6x^2 + 17x + 5}{x(x + 2)^2}$ g) $\frac{-18x + 14}{(2x - 1)^2(x + 2)}$ h) $\frac{8x^2 - x - 5}{x^3 - x^2}$

5) Express the following as partial fractions — they're all improper, so divide them first.

a) $\frac{2x^2 + 18x + 26}{(x + 2)(x + 4)}$ b) $\frac{3x^2 + 9x + 2}{x(x + 1)}$ c) $\frac{24x^2 - 70x + 53}{(2x - 3)^2}$ d) $\frac{3x^3 - 2x^2 - 2x - 3}{(x + 1)(x - 2)}$

6) Give the binomial expansion of $(1 + 2x)^3$.

7) Find the binomial expansion of each of the following, up to and including the term in x^3:

a) $\frac{1}{(1 + x)^4}$ b) $\frac{1}{(1 - 3x)^3}$ c) $\sqrt{1 - 5x}$

8) a) If the full binomial expansion of $(c + dx)^n$ is an infinite series, what values of x is the expansion valid for?

b) What values of x are the expansions from question 7 valid for?

9) Give the binomial expansions of the following, up to and including the term in x^2. State which values of x each expansion is valid for.

a) $\frac{1}{(3 + 2x)^2}$ b) $\sqrt[3]{8 - x}$

Q10 is for AQA C3 & C4, Edexcel C3, OCR C3 & C4 and OCR MEI C3

10) Disprove the following statement: *"$n^2 - n - 1$ is a prime number, for any integer $n > 2$."*

By now all your cares should have floated away on the algebraic breeze.
Time for a bracing dip in an ice-cool bath of exam questions.

Exam Questions

Q1 is for AQA C4, Edexcel C3, OCR C4 and OCR MEI C4

1 Write $\frac{2x^2 - 9x - 35}{x^2 - 49}$ as a fraction in its simplest form.

(3 marks)

Core Section 2 — Practice Questions

Q2-6 are for AQA C4, Edexcel C4, OCR C4 and OCR MEI C4 — but ignore Q4 for OCR

2 $$f(x) = \frac{1}{\sqrt{(9-4x)}}, \text{ for } |x| < \frac{9}{4}.$$

a) Find the binomial expansion of $f(x)$ up to and including the term in x^3. *(5 marks)*

b) Hence find the first three terms in the expansion of $\frac{2-x}{\sqrt{(9-4x)}}$. *(4 marks)*

3 $$\frac{18x^2 - 15x - 62}{(3x+4)(x-2)} \equiv A + \frac{B}{(3x+4)} + \frac{C}{(x-2)}$$

Find the values of the integers A, B and C. *(4 marks)*

4 $$f(x) = \frac{36x^2 + 3x - 10}{(4+3x)(1-3x)^2}$$

a) Given that $f(x)$ can be expressed in the form

$$f(x) = \frac{A}{(4+3x)} + \frac{B}{(1-3x)} + \frac{C}{(1-3x)^2}$$

find the values of A, B and C. *(4 marks)*

b) Find the binomial expansion of $f(x)$, up to and including the term in x^2. *(6 marks)*

c) Find the range of values of x for which the binomial expansion of $f(x)$ is valid. *(2 marks)*

5 a) Find the binomial expansion of $(16+3x)^{\frac{1}{4}}$, for $|x| < \frac{16}{3}$, up to and including the term in x^2. *(5 marks)*

b) (i) Estimate $\sqrt[4]{12.4}$ by substituting a suitable value of x into your expansion from part (a). Give your answer to 6 decimal places. *(2 marks)*

(ii) What is the percentage error in this estimate? Give your answer to 3 s.f. *(2 marks)*

6 a) Find the binomial expansion of $\left(1 - \frac{4}{3}x\right)^{-\frac{1}{2}}$, up to and including the term in x^3. *(4 marks)*

b) Hence find the values of integer constants a, b and c, such that

$$\sqrt{\frac{27}{(3-4x)}} \approx a + bx + cx^2,$$

and state the range of values of x for which this approximation is valid. *(3 marks)*

Core Section 2 — Practice Questions

Congratulations, you've almost achieved A2 algebra nirvana. Just a few more exam question steps and you'll be there...

Q7 is for AQA C4, Edexcel C3, OCR C4 and OCR MEI C4

7 Write $x^3 + 15x^2 + 43x - 30$ in the form $(Ax^2 + Bx + C)(x + 6) + D$, where A, B, C and D are constants to be found.

(3 marks)

Q8 is for AQA C3 & C4, Edexcel C3, OCR C3 & C4 and OCR MEI C3

8 a) Prove the statement below:
For any integer n, $n^2 - n - 1$ is always odd. *(3 marks)*

b) Hence prove that $(n^2 - n - 2)^3$ is always even. *(3 marks)*

Q9-12 are for AQA C4, Edexcel C4, OCR C4 and OCR MEI C4

9 The algebraic fraction $\frac{-80x^2 + 49x - 9}{(5x - 1)(2 - 4x)}$ can be written in the form $4 + \frac{A}{(5x - 1)} + \frac{B}{(2 - 4x)}$,
where A and B are constants. Find the values of A and B.

(4 marks)

10 a) (i) Show that $\sqrt{\frac{1 + 2x}{1 - 3x}} \approx 1 + \frac{5}{2}x + \frac{35}{8}x^2$. *(5 marks)*

(ii) For what values of x is your expansion valid? *(2 marks)*

b) Using the above expansion with $x = \frac{2}{15}$, show that $\sqrt{19} \approx \frac{127}{30}$. *(2 marks)*

11 a) Express the algebraic fraction $\frac{3x^2 + 12x - 11}{(x + 3)(x - 1)}$ in the form $A + \frac{B + Cx}{(x + 3)(x - 1)}$,
where A, B and C are constants.

(4 marks)

b) Express $\frac{3x^2 + 12x - 11}{(x + 3)(x - 1)}$ as partial fractions. *(3 marks)*

12 a) Find the values of A and B such that $\frac{13x - 17}{(5 - 3x)(2x - 1)} \equiv \frac{A}{(5 - 3x)} + \frac{B}{(2x - 1)}$. *(3 marks)*

b) (i) Find the binomial expansion of $(2x - 1)^{-1}$, up to and including the term in x^2. *(2 marks)*

(ii) Show that $\frac{1}{(5 - 3x)} \approx \frac{1}{5} + \frac{3}{25}x + \frac{9}{125}x^2$, for $|x| < \frac{5}{3}$. *(5 marks)*

c) Using your answers to parts a) and b), find the first three terms of the binomial expansion of $\frac{13x - 17}{(5 - 3x)(2x - 1)}$. *(2 marks)*

Arcsine, Arccosine and Arctangent

This page is for AQA C3, Edexcel C3, OCR C3 and OCR MEI C3

So here we go, forging deep into the mathematical jungle that is trigonometry.
Up first it's inverse trig functions — you need to know what they are, and what their graphs look like...

Arcsin, Arccos and Arctan are the Inverses of Sin, Cos and Tan

In Section 1 you saw that some functions have inverses, which reverse the effect of the function.
The trig functions have inverses too.

ARCSINE is the inverse of sine. You might see it written as arcsin or $\sin^{-1}$.

ARCCOSINE is the inverse of cosine. You might see it written as arccos or $\cos^{-1}$.

ARCTANGENT is the inverse of tangent. You might see it written as arctan or $\tan^{-1}$.

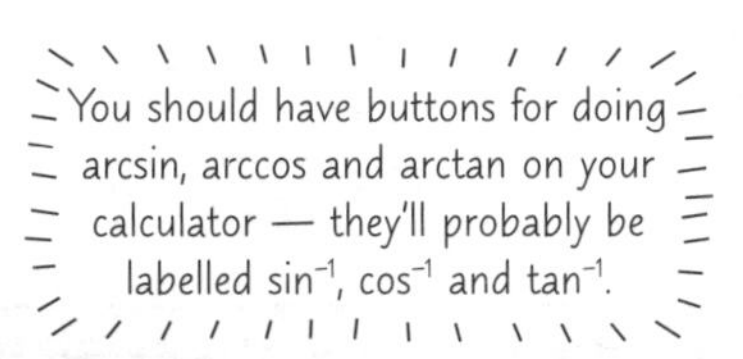

The inverse trig functions reverse the effect of sin, cos and tan.
For example, if sin 30° = 0.5, then arcsin 0.5 = 30°.

To Graph the Inverse Functions you need to Restrict their Domains

1) The functions sine, cosine and tangent are NOT one-to-one mappings (see p.1) — lots of values of x give the same value for $\sin x$, $\cos x$ or $\tan x$. For example: $\cos 0 = \cos 2\pi = \cos 4\pi = 1$, and $\tan 0 = \tan \pi = \tan 2\pi = 0$.
2) If you want the inverses to be functions, you have to restrict their domains. This means that you only plot the graph between certain x values, so that for each x value, you end up with one y value.
3) As the graphs are inverse functions, they're also reflections of the sin, cos and tan functions in the line $y = x$.

ARCSINE

For arcsin, limit the domain of the function to $-1 \leq x \leq 1$.

This limits the range of the function to $-\frac{\pi}{2} \leq \arcsin x \leq \frac{\pi}{2}$.

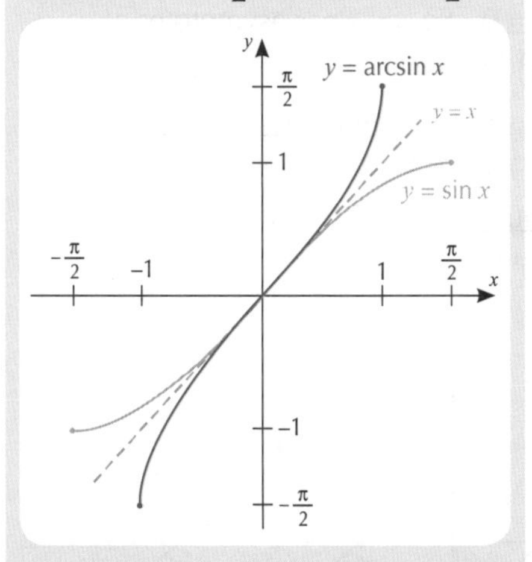

This graph goes through the origin.

The coordinates of its endpoints are $(1, \frac{\pi}{2})$ and $(-1, -\frac{\pi}{2})$.

ARCCOSINE

For arccos, you need to limit the domain of the function to $-1 \leq x \leq 1$.

This limits the range of the function to $0 \leq \arccos x \leq \pi$.

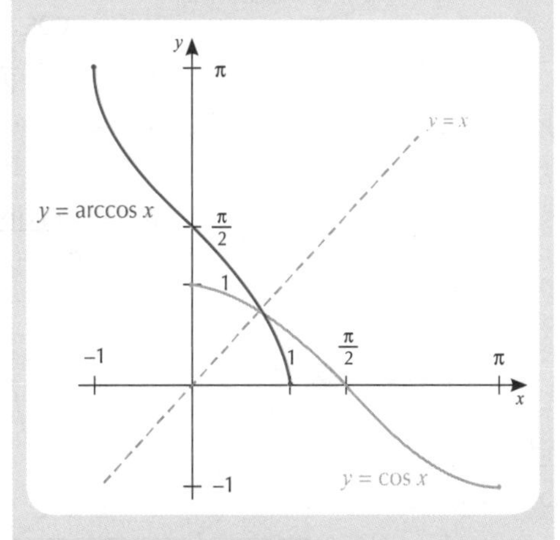

This graph crosses the y-axis at $(0, \frac{\pi}{2})$.

The coordinates of its endpoints are $(-1, \pi)$ and $(1, 0)$.

ARCTANGENT

For arctan, you don't need to limit the domain of the function.

The range of the function is limited to $-\frac{\pi}{2} \leq \arctan x \leq \frac{\pi}{2}$.

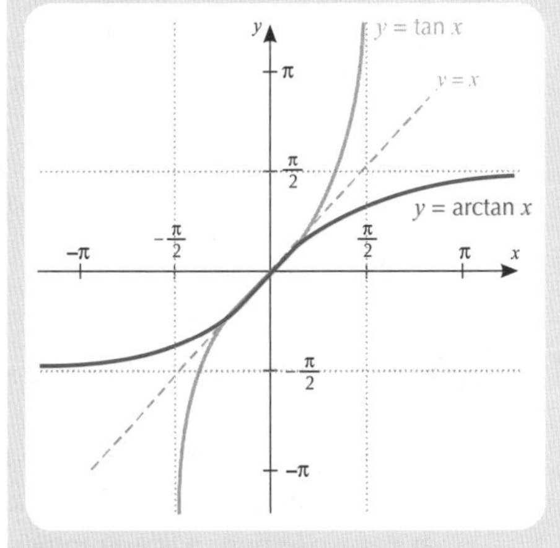

This graph goes through the origin.

It has asymptotes at $y = \frac{\pi}{2}$ and $y = -\frac{\pi}{2}$.

So applying the inverse function reverses everything...

It's really important that you can recognise the graphs of the inverse trig functions — you need to know what shape they are, what restricted domains you need to use to draw them, and any significant points, like where they end or where they cross the axes. You can check the graph by reflecting the curve in the line $y = x$ and seeing if you get the trig function you want.

Secant, Cosecant and Cotangent

This page is for AQA C3, Edexcel C3, OCR C3 and OCR MEI C4

Just when you thought you'd seen all the functions that trigonometry could throw at you, here come three more. These ones are pretty important — they'll come in really handy when you're solving trig equations.

Cosec, Sec and Cot are the Reciprocals of Sin, Cos and Tan

When you take the reciprocal of the three main trig functions, sin, cos and tan, you get three new trig functions — cosecant (or cosec), secant (or sec) and cotangent (or cot).

$$\text{cosec } \theta \equiv \frac{1}{\sin\theta} \qquad \sec \theta \equiv \frac{1}{\cos\theta} \qquad \cot \theta \equiv \frac{1}{\tan\theta}$$

The trick for remembering which is which is to look at the third letter — cosec (1/sin), sec (1/cos) and cot (1/tan).

Since $\tan\theta = \frac{\sin\theta}{\cos\theta}$, you can also think of cot θ as being $\frac{\cos\theta}{\sin\theta}$.

Graphing Cosec, Sec and Cot

COSEC This is the graph of $y = \text{cosec } x$.

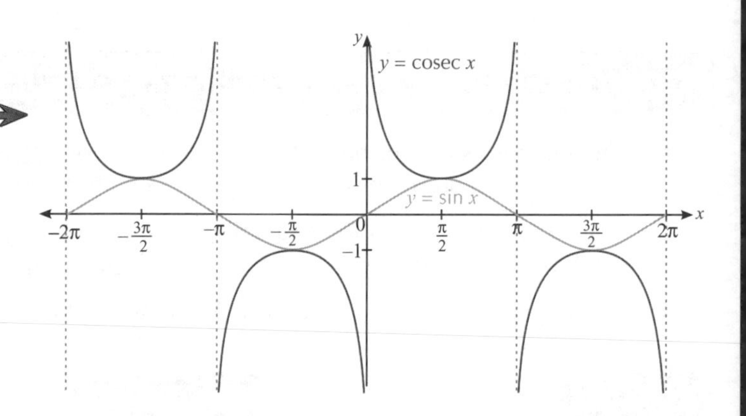

1) Since $\text{cosec } x = \frac{1}{\sin x}$, $y = \text{cosec } x$ is undefined at any point where $\sin x = 0$. So cosec x has asymptotes at $x = n\pi$ (where n is any integer).
2) The graph of cosec x has minimum points at $y = 1$ (wherever the graph of sin x has a maximum).
3) It has maximum points at $y = -1$ (wherever sin x has a minimum).

SEC This is the graph of $y = \sec x$.

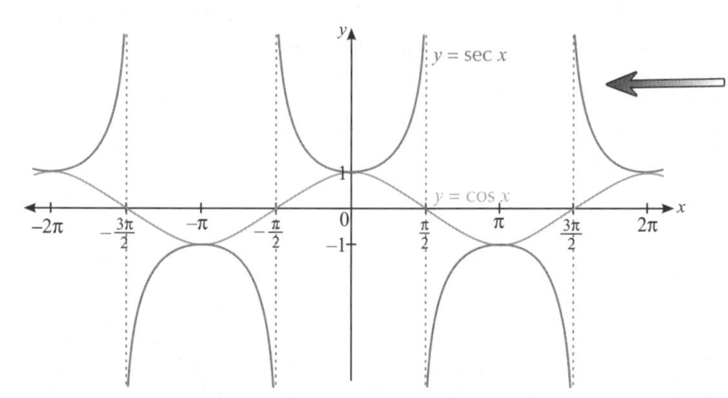

1) As $\sec x = \frac{1}{\cos x}$, $y = \sec x$ is undefined at any point where $\cos x = 0$. So sec x has asymptotes at $x = \left(n\pi + \frac{\pi}{2}\right)$ (where n is any integer).
2) The graph of sec x has minimum points at $y = 1$ (wherever the graph of cos x has a maximum).
3) It has maximum points at $y = -1$ (wherever cos x has a minimum).

COT This is the graph of $y = \cot x$.

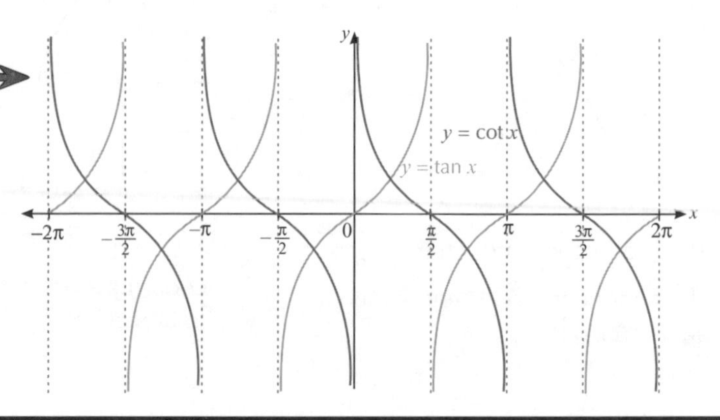

1) Since $\cot x = \frac{1}{\tan x}$, $y = \cot x$ is undefined at any point where $\tan x = 0$. So cot x has asymptotes at $x = n\pi$ (where n is any integer).
2) $y = \cot x$ crosses the x-axis at every place where the graph of tan x has an asymptote — this is any point with the coordinates $\left(\left(n\pi + \frac{\pi}{2}\right), 0\right)$.

Why did I multiply cot x by sin x? Just 'cos...

Remember to look at the third letter to work out which trig function it's the reciprocal of. I'm afraid you do need to be able to sketch the three graphs from memory. Someone in examiner world clearly has a bit of a graph-sketching obsession. You might have to transform a trig graph too — you use the same method as you would for other graphs (see p.7).

Using Trigonometric Identities

This page is for AQA C3, Edexcel C3, OCR C3 and OCR MEI C4

Ahh, trig identities. More useful than a monkey wrench, and more fun than rice pudding. Probably.

Learn these *Three Trig Identities*

Hopefully you remember using this handy little trig identity before:

Identity 1: $\cos^2\theta + \sin^2\theta \equiv 1$

The ≡ sign tells you that this is true for all values of θ, rather than just certain values.

You can use it to produce a couple of other identities that you need to know about...

Identity 2: $\sec^2\theta \equiv 1 + \tan^2\theta$

To get this, you just take everything in Identity 1, and divide it by $\cos^2\theta$:

$$\frac{\cos^2\theta}{\cos^2\theta} + \frac{\sin^2\theta}{\cos^2\theta} \equiv \frac{1}{\cos^2\theta}$$
$$1 + \tan^2\theta \equiv \sec^2\theta$$

Remember that $\cos^2\theta = (\cos\theta)^2$.

Identity 3: $\text{cosec}^2\theta \equiv 1 + \cot^2\theta$

You get this one by dividing everything in Identity 1 by $\sin^2\theta$:

$$\frac{\cos^2\theta}{\sin^2\theta} + \frac{\sin^2\theta}{\sin^2\theta} \equiv \frac{1}{\sin^2\theta}$$
$$\cot^2\theta + 1 \equiv \text{cosec}^2\theta$$

Use the *Trig Identities* to *Simplify Equations*...

You can use identities to get rid of any trig functions that are making an equation difficult to solve.

EXAMPLE Solve the equation $\cot^2 x + 5 = 4\,\text{cosec}\,x$ in the interval $0° \leq x \leq 360°$.

You can't solve this while it has both cot and cosec in it, so use Identity 3 to swap $\cot^2 x$ for $\text{cosec}^2 x - 1$.

$\text{cosec}^2 x - 1 + 5 = 4\,\text{cosec}\,x$

Now rearranging the equation gives: $\text{cosec}^2 x + 4 = 4\,\text{cosec}\,x \Rightarrow \text{cosec}^2 x - 4\,\text{cosec}\,x + 4 = 0$

So you've got a quadratic in cosec x — factorise it like you would any other quadratic equation.

$\text{cosec}^2 x - 4\,\text{cosec}\,x + 4 = 0$
$(\text{cosec}\,x - 2)(\text{cosec}\,x - 2) = 0$

If it helps, think of this as $y^2 - 4y + 4 = 0$. Factorise it, and then replace the y with cosec x.

One of the brackets must be equal to zero — here they're both the same, so you only get one equation:

$(\text{cosec}\,x - 2) = 0 \Rightarrow \text{cosec}\,x = 2$

Now you can convert this into sin x, and solve it easily:

$\text{cosec}\,x = 2 \Rightarrow \sin x = \frac{1}{2}$
$x = 30°$ or $x = 150°$

To find the other values of x, draw a quick sketch of the sin curve: From the graph, you can see that sin x takes the value of ½ twice in the given interval, once at $x = 30°$ and once at $x = 180 - 30 = 150°$.

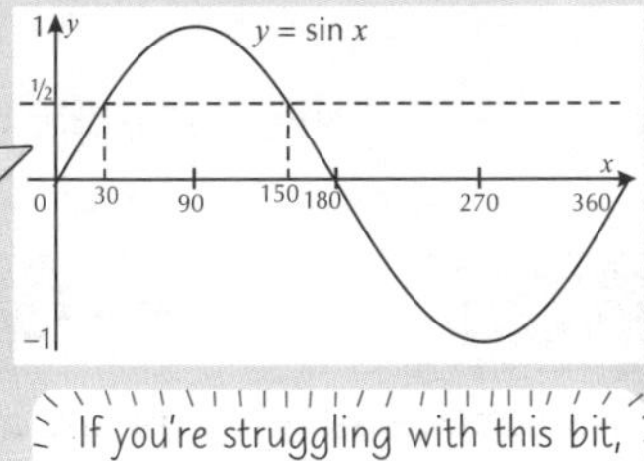

If you're struggling with this bit, have a look back at C2.

...or to *Prove* that two things are *The Same*

You can also use identities to prove that two trig expressions are the same, like this:

EXAMPLE Show that $\frac{\tan^2 x}{\sec x} \equiv \sec x - \cos x$.

You need to take one side of the identity and play about with it until you get the other side. ⟹ Left-hand side: $\frac{\tan^2 x}{\sec x}$.

Try replacing $\tan^2 x$ with $\sec^2 x - 1$: $\equiv \frac{\sec^2 x - 1}{\sec x} \equiv \frac{\sec^2 x}{\sec x} - \frac{1}{\sec x} \equiv \sec x - \cos x$ ...which is the right-hand side.

The Addition Formulas

This page is for AQA C4, Edexcel C3, OCR C3 and OCR MEI C4

You might have noticed that there are quite a lot of formulas lurking in this here trigonometry jungle. Sorry about that.

You can use the *Addition Formulas* to find *Sums of Angles*

You can use the addition formulas to find the sin, cos or tan of the sum or difference of two angles.

When you have an expression like sin $(x + 60°)$ or cos $(n - \frac{\pi}{2})$, you can use these formulas to expand the brackets.

$$\sin(A \pm B) \equiv \sin A \cos B \pm \cos A \sin B$$

$$\cos(A \pm B) \equiv \cos A \cos B \mp \sin A \sin B$$

$$\tan(A \pm B) \equiv \frac{\tan A \pm \tan B}{1 \mp \tan A \tan B}$$

These formulas are given to you on the formula sheet.

Watch out for the $\pm$ and $\mp$ signs in the formulas — especially for cos and tan. If you use the sign on the top on the RHS, you have to use the sign on the top on the left-hand side too — so $\cos(A + B) = \cos A \cos B - \sin A \sin B$.

Use the *Formulas* to find the *Exact Value* of trig expressions

1) You should know the value of sin, cos and tan for common angles (in degrees and radians). These values come from using Pythagoras on right-angled triangles — you did it in C2.
2) In the exam you might be asked to calculate the exact value of sin, cos or tan for another angle using your knowledge of those angles and the addition formulas.
3) Find a pair of angles from the table which add or subtract to give the angle you're after. Then plug them into the addition formula, and work it through.

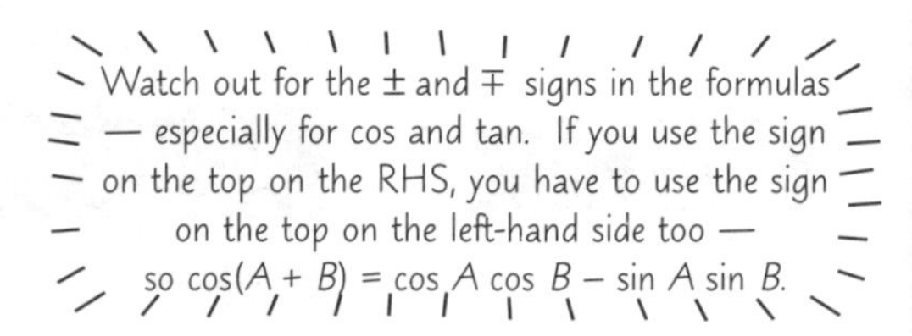

	0°	30°	45°	60°	90°
	0	$\frac{\pi}{6}$	$\frac{\pi}{4}$	$\frac{\pi}{3}$	$\frac{\pi}{2}$
sin	0	$\frac{1}{2}$	$\frac{1}{\sqrt{2}}$	$\frac{\sqrt{3}}{2}$	1
cos	1	$\frac{\sqrt{3}}{2}$	$\frac{1}{\sqrt{2}}$	$\frac{1}{2}$	0
tan	0	$\frac{1}{\sqrt{3}}$	1	$\sqrt{3}$	n/a

EXAMPLE Using the addition formula for tangent, show that $\tan 15° = 2 - \sqrt{3}$.

Pick two angles that add or subtract to give 15°, and put them into the tan addition formula. It's easiest to use tan 60° and tan 45° here, since neither of them are fractions.

$$\tan 15° = \tan(60° - 45°) = \frac{\tan 60° - \tan 45°}{1 + \tan 60° \tan 45°}$$

Using $\tan(A - B) = \frac{\tan A - \tan B}{1 + \tan A \tan B}$

Substitute the values for tan 60° and tan 45° into the equation:

$$= \frac{\sqrt{3} - 1}{1 + (\sqrt{3} \times 1)} = \frac{\sqrt{3} - 1}{\sqrt{3} + 1}$$

Now rationalise the denominator of the fraction to get rid of the $\sqrt{3}$.

$$\frac{\sqrt{3} - 1}{\sqrt{3} + 1} \times \frac{\sqrt{3} - 1}{\sqrt{3} - 1} = \frac{3 - 2\sqrt{3} + 1}{3 - \sqrt{3} + \sqrt{3} - 1}$$

If you can't remember how to rationalise the denominator have a peek at your C1 notes.

Simplify the expression... $= \frac{4 - 2\sqrt{3}}{2} = 2 - \sqrt{3}$ ...and there's the right-hand side.

You can use these formulas to *Prove Identities* too

You might be asked to use the addition formulas to prove an identity. All you need to do is put the numbers and variables from the left-hand side into the addition formulas and simplify until you get the expression you're after.

EXAMPLE Prove that $\cos(a + 60°) + \sin(a + 30°) \equiv \cos a$

Put the numbers from the question into the addition formulas:

Be careful with the + and – signs here.

$$\cos(a + 60°) + \sin(a + 30°) \equiv (\cos a \cos 60° - \sin a \sin 60°) + (\sin a \cos 30° + \cos a \sin 30°)$$

Now substitute in any sin and cos values that you know...

$$= \frac{1}{2}\cos a - \frac{\sqrt{3}}{2}\sin a + \frac{\sqrt{3}}{2}\sin a + \frac{1}{2}\cos a$$

..and simplify:

$$= \frac{1}{2}\cos a + \frac{1}{2}\cos a = \cos a$$

This section's got more identities than Clark Kent...

I was devastated when my secret identity was revealed — I'd been masquerading as a mysterious caped criminal mastermind with an army of minions and a hidden underground lair. It was great fun, but I had to give it all up and write about trig.

The Double Angle Formulas

This page is for AQA C4, Edexcel C3, OCR C3 and OCR MEI C4

Whenever you see a trig expression with an even multiple of x in it, like sin $2x$, you can use one of the double angle formulas to prune it back to an expression just in terms of a single x.

There's a **Double Angle Formula** for **Each Trig Function**

Double angle formulas are just a slightly different kind of identity. They're called "double angle" formulas because they turn any tricky $2x$ type terms in trig equations back into plain x terms.

You need to know the double angle formulas for sin, cos and tan:

$$\sin 2A = 2\sin A\cos A$$

You can use the identity $\cos^2 A + \sin^2 A \equiv 1$ to get the other versions of the cos $2A$ formula.

$$\cos 2A \equiv \cos^2 A - \sin^2 A$$
or $$\equiv 2\cos^2 A - 1$$
or $$\equiv 1 - 2\sin^2 A$$

$$\tan 2A \equiv \frac{2\tan A}{1 - \tan^2 A}$$

You get these formulas by writing $2A$ as $A + A$ and using the addition formulas from the previous page.

Use the **Double Angle Formulas** to **Simplify** and **Solve Equations**

If an equation has a mixture of $\sin x$ and $\sin 2x$ terms in it, there's not much that you can do with it. So that you can simplify it, and then solve it, you have to use one of the double angle formulas.

EXAMPLE Solve the equation $\cos 2x - 5\cos x = 2$ in the interval $0 \le x \le 2\pi$.

First use the double angle formula $\cos 2A \equiv 2\cos^2 A - 1$ to get rid of $\cos 2x$ (use this version so that you don't end up with a mix of sin and cos terms).

$2\cos^2 x - 1 - 5\cos x = 2$

Simplify so you have zero on one side...
...then factorise and solve the quadratic that you've made:

$2\cos^2 x - 5\cos x - 3 = 0$
$(2\cos x + 1)(\cos x - 3) = 0$
So $(2\cos x + 1) = 0$ or $(\cos x - 3) = 0$

The second bracket gives you $\cos x = 3$, which has no solutions since $-1 \le \cos x \le 1$.

So all that's left is to solve the first bracket to find x:

$2\cos x + 1 = 0$
$\cos x = -\frac{1}{2} \Rightarrow x = \frac{2}{3}\pi$ or $x = \frac{4}{3}\pi$.

Sketch the graph of $\cos x$ to find all values of x in the given interval: $\cos x = -\frac{1}{2}$ twice, once at $\frac{2}{3}\pi$ and once at $2\pi - \frac{2}{3}\pi = \frac{4}{3}\pi$.

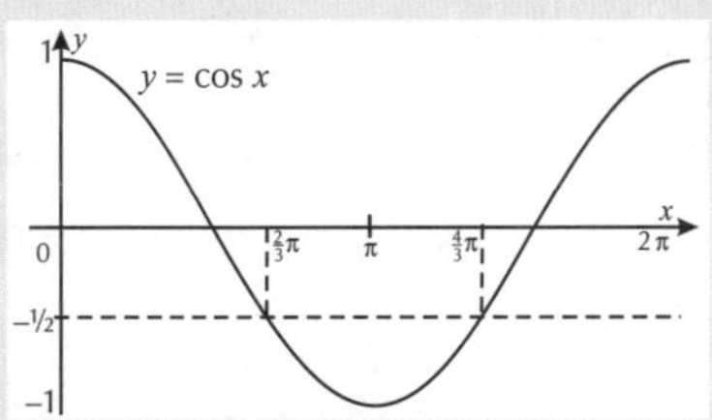

You can use a **Double Angle Formula** even when the x term **Isn't 2x**

Whenever you have an expression that contains any angle that's twice the size of another, you can use the double angle formulas — whether it's $\sin x$ and $\sin 2x$, $\cos 2x$ and $\cos 4x$ or $\tan x$ and $\tan \frac{x}{2}$.

EXAMPLE Prove that $2\cot\frac{x}{2}(1 - \cos^2\frac{x}{2}) \equiv \sin x$

Use the identity $\sin^2\theta + \cos^2\theta \equiv 1$ to replace $1 - \cos^2\frac{x}{2}$ on the left-hand side:

Left-hand side: $2\cot\frac{x}{2}\sin^2\frac{x}{2}$

Now write $\cot\theta$ as $\frac{\cos\theta}{\sin\theta}$:

$2\frac{\cos\frac{x}{2}}{\sin\frac{x}{2}}\sin^2\frac{x}{2} \equiv 2\cos\frac{x}{2}\sin\frac{x}{2}$

Now you can use the sin $2A$ double angle formula to write $\sin x \equiv 2\sin\frac{x}{2}\cos\frac{x}{2}$ (using $A = \frac{x}{2}$).

So using the sin double angle formula... $\equiv \sin x$ ...you get the right-hand side.

You can work out half-angle formulas for cos and tan from the double angle formulas. This example uses the one for sin.

Double the angles, double the fun...

You definitely need to know the double angle formulas off by heart, because they won't be on the exam formula sheet. So it's a case of the old "learn 'em, write 'em out, and keep going until you can do all three perfectly" strategy. And don't forget to be on the lookout for sneaky questions that want you to use a double angle formula but don't contain a "$2x$" bit.

The R Addition Formulas

This page is for AQA C4, Edexcel C3, OCR C3 and OCR MEI C4

A different kind of addition formula this time — one that lets you go from an expanded expression to one with brackets...

Use the R Formulas when you've got a Mix of Sin and Cos

If you're solving an equation that contains both $\sin\theta$ and $\cos\theta$ terms, e.g. $3\sin\theta + 4\cos\theta = 1$, you need to rewrite it so that it only contains one trig function. The formulas that you use to do that are known as the *R* formulas:

One set for sine: $a\sin\theta \pm b\cos\theta \equiv R\sin(\theta \pm \alpha)$

And one set for cosine: $a\cos\theta \pm b\sin\theta \equiv R\cos(\theta \mp \alpha)$

where a and b are positive. Again, you need to be careful with the + and – signs here — see p.28.

Using the R Formulas

1) **You'll start with an identity like $2\sin x + 5\cos x \equiv R\sin(x + \alpha)$, where R and α need to be found.**
2) **First, expand the RHS using the addition formulas (see p.28): $2\sin x + 5\cos x \equiv R\sin x\cos\alpha + R\cos x\sin\alpha$.**
3) **Equate the coefficients of $\sin x$ and $\cos x$. You'll get two equations: ① $R\cos\alpha = 2$ and ② $R\sin\alpha = 5$.**
4) **To find α, divide equation ② by equation ①, then take $\tan^{-1}$ of the result.**
5) **To find R, square equations ① and ② and add them together, then take the square root of the answer.**

This is because $\frac{R\sin\alpha}{R\cos\alpha} = \tan\alpha$.

$(R\sin\alpha)^2 + (R\cos\alpha)^2 \equiv R^2(\sin^2\alpha + \cos^2\alpha) \equiv R^2$ (using the identity $\sin^2\alpha + \cos^2\alpha \equiv 1$).

This method looks a bit scary, but follow the example below through and it should make more sense.

Solve the equation in Stages

You'll almost always be asked to solve equations like this in different stages — first writing out the equation in the form of one of the *R* formulas, then solving it. You might also have to find the maximum or minimum value.

EXAMPLE (Part 1): Express $2\sin x - 3\cos x$ in the form $R\sin(x - \alpha)$, given that $R > 0$ and $0 \le \alpha \le 90°$.

$2\sin x - 3\cos x \equiv R\sin(x - \alpha)$, so expand the RHS to get $2\sin x - 3\cos x \equiv R(\sin x\cos\alpha - \cos x\sin\alpha)$.

Equating coefficients gives the equations $R\cos\alpha = 2$ and $R\sin\alpha = 3$.

Look at the coefficients of $\sin x$ on each side of the equation — on the LHS it's 2 and on the RHS it's $R\cos\alpha$, so $2 = R\cos\alpha$. You find the coefficient of $\cos x$ in the same way.

Solving for α: $\frac{R\sin\alpha}{R\cos\alpha} = \frac{3}{2} = \tan\alpha$

$\alpha = \tan^{-1} 1.5 = 56.31°$

This value fits into the correct range so you can leave it as it is.

Solving for R: $(R\cos\alpha)^2 + (R\sin\alpha)^2 = 2^2 + 3^2 = R^2$

$R = \sqrt{2^2 + 3^2} = \sqrt{13}$

So $2\sin x - 3\cos x = \sqrt{13}\sin(x - 56.31°)$

EXAMPLE (Part 2): Hence solve $2\sin x - 3\cos x = 1$ in the interval $0 \le x \le 360°$.

If $2\sin x - 3\cos x = 1$, that means $\sqrt{13}\sin(x - 56.31°) = 1$, so $\sin(x - 56.31°) = \frac{1}{\sqrt{13}}$.

$0 \le x \le 360°$, so $-56.31° \le x - 56.31° \le 303.69°$.

Careful — you're looking for solutions between –56.31° and 303.69° here.

Solve the equation using arcsin:

$x - 56.31° = \sin^{-1}\left(\frac{1}{\sqrt{13}}\right) = 16.10°$ or $180 - 16.10 = 163.90°$.

So $x = 16.10 + 56.31 = 72.4°$ or $x = 163.90 + 56.31 = 220.2°$

EXAMPLE (Part 3): What are the max and min values of $2\sin x - 3\cos x$?

The maximum and minimum values of sin (and cos) are ±1, so the maximum and minimum values of $R\sin(x - \alpha)$ are $\pm R$.

As $2\sin x - 3\cos x = \sqrt{13}\sin(x - 56.31°)$, $R = \sqrt{13}$, so the maximum and minimum values are $\pm\sqrt{13}$.

A pirate's favourite trigonometry formula...

The *R* formulas might look a bit scary, but they're OK really — just do lots of examples until you're happy with the method. Careful with the adjusting the interval bit that came up in part 2 above — it's pretty fiddly and easy to get muddled over.

More Trigonometry Stuff

This page is for AQA C4, Edexcel C3, OCR C3 and OCR MEI C4

And here we have the final trig page... a collection of random bits that didn't really fit on the other pages. That's one of the scary things about trig questions — you never know what you're going to get.

The Factor Formulas come from the Addition Formulas

As if there weren't enough trig formulas already, here come a few more. Don't worry though — these ones are given to you on the exam formula sheet so you don't need to learn them off by heart.

$$\sin A + \sin B = 2\sin\left(\frac{A+B}{2}\right)\cos\left(\frac{A-B}{2}\right)$$

$$\sin A - \sin B = 2\cos\left(\frac{A+B}{2}\right)\sin\left(\frac{A-B}{2}\right)$$

$$\cos A + \cos B = 2\cos\left(\frac{A+B}{2}\right)\cos\left(\frac{A-B}{2}\right)$$

$$\cos A - \cos B = -2\sin\left(\frac{A+B}{2}\right)\sin\left(\frac{A-B}{2}\right)$$

These are the factor formulas, and they come from the addition formulas (see below). They come in handy for some integrations — it's a bit tricky to integrate $2\cos 3\theta \cos\theta$, but integrating $\cos 4\theta + \cos 2\theta$ is much easier.

EXAMPLE Use the addition formulas to show that $\cos A + \cos B \equiv 2\cos\left(\frac{A+B}{2}\right)\cos\left(\frac{A-B}{2}\right)$

You can derive the other formulas using the same method.

Use the cos addition formulas: $\cos(x+y) \equiv \cos x \cos y - \sin x \sin y$ and $\cos(x-y) \equiv \cos x \cos y + \sin x \sin y$.

Add them together to get: $\cos(x+y) + \cos(x-y) \equiv \cos x \cos y - \sin x \sin y + \cos x \cos y + \sin x \sin y$
$\equiv 2\cos x \cos y$.

Now substitute in $A = x + y$ and $B = x - y$.

Subtracting these gives $A - B = x + y - (x - y) = 2y$, so $y = \frac{A-B}{2}$.

Adding gives $A + B = x + y + (x - y) = 2x$, so $x = \frac{A+B}{2}$.

So $\cos A + \cos B = 2\cos\left(\frac{A+B}{2}\right)\cos\left(\frac{A-B}{2}\right)$.

You might have to use Different Bits of Trig in the Same Question

Some exam questions might try and catch you out by making you use more than one identity to show that two things are equal...

EXAMPLE Show that $\cos 3\theta \equiv 4\cos^3\theta - 3\cos\theta$.

You have to use both the addition formula and the double angle formulas in this question.

First, write $\cos 3\theta$ as $\cos(2\theta + \theta)$, then you can use the cos addition formula:
$\cos(3\theta) \equiv \cos(2\theta + \theta) \equiv \cos 2\theta \cos\theta - \sin 2\theta \sin\theta$.

Now you can use the cos and sin double angle formulas to get rid of the 2θ:
$\cos 2\theta \cos\theta - \sin 2\theta \sin\theta \equiv (2\cos^2\theta - 1)\cos\theta - (2\sin\theta\cos\theta)\sin\theta$

$\equiv 2\cos^3\theta - \cos\theta - 2\sin^2\theta\cos\theta \equiv 2\cos^3\theta - \cos\theta - 2(1 - \cos^2\theta)\cos\theta$

$\equiv 2\cos^3\theta - \cos\theta - 2\cos\theta + 2\cos^3\theta \equiv 4\cos^3\theta - 3\cos\theta$.

This uses the identity $\sin^2\theta + \cos^2\theta \equiv 1$ in the form $\sin^2\theta \equiv 1 - \cos^2\theta$.

...or even drag up trig knowledge from C2 or even GCSE. This question looks short and sweet, but it's actually pretty nasty — you need to know a sneaky conversion between sin and cos.

EXAMPLE If $y = \arcsin x$ for $-1 \le x \le 1$ and $-\frac{\pi}{2} \le y \le \frac{\pi}{2}$, show that $\arccos x = \frac{\pi}{2} - y$.

$y = \arcsin x$, so $x = \sin y$ (as arcsin is the inverse of sin).

Now the next bit isn't obvious — you need to use an identity to switch from sin to cos. This gives... $x = \cos(\frac{\pi}{2} - y)$.

Now, taking inverses gives $\arccos x = \arccos(\cos(\frac{\pi}{2} - y))$, so $\arccos x = \frac{\pi}{2} - y$.

Converting Sin to Cos (and back):
$\sin t \equiv \cos(\frac{\pi}{2} - t)$
and $\cos t \equiv \sin(\frac{\pi}{2} - t)$.
Remember sin is just cos shifted by $\frac{\pi}{2}$ and vice versa.

Trig is like a box of chocolates...

You'll be pleased to know that you've seen all the trig formulas you need for Core. I know there are about 1000 of them (N.B. exaggerations like this may lose you marks in the exam), but any of them could pop up. Examiners particularly like it when you have to use one identity or formula to prove or derive another, so get practising. Then have a nice cup of tea.

Core Section 3 — Practice Questions

There are a lot of formulas in this section — try writing them all out and sticking them somewhere so you can learn them. The best way to get to grips with them is to practise using them — so here are some questions for you to have a go at.

Warm-up Questions

Q1-2 are for AQA C3, Edexcel C3, OCR C3 and OCR MEI C3

1) Using your vast knowledge of trig values for common angles, evaluate these (in radians, between 0 and $\frac{\pi}{2}$):
   a) $\sin^{-1}\frac{1}{\sqrt{2}}$ b) $\cos^{-1}0$ c) $\tan^{-1}\sqrt{3}$
2) Sketch the graphs of arcsine, arccosine and arctangent. Make sure you show their domains and ranges.

Q3-6 are for AQA C3, Edexcel C3, OCR C3 and OCR MEI C4

3) For $\theta = 30°$, find the exact values of:
   a) $\operatorname{cosec}\theta$ b) $\sec\theta$ c) $\cot\theta$
4) Sketch the graphs of cosecant, secant and cotangent for $-2\pi \leq x \leq 2\pi$.
5) Use the identity $\cos^2\theta + \sin^2\theta \equiv 1$ to produce the identity $\sec^2\theta \equiv 1 + \tan^2\theta$.
6) Use the trig identities to show that $\cot^2\theta + \sin^2\theta \equiv \operatorname{cosec}^2\theta - \cos^2\theta$.

Q7-14 are for AQA C4, Edexcel C3, OCR C3 and OCR MEI C4

7) Using the addition formula for cos, find the exact value of $\cos\frac{\pi}{12}$.
8) Find $\sin(A + B)$, given that $\sin A = \frac{4}{5}$ and $\sin B = \frac{7}{25}$.
   You might find these triangles useful:

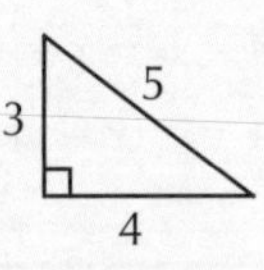

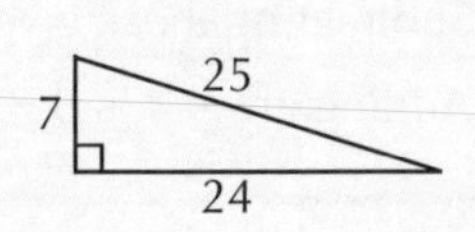

9) State the three different versions of the double angle formula for cos.
10) Use the double angle formula to solve the equation: $\sin 2\theta = -\sqrt{3}\sin\theta,\ 0 \leq \theta \leq 360°$.
11) Which two R formulas could you use to write $a\cos\theta + b\sin\theta$ ($a, b > 0$) in terms of just sin or just cos?
12) Write $5\sin\theta - 6\cos\theta$ in the form $R\sin(\theta - \alpha)$, where $R > 0$ and $0 \leq \alpha \leq 90°$.
13) Use the addition formulas to show that $\sin A - \sin B = 2\cos\left(\frac{A+B}{2}\right)\sin\left(\frac{A-B}{2}\right)$.
14) Show that $\frac{\cos\theta}{\sin\theta} + \frac{\sin\theta}{\cos\theta} \equiv 2\operatorname{cosec}2\theta$.

Here is a selection of the finest trigonometry exam questions available, matured for 21 days and served with a delicious peppercorn sauce.

Exam Questions

Q1-3 are for AQA C3, Edexcel C3, OCR C3 and OCR MEI C4

1 a) Sketch the graph of $y = \operatorname{cosec} x$ for $-\pi \leq x \leq \pi$. *(3 marks)*

Don't forget to put your calculator in RAD mode when you're using radians (and DEG mode when you're using degrees)...

b) Solve the equation $\operatorname{cosec} x = \frac{5}{4}$ for $-\pi \leq x \leq \pi$.
Give your answers correct to 3 significant figures. *(3 marks)*

c) Solve the equation $\operatorname{cosec} x = 3\sec x$ for $-\pi \leq x \leq \pi$.
Give your answers correct to 3 significant figures. *(3 marks)*

Core Section 3 — Practice Questions

Take a deep breath and get ready to dive in again — here come some more lovely trig questions...

2 a) Show that $\frac{2\sin x}{1-\cos x} - \frac{2\cos x}{\sin x} \equiv 2\operatorname{cosec} x$ *(4 marks)*

b) Use this result to find all the solutions for which

$$\frac{2\sin x}{1-\cos x} - \frac{2\cos x}{\sin x} = 4 \qquad 0 < x < 2\pi.$$

(3 marks)

3 a) (i) Using an appropriate identity, show that $3\tan^2\theta - 2\sec\theta = 5$ can be written as $3\sec^2\theta - 2\sec\theta - 8 = 0$. *(2 marks)*

(ii) Hence or otherwise show that $\cos\theta = -\frac{3}{4}$ or $\cos\theta = \frac{1}{2}$. *(3 marks)*

b) Use your results from part a) above to solve the equation $3\tan^2 2x - 2\sec 2x = 5$ for $0 \leq x \leq 180°$. Give your answers to 2 decimal places. *(3 marks)*

Q4 is for AQA C3, Edexcel C3, OCR C3 and OCR MEI C3

4 **Figure 1** shows the graph of $y = \arccos x$, where y is in radians. A and B are the end points of the graph.

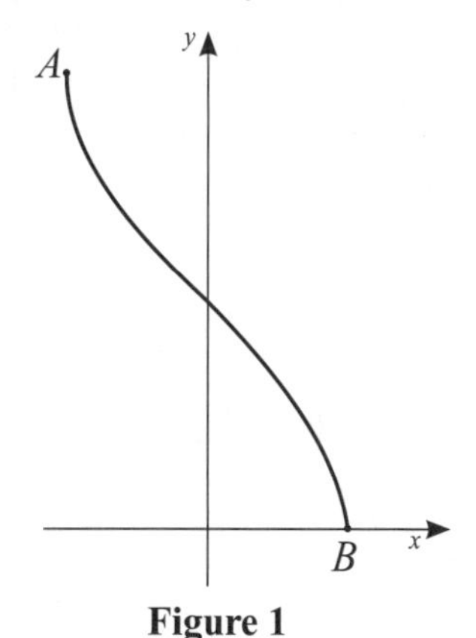

Figure 1

a) Write down the coordinates of A and B. *(2 marks)*

b) Express x in terms of y. *(1 mark)*

c) Solve, to 3 significant figures, the equation $\arccos x = 2$ for the interval shown on the graph. *(2 marks)*

Q5-7 are for AQA C4, Edexcel C3, OCR C3 and OCR MEI C4

5 a) Write $9\sin\theta + 12\cos\theta$ in the form $R\sin(\theta + \alpha)$, where $R > 0$ and $0 \leq \alpha \leq \frac{\pi}{2}$. *(3 marks)*

b) Using the result from part (a) solve $9\sin\theta + 12\cos\theta = 3$, giving all solutions for θ in the range $0 \leq \theta \leq 2\pi$. *(5 marks)*

6 Using the double angle and addition identities for sin and cos, find an expression for $\sin 3x$ in terms of $\sin x$ only. *(4 marks)*

7 a) Write $5\cos\theta + 12\sin\theta$ in the form $R\cos(\theta - \alpha)$, where $R > 0$ and $0 \leq \alpha \leq 90°$. *(4 marks)*

b) Hence solve $5\cos\theta + 12\sin\theta = 2$ for $0 \leq \theta \leq 360°$, giving your answers to 2 decimal places. *(5 marks)*

c) Use your results from part a) above to find the minimum value of $(5\cos\theta + 12\sin\theta)^3$. *(2 marks)*

e^x, ln x and Graphs

This page is for AQA C3, Edexcel C3, OCR C3 and OCR MEI C3

This section is useful 'cos lots of 'real' things increase (or decrease) exponentially — student debts, horrible diseases... We'll start off with a quick recap of some things from C2, then I'll introduce you to some very special functions...

Graphs of $y = a^x$ and $y = a^{-x}$ show Exponential Growth and Decay

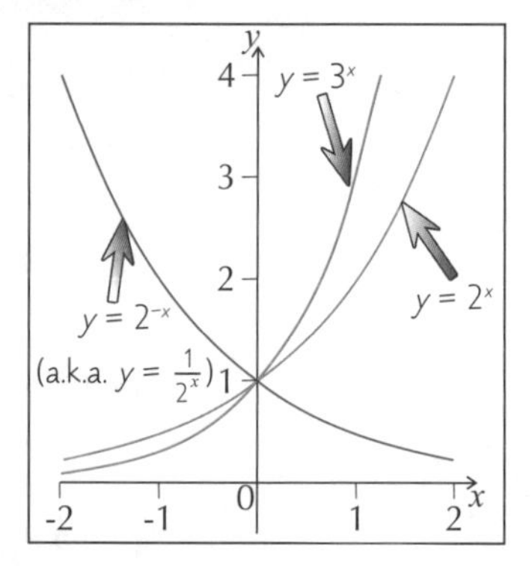

You should be familiar with these graphs from C2.
The main feature of exponential growth / decay is that the rate of increase / decrease of the function is proportional to the function itself.
So if we plotted the gradient of $y = a^x$, it would have the same shape as $y = a^x$.

The main points to remember for $y = a^x$ functions ($a > 0$) are:

1) As $x \to \infty$, $y \to \infty$ (and the gradient also $\to \infty$).
2) As $x \to -\infty$, $y \to 0$ (which means that a^x is always positive).
3) When $x = 0$, $y = 1$ (so they all pass through (0, 1) on the y-axis).

$\to$ means 'tends to'.

The Gradient of the Exponential Function $y = e^x$ is e^x

There is a value of 'a' for which the gradient of $y = a^x$ is exactly the same as a^x. That value is known as e, an irrational number around 2.7183 (it's stored in your calculator just like π). Because e is just a number, the graph of $y = e^x$ has all the properties of $y = a^x$...

1) $y = e^x$ cuts the y-axis at (0, 1).
2) As $x \to \infty$, $e^x \to \infty$ and as $x \to -\infty$, $e^x \to 0$.
3) $y = e^x$ does not exist for $y \leq 0$ (i.e. e^x can't be zero or –ve).

The disturbingly interesting fact that e^x doesn't change when you differentiate is used lots in the differentiation section — see p.46.

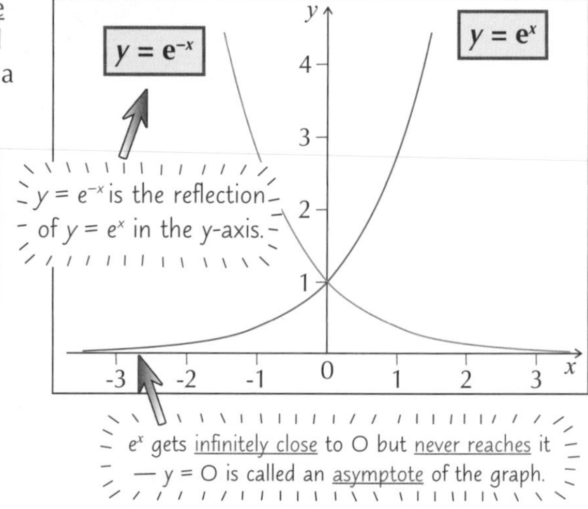

e^x gets infinitely close to 0 but never reaches it — $y = 0$ is called an asymptote of the graph.

ln x is the Inverse Function of e^x

$y = \ln x$ has an asymptote at $x = 0$.

ln x (also known as $\log_e x$, or 'natural log'*) is the inverse function of e^x (see p.3):

1) $y = \ln x$ is the reflection of $y = e^x$ in the line $y = x$.
2) It cuts the x-axis at (1, 0) (so $\ln 1 = 0$).
3) As $x \to \infty$, $\ln x \to \infty$ (but 'slowly'), and as $x \to 0$, $\ln x \to -\infty$.
4) $\ln x$ does not exist for $x \leq 0$ (i.e. x can't be zero or negative).

Because ln x is a logarithmic function and the inverse of e^x, we get these juicy formulas and log laws...

These formulas are extremely useful for dealing with equations containing 'e^x's or 'ln x's, as you'll see on the next page...

'Log laws' for ln x

$\ln x + \ln y = \ln (xy)$

$\ln x - \ln y = \ln \left(\frac{x}{y}\right)$

$\ln x^k = k \ln x$

These are the same old log laws you saw in C2, applied to ln x.

*Certified organic

'e' is for exponential, but also for easy exam questions — no excuses...

When it comes to logs, I prefer the natural look. Remember the limits of $y = e^x$, $y = e^{-x}$ and $y = \ln x$ from the graphs, and polish up your skills with the log laws from C2, and the rest of the section should be a breeze. Naturally.

Using e^x and ln x — Solving Equations

This page is for AQA C3, Edexcel C3, OCR C3 and OCR MEI C3

Now what makes e^x and ln x so clever is that you can use one to cancel out the other, which comes in very handy for solving equations. You'll need all those fruity formulas from the previous page to get through this one...

Use the *Inverse Functions* and *Log Laws* to *Solve Equations*

EXAMPLES

a) Solve the equation $2\ln x - \ln 2x = 6$, giving your answer as an exact value of x.

1) Use the log laws (see previous page) to simplify $2\ln x - \ln 2x = 6$ into:
$\ln x^2 - \ln 2x = 6 \Rightarrow \ln (x^2 \div 2x) = 6 \Rightarrow \ln (\frac{x}{2}) = 6$.

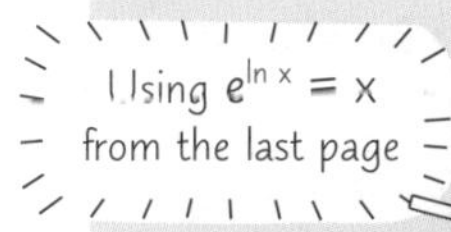

2) Now apply the inverse function e^x to both sides — this will remove the ln $(\frac{x}{2})$:
$e^{\ln(\frac{x}{2})} = e^6 \Rightarrow \frac{x}{2} = e^6 \Rightarrow x = 2e^6$. And since we need an exact value, leave it as that.

b) Find the exact solutions of the equation $e^x + 5e^{-x} = 6$.

1) A big clue here is that you're asked for more than one solution. Think quadratics...
2) Multiply each part of the equation by e^x to get rid of that e^{-x}:
$e^x + 5e^{-x} = 6 \Rightarrow e^{2x} + 5 = 6e^x \Rightarrow e^{2x} - 6e^x + 5 = 0$.
3) It starts to look a bit nicer if you substitute y for e^x: $y^2 - 6y + 5 = 0$.
4) Since we're asked for exact solutions, it will probably factorise:
$(y - 1)(y - 5) = 0 \Rightarrow y = 1$ and $y = 5$.
5) Put e^x back in: $e^x = 1$ and $e^x = 5$.
6) Take 'ln' of both sides to solve: $\ln e^x = \ln 1 \Rightarrow x = \ln 1 = 0$ and $\ln e^x = \ln 5 \Rightarrow x = \ln 5$.

Basic power laws — $(e^x)^2 = e^{2x}$ and $e^{-x} \times e^x = e^0 = 1$.

Using $\ln e^x = x$

Real-Life functions look like $y = e^{ax+b} + c$ and $y = \ln (ax + b)$

You should be familiar with the shape of the bog-standard exponential graphs, but most exponential functions will be transformed in some way. You need to know how the key features of the graph change depending on the function.

EXAMPLES Sketch the graphs of the following functions, labelling any key points and stating the value of 'a':
a) $y = e^{-7x+1} - 5$ ($x \in \mathbb{R}, y > a$) and b) $y = \ln (2x + 4)$ ($x \in \mathbb{R}, x > a$).

$y = e^{-7x+1} - 5$

1) 'Key points' usually means where the graph crosses the axes, i.e. where x and y are 0:
When $x = 0$, $y = e^1 - 5 = -2.28$. When $y = 0$, $e^{-7x+1} = 5 \Rightarrow -7x + 1 = \ln 5 \Rightarrow x = -0.0871$.
2) Next see what happens as x goes to $\pm\infty$ to find any asymptotes:
As $x \to \infty$, $e^{-7x+1} \to 0$, so $y \to -5$. As $x \to -\infty$, $e^{-7x+1} \to \infty$, so $y \to \infty$.
3) Now use this information to sketch out a graph. y can't go below -5, so if $y > a$, $a = -5$.

This tells you the range of values for the function (see p.1).

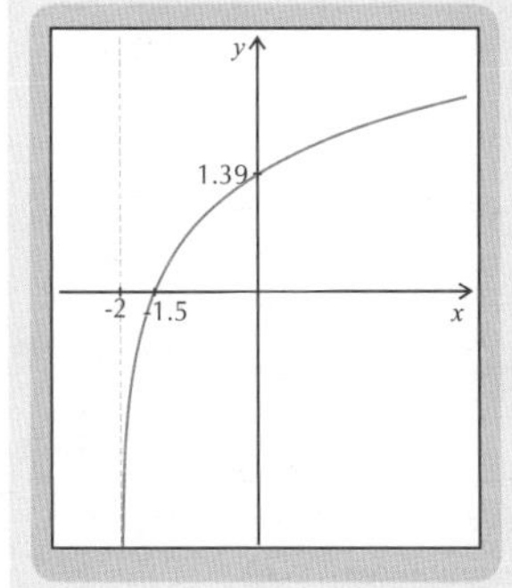

$y = \ln (2x + 4)$

1) First the intercepts: When $x = 0$, $y = \ln 4 = 1.39$. When $y = 0$, $2x + 4 = e^0 = 1 \Rightarrow x = -1.5$.
2) As $x \to \infty$, $y \to \infty$ (gradually).
3) As $x \to -\infty$, y decreases up to the point where $2x + 4 = 0$, at which it can no longer exist (since ln x can only exist for $x > 0$). This gives an asymptote at $2x + 4 = 0$, i.e. $x = -2$.
4) Sketch the graph using this information. x can't go below -2, so if $x > a$, $a = -2$.

This tells you the domain (see p.1).

No problems — only solutions...

All the individual steps to solving these equations are easy — the hard bit is spotting what combination of things to try. A good thing to look for is hidden quadratics, so try and substitute for e^x or ln x to make things look a bit nicer.

Using e^x and ln x — Solving Equations

This page is for AQA C4, Edexcel C3, OCR C3 and OCR MEI C3

This page is all about models. Except they're modelling exponential growth and decay in real-world applications rather than the Chanel Autumn/Winter collection. Sorry.

Use the Exponential Functions to Model real-life Growth and Decay

In the exam you'll usually be given a background story to an exponential equation.
They may then ask you to find some values, work out a missing part of the equation, or even sketch a graph.
There's nothing here you haven't seen before — you just need to know how to deal with all the wordy bits.

EXAMPLE The exponential growth of a colony of bacteria can be modelled by the equation $B = 60e^{0.03t}$, where B is the number of bacteria, and t is the time in hours from the point at which the colony is first monitored ($t \geq 0$). Use the model to predict:

a) the number of bacteria after 4 hours.

You need to find B when $t = 4$, so put the numbers into the equation:

$B = 60 \times e^{(0.03 \times 4)}$
$= 60 \times 1.1274...$
$= 67.6498...$

So B = 67 bacteria.

You shouldn't round up here — there are only 67 whole bacteria, not 68.

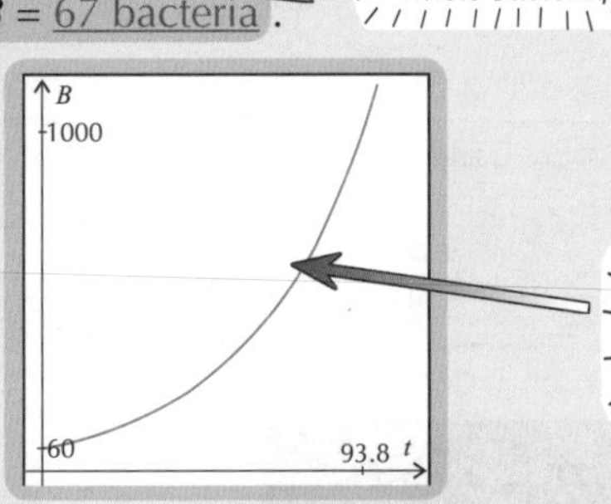

b) the time taken for the colony to grow to 1000.

1) You need to find t when B = 1000, so put the numbers into the equation:
$1000 = 60e^{0.03t}$
$\Rightarrow e^{0.03t} = 1000 \div 60 = 16.6666...$

2) Now take 'ln' of both sides as usual:
$\ln e^{0.03t} = \ln (16.6666...)$
$\Rightarrow 0.03t = 2.8134...$
$\Rightarrow t = 2.8134... \div 0.03 =$ 93.8 hours to 3 s.f.

Even if the question doesn't ask for a sketch of the equation, you may still find it useful to do one to give you an idea of what's going on.

EXAMPLE The concentration (C) of a drug in the bloodstream, t hours after taking an initial dose, decreases exponentially according to $C = Ae^{-kt}$, where k is a constant. If the initial concentration is 0.72, and this halves after 5 hours, find the values of A and k and sketch a graph of C against t.

1) The 'initial concentration' is 0.72 when $t = 0$, so put this information in the equation to find the missing constant A:
$0.72 = A \times e^0 \Rightarrow 0.72 = A \times 1 \Rightarrow A = 0.72$.

2) The question also says that when $t = 5$ hours, C is half of 0.72. So using the value for A found above:
$C = 0.72e^{-kt}$
$0.72 \div 2 = 0.72 \times e^{(-k \times 5)}$
$\Rightarrow 0.36 = 0.72 \times e^{-5k} \Rightarrow 0.36 = \frac{0.72}{e^{5k}} \Rightarrow e^{5k} = \frac{0.72}{0.36} = 2.$

3) Now take 'ln' of both sides to solve:
$\ln e^{5k} = \ln 2 \Rightarrow 5k = \ln 2$
$\Rightarrow k = \ln 2 \div 5 = 0.139$ to 3 s.f.

4) So the equation is $C = 0.72e^{-0.139t}$.
You still need to do a sketch though, so find the intercepts and asymptotes as you did on the last page:
When $t = 0$, $C = 0.72$. As $t \to \infty$, $e^{-0.139t} \to 0$, so C → 0.

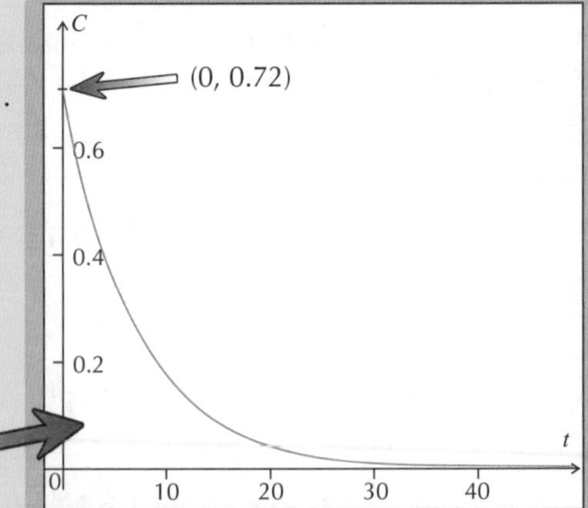

The sketch should make sense for the situation in the question — here t can only be positive as it is the time after an event, so only sketch the graph for $t \geq 0$.

Learn this and watch your knowledge grow exponentially...

For these wordy problems the key is just to extract the relevant information and solve like you did on the last page. The more you practise, the more familiar they'll become — fortunately there's a fair bit of practice on the next two pages.

Core Section 4 — Practice Questions

Well that section was short and sweet, rather like that lovely Richard Hammond.
While it's all fresh in your mind, have a go at these little hamsters...

Warm-up Questions

Q1-4 are for AQA C3, Edexcel C3, OCR C3 and OCR MEI C3

1) Plot the following graphs on the same axes, for $-2 \leq x \leq 2$:
   a) $y = 4e^x$ b) $y = 4e^{-x}$ c) $y = 4 \ln x$ d) $y = \ln 4x$.

2) Find the value of x, to 4 decimal places, when:
   a) $e^{2x} = 6$ b) $\ln (x + 3) = 0.75$ c) $3e^{-4x+1} = 5$ d) $\ln x = \ln 5 - \ln 4$.

3) Solve the following equations, giving your solutions as exact values:
   a) $\ln (2x - 7) + \ln 4 = -3$ b) $2e^{2x} + e^x = 3$.

4) Sketch graphs of the following, labelling key points and asymptotes:
   a) $y = 2 - e^{x+1}$ b) $y = 5e^{0.5x} + 5$ c) $y = \ln (2x) + 1$ d) $y = \ln (x + 5)$

Q5 is for AQA C4, Edexcel C3, OCR C3 and OCR MEI C3

5) The value of a motorbike (£V) varies with age (in t years from new) according to $V = 7500e^{-0.2t}$.
   a) How much did it originally cost?
   b) What is its value after 10 years (to the nearest £)?
   c) After how many years will the motorbike's value have fallen below £500?
   d) Sketch a graph showing how the value of the motorbike varies with age, labelling all key points.

Feeling confident? Thought so.
Let's see how you handle these exam-style problems — they're a wee bit more problematic...

Exam Questions

Q1-4 are for AQA C3, Edexcel C3, OCR C3 and OCR MEI C3

1 a) Given that $6e^x = 3$, find the exact value of x. *(2 marks)*

b) Find the exact solutions to the equation:

$$e^{2x} - 8e^x + 7 = 0.$$

(4 marks)

c) Given that $4 \ln x = 3$, find the exact value of x. *(2 marks)*

d) Solve the equation:

$$\ln x + \frac{24}{\ln x} = 10$$

giving your answers as exact values of x. *(4 marks)*

2 The sketch below shows the function $y = e^{ax} + b$, where a and b are constants.

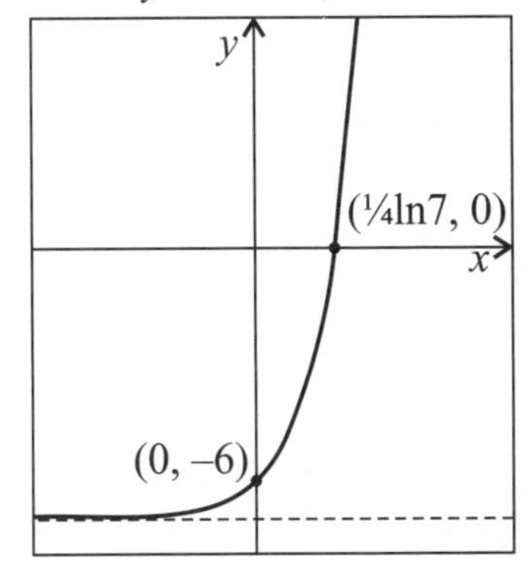

Find the values of a and b, and the equation of the asymptote shown on the sketch. *(5 marks)*

Core Section 4 — Practice Questions

3 Solve the following equations, giving your answers as exact values of x.

a) $2e^x + 18e^{-x} = 20$ *(4 marks)*

b) $2 \ln x - \ln 3 = \ln 12$ *(3 marks)*

4 A curve has the equation $y = \ln(4x - 3)$.

a) The point A with coordinate $(a, 1)$ lies on the curve. Find a to 2 decimal places. *(2 marks)*

b) The curve is only defined for $x > b$. State the value of b. *(2 marks)*

c) Sketch the curve, labelling any important points. *(2 marks)*

Q5-6 are for AQA C4, Edexcel C3, OCR C3 and OCR MEI C3

5 A breed of mink is introduced to a new habitat.
The number of mink, M, after t years in the habitat, is modelled by:

$$M = 74e^{0.6t} \quad (t \geq 0)$$

a) State the number of mink that were introduced to the new habitat originally. *(1 mark)*

b) Predict the number of mink after 3 years in the habitat. *(2 marks)*

c) Predict the number of complete years it would take for the population of mink to exceed 10 000. *(2 marks)*

d) Sketch a graph to show how the mink population varies with time in the new habitat. *(2 marks)*

6 A radioactive substance decays exponentially so that its activity, A, can be modelled by

$$A = Be^{-kt}$$

where t is the time in days, and $t \geq 0$. Some experimental data is shown below.

t	0	5	10
A	50	42	

a) State the value of B. *(1 mark)*

b) Find the value of k, to 3 significant figures. *(2 marks)*

c) Find the missing value from the table, to the nearest whole number. *(2 marks)*

d) The half-life of a substance is the time it takes for the activity to halve.
Find the half-life of this substance, in days. Give your answer to the nearest day. *(3 marks)*

Parametric Equations of Curves

This page is for AQA C4, Edexcel C4, OCR C4 and OCR MEI C4

Parametric equations seem kinda weirdy to start with, but they're actually pretty clever. You can use them to replace one horrifically complicated equation with two relatively normal-looking ones. I bet that's just what you always wanted...

Parametric Equations split up x and y into Separate Equations

1) Normally, graphs in the (x, y) plane are described using a Cartesian equation — a single equation linking x and y.
2) Sometimes, particularly for more complicated graphs, it's easier to have two linked equations, called parametric equations.
3) In parametric equations x and y are each defined separately in terms of a third variable, called a parameter. The parameter is usually either t or θ.

EXAMPLE

This graph is given by the parametric equations $y = t^2 - 1$ and $x = t + 1$:

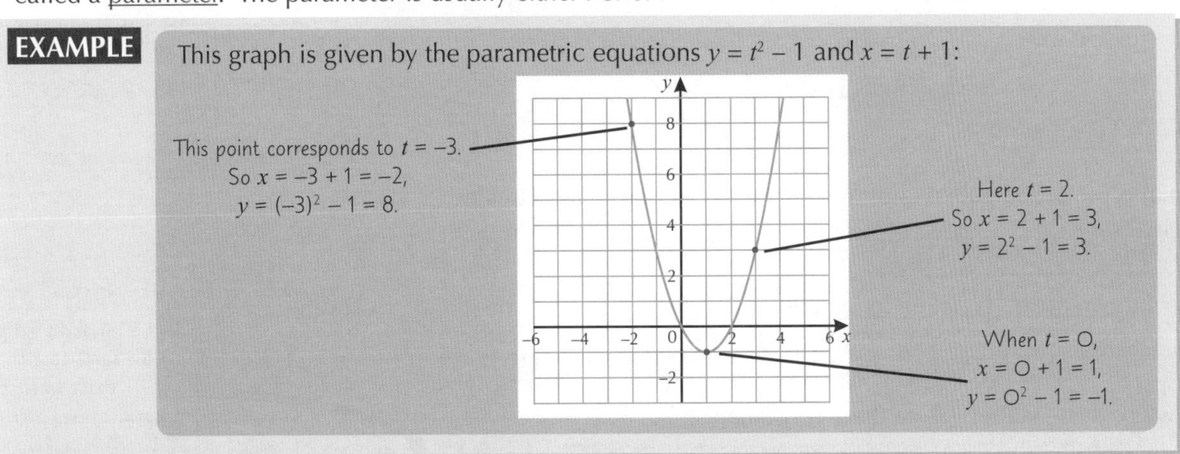

You can use the parametric equations of a graph to find coordinates of points on the graph, and to find the value of the parameter for given x- or y-coordinates.

EXAMPLE

A curve is defined by the parametric equations $y = \frac{1}{3t}$ and $x = 2t - 3$, $t \neq 0$.

a) Find the x- and y- values of the point the curve passes through when $t = 4$.
b) What value of t corresponds to the point where $y = 9$?
c) What is the value of y when $x = -15$?

Nothing to this question — just sub the right values into the right equations and you're away:

a) When $t = 4$, $x = 8 - 3 = 5$, and $y = \frac{1}{12}$

b) $9 = \frac{1}{3t} \Rightarrow t = \frac{1}{27}$

c) $-15 = 2t - 3 \Rightarrow t = -6 \Rightarrow y = -\frac{1}{18}$

Use the equation for x to find t first, then use that value of t in the other equation to find y.

Circles can be given by Parametric Equations too

This bit is just for OCR MEI C4

You saw the Cartesian equations of circles way back at AS-level, but now you get to use their parametric equations.

A circle with centre (0, 0) and radius r is defined by the parametric equations $x = r\cos\theta$ and $y = r\sin\theta$...

...and a circle with centre (a, b) and radius r is defined by the parametric equations $x = r\cos\theta + a$ and $y = r\sin\theta + b$.

EXAMPLES

This is the circle given by the equations $x = 4\cos\theta$ and $y = 4\sin\theta$. It has radius 4 and centre (0, 0).

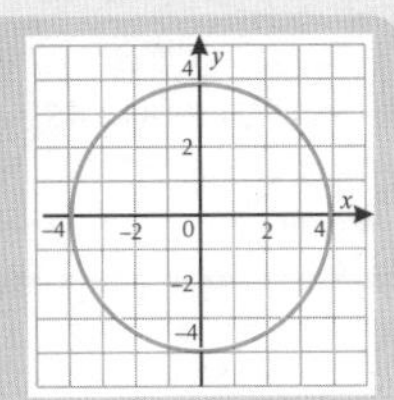

And this is the circle given by the equations $x = 3\cos\theta + 1$ and $y = 3\sin\theta + 2$. It has radius 3 and centre (1, 2).

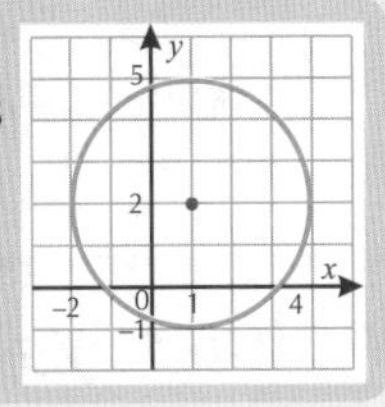

If the values of r are different in each equation (e.g. $x = 2\cos\theta$ and $y = 3\sin\theta$), you'll get an ellipse instead of a circle. For this example, the ellipse will be 4 units wide and 6 units tall.

Using Parametric Equations

This page is for AQA C4, Edexcel C4, OCR C4 and OCR MEI C4

There's plenty of tinkering around with equations to be done in this topic, so get your rearranging hat on. My rearranging hat is a jaunty straw boater.

Use *Parametric Equations* to find where graphs *Intersect*

A lot of parametric equations questions involve identifying points on the curve defined by the equations.

EXAMPLE

The curve shown in this sketch has the parametric equations $y = t^3 - t$ and $x = 4t^2 - 1$.

Find the coordinates of the points where the graph crosses:
a) the x-axis,
b) the y-axis,
c) the line $8y = 3x + 3$.

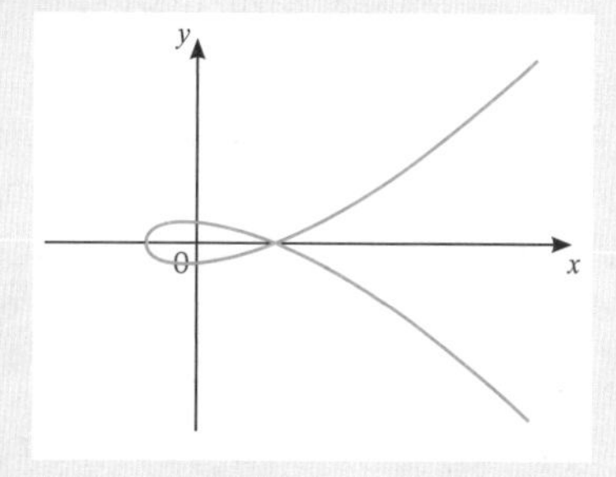

Part a) is pretty straightforward. You've got the y-coordinates already:

a) On the x-axis, $y = 0$.

Use the parametric equation for y to find the values of t where the graph crosses the x-axis:

So $0 = t^3 - t \Rightarrow t(t^2 - 1) = 0 \Rightarrow t(t + 1)(t - 1) = 0 \Rightarrow t = 0, t = -1, t = 1$

$t = -1$ and $t = 1$ give the same coordinates — that's where the curve crosses over itself.

Now use those values to find the x-coordinates:

$t = 0 \Rightarrow x = 4(0)^2 - 1 = -1$ $\quad t = -1 \Rightarrow x = 4(-1)^2 - 1 = 3$ $\quad t = 1 \Rightarrow x = 4(1)^2 - 1 = 3$

So the graph crosses the x-axis at the points $(-1, 0)$ and $(3, 0)$.

The sketch shows there are two points where the graph crosses each axis.

And b) is very similar:

b) On the y-axis, $x = 0$.

So $0 = 4t^2 - 1 \Rightarrow t^2 = \frac{1}{4} \Rightarrow t = \pm\frac{1}{2}$

$t = \frac{1}{2} \Rightarrow y = \left(\frac{1}{2}\right)^3 - \frac{1}{2} = -\frac{3}{8}$ $\qquad t = -\frac{1}{2} \Rightarrow y = \left(-\frac{1}{2}\right)^3 - \left(-\frac{1}{2}\right) = \frac{3}{8}$

So the graph crosses the y-axis at the points $(0, -\frac{3}{8})$ and $(0, \frac{3}{8})$.

Part c) is just a little trickier. First, sub the parametric equations into $8y = 3x + 3$:

c) $8y = 3x + 3 \Rightarrow 8(t^3 - t) = 3(4t^2 - 1) + 3$

Rearrange and factorise to find the values of t you need:

$\Rightarrow 8t^3 - 8t = 12t^2 \Rightarrow 8t^3 - 12t^2 - 8t = 0 \Rightarrow t(2t + 1)(t - 2) = 0 \Rightarrow t = 0, t = -\frac{1}{2}, t = 2$

Go back to the parametric equations to find the x- and y-coordinates:

$t = 0 \Rightarrow x = -1, y = 0$
$t = -\frac{1}{2} \Rightarrow x = 4(\frac{1}{4}) - 1 = 0, y = (-\frac{1}{2})^3 + \frac{1}{2} = \frac{3}{8}$
$t = 2 \Rightarrow x = 4(4) - 1 = 15, y = 2^3 - 2 = 6$

So the graph crosses the line $4y = 3x + 3$ at the points $(-1, 0)$, $(0, \frac{3}{8})$, $(15, 6)$.

You can check the answers by sticking these values back into $8y = 3x + 3$.

y-coordinates? y not...

You quite often get given a sketch of the curve that the parametric equations define. Don't forget that the sketch can be useful for checking your answers — if the curve crosses the x-axis twice, and you've only found one x-coordinate for when $y = 0$, you know something's gone a bit pear-shaped and you should go back and sort it out, sunshine.

Parametric and Cartesian Equations

This page is for AQA C4, Edexcel C4, OCR C4 and OCR MEI C4

If you've been dying for θ to put in another appearance since it first popped up on page 39, then I've got good news. If, on the other hand, you're bored of parametric equations already... I'm sorry.

Rearrange Parametric Equations to get the Cartesian Equation

Some parametric equations can be converted into Cartesian equations. There are two main ways to do this:

To convert Parametric Equations to a Cartesian Equation:

1. **Rearrange one of the equations to make the parameter the subject, then substitute the result into the other equation.**

or

2. **If your equations involve trig functions, use trig identities (see Core Section 3) to eliminate the parameter.**

You can use the first method to combine the parametrics used in the examples on p39:

EXAMPLE Give the Cartesian equations, in the form $y = \mathrm{f}(x)$, of the curves represented by the following pairs of parametric equations:

a) $y = t^2 - 1$ and $x = t + 1$, b) $y = \frac{1}{3t}$ and $x = 2t - 3$, $t \neq 0$.

You want the answer in the form $y = \mathrm{f}(x)$, so leave y alone for now, and rearrange the equation for x to make t the subject:

a) $x = t + 1 \Rightarrow t = x - 1$

Now you can eliminate t from the equation for y:

$$y = t^2 - 1 \quad \Rightarrow y = (x-1)^2 - 1 = x^2 - 2x + 1 - 1$$
$$\Rightarrow y = x^2 - 2x$$

b) $x = 2t - 3 \Rightarrow t = \frac{x+3}{2}$

So $y = \frac{1}{3t} \Rightarrow y = \frac{1}{3\left(\frac{x+3}{2}\right)} \Rightarrow y = \frac{1}{\frac{3(x+3)}{2}} \Rightarrow y = \frac{2}{3x+9}$

If there are Trig Functions... use Trig Identities

Things get a little trickier when the likes of sin and cos decide to put in an appearance:

EXAMPLE A curve has parametric equations

$$x = 1 + \sin\theta, \quad y = 1 - \cos 2\theta$$

Give the Cartesian equation of the curve in the form $y = \mathrm{f}(x)$.

If you try to make θ the subject of these equations, things will just get messy. The trick is to find a way to get both x and y in terms of the same trig function.

You can get $\sin\theta$ into the equation for y using the identity $\cos 2\theta = 1 - 2\sin^2\theta$:

$$y = 1 - \cos 2\theta = 1 - (1 - 2\sin^2\theta) = 2\sin^2\theta$$

Rearranging the equation for x gives:

$\sin\theta = x - 1$, so $y = 2\sin^2\theta$
$\Rightarrow y = 2(x-1)^2 = 2x^2 - 4x + 2$

If one of the parametric equations includes $\cos 2\theta$ or $\sin 2\theta$, that's probably the one you need to substitute — so make sure you know the double angle formulas.

Cartesy peasy, lemon squeezy...

Sometimes you'll get a nasty question where it's really difficult to get the parameter on its own — in that case you might have to do something clever like think about multiplying x and y or dividing y by x. If something like that comes up in an exam, they'll usually give you a hint — but be aware that you might need to think outside the box.

Areas Under Parametric Curves

This page is just for Edexcel C4

Did I ever tell you about the time I spent the night under a parametric curve... or was it a bridge...

Learn the method for **Integrating Parametrics**

1) Normally, to find the area under a graph, you can do a simple integration.
   But if you've got parametric equations, things are more difficult — you can't find $\int y\,dx$ if y isn't written in terms of x.
2) There's a sneaky way to get around this. Suppose your parameter's t, then $\int y\,dx = \int y\frac{dx}{dt}\,dt$
3) Both y and $\frac{dx}{dt}$ are written in terms of t, so you can multiply them together to get an expression you can integrate with respect to t.

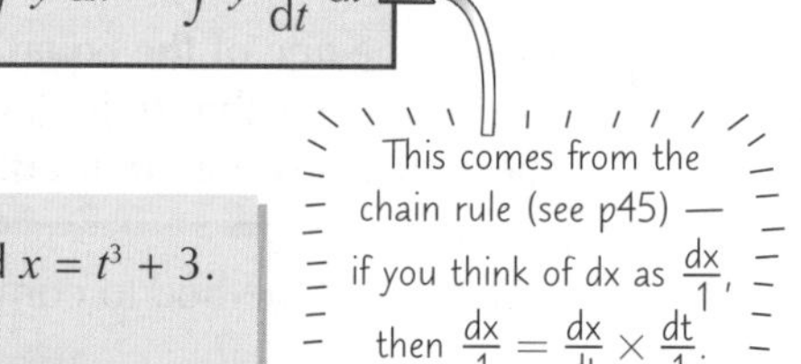

EXAMPLE A curve is defined by the parametric equations $y = t^2 + 2t + 3$ and $x = t^3 + 3$.
Show that $\int y\,dx = \int 3t^4 + 6t^3 + 9t^2\,dt$.

$\frac{dx}{dt} = 3t^2$, so using the formula above:

$$\int y\,dx = \int y\frac{dx}{dt}\,dt = \int (t^2 + 2t + 3)(3t^2)\,dt = \int 3t^4 + 6t^3 + 9t^2\,dt.$$

Remember to Convert the **Limits of Definite Integrals**

With a definite integral, you need to alter the limits as well.

EXAMPLE The shaded region marked A on this sketch is bounded by the lines $y = 0$ and $x = 2$, and by the curve with parametric equations $x = t^2 - 2$ and $y = t^2 - 9t + 20$, $t \geq 0$, which crosses the x-axis at $x = 14$.

Find the area of A.

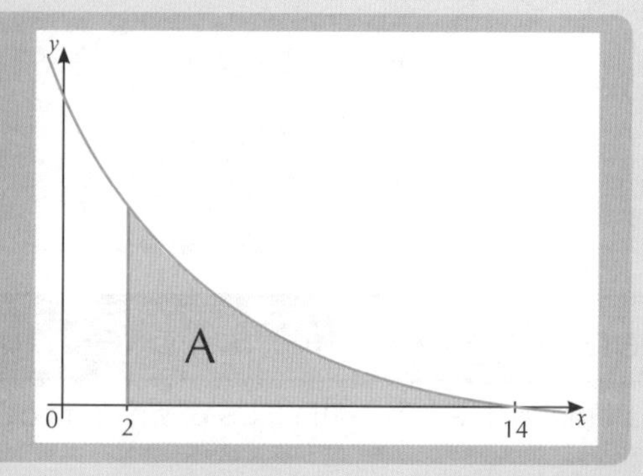

The area of A is $\int_2^{14} y\,dx$. We're going to use $\int y\,dx = \int y\frac{dx}{dt}\,dt$, so first we need to find $\frac{dx}{dt} = \frac{d}{dt}(t^2 - 2) = 2t$

Now we need to do something about those limits.
14 and 2 are the limits for integrating with respect to x.
We need to find the corresponding values of t.

$x = 2 \Rightarrow t^2 - 2 = 2 \Rightarrow t^2 = 4 \Rightarrow t = 2$
$x = 14 \Rightarrow t^2 - 2 = 14 \Rightarrow t^2 = 16 \Rightarrow t = 4$

Now we can integrate:

$$\begin{aligned} A &= \int_2^{14} y\,dx = \int_2^4 y\frac{dx}{dt}\,dt \\ &= \int_2^4 (t^2 - 9t + 20)(2t)\,dt \\ &= \int_2^4 2t^3 - 18t^2 + 40t\,dt \\ &= [\tfrac{1}{2}t^4 - 6t^3 + 20t^2]_2^4 \\ &= (\tfrac{1}{2}(4)^4 - 6(4)^3 + 20(4)^2) - (\tfrac{1}{2}(2)^4 - 6(2)^3 + 20(2)^2) \\ &= 64 - 40 = 24 \end{aligned}$$

You might be having problems with premature integration...

There's lots more integration in Core Sections 9 and 10 — this is just an appetiser. If you can't wait that long, flick ahead and check it out. But I'll know you're only doing it to put off getting to the Practice Questions...

Core Section 5 — Practice Questions

Before it became famous in the world of maths, the word 'parametric' had several other jobs. For example, it once starred as the last name of a Bond villain from the former Yugoslavia. Here are some questions. Enjoy.

Warm-up Questions

Q1-4 are for AQA C4, Edexcel C4, OCR C4 and OCR MEI C4

1) A curve is defined by the parametric equations $y = 2t^2 + t + 4$ and $x = \frac{6-t}{2}$.
   a) Find the values of x and y when $t = 0, 1, 2$ and 3.
   b) What are the values of t when: (i) $x = -7$ (ii) $y = 19$?
   c) Find the Cartesian equation of the curve, in the form $y = f(x)$.

2) The parametric equations of a curve are $x = 2\sin\theta$ and $y = \cos^2\theta + 4$, $-\frac{\pi}{2} \leq \theta \leq \frac{\pi}{2}$.
   a) What are the coordinates of the points where: (i) $\theta = \frac{\pi}{4}$ (ii) $\theta = \frac{\pi}{6}$
   b) What is the Cartesian equation of the curve?
   c) What restrictions are there on the values of x for this curve?

3) The curve C is defined by the parametric equations $x = \frac{\sin\theta}{3}$ and $y = 3 + 2\cos 2\theta$.
   Find the Cartesian equation of C.

4) A curve has parametric equations $y = 4 + \frac{3}{t}$ and $x = t^2 - 1$.
   a) What are the coordinates of the points where this curve crosses
      (i) the y-axis (ii) the line $x + 2y = 14$?
   b) ***[just Edexcel]*** Write the integral $\int y\,dx$ in the form $\int f(t)\,dt$. (You don't need to do the integration.)

Q5 is just for OCR MEI C4

5) For the following circles, write down the radius and the coordinates of the centre:
   a) The circle given by the parametric equations $x = 7\cos\theta$ and $y = 7\sin\theta$.
   b) The circle given by the parametric equations $x = 5\cos\theta + 2$ and $y = 5\sin\theta - 1$.

Former career of the word 'parametric' #2 — stand-in for the word 'hallelujah' in an early draft of Handel's Messiah. Meanwhile, back at the practice questions...

Exam Questions

Q1-3 are for AQA C4, Edexcel C4, OCR C4 and OCR MEI C4

1 The curve C is defined by the parametric equations

$$x = 1 - \tan\theta, \quad y = \frac{1}{2}\sin 2\theta, \quad -\frac{\pi}{2} < \theta < \frac{\pi}{2}.$$

a) P is the point on curve C where $\theta = \frac{\pi}{3}$. Find the exact coordinates of P.
(2 marks)

b) Point Q on curve C has coordinates $(2, -\frac{1}{2})$. Find the value of θ at Q.
(2 marks)

c) Using the identity $\sin 2\theta \equiv \frac{2\tan\theta}{1+\tan^2\theta}$, show that the Cartesian equation of C is $y = \frac{1-x}{x^2 - 2x + 2}$.
(3 marks)

2

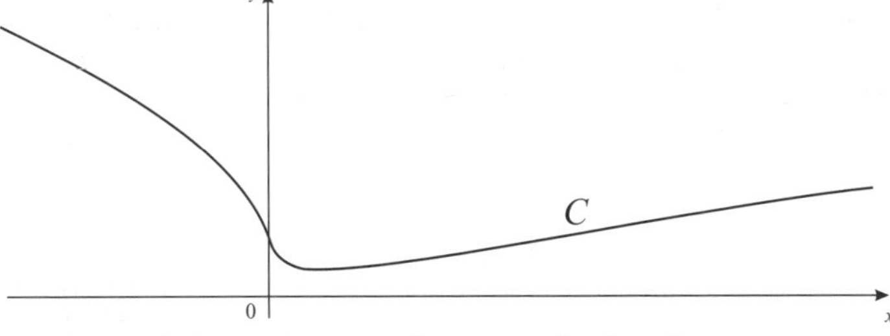

Curve C has parametric equations $x = t^3 + t$, $y = t^2 - 2t + 2$.

a) K is a point on C, and has the coordinates $(a, 1)$. Find the value of a.
(2 marks)

b) The line $8y = x + 6$ passes through C at points K, L and M.
Find the coordinates of L and M, given that the x-coordinate of M is greater than the x-coordinate of L.
(6 marks)

Core Section 5 — Practice Questions

Former career of the word 'parametric' #3 — proposed name for the next ocean to be discovered. Unfortunately it turned out all the oceans have already been discovered. Ooh look, more questions...

Part b) is just for Edexcel C4

3

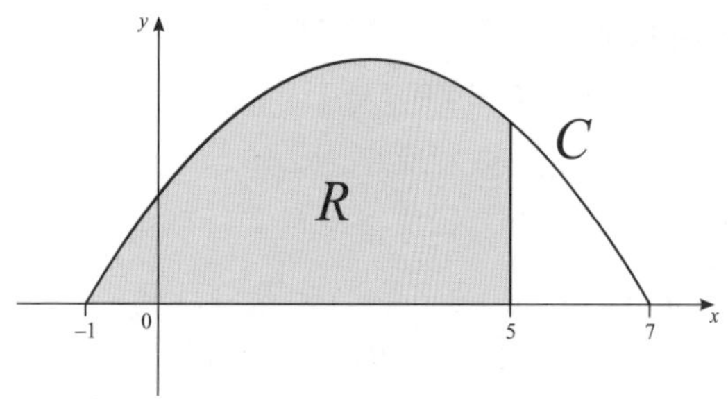

The parametric equations of curve C are

$$x = 3 + 4\sin\theta, \qquad y = \frac{1 + \cos 2\theta}{3}, \qquad -\frac{\pi}{2} \leq \theta \leq \frac{\pi}{2}.$$

Point H on C has coordinates $(5, \frac{1}{2})$.

a) Find the value of θ at point H. *(2 marks)*

The region R is enclosed by C, the line $x = 5$, and the x-axis, as shown.

b) Show that the area of R is given by the integral $\frac{8}{3}\int_{-\frac{\pi}{2}}^{\frac{\pi}{6}} \cos^3\theta \, d\theta$. *(5 marks)*

c) Show that the Cartesian equation of C can be written $y = \frac{-x^2 + 6x + 7}{24}$. *(4 marks)*

d) State the domain of values of x for the curve C. *(1 mark)*

Q4 is just for Edexcel C4

4

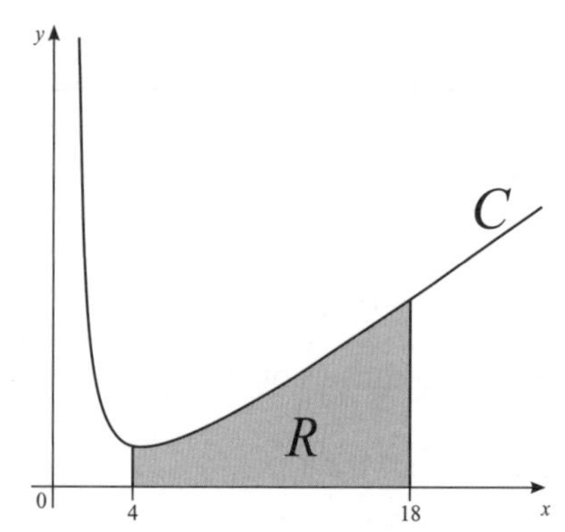

The curve C has parametric equations

$$x = t^2 + 3t, \qquad y = t^2 + \frac{1}{t^3}, \qquad t > 0.$$

The shaded region marked R is enclosed by C, the x-axis and the lines $x = 4$ and $x = 18$.

a) Show that the area of R is given by $\int_1^3 \frac{(t^5 + 1)(2t + 3)}{t^3} \, dt$. *(4 marks)*

b) Find the area of R. *(4 marks)*

Q5 is just for OCR MEI C4

5 The circle on the right is given by the parametric equations
$x = 5\cos\theta + a$ and $y = 5\sin\theta + b$.
Work out the values of a and b. *(2 marks)*

Chain Rule

This page is for AQA C3, Edexcel C3, OCR C3 and OCR MEI C3

That's right — our old friend differentiation is back again, this time with some new exciting features. Before you start panicking about how much you've already forgotten, all you need for now is: $\frac{d}{dx}(x^n) = nx^{n-1}$

The Chain Rule is used for Functions of Functions

The chain rule is a nifty little tool that allows you to differentiate complicated functions by splitting them up into easier ones. The trick is spotting how to split them up, and choosing the right bit to substitute.

Chain Rule Method

- **Pick a suitable function of x for 'u' and rewrite y in terms of u.**
- **Differentiate u (with respect to x) to get $\frac{du}{dx}$, and differentiate y (with respect to u) to get $\frac{dy}{du}$.**
- **Stick it all in the formula.**

If $y = f(u)$ and $u = g(x)$ then:

$$\frac{dy}{dx} = \frac{dy}{du} \times \frac{du}{dx}$$

EXAMPLE Find the exact value of $\frac{dy}{dx}$ when $x = 1$ for $y = \frac{1}{\sqrt{x^2 + 4x}}$.

1) First, write y in terms of powers to make it easier to differentiate: $y = (x^2 + 4x)^{-\frac{1}{2}}$.
2) Pick a chunk of the equation to call 'u', and rewrite y in terms of u: e.g. in this case let $u = x^2 + 4x$, so $y = u^{-\frac{1}{2}}$.
3) Now differentiate both bits separately: $u = x^2 + 4x$, so $\frac{du}{dx} = 2x + 4$ and $y = u^{-\frac{1}{2}}$, so $\frac{dy}{du} = -\frac{1}{2}u^{-\frac{3}{2}}$.
4) Use the chain rule to find $\frac{dy}{dx}$: $\frac{dy}{dx} = \frac{dy}{du} \times \frac{du}{dx} = -\frac{1}{2}u^{-\frac{3}{2}} \times (2x + 4)$.
5) Substitute in for u and rearrange: $u = x^2 + 4x$, so $\frac{dy}{dx} = -\frac{1}{2}(x^2 + 4x)^{-\frac{3}{2}}(2x + 4) = -\frac{x + 2}{(\sqrt{x^2 + 4x})^3}$.
6) Finally, put in $x = 1$ to answer the question: $\frac{dy}{dx} = -\frac{1 + 2}{(\sqrt{1^2 + (4 \times 1)})^3} = \frac{-3}{5\sqrt{5}} = \frac{-3\sqrt{5}}{25}$.

Write down all the steps — it'll help you avoid small mistakes that could affect your final answer.

'Exact' means leave in surd form where necessary.

Use dy/dx = 1 ÷ dx/dy for x = f(y)

For $x = f(y)$, use $\frac{dy}{dx} = \frac{1}{\left(\frac{dx}{dy}\right)}$

The principle of the chain rule can also be used where x is given in terms of y (i.e. $x = f(y)$). This comes from a bit of mathematical fiddling, but it's quite useful:

$\frac{dy}{dx} \times \frac{dx}{dy} = \frac{dy}{dy} = 1$, so rearranging gives $\frac{dy}{dx} = \frac{1}{\left(\frac{dx}{dy}\right)}$. Here's how to use it...

EXAMPLE A curve has the equation $x = y^3 + 2y - 7$. Find $\frac{dy}{dx}$ at the point $(-4, 1)$.

1) Forget that the xs and ys are in the 'wrong' places and differentiate as usual: $x = y^3 + 2y - 7$, so $\frac{dx}{dy} = 3y^2 + 2$.
2) Use $\frac{dy}{dx} = \frac{1}{\left(\frac{dx}{dy}\right)}$ to find $\frac{dy}{dx}$: $\frac{dy}{dx} = \frac{1}{3y^2 + 2}$.
3) $y = 1$ at the point $(-4, 1)$, so put this in the equation: $\frac{dy}{dx} = \frac{1}{3(1)^2 + 2} = \frac{1}{5} = 0.2$, so $\frac{dy}{dx} = 0.2$ at the point $(-4, 1)$.

You'll be using this again on the next page so make sure you've learnt it now.

I'm in the middle of a chain rule differentiation...

You know, I'm not sure I've stressed enough just how important differentiation is. It's one of those bits of maths that examiners can tag on to almost any other A-Level topic. It's almost like they have a mantra: 'Give me ANY function and I will ask you to differentiate it, in a multitude of intricate ways'. To which you should respond: 'Bring. It. On.'

Differentiation of e^x and ln x

This page is for AQA C3, Edexcel C3, OCR C3 and OCR MEI C3

Remember those special little functions from Section 4? Well you're about to find out just how special they are as we take a look at how to differentiate them. I can tell you're overcome by excitement so I'll not keep you waiting...

The **Gradient** of $y = e^x$ is e^x by **Definition**

$$y = e^x \qquad \frac{dy}{dx} = e^x$$

OR

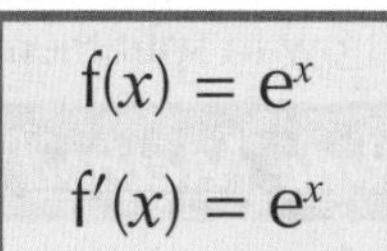

Get used to using both types of function notation. You should remember from AS that $f'(x)$ means the same as dy/dx.

In Section 4 (see p.34) we saw that 'e' was just a number for which the gradient of e^x was e^x. Which makes it pretty simple to differentiate.

EXAMPLE If $f(x) = e^{x^2} + 2e^x$, find $f'(x)$ for $x = 0$.

1) Let's break down the function into its two bits and differentiate them separately:

$y = e^{x^2}$ and $y = 2e^x$

2) This is the tricky bit.
Use the chain rule from the last page:
$u = x^2$ and $y = e^u$

3) Both u and y are now easy to differentiate:
$\frac{du}{dx} = 2x$ and $\frac{dy}{du} = e^u$

4) $\frac{dy}{dx} = \frac{du}{dx} \times \frac{dy}{du} = 2x \cdot e^u = 2x \cdot e^{x^2}$

5) This bit's easy.
If $y = 2e^x$ then $\frac{dy}{dx} = 2e^x$ too.

When $y = kf(x)$ where k is a constant, then dy/dx is just $kf'(x)$.

6) Put the bits back together and you end up with $f'(x) = 2xe^{x^2} + 2e^x$.

7) So when $x = 0$, $f'(x) = 0 + 2e^0 = 2$.

Turn $y = \ln x$ into $x = e^y$ to Differentiate

$$y = \ln x \qquad \frac{dy}{dx} = \frac{1}{x}$$

This result you can just learn, but it comes from another bit of mathematical fiddling:

If $y = \ln x$, then $x = e^y$ (see p.34).

Differentiating gives $\frac{dx}{dy} = e^y$, and $\frac{dy}{dx} = \frac{1}{\left(\frac{dx}{dy}\right)} = \frac{1}{e^y} = \frac{1}{x}$ (since $x = e^y$). Nice eh.

EXAMPLE Find $\frac{dy}{dx}$ if $y = \ln(x^2 + 3)$.

1) Use the chain rule again for this one: $y = \ln u$ and $u = x^2 + 3$.

2) $\frac{dy}{du} = \frac{1}{u}$ (from above) and $\frac{du}{dx} = 2x$.

3) So $\frac{dy}{dx} = \frac{dy}{du} \times \frac{du}{dx} = \frac{1}{u} \times 2x = \frac{2x}{x^2+3}$.

Look again at your final answer. It comes out to $\frac{f'(x)}{f(x)}$.

This will always be the case for $y = \ln(f(x))$ so you can just learn this result:

$$y = \ln(f(x)) \qquad \frac{dy}{dx} = \frac{f'(x)}{f(x)}$$

These functions pop up everywhere in the e^xams...

There's nothing too tough on this page, so you have no excuse for not getting a good grasp of the basics while you can. The derivatives of e^x and $\ln x$ are a couple more of those essential little things you've just got to learn. If you don't, you could get stumped by a fairly easy exam question. I know I'd gladly spend every waking hour learning this stuff if I could...

Differentiation of Sin, Cos and Tan

This page is for AQA C3, Edexcel C3, OCR C4 and OCR MEI C3

So you think you know all there is to know about trigonometry. Well think again, 'cos here it comes again. (You see what I did there with the 'cos'? Pun #27 from 'Ye Olde Booke of Maths Punnes'...)

The *Rules* for *dy/dx* of *Sin*, *Cos* and *Tan* only work in *Radians*

For trigonometric functions, where the angle is measured in radians, the following rules apply:

If $y =$	$\frac{dy}{dx} =$
$\sin x$	$\cos x$
$\cos x$	$-\sin x$
$\tan x$	$\sec^2 x$

There's loads more about sec (and cosec and cot) on p.26.

Use the *Chain Rule* with *Sin/Cos/Tan (f(x))*

If you can't follow what's happening here, go back to p.45 and brush up on the chain rule.

EXAMPLE: Differentiate $y = \cos 2x + \sin(x + 1)$ with respect to x.

It's the chain rule (again) for both parts of this equation:

1) Differentiate '$y = \cos 2x$': $y = \cos u$, $u = 2x$,

   so $\frac{dy}{du} = -\sin u$ (see above) and $\frac{du}{dx} = 2 \Rightarrow \frac{dy}{dx} = -2\sin 2x$.

2) Differentiate '$y = \sin(x + 1)$': $y = \sin u$, $u = x + 1$,

   so $\frac{dy}{du} = \cos u$ (see above) and $\frac{du}{dx} = 1 \Rightarrow \frac{dy}{dx} = \cos(x + 1)$.

3) Put it all together to get $\frac{dy}{dx} = -2\sin 2x + \cos(x + 1)$.

EXAMPLE: Find $\frac{dy}{dx}$ when $x = \tan 3y$.

1) First find $\frac{dx}{dy}$ using the chain rule: $x = \tan u$, $u = 3y$, $\frac{dx}{du} = \sec^2 u$, $\frac{du}{dy} = 3$, so $\frac{dx}{dy} = 3\sec^2 3y$.

2) Then use $\frac{dy}{dx} = \frac{1}{\left(\frac{dx}{dy}\right)}$ to get the final answer: $\frac{dy}{dx} = \frac{1}{3\sec^2 3y} = \frac{1}{3}\cos^2 3y$.

See p.45 if you can't remember this.

Remember to use *Trig Identities* where *Necessary*

OCR MEI can skip this bit

EXAMPLE For $y = 2\cos^2 x + \sin 2x$, show that $\frac{dy}{dx} = 2(\cos 2x - \sin 2x)$.

1) Writing out the equation in a slightly different way helps with the chain rule: $y = 2(\cos x)^2 + \sin 2x$.

2) For the first bit, $y = 2u^2$, $u = \cos x$, so $\frac{dy}{du} = 4u$ and $\frac{du}{dx} = -\sin x$.

   For the second bit, $y = \sin u$, $u = 2x$, so $\frac{dy}{du} = \cos u$ and $\frac{du}{dx} = 2$.

3) Putting it all in the chain rule formula gives $\frac{dy}{dx} = -4\sin x \cos x + 2\cos 2x$.

4) From the target answer in the question it looks like we need a $\sin 2x$ from somewhere, so use the 'double angle' formula (see p.29) $\sin 2x \equiv 2\sin x \cos x$:

   $\frac{dy}{dx} = -2\sin 2x + 2\cos 2x$, which rearranges nicely to give $\frac{dy}{dx} = 2(\cos 2x - \sin 2x)$. Et voilà.

You could also use the identity $\cos 2x \equiv 2\cos^2 x - 1$ before differentiating. You'll get the same answer.

I'm having an identity crisis — I can't differentiate between sin and cos...

Don't get tied down by the chain rule (pun #28...). After a bit of practice you'll be able to do it a lot quicker in one step — just say in your working 'using the chain rule...' so the examiner can see how clever you are.

Product Rule

This page is for AQA C3, Edexcel C3, OCR C3 and OCR MEI C3

In maths-speak, a 'product' is what you get when you multiply things together. So the 'product rule' is a rule about differentiating things that are multiplied together. And it's yet another rule you have to learn I'm afraid.

Use the **Product Rule** to differentiate **Two Functions Multiplied Together**

This is what it looks like:

If $y = u(x)v(x)$

$$\frac{dy}{dx} = u\frac{dv}{dx} + v\frac{du}{dx}$$

(u and v are functions of x.)

And here's how to use it:

EXAMPLES Differentiate the following with respect to x: a) $x^3 \tan x$ and b) $e^{2x}\sqrt{2x-3}$.

a) $x^3 \tan x$

1) The crucial thing is to write down everything in steps. Start with identifying 'u' and 'v':
$u = x^3$ and $v = \tan x$.

2) Now differentiate these two separately, with respect to x:
$\frac{du}{dx} = 3x^2$ and $\frac{dv}{dx} = \sec^2 x$.

3) Very carefully put all the bits into the formula:
$\frac{dy}{dx} = u\frac{dv}{dx} + v\frac{du}{dx} = (x^3 \cdot \sec^2 x) + (\tan x \cdot 3x^2)$

4) Finally, rearrange to make it look nicer:
$\frac{dy}{dx} = x^3 \sec^2 x + 3x^2 \tan x$.

b) $e^{2x}\sqrt{2x-3}$

1) Again, start with identifying 'u' and 'v':
$u = e^{2x}$ and $v = \sqrt{2x-3}$.

2) Each of these needs the chain rule to differentiate:
$\frac{du}{dx} = 2e^{2x}$ and $\frac{dv}{dx} = \frac{1}{\sqrt{2x-3}}$ (do it in steps if you need to...)

3) Put it all into the product rule formula:
$\frac{dy}{dx} = u\frac{dv}{dx} + v\frac{du}{dx} = \left(e^{2x} \cdot \frac{1}{\sqrt{2x-3}}\right) + \left(\sqrt{2x-3} \cdot 2e^{2x}\right)$

4) Rearrange and simplify:
$\frac{dy}{dx} = e^{2x}\left(\frac{1}{\sqrt{2x-3}} + 2\sqrt{2x-3}\right) = e^{2x}\left(\frac{1 + 2(2x-3)}{\sqrt{2x-3}}\right)$
$= \frac{e^{2x}(4x-5)}{\sqrt{2x-3}}$.

Use the Rules **Together** to differentiate **Complicated Functions**

In the exam they might tell you which rules to use, but chances are they won't.
And you'll probably have to throw a whole load of rules at any one question.

EXAMPLE Solve the equation $\frac{d}{dx}((x^3 + 3x^2)\ln x) = 2x^2 + 5x$, leaving your answer as an exact value of x.

1) The $\frac{d}{dx}$ just tells you to differentiate the bit in brackets first.

And since $(x^3 + 3x^2)\ln x$ is a product of two functions, use the product rule:
$u = x^3 + 3x^2 \Rightarrow \frac{du}{dx} = 3x^2 + 6x$ and $v = \ln x \Rightarrow \frac{dv}{dx} = \frac{1}{x}$ (see p.46)

So $\frac{d}{dx}((x^3 + 3x^2)\ln x) = [(x^3 + 3x^2) \cdot \frac{1}{x}] + [\ln x \cdot (3x^2 + 6x)] = x^2 + 3x + (3x^2 + 6x)\ln x$.

You should be well up on $\ln x$ and e^x after Section 4, but glance back at pages 34-36 if you need to.

2) Now put this into the equation from the question in place of $\frac{d}{dx}((x^3 + 3x^2)\ln x)$:
$x^2 + 3x + (3x^2 + 6x)\ln x = 2x^2 + 5x$

You're asked for an exact value so leave in terms of e.

3) Rearrange and solve as follows:
$(3x^2 + 6x)\ln x = 2x^2 + 5x - x^2 - 3x \Rightarrow (3x^2 + 6x)\ln x = x^2 + 2x \Rightarrow \ln x = \frac{x^2 + 2x}{3(x^2 + 2x)} = \frac{1}{3} \Rightarrow x = e^{\frac{1}{3}}$.

The first rule of maths club is — you do not talk about maths club...

These rules are supposed to make your life easier when differentiating. Learning them means you don't have to do everything from first principles every time. Try not to get the product rule mixed up with the chain rule. Repeat after me: 'The chain rule is for functions of functions but the product rule is for products of functions'. Snappy, I know...

Quotient Rule

This page is for AQA C3, Edexcel C3, OCR C3 & C4 and OCR MEI C3

The world is a beautiful, harmonious place full of natural symmetry. So of course, if we have a 'product rule' to differentiate products, we must also have a 'quotient rule' to differentiate... er... quotients. Read on and learn.

Use the **Quotient Rule** for one function **Divided By** another

(OCR — this bit's C3)

A quotient is one function divided by another one.
The rule for differentiating quotients looks like this:

$$\text{If } y = \frac{u(x)}{v(x)}$$

$$\frac{dy}{dx} = \frac{v\frac{du}{dx} - u\frac{dv}{dx}}{v^2}$$

You could, if you wanted to, just use the product rule on $y = uv^{-1}$ (try it — you'll get the same answer).
This way is so much quicker and easier though — and it's on the formula sheet.

EXAMPLE:

Find the gradient of the tangent to the curve with equation $y = \frac{(2x^2 - 1)}{(3x^2 + 1)}$, at the point (1, 0.25).

1) 'Gradient of tangent' means differentiate.

2) First identify u and v for the quotient rule, and differentiate separately:

$$u = 2x^2 - 1 \Rightarrow \frac{du}{dx} = 4x \quad \text{and} \quad v = 3x^2 + 1 \Rightarrow \frac{dv}{dx} = 6x.$$

This bit's just like the product rule from the last page.

3) It's very important that you get things in the right order, so concentrate on what's going where:

$$\frac{dy}{dx} = \frac{v\frac{du}{dx} - u\frac{dv}{dx}}{v^2} = \frac{(3x^2+1)(4x) - (2x^2-1)(6x)}{(3x^2+1)^2}$$

Don't try and simplify straight away or you'll get things mixed up.

4) Now you can simplify things:

$$\frac{dy}{dx} = \frac{x[4(3x^2+1) - 6(2x^2-1)]}{(3x^2+1)^2} = \frac{x[12x^2 + 4 - 12x^2 + 6]}{(3x^2+1)^2} = \frac{10x}{(3x^2+1)^2}.$$

5) Finally, put in $x = 1$ to find the gradient at (1, 0.25): $\frac{dy}{dx} = \frac{10}{(3+1)^2} = 0.625.$

If it's a 'normal' rather than a 'tangent' do $-1 \div$ gradient.

Find **Further Rules** using the **Quotient Rule**

(OCR — this bit's C4)

EXAMPLE Use the quotient rule to differentiate $y = \frac{\cos x}{\sin x}$, and hence show that for $y = \cot x$, $\frac{dy}{dx} = -\text{cosec}^2\, x$.

1) Start off identifying $u = \cos x$ and $v = \sin x$.

2) Differentiating separately gives: $\frac{du}{dx} = -\sin x$, and $\frac{dv}{dx} = \cos x$ (see p.47).

3) Putting everything in the quotient rule formula gives:

$$\frac{dy}{dx} = \frac{(\sin x \times -\sin x) - (\cos x \times \cos x)}{(\sin x)^2} = \frac{-\sin^2 x - \cos^2 x}{\sin^2 x}.$$

Don't forget your easy AS trig identities as well as the ones covered in Section 3.

4) Use a trig identity to simplify this ($\sin^2 x + \cos^2 x \equiv 1$ should do the trick...):

$$\frac{dy}{dx} = \frac{-(\sin^2 x + \cos^2 x)}{\sin^2 x} = \frac{-1}{\sin^2 x}.$$

5) Linking this back to the question, since $\tan x = \frac{\sin x}{\cos x}$, and $\cot x = \frac{1}{\tan x}$, then $y = \frac{\cos x}{\sin x} = \cot x$.

And since $\text{cosec}\, x = \frac{1}{\sin x}$, then $\frac{dy}{dx} = \frac{-1}{\sin^2 x} = -\text{cosec}^2\, x$. QED*

There's more of this trig stuff on the next page. This was just a taste of things to come...

*Quite Exciting Differentiation

The second rule of maths club is — you do not talk about maths club...

Confused yet? Yes I know, there are three very similar looking rules in this section, all using us and vs and xs and ys all over the shop. You won't remember them by reading them over and over again like some mystical code. You will remember them by using them lots and lots in practice questions. Plain and simple — just how I like my men...

More Trig Differentiation

This page is for AQA C3, Edexcel C3 and OCR C4

After whetting your appetite with the little proof on the last page, let's have a gander at some more trig differentiation. Namely, the rules for differentiating cosec x, sec x and cot x, and the vast array of things you can do with them.

d/dx of **Cosec**, **Sec** and **Cot** come from the **Quotient Rule**

Since cosec, sec and cot are just the reciprocals of sin, cos and tan, the quotient rule can be used to differentiate them.

$$y = \text{cosec}\, x = \frac{1}{\sin x}$$

1) For the quotient rule:
   $u = 1 \Rightarrow \frac{du}{dx} = 0$ and $v = \sin x \Rightarrow \frac{dv}{dx} = \cos x$
2) $\frac{dy}{dx} = \frac{v\frac{du}{dx} - u\frac{dv}{dx}}{v^2} = \frac{(\sin x \cdot 0) - (1 \cdot \cos x)}{\sin^2 x} = -\frac{\cos x}{\sin^2 x}$
3) Since $\cot x = \frac{\cos x}{\sin x}$, and $\text{cosec}\, x = \frac{1}{\sin x}$,
   $\frac{dy}{dx} = -\frac{\cos x}{\sin x} \times \frac{1}{\sin x} = -\text{cosec}\, x \cot x.$

$$y = \sec x = \frac{1}{\cos x}$$

1) For the quotient rule:
   $u = 1 \Rightarrow \frac{du}{dx} = 0$ and $v = \cos x \Rightarrow \frac{dv}{dx} = -\sin x$
2) $\frac{dy}{dx} = \frac{v\frac{du}{dx} - u\frac{dv}{dx}}{v^2} = \frac{(\cos x \cdot 0) - (1 \cdot -\sin x)}{\cos^2 x} = \frac{\sin x}{\cos^2 x}$
3) Since $\tan x = \frac{\sin x}{\cos x}$, and $\sec x = \frac{1}{\cos x}$,
   $\frac{dy}{dx} = \frac{\sin x}{\cos x} \times \frac{1}{\cos x} = \sec x \tan x.$

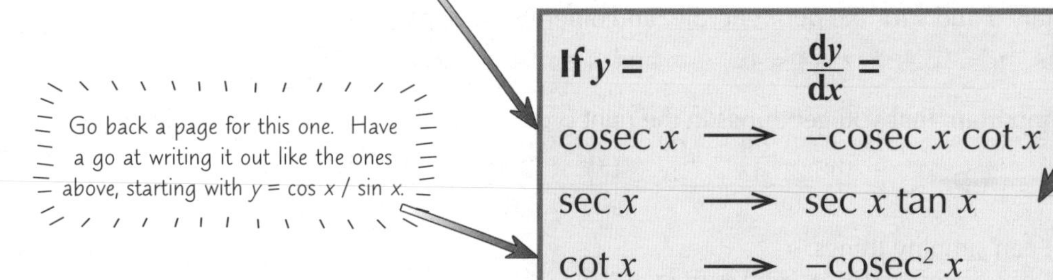

If you can't remember which trig functions give a negative result when you differentiate them, just remember it's all the ones that begin with c — cos, cosec and cot.

Use the **Chain**, **Product** and **Quotient Rules** with **Cosec**, **Sec** and **Cot**

So once you're familiar with the three rules in the box above you can use them with the chain, product and quotient rules and in combination with all the other functions we've seen so far.

EXAMPLES Find $\frac{dy}{dx}$ for the following functions: a) $y = \sec(2x^2)$ and b) $y = e^x \cot x$.

a) $y = \sec(2x^2)$

1) This is a function of a function, so think 'chain rule':
   $y = \sec u$ and $u = 2x^2$
2) $\frac{dy}{du} = \sec u \tan u$ (see above) $= \sec(2x^2)\tan(2x^2)$
3) $\frac{du}{dx} = 4x$
4) So $\frac{dy}{dx} = \frac{dy}{du} \times \frac{du}{dx} = 4x\sec(2x^2)\tan(2x^2).$

b) $y = e^x \cot x$

1) This is a product of two functions, so think 'product rule':
   $u = e^x$ and $v = \cot x$
2) $\frac{du}{dx} = e^x$
3) $\frac{dv}{dx} = -\text{cosec}^2 x$ (see above)
4) So $\frac{dy}{dx} = u\frac{dv}{dx} + v\frac{du}{dx} = (e^x \cdot -\text{cosec}^2 x) + (\cot x \cdot e^x)$
   $= e^x(\cot x - \text{cosec}^2 x).$

Get it? Got it? Good.

I'm co-sec-sy for my shirt — co-sec-sy it hurts...

I have good news — the formulas above will be on the formula sheet in the exam, so you only need to know how to use them. So there's no excuse for anything less than excellence. Also acceptable: perfection, full marks and sheer brilliance.

More Differentiation

This page is for AQA C3, Edexcel C3, OCR C3 and OCR MEI C3

What?! More differentiation?! Surely not. This page is all about using what you know.

Finding the **Gradient**, **Tangent**, **dy/dx**, **f'(x)**, **d/dx(f(x))** — all mean **'Differentiate'**

Usually in exams, differentiation will be disguised as something else — either through different notation ($f'(x)$, $\frac{dy}{dx}$ etc.) or by asking for the gradient or rate of change of something.
You could also be asked to find the equation of a tangent or normal to a curve at a given point:

EXAMPLE Find the equation of the tangent to the curve $y = \frac{5x+2}{3x-2}$ at the point (1, 7), in the form $y = mx + c$.

1) The gradient of the tangent is just the gradient of the curve at that point. So differentiate...
2) Use the quotient rule: $u = 5x + 2 \Rightarrow \frac{du}{dx} = 5$ and $v = 3x - 2 \Rightarrow \frac{dv}{dx} = 3$.
   So $\frac{dy}{dx} = \frac{5(3x-2) - 3(5x+2)}{(3x-2)^2} = -\frac{16}{(3x-2)^2}$.
3) Gradient of tangent at (1, 7) is $\frac{dy}{dx}$ at $x = 1$, which is $-\frac{16}{(3-2)^2} = -16$.
4) Use the equation of a straight line $y - y_1 = m(x - x_1)$ with $m = -16$, $y_1 = 7$ and $x_1 = 1$, to give:
   $y - 7 = -16(x - 1) \Rightarrow y = -16x + 23$ is the equation of the tangent.

If you're asked for a 'normal', do $-1 \div$ gradient of tangent here — then the rest is the same.

The **Rules** might need to be used **Twice**

OCR can skip this section

Some questions will really stretch your alphabet with a multitude of us and vs:

EXAMPLE Differentiate $y = e^x \tan^2(3x)$

1) First off, this is product rule: $u = e^x$ (so $\frac{du}{dx} = e^x$) and $v = \tan^2(3x)$.
2) To find $\frac{dv}{dx}$ for the product rule, we need the chain rule twice:
   $v = u_1^2$, where $u_1 = \tan(3x)$.
   $\frac{dv}{du_1} = 2u_1 = 2\tan(3x)$, and $\frac{du_1}{dx} = 3\sec^2(3x)$ (which is an easy chain rule solution itself). So $\frac{dv}{dx} = 6\tan(3x)\sec^2(3x)$.
3) Now we can put this result in the product rule formula to get $\frac{dy}{dx}$:
   $\frac{dy}{dx} = (e^x \cdot 6\tan(3x)\sec^2(3x)) + (\tan^2(3x) \cdot e^x) = e^x \tan(3x)[6\sec^2(3x) + \tan(3x)]$. Job done.
   (u, $\frac{dv}{dx}$, v, $\frac{du}{dx}$)

Differentiate **Again** for **d²y/dx²**, **Turning Points**, **Stationary Points** etc.

Refresh your memory of AS Maths, where you learnt all about maximums and minimums...

EXAMPLE Determine the nature of the stationary point of the curve $y = \frac{\ln x}{x^2}$ $(x > 0)$.

1) First use the quotient rule to find $\frac{dy}{dx}$: $u = \ln x \Rightarrow \frac{du}{dx} = \frac{1}{x}$, $v = x^2 \Rightarrow \frac{dv}{dx} = 2x$. So $\frac{dy}{dx} = \frac{1 - 2\ln x}{x^3}$.
2) The stationary points occur where $\frac{dy}{dx} = 0$ (i.e. zero gradient) so this is when:
   $\frac{1 - 2\ln x}{x^3} = 0 \Rightarrow \ln x = \frac{1}{2} \Rightarrow x = e^{\frac{1}{2}}$.
3) To find out whether it's a maximum or minimum, differentiate $\frac{dy}{dx}$ to get $\frac{d^2y}{dx^2}$:
   $u = 1 - 2\ln x \Rightarrow \frac{du}{dx} = -\frac{2}{x}$, $v = x^3 \Rightarrow \frac{dv}{dx} = 3x^2$. So $\frac{d^2y}{dx^2} = \frac{6\ln x - 5}{x^4}$.
4) When $x = e^{\frac{1}{2}}$, $\frac{d^2y}{dx^2} < 0$ (i.e. negative), which means it's a maximum point.

Positive means minimum, negative means maximum — it's all there in your AS notes.

Parlez vous exam?

It's often noted that mathematics has its own language — you need to make sure you're fluent or all your hard work will go to waste. Become an expert in deciphering exam questions so you do exactly what's expected with the minimum of fuss.

Core Section 6 — Practice Questions

Those who know it, know they know it. Those who think they know it, need to know they know it.
So, you think you know it, no? Try these to make sure.

Warm-up Questions

Q1-7 are for AQA C3, Edexcel C3, OCR C3 [Q1-2, 4a, 5, 7], OCR C4 [Q3, 4b, 6] and OCR MEI C3 [skip Q6]

1) Differentiate with respect to x:
   a) $y = \sqrt{x^3 + 2x^2}$ b) $y = \frac{1}{\sqrt{x^3 + 2x^2}}$ c) $y = e^{5x^2}$ d) $y = \ln(6 - x^2)$
2) Find $\frac{dy}{dx}$ when a) $x = 2e^y$ b) $x = \ln(2y + 3)$

Assume that questions involving trig are using radians unless stated otherwise.

3) Find $f'(x)$ for the following functions:
   a) $f(x) = \sin^2(x + 2)$ b) $f(x) = 2\cos 3x$ c) $f(x) = \sqrt{\tan x}$
4) Find the value of the gradient for:
   a) $y = e^{2x}(x^2 - 3)$ when $x = 0$ b) $y = \ln x \sin x$ when $x = 1$
5) Find the equation of the tangent to the curve $y = \frac{6x^2 + 3}{4x^2 - 1}$ at the point (1, 3).
6) Find $\frac{dy}{dx}$ when $x = 0$ for $y = \operatorname{cosec}(3x - 2)$.

And next — a megabeast of a question. You probably won't get anything as involved as this in the exam, but if you think you're hard enough...

7) Find the stationary point on the curve $y = \frac{e^x}{\sqrt{x}}$, and say whether it is a maximum or minimum.

Well that's put some colour in your cheeks. Now to really excel yourself on the exam practice, but try not to pull a muscle — you need to be match fit for the real thing.

Exam Questions

Q1-4 are for AQA C3, Edexcel C3, OCR C3 and OCR MEI C3

1 The curve shown below has the equation $x = \sqrt{y^2 + 3y}$.

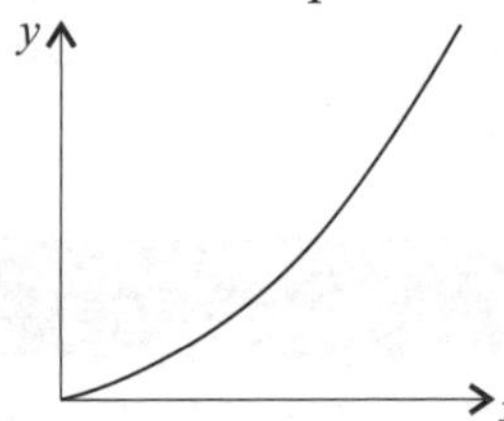

a) Find $\frac{dy}{dx}$ at the point (2, 1). *(5 marks)*

b) Hence find the equation of the tangent to the curve at (2, 1), in the form $y = ax + b$, where a and b are constants. *(2 marks)*

2 Differentiate the following with respect to x.

a) $\sqrt{(e^x + e^{2x})}$. *(3 marks)*

b) $3e^{2x+1} - \ln(1 - x^2) + 2x^3$. *(3 marks)*

3 Given that $y = \frac{e^x + x}{e^x - x}$, find $\frac{dy}{dx}$ when $x = 0$. *(3 marks)*

Core Section 6 — Practice Questions

Some see beauty in lakes and mountains, others in Michelangelo's sculptures and others in Mozart's operas. For me, it's differentiation. Pure, unspoiled differentiation.

4 A sketch of the function $f(x) = 4\ln 3x$ is shown in the diagram.

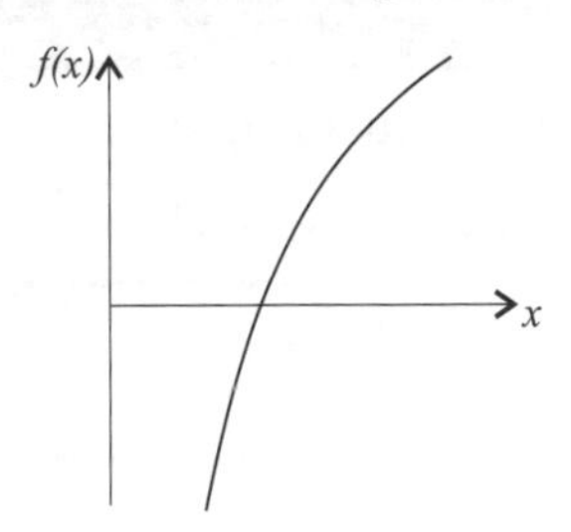

a) Find $f'(x)$ at the point where $x = 1$. *(3 marks)*

b) Find the equation of the tangent to the curve at the point $x = 1$. *(3 marks)*

Q5-9 are for AQA C3, Edexcel C3, OCR C4 and OCR MEI C3 [MEI can skip Q5 and 6d]

5 Use the quotient rule to show that, for the function $f(x) = \sec x$:

$$f'(x) = \sec x \tan x.$$

(4 marks)

6 Find $\frac{dy}{dx}$ for each of the following functions. Simplify your answer where possible.

a) $y = \ln(3x + 1)\sin(3x + 1)$. *(4 marks)*

b) $y = \frac{\sqrt{x^2 + 3}}{\cos 3x}$. *(4 marks)*

c) $y = \sin^3(2x^2)$ *(3 marks)*

d) $y = 2\ \mathrm{cosec}\ (3x)$ *(2 marks)*

7 Find the gradient of the tangent to the curve:

$$y = \sin^2 x - 2\cos 2x$$

at the point where $x = \frac{\pi}{12}$ radians. *(4 marks)*

8 Find the equation of the normal to the curve $x = \sin 4y$ that passes through the point $\left(0, \frac{\pi}{4}\right)$.

Give your answer in the form $y = mx + c$, where m and c are constants to be found. *(6 marks)*

9 A curve with equation $y = e^x \sin x$ has 2 turning points in the interval $-\pi \leq x \leq \pi$.

a) Find the value of x at each of these turning points. *(6 marks)*

b) Determine the nature of each of the turning points. *(5 marks)*

Differentiation with Parametric Equations

This page is for AQA C4, Edexcel C4, OCR C4 and OCR MEI C4

Another shiny new section, and it starts with the return of an old friend. If you've forgotten what parametric equations are already, go back to Section 5. Go on, I'll wait for you... ... ...OK, are you back now? Ready? Right, on we go...

Differentiating Parametric Equations is a lot Simpler than you might expect

Just suppose you've got a curve defined by two parametric equations, with the parameter t: $y = f(t)$ and $x = g(t)$.

If you can't find the Cartesian equation, it seems like it would be a bit tricky to find the gradient, $\frac{dy}{dx}$.

Luckily the chain rule (see p.45) is on hand to help out:

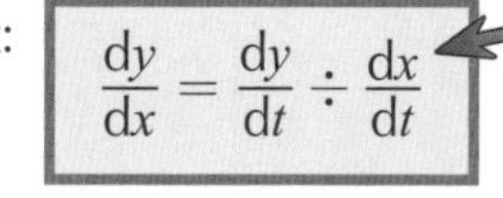

$$\frac{dy}{dx} = \frac{dy}{dt} \div \frac{dx}{dt}$$

This is exactly the same as on p45, except we've replaced '$\times \frac{dt}{dx}$' with '$\div \frac{dx}{dt}$'

EXAMPLE The curve C is defined by the parametric equations $y = t^3 - 2t + 4$ and $x = t^2 - 1$.
Find: a) $\frac{dy}{dx}$ in terms of t, b) the gradient of C when $t = -1$.

Start by differentiating the two parametric equations with respect to t:

a) $\frac{dy}{dt} = 3t^2 - 2$, $\frac{dx}{dt} = 2t$

Now use the chain rule to combine them:

$\frac{dy}{dx} = \frac{dy}{dt} \div \frac{dx}{dt} = \frac{3t^2 - 2}{2t}$

Use the answer to a) to find the gradient for a specific value of t:

b) When $t = -1$, $\frac{dy}{dx} = \frac{3(-1)^2 - 2}{2(-1)} = \frac{3-2}{-2} = -\frac{1}{2}$

Use the Gradient to find Tangents and Normals

Of course, it's rarely as straightforward as just finding the gradient. A lot of the time, you'll then have to use it in the equation of a tangent or normal to the parametric curve.

EXAMPLE For the curve C in the example above, find:
a) the equation of the tangent to the curve when $t = 2$,
b) the equation of the normal to the curve when $t = 2$.

First you need the coordinates of the point where $t = 2$:

a) When $t = 2$, $x = (2)^2 - 1 = 3$ and $y = (2)^3 - 2(2) + 4 = 8 - 4 + 4 = 8$.

You also need the gradient at that point:

When $t = 2$, $\frac{dy}{dx} = \frac{3(2)^2 - 2}{2(2)} = \frac{10}{4} = \frac{5}{2}$

Now use that information to find the equation of the tangent:

The tangent to C at (3, 8) has an equation of the form $y = mx + c$.

So $8 = \frac{5}{2}(3) + c \Rightarrow c = \frac{1}{2}$.

The tangent to curve C when $t = 2$ is $y = \frac{5}{2}x + \frac{1}{2}$.

You could also use $y - y_1 = m(x - x_1)$ to get the equation.

You can find the normal in a similar way:

b) The normal to C at (3, 8) has gradient $-\frac{1}{\left(\frac{5}{2}\right)} = -\frac{2}{5}$.

So $8 = -\frac{2}{5}(3) + c \Rightarrow c = \frac{46}{5}$.

The normal to curve C when $t = 2$ is $y = -\frac{2}{5}x + \frac{46}{5}$.

If you're not quite following all this tangents and normals business, take a look back at your AS notes to refresh your memory.

And now, yet another chocolate biscuit reference...

To an examiner, adding a 'find the tangent' or 'find the normal' part to a parametric equations question is like adding chocolate to a digestive biscuit — it makes it at least 4 times better. In other words: this is very likely to show up in your A2 exam, so be ready for it. And in case you were wondering, tangent = milk chocolate, normal = dark chocolate.

Implicit Differentiation

This page is for AQA C4, Edexcel C4, OCR C4 and OCR MEI C3

This really isn't as complicated as it looks... in fact, I think you'll find that if something's implicit between x and y, it can be ximplicity itself. No, that's not a typo, it's a hilarious joke... 'implicit' between 'x' and 'y'... do you see?...

You need *Implicit Differentiation* if you can't write the *Equation* as $y = f(x)$

1) An 'implicit relation' is the maths name for any equation in x and y that's written in the form $f(x, y) = g(x, y)$ instead of $y = f(x)$.

   f(x, y) and g(x, y) don't actually both have to include x and y — one of them could even be a constant.

2) Some implicit relations are either awkward or impossible to rewrite in the form $y = f(x)$. This can happen, for example, if the equation contains a number of different powers of y, or terms where x is multiplied by y.
3) This can make implicit relations tricky to differentiate — the solution is implicit differentiation:

Implicit Differentiation

To find $\frac{dy}{dx}$ for an implicit relation between x and y:

1) **Differentiate terms in x^n only (and constant terms) with respect to x, as normal.**
2) **Use the chain rule to differentiate terms in y^m only:**

$$\frac{d}{dx}f(y) = \frac{d}{dy}f(y)\frac{dy}{dx}$$

In other words, 'differentiate with respect to y, and stick a $\frac{dy}{dx}$ on the end'.

3) **Use the product rule to differentiate terms in both x and y:**

$$\frac{d}{dx}u(x)v(y) = u(x)\frac{d}{dx}v(y) + v(y)\frac{d}{dx}u(x)$$

This version of the product rule is slightly different from the one on p48 — it's got v(y) instead of v(x).

4) **Rearrange the resulting equation in x, y and $\frac{dy}{dx}$ to make $\frac{dy}{dx}$ the subject.**

EXAMPLE Use implicit differentiation to find $\frac{dy}{dx}$ if $2x^2y + y^3 = 6x^2 + 5$.

We need to differentiate each term of the equation with respect to x.

Start by sticking '$\frac{d}{dx}$' in front of each term:

$$\frac{d}{dx}2x^2y + \frac{d}{dx}y^3 = \frac{d}{dx}6x^2 + \frac{d}{dx}5$$

First, deal with the terms in x and constant terms — in this case that's the two terms on the RHS:

$$\Rightarrow \frac{d}{dx}2x^2y + \frac{d}{dx}y^3 = 12x + 0$$

Now use the chain rule on the term in y:

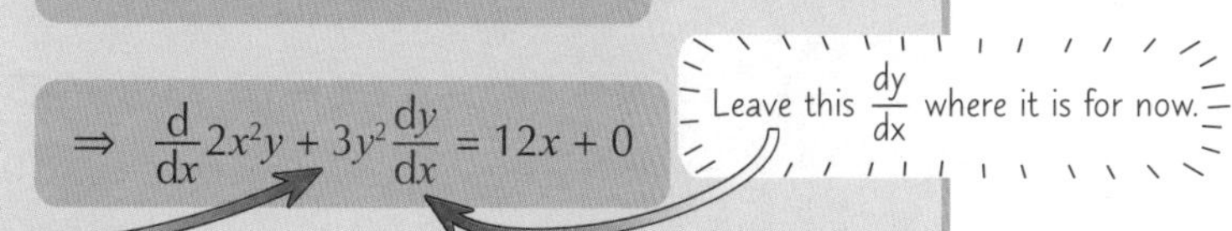

$$\Rightarrow \frac{d}{dx}2x^2y + 3y^2\frac{dy}{dx} = 12x + 0$$

Using the chain rule from the box above, $f(y) = y^3$.

Leave this $\frac{dy}{dx}$ where it is for now.

And use the product rule on the term in x and y:

$$\Rightarrow 2x^2\frac{d}{dx}(y) + y\frac{d}{dx}(2x^2) + 3y^2\frac{dy}{dx} = 12x + 0$$

$$\Rightarrow 2x^2\frac{dy}{dx} + y4x + 3y^2\frac{dy}{dx} = 12x + 0$$

So in terms of the box above, $u(x) = 2x^2$ and $v(y) = y$.

You get a $\frac{dy}{dx}$ term here too (from the '$\frac{d}{dx}v(y)$' bit).

Finally, rearrange to make $\frac{dy}{dx}$ the subject:

$$\Rightarrow \frac{dy}{dx}(2x^2 + 3y^2) = 12x - 4xy$$

$$\Rightarrow \frac{dy}{dx} = \frac{12x - 4xy}{2x^2 + 3y^2}$$

If an imp asks to try your ice lolly, don't let the imp lick it...

Learn the versions of the chain rule and product rule from the box above. All the different bits of the method for implicit differentiation can make it confusing — read the example carefully and make sure you understand every little bit of it.

Implicit Differentiation

This page is for AQA C4, Edexcel C4, OCR C4 and OCR MEI C3

If you've gone to all the hard work of differentiating an implicit relation, it would be a shame not to use it. It'd be like a shiny toy that's been kept in its box and never played with. Don't make the maths sad — play with it.

Implicit Differentiation still gives you an expression for the Gradient

Most implicit differentiation questions aren't really that different at heart to any other differentiation question. Once you've got an expression for the gradient, you'll have to use it to do the sort of stuff you'd normally expect.

EXAMPLE Curve A has the equation $x^2 + 2xy - y^2 = 10x + 4y - 21$

a) Show that when $\frac{dy}{dx} = 0$, $y = 5 - x$.

b) Find the coordinates of the stationary points of A.

For starters, we're going to need to find $\frac{dy}{dx}$ by implicit differentiation:

a)
$$\frac{d}{dx}x^2 + \frac{d}{dx}2xy - \frac{d}{dx}y^2 = \frac{d}{dx}10x + \frac{d}{dx}4y - \frac{d}{dx}21$$
$$\Rightarrow 2x + \frac{d}{dx}2xy - \frac{d}{dx}y^2 = 10 + \frac{d}{dx}4y - 0$$
$$\Rightarrow 2x + \frac{d}{dx}2xy - 2y\frac{dy}{dx} = 10 + 4\frac{dy}{dx}$$
$$\Rightarrow 2x + 2x\frac{dy}{dx} + y\frac{d}{dx}2x - 2y\frac{dy}{dx} = 10 + 4\frac{dy}{dx}$$
$$\Rightarrow 2x + 2x\frac{dy}{dx} + 2y - 2y\frac{dy}{dx} = 10 + 4\frac{dy}{dx}$$
$$\Rightarrow 2x\frac{dy}{dx} - 2y\frac{dy}{dx} - 4\frac{dy}{dx} = 10 - 2x - 2y$$
$$\Rightarrow \frac{dy}{dx} = \frac{10 - 2x - 2y}{2x - 2y - 4}$$

Differentiate x^2, 10x and 21 with respect to x.

Use the chain rule to differentiate y^2 and 4y.

Use the product rule to differentiate 2xy.

Collect '$\frac{dy}{dx}$' terms on one side, and everything else on the other side.

So when $\frac{dy}{dx} = 0$, $\frac{10 - 2x - 2y}{2x - 2y - 4} = 0 \Rightarrow 10 - 2x - 2y = 0 \Rightarrow y = 5 - x$

Now we can use the answer to part a) in the equation of the curve to find the points where $\frac{dy}{dx} = 0$.

b) When $\frac{dy}{dx} = 0$, $y = 5 - x$. So at the stationary points,
$$x^2 + 2xy - y^2 = 10x + 4y - 21$$
$$\Rightarrow x^2 + 2x(5 - x) - (5 - x)^2 = 10x + 4(5 - x) - 21$$
$$\Rightarrow x^2 + 10x - 2x^2 - 25 + 10x - x^2 = 10x + 20 - 4x - 21$$
$$\Rightarrow -2x^2 + 20x - 25 = 6x - 1$$
$$\Rightarrow -2x^2 + 14x - 24 = 0$$
$$\Rightarrow x^2 - 7x + 12 = 0$$
$$\Rightarrow (x - 3)(x - 4) = 0$$
$$\Rightarrow x = 3 \text{ or } x = 4$$
$x = 3 \Rightarrow y = 5 - 3 = 2 \quad x = 4 \Rightarrow y = 5 - 4 = 1$

So the stationary points of A are (3, 2) and (4, 1).

Substitute y = 5 – x into the original equation to find the values of x at the stationary points.

Pah, differentiation? They should have called it same-iation...

...you know, cos all the questions basically end up asking for the same thing. Other familiar faces that are likely to show up in implicit differentiation questions include finding tangents and normals to implicitly defined curves. All these differentiation questions set off in different ways to end up asking you the same thing, so make sure you know the basics.

Differentiation of a^x

This page is for Edexcel C4 only.

And so, in our never-ending quest to find yet more stuff to differentiate, we arrive in this peculiar and uncharted corner of maths. Keep your wits about you, and you'll probably make it through this page alive...

Learn the rule for Differentiating a^x

Here's another little rule you need to learn:

For any constant a,

$$\frac{d}{dx}(a^x) = a^x \ln a$$

The rule $\frac{d}{dx}(e^x) = e^x$ (see p46) is actually just a special case of this rule — $\frac{d}{dx}(e^x) = e^x \ln e = e^x \times 1 = e^x$

With a little bit of implicit differentiation, you can prove this rule:

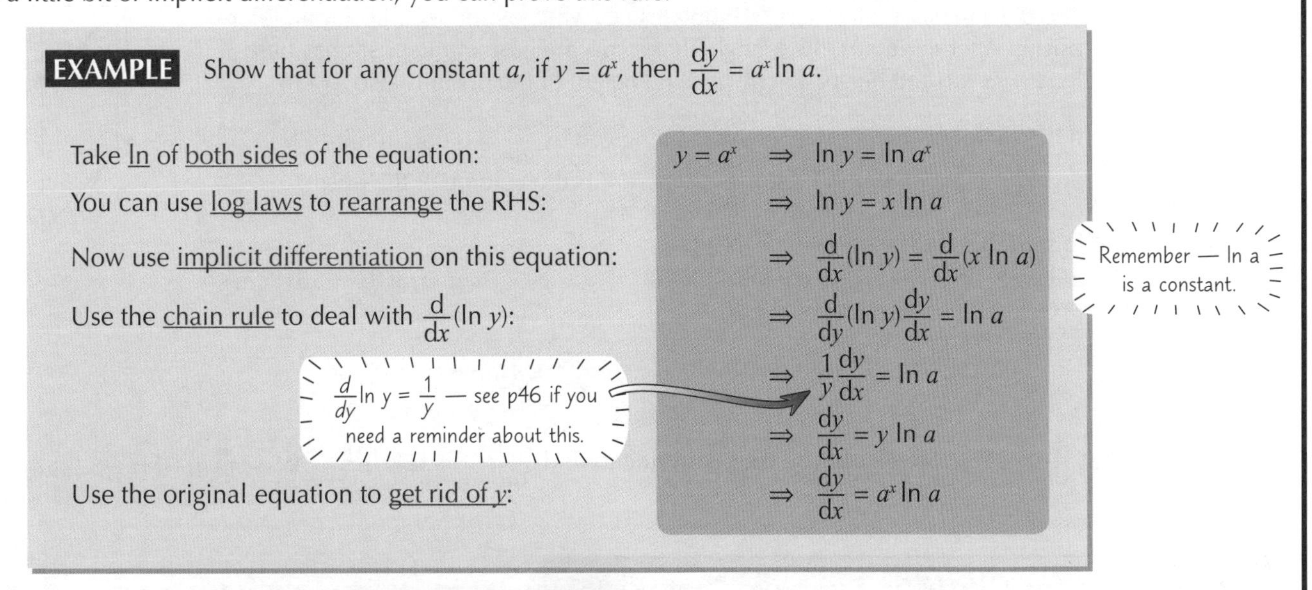

EXAMPLE Show that for any constant a, if $y = a^x$, then $\frac{dy}{dx} = a^x \ln a$.

Take ln of both sides of the equation: $y = a^x \Rightarrow \ln y = \ln a^x$

You can use log laws to rearrange the RHS: $\Rightarrow \ln y = x \ln a$

Now use implicit differentiation on this equation: $\Rightarrow \frac{d}{dx}(\ln y) = \frac{d}{dx}(x \ln a)$

Use the chain rule to deal with $\frac{d}{dx}(\ln y)$: $\Rightarrow \frac{d}{dy}(\ln y)\frac{dy}{dx} = \ln a$

$\frac{d}{dy}\ln y = \frac{1}{y}$ — see p46 if you need a reminder about this.

$\Rightarrow \frac{1}{y}\frac{dy}{dx} = \ln a$

$\Rightarrow \frac{dy}{dx} = y \ln a$

Use the original equation to get rid of y: $\Rightarrow \frac{dy}{dx} = a^x \ln a$

Remember — ln a is a constant.

Differentiate $a^{f(x)}$ using the Chain Rule

Once you've got the basic rule sorted, you can apply it to more complicated examples:

EXAMPLE Find the equation of the tangent to the curve $y = 3^{-2x}$ at the point $\left(\frac{1}{2}, \frac{1}{3}\right)$.

Use the 'normal' chain rule (from p45 — $\frac{dy}{dx} = \frac{dy}{du} \times \frac{du}{dx}$) to find $\frac{dy}{dx}$:

Let $u = -2x$, then $y = 3^u$ and $\frac{dy}{dx} = \frac{d}{du}(3^u)\,\frac{d}{dx}(-2x) = 3^u \ln 3 \times -2 = -2(3^{-2x} \ln 3)$

Now we can find the gradient and y-intercept of the tangent:

So at $\left(\frac{1}{2}, \frac{1}{3}\right)$, $\frac{dy}{dx} = -2(3^{-1} \ln 3) = -\frac{2}{3} \ln 3 = -0.732$ (to 3 s.f.)

So if the equation of the tangent at $\left(\frac{1}{2}, \frac{1}{3}\right)$ has the form $y = mx + c$, then

$\frac{1}{3} = (-\frac{2}{3} \ln 3)\frac{1}{2} + c \quad \Rightarrow \quad c = \frac{1}{3} + \frac{1}{3} \ln 3 = 0.700$ (to 3 s.f.)

So the equation of the tangent to $y = 3^{-2x}$ at $\left(\frac{1}{2}, \frac{1}{3}\right)$ is $y = -0.732x + 0.700$

This is a topic for lumberjacks — it's all about logs and a^xes...

Make sure you understand and learn the method used in the first example on this page. It's a prime candidate for a 'show that' exam question, so if you can regurgitate it at will, it just might net you a couple more precious marks.

Relating Rates of Change

This page is for Edexcel C4, OCR C3 and OCR MEI C3

This is one of those topics where the most awkward bit is getting your head round the information in the question. The actual maths is nothing like as bad as the questions usually make it sound. Honest.

The Chain Rule lets you Connect different Rates of Change

1) Some situations have a number of linked variables, like length, surface area and volume, or distance, speed and acceleration.
2) If you know the rate of change of one of these linked variables, and the equations that connect the variables, you can use the chain rule to help you find the rate of change of the other variables.

EXAMPLE A scientist is testing how a new material expands when it is gradually heated. The diagram shows the sample being tested, which is shaped like a triangular prism. After t minutes, the triangle that forms the base of the prism has base length $7x$ cm and height $4x$ cm, and the height of the prism is also $4x$ cm.

If the sample expands at a constant rate, given by $\frac{dx}{dt} = 0.05$ cm min^{-1}, find an expression in terms of x for $\frac{dV}{dt}$, where V is the volume of the prism.

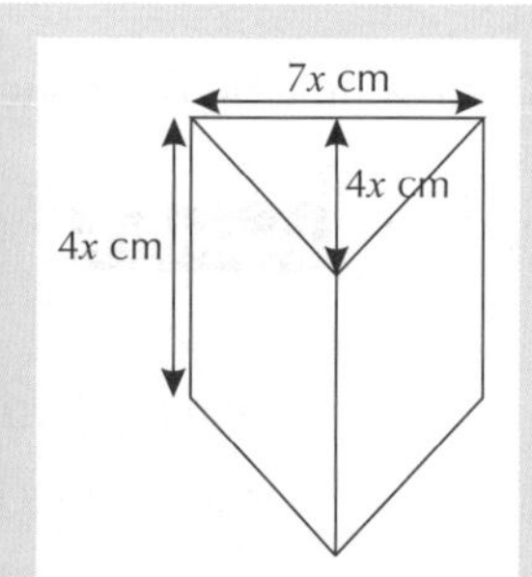

The best way to start this kind of question is to write down what you know. We've got enough information to write an expression for the volume of the prism:

$$V = (\tfrac{1}{2} \times 7x \times 4x) \times 4x = 56x^3 \text{ cm}^3$$

Differentiate this with respect to x:

$$\frac{dV}{dx} = 168x^2$$

We know that $\frac{dx}{dt} = 0.05$. So we can use the chain rule to find $\frac{dV}{dt}$:

$$\frac{dV}{dt} = \frac{dV}{dx} \times \frac{dx}{dt} = 168x^2 \times 0.05 = 8.4x^2$$

Watch out for Slightly Trickier questions

1) There are a couple of sneaky tricks in this type of question that could catch you out if you're not prepared for them.
2) In this next example, you have to spot that there's a hidden derivative described in words.
3) You also need to remember the rule $\frac{dy}{dx} = \frac{1}{\left(\frac{dx}{dy}\right)}$ (see p.45).

EXAMPLE A giant metal cube from space is cooling after entering the Earth's atmosphere. As it cools, the surface area of the cube decreases at a constant rate of 0.027 m^2 s^{-1}. If the side length of the cube after t seconds is x m, find $\frac{dx}{dt}$ at the point when $x = 15$ m.

Start with what you know:

We use $\frac{d}{dt}$ because it's a rate of time.

The cube has side length x m, so the surface area of the cube is $A = 6x^2 \Rightarrow \frac{dA}{dx} = 12x$

A decreases at a constant rate of 0.027 m^2 s^{-1} — we can write this as $\frac{dA}{dt} = -0.027$

This value is negative because A is decreasing.

Now use the chain rule to find $\frac{dx}{dt}$:

$$\frac{dx}{dt} = \frac{dx}{dA} \times \frac{dA}{dt} = \frac{1}{\left(\frac{dA}{dx}\right)} \times \frac{dA}{dt} = \frac{1}{12x} \times -0.027 = -\frac{0.00225}{x}$$

So when $x = 15$, $\frac{dx}{dt} = -\frac{0.00225}{x} = -\frac{0.00225}{15} = -0.00015$ m s^{-1}

I'd rate this page 10 out of 10 — if I do say so myself...

If you get stuck on a question like this, don't panic. Somewhere in the question there'll be enough information to write at least one equation linking some of the variables. If in doubt, write down any equations you can make, differentiate them all, and then see which of the resulting expressions you can link using the chain rule to make the thing you're looking for.

Core Section 7 — Practice Questions

If you think that was a lot of differentiation, be thankful you didn't live in Ancient Molgarahenia, where differentiation was the only maths permitted. Try these tasty warm-up questions for an authentic taste of Molgarahenian life.

Warm-up Questions

Q1 is for AQA C4, Edexcel C4, OCR C4 and OCR MEI C4

1) A curve is defined by the parametric equations $x = t^2$, $y = 3t^3 - 4t$.
   a) Find $\frac{dy}{dx}$ for this curve.
   b) Find the coordinates of the stationary points of the curve.

Q2-3 are for AQA C4, Edexcel C4, OCR C4 and OCR MEI C3

2) Use implicit differentiation to find $\frac{dy}{dx}$ for each of the following equations:
   a) $4x^2 - 2y^2 = 7x^2y$ b) $3x^4 - 2xy^2 = y$ c) $\cos x \sin y = xy$

3) Using your answers to question 2, find:
   a) the gradient of the tangent to the graph of $4x^2 - 2y^2 = 7x^2y$ at $(1, -4)$,
   b) the gradient of the normal to the graph of $3x^4 - 2xy^2 = y$ at $(1, 1)$.

Q4 is just for Edexcel C4

4) Write down the proof that $\frac{d}{dx}a^x = a^x \ln a$, where a is a constant.

Q5 is for Edexcel C4, OCR C3 and OCR MEI C3

5) A cuboid has length x cm, width $2x$ cm and height $3x$ cm.
   The cuboid is expanding, for some unexplained reason.
   If A is the surface area of the cuboid and V is its volume, find $\frac{dA}{dx}$ and $\frac{dV}{dx}$,
   and use them to show that if $\frac{dV}{dt} = 3$, then $\frac{dA}{dt} = \frac{22}{3x}$.

It is said that the Great Molgarahenian Plain was carpeted with differentiation as far as the eye could see. Your exams won't be quite that bad, but there will be some differentiation in there, so get practising...

Exam Questions

Q1-2 are for AQA C4, Edexcel C4, OCR C4 and OCR MEI C3

1 The equation of curve C is $6x^2y - 7 = 5x - 4y^2 - x^2$.

a) The line T has the equation $y = c$ and passes through a point on C where $x = 2$.
Find c, given that $c > 0$. *(2 marks)*

b) T also crosses C at point Q.

(i) Find the coordinates of Q. *(2 marks)*

(ii) Find the gradient of C at Q. *(6 marks)*

2 The curve C has the equation $3e^x + 6y = 2x^2y$.

a) (i) Use implicit differentiation to find an expression for $\frac{dy}{dx}$. *(3 marks)*

(ii) Show that at the stationary points of C, $y = \frac{3e^x}{4x}$. *(2 marks)*

b) Hence find the exact coordinates of the two stationary points of C. *(4 marks)*

Core Section 7 — Practice Questions

In 272 BC, the famous Molgarahenian philosopher, Bobby the Wise, was put to death for straying from the path of differentiation and doing some simultaneous equations. Don't be like Bobby, stick with these questions (for now)...

Exam Questions

Q3 is for just for Edexcel C4

3 a) Curve A has the equation $y = 4^x$.
What are the coordinates of the point on A where $\frac{dy}{dx} = \ln 4$? *(2 marks)*

b) Curve B has the equation $y = 4^{(x-4)^3}$. Find the gradient of B at the point $(3, \frac{1}{4})$. *(4 marks)*

Q4-5 are for AQA C4, Edexcel C4, OCR C4 and OCR MEI C4

4 The curve C is defined by the parametric equations

$$x = 3\theta - \cos 3\theta, \quad y = 2\sin\theta, \quad -\pi \leq \theta \leq \pi.$$

a) Find an expression for $\frac{dy}{dx}$. *(3 marks)*

b) (i) Show that the gradient of C at the point $(\pi + 1, \sqrt{3})$ is $\frac{1}{3}$. *(3 marks)*

(ii) Find the equation of the normal to C when $\theta = \frac{\pi}{6}$. *(4 marks)*

5 A curve, C, has parametric equations
$x = t^2 + 2t - 3, \qquad y = 2 - t^3.$

a) The line L is the tangent to C at $y = -6$. Show that the equation of L is $y = -2x + 4$. *(4 marks)*

b) L also meets C at point P.

(i) Find the coordinates of P. *(4 marks)*

(ii) Find the equation of the normal to the curve at P. *(3 marks)*

Q6 is for Edexcel C4, OCR C3 and OCR MEI C3

6

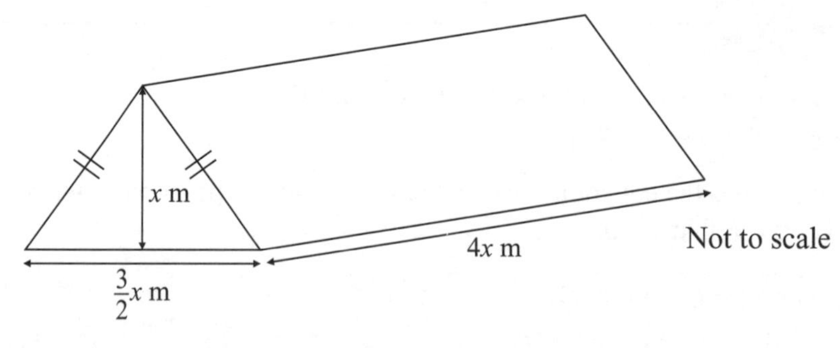

The triangular prism shown in the diagram is expanding.
The dimensions of the prism after t seconds are given in terms of x.
The prism is $4x$ m long, and its cross-section is an isosceles triangle with base $\frac{3}{2}x$ m and height x m.

a) Show that, if the surface area of the prism after t seconds is A m², then $A = \frac{35}{2}x^2$. *(3 marks)*

The surface area of the prism is increasing at a constant rate of $0.07 \text{ m}^2 \text{ s}^{-1}$.

b) Find $\frac{dx}{dt}$ when $x = 0.5$. *(3 marks)*

c) If the volume of the prism is V m³, find the rate of change of V when $x = 1.2$. *(4 marks)*

Location of Roots

This page is for AQA C3, Edexcel C3, OCR C3 and OCR MEI C3 (coursework only)

Small but perfectly formed, this section will tell you everything you need to know (for now) about finding approximations of roots. Oh the thrills.

A Change of Sign from f(a) to f(b) means a Root Between a and b

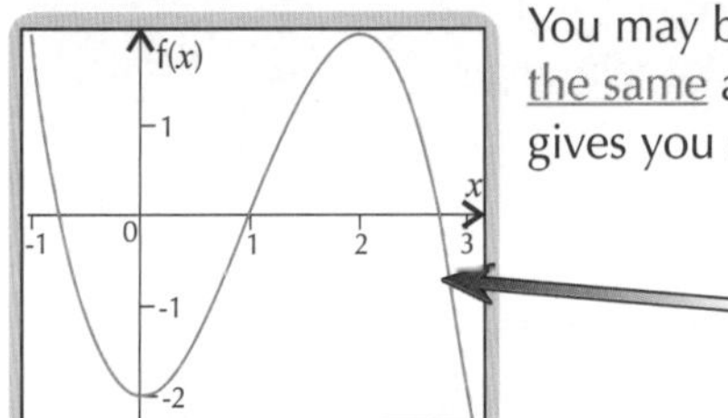

You may be asked to 'solve' or 'find the roots of' an equation (where $f(x) = 0$). This is exactly the same as finding the value of x where the graph crosses the x-axis. The graph of the function gives you a rough idea how many roots there are (if any) and where.

E.g. the function $f(x) = 3x^2 - x^3 - 2$ (shown here) has 3 roots in the interval $-1 \leq x \leq 3$, since it crosses the x-axis three times (i.e. there are 3 solutions to the equation $3x^2 - x^3 - 2 = 0$). You can also see from the graph that $x = 1$ is a root, and the other roots are close to $x = -1$ and $x = 3$.

Look at the graph above at the root $x = 1$. For x-values just before the root, $f(x)$ is negative, and just after the root, $f(x)$ is positive. It's the other way around for the other two roots, but either way:

> $f(x)$ changes sign as it passes through a root.

This is only true for continuous functions — ones that are joined up all the way along with no 'jumps' or gaps.

> To show that a root lies in the interval between two values 'a' and 'b':
>
> 1) Find f(a) and f(b).
> 2) If the two answers have different signs, and the function is continuous, there's a root somewhere between 'em.

$f(x) = \tan x$ is an example of a non-continuous function — it has gaps where $f(x)$ changes sign even though there's no root.

EXAMPLE Show that $x^4 + 3x - 5 = 0$ has a root in the interval $1.1 \leq x \leq 1.2$.

1) Put both 1.1 and 1.2 into the expression:
   $f(1.1) = (1.1)^4 + (3 \times 1.1) - 5 = -0.2359$. $f(1.2) = (1.2)^4 + (3 \times 1.2) - 5 = 0.6736$.
2) $f(1.1)$ and $f(1.2)$ have different signs, and $f(x)$ is continuous, so there's a root in the interval $1.1 \leq x \leq 1.2$.

Use an Iteration Formula to find Approximations of Roots

Some equations are just too darn tricky to solve properly. For these, you need to find approximations to the roots, to a certain level of accuracy. You'll usually be told the value of x that a root is close to, and then iteration does the rest.

Iteration is like fancy trial and improvement. You put an approximate value of a root x into an iteration formula, and out pops a slightly more accurate value. Then repeat as necessary until you have an accurate enough answer.

EXAMPLE Use the iteration formula $x_{n+1} = \sqrt[3]{x_n + 4}$ to solve $x^3 - 4 - x = 0$, to 2 d.p. Start with $x_0 = 2$.

1) The notation x_n just means the approximation of x at the nth iteration.
   So putting x_0 in the formula for x_n, gives you x_{n+1}, which is x_1, the first iteration.
2) $x_0 = 2$, so $x_1 = \sqrt[3]{x_0 + 4} = \sqrt[3]{2 + 4} = 1.8171...$ ← Leave this in your calculator for accuracy.
3) This value now gets put back into the formula to find x_2:
   $x_1 = 1.8171...$, so $x_2 = \sqrt[3]{x_1 + 4} = \sqrt[3]{1.8171... + 4} = 1.7984...$ ← You should now just be able to type '$\sqrt[3]{(\text{ANS} + 4)}$' in your calculator and keep pressing enter for each iteration.
4) Carry on until you get answers that are the same when rounded to 2 d.p:
   $x_2 = 1.7984...$, so $x_3 = \sqrt[3]{x_2 + 4} = \sqrt[3]{1.7984... + 4} = 1.7965...$
5) x_2, x_3, and all further iterations are the same when rounded to 2 d.p., so the root is $x = 1.80$ to 2 d.p.

The hat — an approximate solution to root problems...

Just to re-iterate (ho ho), the main ways to find those roots are sign changes and iteration formulas. It's a doddle. Don't get confused and go looking for tree roots — that involves a lot of digging, and you'll end up getting all muddy.

Iterative Methods

This page is for AQA C3, Edexcel C3, OCR C3 and OCR MEI C3 (coursework only)

Now we come to the trickier bits. It's all well and good being able to plug numbers into a formula, but where do those formulas come from? And why don't they always work? Read on to find out...

Rearrange the Equation to get the Iteration Formula

The iteration formula is just a rearrangement of the equation, leaving a single 'x' on one side.

There are often lots of different ways to rearrange the equation, so in the exam you'll usually be asked to 'show that' it can be rearranged in a certain way, rather than starting from scratch.

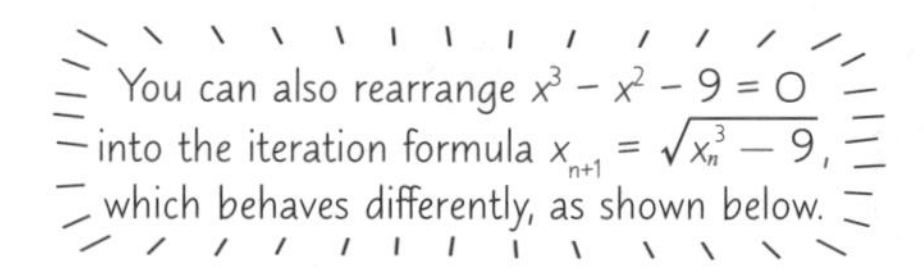

EXAMPLE Show that $x^3 - x^2 - 9 = 0$ can be rearranged into $x = \sqrt{\frac{9}{x-1}}$.

1) The '9' is on its own in the fraction so try:
   $x^3 - x^2 - 9 = 0 \Rightarrow x^3 - x^2 = 9$
2) The LHS can be factorised now: $x^2(x - 1) = 9$
3) Get the x^2 on its own by dividing by $x - 1$: $x^2 = \frac{9}{x-1}$
4) Finally square root both sides: $x = \sqrt{\frac{9}{x-1}}$

You can now use the iteration formula $x_{n+1} = \sqrt{\frac{9}{x_n - 1}}$ to find approximations of the roots.

Sometimes an iteration formula just will not find a root. In these cases, no matter how close to the root you have x_0, the iteration sequence diverges — the numbers get further and further apart. The iteration also might stop working — e.g. if you have to take the square root of a negative number.

EXAMPLE The equation $x^3 - x^2 - 9 = 0$ has a root close to $x = 2.5$.
What is the result of using $x_{n+1} = \sqrt{x_n^3 - 9}$ with $x_0 = 2.5$ to find this root?

1) Start with $x_1 = \sqrt{2.5^3 - 9} = 2.5739...$ (seems okay so far...)
2) Subsequent iterations give: $x_2 = 2.8376...$, $x_3 = 3.7214...$, $x_4 = 6.5221...$ — so the sequence diverges.

Usually though, in an exam question, you'll be given a formula that converges to a certain root — otherwise there's not much point in using it. If your formula diverges when it shouldn't, go back and check you've not made a mistake.

Use Upper and Lower Bounds to 'Show that' a root is correct

Quite often you'll be given an approximation to a root and be asked to show that it's correct to a certain accuracy. This is a lot like showing that the root lies in a certain interval (on the last page) — the trick is to work out the right interval.

Upper and lower bounds are sometimes known as error bounds.

EXAMPLE Show that $x = 2.472$ is a root of the equation $x^3 - x^2 - 9 = 0$ to 3 d.p.

1) If $x = 2.472$ is a root rounded to 3 decimal places, the exact root must lie between the upper and lower bounds of this value — 2.4715 and 2.4725. Any value in this interval would be rounded to 2.472 to 3 d.p.

2.471 2.4715 2.472 2.4725 2.473

2) The function $f(x) = x^3 - x^2 - 9$ is continuous, so you know the root lies in the interval $2.4715 \leq x \leq 2.4725$ if f(2.4715) and f(2.4725) have different signs.
3) $f(2.4715) = 2.4715^3 - 2.4715^2 - 9 = -0.0116...$
   and $f(2.4725) = 2.4725^3 - 2.4725^2 - 9 = 0.0017...$
4) f(2.4715) and f(2.4725) have different signs, so the root must lie in between them. Since any value between would be rounded to 2.472 to 3 d.p. this answer must be correct.

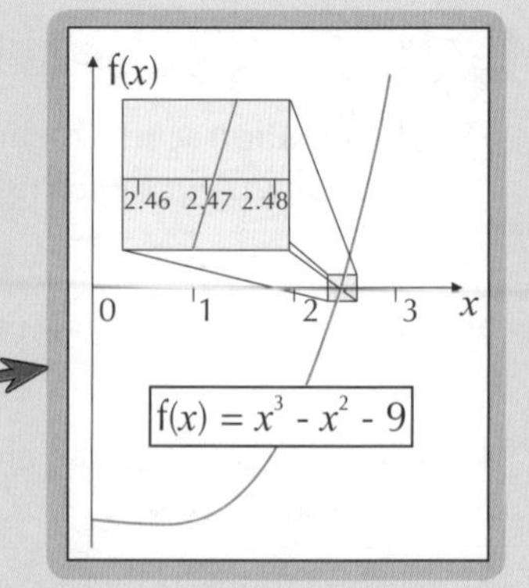

You're bound to be asked questions on this...

There are usually several parts to an exam question on iteration, but it's all pretty standard stuff. I'd put good money on you having to rearrange an equation to get an iteration formula, or show that an approximation to a root is correct.

Iterative Methods

This page is for AQA C3 and OCR MEI C3 (coursework only)

Whilst iterations are pretty thrilling on their own, put them into a diagram and the world's your lobster. Well, that might be a bit of an exaggeration, but you can show convergence or divergence easily on a diagram. And it looks pretty.

You can show Iterations on a Diagram

Once you've calculated a sequence of iterations, you can plot the points on a diagram and use it to show whether your sequence converges or diverges.

Sketching Iterations

1) **First, sketch the graphs of $y = x$ and $y = f(x)$ (where f(x) is the iterative formula). The point where the two graphs meet is the root you're aiming for.**
2) **Draw a vertical line from the x-value of your starting point (x_0) until it meets the curve $y = f(x)$.**
3) **Now draw a horizontal line from this point to the line $y = x$. At this point, the x-value is x_1, the value of your first iteration. This is one step.**
4) **Draw another step — a vertical line from this point to the curve, and a horizontal line joining it to the line $y = x$.**
5) **Repeat step 4) for each of your iterations.**
6) **If your steps are getting closer and closer to the root, the sequence of iterations is converging. If the steps are moving further and further away from the root, the sequence is diverging.**

This method produces two different types of diagrams — cobweb diagrams and staircase diagrams.

Convergent iterations Home In on the Root

It's probably easiest to follow the method by looking at a few examples:

EXAMPLES

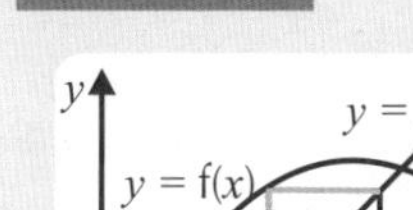

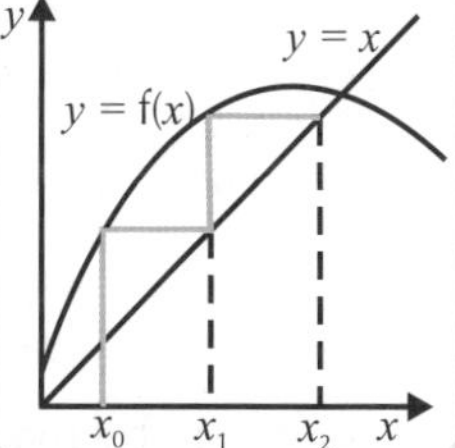

This is an example of a convergent staircase diagram. Starting at x_0, the next iterations x_1 and x_2 are getting closer to the point where the two graphs intersect (the root).

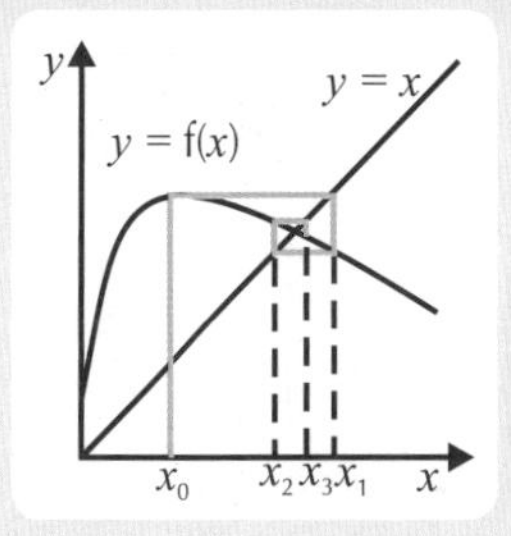

This is an example of a convergent cobweb diagram. In this case, the iterations alternate between being below the root and above the root, but are still getting closer each time.

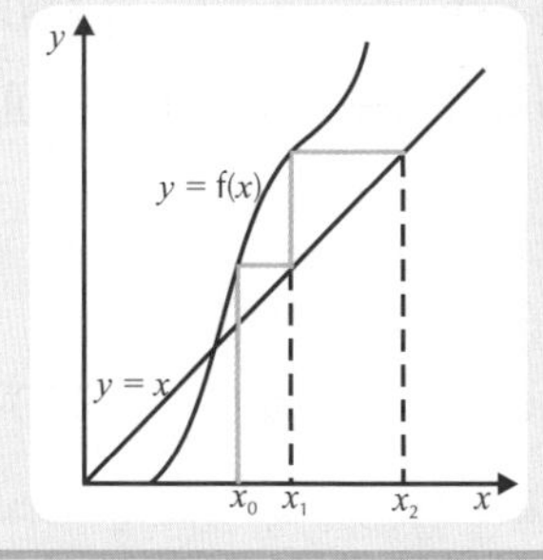

This is an example of a divergent staircase diagram. Starting at x_0, the iterations x_1 and x_2 are getting further away from the root.

In each case, the diagram will look different depending on where your starting point is.

There are cobwebs on my staircase...

Drawing diagrams is actually pretty easy — it's mainly just a case of drawing straight lines. If your iterative function is a bit nasty (which it probably will be), you'll be given the graph of it in the exam and you'll just have to add on the iteration steps.

Iterative Methods

This page is for AQA C3, Edexcel C3, OCR C3 and OCR MEI C3 (coursework only)

So now that you know all you need to know to be able to tackle the exam questions, let's have a look at how it all fits together in a worked example. Brace yourself...

Questions on Locating Roots combine all the Different Methods

Obviously, the questions you come across in the exam won't be identical to the one below (if only...), but there are, after all, only a limited number of ways you can be asked to find a root using the numerical methods in this section. If you can follow the steps shown below you won't go far wrong.

EXAMPLE The graph below shows both roots of the continuous function $f(x) = 6x - x^2 + 13$.

a) Show that the positive root, α, lies in the interval $7 < x < 8$.

b) Show that $6x - x^2 + 13 = 0$ can be rearranged into the formula: $x = \sqrt{6x + 13}$.

c) Use the iteration formula $x_{n+1} = \sqrt{6x_n + 13}$ and $x_0 = 7$ to find α to 1 d.p.

d) Sketch a diagram to show the convergence of the sequence for x_1, x_2 and x_3.

e) Show that the negative root, β, is -1.690 to 3 d.p.

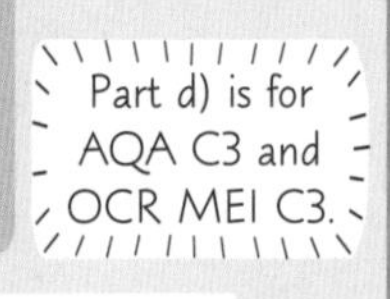

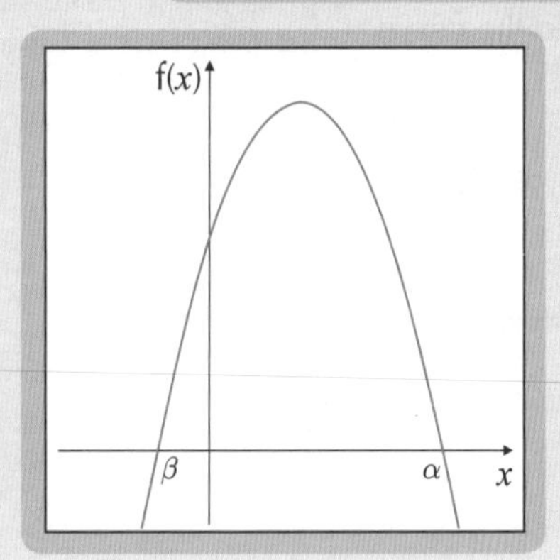

a) f(x) is a continuous function, so if f(7) and f(8) have different signs then there is a root in the interval $7 < x < 8$:

$f(7) = (6 \times 7) - 7^2 + 13 = 6.$
$f(8) = (6 \times 8) - 8^2 + 13 = -3.$

There is a change of sign so $7 < \alpha < 8$.

b) Get the x^2 on its own to make: $6x + 13 = x^2$

Now take the (positive) square root to leave: $x = \sqrt{6x + 13}$.

c) Using $x_{n+1} = \sqrt{6x_n + 13}$ with $x_0 = 7$, gives $x_1 = \sqrt{6 \times 7 + 13} = 7.4161...$

Continuing the iterations:

$x_2 = \sqrt{6 \times 7.4161... + 13} = 7.5826...$ $\quad x_3 = \sqrt{6 \times 7.5826... + 13} = 7.6482...$
$x_4 = \sqrt{6 \times 7.6482... + 13} = 7.6739...$ $\quad x_5 = \sqrt{6 \times 7.6739... + 13} = 7.6839...$
$x_6 = \sqrt{6 \times 7.6839... + 13} = 7.6879...$ $\quad x_7 = \sqrt{6 \times 7.6879... + 13} = 7.6894...$

x_4 to x_7 all round to 7.7 to 1 d.p., so to 1 d.p. $\alpha = 7.7$.

The list of results from each iteration x_1, x_2, x_3... is called the iteration sequence.

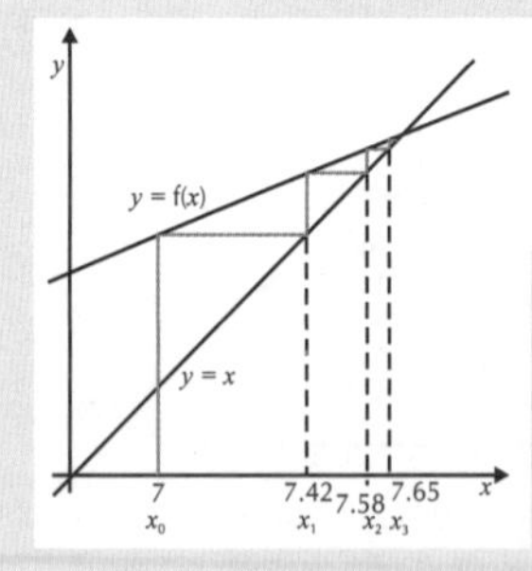

d) $y = f(x)$ and $y = x$ will be drawn for you, and the position of x_0 will be marked. All you have to do is draw on the lines and label the values of x_1, x_2 and x_3. You can see from the diagram that the sequence is a convergent staircase.

e) If $\beta = -1.690$ to 3 d.p. the upper and lower bounds are -1.6895 and -1.6905. The root must lie between these values in order to be rounded to -1.690.

As the function is continuous, if $f(-1.6895)$ and $f(-1.6905)$ have different signs then $-1.6905 \le \beta \le -1.6895$:

$f(-1.6895) = (6 \times -1.6895) - (-1.6895)^2 + 13 = 0.00858...$
$f(-1.6905) = (6 \times -1.6905) - (-1.6905)^2 + 13 = -0.00079...$

There is a change of sign, so $-1.6905 \le \beta \le -1.6895$, and so $\beta = -1.690$ to 3 d.p.

Trouble finding a root? Try sat-nav...

And that's your lot — wasn't so bad, was it? All done and dusted, except for those practice questions you've come to know and love so well. So calculators at the ready, grab your lucky pen and prepare to iterate your heart out...

Core Section 8 — Practice Questions

To make up for a short section I'm giving you lots of lovely practice. Stretch those thinking muscles with this warm-up:

Warm-up Questions

Q1-6 are for AQA C3, Edexcel C3, OCR C3 and OCR MEI C3 (coursework only)

1) The graph shows the function $f(x) = e^x - x^3$ for $0 \le x \le 5$.
How many roots does the equation $e^x - x^3 = 0$ have in the interval $0 \le x \le 5$?

2) Show that there is a root in the interval:
   a) $3 < x < 4$ for $\sin(2x) = 0$, Don't forget to use radians when you're given trig functions.
   b) $2.1 < x < 2.2$ for $\ln(x - 2) + 2 = 0$,
   c) $4.3 < x < 4.5$ for $x^3 - 4x^2 = 7$.

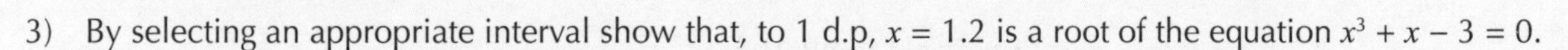

3) By selecting an appropriate interval show that, to 1 d.p, $x = 1.2$ is a root of the equation $x^3 + x - 3 = 0$.

4) Use the formula $x_{n+1} = -\frac{1}{2}\cos x_n$, with $x_0 = -1$, to find a root of $\cos x + 2x = 0$ to 2 d.p.

5) Use the formula $x_{n+1} = \sqrt{\ln x_n + 4}$, with $x_0 = 2$, to find a root of $x^2 - \ln x - 4 = 0$ to 3 d.p.

6) a) Show that the equation $2x^2 - x^3 + 1 = 0$ can be written in the form:
   i) $x = \sqrt{\frac{-1}{2-x}}$ ii) $x = \sqrt[3]{2x^2 + 1}$ iii) $x = \sqrt{\frac{x^3 - 1}{2}}$
   b) Use iteration formulas based on each of the above rearrangements with $x_0 = 2.3$ to find a root of $2x^2 - x^3 + 1 = 0$ to 2 d.p.
   Which of the three formulas converge to a root?

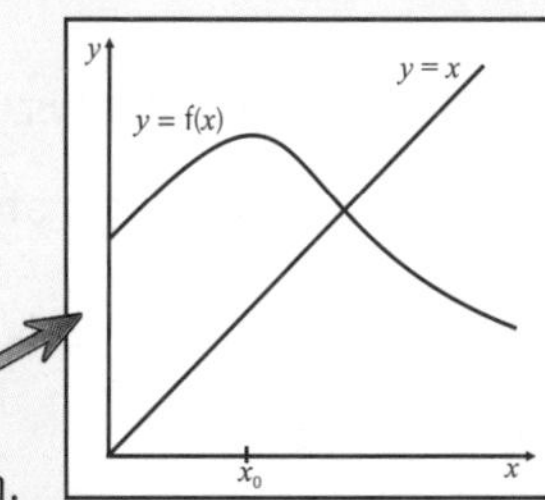

Q7 is for AQA C3 and OCR MEI C3 (coursework only)

7) Using the position of x_0 as given on the graph, draw a staircase or cobweb diagram showing how the sequence converges. Label x_1 and x_2 on the diagram.

And for my next trick... Sadly no magic here, but all the right kinds of questions to prepare you for the exam. Which may not be what you want, but it's definitely what you need.

Exam Questions

Q1-3 are for AQA C3, Edexcel C3 and OCR C3

1 The sketch below shows part of the graph of the function $f(x) = 2xe^x - 3$.
The curve crosses the x-axis at the point $P\ (p, 0)$, as shown, so p is a root of the equation $f(x) = 0$.

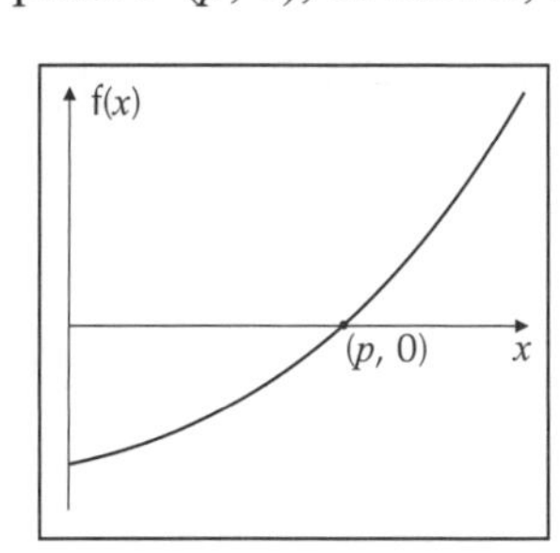

a) Show that $0.7 < p < 0.8$. *(3 marks)*

b) Show that $f(x) = 0$ can be rewritten as:

$$x = \frac{3}{2}e^{-x}.$$

(2 marks)

c) Starting with $x_0 = 0.7$, use the iteration

$$x_{n+1} = \frac{3}{2}e^{-x_n}$$

to find x_1, x_2, x_3 and x_4 to 4 d.p. *(3 marks)*

d) Show that $p = 0.726$, to 3 d.p. *(3 marks)*

Core Section 8 — Practice Questions

2 The graph of the function $y = \sin 3x + 3x$, $0 < x < \pi$, meets the line $y = 1$ when $x = a$.

a) Show that $0.1 < a < 0.2$. *(4 marks)*

b) Show that the equation:

$$\sin 3x + 3x = 1$$

can be written as:

$$x = \tfrac{1}{3}(1 - \sin 3x).$$

(2 marks)

c) Starting with $x_0 = 0.2$, use the iteration:

$$x_{n+1} = \tfrac{1}{3}(1 - \sin 3x_n)$$

to find x_4, to 3 d.p. *(2 marks)*

3 The sequence given by:

$$x_{n+1} = \sqrt[3]{x_n^2 - 4}, \quad x_0 = -1$$

converges to a number 'b'.

a) Find the values of x_1, x_2, x_3 and x_4 correct to 4 decimal places. *(3 marks)*

b) Show that $x = b$ is a root of the equation:

$$x^3 - x^2 + 4 = 0$$

(2 marks)

c) Show that $b = -1.315$ to 3 decimal places, by choosing an appropriate interval. *(3 marks)*

Q4 a)-c) are for AQA C3, Edexcel C3 and OCR C3
Q4 d) is just for AQA C3

4 The function:

$$f(x) = \ln(x + 3) - x + 2, \quad x > -3$$

has a root at $x = m$.

a) Show that m lies between 3 and 4. *(3 marks)*

b) Find, using iteration, the value of m correct to 2 decimal places.
Use the iteration formula: $x_{n+1} = \ln(x_n + 3) + 2$
with $x_0 = 3$. *(3 marks)*

c) Use a suitable interval to verify that your answer to part b) is correct to 2 decimal places. *(3 marks)*

d) The graph below shows part of the curve $y = \ln(x + 3) + 2$, the line $y = x$ and the position of x_0. Complete the diagram showing the convergence of the iteration sequence, showing the locations of x_1 and x_2 on the graph.

You can trace the graph if you don't want to draw in your book.

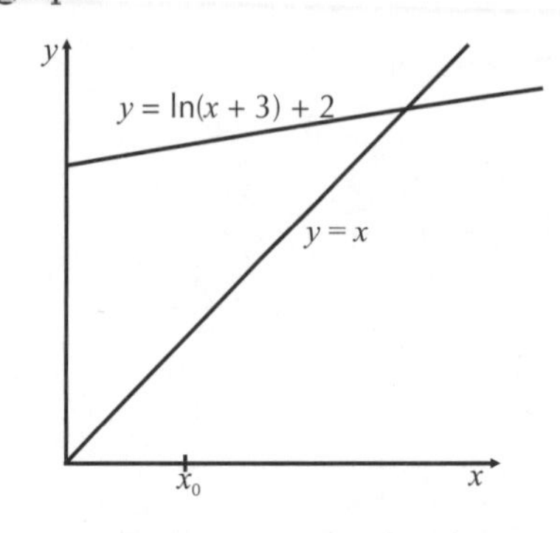

(2 marks)

Integration of e^x and 1/x

This page is for AQA C3, Edexcel C4, OCR C3 and OCR MEI C3

Although it was many moons ago that you last encountered integration, way back in AS Maths, it's an integral part of A2 Maths. It does the opposite of differentiation, so some of this stuff should look familiar to you.

e^x integrates to give e^x (+ C)

As e^x differentiates to give e^x (see p.46), it makes sense that

$$\int e^x dx = e^x + C$$

Don't forget the constant of integration.

Once you're happy with that, you can use it to solve lots of integrations that have an e^x term in them.
If the coefficient of x isn't 1, you need to divide by that coefficient when you integrate — so $\int e^{kx} dx = \frac{1}{k}e^{kx} + C$

EXAMPLES Integrate the following: a) e^{7x} b) $2e^{4-3x}$ c) $e^{\frac{x}{2}}$.

a) $\int e^{7x} dx = \frac{1}{7}e^{7x} + C$ If you differentiated e^{7x} using the chain rule, you'd get $7e^{7x}$. So when you integrate, you need to divide by 7 (the coefficient of x). This is so that if you differentiated your answer you'd get back to e^{7x}.

b) $\int 2e^{4-3x} dx = -\frac{2}{3}e^{4-3x} + C$ This one isn't as bad as it looks — if you differentiated $2e^{4-3x}$, you'd get $-6e^{4-3x}$, so you need to divide by -3 (the coefficient of x) when you integrate. Differentiating your answer gives you $2e^{4-3x}$.

c) $\int e^{\frac{x}{2}} dx = \int e^{\frac{1}{2}x} dx = 2e^{\frac{x}{2}} + C$ If you differentiated this one using the chain rule, you'd get $\frac{1}{2}e^{\frac{x}{2}}$, so you need to multiply by 2 when you integrate.

Whenever you integrate,

ALWAYS DIFFERENTIATE YOUR ANSWER TO CHECK IT WORKS

— you should end up with the thing you integrated in the first place. It's the best way to check that you divided or multiplied by the right number.

1/x integrates to ln |x| (+ C)

When you first came across integration, you couldn't integrate $\frac{1}{x}$ ($= x^{-1}$) by increasing the power by 1 and dividing by it, as you ended up dividing by 0 (which is baaaaad).

However, on p.46, you saw that $\ln x$ differentiates to give $\frac{1}{x}$, so

$$\int \frac{1}{x} dx = \ln|x| + C$$

Don't worry about where the modulus sign (see p.4) comes from — using $|x|$ just means that there isn't a problem when x is negative.

EXAMPLES Integrate the following: a) $\frac{5}{x}$ b) $\frac{1}{3x}$ c) $\frac{1}{4x+5}$.

There are more examples like this one on p.69.

a) $\int \frac{5}{x} dx = 5\int \frac{1}{x} dx = 5\ln|x| + C$ 5 is a constant coefficient — you can take it outside the integral if you want.

You could also write $5\ln|x|$ as $\ln|x^5|$.

b) $\int \frac{1}{3x} dx = \frac{1}{3}\int \frac{1}{x} dx = \frac{1}{3}\ln|x| + C$ Be careful with ones like this — 1/3 is just the coefficient, so it goes outside $\ln|x|$. Don't make the mistake of putting $\ln|3x|$ — this would differentiate to give $1/x$ (as $\ln 3x = \ln 3 + \ln x$, so when you differentiate, $\ln 3$ disappears).

c) $\int \frac{1}{4x+5} dx = \frac{1}{4}\ln|4x+5| + C$ However, for this one you have to leave the coefficient (4) inside ln because it's part of the function $4x + 5$. You still have to divide by 4 though (again, try differentiating it to see why).

Integration feels pretty constant to me...

These integrations are pretty easy — the only thing you have to worry about is if x has a coefficient that isn't 1. When this happens, work out what you think the answer will look like (e.g. e^x, $\ln|x|$, etc.), then differentiate to see what you get. Then you might have to adjust your answer, usually by dividing or multiplying by the coefficient, to get back to what you started with.

Integration of Sin and Cos

This page is for AQA C3, Edexcel C4, OCR C4 and OCR MEI C3

If you thought you'd killed off the dragon that is trigonometry, you're sadly mistaken. It rears its ugly head again, and now you need to know how to integrate trig functions. Find your most trusty dragon-slaying sword and read on...

Sin and Cos are Easy to integrate

From Section 6, you know that $\sin x$ differentiates to give $\cos x$, $\cos x$ differentiates to give $-\sin x$ and $\tan x$ differentiates to give $\sec^2 x$ (where the angle x is in radians). So it's pretty obvious that:

$$\int \sin x \, dx = -\cos x + C$$
$$\int \cos x \, dx = \sin x + C$$
$$\int \sec^2 x \, dx = \tan x + C$$

Integrating $\tan x$ is a bit different — see p.69.

If x has a coefficient that isn't 1 (e.g. $\sin 3x$), you just divide by the coefficient when you integrate — just like on the previous page.

EXAMPLE Find $\int \cos 4x - 2\sin 2x + \sec^2 \frac{1}{2}x \, dx$.

Integrate each term separately using the results from above:

$\int \cos 4x \, dx = \frac{1}{4}\sin 4x \qquad \int -2\sin 2x \, dx = -2\left(-\frac{1}{2}\cos 2x\right) = \cos 2x \qquad \int \sec^2 \frac{1}{2}x \, dx = \frac{1}{\frac{1}{2}}\tan \frac{1}{2}x = 2\tan\frac{1}{2}x$

Putting these terms together and adding the constant gives:

$$\int \cos 4x - 2\sin 2x + \sec^2 \tfrac{1}{2}x \, dx = \frac{1}{4}\sin 4x + \cos 2x + 2\tan\tfrac{1}{2}x + C$$

There are some Results you can just Learn

OCR MEI can just ignore this bit

There are a list of trig integrals that you can just learn — you don't need to know where they came from, you can just use them. You've met these ones before — they're the results of differentiating $\text{cosec}\, x$, $\sec x$ and $\cot x$ (see p.50).

There are some more trig integrals on pages 69 and 71.

$$\int \text{cosec}\, x \cot x \, dx = -\text{cosec}\, x + C$$
$$\int \sec x \tan x \, dx = \sec x + C$$
$$\int \text{cosec}^2 x \, dx = -\cot x + C$$

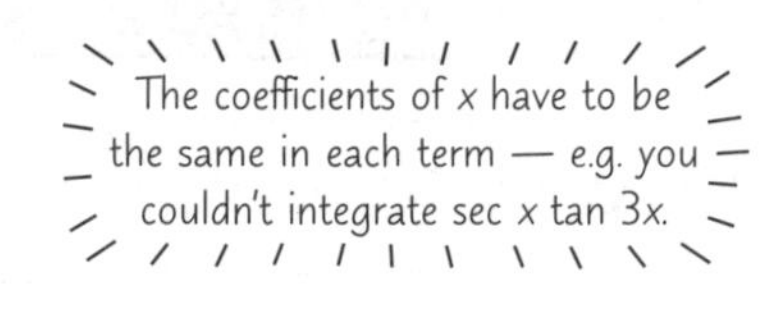

As usual, you need to divide by the coefficient of x when you integrate.

EXAMPLE Find $\int 10\sec 5x \tan 5x + \frac{1}{2}\text{cosec}\, 3x \cot 3x - \text{cosec}^2(6x+1) \, dx$.

This one looks a bit scary, but take it one step at a time. Integrate each bit in turn to get:

1. $\int 10\sec 5x \tan 5x \, dx = \frac{1}{5}\cdot 10\sec 5x = 2\sec 5x$

2. $\int \frac{1}{2}\text{cosec}\, 3x \cot 3x \, dx = -\frac{1}{3}\cdot\frac{1}{2}\text{cosec}\, 3x = -\frac{1}{6}\text{cosec}\, 3x$

3. $\int -\text{cosec}^2(6x+1) \, dx = -\frac{1}{6}(-\cot(6x+1)) = \frac{1}{6}\cot(6x+1)$

The + 1 inside the brackets has no effect on the integration — differentiate to see why.

Putting these terms together and adding the constant gives:

$$\int 10\sec 5x \tan 5x + \tfrac{1}{2}\text{cosec}\, 3x \cot 3x - \text{cosec}^2(6x+1) \, dx = 2\sec 5x - \tfrac{1}{6}\text{cosec}\, 3x + \tfrac{1}{6}\cot(6x+1) + C$$

This is starting to grate on me now...

Although you're not given all of these integrals on the formula sheet, some of the trickier ones (e.g. $\sec x \tan x$ and $\text{cosec}^2 x$) are on the list of differentiation formulas. As long as you work backwards (i.e. from $f'(x)$ to $f(x)$) you can just use these results without having to remember them all. Be careful with the coefficients — don't forget to divide by them when you integrate.

Integration of f′(x)/f(x)

This page is for AQA C3, Edexcel C4, OCR C4 and OCR MEI C3

Sometimes you get integrals that look really nasty — like fractions. However, there are a couple of clever tricks that can make them easy to integrate.

Some Fractions integrate to ln

If you have a fraction that has a function of x on the numerator and a different function of x on the denominator (e.g. $\frac{x-2}{x^3+1}$), you'll probably struggle to integrate it. However, if you have a fraction where the numerator is the derivative of the denominator (e.g. $\frac{3x^2}{x^3+1}$), it integrates to give ln of whatever the denominator is (in this case, $x^3 + 1$).

In general terms, this is written as:

$$\int \frac{f'(x)}{f(x)}\,dx = \ln|f(x)| + C$$

This is another one that comes from the chain rule (p.45) — if you differentiated $\ln|f(x)|$, you'd end up with the fraction on the left.

The hardest bit about questions like this is recognising that the denominator differentiates to give the numerator. Once you've spotted that, it's dead easy. They might make the numerator a multiple of the denominator just to confuse things, so watch out for that.

Trig identities can even sneak into questions like this, but you probably won't get anything too nasty.

EXAMPLES

Find a) $\int \frac{8x^3-4}{x^4-2x}\,dx$ and b) $\int \frac{3\sin 3x}{\cos 3x+2}\,dx$.

a) $\frac{d}{dx}(x^4-2x) = 4x^3-2$

and $8x^3-4 = 2(4x^3-2)$

The numerator is 2 × the derivative of the denominator, so

$\int \frac{8x^3-4}{x^4-2x}\,dx = 2\ln|x^4-2x| + C$

b) $\frac{d}{dx}(\cos 3x+2) = -3\sin 3x$

The numerator is minus the derivative of the denominator, so

$\int \frac{3\sin 3x}{\cos 3x+2}\,dx = -\ln|\cos 3x+2| + C$

$= -\ln|\cos 3x+2| + \ln k = -\ln|k(\cos 3x+2)|$

Using $C = \ln k$, you can combine all the terms into one using the laws of logs. $\ln k$ is just a constant.

You can get ln of Trig Functions too

OCR MEI can just ignore this bit

You might have noticed from part (b) above that you can work out the integral of $\tan x$ using this method:

$\tan x = \frac{\sin x}{\cos x}$, and $\frac{d}{dx}(\cos x) = -\sin x$

The numerator is minus the derivative of the denominator, so

$\int \tan x\,dx = \int \frac{\sin x}{\cos x}\,dx = -\ln|\cos x| + C$

$-\ln|\cos x|$ is the same as $\ln|\sec x|$ — this comes from the laws of logs on p.34.

There are some other trig functions that you can integrate in the same way:

$$\int \cot x\,dx = \ln|\sin x| + C$$

$$\int \operatorname{cosec} x\,dx = -\ln|\operatorname{cosec} x + \cot x| + C$$

$$\int \sec x\,dx = \ln|\sec x + \tan x| + C$$

This list is given in the formula booklet — so you don't need to learn them (just be able to use them).

EXAMPLE

Find $\int \frac{1}{2}\operatorname{cosec} 2x\,dx$.

You can just use the result above — so all you have to do is work out what happens to the coefficient. The coefficient is 2, so you need to divide by 2 when you integrate:

$\int \frac{1}{2}\operatorname{cosec} 2x\,dx = -\frac{1}{4}\ln|\operatorname{cosec} 2x + \cot 2x| + C$

Check this by differentiating (using the chain rule with $u = \operatorname{cosec} 2x + \cot 2x$).

3 pages in and I've run out of jokes on integration. Please help...

If you come across an integration question with a fraction that doesn't seem to integrate easily, have a quick look and see if one bit is the derivative of the other. If it is, use the rule above and you'll be as happy as Larry (and Larry's always happy).

Integration Using the Chain Rule Backwards

This page is for AQA C3, Edexcel C4, OCR C4 and OCR MEI C3

Most integrations aren't as bad as they look — on the previous page, you saw how to integrate special fractions, and now it's time for certain products. There are some things you can look out for when you're integrating...

You can use the *Chain Rule* in *Reverse*

You came across the chain rule in Section 6 (back on p.45) — it's where you write the thing you're differentiating in terms of u (and u is a function of x). You end up with the product of two derivatives ($\frac{dy}{du}$ and $\frac{du}{dx}$).

When it comes to integrating, if you spot that your integral is a product where one bit is the derivative of part of the other bit, you can use this rule:

$$\int \frac{du}{dx} f'(u)\,dx = f(u) + C$$

where u is a function of x.

EXAMPLE

Find a) $\int 6x^5 e^{x^6}\,dx$ and b) $\int e^{\sin x}\cos x\,dx$.

a) $\int 6x^5 e^{x^6}\,dx = e^{x^6} + C$ — If you differentiated $y = e^{x^6}$ using the chain rule, you'd get $6x^5 e^{x^6}$. This is the function you had to integrate.

b) $\int e^{\sin x}\cos x\,dx = e^{\sin x} + C$ — If you differentiated $y = e^{\sin x}$ using the chain rule, you'd get $e^{\sin x}\cos x$. This is the function you had to integrate.

Some *Products* are made up of a *Function* and its *Derivative*

Similarly, if you spot that part of a product is the derivative of the other part of it (which is raised to a power), you can integrate it using this rule:

$$\int (n+1)f'(x)[f(x)]^n\,dx = [f(x)]^{n+1} + C$$

Remember that the derivative will be a multiple of $n + 1$ (not n) — watch out for any other multiples too. This will probably make more sense if you have a look at an example:

EXAMPLE

Find a) $\int 12x^3(2x^4-5)^2\,dx$ and b) $\int 8\,\text{cosec}^2 x\cot^3 x\,dx$.

This one looks pretty horrific, but it isn't too bad once you spot that $-\text{cosec}^2 x$ is the derivative of cot x.

a) Here, $f(x) = 2x^4 - 5$, so differentiating gives $f'(x) = 8x^3$. $n = 2$, so $n + 1 = 3$. Putting all this into the rule above gives:

$$\int 3(8x^3)(2x^4-5)^2\,dx = \int 24x^3(2x^4-5)^2\,dx = (2x^4-5)^3 + C$$

Divide everything by 2 to match the original integral:

$$\int 12x^3(2x^4-5)^2\,dx = \frac{1}{2}(2x^4-5)^3 + C.$$

b) For this one, $f(x) = \cot x$, so differentiating gives $f'(x) = -\text{cosec}^2 x$. $n = 3$, so $n + 1 = 4$. Putting all this into the rule gives:

$$\int -4\,\text{cosec}^2 x\cot^3 x\,dx = \cot^4 x + C$$

Multiply everything by –2 to match the original integral:

$$\int 8\,\text{cosec}^2 x\cot^3 x\,dx = -2\cot^4 x + C$$

To get rid of hiccups, drink a glass of water backwards...

It seems to me that most of this section is about reversing the things you learnt in Section 6. I don't know why they ask you to differentiate stuff if you're just going to have to integrate it again and end up where you started. At least it keeps you busy.

Integrating Trig Things Using Trig Identities

This page is for AQA C4, Edexcel C4 and OCR C4

Examiners have a nasty habit of expecting you to remember things from previous modules — they just can't let go of the past. In this case, it's the trig identities that popped up in Section 3 (see p.27-29 if you need a reminder).

*The **Double Angle Formulas** are useful for **Integration***

If you're given a tricky trig function to integrate, see if you can simplify it using one of the double angle formulas. They're especially useful for things like $\cos^2 x$, $\sin^2 x$ and $\sin x \cos x$. Here are the double angle formulas (see p.29):

$$\sin 2x \equiv 2\sin x \cos x$$

$$\tan 2x \equiv \frac{2\tan x}{1-\tan^2 x}$$

$$\cos 2x \equiv \cos^2 x - \sin^2 x$$

$$\cos 2x \equiv 2\cos^2 x - 1$$

$$\cos 2x \equiv 1 - 2\sin^2 x$$

You can rearrange the second two cos $2x$ formulas to get expressions for $\cos^2 x$ and $\sin^2 x$: $\cos^2 x = \frac{1}{2}(\cos 2x + 1)$
$\sin^2 x = \frac{1}{2}(1 - \cos 2x)$

Once you've replaced the original function with one of the double angle formulas, you can just integrate as normal.

EXAMPLE Find a) $\int \sin^2 x\, dx$ b) $\int \cos^2 5x\, dx$ c) $\int \sin x \cos x\, dx$.

a) Using the double angle formula above, write $\sin^2 x$ as $\frac{1}{2}(1 - \cos 2x)$, then integrate.

$$\int \sin^2 x\, dx = \int \tfrac{1}{2}(1-\cos 2x)dx$$
$$= \tfrac{1}{2}\left(x - \tfrac{1}{2}\sin 2x\right) + C = \tfrac{1}{2}x - \tfrac{1}{4}\sin 2x + C$$

b) Using the double angle formula above, write $\cos^2 5x$ as $\frac{1}{2}(\cos 10x + 1)$, then integrate.

$$\int \cos^2 5x\, dx = \int \tfrac{1}{2}(\cos 10x + 1)dx$$
$$= \tfrac{1}{2}\left(\tfrac{1}{10}\sin 10x + x\right) + C = \tfrac{1}{20}\sin 10x + \tfrac{1}{2}x + C$$

Don't forget to double the coefficient of x here. You'll also need to divide by 10 when you integrate.

c) Using the double angle formula above, write $\sin x \cos x$ as $\frac{1}{2}\sin 2x$, then integrate.

$$\int \sin x \cos x\, dx = \int \tfrac{1}{2}\sin 2x\, dx = \tfrac{1}{2}\left(-\tfrac{1}{2}\cos 2x\right) + C = -\tfrac{1}{4}\cos 2x + C$$

*Use the **Identities** to get a function you **Know** how to **Integrate***

There are a couple of other identities you can use to simplify trig functions (see p.27):

$$\sec^2\theta \equiv 1 + \tan^2\theta \qquad \operatorname{cosec}^2\theta \equiv 1 + \cot^2\theta$$

These two identities are really useful if you have to integrate $\tan^2 x$ or $\cot^2 x$, as you already know how to integrate $\sec^2 x$ and $\operatorname{cosec}^2 x$ (see p.68). Don't forget the stray 1s flying around — they'll just integrate to x.

EXAMPLE Find a) $\int \tan^2 x - 1\, dx$ b) $\int \cot^2 3x\, dx$.

a) Rewrite the function in terms of $\sec^2 x$:
$\tan^2 x - 1 \equiv \sec^2 x - 1 - 1 \equiv \sec^2 x - 2$.
Now integrate:
$$\int \sec^2 x - 2\, dx = \tan x - 2x + C$$

b) Get the function in terms of $\operatorname{cosec}^2 x$:
$\cot^2 3x \equiv \operatorname{cosec}^2 3x - 1$.
Now integrate:
$$\int \operatorname{cosec}^2 3x - 1\, dx = -\tfrac{1}{3}\cot 3x - x + C$$

Remember to divide by 3, the coefficient of x, when you integrate.

EXAMPLE Evaluate $\int_0^{\frac{\pi}{3}} 6\sin 3x \cos 3x + \tan^2 \tfrac{1}{2}x + 1\, dx$.

Using the identities, $6\sin 3x \cos 3x \equiv 3\sin 6x$ and $\tan^2 \frac{1}{2}x + 1 \equiv \sec^2 \frac{1}{2}x$ gives:

$$\int_0^{\frac{\pi}{3}} 3\sin 6x + \sec^2 \tfrac{1}{2}x\, dx = \left[-\tfrac{3}{6}\cos 6x + 2\tan \tfrac{1}{2}x\right]_0^{\frac{\pi}{3}}$$
$$= \left[-\tfrac{1}{2}\cos 6\left(\tfrac{\pi}{3}\right) + 2\tan \tfrac{1}{2}\left(\tfrac{\pi}{3}\right)\right] - \left[-\tfrac{1}{2}\cos 6(0) + 2\tan \tfrac{1}{2}(0)\right]$$
$$= \left[-\tfrac{1}{2}\cos(2\pi) + 2\tan\left(\tfrac{\pi}{6}\right)\right] - \left[-\tfrac{1}{2}\cos(0) + 2\tan(0)\right]$$
$$= \left[-\tfrac{1}{2}(1) + 2\left(\tfrac{1}{\sqrt{3}}\right)\right] - \left[-\tfrac{1}{2}(1) + 2(0)\right] = -\tfrac{1}{2} + \tfrac{2}{\sqrt{3}} + \tfrac{1}{2} = \tfrac{2}{\sqrt{3}}$$

Use the table of common trig angles on p.28 to help you here.

I can't help feeling I've seen these somewhere before...

If you're given a trig function that you don't know how to integrate, play around with these identities and see if you can turn it into something you can integrate. Watch out for coefficients though — they can trip you up if you're not careful.

Core Section 9 — Practice Questions

That was a fairly small section to ease you into the magical world of integration — and to test your powers, here are some questions for you to have a go at.

Warm-up Questions

Q1-3 are for AQA C3, Edexcel C4, OCR C3 and OCR MEI C3

1) Find $\int 4e^{2x}\,dx$.

2) Find $\int e^{3x-5}\,dx$.

3) Find $\int \frac{2}{3x}\,dx$.

Q4-9 are for AQA C3, Edexcel C4, OCR C4 and OCR MEI C3

4) Find $\int \frac{2}{2x+1}\,dx$.

5) Find

   a) $\int \cos 4x - \sec^2 7x\,dx$,

   b) ***[Not OCR MEI]*** $\int 6\sec 3x \tan 3x - \operatorname{cosec}^2 \frac{x}{5}\,dx$.

6) Integrate $\int \frac{\cos x}{\sin x}\,dx$.

7) Integrate $\int 3x^2 e^{x^3}\,dx$.

8) Integrate $\int \frac{20x^4 + 12x^2 - 12}{x^5 + x^3 - 3x}\,dx$.

9) ***[Not OCR MEI]*** Use the trig identity $\sec^2 x \equiv 1 + \tan^2 x$ to find $\int 2\tan^2 3x + 2\,dx$.

Q10 is for AQA C4, Edexcel C4 and OCR C4

10) Use the appropriate trig identity to find $\int \frac{2\tan 3x}{1-\tan^2 3x}\,dx$.

Most of the exam questions on integration will ask you to do more than simply integrate — there's more on this in the next section. But there are some that are bog-forward, straight-standard, mill-of-the-run questions like the ones below.

Exam Questions

Q1 is for AQA C3, Edexcel C4, OCR C3 [part a)], OCR C4 [part b)] and OCR MEI C3 [part a)]

1 Find

a) $\int 3e^{(5-6x)}\,dx$. *(2 marks)*

b) $\int \frac{\operatorname{cosec}^2 x - 2}{\cot x + 2x}\,dx$. *(3 marks)*

Q2 is for AQA C3, Edexcel C4 and OCR C4

2 Use an appropriate identity to find $\int 2\cot^2 x\,dx$. *(3 marks)*

Integration by Substitution

This page is for AQA C3, Edexcel C4, OCR C3 & C4 and OCR MEI C3

I know, I know — you've already done one integration section, surely there can't be another? Well, I'm afraid there is. And this time, there's nowhere to run...

Use *Integration by Substitution* on *Products* of *Two Functions*

On p.45, you saw how to differentiate functions of functions using the chain rule. Integration by substitution is a bit like doing the reverse of the chain rule — it lets you integrate functions of functions by simplifying the integral. Like the chain rule, you have to write part of the function in terms of u, where u is some function of x.

Integration by Substitution

1) **You'll be given an integral that's made up of two functions of x (one is often just x) — e.g. $x(3x + 2)^3$.**
2) **Substitute u for one of the functions of x (to give a function that's easier to integrate) — e.g. $u = 3x + 2$.**
3) **Next, find $\frac{du}{dx}$, and rewrite it so that dx is on its own — e.g. $\frac{du}{dx} = 3$, so $dx = \frac{1}{3}du$.**
4) **Rewrite the original integral in terms of u and du — e.g. $\int x(3x + 2)^3\,dx$ becomes $\int \left(\frac{u-2}{3}\right)u^3\frac{1}{3}du = \int \frac{u^4 - 2u^3}{9}\,du$.**
5) **You should now be left with something that's easier to integrate — just integrate as normal, then at the last step replace u with the original substitution (so for this one, replace u with $3x + 2$).**

$\frac{du}{dx}$ isn't really a fraction, but you can treat it as one for this bit.

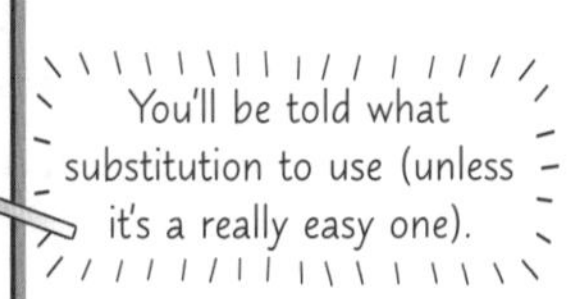

EXAMPLE

Use the substitution $u = x^2 - 2$ to find $\int 4x(x^2 - 2)^4\,dx$.

As $u = x^2 - 2$, $\frac{du}{dx} = 2x$, so $dx = \frac{1}{2x}du$.

Substituting gives $\int 4x(x^2 - 2)^4\,dx = \int 4xu^4\frac{1}{2x}du = \int 2u^4du$.

The x's cancel, making it a lot easier to integrate — this often happens.

Integrate... $\int 2u^4du = \frac{2}{5}u^5 + C$.

...and substitute x back in:

$= \frac{2}{5}(x^2 - 2)^5 + C$.

For *Definite Integrals*, you have to *Change* the *Limits*

If you're given a definite integral, it's really important that you remember to change the limits to u. Doing it this way means you don't have to put x back in at the last step — just put the numbers into the integration for u.

EXAMPLE

Use the substitution $u = \cos x$ to find $\int_{\frac{\pi}{2}}^{2\pi} -12\sin x\cos^3 x\,dx$.

Trigonometry can pop up here as well.

As $u = \cos x$, $\frac{du}{dx} = -\sin x$, so $dx = -\frac{1}{\sin x}du$.

Find the limits of u:

when $x = \frac{\pi}{2}$, $u = \cos\frac{\pi}{2} = 0$,

when $x = 2\pi$, $u = \cos 2\pi = 1$.

So the limits of u are 0 and 1.

Substituting all this gives:

$$\int_{\frac{\pi}{2}}^{2\pi} -12\sin x\cos^3 x\,dx = \int_0^1 -12\sin x\,u^3\frac{-1}{\sin x}du = \int_0^1 12u^3du$$

Integrating and putting in the values of the limits gives:

$[3u^4]_0^1 = [3(1)^4] - [3(0)^4] = 3$

You could also have solved this one using the method on p.70.

Never substitute salt for sugar...

Life is full of limits — age limits, time limits, height limits, limits of how many times I can gaze at my Hugh Jackman poster while still getting my work done... But at least limits of integration will get you exam marks, so it's worth practising them.

Integration by Parts

This page is for AQA C3, Edexcel C4, OCR C4 and OCR MEI C3

Just like you can differentiate products using the product rule (see p.48), you can integrate products using the... er... integration by parts. Not quite as catchy I know, but just as thrilling.

Integration by Parts is the Reverse of the Product Rule

If you have to integrate a product but can't use integration by substitution (see previous page), you might be able to use integration by parts. The formula for integrating by parts is:

$$\int u\frac{dv}{dx}\,dx = uv - \int v\frac{du}{dx}\,dx$$

where u and v are both functions of x.

The hardest thing about integration by parts is deciding which bit of your product should be u and which bit should be $\frac{dv}{dx}$. There's no set rule for this — you just have to look at both parts and see which one differentiates to give something nice, then set that one as u. For example, if you have a product that has a single x as one part of it, choose this to be u. It differentiates to 1, which makes integrating $v\frac{du}{dx}$ dead easy.

EXAMPLE Find $\int 2xe^x\,dx$.

Let $u = 2x$ and let $\frac{dv}{dx} = e^x$. Then u differentiates to give $\frac{du}{dx} = 2$ and $\frac{dv}{dx}$ integrates to give $v = e^x$.

Putting these into the formula gives: $\int 2xe^x\,dx = 2xe^x - \int 2e^x\,dx$

$= 2xe^x - 2e^x + C$

If you have a product that has ln x as one of its factors, let $u = \ln x$, as ln x is easy to differentiate but quite tricky to integrate (see below).

You can integrate ln x using Integration by Parts

Up till now, you haven't been able to integrate ln x, but all that is about to change.
There's a little trick you can use — write ln x as $1 \cdot \ln x$ then integrate by parts.

To find $\int \ln x\,dx$, write $\ln x = 1 \cdot \ln x$.

Let $u = \ln x$ and let $\frac{dv}{dx} = 1$. Then u differentiates to give $\frac{du}{dx} = \frac{1}{x}$ and $\frac{dv}{dx}$ integrates to give $v = x$.

$$\int \ln x\,dx = x\ln x - \int x\frac{1}{x}\,dx = x\ln x - \int 1\,dx = x\ln x - x + C$$

You might have to integrate by parts More Than Once

Not OCR MEI C3

If you have an integral that doesn't produce a nice, easy-to-integrate function for $v\frac{du}{dx}$, you might have to carry out integration by parts more than once.

EXAMPLE Find $\int x^2\sin x\,dx$.

Let $u = x^2$ and let $\frac{dv}{dx} = \sin x$.
Then u differentiates to give $\frac{du}{dx} = 2x$
and $\frac{dv}{dx}$ integrates to give $v = -\cos x$.
Putting these into the formula gives:

$$\int x^2\sin x\,dx = -x^2\cos x - \int -2x\cos x\,dx$$
$$= -x^2\cos x + \int 2x\cos x\,dx$$

$2x\cos x$ isn't very easy to integrate, so integrate by parts again (the $-x^2\cos x$ at the front just stays as it is):

Let $u = 2x$ and let $\frac{dv}{dx} = \cos x$. Then u differentiates to give $\frac{du}{dx} = 2$ and $\frac{dv}{dx}$ integrates to give $v = \sin x$.

Putting these into the formula gives:

$$\int 2x\cos x\,dx = 2x\sin x - \int 2\sin x\,dx = 2x\sin x + 2\cos x$$

So $\int x^2\sin x\,dx = -x^2\cos x + 2x\sin x + 2\cos x + C$.

Every now and then I fall apart...

After you've had a go at some examples, you'll probably realise that integrals with e^x, $\sin x$ or $\cos x$ in them are actually quite easy, as all three are really easy to integrate and differentiate. Fingers crossed you get one of them in the exam.

Tough Integrals

This page is for AQA C3 & C4, Edexcel C4, OCR C4 and OCR MEI C4

With a name like 'Tough Integrals', it doesn't sound like it's going to be a very nice page. However, names can be deceiving. Maybe not in this case, but they can be.

You can integrate **Partial Fractions**

This bit is for AQA C4...

In Section 2 (pages 13-14), you saw how to break down a scary-looking algebraic fraction into partial fractions. This comes in pretty handy when you're integrating — you could try integration by parts on the original fraction, but it would get messy and probably end in tears. Fortunately, once you've split it up into partial fractions, it's much easier to integrate, using the methods on p.67 and p.69.

EXAMPLE

Find $\int \frac{9x^2 + x + 16}{(x+2)(2x-1)(x-3)}\,dx$.

This is the example from p.13, and it can be written as partial fractions like this: $\frac{2}{(x+2)} - \frac{3}{(2x-1)} + \frac{4}{(x-3)}$

Integrating the partial fractions is much easier: $\int \frac{2}{(x+2)} - \frac{3}{(2x-1)} + \frac{4}{(x-3)}\,dx = 2\ln|x+2| - \frac{3}{2}\ln|2x-1| + 4\ln|x-3| + C$

$$= \ln\left|\frac{(x+2)^2(x-3)^4}{(2x-1)^{\frac{3}{2}}}\right| + C$$

Don't forget the coefficients here. Have a look back at p.67 if you can't remember how to do this.

EXAMPLE

Find $\int \frac{x^2 + 17x + 16}{(x+2)^2(3x-1)}\,dx$.

This is the example from p.14. It's a bit trickier because it has a repeated factor. Written in partial fractions, it looks like this: $\frac{2}{(x+2)^2} - \frac{1}{(x+2)} + \frac{4}{(3x-1)}$

You might find it easiest to use integration by substitution (p.73) on the first fraction: Let $u = x + 2$, then $\frac{du}{dx} = 1$, so $du = dx$. Substituting gives:

$$\int \frac{2}{(x+2)^2}\,dx = \int \frac{2}{u^2}\,du = -\frac{2}{u} = -\frac{2}{(x+2)}$$

Putting it all together: $\int \frac{2}{(x+2)^2} - \frac{1}{(x+2)} + \frac{4}{(3x-1)}\,dx = -\frac{2}{x+2} - \ln|x+2| + \frac{4}{3}\ln|3x-1| + C$

$$= -\frac{2}{x+2} + \ln\left|\frac{(3x-1)^{\frac{4}{3}}}{x+2}\right| + C$$

Some **Trig Integrals** can be really **Nasty**

...and this bit is for AQA C3

Unfortunately, the vast range of trig identities and formulas you've seen, as well as lots of different rules for integration, mean that there's no end of evil integration questions they can ask you. Here's a particularly nasty example:

EXAMPLE

Use the substitution $u = \tan x$ to find $\int \frac{\sec^4 x}{\sqrt{\tan x}}\,dx$.

First, work out what all the substitutions will be:
If $u = \tan x$, then $\frac{du}{dx} = \sec^2 x$, so $dx = \frac{du}{\sec^2 x}$.
This will leave $\sec^2 x$ on the numerator — you need to find this in terms of u:
From the identity $\sec^2 x \equiv 1 + \tan^2 x$, you get $\sec^2 x \equiv 1 + u^2$,

Then substitute all these bits into the integral:

$$\int \frac{\sec^4 x}{\sqrt{\tan x}}\,dx = \int \frac{1+u^2}{\sqrt{u}}\,du$$

$$= \int \frac{1}{\sqrt{u}} + \frac{u^2}{\sqrt{u}}\,du = \int u^{-\frac{1}{2}} + u^{\frac{3}{2}}\,du$$

$$= 2u^{\frac{1}{2}} + \frac{2}{5}u^{\frac{5}{2}} + C$$

Remember to stick $u = \tan x$ back into the equation.

$$= 2\sqrt{\tan x} + \frac{2}{5}\sqrt{\tan^5 x} + C$$

I'm partial to a cup of tea...

Partial fractions quite often pop up in a two-part question — for the first part, you'll have to write a tricky fraction in partial fractions, and in the second part you'll have to integrate it. It's a good job you're such a whizz at integrating.

Volumes of Revolution

This page is for AQA C3, Edexcel C4, OCR C3 and OCR MEI C4

Sadly, volumes of revolution isn't to do with plotting your own revolution — it calculates the volumes of weird solids.

You have to find the *Volume* of an area *Rotated About the X- or Y- Axis*

If you're given a definite integral, the solution you come up with is the area under the graph between the two limits (you did this back in AS). If you now rotate that area 2π radians about the x-axis, you'll come up with a solid — and this is what you want to find the volume of. The formula for finding the volume of revolution is:

$$V = \pi \int_{x=x_1}^{x=x_2} y^2 \, dx$$

where y is a function of x (i.e. $y = f(x)$) and x_1 and x_2 are the limits of x.

If you wanted to rotate an area about the y-axis, you'd use the formula $V = \pi \int_{y_1}^{y_2} x^2 \, dy$ — you'd have to write x^2 as a function of y first.

EXAMPLE Find the volume, V, of the solid formed when the area enclosed by the curve $y = \sqrt{6x^2 - 3x + 2}$, the x-axis and the lines $x = 1$ and $x = 2$ is rotated 2π radians about the x-axis.

If $y = \sqrt{6x^2 - 3x + 2}$, then $y^2 = 6x^2 - 3x + 2$.
Putting this into the formula gives:

$$V = \pi \int_1^2 6x^2 - 3x + 2 \, dx = \pi\left[2x^3 - \frac{3}{2}x^2 + 2x\right]_1^2$$
$$= \pi\left(\left[2(2)^3 - \frac{3}{2}(2)^2 + 2(2)\right] - \left[2(1)^3 - \frac{3}{2}(1)^2 + 2(1)\right]\right)$$
$$= \pi\left([16 - 6 + 4] - \left[2 - \frac{3}{2} + 2\right]\right) = \pi\left(14 - 2\frac{1}{2}\right) = 11\frac{1}{2}\pi$$

Don't forget to square y — you might think it's obvious, but it's easily done.

EXAMPLE Find the volume, V, of the solid formed when the area enclosed by the curve $y = \sqrt{x^2 + 5}$, the y-axis and the lines $y = 3$ and $y = 6$ is rotated 2π radians about the y-axis.

First, rearrange the equation to get x^2 on its own:
$y = \sqrt{x^2 + 5} \Rightarrow y^2 = x^2 + 5$
so $x^2 = y^2 - 5$.

You don't need rotation about the y-axis for Edexcel.

Now integrate: $V = \pi \int_3^6 y^2 - 5 \, dy = \pi\left[\frac{1}{3}y^3 - 5y\right]_3^6$
$$= \pi\left(\left[\frac{1}{3}(6)^3 - 5(6)\right] - \left[\frac{1}{3}(3)^3 - 5(3)\right]\right)$$
$$= \pi(42 - (-6)) = 48\pi$$

You can use *Parametric Equations* to find a *Volume*

This bit is just for Edexcel C4

In Section 5 (pages 39-42), you met curves with parametric equations — where x and y are functions of t.
You saw how to integrate them on p.42 — now you need to find the volume of revolution.
Here is the formula you need, for a curve with parametric equations $x = f(t)$ and $y = g(t)$ and limits t_1 and t_2.

Don't forget to change the limits from x to t.

$$V = \pi \int_{t=t_1}^{t=t_2} y^2 \frac{dx}{dt} \, dt$$

You need the dx/dt to get the whole thing in terms of t.

EXAMPLE A curve is given by the parametric equations $x = 2t$, $y = \sin 3t$. Find the volume formed when the area bounded by the curve, the x-axis and the lines $x = 0$ and $x = 2\pi$ is rotated 2π radians about the x-axis.

First of all, $\frac{dx}{dt} = 2$.
Change the limits: $x = 0$, so $2t = 0 \Rightarrow t = 0$
$x = 2\pi$, so $2t = 2\pi \Rightarrow t = \pi$.

Squaring y gives $\sin^2 3t$.

Put the expressions for y^2, $\frac{dx}{dt}$ and the new limits into the formula:

$$V = \pi \int_0^\pi 2\sin^2 3t \, dt = \pi \int_0^\pi 1 - \cos 6t \, dt$$
$$= \pi\left[t - \frac{1}{6}\sin 6t\right]_0^\pi$$
$$= \pi\left[\pi - \frac{1}{6}\sin 6\pi\right] - \pi\left[0 - \frac{1}{6}\sin 0\right]$$
$$= \pi\left[\pi - \frac{1}{6}(0)\right] - 0 = \pi^2$$

This bit uses the identity $\cos 2t \equiv 1 - 2\sin^2 t$ — don't forget to double the coefficient of t.

Don't be put off if the parametric equations are written in terms of something other than t (trig equations will sometimes be given in terms of θ) — just change $\frac{dx}{dt}$ to $\frac{dx}{d\theta}$ and dt to $d\theta$ (or to whatever the variable is).

Come the revolution, I will have to kill you all...

For OCR, you might be given an area between two curves and asked to work out the volume when this region is rotated. To do this, you'd just work out the volume for each curve separately and subtract to find the volume you want.

Differential Equations

This page is for AQA C4, Edexcel C4, OCR C4 and OCR MEI C4

Differential equations are tricky little devils that have a lot to do with differentiation as well as integration. They're often about rates of change, so the variable t pops up quite a lot.

Differential Equations have a dy/dx Term

(or $\frac{dP}{dt}, \frac{ds}{dt}, \frac{dV}{dr}$, etc. — depending on the variables)

1) A differential equation is an equation that includes a derivative term (such as $\frac{dy}{dx}$), as well as other variables (like x and y).
2) Before you even think (or worry) about solving them, you have to be able to set up ('formulate') differential equations.
3) Differential equations tend to involve a rate of change (giving a derivative term) and a proportion relation. Remember — if $a \propto b$, then $a = kb$ for some constant k.

EXAMPLE The number of bacteria in a petri dish is increasing over time, t, at a rate directly proportional to the number of bacteria at a given time, b. Formulate a differential equation that shows this information.

The rate of change, $\frac{db}{dt}$, is proportional to b, so $\frac{db}{dt} \propto b$. This means that $\frac{db}{dt} = kb$ for some constant k, $k > 0$.

EXAMPLE The volume of interdimensional space jelly, V, in a container is decreasing over time, t, at a rate directly proportional to the square of its volume. Show this as a differential equation.

The rate of change, $\frac{dV}{dt}$, is proportional to V^2, so $\frac{dV}{dt} \propto -V^2$. $\frac{dV}{dt} = -kV^2$ for some constant k, $k > 0$.

V is decreasing, so don't forget the −.

Solve differential equations by Integrating

Now comes the really juicy bit — solving differential equations. It's not as bad as it looks (honest).

Solving Differential Equations

1) **You can only solve differential equations if they have separable variables — where x and y can be separated into functions $f(x)$ and $g(y)$.**
2) **Write the differential equation in the form $\frac{dy}{dx} = f(x)g(y)$.**
3) **Then rearrange the equation to get all the terms with y on the LHS and all the terms with x on the RHS. It'll look something like this: $\frac{1}{g(y)}dy = f(x)dx$.**
4) **Now integrate both sides: $\int \frac{1}{g(y)}dy = \int f(x)dx$.**
5) **Rearrange your answer to get it in a nice form — you might be asked to find it in the form $y = h(x)$. Don't forget the constant of integration (you only need one — not one on each side). It might be useful to write the constant as $\ln k$ rather than C (see p.69).**
6) **If you're asked for a general solution, leave C (or k) in your answer. If they want a particular solution, they'll give you x and y values for a certain point. All you do is put these values into your equation and use them to find C (or k).**

Remember — it might not be in terms of x and y.

Like in integration by substitution, you can treat dy/dx as a fraction here.

EXAMPLE Find the particular solution of $\frac{dy}{dx} = 2y(1 + x)^2$ when $x = -1$ and $y = 4$.

This equation has separable variables: $f(x) = 2(1 + x)^2$ and $g(y) = y$.

Rearranging this equation gives: $\frac{1}{y}dy = 2(1 + x)^2 dx$

And integrating: $\int \frac{1}{y}dy = \int 2(1 + x)^2 dx$

$\Rightarrow \ln|y| = \frac{2}{3}(1 + x)^3 + C$

If you were asked for a general solution, you could just leave it in this form.

Now put in the values of x and y to find the value of C:

$\ln 4 = \frac{2}{3}(1 + (-1))^3 + C \Rightarrow \ln 4 = C$

so $\ln|y| = \frac{2}{3}(1 + x)^3 + \ln 4$

I will formulate a plan to take over the world...

...starting with Cumbria. I've always liked Cumbria. You're welcome to join my army of minions, but first you'll have to become an expert on solving differential equations. Do that, and I'll give you Grasmere — or name a mountain after you.

Differential Equations

This page is for AQA C4, Edexcel C4, OCR C4 and OCR MEI C4

One of the most exciting things about differential equations is that you can apply them to real-life situations. Well, I say exciting, but perhaps I should say 'mildly interesting', or maybe just 'more stuff for you to learn'.

You might be given **Extra Information**

1) In the exam, you might be given a question that takes a real-life problem and uses differential equations to model it.
2) Population questions come up quite often — the population might be increasing or decreasing, and you have to find and solve differential equations to show it. In cases like this, one of your variables will usually be t, time.
3) You might be given a starting condition — e.g. the initial population. The important thing to remember is that:

   the starting condition occurs when $t = 0$.

   This is pretty obvious, but it's really important.

4) You might also be given extra information — e.g. the population after a certain number of years (where you have to figure out what t is), or the number of years it takes to reach a certain population (where you have to work out what the population will be). Make sure you always link the numbers you get back to the situation.

Exam Questions are often **Broken Down** into lots of **Parts**

Questions like the one below can be a bit overwhelming, but follow it through step by step and it shouldn't be too bad.

EXAMPLE

The population of rabbits in a park is decreasing as winter approaches.
The rate of decrease is directly proportional to the current number of rabbits (P).

a) Formulate a differential equation to model the rate of decrease in terms of the variables P, t (time in days) and k, a positive constant.

b) If the initial population is P_0, solve your differential equation to find P in terms of P_0, k and t.

c) Given that $k = 0.1$, find the time at which the population of rabbits will have halved, to the nearest day.

a) If the rate of decrease is proportional to the number of rabbits, then $\frac{dP}{dt} = -kP$ (it's negative because the population is decreasing).

b) First, solve the differential equation to find the general solution: $\frac{dP}{dt} = -kP \Rightarrow \frac{1}{P}\,dP = -k\,dt$

Integrating this gives: $\int \frac{1}{P}\,dP = \int -k\,dt$

$\Rightarrow \ln P = -kt + C$

You don't need modulus signs for ln P as $P \geq 0$ — you can't have a negative population.

At $t = 0$, $P = P_0$. Putting these values into the equation gives: $\ln P_0 = -k(0) + C$

$\Rightarrow \ln P_0 = C$

So the differential equation becomes: $\ln P = -kt + \ln P_0$

$\Rightarrow P = e^{(-kt + \ln P_0)} = e^{-kt}e^{\ln P_0}$

$\Rightarrow P = P_0 e^{-kt}$

Remember that $e^{\ln x} = x = \ln e^x$.

c) When the population of rabbits has halved, $P = \frac{1}{2}P_0$. You've been told that $k = 0.1$, so substitute these values into the equation above and solve for t:

$\frac{1}{2}P_0 = P_0 e^{-0.1t}$

$\frac{1}{2} = e^{-0.1t}$

$\ln\frac{1}{2} = -0.1t$

$-0.6931 = -0.1t \Rightarrow t = 6.931$

So, to the nearest day, $t = 7$. This means that it will take 7 days for the population of rabbits to halve.

At t = 10, we kill all the bunnies...

These questions can get a bit morbid — just how I like them. They might look a bit scary, as they throw a lot of information at you in one go, but once you know how to solve them, they're a walk in the park. Rabbit traps optional.

Numerical Integration

This page is for Edexcel C4 and OCR MEI C4

In the interests of avoiding cruelty to animals, I think it's best to move on. Next up, numerical integration — I promise you it's not as bad as it sounds. Most of it's stuff you've done before — like the Trapezium Rule.

Estimate** the area using the **Trapezium Rule

Here's a quick reminder of the Trapezium Rule — look back over your AS notes if you can't remember how to do it.

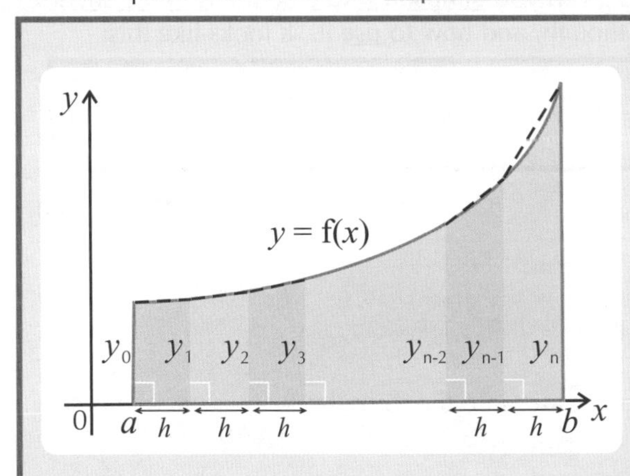

The area of each trapezium is $A_n = \frac{h}{2}(y_n + y_{n+1})$

The area represented by $\int_a^b y\,dx$ is approximately:

$$\int_a^b y\,dx \approx \frac{h}{2}[y_0 + 2(y_1 + y_2 + \ldots + y_{n-1}) + y_n]$$

Remember that 5 ordinates is the same as 4 strips.

where n is the number of strips or intervals and h is the width of each strip.

You can find the width of each strip using $h = \frac{(b-a)}{n}$

$y_0, y_1, y_2, \ldots, y_n$ are the heights of the sides of the trapeziums — you get these by putting the x-values into the equation of the curve.

*Use **More Strips** to get a **More Accurate** answer*

Using more strips (i.e. increasing n) gives you a more accurate approximation. You can check how accurate your answer is by working out the percentage error (see below).

EXAMPLE Use the Trapezium Rule to approximate the area of $\int_0^4 \frac{6x^2}{x^3+2}\,dx$, using a) $n = 2$ and b) $n = 4$.

a) For 2 strips, the width of each strip is $h = \frac{4-0}{2} = 2$, so the x-values are 0, 2 and 4.

x	$y = \frac{6x^2}{x^3+2}$
$x_0 = 0$	$y_0 = 0$
$x_1 = 2$	$y_1 = 2.4$
$x_2 = 4$	$y_2 = 1.455$

(3 d.p.)

Putting these values into the formula gives:

$\int_0^4 \frac{6x^2}{x^3+2}\,dx$

$\approx \frac{2}{2}[0 + 2(2.4) + 1.455]$

$= [4.8 + 1.455] = 6.255$ (3 d.p.)

b) For 4 strips, the width of each strip is $h = \frac{4-0}{4} = 1$, so the x-values are 0, 1, 2, 3 and 4.

x	$y = \frac{6x^2}{x^3+2}$
$x_0 = 0$	$y_0 = 0$
$x_1 = 1$	$y_1 = 2$
$x_2 = 2$	$y_2 = 2.4$
$x_3 = 3$	$y_3 = 1.862$
$x_4 = 4$	$y_4 = 1.455$

(3 d.p.)

Putting these values into the formula gives:

$\int_0^4 \frac{6x^2}{x^3+2}\,dx$

$\approx \frac{1}{2}[0 + 2(2 + 2.4 + 1.862) + 1.455]$

$= \frac{1}{2}[12.524 + 1.455] = 6.990$ (3 d.p.)

*You need the **Exact Answer** to work out the **Percentage Error***

To work out the percentage error, calculate or use the exact value of the integral, then use this formula:

$$\% \text{ Error} = \frac{\text{exact value} - \text{approximate value}}{\text{exact value}} \times 100$$

EXAMPLE Calculate the percentage error for a) and b) above to 2 d.p.

First, work out the exact value of the integral: $\int_0^4 \frac{6x^2}{x^3+2}\,dx = [2\ln|x^3+2|]_0^4 = [2\ln 66] - [2\ln 2] = 6.993$ (3 d.p.)

This was calculated using the formula on p.69.

For part a), the percentage error is $\frac{6.993 - 6.255}{6.993} \times 100 = 10.55\%$, and for part b), $\frac{6.993 - 6.990}{6.993} \times 100 = 0.04\%$.

The approximation with more strips has a lower percentage error — so it's a more accurate approximation.

I lost my notes in the Bermuda Trapezium...

The key thing to remember about the Trapezium Rule (well, apart from the rule itself) is that the more strips you have, the more accurate your answer (and so the lower the % error). More strips = more accurate = smaller % error. Got it?

Numerical Integration

This page is for AQA C3 and OCR C3

There are a few more ways of estimating areas under graphs — unfortunately looking at the picture and saying "I reckon that's about 5 cm²" isn't one of them. The ones you need are Simpson's Rule and the Mid-Ordinate Rule.

Simpson's Rule works for an *Even Number* of *Strips*

Gosh, the examiners must have been having a good day — they've given you Simpson's Rule on the formula sheet, so you don't have to learn it. You do need to know what it means though, and how to use it. It looks like this:

$$\int_a^b y\,dx \approx \frac{1}{3}h[(y_0 + y_n) + 4(y_1 + y_3 + \ldots + y_{n-1}) + 2(y_2 + y_4 + \ldots + y_{n-2})]$$

The y-values are found in the same way as for the Trapezium Rule (have a look at the diagram at the top of previous page).

where $h = \frac{b-a}{n}$ and n is even.

EXAMPLE

Use Simpson's Rule with 4 strips to approximate the area of $\int_0^1 e^{2x^2-1}\,dx$.

The width of each strip is $h = \frac{1-0}{4} = 0.25$, so the x-values are 0, 0.25, 0.5, 0.75 and 1.
Using these, calculate the y-values (to 3 d.p.):

x	$y = e^{2x^2-1}$
$x_0 = 0$	$y_0 = 0.368$
$x_1 = 0.25$	$y_1 = 0.417$
$x_2 = 0.5$	$y_2 = 0.607$
$x_3 = 0.75$	$y_3 = 1.133$
$x_4 = 1$	$y_4 = 2.718$

Now put the y-values into the formula:

$$\int_0^1 e^{2x^2-1}\,dx \approx \frac{1}{3}(0.25)[(0.368 + 2.718) + 4(0.417 + 1.133) + 2(0.607)]$$

$$= \frac{1}{12}[3.086 + 6.2 + 1.214] = 0.875 \text{ (3 d.p.)}$$

Like the Trapezium Rule, using more strips will give you a more accurate approximation. You can calculate the % error in the same way as for the Trapezium Rule too — have a look at the previous page to see how to work out the % error.

The *Mid-Ordinate Rule* adds up the *Areas* of lots of *Rectangles*

The Mid-Ordinate Rule is for AQA C3 only

The Mid-Ordinate Rule works in a similar way to the Trapezium Rule — except that instead of adding up the areas of lots of trapeziums (or trapezia if you want to be fancy), it uses the areas of lots of rectangles (rectanglia?) instead. The height of the rectangle is given by the y-value at the midpoint of the strip.

Here's the Mid-Ordinate Rule:

$$\int_a^b y\,dx \approx h(y_{0.5} + y_{1.5} + \ldots + y_{n-1.5} + y_{n-0.5})$$ where $h = \frac{b-a}{n}$.

This rule is also given on the formula sheet.

EXAMPLE

Use the Mid-Ordinate Rule to approximate the area of $\int_1^4 \ln(x^3 + 2x)\,dx$, where $n = 3$.

The width of each strip is $h = \frac{4-1}{3} = 1$, so the normal x-values would be $x_0 = 1$, $x_1 = 2$, $x_2 = 3$ and $x_3 = 4$. But for this rule, you want the midpoints of the strips, so instead you need to find the y-values at $x_{0.5} = 1.5$, $x_{1.5} = 2.5$ and $x_{2.5} = 3.5$:

x	$y = \ln(x^3 + 2x)$
$x_{0.5} = 1.5$	$y_{0.5} = 1.852$
$x_{1.5} = 2.5$	$y_{1.5} = 3.027$
$x_{2.5} = 3.5$	$y_{2.5} = 3.910$

(3 d.p.)

Putting these values into the formula gives:

$$\int_1^4 \ln(x^3 + 2x)\,dx \approx 1(1.852 + 3.027 + 3.910)$$

$$= 8.789 \text{ (3 d.p.)}$$

I'd rather have subordinates than mid-ordinates...

There's nothing too tricky here — and it helps that both rules are on the formula sheet. Just remember that you must have an even number of strips for Simpson's Rule, and you use the midpoint of the strip to find the y-value for the Mid-Ordinate Rule.

Core Section 10 — Practice Questions

Phew, that was a whopper of a section. I bet you could do with a break. Well, hold on just a minute — here are some practice questions to do first to check you know your stuff. Let's start with a gentle warm-up.

Warm-up Questions

Questions 1-4 are for AQA C3, Edexcel C4, OCR C4 and OCR MEI C3

1) Use the substitution $u = e^x - 1$ to find $\int e^x(e^x + 1)(e^x - 1)^2\,dx$.

2) Find the exact value of $\int_{\frac{\pi}{4}}^{\frac{\pi}{3}} \sec^4 x \tan x\,dx$, using the substitution $u = \sec x$.

3) Use integration by parts to solve $\int 3x^2 \ln x\,dx$.

4) Use integration by parts to solve $\int 4x \cos 4x\,dx$.

Questions 5-6 are for AQA C3, Edexcel C4, OCR C3 and OCR MEI C4

5) Find the volume of the solid formed when the area bounded by the curve $y = \frac{1}{x}$, the x-axis and the lines $x = 2$ and $x = 4$ is rotated 2π radians about the x-axis.

6) ***[Not Edexcel]*** Find the volume of the solid formed when the area bounded by the curve $y = x^2 + 1$, the y-axis and the lines $y = 1$ and $y = 3$ is rotated 2π radians about the y-axis.

Question 7 is just for Edexcel C4

7) A curve is given by the parametric equations $x = t^2$ and $y = \frac{1}{t}$, $(t > 0)$. Find the volume formed when the area bounded by the curve, the x-axis and the lines $x = 4$ and $x = 9$ is rotated 2π radians about the x-axis.

Questions 8-10 are for AQA C4, Edexcel C4, OCR C4 and OCR MEI C4

8) Use $\frac{3x + 10}{(2x + 3)(x - 4)} \equiv \frac{A}{2x + 3} + \frac{B}{x - 4}$ to find $\int \frac{3x + 10}{(2x + 3)(x - 4)}\,dx$.

9) Find the general solution to the differential equation $\frac{dy}{dx} = \frac{1}{y}\cos x$. Give your answer in the form $y^2 = f(x)$.

10) The population of squirrels is increasing suspiciously quickly. The rate of increase is directly proportional to the current number of squirrels, S.
    a) Formulate a differential equation to model the rate of increase in terms of S, t (time in weeks) and k, a positive constant.
    b) The squirrels need a population of 150 to successfully take over the forest. If the initial population is 30 and the value of k is 0.2, how long (to the nearest week) will it take before they can overthrow the evil hedgehogs?

Question 11 is for Edexcel C4 and OCR MEI C4

11) Use the Trapezium Rule to estimate the value of $\int_0^6 (6x - 12)(x^2 - 4x + 3)^2\,dx$, first using 4 strips and then again with 6 strips. Calculate the percentage error for each answer.

Question 12 is for AQA C3 and OCR C3

12) a) Use Simpson's Rule with 6 strips to approximate the area of $\int_1^4 \ln(\sqrt{x} + 2)\,dx$.
    b) Why can't you use Simpson's Rule with 5 strips to find an estimate?

For extra practice, have a go at q12 using the Mid-Ordinate Rule and q13 using Simpson's Rule.

Question 13 is just for AQA C3

13) Use the Mid-Ordinate Rule to estimate the value of $\int_2^4 x^2 e^{x-2}\,dx$, using $n = 4$.

Core Section 10 — Practice Questions

Unfortunately the exam questions are less likely to be about rebel squirrels, as the examiners tend to be on the hedgehogs' side. If you ever meet an examiner, look closely to make sure he's not a hedgehog in disguise.

Exam Questions

Question 1 is for AQA C3, Edexcel C4 and OCR C3

1 Find the volume of the solid formed when the region R, bounded by the curve $y = \text{cosec } x$, the x-axis and the lines $x = \frac{\pi}{4}$ and $x = \frac{\pi}{3}$, is rotated 2π radians about the x-axis.
Give your answer to 3 decimal places.

(3 marks)

Question 2 is for Edexcel C4 and OCR MEI C4

2 **Figure 1** shows the graph of $y = x \sin x$. The region R is bounded by the curve and the x-axis ($0 \leq x \leq \pi$).

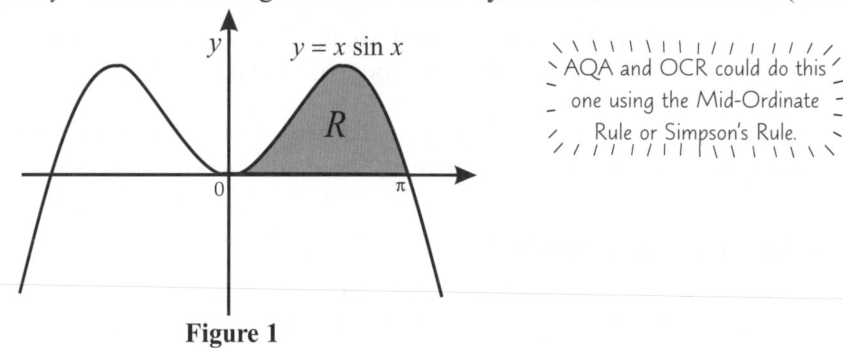

Figure 1

a) Fill in the missing values of y in the table below. Give your answers to 4 decimal places.

x	0	$\frac{\pi}{4}$	$\frac{\pi}{2}$	$\frac{3\pi}{4}$	π
y	0	0.5554			0

(2 marks)

b) Hence find an approximation for the area of R, using the Trapezium Rule.
Give your answer to 3 decimal places.

(4 marks)

c) Find the exact area of R using integration by parts.

(4 marks)

d) Hence find the percentage error of the approximation.

(2 marks)

Question 3 is for AQA C3, Edexcel C4, OCR C4 and OCR MEI C3

3 Find the value of $\int_1^2 \frac{8}{x}(\ln x + 2)^3 \, dx$ using the substitution $u = \ln x$. Give your answer to 4 s.f.

(6 marks)

Question 4 is just for AQA C3

4 Use the Mid-Ordinate Rule to approximate the area of $\int_1^3 2^x \, dx$, with 4 strips.
Give your answer to 4 s.f.

(3 marks)

Core Section 10 — Practice Questions

One more page of questions, then you're onto the final section of Core. That's right, the last one.

Question 5 is for AQA C3, OCR C3 and OCR MEI C4

5 The graph below shows the curve of $y = \frac{1}{x^2}$ for $x > 0$.

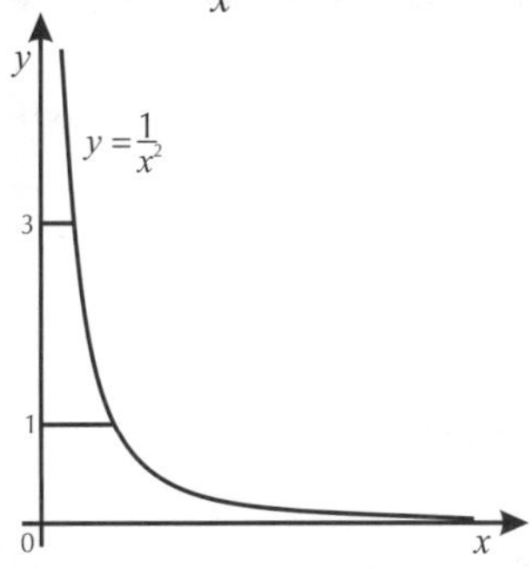

a) The region bounded by the curve, the y-axis and the lines $y = 1$ and $y = 3$ is rotated 2π radians about the y-axis. Calculate the exact volume of the solid formed. *(5 marks)*

b) *[AQA and OCR only]* Using Simpson's Rule with 4 strips, approximate the area of $\int_1^5 \frac{1}{x^2 + 3x}\,dx$. Give your answer to 3 s.f. *(4 marks)*

Question 6 is just for AQA C3, Edexcel C4, OCR C3 and OCR MEI C4

6

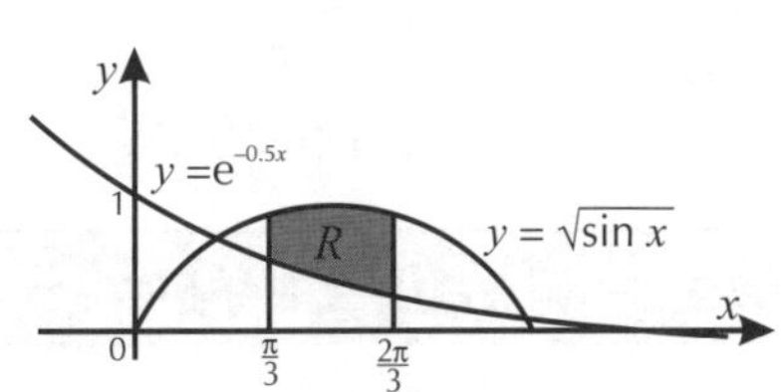

The region R above is formed by the curves $y = \sqrt{\sin x}$, $y = e^{-0.5x}$, $x = \frac{\pi}{3}$ and $x = \frac{2\pi}{3}$. Calculate the volume of the solid formed when R is rotated completely about the x-axis. Give your answer to 4 s.f. *(7 marks)*

Questions 7-8 are for AQA C4, Edexcel C4, OCR C4 and OCR MEI C4

7 a) Find the general solution to the differential equation

$$\frac{dy}{dx} = \frac{\cos x \cos^2 y}{\sin x}.$$

(4 marks)

b) Given that $y = \pi$ when $x = \frac{\pi}{6}$, solve the differential equation above. *(2 marks)*

8 A company sets up an advertising campaign to increase sales of margarine. After the campaign, the number of tubs of margarine sold each week, m, increases over time, t weeks, at a rate that is directly proportional to the square root of the number of tubs sold.

a) Formulate a differential equation in terms of t, m and a constant k. *(2 marks)*

b) At the start of the campaign, the company was selling 900 tubs of margarine a week. Use this information to solve the differential equation, giving m in terms of k and t. *(4 marks)*

c) Hence calculate the number of tubs sold in the fifth week after the campaign, given that $k = 2$. *(3 marks)*

Vectors

This page is for AQA C4, Edexcel C4, OCR C4 and OCR MEI C4

If you did M1, then you've probably seen some of this vector stuff before. If not, you've got lots to look forward to. In any case, we're going to start with the basics — like what vectors are.

Vectors have Magnitude and Direction — Scalars Don't

1) Vectors have both size and direction — e.g. a velocity of 2 m/s on a bearing of 050°, or a displacement of 3 m north. Scalars are just quantities without a direction, e.g. a speed of 2 m/s, a distance of 3 m.
2) Vectors are drawn as lines with arrowheads on them.
   - The length of the line represents the magnitude (size) of the vector (e.g. the speed component of velocity). Sometimes vectors are drawn to scale.
   - The direction of the arrowhead shows the direction of the vector.

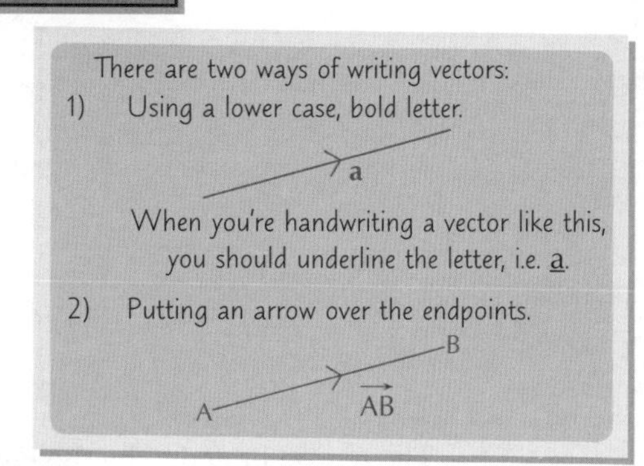

Find the Resultant by Drawing Vectors Nose to Tail

You can add vectors together by drawing the arrows nose to tail.
The single vector that goes from the start to the end of the vectors is called the resultant vector.

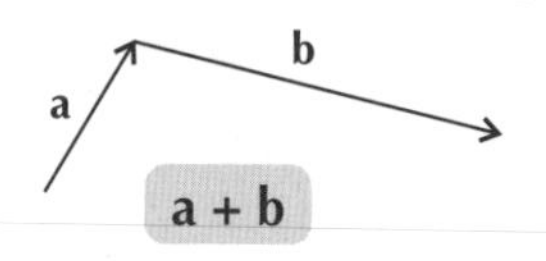

$\mathbf{a} + \mathbf{b}$

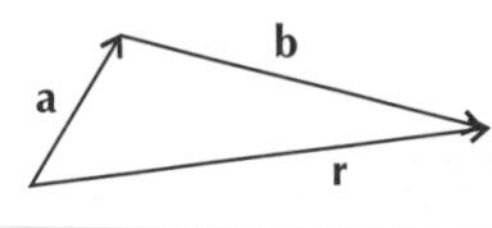

Resultant: $\mathbf{r} = \mathbf{a} + \mathbf{b}$

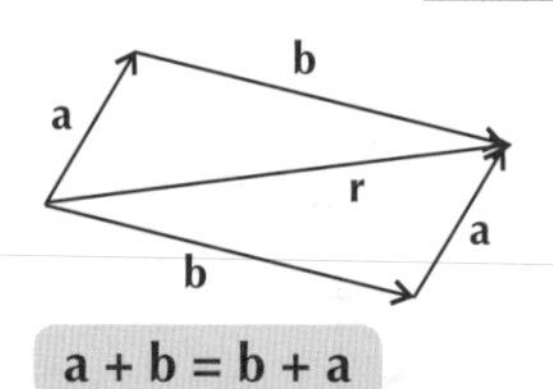

$\mathbf{a} + \mathbf{b} = \mathbf{b} + \mathbf{a}$

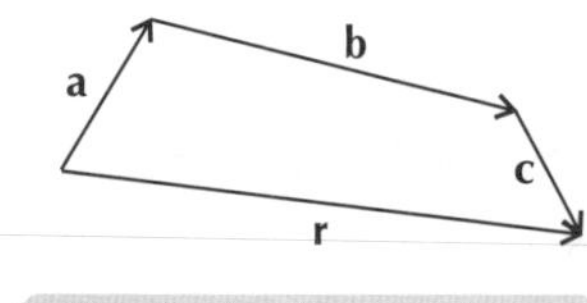

Resultant: $\mathbf{r} = \mathbf{a} + \mathbf{b} + \mathbf{c}$

Subtracting a Vector is the Same as Adding a Negative Vector

1) The vector $-\mathbf{a}$ is in the opposite direction to the vector $\mathbf{a}$. They're both exactly the same size.

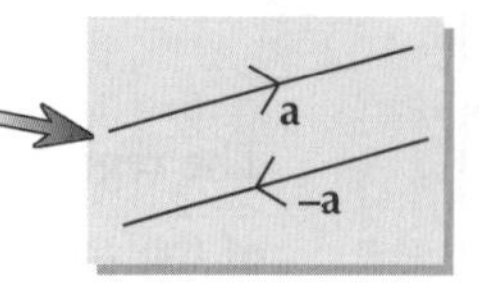

2) So subtracting a vector is the same as adding the negative vector:

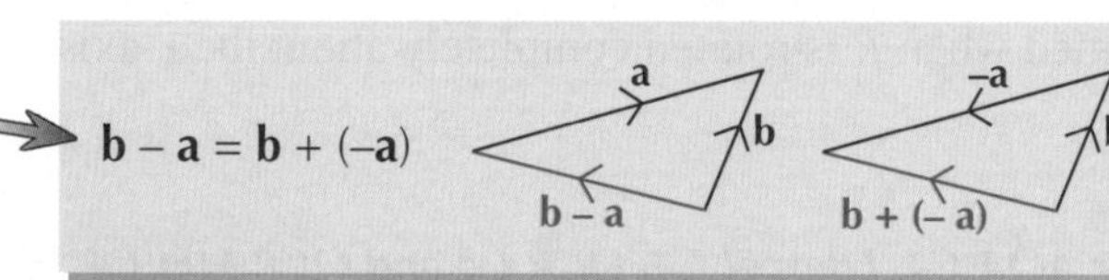

3) You can use the adding and subtracting rules to find a vector in terms of other vectors.

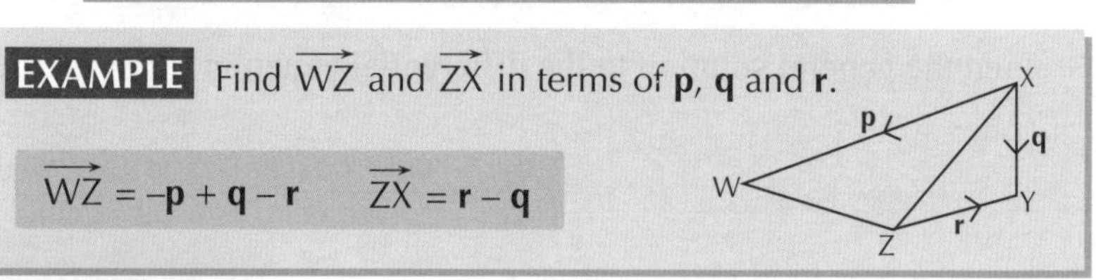

Vectors a, 2a and 3a are all Parallel

You can multiply a vector by a scalar (just a number, remember) — the length changes but the direction stays the same.

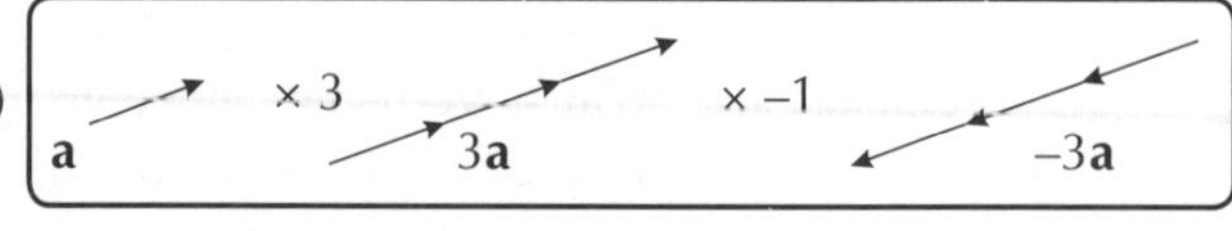

Multiplying a vector by a non-zero scalar always produces a parallel vector.

All these vectors are parallel: $9\mathbf{a} + 15\mathbf{b}$ | $18\mathbf{a} + 30\mathbf{b}$ | $6\mathbf{a} + 10\mathbf{b}$ | $3\mathbf{a} + 5\mathbf{b}$

This is $2(9\mathbf{a} + 15\mathbf{b})$. This is $\frac{2}{3}(9\mathbf{a} + 15\mathbf{b})$. This is $\frac{1}{3}(9\mathbf{a} + 15\mathbf{b})$.

Eating pasta = buying anti-pasta?...

If an exam question asks you to show that two lines are parallel, you just have to show that one vector's a multiple of the other. By the way — exam papers often use λ and μ as scalars in vector questions (so you don't confuse them with vectors).

Vectors

This page is for AQA C4, Edexcel C4, OCR C4 and OCR MEI C4

There are a few more ways of representing vectors that you need to know about. Then it's off to the third dimension...

Position Vectors Describe Where a Point Lies

You can use a vector to describe the position of a point, in relation to the origin, O.

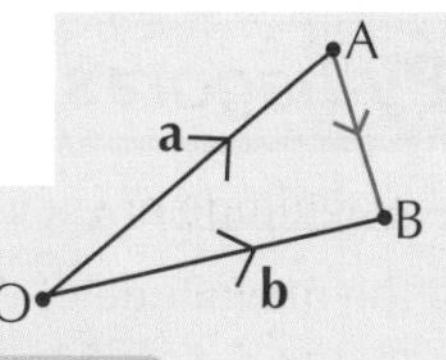

The position vector of point A is $\overrightarrow{OA}$. It's usually called vector **a**.
The position vector of point B is $\overrightarrow{OB}$. It's usually called vector **b**.

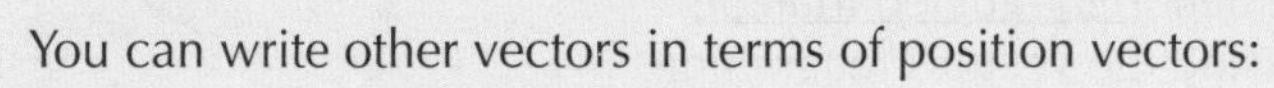

You can write other vectors in terms of position vectors: $\overrightarrow{AB} = -\overrightarrow{OA} + \overrightarrow{OB} = \overrightarrow{OB} - \overrightarrow{OA} = -\mathbf{a} + \mathbf{b} = \mathbf{b} - \mathbf{a}$

Vectors can be described using i + j Units

1) A unit vector is any vector with a magnitude of 1 unit. ← There's more on unit vectors on the next page.
2) The vectors **i** + **j** are standard unit vectors. **i** is in the direction of the x-axis, and **j** is in the direction of the y-axis. They each have a magnitude of 1 unit, of course.
3) They're a dead handy way of describing any vector. You use them to say how far horizontally and vertically you have to go to get from the start of the vector to the end.

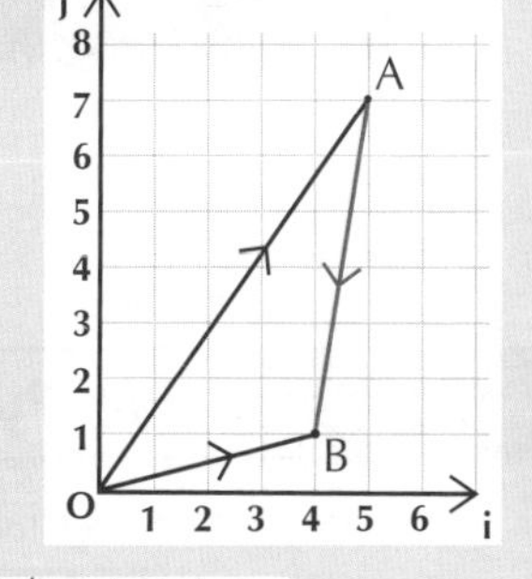

The position vector of point A = $\mathbf{a} = 5\mathbf{i} + 7\mathbf{j}$
The position vector of point B = $\mathbf{b} = 4\mathbf{i} + \mathbf{j}$

This tells you that point B lies 4 units to the right and 1 unit above the origin — it's just like coordinates.

Vector $\overrightarrow{AB} = \mathbf{b} - \mathbf{a} = (4\mathbf{i} + \mathbf{j}) - (5\mathbf{i} + 7\mathbf{j}) = -\mathbf{i} - 6\mathbf{j}$

Add/subtract the **i** and **j** components separately.

To go from A to B, you go 1 unit left and 6 units down. It's just like a translation.

And then there are Column Vectors

1) If writing i's and j's gets a bit much for your wrists, you can use column vectors instead.

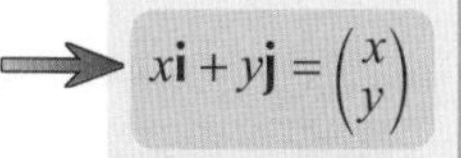

$x\mathbf{i} + y\mathbf{j} = \begin{pmatrix} x \\ y \end{pmatrix}$

2) Calculating with them is a breeze. Just add or subtract the top row, then add or subtract the bottom row separately.
3) When you're multiplying a column vector by a scalar, you multiply each number in the column vector by the scalar.

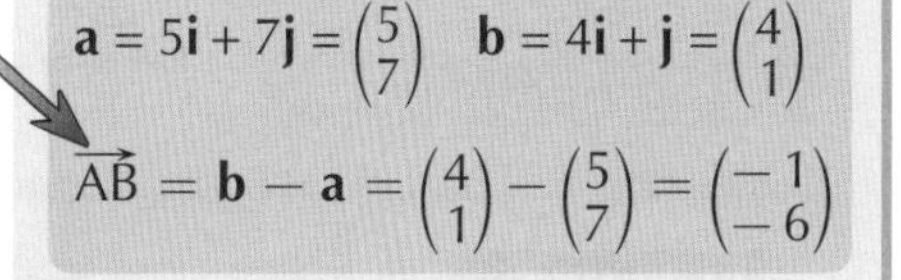

$\mathbf{a} = 5\mathbf{i} + 7\mathbf{j} = \begin{pmatrix} 5 \\ 7 \end{pmatrix}$ $\quad \mathbf{b} = 4\mathbf{i} + \mathbf{j} = \begin{pmatrix} 4 \\ 1 \end{pmatrix}$

$\overrightarrow{AB} = \mathbf{b} - \mathbf{a} = \begin{pmatrix} 4 \\ 1 \end{pmatrix} - \begin{pmatrix} 5 \\ 7 \end{pmatrix} = \begin{pmatrix} -1 \\ -6 \end{pmatrix}$

$$2\mathbf{b} - 3\mathbf{a} = 2\begin{pmatrix} 4 \\ 1 \end{pmatrix} - 3\begin{pmatrix} 5 \\ 7 \end{pmatrix} = \begin{pmatrix} 8 \\ 2 \end{pmatrix} - \begin{pmatrix} 15 \\ 21 \end{pmatrix} = \begin{pmatrix} -7 \\ -19 \end{pmatrix}$$

You Can Have Vectors in Three Dimensions Too

1) Imagine that the x- and y-axes lie flat on the page. Then imagine a third axis sticking straight through the page at right angles to it — this is the z-axis.
2) The points in three dimensions are given (x, y, z) coordinates.
3) When you're talking vectors, **k** is the unit vector in the direction of the z-axis.
4) You can write three-dimensional vectors as column vectors like this:

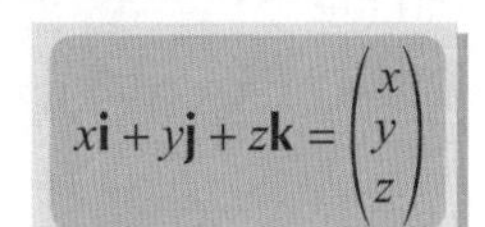

$x\mathbf{i} + y\mathbf{j} + z\mathbf{k} = \begin{pmatrix} x \\ y \\ z \end{pmatrix}$

5) So the position vector of point Q is:

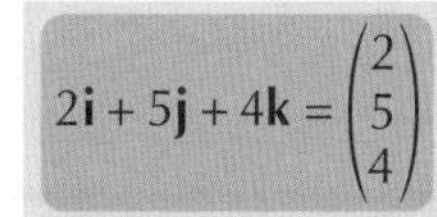

$2\mathbf{i} + 5\mathbf{j} + 4\mathbf{k} = \begin{pmatrix} 2 \\ 5 \\ 4 \end{pmatrix}$

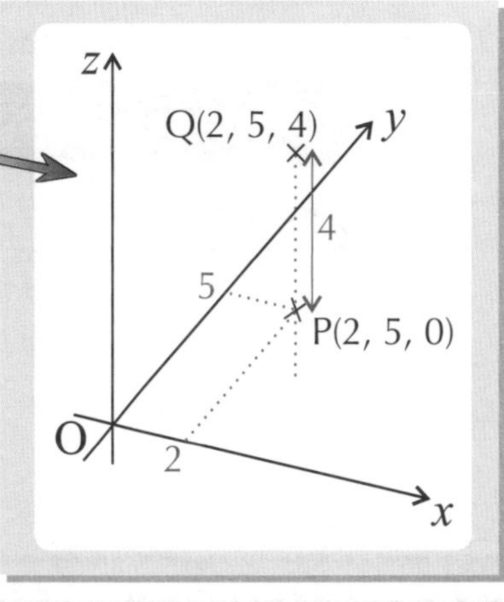

I've got B + Q units in my kitchen...

Three dimensions doesn't really make things much more difficult — it just gives you an extra number to calculate with. You add, subtract and multiply 3D column vectors in the same way as 2D ones — you just have three rows to deal with.

Vectors

This page is for AQA C4, Edexcel C4, OCR C4 and OCR MEI C4

Pythagoras pops up all over the place, and here he is again. Fascinating fact — Pythagoras refused to say words containing the Greek equivalent of the letter c. I read it on the internet, so it has to be true.

Use *Pythagoras' Theorem* to Find Vector *Magnitudes*

1) The magnitude of vector **a** is written as $|\mathbf{a}|$, and the magnitude of $\overrightarrow{AB}$ is written as $|\overrightarrow{AB}|$.
2) The **i** and **j** components of a vector form a convenient right-angled triangle, so just bung them into the Pythagoras formula to find the vector's magnitude.
3) You might be asked to find a unit vector in the direction of a particular vector. Remember — a unit vector has a magnitude of 1 (see the previous page).

A vector's magnitude is sometimes called its modulus.

EXAMPLE

$\mathbf{a} = 5\mathbf{i} + 3\mathbf{j}$

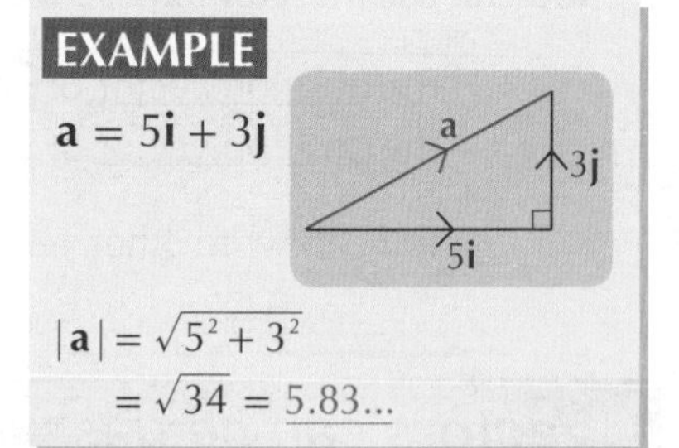

$|\mathbf{a}| = \sqrt{5^2 + 3^2}$
$= \sqrt{34} = 5.83...$

A unit vector in the direction of vector $\mathbf{a} = \dfrac{\mathbf{a}}{|\mathbf{a}|}$

EXAMPLE If vector **p** has a magnitude of 12 units, find a unit vector parallel to **p**.

$$\frac{\mathbf{p}}{|\mathbf{p}|} = \frac{\mathbf{p}}{12} = \frac{1}{12}\mathbf{p}$$

You Can Use *Pythagoras* in *Three Dimensions* Too

1) You can use a variation of Pythagoras' theorem to find the distance of any point in 3 dimensions from the origin, O.

The distance of point (x, y, z) from the origin is $\sqrt{x^2 + y^2 + z^2}$

EXAMPLE 1

Find $|\overrightarrow{OQ}|$.

$|\overrightarrow{OQ}| = \sqrt{x^2 + y^2 + z^2}$
$= \sqrt{2^2 + 5^2 + 4^2}$
$= \sqrt{45}$
$= 6.7$ units

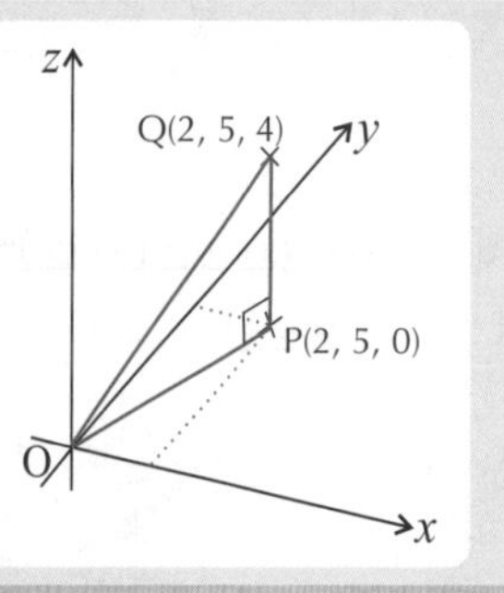

Here's where this formula comes from:
$OP = \sqrt{x^2 + y^2}$
$OP^2 = x^2 + y^2$
$OQ = \sqrt{OP^2 + z^2}$
$OQ = \sqrt{x^2 + y^2 + z^2}$

EXAMPLE 2 Find the magnitude of the vector $\mathbf{r} = 5\mathbf{i} + 7\mathbf{j} + 3\mathbf{k}$.

$|\mathbf{r}| = \sqrt{5^2 + 7^2 + 3^2}$
$= \sqrt{83} = 9.1$ units

2) There's also a Pythagoras-based formula for finding the distance between any two points.

The distance between points (x_1, y_1, z_1) and (x_2, y_2, z_2) is $\sqrt{(x_1 - x_2)^2 + (y_1 - y_2)^2 + (z_1 - z_2)^2}$

EXAMPLE

The position vector of point A is $3\mathbf{i} + 2\mathbf{j} + 4\mathbf{k}$, and the position vector of point B is $2\mathbf{i} + 6\mathbf{j} - 5\mathbf{k}$. Find $|\overrightarrow{AB}|$.

A has the coordinates (3, 2, 4),
B has the coordinates (2, 6, –5).

$|\overrightarrow{AB}| = \sqrt{(x_1 - x_2)^2 + (y_1 - y_2)^2 + (z_1 - z_2)^2}$
$= \sqrt{(3-2)^2 + (2-6)^2 + (4-(-5))^2}$
$= \sqrt{1 + 16 + 81} = 9.9$ units

You can play Battleships with 3D coordinates too — but you don't have to...

The magnitude is just a scalar, so it doesn't have a direction — the magnitude of $\overrightarrow{AB}$ is the same as the magnitude of $\overrightarrow{BA}$. Squaring the numbers in the formulas gets rid of any minus signs, so you don't have to worry about which way round you subtract the coordinates (phew). There's not a lot new on this page — in fact, it's mostly just good old Pythagoras.

Vector Equations of Lines

This page is for AQA C4, Edexcel C4, OCR C4 and OCR MEI C4

At first glance, vector equations of straight lines don't look much like normal straight-line equations.
But they're pretty similar if you look closely. In any case, just learn the formulas really well and you'll be fine.

Learn the Equation of the Line **Through a Point** *and* **Parallel to Another Vector**

A straight line which goes through point A, and is parallel to vector **b**, has the vector equation: $\mathbf{r} = \mathbf{a} + t\mathbf{b}$

a = position vector of point A
r = position vector of a point on the line, and t = a scalar.

A is a fixed point.

This is pretty much a 3D version of the old $y = mx + c$ equation.
b is similar to the gradient, m, and **a** gives a point that the line passes through, just like c gives the y-axis intercept.

Each different value you stick in for t in the vector equation gives you the position vector, **r**, of a different point on the line.

EXAMPLE A straight line is parallel to the vector $\mathbf{i} + 3\mathbf{j} - 2\mathbf{k}$. It passes through a point with the position vector $3\mathbf{i} + 2\mathbf{j} + 6\mathbf{k}$. Find its vector equation.

$$\mathbf{r} = \mathbf{a} + t\mathbf{b} = (3\mathbf{i} + 2\mathbf{j} + 6\mathbf{k}) + t(\mathbf{i} + 3\mathbf{j} - 2\mathbf{k})$$

Alternative ways of writing this are:
$\mathbf{r} = (3 + t)\mathbf{i} + (2 + 3t)\mathbf{j} + (6 - 2t)\mathbf{k}$,
$\mathbf{r} = \begin{pmatrix}3\\2\\6\end{pmatrix} + t\begin{pmatrix}1\\3\\-2\end{pmatrix}$ and $\mathbf{r} = \begin{pmatrix}3+t\\2+3t\\6-2t\end{pmatrix}$

And the Equation of the Line Passing **Through Two Known Points**

A straight line through points C and D, with position vectors **c** and **d**, has the vector equation:

$$\mathbf{r} = \mathbf{c} + t(\mathbf{d} - \mathbf{c})$$

r = position vector of a point on the line, t = a scalar.

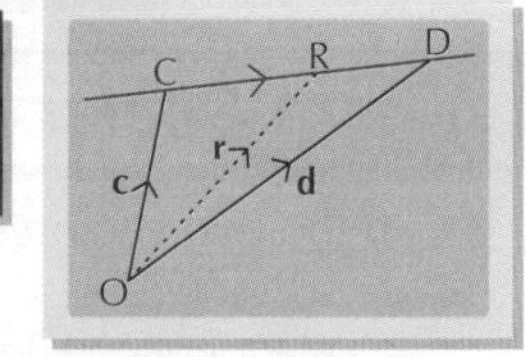

This is basically the same as the vector equation above. You just have to find a vector in the direction of CD first (i.e. $\mathbf{d} - \mathbf{c}$).

EXAMPLE
Find the vector equation of the line through (3, 2, 4) and (–1, 3, 0).

If $\mathbf{c} = \begin{pmatrix}3\\2\\4\end{pmatrix}$, and $\mathbf{d} = \begin{pmatrix}-1\\3\\0\end{pmatrix}$, then $\mathbf{r} = \begin{pmatrix}3\\2\\4\end{pmatrix} + t\left(\begin{pmatrix}-1\\3\\0\end{pmatrix} - \begin{pmatrix}3\\2\\4\end{pmatrix}\right) \Rightarrow \mathbf{r} = \begin{pmatrix}3\\2\\4\end{pmatrix} + t\begin{pmatrix}-4\\1\\-4\end{pmatrix}$

Find the **Point of Intersection** *of two Lines with* **Simultaneous Equations**

If l_1, $\mathbf{r} = \begin{pmatrix}5\\2\\-1\end{pmatrix} + \mu\begin{pmatrix}1\\-2\\-3\end{pmatrix}$, and l_2, $\mathbf{r} = \begin{pmatrix}2\\0\\4\end{pmatrix} + \lambda\begin{pmatrix}1\\2\\-1\end{pmatrix}$, intersect, there'll be a value for μ and a value for λ that result in the same point for both lines. This is the point of intersection.

If you can't find values that work for both lines, it means the lines don't intersect. Two lines that aren't parallel and don't intersect are called skew lines.

EXAMPLE Determine whether Line 1 and Line 2 (above) intersect. If they do, find the point of intersection.

At the point of intersection, $\begin{pmatrix}5\\2\\-1\end{pmatrix} + \mu\begin{pmatrix}1\\-2\\-3\end{pmatrix} = \begin{pmatrix}2\\0\\4\end{pmatrix} + \lambda\begin{pmatrix}1\\2\\-1\end{pmatrix}$. You can get 3 equations from this:
① $5 + \mu = 2 + \lambda$
② $2 - 2\mu = 0 + 2\lambda$
③ $-1 - 3\mu = 4 - \lambda$

Solve the first two simultaneously: $2 \times$①: $10 + 2\mu = 4 + 2\lambda$ ④
④ – ②: $8 + 4\mu = 4 \Rightarrow \mu = -1$
sub. in ②: $2 - 2(-1) = 0 + 2\lambda \Rightarrow \lambda = 2$

Substitute the values for μ and λ into equation ③. If they make the equation true, then the lines do intersect:
$-1 - (3 \times -1) = 4 - 2 \Rightarrow 2 = 2$ — True, so they do intersect.

Now find the intersection point: $\mathbf{r} = \begin{pmatrix}5\\2\\-1\end{pmatrix} + \mu\begin{pmatrix}1\\-2\\-3\end{pmatrix} = \begin{pmatrix}5\\2\\-1\end{pmatrix} - 1\begin{pmatrix}1\\-2\\-3\end{pmatrix} \Rightarrow \mathbf{r} = \begin{pmatrix}4\\4\\2\end{pmatrix} = 4\mathbf{i} + 4\mathbf{j} + 2\mathbf{k}$

This is the position vector of the intersection point. The coordinates are (4, 4, 2).

*Stardate 45283.5, position vector 20076**i** + 23485**j** + 48267**k**...*

You might be given vector equations in **i**, **j**, **k** form or in column form, so practise these examples using each vector form.

Scalar Product

This page is for AQA C4, Edexcel C4, OCR C4 and OCR MEI C4

The scalar product of two vectors is kind of what it says on the tin — two vectors multiplied together to give a scalar result. But this is A2, so it's going to be trickier than simple multiplying. It even involves a bit of cos-ing.

Learn the Definition of the **Scalar Product of Two Vectors**

Scalar Product of Two Vectors

$$\mathbf{a}.\mathbf{b} = |\mathbf{a}||\mathbf{b}|\cos\theta$$

θ is the angle between position vectors **a** and **b**.

Both vectors have to be directed away from the intersection point.

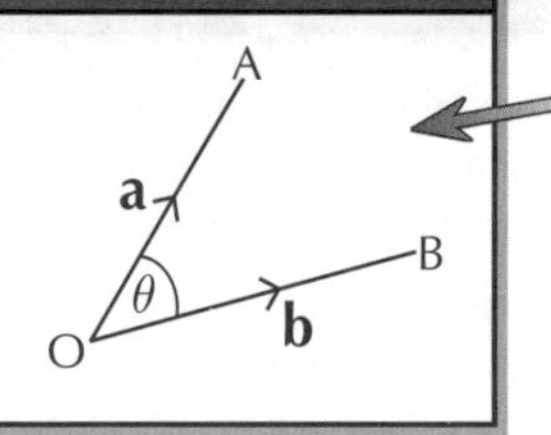

Watch out — the correct angle might not always be obvious. θ is the angle in the definition. Here you have to continue **b** on so that it's also directed away from the intersection point.

1) The scalar product of two vectors is always a scalar quantity — it's never a vector.
2) The scalar product can be used to calculate the angle between two lines (see the next page) — $\mathbf{a}.\mathbf{b} = |\mathbf{a}||\mathbf{b}|\cos\theta$ rearranges to $\cos\theta = \frac{\mathbf{a}.\mathbf{b}}{|\mathbf{a}||\mathbf{b}|}$.
3) The scalar product $\mathbf{a}\,.\,\mathbf{b}$ is read '**a** dot **b**'. It's really, really important to put the dot in, as it shows you mean the scalar product (rather than a different sort of vector product that you don't have to worry about in C4).

A **Zero Scalar Product** Means the Vectors are **Perpendicular**

1) If the two vectors are perpendicular, they're at 90° to each other.
2) Cos 90° = 0, so the scalar product of the two vectors is 0.

Scalar Product of Two Perpendicular Vectors

$$\mathbf{a}.\mathbf{b} = |\mathbf{a}||\mathbf{b}|\cos 90^\circ = 0$$

3) The unit vectors **i**, **j** and **k** are all perpendicular to each other.

So, $\mathbf{i}\,.\,\mathbf{j} = 1 \times 1 \times 0 = 0$ and $3\mathbf{j}\,.\,4\mathbf{k} = 3 \times 4 \times 0 = 0$

4) This all assumes that the vectors are non-zero. Because if either vector was 0, you'd always get a scalar product of 0, regardless of the angle between them.

The Scalar Product of **Parallel Vectors** is just the **Product of the Magnitudes**

1) If two vectors are parallel, the angle between them is 0°. And cos 0° = 1, so...

Scalar Product of Two Parallel Vectors

$$\mathbf{a}.\mathbf{b} = |\mathbf{a}||\mathbf{b}|\cos 0^\circ = |\mathbf{a}||\mathbf{b}|$$

2) Two **i** unit vectors are parallel to each other (as are two **j**s or two **k**s).

So, $\mathbf{j}\,.\,\mathbf{j} = 1 \times 1 \times 1 = 1$ and $3\mathbf{k}\,.\,4\mathbf{k} = 3 \times 4 \times 1 = 12$

3) Again, this all assumes that the vectors are non-zero.

Scaly product — a lizard-skin handbag...

The fact that two perpendicular vectors have a zero scalar product is the key to loads of vector exam questions. E.g. you might be asked to show two vectors are perpendicular, or told that two vectors are perpendicular and asked to find a missing vector. Whatever they ask, you'll definitely have to multiply the two vectors — and you're about to learn how.

Scalar Product

This page is for AQA C4, Edexcel C4, OCR C4 and OCR MEI C4

Finding the scalar product of two vectors is super quick and easy once you know how to do it.

*Learn This Result for the **Scalar Product***

1) You can use this result to find the scalar product of two known vectors:

If $\mathbf{a} = a_1\mathbf{i} + a_2\mathbf{j} + a_3\mathbf{k}$, and $\mathbf{b} = b_1\mathbf{i} + b_2\mathbf{j} + b_3\mathbf{k}$, then $\mathbf{a.b} = a_1b_1 + a_2b_2 + a_3b_3$

2) The normal laws of multiplication apply to scalar products too — e.g. the commutative law ($\mathbf{a}.\mathbf{b} = \mathbf{b}.\mathbf{a}$) and the distributive law ($\mathbf{a}.(\mathbf{b} + \mathbf{c}) = \mathbf{a}.\mathbf{b} + \mathbf{a}.\mathbf{c}$).

3) By applying these laws, you can derive the result above...

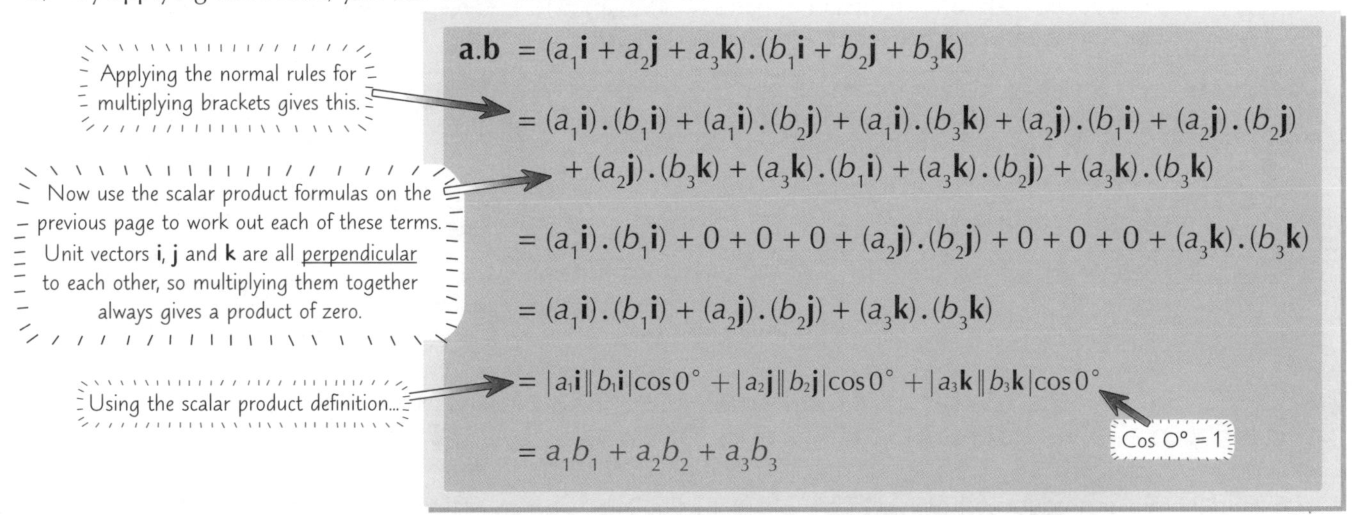

$$\mathbf{a.b} = (a_1\mathbf{i} + a_2\mathbf{j} + a_3\mathbf{k}).(b_1\mathbf{i} + b_2\mathbf{j} + b_3\mathbf{k})$$

$$= (a_1\mathbf{i}).(b_1\mathbf{i}) + (a_1\mathbf{i}).(b_2\mathbf{j}) + (a_1\mathbf{i}).(b_3\mathbf{k}) + (a_2\mathbf{j}).(b_1\mathbf{i}) + (a_2\mathbf{j}).(b_2\mathbf{j}) + (a_2\mathbf{j}).(b_3\mathbf{k}) + (a_3\mathbf{k}).(b_1\mathbf{i}) + (a_3\mathbf{k}).(b_2\mathbf{j}) + (a_3\mathbf{k}).(b_3\mathbf{k})$$

$$= (a_1\mathbf{i}).(b_1\mathbf{i}) + 0 + 0 + 0 + (a_2\mathbf{j}).(b_2\mathbf{j}) + 0 + 0 + 0 + (a_3\mathbf{k}).(b_3\mathbf{k})$$

$$= (a_1\mathbf{i}).(b_1\mathbf{i}) + (a_2\mathbf{j}).(b_2\mathbf{j}) + (a_3\mathbf{k}).(b_3\mathbf{k})$$

$$= |a_1\mathbf{i}||b_1\mathbf{i}|\cos 0° + |a_2\mathbf{j}||b_2\mathbf{j}|\cos 0° + |a_3\mathbf{k}||b_3\mathbf{k}|\cos 0°$$

$$= a_1b_1 + a_2b_2 + a_3b_3$$

*Use the **Scalar Product** to Find the **Angle** Between Two Vectors*

Finding the angle between two vectors often crops up in vector exam questions.
It's just a matter of using the above result to find the scalar product of the two vectors, then popping it into the scalar product definition, $\cos\theta = \frac{\mathbf{a.b}}{|\mathbf{a}||\mathbf{b}|}$, to find the angle.

EXAMPLE Find the angle between the vectors $-\mathbf{i} - 6\mathbf{j}$ and $4\mathbf{i} + 2\mathbf{j} + 8\mathbf{k}$.

$\cos\theta = \frac{\mathbf{a.b}}{|\mathbf{a}||\mathbf{b}|}$. Let $\mathbf{a} = -\mathbf{i} - 6\mathbf{j}$ and $\mathbf{b} = 4\mathbf{i} + 2\mathbf{j} + 8\mathbf{k}$.

1) Find the scalar product of the vectors.

$$\mathbf{a.b} = (-1 \times 4) + (-6 \times 2) + (0 \times 8) = -4 - 12 + 0 = -16$$

This uses the result above.

2) Find the magnitude of each vector (see page 86).

$$|\mathbf{a}| = \sqrt{(-1)^2 + (-6)^2 + (0)^2} = \sqrt{37} \qquad |\mathbf{b}| = \sqrt{(4)^2 + (2)^2 + (8)^2} = \sqrt{84}$$

3) Now plug these values into the equation and find the angle.

$$\cos\theta = \frac{\mathbf{a.b}}{|\mathbf{a}||\mathbf{b}|} = \frac{-16}{\sqrt{37}\sqrt{84}} \Rightarrow \theta = 106.7°$$

Scalar product — Ooops. I'd best stop eating chips every day...

So when you scalar multiply two vectors, you basically multiply the **i** components together, multiply the **j** components together, multiply the **k** components together, then add up all the products. You end up with just a number, with no **i**s, **j**s or **k**s attached to it. You'll see this more in the examples on the next page, so don't worry if it seems a bit strange at the mo.

Scalar Product

This page is for AQA C4, Edexcel C4, OCR C4 and OCR MEI C4

Right, you've learnt the definitions and the facts. Now it's time to put them to good use.

You Might have to Find the Angle from **Vector Equations** or from **Two Points**

1) If you're given the vector equations for lines that you're finding the angle between, it's important to use the correct bits of the vector equations.
2) You use the **b** bit in $\mathbf{r} = \mathbf{a} + t\mathbf{b}$ (the 'parallel to' or the 'direction' bit).

EXAMPLE Line l has the equation $\mathbf{r} = \begin{pmatrix} 2 \\ 0 \\ 4 \end{pmatrix} + \lambda \begin{pmatrix} 1 \\ 2 \\ -1 \end{pmatrix}$.

Point A and point B have the coordinates (4, 4, 2) and (1, 0, 3) respectively. Point A lies on l. Find the acute angle between l and line segment AB.

1) First draw a diagram — it'll make everything clearer.

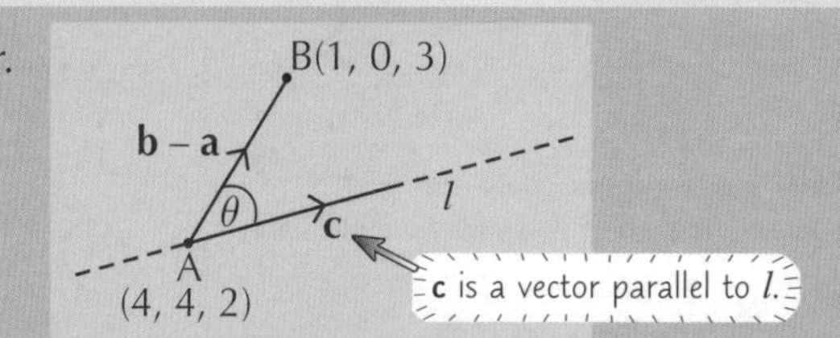

2) Find the vectors that you want to know the angle between.

$$\overrightarrow{AB} = \mathbf{b} - \mathbf{a} = \begin{pmatrix} 1 \\ 0 \\ 3 \end{pmatrix} - \begin{pmatrix} 4 \\ 4 \\ 2 \end{pmatrix} = \begin{pmatrix} -3 \\ -4 \\ 1 \end{pmatrix}$$ and the 'parallel to' bit of l (which we've called **c**): $\mathbf{c} = \begin{pmatrix} 1 \\ 2 \\ -1 \end{pmatrix}$

3) Find the scalar product of these vectors. $\overrightarrow{AB} \cdot \mathbf{c} = (-3 \times 1) + (-4 \times 2) + (1 \times -1) = -3 - 8 - 1 = -12$

4) Find the magnitude of each vector.

$$|\overrightarrow{AB}| = \sqrt{(-3)^2 + (-4)^2 + (1)^2} = \sqrt{26} \quad |\mathbf{c}| = \sqrt{(1)^2 + (2)^2 + (-1)^2} = \sqrt{6}$$

5) Now plug these values into the equation and find the angle.

$$\cos\theta = \frac{\overrightarrow{AB}.\mathbf{c}}{|\overrightarrow{AB}||\mathbf{c}|} = \frac{-12}{\sqrt{26}\sqrt{6}} \Rightarrow \theta = 164°\,(3\text{ s.f.})$$

6) Whoops. The formula gives the non-acute angle — the situation must have been more like this:

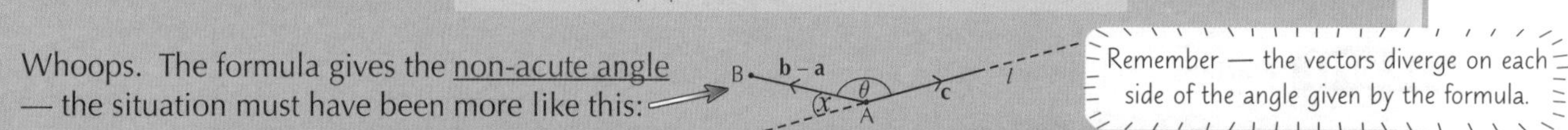

Don't panic — just subtract this angle from 180° to get the acute angle, x, between the lines.

$180° - 164° = 16°$

Prove Lines are **Perpendicular** by Showing that the **Scalar Product = 0**

EXAMPLE Show that the lines $\mathbf{r}_1 = (\mathbf{i} + 6\mathbf{j} + 2\mathbf{k}) + \lambda(\mathbf{i} + 2\mathbf{j} + 2\mathbf{k})$ and $\mathbf{r}_2 = (3\mathbf{i} - \mathbf{j} + \mathbf{k}) + \mu(4\mathbf{i} - 3\mathbf{j} + \mathbf{k})$ are perpendicular.

1) Make sure you've got the right bit of each vector equation — it's the direction you're interested in, so it's **b** in $\mathbf{r} = \mathbf{a} + t\mathbf{b}$. $\mathbf{i} + 2\mathbf{j} + 2\mathbf{k}$ and $4\mathbf{i} - 3\mathbf{j} + \mathbf{k}$
2) Find the scalar product of the vectors. $(\mathbf{i} + 2\mathbf{j} + 2\mathbf{k}).(4\mathbf{i} - 3\mathbf{j} + \mathbf{k}) = 4 - 6 + 2 = 0$
3) Draw the correct conclusion. The scalar product is 0, so the vectors are perpendicular.

P...P...P... — prove perpendicularity using products...

They'll word these questions in a zillion different ways. Drawing a diagram can often help you figure out what's what.

Equations of Planes

This page is for OCR MEI C4

Planes might seem a bit complicated at first glance — but most of the maths you need to do with them is surprisingly similar to stuff you've already seen with lines. Plus, the word 'plane' offers almost endless potential for dodgy puns.

Learn how to find **Equations of Planes**

1) A plane is a flat 2D surface. You can define a plane with a Cartesian equation in three coordinates (normally x, y and z) or using a vector equation.
2) The Cartesian equation of a plane will be in the form $ax + by + cz + d = 0$. The vector equation of a plane will be in the form $\mathbf{r} = \mathbf{a} + \lambda\mathbf{b} + \mu\mathbf{c}$.
3) A normal to a plane is a line or vector that's perpendicular to the plane — so it meets the plane at a right angle.

Finding the **Cartesian Equation** from a **Point** and a **Normal** is **Easy**

If $\mathbf{n}$ is a normal vector to a plane, and $\mathbf{a}$ is the position vector of a point on the plane, then the Cartesian equation of the plane is $n_1x + n_2y + n_3z + d = 0$, where $d = -\mathbf{a}\cdot\mathbf{n}$.

$\mathbf{n} = \begin{pmatrix} n_1 \\ n_2 \\ n_3 \end{pmatrix}$

EXAMPLE The point A has coordinates (10, –5, 1) and lies on a plane, P. The vector $3\mathbf{i} + 2\mathbf{j} - \mathbf{k}$ is perpendicular to P. Find the Cartesian equation of P.

Remember, if a vector's perpendicular to a plane (or line), that means it's a normal.

The position vector of A is $10\mathbf{i} - 5\mathbf{j} + \mathbf{k}$.

So the constant term d is: $d = -\mathbf{a}\cdot\mathbf{n} = -[a_1n_1 + a_2n_2 + a_3n_3] = -[(10 \times 3) + (-5 \times 2) + (1 \times -1)] = -19$

So the Cartesian equation of P is $3x + 2y - z - 19 = 0$.

You might have to prove that your normal is actually a normal before you find the equation of the plane. A vector is a normal to a plane if it's perpendicular to any two non-parallel vectors on the plane — so to prove it, get two non-parallel vectors on the plane (e.g. by finding the vectors joining given points on the plane) and show their scalar products with the normal are both 0 (see p90).

You can find the **Vector Equation** from a **Point** and **Two Vectors**

The vector equation of a plane is $\mathbf{r} = \mathbf{a} + \lambda\mathbf{b} + \mu\mathbf{c}$, where $\mathbf{a}$ is the position vector of a point on the plane, and $\mathbf{b}$ and $\mathbf{c}$ are two non-parallel vectors that lie on the plane.

EXAMPLE Find the vector equation of the plane passing through A = (2, 9, 2), B = (0, 6, 3) and C = (8, 4, 1).

The vectors $\overrightarrow{AB}$ and $\overrightarrow{AC}$ lie on the plane we're looking for, and they're pretty much guaranteed not to be parallel, so we can use the position vector of A as $\mathbf{a}$, the vector $\overrightarrow{AB}$ as $\mathbf{b}$ and the vector $\overrightarrow{AC}$ as $\mathbf{c}$:

$$\mathbf{a} = \begin{pmatrix} 2 \\ 9 \\ 2 \end{pmatrix}, \quad \mathbf{b} = \overrightarrow{AB} = -\begin{pmatrix} 2 \\ 9 \\ 2 \end{pmatrix} + \begin{pmatrix} 0 \\ 6 \\ 3 \end{pmatrix} = \begin{pmatrix} -2 \\ -3 \\ 1 \end{pmatrix}, \quad \mathbf{c} = \overrightarrow{AC} = -\begin{pmatrix} 2 \\ 9 \\ 2 \end{pmatrix} + \begin{pmatrix} 8 \\ 4 \\ 1 \end{pmatrix} = \begin{pmatrix} 6 \\ -5 \\ -1 \end{pmatrix}$$

So the vector equation of the plane is $\mathbf{r} = \mathbf{a} + \lambda\mathbf{b} + \mu\mathbf{c} = \begin{pmatrix} 2 \\ 9 \\ 2 \end{pmatrix} + \lambda\begin{pmatrix} -2 \\ -3 \\ 1 \end{pmatrix} + \mu\begin{pmatrix} 6 \\ -5 \\ -1 \end{pmatrix}$

Time for some plane speaking...

Check out how similar the second example on this page is to the examples for finding the vector equation of a line on p87. I told you this stuff with planes was really similar to the stuff with lines... I reckon this page has been plane sailing...

Lines and Planes

This page is for OCR MEI C4

It should be as plane as the nose on your face that there's more to planes than just finding their equations. The next thrill that awaits is the intersection of lines and planes.

You can find the Point where a Plane meets a Line

Finding the point of intersection between a line and a plane is, unsurprisingly, pretty similar to finding the point of intersection between two lines. In fact, when you've got the Cartesian equation of the plane, it's easier.

EXAMPLE Find the point where the line $\mathbf{r} = \begin{pmatrix} 6 \\ 0 \\ 1 \end{pmatrix} + t\begin{pmatrix} 3 \\ 1 \\ 10 \end{pmatrix}$ meets the plane $3x + 5y - z - 5 = 0$.

Write the components of the vector equation of the line as three equations:

$$x = 6 + 3t, \quad y = t, \quad z = 1 + 10t$$

Now stick those into the equation of the plane and solve for t:

$3x + 5y - z - 5 = 0$

$\Rightarrow \ 3(6 + 3t) + 5(t) - (1 + 10t) - 5 = 0$

$\Rightarrow \ 18 + 9t + 5t - 1 - 10t - 5 = 0$

$\Rightarrow \ 12 + 4t = 0$

$\Rightarrow \ t = -3$

So at the point where the line and plane meet, $t = -3$.
Put $t = -3$ back into the equation of the line to get the coordinates of this point:

$$\mathbf{r} = \begin{pmatrix} 6 \\ 0 \\ 1 \end{pmatrix} + t\begin{pmatrix} 3 \\ 1 \\ 10 \end{pmatrix} = \begin{pmatrix} 6 \\ 0 \\ 1 \end{pmatrix} - 3\begin{pmatrix} 3 \\ 1 \\ 10 \end{pmatrix} = \begin{pmatrix} 6 - 9 \\ 0 - 3 \\ 1 - 30 \end{pmatrix} = \begin{pmatrix} -3 \\ -3 \\ -29 \end{pmatrix}$$

So the line and plane meet at $(-3, -3, -29)$.

Use Normal Vectors to find the Angle Between Two Planes

The angle between two planes is the same as the angle between their normals.

To find a normal to a plane from its Cartesian equation, use the coefficients of x, y and z (when they're all on the same side of the equation) as the components of the normal vector.

This is almost the reverse of the method for finding the Cartesian equation from the previous page.

When you've got your two normals, you can use the scalar product to find the angle between them (see p89).

EXAMPLE Find the angle between the planes $5x + 2y + 3z = 9$ and $x - 3y + z = 7$.

Use the coefficients of x, y and z in the equations to find a normal to each plane:

$\mathbf{n}_1 = \begin{pmatrix} 5 \\ 2 \\ 3 \end{pmatrix}$ is a normal to $5x + 2y + 3z = 9$ and $\mathbf{n}_2 = \begin{pmatrix} 1 \\ -3 \\ 1 \end{pmatrix}$ is a normal to $x - 3y + z = 7$.

The angle between the planes is equal to the angle between the normals, so

$$\cos\theta = \frac{\mathbf{n}_1.\mathbf{n}_2}{|\mathbf{n}_1||\mathbf{n}_2|} = \frac{(5 \times 1) + (2 \times -3) + (3 \times 1)}{\sqrt{5^2 + 2^2 + 3^2}\sqrt{1^2 + (-3)^2 + 1^2}} = \frac{2}{\sqrt{38}\sqrt{11}} = 0.0978$$

$\Rightarrow \theta = \cos^{-1} 0.0978 = 84.4°$

I hope that ex-planes everything...

Most exam questions about planes will be a combination of the examples on the last two pages, with some bits from the rest of this section thrown in for good measure. Make sure you practise lots of exam-style questions so you can see how all this vector stuff fits together. And if that gives you a headache, take a couple of planekillers... ho ho... ... ...I'll get my coat.

Core Section 11 — Practice Questions

Vectors might cause some mild vexation. It's not the simplest of topics, but practising does help. Try these warm-up questions and see if you can remember what you've just read.

Warm-up Questions

Q1-10 are for AQA C4, Edexcel C4, OCR C4 and OCR MEI C4

1) Give two vectors that are parallel to each of the following: a) $2\mathbf{a}$ b) $3\mathbf{i} + 4\mathbf{j} - 2\mathbf{k}$ c) $\begin{pmatrix} 1 \\ 2 \\ -1 \end{pmatrix}$

2) Using the diagram on the right, find these vectors in terms of vectors **a**, **b** and **c**.
a) $\overrightarrow{AB}$ b) $\overrightarrow{BA}$ c) $\overrightarrow{CB}$ d) $\overrightarrow{AC}$

3) Give the position vector of point P, which has the coordinates (2, –4, 5).
Give your answer in unit vector form.

4) Find the magnitudes of these vectors: a) $3\mathbf{i} + 4\mathbf{j} - 2\mathbf{k}$ b) $\begin{pmatrix} 1 \\ 2 \\ -1 \end{pmatrix}$

5) If A(1, 2, 3) and B(3, –1, –2), find: a) $|\overrightarrow{AB}|$ b) $|\overrightarrow{OA}|$ c) $|\overrightarrow{OB}|$

6) Find the vector equations of the following lines.
Give your answer in **i**, **j**, **k** form and in column vector form.
a) a straight line through (4, 1, 2), parallel to vector $3\mathbf{i} + \mathbf{j} - \mathbf{k}$.
b) a straight line through (2, –1, 1) and (0, 2, 3).

7) Find three points that lie on the line with vector equation $\mathbf{r} = \begin{pmatrix} 3 \\ 2 \\ 4 \end{pmatrix} + t\begin{pmatrix} -1 \\ 3 \\ 0 \end{pmatrix}$.

8) Find **a.b** if: a) $\mathbf{a} = 3\mathbf{i} + 4\mathbf{j}$ and $\mathbf{b} = \mathbf{i} - 2\mathbf{j} + 3\mathbf{k}$ b) $\mathbf{a} = \begin{pmatrix} 4 \\ 2 \\ 1 \end{pmatrix}$ and $\mathbf{b} = \begin{pmatrix} 3 \\ -4 \\ -3 \end{pmatrix}$

9) $\mathbf{r}_1 = \begin{pmatrix} 2 \\ -1 \\ 2 \end{pmatrix} + t\begin{pmatrix} -4 \\ 6 \\ -2 \end{pmatrix}$ and $\mathbf{r}_2 = \begin{pmatrix} 3 \\ 2 \\ 4 \end{pmatrix} + u\begin{pmatrix} -1 \\ 3 \\ 0 \end{pmatrix}$
a) Show that these lines intersect and find the position vector of their intersection point.
b) Find the angle between these lines.

10) Find a vector that is perpendicular to $3\mathbf{i} + 4\mathbf{j} - 2\mathbf{k}$.

Q11-12 are just for OCR MEI C4

11) Find the Cartesian equation of the plane perpendicular to $\mathbf{i} + 3\mathbf{j} - 3\mathbf{k}$ which contains the point (2, 2, 4).

12) Find a vector equation for the plane containing the points (1, –2, 5), (6, 2, –3) and (4, 0, 2).

You might look at an exam question and think that it's complete gobbledegook. But chances are, when you look at it carefully, you can use what you know to solve it. If you don't know what you need to, you can peek back while doing these questions. You won't be able to in the proper exam, so all the more reason to practise on these.

Exam Questions

Q1-4 are for AQA C4, Edexcel C4, OCR C4 and OCR MEI C4

1 The quadrilateral ABCD has vertices A(1, 5, 9), B(3, 2, 1), C(–2, 4, 3) and D(5, –1, –7).

a) Find the vector $\overrightarrow{AB}$. *(2 marks)*

b) C and D lie on line l_1. Using the parameter μ, find the vector equation of l_1. *(2 marks)*

c) Find the coordinates of the intersection point of l_1 and the line that passes through AB. *(5 marks)*

d) (i) Find the acute angle between l_1 and AB. Give your answer to 1 decimal place. *(4 marks)*

(ii) Find the shortest distance from point A to l_1. *(4 marks)*

Core Section 11 — Practice Questions

2 The lines l_1 and l_2 are given by the vector equations:

l_1: $\mathbf{r} = (3\mathbf{i} - 3\mathbf{j} - 2\mathbf{k}) + \mu(\mathbf{i} - 4\mathbf{j} + 2\mathbf{k})$ l_2: $\mathbf{r} = (10\mathbf{i} - 21\mathbf{j} + 11\mathbf{k}) + \lambda(-3\mathbf{i} + 12\mathbf{j} - 6\mathbf{k})$

a) Show that l_1 and l_2 are parallel. *(1 mark)*

b) Show that point A(2, 1, –4) lies on l_1. *(2 marks)*

c) Point B lies on l_2 and is such that the line segment AB is perpendicular to l_1 and l_2. Find the position vector of point B. *(6 marks)*

d) Find $|\overrightarrow{AB}|$. *(2 marks)*

3 The lines l_1 and l_2 are given by the equations: l_1: $\mathbf{r} = \begin{pmatrix} 3 \\ 0 \\ -2 \end{pmatrix} + \lambda \begin{pmatrix} 1 \\ 3 \\ -2 \end{pmatrix}$ l_2: $\mathbf{r} = \begin{pmatrix} 0 \\ 2 \\ 1 \end{pmatrix} + \mu \begin{pmatrix} 2 \\ -5 \\ -3 \end{pmatrix}$

a) Show that l_1 and l_2 do not intersect. *(4 marks)*

b) Point P has position vector $\begin{pmatrix} 5 \\ 8 \\ -3 \end{pmatrix}$. Point Q is the image of point P after reflection in line l_1.

Point P and Q both lie on the line with equation $\mathbf{r} = \begin{pmatrix} 5 \\ 4 \\ -9 \end{pmatrix} + t \begin{pmatrix} 0 \\ 2 \\ 3 \end{pmatrix}$.

(i) Find the intersection point of line segment PQ and line l_1. *(4 marks)*

(ii) Show that the line segment PQ and line l_1 are perpendicular. *(2 marks)*

(iii) Find the position vector of point Q. *(3 marks)*

4 Point A has the position vector $3\mathbf{i} + 2\mathbf{j} + \mathbf{k}$ and point B has position vector $3\mathbf{i} - 4\mathbf{j} - \mathbf{k}$.

a) Show that AOB is a right-angled triangle. *(3 marks)*

b) Find angle ABO in the triangle using the scalar product definition. *(5 marks)*

c) (i) Point C has the position vector $3\mathbf{i} - \mathbf{j}$. Show that triangle OAC is isosceles. *(3 marks)*

(ii) Calculate the area of triangle OAC. *(4 marks)*

d) (i) Find the vector equation for line l, which passes through points A and B. *(2 marks)*

(ii) The point D lies on line l and has the position vector $a\mathbf{i} + b\mathbf{j} + \mathbf{k}$. Find a and b. *(3 marks)*

Q5 is just for OCR MEI C4

5 The equation of plane A is $3x + 4y + 2z = 1$. The equation of plane B is $x - y + 6z = -5$.

a) Write down normal vectors to each plane. *(1 mark)*

b) The acute angle between planes A and B measures $\theta°$. Find the value of θ. *(3 marks)*

c) Find the coordinates of the point of intersection between the line $\mathbf{r} = \begin{pmatrix} 5 \\ 1 \\ 3 \end{pmatrix} + \lambda \begin{pmatrix} -3 \\ 2 \\ 0 \end{pmatrix}$ and plane A. *(4 marks)*

Probability Distributions

This page is for AQA S2. Note — the whole of Section 1 is for AQA S2 only.

This page might seem familiar from S1. I wish I could say the same for the rest of the book, but alas it's not true.

Getting your head round this **Basic Stuff** is Important

This first bit isn't particularly interesting. But understanding the difference between X and x (bear with me) might make the later stuff a bit less confusing. Might.

1) X (upper case) is just the name of a random variable. So X could be 'score on a dice' — it's just a name.
2) A random variable doesn't have a fixed value. Like with a dice score — the value on any 'roll' is all down to chance.
3) x (lower case) is a particular value that X can take. So for one roll of a dice, x could be 1, 2, 3, 4, 5 or 6.
4) Discrete random variables only have a certain number of possible values. Often these values are whole numbers, but they don't have to be. Usually there are only a few possible values (e.g. the possible scores with one roll of a dice).
5) A probability distribution is a table showing the possible values of x, plus the probability for each one.
6) A probability function is a formula that generates the probabilities for different values of x.

All the Probabilities **Add up to 1**

For a discrete random variable X:

$$\sum_{\text{all } x} P(X = x) = 1$$

EXAMPLE The random variable X has probability function $P(X = x) = kx$ for $x = 1, 2, 3$. Find the value of k.

So X has three possible values (x = 1, 2 and 3), and the probability of each is kx (where you need to find k).

It's easier to understand with a table:

x	1	2	3
$P(X = x)$	$k \times 1 = k$	$k \times 2 = 2k$	$k \times 3 = 3k$

Now just use the formula: $\sum_{\text{all } x} P(X = x) = 1$ Here, this means: $k + 2k + 3k = 6k = 1$

i.e. $k = \frac{1}{6}$ Piece of cake.

The **mode** is the most likely value — so it's the value with the biggest probability.

EXAMPLE The discrete random variable X has the probability distribution shown below.

x	0	1	2	3	4
$P(X = x)$	0.1	0.2	0.3	0.2	a

Find: (i) the value of a, (ii) $P(2 \leq X < 4)$, (iii) the mode.

(i) Use the formula $\sum_{\text{all } x} P(X = x) = 1$ again.

From the table: $0.1 + 0.2 + 0.3 + 0.2 + a = 1$
$0.8 + a = 1$
$a = 0.2$

Careful with the inequality signs — you need to include $x = 2$ but not $x = 4$.

(ii) This is asking for the probability that 'X is greater than or equal to 2, but less than 4'. Easy — just add up the probabilities.

$P(2 \leq X < 4) = P(X = 2) + P(X = 3) = 0.3 + 0.2 = 0.5$

(iii) The mode is the value of x with the biggest probability — so mode = 2.

Probability Distributions

This page is for AQA S2

EXAMPLE An unbiased six-sided dice has faces marked 1, 1, 1, 2, 2, 3.
The dice is rolled twice. Let X be the random variable "sum of the two scores on the dice".
Show that $P(X = 4) = \frac{5}{18}$. Find the probability distribution of X.

① Make a table showing the 36 possible outcomes.
You can see from the table that 10 of these have the outcome $X = 4$

... so $P(X = 4) = \frac{10}{36} = \frac{5}{18}$

Score on roll 1 (columns); Score on roll 2 (rows)

+	1	1	1	2	2	3
1	2	2	2	3	3	4
1	2	2	2	3	3	4
1	2	2	2	3	3	4
2	3	3	3	4	4	5
2	3	3	3	4	4	5
3	4	4	4	5	5	6

Don't forget to change the fractions into their simplest form.

② Use the table to work out the probabilities for the other outcomes and then fill in a table summarising the probability distribution. So...

... $\frac{9}{36}$ of the outcomes are a score of 2

... $\frac{12}{36}$ of the outcomes are a score of 3

... $\frac{4}{36}$ of the outcomes are a score of 5

... $\frac{1}{36}$ of the outcomes are a score of 6

x	2	3	4	5	6
$P(X = x)$	$\frac{1}{4}$	$\frac{1}{3}$	$\frac{5}{18}$	$\frac{1}{9}$	$\frac{1}{36}$

Do Complicated questions *Bit by bit*

EXAMPLE A game involves rolling two fair dice. If the sum of the scores is greater than 10 then the player wins 50p.
If the sum is between 8 and 10 (inclusive) then they win 20p. Otherwise they get nothing.
If X is the random variable "amount player wins", find the probability distribution of X.

There are 3 possible values for X (0, 20 and 50) and you need the probability of each.
To work these out, you need the probability of getting various totals on the dice.

① You need to know $P(8 \leq \text{score} \leq 10)$ — the probability that the score is between 8 and 10 inclusive (i.e. including 8 and 10) and $P(11 \leq \text{score} \leq 12)$ — the probability that the score is greater than 10.

This means working out: P(score = 8), P(score = 9), P(score = 10), P(score = 11) and P(score = 12). Use a table...

② Score on dice 1 (columns); Score on dice 2 (rows)

+	1	2	3	4	5	6
1	2	3	4	5	6	7
2	3	4	5	6	7	8
3	4	5	6	7	8	9
4	5	6	7	8	9	10
5	6	7	8	9	10	11
6	7	8	9	10	11	12

There are 36 possible outcomes...

...5 of these have a total of 8 — so the probability of scoring 8 is $\frac{5}{36}$

...4 have a total of 9 — so the probability of scoring 9 is $\frac{4}{36}$

...the probability of scoring 10 is $\frac{3}{36}$

...the probability of scoring 11 is $\frac{2}{36}$

...the probability of scoring 12 is $\frac{1}{36}$

③ To find the probabilities you need, you just add the right bits together:

$P(X = 20p) = P(8 \leq \text{score} \leq 10) = \frac{5}{36} + \frac{4}{36} + \frac{3}{36} = \frac{12}{36} = \frac{1}{3}$

$P(X = 50p) = P(11 \leq \text{score} \leq 12) = \frac{2}{36} + \frac{1}{36} = \frac{3}{36} = \frac{1}{12}$

To find $P(X = 0)$ just take the total of the two probabilities above from 1 (since $X = 0$ is the only other possibility).

$P(X = 0) = 1 - \left[\frac{12}{36} + \frac{3}{36}\right] = 1 - \frac{15}{36} = \frac{21}{36} = \frac{7}{12}$

④ Now just stick all this info in a table (and check that the probabilities all add up to 1):

x	0	20	50
$P(X = x)$	$\frac{7}{12}$	$\frac{1}{3}$	$\frac{1}{12}$

Useful quotes: All you need in life is ignorance and confidence, then success is sure*...

With all this working out of probabilities, you could be forgiven for thinking you're still doing S1. But I can confirm this is S2 alright. And it seems like your probability skills are as in demand as ever — if you can find the probabilities of the different outcomes, you can find the probability distribution. And remember — ALL THE PROBABILITIES SHOULD ADD UP TO 1.

* Mark Twain

Expected Values, Mean & Variance

This page is for AQA S2

This is all about the mean and variance of random variables — not a load of data. It's a tricky concept, but bear with it.

Discrete Random Variables have an 'Expected Value' or 'Mean'

You can work out the expected value (or 'mean') $E(X)$ for a discrete random variable X. $E(X)$ is a kind of 'theoretical mean' — it's what you'd expect the mean of X to be if you took loads of readings. In practice, the mean of your results is unlikely to match the theoretical mean exactly, but it should be pretty near.

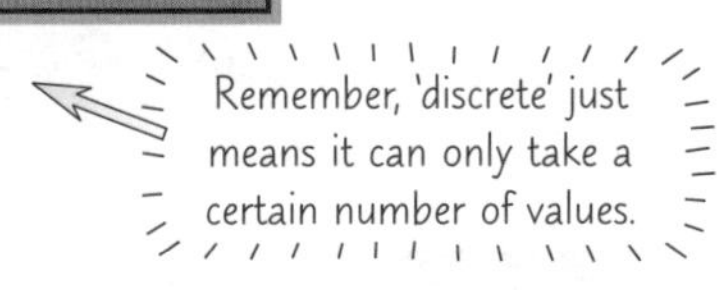

If the possible values of X are $x_1, x_2, x_3, \ldots$ then the expected value of X is:

$$\text{Mean} = \text{Expected Value } E(X) = \sum x_i P(X = x_i) = \sum x_i p_i$$

$p_i = P(X = x_i)$

EXAMPLE The probability distribution of X, the number of daughters in a family of 3 children, is shown in the table. Find the expected number of daughters.

x_i	0	1	2	3
p_i	$\frac{1}{8}$	$\frac{3}{8}$	$\frac{3}{8}$	$\frac{1}{8}$

$$\text{Mean} = \sum x_i p_i = \left[0 \times \tfrac{1}{8}\right] + \left[1 \times \tfrac{3}{8}\right] + \left[2 \times \tfrac{3}{8}\right] + \left[3 \times \tfrac{1}{8}\right] = 0 + \tfrac{3}{8} + \tfrac{6}{8} + \tfrac{3}{8} = \tfrac{12}{8} = 1.5$$

So the expected number of daughters is 1.5 — which sounds a bit weird.
But all it means is that if you check a large number of 3-child families, the mean will be close to 1.5.

The Variance measures how Spread Out the distribution is

You can also find the variance of a random variable. It's the 'expected variance' of a large number of readings.

$$\text{Var}(X) = E(X^2) - [E(X)]^2 = \sum x_i^2 p_i - \left[\sum x_i p_i\right]^2$$

This formula needs $E(X^2) = \sum x_i^2 p_i$ — take each possible value of x, square it, multiply it by its probability and then add up all the results. There's more on this on the next page.

EXAMPLE Work out the variance for the '3 daughters' example above:

First work out $E(X^2)$: $E(X^2) = \sum x_i^2 p_i = \left[0^2 \times \tfrac{1}{8}\right] + \left[1^2 \times \tfrac{3}{8}\right] + \left[2^2 \times \tfrac{3}{8}\right] + \left[3^2 \times \tfrac{1}{8}\right]$

$$= 0 + \tfrac{3}{8} + \tfrac{12}{8} + \tfrac{9}{8} = \tfrac{24}{8} = \underline{3}$$

As always, if you're asked for the standard deviation of X, you just take the square root of Var(X).

Now you take away the mean squared: $\text{Var}(X) = E(X^2) - [E(X)]^2 = 3 - 1.5^2 = 3 - 2.25 = \underline{0.75}$

EXAMPLE X has the probability function $P(X = x) = k(x + 1)$ for $x = 0, 1, 2, 3, 4$.
Find the mean and variance of X.

(1) First you need to find k — work out all the probabilities and make sure they add up to 1.

$P(X = 0) = k \times (0 + 1) = k$. Similarly, $P(X = 1) = 2k$, $P(X = 2) = 3k$, $P(X = 3) = 4k$, $P(X = 4) = 5k$.

So $k + 2k + 3k + 4k + 5k = 1$, i.e. $15k = 1$, and so $k = \frac{1}{15}$

Now you can work out $p_1, p_2, p_3, \ldots$ where $p_1 = P(X = 1)$ etc.

(2) Now use the formulas — find the mean $E(X)$ first:

$$E(X) = \sum x_i p_i = \left[0 \times \tfrac{1}{15}\right] + \left[1 \times \tfrac{2}{15}\right] + \left[2 \times \tfrac{3}{15}\right] + \left[3 \times \tfrac{4}{15}\right] + \left[4 \times \tfrac{5}{15}\right] = \tfrac{40}{15} = \tfrac{8}{3}$$

For the variance you need $E(X^2)$:

$$E(X^2) = \sum x_i^2 p_i = \left[0^2 \times \tfrac{1}{15}\right] + \left[1^2 \times \tfrac{2}{15}\right] + \left[2^2 \times \tfrac{3}{15}\right] + \left[3^2 \times \tfrac{4}{15}\right] + \left[4^2 \times \tfrac{5}{15}\right] = \tfrac{130}{15} = \tfrac{26}{3}$$

And finally: $\text{Var}(X) = E(X^2) - [E(X)]^2 = \frac{26}{3} - \left[\frac{8}{3}\right]^2 = \frac{14}{9}$

Expected Values, Mean and Variance

This page is for AQA S2

You can find the **Expected Value** for a **Function of X**

A function of a random variable X, $g(X)$, is an expression that takes X and does something to it. For example, $g(X) = X^3$, or $g(X) = \frac{12}{X}$, etc. If you're asked to find the expected value of a function of X, just use the following formula:

$$E(g(X)) = \sum g(x_i)p_i$$

EXAMPLE A discrete random variable X has the probability function $P(X = x) = \frac{x}{10}$, for $x = 1, 2, 3, 4$.
Find: a) $E(X^2)$ b) $E\left(\frac{1}{X}\right)$ c) $\text{Var}\left(\frac{1}{X}\right)$

You did this on the last page when you found the variance of X. But now you can see the same method works for functions of X in general, not just X^2.

a) Here $g(X) = X^2$, so $E(X^2) = \sum x_i^2 p_i = \left(1^2 \times \frac{1}{10}\right) + \left(2^2 \times \frac{2}{10}\right) + \left(3^2 \times \frac{3}{10}\right) + \left(4^2 \times \frac{4}{10}\right)$
$= \frac{1}{10} + \frac{8}{10} + \frac{27}{10} + \frac{64}{10} = \frac{100}{10} = 10$

b) Now $g(X) = \frac{1}{X}$, so $E\left(\frac{1}{X}\right) = \sum \frac{1}{x_i} \times p_i = \left(\frac{1}{1} \times \frac{1}{10}\right) + \left(\frac{1}{2} \times \frac{2}{10}\right) + \left(\frac{1}{3} \times \frac{3}{10}\right) + \left(\frac{1}{4} \times \frac{4}{10}\right)$
$= \frac{1}{10} + \frac{1}{10} + \frac{1}{10} + \frac{1}{10} = \frac{4}{10} = \frac{2}{5}$

c) Replace X with $\frac{1}{X}$ in the variance formula on p.97 to get:

$$\text{Var}\left(\frac{1}{X}\right) = E\left[\left(\frac{1}{X}\right)^2\right] - \left[E\left(\frac{1}{X}\right)\right]^2 = E\left(\frac{1}{X^2}\right) - \left[E\left(\frac{1}{X}\right)\right]^2 = \sum \frac{1}{x_i^2} \times p_i - \left(\frac{2}{5}\right)^2$$
$$= \left[\left(\frac{1}{1}\cdot\frac{1}{10}\right) + \left(\frac{1}{4}\cdot\frac{2}{10}\right) + \left(\frac{1}{9}\cdot\frac{3}{10}\right) + \left(\frac{1}{16}\cdot\frac{4}{10}\right)\right] - \frac{4}{25} = \frac{5}{24} - \frac{4}{25} = \frac{29}{600}$$

A function of X.

From part b) above.

For **Linear** Functions of X, there are **Two Formulas** to **Make Life Easier**

These formulas really will save you lots of time in the exam, so it's well worth your while learning them.

$$E(aX + b) = aE(X) + b \qquad \text{Var}(aX + b) = a^2\text{Var}(X)$$

Here a and b are any numbers.

EXAMPLE If $E(X) = 3$ and $\text{Var}(X) = 7$, find $E(2X + 5)$ and $\text{Var}(2X + 5)$.

Easy. $E(2X + 5) = 2E(X) + 5 = (2 \times 3) + 5 = 11$
$\text{Var}(2X + 5) = 2^2\text{Var}(X) = 4 \times 7 = 28$

EXAMPLE The discrete random variable X has the following probability distribution:

x	2	3	4	5	6
$P(X = x)$	0.1	0.2	0.3	0.2	k

Find: a) k, b) $E(X)$, c) $\text{Var}(X)$, d) $E(3X - 1)$, e) $\text{Var}(3X - 1)$.

Slowly, slowly — one bit at a time...

a) Remember the probabilities add up to 1 — $0.1 + 0.2 + 0.3 + 0.2 + k = 1$, and so $k = 0.2$

b) Now you can use the formula to find $E(X)$: $E(X) = \sum x_i p_i = (2 \times 0.1) + (3 \times 0.2) + (4 \times 0.3) + (5 \times 0.2) + (6 \times 0.2) = 4.2$

c) Next work out $E(X^2)$: $E(X^2) = \sum x_i^2 p_i = [2^2 \times 0.1] + [3^2 \times 0.2] + [4^2 \times 0.3] + [5^2 \times 0.2] + [6^2 \times 0.2] = 19.2$
and then the variance is easy: $\text{Var}(X) = E(X^2) - [E(X)]^2 = 19.2 - 4.2^2 = 1.56$

d) You'd expect the question to get harder but it doesn't: $E(3X - 1) = 3E(X) - 1 = 3 \times 4.2 - 1 = 11.6$

e) And finally: $\text{Var}(3X - 1) = 3^2\text{Var}(X) = 9 \times 1.56 = 14.04$

Statisticians say: E(Bird in hand) = E(2 Birds in bush)...

The mean and variance here are theoretical values — don't get them confused with the mean and variance of a load of practical observations. And remember, to find $E(g(X))$, you just replace x_i with $g(x_i)$. You can also work out linear functions like this if you want — but since the statistics gods have provided 2 easy-to-learn formulas, it'd be a shame not to use them.

S2 Section 1 — Practice Questions

Probability distribution and probability function are fancy-looking names for things that are actually quite straightforward...ish. Have a go at these to make sure you know who's who in the glitzy world of discrete random variables.

Warm-up Questions

All questions are for AQA S2

1) The probability distribution of Y is:

y	0	1	2	3
$P(Y = y)$	0.5	k	k	3k

a) Find the value of k. b) Find $P(Y < 2)$.

2) The discrete random variable X has the probability function $P(X = x) = k$ for $x = 0, 1, 2, 3$ and 4. Find the value of k, and then find the mean and variance of X.

3) A discrete random variable X has the probability distribution shown in the table, where k is a constant.

x_i	1	2	3	4
p_i	$\frac{1}{6}$	$\frac{1}{2}$	k	$\frac{5}{24}$

a) Find k.
b) Find E(X) and show that Var(X) = 63/64.
c) Find E($2X - 1$) and Var($2X - 1$).

4) A discrete random variable X has the probability distribution shown in the table.

x_i	1	2	3	4	5	6
p_i	0.1	0.2	0.25	0.2	0.1	0.15

a) Find E(X).
b) Find Var(X).
c) Show that $E\left(\frac{2}{X}\right) = 0.757$, correct to 3 decimal places.

Exam Questions

All questions are for AQA S2

1 In a game a player tosses three fair coins.
If three heads occur then the player gets 20p; if two heads occur then the player gets 10p.
For any other outcome, the player gets nothing.

(a) If X is the random variable 'amount received', tabulate the probability distribution of X.
(4 marks)

The player pays 10p to play one game.

(b) Use the probability distribution to find the probability that the player wins (i.e. gets more money than they pay to play) in one game.
(2 marks)

S2 Section 1 — Practice Questions

There's nothing I enjoy more than pretending I'm in an exam. An eerie silence, sweaty palms, having to be escorted to the toilet by a responsible adult... and all these lovely maths questions too. Just like the real thing.

Exam Questions

2 A discrete random variable X can only take values 0, 1, 2 and 3.
Its probability distribution is shown below.

x	0	1	2	3
$P(X = x)$	$2k$	$3k$	k	k

a) Find the value of k. *(1 mark)*

b) Find $P(X > 2)$. *(1 mark)*

3 The random variable X has the probability function $P(X = x) = k$ for $x = 0, 1, ..., 9$, where k is a constant.

a) Write down the probability distribution of X. *(1 mark)*

b) Find the mean and variance of X. *(3 marks)*

c) Calculate the probability that X is less than the mean. *(2 marks)*

4 A discrete random variable X has the probability function:
$P(X = x) = ax$ for $x = 1, 2, 3$, where a is a constant.

a) Show $a = \frac{1}{6}$. *(1 mark)*

b) Find $E(X)$. *(2 marks)*

c) If $Var(X) = \frac{5}{9}$ find $E(X^2)$. *(2 marks)*

d) Find $E(3X + 4)$ and $Var(3X + 4)$. *(3 marks)*

5 The number of points awarded to each contestant in a talent competition is given by the discrete random variable X with the following probability distribution:

x	0	1	2	3
$P(X = x)$	0.4	0.3	0.2	0.1

a) Find $E(X)$. *(2 marks)*

b) Find $E(6X + 8)$. *(2 marks)*

c) Show that $Var(X) = 1$. *(4 marks)*

d) Find $Var(5 - 3X)$. *(2 marks)*

Binomial Coefficients

This page is for Edexcel S2. Note — the whole of Section 2 is for Edexcel S2 only.

Welcome to Statistics 2, Edexcel people. Nice of you to join us.
It's a bit of a gentle introduction, to be honest, because this page is basically about counting things.

n Different Objects can be Arranged in *n!* Different Ways...

There are $n!$ ("n factorial") ways of arranging n different objects, where $n! = n \times (n-1) \times (n-2) \times ... \times 3 \times 2 \times 1$.

EXAMPLE a) In how many ways can 4 different ornaments be arranged on a shelf?
b) In how many ways can 8 different objects be arranged?

a) You have 4 choices for the first ornament, 3 choices for the second ornament, 2 choices for the third ornament, and 1 choice for the last ornament. So there are $4! = 4 \times 3 \times 2 \times 1 = 24$ arrangements.

b) There are $8! = 40\,320$ arrangements.

...but **Divide by r!** if **r** of These Objects are the **Same**

If r of your n objects are identical, then the total number of possible arrangements is $n! \div r!$.

EXAMPLE a) In how many different ways can 5 objects be arranged if 2 of those objects are identical?
b) In how many different ways can 7 objects be arranged if 4 of those objects are identical?

a) Imagine those 2 identical objects were different. Then there would be $5! = 120$ possible arrangements. But because those 2 objects are actually identical, you can always swap them round without making a different arrangement. So there are really only $120 \div 2 = 60$ different ways to arrange the objects.

b) There are $\frac{n!}{r!} = \frac{7!}{4!} = \frac{5040}{24} = 210$ different ways to arrange the objects.

Use **Binomial Coefficients** if There are **Only Two Types** of Object

Binomial Coefficients

$$\binom{n}{r} = {}^nC_r = \frac{n!}{r!(n-r)!}$$

${}^nC_r$ and $\binom{n}{r}$ both mean $\frac{n!}{r!(n-r)!}$

EXAMPLE a) In how many different ways can n objects of two types be arranged if r are of the first type?
b) How many ways are there to select 11 players from a squad of 16?
c) How many ways are there to pick 6 lottery numbers from 49?

a) If the objects were all different, there would be $n!$ ways to arrange them. But r of the objects are of the same type and could be swapped around, so divide by $r!$. Since there are only two types, the other $(n-r)$ could also be swapped around — so divide by $(n-r)!$. This means there are $\frac{n!}{r!(n-r)!}$ arrangements.

b) This is basically a 'number of different arrangements' problem. Imagine the 16 players are lined up — then you could 'pick' or 'not pick' players by giving each of them a sign marked with a tick or a cross. So just find the number of ways to arrange 11 ticks and 5 crosses — this is $\binom{16}{11} = \frac{16!}{11!5!} = 4368$.

c) Again, numbers are either 'picked' or 'unpicked', so there are $\binom{49}{6} = \frac{49!}{6!43!} = 13\,983\,816$ possibilities.

You can use your fingers and toes for counting up to 5! ÷ 3!...

So there you go — your first taste of S2 and hopefully it didn't seem too bad. But statistics (like maths generally) is one of those subjects where everything builds on what you've just learnt. So you need to make really sure you commit all this to memory, and (preferably) understand why it's true too — which is what the three part a)'s are about in the above examples.

The Binomial Probability Function

This page is for Edexcel S2

Being able to count the number of different arrangements of things is a big help when it comes to finding probabilities. This is because the probability of something depends on the number of different ways things could turn out.

Use Binomial Coefficients to Count Arrangements of 'Successes' and 'Failures'

Ages ago, you probably learnt that if p = P(something happens), then $1 - p$ = P(that thing doesn't happen). You'll need that fact now.

EXAMPLE I toss a fair coin 5 times. Find the probability of: a) 0 heads, b) 1 head, c) 2 heads.

First, note that each coin toss is independent of the others. That means you can multiply individual probabilities together.

a) P(0 heads) = P(tails) × P(tails) × P(tails) × P(tails) × P(tails) = $0.5^5 = 0.03125$

P(tails) = P(heads) = 0.5.

b) P(1 head) = P(heads) × P(tails) × P(tails) × P(tails) × P(tails)
+ P(tails) × P(heads) × P(tails) × P(tails) × P(tails)
+ P(tails) × P(tails) × P(heads) × P(tails) × P(tails)
+ P(tails) × P(tails) × P(tails) × P(heads) × P(tails)
+ P(tails) × P(tails) × P(tails) × P(tails) × P(heads)

These are the $\binom{5}{1} = 5$ ways to arrange 1 head and 4 tails.

So $P(1 \text{ head}) = 0.5 \times (0.5)^4 \times \binom{5}{1} = 0.03125 \times \frac{5!}{1!4!} = 0.15625$

= P(heads) × $[\text{P(tails)}]^4$ × ways to arrange 1 head and 4 tails.

c) P(2 heads) = $[\text{P(heads)}]^2 \times [\text{P(tails)}]^3$ × ways to arrange 2 heads and 3 tails = $(0.5)^2 \times (0.5)^3 \times \binom{5}{2} = 0.3125$

The Binomial Probability Function gives P(r successes out of n trials)

The previous example really just shows why this thing in a box must be true.

Binomial Probability Function

$$P(r \text{ successes in } n \text{ trials}) = \binom{n}{r} \times [\text{P(success)}]^r \times [\text{P(failure)}]^{n-r}$$

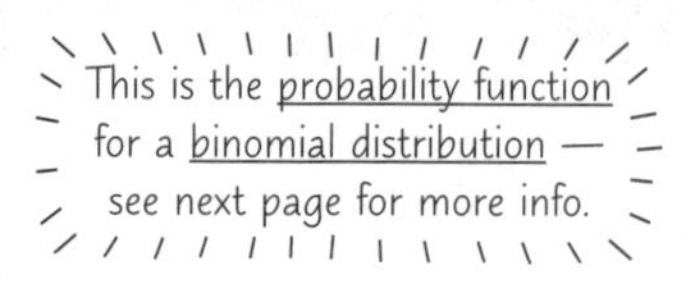

EXAMPLE I roll a fair dice 5 times. Find the probability of rolling: a) 2 sixes, b) 3 sixes, c) 4 numbers less than 3.

Again, note that each roll of a dice is independent of the other rolls.

a) For this part, call "roll a 6" a success, and "roll anything other than a 6" a failure.

Then $P(\text{roll 2 sixes}) = \binom{5}{2} \times \left(\frac{1}{6}\right)^2 \times \left(\frac{5}{6}\right)^3 = \frac{5!}{2!3!} \times \frac{1}{36} \times \frac{125}{216} = 0.161$ (to 3 d.p.).

b) Again, call "roll a 6" a success, and "roll anything other than a 6" a failure.

Then $P(\text{roll 3 sixes}) = \binom{5}{3} \times \left(\frac{1}{6}\right)^3 \times \left(\frac{5}{6}\right)^2 = \frac{5!}{3!2!} \times \frac{1}{216} \times \frac{25}{36} = 0.032$ (to 3 d.p.).

Notice how $\binom{5}{2} = \binom{5}{3}$. In fact, $\binom{n}{r} = \binom{n}{n-r}$.

c) This time, success means "roll a 1 or a 2", while failure is now "roll a 3, 4, 5 or 6".

Then $P(\text{roll 4 numbers less than 3}) = \binom{5}{4} \times \left(\frac{1}{3}\right)^4 \times \left(\frac{2}{3}\right) = \frac{5!}{4!1!} \times \frac{1}{81} \times \frac{2}{3} = 0.041$ (to 3 d.p.).

Let this formula for success go to your head — and then keep it there...

This page is all about finding the probabilities of different numbers of successes in n trials. Now then... if you carry out n trials, there are $n + 1$ possibilities for the number of successes (0, 1, 2, ..., n). This 'family' of possible results along with their probabilities is sounding suspiciously like a probability distribution. Oh rats... I've given away what's on the next page.

The Binomial Distribution

This page is for Edexcel S2

Remember the fun you had in S1 when you learnt all about random variables... well, happy days are here again. This page is about random variables following a binomial distribution (whose probability function you saw on p102).

There are 5 Conditions for a Binomial Distribution

Binomial Distribution: B(n, p)

A random variable X follows a Binomial Distribution as long as these 5 conditions are satisfied:

1) There is a fixed number (n) of trials.
2) Each trial involves either "success" or "failure".
3) All the trials are independent.
4) The probability of "success" (p) is the same in each trial.
5) The variable is the total number of successes in the n trials.

Binomial random variables are discrete, since they only take values 0, 1, 2... n.

n and p are the two parameters of the binomial distribution. (Or n is sometimes called the 'index'.)

In this case, $P(X = x) = \binom{n}{x} \times p^x \times (1-p)^{n-x}$ for $x = 0, 1, 2,..., n$, and you can write $X \sim B(n, p)$.

EXAMPLE: Which of the random variables described below would follow a binomial distribution? For those that do, state the distribution's parameters.

a) **The number of faulty items (T) produced in a factory per day, if the probability of each item being faulty is 0.01 and there are 10 000 items produced every day.**
Binomial — there's a fixed number (10 000) of trials with two possible results ('faulty' or 'not faulty'), a constant probability of 'success', and T is the total number of 'faulty' items.
So (as long as faulty items occur independently) $T \sim B(10\,000, 0.01)$.

b) **The number of red cards (R) drawn from a standard 52-card deck in 10 picks, not replacing the cards each time.**
Not binomial, since the probability of 'success' changes each time (as I'm not replacing the cards).

c) **The number of red cards (R) drawn from a standard 52-card deck in 10 picks, replacing the cards each time.**
Binomial — there's a fixed number (10) of independent trials with two possible results ('red' or 'black/not red'), a constant probability of success (I'm replacing the cards), and R is the number of red cards drawn. $R \sim B(10, 0.5)$.

d) **The number of times (T) I have to toss a coin before I get heads.**
Not binomial, since the number of trials isn't fixed.

e) **The number of left-handed people (L) in a sample of 500 randomly chosen people, if the proportion of left-handed people in the population as a whole is 0.13.**
Binomial — there's a fixed number (500) of independent trials with two possible results ('left-handed' or 'not left-handed'), a constant probability of success (0.13), and L is the number of left-handers. $L \sim B(500, 0.13)$.

EXAMPLE: When I toss a grape in the air and try to catch it in my mouth, my probability of success is always 0.8. The number of grapes I catch in 10 throws is described by the discrete random variable X.

a) How is X distributed? Name the type of distribution, and give the values of any parameters.
b) Find the probability of me catching at least 9 grapes.

a) There's a fixed number (10) of independent trials with two possible results ('catch' and 'not catch'), a constant probability of success (0.8), and X is the total number of catches.
Therefore X follows a binomial distribution, $X \sim B(10, 0.8)$.

b) P(at least 9 catches) $= P(9 \text{ catches}) + P(10 \text{ catches})$

$$= \left\{\binom{10}{9} \times 0.8^9 \times 0.2^1\right\} + \left\{\binom{10}{10} \times 0.8^{10} \times 0.2^0\right\}$$

$= 0.268435... + 0.107374... = 0.376$ (to 3 d.p.).

Binomial distributions come with 5 strings attached...

There's a big, boring box at the top of the page with a list of 5 conditions in — and you do need to know it, unfortunately. There's only one way to learn it — keep trying to write down the 5 conditions until you can do it in your sleep.

Using Binomial Tables

This page is for Edexcel S2

Your life is just about to be made a whole lot easier. So smile sweetly and admit that statistics isn't all bad.

Look up Probabilities in **Binomial Tables**

EXAMPLE I have an unfair coin. When I toss this coin, the probability of getting heads is 0.35. Find the probability that it will land on heads fewer than 3 times when I toss it 12 times in total.

If the random variable X represents the number of heads I get in 12 tosses, then $X \sim B(12, 0.35)$.
You need to find $P(X \leq 2)$.

① You could work this out 'manually'...

$$P(\text{0 heads}) + P(\text{1 head}) + P(\text{2 heads}) = \left\{\binom{12}{0} \times 0.35^0 \times 0.65^{12}\right\} + \left\{\binom{12}{1} \times 0.35^1 \times 0.65^{11}\right\} + \left\{\binom{12}{2} \times 0.35^2 \times 0.65^{10}\right\}$$

$$= 0.0057 + 0.0368 + 0.1088 = 0.1513$$

See p161 for more binomial tables.

② But it's much quicker to use tables of the binomial cumulative distribution function (c.d.f.).

These show $P(X \leq x)$, for $X \sim B(n, p)$.

- First find the table for the correct values of n and p. Then the table gives you a value for $P(X \leq x)$.
- So here, $n = 12$ and $p = 0.35$, and you need $P(X \leq 2)$. The table tells you this is 0.1513.

Binomial Cumulative Distribution Function
Values show P(X ≤ x), where X~B(n, p)

	p =	0.05	0.10	0.15	0.20	0.25	0.30	0.35	0.40	0.45	0.50
n = 12	x = 0	0.5404	0.2824	0.1422	0.0687	0.0317	0.0138	0.0057	0.0022	0.0008	0.0002
	1	0.8816	0.6590	0.4435	0.2749	0.1584	0.0850	0.0424	0.0196	0.0083	0.0032
	2	0.9804	0.8891	0.7358	0.5583	0.3907	0.2528	0.1513	0.0834	0.0421	0.0193
	3	0.9978	0.9744	0.9078	0.7946	0.6488	0.4925	0.3467	0.2253	0.1345	0.0730
	4	0.9998	0.9957	0.9761	0.9274	0.8424	0.7237	0.5833	0.4382	0.3044	0.1938
	5	1.0000	0.9995	0.9954	0.9806	0.9456	0.8822	0.7873	0.6652	0.5269	0.3872
	6	1.0000	0.9999	0.9993	0.9961	0.9857	0.9614	0.9154	0.8418	0.7393	0.6128
	7	1.0000	1.0000	0.9999	0.9994	0.9972	0.9905	0.9745	0.9427	0.8883	0.8062
	8	1.0000	1.0000	1.0000	0.9999	0.9996	0.9983	0.9944	0.9847	0.9644	0.9270
	9	1.0000	1.0000	1.0000	1.0000	1.0000	0.9998	0.9992	0.9972	0.9921	0.9807
	10	1.0000	1.0000	1.0000	1.0000	1.0000	1.0000	0.9999	0.9997	0.9989	0.9968
	11	1.0000	1.0000	1.0000	1.0000	1.0000	1.0000	1.0000	1.0000	0.9999	0.9998

Binomial Tables Tell You **More** Than You Might Think

With a bit of cunning, you can get binomial tables to tell you anything you want to know...

EXAMPLE I have a different unfair coin. When I toss this coin, the probability of getting tails is 0.6. The random variable X represents the number of tails in 12 tosses, so $X \sim B(12, 0.6)$.
If I toss this coin 12 times, use the table above to find the probability that:

a) it will land on tails more than 8 times,
b) it will land on heads at least 6 times,
c) it will land on heads exactly 9 times,
d) it will land on heads more than 3 but fewer than 6 times.

a) The above table only goes up to $p = 0.5$.
So switch things round... if P(tails) = 0.6, then P(heads) = 1 – 0.6 = 0.4 (and $p = 0.4$ is in the table).
So define a new random variable, Y, representing the number of heads in 12 throws — then $Y \sim B(12, 0.4)$.
This means $P(X > 8) = P(Y \leq 3) = 0.2253$

b) $P(Y \geq 6) = 1 - P(Y < 6) = 1 - P(Y \leq 5) = 1 - 0.6652 = 0.3348$

1) P(event happens) = 1 – P(event doesn't happen),
2) P(Y < 6) = P(Y ≤ 5), as Y takes whole number values.

c) $P(Y = 9) = P(Y \leq 9) - P(Y \leq 8) = 0.9972 - 0.9847 = 0.0125$

Use P(A or B) = P(A) + P(B) with the mutually exclusive events "Y ≤ 8" and "Y = 9" to get P(Y ≤ 9) = P(Y ≤ 8) + P(Y = 9).
Or you can think of it as "subtracting P(Y ≤ 8) from P(Y ≤ 9) leaves just P(Y = 9)".

d) $P(3 < Y < 6) = P(Y \leq 5) - P(Y \leq 3) = 0.6652 - 0.2253 = 0.4399$

Statistical tables are the original labour-saving device...

...as long as you know what you're doing. Careful, though — it's easy to trip yourself up. Basically, as long as you can find the right value of n and p (or $1 - p$, if necessary) in a table, you can use those tables to work out anything you might need. Work through those last examples *really slowly* and make sure you can follow them — they're pretty vital.

Mean and Variance of B(n, p)

This page is for Edexcel S2

You know from S1 what the mean (or expected value) and variance of a random variable are.
And you also know what the binomial distribution is. Put those things together, and you get this page.

For a Binomial Distribution: Mean = np

This formula will be in your formula booklet, but it's worth committing to memory anyway.

Mean of a Binomial Distribution

If $X \sim B(n, p)$, then:

Mean (or Expected Value) = μ = E(X) = np

Greek letters (e.g. μ) often show something based purely on theory rather than experimental results.

Remember... the expected value is the value you'd expect the random variable to take on average if you took loads and loads of readings. It's a "theoretical mean" — the mean of experimental results is unlikely to match it exactly.

EXAMPLE If $X \sim B(20, 0.2)$, what is E(X)?

Just use the formula: $E(X) = np = 20 \times 0.2 = 4$

EXAMPLE What's the expected number of sixes when I roll a fair dice 30 times? Interpret your answer.

If the random variable X represents the number of sixes in 30 rolls, then $X \sim B(30, \frac{1}{6})$.

So the expected value of X is $E(X) = 30 \times \frac{1}{6} = 5$

If I were to repeatedly throw the dice 30 times, and find the average number of sixes in each set of 30 throws, then I would expect it to end up pretty close to 5.
And the more sets of 30 throws I did, the closer to 5 I'd expect the average to be.

Notice that the probability of getting exactly 5 sixes on my next set of 30 throws = $\binom{30}{5} \times \left(\frac{1}{6}\right)^5 \times \left(\frac{5}{6}\right)^{25} = 0.192$
So I'm much more likely not to get exactly 5 sixes (= 1 – 0.192 = 0.808).
This is why it only makes sense to talk about the mean as a "long-term average", and not as "what I expect to happen next".

For a Binomial Distribution: Variance = npq

Variance of a Binomial Distribution

If $X \sim B(n, p)$, then:

Variance = Var(X) = $\sigma^2 = np(1 - p) = npq$

Standard Deviation = $\sigma = \sqrt{np(1-p)} = \sqrt{npq}$

For a binomial distribution, P(success) is usually called p, and P(failure) is sometimes called q (= 1 – p).

EXAMPLE If $X \sim B(20, 0.2)$, what is Var(X)?

Just use the formula: $\text{Var}(X) = np(1-p) = 20 \times 0.2 \times 0.8 = 3.2$

EXAMPLE If $X \sim B(25, 0.2)$, find: a) $P(X \leq \mu)$, b) $P(X \leq \mu - \sigma)$, c) $P(X \leq \mu - 2\sigma)$

$E(X) = \mu = 25 \times 0.2 = 5$, and $\text{Var}(X) = \sigma^2 = 25 \times 0.2 \times (1 - 0.2) = 4$, which gives $\sigma = 2$.

So, using tables (for $n = 25$ and $p = 0.2$): a) $P(X \leq \mu) = P(X \leq 5) = 0.6167$

b) $P(X \leq \mu - \sigma) = P(X \leq 3) = 0.2340$

c) $P(X \leq \mu - 2\sigma) = P(X \leq 1) = 0.0274$

See page 163.

For B(n, p) — the variance is always less than the mean...

Nothing too fancy there really. A couple of easy-to-remember formulas, and some stuff about how to interpret these figures which you've seen before anyway. So learn the formulas, put the kettle on, and have a cup of tea while the going's good.

Binomial Distribution Problems

This page is for Edexcel S2

That's everything you need to know about binomial distributions (for now).
So it's time to put it all together and have a look at the kind of thing you might get asked in the exam.

EXAMPLE 1: Selling Double Glazing

A double-glazing salesman is handing out leaflets in a busy shopping centre. He knows that the probability of each passing person taking a leaflet is always 0.3. During a randomly chosen one-minute interval, 30 people passed him.

a) Suggest a suitable model to describe the number of people (X) who take a leaflet.
b) What is the probability that more than 10 people take a leaflet?
c) How many people would the salesman expect to take a leaflet?
d) Find the variance and standard deviation of X.

a) During this one-minute interval, there's a fixed number (30) of independent trials with two possible results ("take a leaflet" and "do not take a leaflet"), a constant probability of success (0.3), and X is the total number of people taking leaflets. So $X \sim B(30, 0.3)$.

b) $P(X > 10) = 1 - P(X \leq 10) = 1 - 0.7304 = 0.2696$

Use binomial tables for this — see p161.

c) The number of people the salesman could expect to take a leaflet is $E(X) = np = 30 \times 0.3 = 9$

d) Variance $= np(1-p) = 30 \times 0.3 \times (1 - 0.3) = 6.3$ Standard deviation $= \sqrt{6.3} = 2.51$ (to 2 d.p.)

EXAMPLE 2: Multiple-Choice Guessing

A student has to take a 50-question multiple-choice exam, where each question has five possible answers of which only one is correct. He believes he can pass the exam by guessing answers at random.

a) How many questions could the student be expected to guess correctly?
b) If the pass mark is 15, what is the probability that the student will pass the exam?
c) The examiner decides to set the pass mark so that it is at least 3 standard deviations above the expected number of correct guesses. What should the minimum pass mark be?

Let X be the number of correct guesses over the 50 questions. Then $X \sim B(50, 0.2)$.

Define your random variable first, and say how it will be distributed.

a) $E(X) = np = 50 \times 0.2 = 10$

b) $P(X \geq 15) = 1 - P(X < 15) = 1 - P(X \leq 14) = 1 - 0.9393 = 0.0607$

c) $\text{Var}(X) = np(1-p) = 50 \times 0.2 \times 0.8 = 8$ — so the standard deviation $= \sqrt{8} = 2.828$ (to 3 d.p.).
So the pass mark needs to be at least $10 + (3 \times 2.828) \approx 18.5$ — i.e. the minimum pass mark should be 19.

EXAMPLE 3: An unfair coin (again)

I am spinning a coin that I know is three times as likely to land on heads as it is on tails.

a) What is the probability that it lands on tails for the first time on the third spin?
b) What is the probability that in 10 spins, it lands on heads at least 7 times?

You know that P(heads) = 3 × P(tails), and that P(heads) + P(tails) = 1.
This means that P(heads) = 0.75 and P(tails) = 0.25.

Careful... this doesn't need you to use one of the binomial formulas.

a) P(lands on tails for the first time on the third spin) $= 0.75 \times 0.75 \times 0.25 = 0.141$ (to 3 d.p.).

b) If X represents the number of heads in 10 spins, then $X \sim B(10, 0.75)$.
This means the number of tails in 10 spins can be described by the random variable Y, where $Y \sim B(10, 0.25)$.
$P(X \geq 7) = P(Y \leq 3) = 0.7759$

$p = 0.75$ isn't in your tables, so define a new binomial random variable Y with probability of success $p = 0.25$.

Proof that you shouldn't send a monkey to take your multi-choice exams...

You can see now how useful a working knowledge of statistics is. Ever since you first started using CGP books, I've been banging on about how hard it is to pass an exam without revising. Well, now you can prove I was correct using a bit of knowledge and binomial tables. Yup... statistics can help out with some of those tricky situations you face in life.

S2 Section 2 — Practice Questions

Hopefully, everything you've just read will already be stuck in your brain. But if you need a bit of help to wedge it in place, then try these questions. Actually... I reckon you'd best try them anyway — a little suffering is good for the soul.

Warm-up Questions

All questions are for Edexcel S2

1) In how many different orders can the following be arranged?
   a) 15 identical red balls, plus 6 other balls, all of different colours.
   b) 4 red counters, 4 blue counters, 4 yellow counters and 4 green counters.

2) What is the probability of the following?
   a) Getting exactly 5 heads when you spin a fair coin 10 times.
   b) Getting exactly 9 heads when you spin a fair coin 10 times.

3) Which of the following would follow a binomial distribution? Explain your answers.
   a) The number of prime numbers you throw in 30 throws of a standard dice.
   b) The number of people in a particular class at a school who get 'heads' when they flip a coin.
   c) The number of aces in a 7-card hand dealt from a standard deck of 52 cards.
   d) The number of shots I have to take before I score from the free-throw line in basketball.

4) What is the probability of the following?
   a) Getting at least 5 heads when you spin a fair coin 10 times.
   b) Getting at least 9 heads when you spin a fair coin 10 times.

5) If $X \sim B(14, 0.27)$, find:
   a) $P(X = 4)$ b) $P(X < 2)$ c) $P(5 < X \leq 8)$

6) If $X \sim B(25, 0.15)$ and $Y \sim B(15, 0.65)$ find:
   a) $P(X \leq 3)$ b) $P(X \leq 7)$
   c) $P(X \leq 15)$ d) $P(Y \leq 3)$
   e) $P(Y \leq 7)$ f) $P(Y \leq 15)$

7) Find the required probability for each of the following binomial distributions.
   a) $P(X \leq 15)$ if $X \sim B(20, 0.4)$ b) $P(X < 4)$ if $X \sim B(40, 0.15)$
   c) $P(X > 7)$ if $X \sim B(25, 0.45)$ d) $P(X \geq 40)$ if $X \sim B(50, 0.8)$
   e) $P(X = 20)$ if $X \sim B(30, 0.7)$ f) $P(X = 7)$ if $X \sim B(10, 0.75)$

8) Find the mean and variance of the following random variables.
   a) $X \sim B(20, 0.4)$ b) $X \sim B(40, 0.15)$
   c) $X \sim B(25, 0.45)$ d) $X \sim B(50, 0.8)$
   e) $X \sim B(30, 0.7)$ f) $X \sim B(45, 0.012)$

S2 Section 2 — Practice Questions

Right then... you're nearly at the end of the section, and with any luck your tail is up, the wind is in your sails and the going is good. But the real test of whether you're ready for the exam is some exam questions. And as luck would have it, there are some right here. So give them a go and test your mettle, see if you can walk the walk... and so on.

Exam Questions

All questions are for Edexcel S2

1 a) The random variable X follows the binomial distribution B(12, 0.6). Find:

(i) $P(X < 8)$,

(2 marks)

(ii) $P(X = 5)$,

(2 marks)

(iii) $P(3 < X \leq 7)$.

(3 marks)

b) If $Y \sim B(11, 0.8)$, find:

(i) $P(Y = 4)$,

(2 marks)

(ii) $E(Y)$,

(1 mark)

(iii) $Var(Y)$.

(1 mark)

2 The probability of an apple containing a maggot is 0.15.

a) Find the probability that in a random sample of 40 apples there are:

(i) fewer than 6 apples containing maggots,

(2 marks)

(ii) more than 2 apples containing maggots,

(2 marks)

(iii) exactly 12 apples containing maggots.

(2 marks)

b) These apples are sold in crates of 40. Ed buys 3 crates.
Find the probability that more than 1 crate contains more than 2 apples with maggots.

(3 marks)

3 Simon tries to solve the crossword puzzle in his newspaper every day for two weeks.
He either succeeds in solving the puzzle, or he fails to solve it.

a) Simon believes that this situation can be modelled by a random variable following a binomial distribution.

(i) State two conditions needed for a binomial distribution to arise here.

(2 marks)

(ii) State which quantity would follow a binomial distribution (assuming the above conditions are satisfied).

(1 mark)

b) Simon believes a random variable X follows the distribution $B(18, p)$.
If $P(X = 4) = P(X = 5)$, find p.

(5 marks)

The Poisson Distribution

This page is for AQA S2, Edexcel S2, OCR S2 and OCR MEI S2

Welcome to S2 — those of you I've not met already. You know who you are. It's time to introduce the Poisson distribution. If you speak French, you'll know that poisson means fish. I think.

A Poisson Distribution has **Only One Parameter**

A Poisson Distribution has just one parameter: λ.
If the random variable X follows a Poisson distribution, then you can write $X \sim \text{Po}(\lambda)$.

The Greek letter lambda is often used for the Poisson parameter. You might also see the Greek letter mu (μ) used.

Poisson Probability Distribution Po(λ)

If $X \sim \text{Po}(\lambda)$, then X can take values 0, 1, 2, 3... with probability:

$$P(X = x) = \frac{e^{-\lambda}\lambda^x}{x!}$$

Random variables following a Poisson distribution are discrete — there are 'gaps' between the possible values.

EXAMPLE If $X \sim \text{Po}(2.8)$, find:
a) $P(X = 0)$, b) $P(X = 1)$, c) $P(X = 2)$, d) $P(X < 3)$, e) $P(X \geq 3)$

Use the formula:

a) $P(X = 0) = \frac{e^{-2.8} \times 2.8^0}{0!} = e^{-2.8} = 0.061$ (to 3 d.p.). ← Remember... 0! = 1.

b) $P(X = 1) = \frac{e^{-2.8} \times 2.8^1}{1!} = e^{-2.8} \times 2.8 = 0.170$ (to 3 d.p.).

c) $P(X = 2) = \frac{e^{-2.8} \times 2.8^2}{2!} = \frac{e^{-2.8} \times 2.8^2}{2 \times 1} = 0.238$ (to 3 d.p.).

If $X \sim \text{Po}(\lambda)$, then it can only take whole number values, so $P(X < 3)$ is the same as $P(X \leq 2)$.

d) $P(X < 3) = P(X \leq 2) = P(X = 0) + P(X = 1) + P(X = 2) = 0.061 + 0.170 + 0.238 = 0.469.$

e) $P(X \geq 3) = 1 - P(X < 3) = 1 - 0.469 = 0.531.$ ← All the normal probability rules apply.

For a Poisson Distribution: **Mean = Variance**

For a Poisson distribution, the mean and the variance are the same — and they both equal λ, the Poisson parameter. Remember that and you've probably learnt the most important Poisson fact. Ever.

Poisson Mean and Variance

If $X \sim \text{Po}(\lambda)$: **Mean ($\mu$) of X = E(X) = λ**

Variance (σ^2) of X = Var(X) = λ

So the standard deviation is: $\sigma = \sqrt{\lambda}$

EXAMPLE If $X \sim \text{Po}(7)$, find: a) E(X), b) Var(X).
It's Poisson, so $E(X) = \text{Var}(X) = \lambda = 7$.

This is the easiest question ever. So enjoy it while it lasts.

EXAMPLE If $X \sim \text{Po}(1)$, find: a) $P(X \leq \mu)$, b) $P(X \leq \mu - \sigma)$

$E(X) = \mu = 1$, and $\text{Var}(X) = \sigma^2 = 1$, and so $\sigma = 1$.

a) $P(X \leq \mu) = P(X \leq 1) = P(0) + P(1) = \frac{e^{-1} \times 1^0}{0!} + \frac{e^{-1} \times 1^1}{1!} = 0.736$ (to 3 d.p.).

b) $P(X \leq \mu - \sigma) = P(X \leq 0) = P(0) = \frac{e^{-1} \times 1^0}{0!} = 0.368$ (to 3 d.p.).

The Poisson distribution is named after its inventor...

...the great French mathematician Monsieur Siméon-Denis Distribution. Boom boom. I always tell that joke at parties (which probably explains why I don't get to go to many parties these days). Most important thing here is that bit about the mean and variance being equal... so if you ever come across a distribution where $\mu = \sigma^2$, think 'Poisson' immediately.

The Poisson Parameter

This page is for AQA S2, Edexcel S2, OCR S2 and OCR MEI S2

I know what you're thinking... if only everything could be as accommodating as the Poisson distribution, with only one parameter and most things of interest being equal to it, then life would be so much easier. (Sigh.)

The Poisson Parameter is a **Rate**

The number of events/things that occur/are present in a particular period often follows a Poisson distribution. It could be a period of: time (e.g. minute/hour etc.), or space (e.g. litre/kilometre etc.).

Poisson Probability Distribution: Po(λ)

If X represents the number of events that occur in a particular space or time, then X will follow a Poisson distribution as long as:

1) The events occur randomly, and are all independent of each other.
2) The events happen singly (i.e. "one at a time").
3) The events happen (on average) at a constant rate (either in space or time).

The Poisson parameter λ is then the average rate at which these events occur (i.e. the average number of events in a given interval of space or time).

So the expected number of events that occur is proportional to the length of the period.

EXAMPLE The random variable X represents the number of a certain type of cell in a particular volume of a blood sample. Assuming that the blood sample has been stirred, and that a given volume of blood always contains the same number of cells, show that X follows a Poisson distribution.

The sample has been stirred, so that should mean the cells of interest aren't all clustered together. This should ensure the 'events' (i.e. the cells you're interested in) occur randomly and singly. And since the total number of cells in a given volume is constant, the cells of interest should occur (on average) at a constant rate. Since X is the total number of 'events' in a given volume, X must follow a Poisson distribution.

The Poisson Parameter is **Additive**

Additive Property of the Poisson Distribution

- If X represents the number of events in 1 unit of time/space (e.g. 1 minute / hour / m^2 / m^3), and $X \sim \text{Po}(\lambda)$, then the number of events in x units of time/space follows the distribution $\text{Po}(x\lambda)$.
- If X and Y are independent variables with $X \sim \text{Po}(\lambda)$ and $Y \sim \text{Po}(\mu)$, then $X + Y \sim \text{Po}(\lambda + \mu)$.

EXAMPLE Sunflowers grow singly and randomly in a field with an average of 10 sunflowers per square metre. What is the probability that a randomly chosen area of 0.25 m^2 contains no sunflowers?

The number of sunflowers in 1 m^2 follows the distribution Po(10).

So the number of sunflowers in 0.25 m^2 must follow the distribution Po(2.5).

This means P(no sunflowers) $= \frac{e^{-2.5} \times 2.5^0}{0!} = e^{-2.5} = 0.082$ (to 3 d.p.).

X ~ Po(10)

EXAMPLE The number of radioactive atoms that decay per second follows the Poisson distribution Po(5). If the probability of no atoms decaying in t seconds is 0.5, verify that $t = 0.1386$.

If the random variable X represents the number of radioactive atoms that decay in t seconds, then $X \sim \text{Po}(5t)$.

This means $P(X = 0) = \frac{e^{-5t}(5t)^0}{0!} = e^{-5t} = 0.5$.

This equation is satisfied by $t = 0.1386$, since $e^{-5 \times 0.1386} = e^{-0.693} = 0.500$ (to 3 d.p.).

If events happen randomly, singly and at a constant rate, it's Poisson...

Lots of things follow a Poisson distribution — e.g. the number of radioactive atoms that decay in a given time, the number of sixes in 5 minutes of dice-throwing, the number of raindrops per minute that hit a bit of your tongue as you stare open-mouthed at the sky on a rainy day. Think of a few others... make sure events happen randomly, singly and at a constant rate.

Using Poisson Tables

This page is for AQA S2, Edexcel S2, OCR S2 and OCR MEI S2

You've seen statistical tables before — for example, you've seen how great they are for working out probabilities for the binomial distribution. So this should all seem eerily familiar.

*Look up Probabilities in **Poisson Tables***

Going back to the sunflowers example near the bottom of the previous page...

EXAMPLE Sunflowers grow singly and randomly in a field with an average of 10 sunflowers per square metre. Find the probability that a randomly chosen square metre contains no more than 8 sunflowers.

If the random variable X represents the number of sunflowers in 1 m², then $X \sim \text{Po}(10)$. You need to find $P(X \leq 8)$.

① You could do this 'manually': $P(X=0) + P(X=1) + \ldots + P(X=8) = \frac{e^{-10} \times 10^0}{0!} + \frac{e^{-10} \times 10^1}{1!} + \ldots + \frac{e^{-10} \times 10^8}{8!}$

② But it's much quicker and easier to use tables of the Poisson cumulative distribution function (c.d.f.).

These show $P(X \leq x)$ if $X \sim \text{Po}(\lambda)$.

Here's a bit of a Poisson table:

- Find your value of λ (here, 10), and the value of x (here, 8).
- You can quickly see that $P(X \leq 8) = 0.3328$.

Poisson Cumulative Distribution Function

Values show $P(X \leqslant x)$, where $X \sim \text{Po}(\lambda)$

λ =	5.5	6.0	6.5	7.0	7.5	8.0	8.5	9.0	9.5	10.0
x = 0	0.0041	0.0025	0.0015	0.0009	0.0006	0.0003	0.0002	0.0001	0.0001	0.0000
1	0.0266	0.0174	0.0113	0.0073	0.0047	0.0030	0.0019	0.0012	0.0008	0.0005
2	0.0884	0.0620	0.0430	0.0296	0.0203	0.0138	0.0093	0.0062	0.0042	0.0028
3	0.2017	0.1512	0.1118	0.0818	0.0591	0.0424	0.0301	0.0212	0.0149	0.0103
4	0.3575	0.2851	0.2237	0.1730	0.1321	0.0996	0.0744	0.0550	0.0403	0.0293
5	0.5289	0.4457	0.3690	0.3007	0.2414	0.1912	0.1496	0.1157	0.0885	0.0671
6	0.6860	0.6063	0.5265	0.4497	0.3782	0.3134	0.2562	0.2068	0.1649	0.1301
7	0.8095	0.7440	0.6728	0.5987	0.5246	0.4530	0.3856	0.3239	0.2687	0.2202
8	0.8944	0.8472	0.7916	0.7291	0.6620	0.5925	0.5231	0.4557	0.3918	0.3328
9	0.9462	0.9161	0.8774	0.8305	0.7764	0.7166	0.6530	0.5874	0.5218	0.4579
10	0.9747	0.9574	0.9332	0.9015	0.8622	0.8159	0.7634	0.7060	0.6453	0.5830
11	0.9890	0.9799	0.9661	0.9467	0.9208	0.8881	0.8487	0.8030	0.7520	0.6968

*You Need to Use **Poisson Tables** with a Bit of **Cunning***

See p166 for the full set of Poisson tables.

This is exactly the same as you've already seen for binomial tables.

EXAMPLE When cloth is manufactured, faults occur randomly in the cloth at a rate of 8 faults per square metre. Use the above Poisson table to find:

a) The probability of 7 or fewer faults in a square metre of cloth.
b) The probability of more than 4 faults in a square metre of cloth.
c) The probability of exactly 10 faults in a square metre of cloth.
d) The probability of at least 9 faults in a square metre of cloth.
e) The probability of exactly 4 faults in 0.75 m² of cloth.

The faults occur randomly, singly and at a constant rate (= 8 faults per square metre). So if X represents the number of faults in a square metre, then $X \sim \text{Po}(8)$.

So use the column showing $\lambda = 8$.

a) $P(X \leq 7) = 0.4530$

b) $P(X > 4) = 1 - P(X \leq 4) = 1 - 0.0996 = 0.9004$

c) $P(X = 10) = P(X \leq 10) - P(X \leq 9) = 0.8159 - 0.7166 = 0.0993$

d) $P(X \geq 9) = 1 - P(X < 9) = 1 - P(X \leq 8) = 1 - 0.5925 = 0.4075$

e) Let the random variable Y represent the number of faults in 0.75 m² of cloth. If the number of faults in 1 m² of cloth ~ Po(8), then $Y \sim \text{Po}(0.75 \times 8) = \text{Po}(6)$.

Now use the column showing $\lambda = 6$.

So P(exactly 4 faults in 0.75 m² of cloth) $= P(Y \leq 4) - P(Y \leq 3) = 0.2851 - 0.1512 = 0.1339$

Poisson tables — the best thing since binomial tables...

Learn the ways of the Poisson tables, and you shall prove your wisdom. In the exam, you'll be given a big booklet of fun containing all the statistical tables you could ever want. You need to think carefully about how to use them though — e.g. you might have to subtract one figure from another, or subtract one of the figures from 1. Or something else similar.

Po(λ) as an Approximation to B(n, p)

This page is for Edexcel S2, OCR S2 and OCR MEI S2

This page is a bit like a buy-one-get-one-free offer — it's in the section about Poisson distributions, but it's actually about binomial distributions. This really is your lucky day...

For Big n and Small p — Po(np) Approximates a Binomial Distribution

Sometimes, a Poisson distribution can be used as an approximation to a binomial distribution.

If you're struggling to remember much about the binomial distribution, this would be a good time to brush up on your knowledge.

Po(np) as an Approximation to B(n, p)

If $X \sim B(n, p)$, and: 1) **n is large**, 2) **p is small**,

then X can be approximated by **Po(np)**.

The mean of the binomial distribution is np, so use that as the mean of your Poisson approximation.

EXAMPLE In a school of 1825 students, what is the probability that at least 6 of them were born on June 21st? Use a suitable approximation to find your answer.
(You may assume that all birthdays are independent, and are distributed evenly throughout the year.)

If X represents the number of children in the school born on June 21st, then $X \sim B(1825, \frac{1}{365})$.

You need to find $P(X \geq 6)$.

So far so good. However, your binomial tables don't go past n = 50. And working this out 'by hand' isn't easy. But look at those values of n and p...

Since n is large and p is small, $B(1825, \frac{1}{365})$ can be approximated by $Po(1825 \times \frac{1}{365}) = Po(5)$.

So $P(X \geq 6) = 1 - P(X < 6) = 1 - P(X \leq 5) = 1 - 0.6160 = 0.3840$.

From Poisson tables — see p166.

If you work it out using $B(1825, \frac{1}{365})$, you also get 0.3840 — so this is a very good approximation.

The Smaller the Value of p, the Better

1) To use the Poisson approximation to B(n, p), you ideally want n "as large as possible" and p "as small as possible". The bigger n is and the smaller p is, the better the approximation will be.
2) It's important p is small because then the mean and the variance of B(n, p) are approximately equal — something you need if Po(np) is going to be a good approximation.

   If $X \sim B(n, p)$, then $E(X) = np$.
   And if p is small, $(1 - p) \approx 1$ — this means $Var(X) = np(1 - p) \approx np \times 1 = np$.
3) In your exam, you'll usually be told when to use an approximation.

EXAMPLES:

① Factory A forgets to add icing to its chocolate cakes with a uniform probability of 0.02. Use a suitable approximation to find the probability that fewer than 6 of the next 100 cakes made will not be iced.

If X represents the number of "un-iced" cakes, then $X \sim B(100, 0.02)$.
Since n is quite large and p is quite small, $X \sim Po(100 \times 0.02) = Po(2)$.
So $P(X < 6) = P(X \leq 5) = 0.9834$

If you work it out using B(100, 0.02), you get 0.9845.

Sometimes you can still use the approximation if p is very close to 1.

② Factory B adds icing to its chocolate cakes with a uniform probability of 0.99. Use a suitable approximation to find the probability that more than 95 of the next 100 cakes made will be iced.

- If Y represents the number of iced cakes produced by Factory B, then $Y \sim B(100, 0.99)$. Here, n is quite large, but p is not small.
- However, if you let W represent the number of "un-iced" cakes made, then $W \sim B(100, 0.01)$. Now you can use a Poisson approximation: $W \sim Po(100 \times 0.01) = Po(1)$.
  So $P(Y > 95) = P(W < 5) = P(W \leq 4) = 0.9963$

Using B(100, 0.01), you get 0.9966.

Remember — you need a small p...

This approximation only works if p is very small (although in the right circumstances, it'll also work if p is very close to 1). On a practical note... check what value of λ your tables go up to — you can only use the approx. for $np \leq$ max value of λ.

Worked Problems

This page is for AQA S2, Edexcel S2, OCR S2 and OCR MEI S2

Make sure you understand what's going on in these examples.

EXAMPLE 1: A breaking-down car

A car randomly breaks down twice a week on average.
The random variable X represents the number of times the car will break down next week.

a) What probability distribution could be used to model X? Explain your answer.
b) Find the probability that the car breaks down fewer than 3 times next week.
c) Find the probability that the car breaks down more than 4 times next week.
d) Find the probability that the car breaks down exactly 6 times in the next fortnight.

a) Since the breakdowns occur randomly, singly and (on average) at a constant rate, and X is the total number of breakdowns in one week, X follows a Poisson distribution: $X \sim \text{Po}(2)$
b) Using tables for $\lambda = 2$: $P(X < 3) = P(X \leq 2) = 0.6767$
c) Again, using tables for $\lambda = 2$: $P(X > 4) = 1 - P(X \leq 4) = 1 - 0.9473 = 0.0527$
d) If the random variable Y represents the number of breakdowns in the next fortnight, then $Y \sim \text{Po}(2 \times 2) = \text{Po}(4)$.
So using tables for $\lambda = 4$: $P(Y = 6) = P(Y \leq 6) - P(Y \leq 5) = 0.8893 - 0.7851 = 0.1042$

EXAMPLE 2: Bad apples

A restaurant owner needs to buy several crates of apples, so she visits a farm that sells apples by the crate. Each crate contains 150 apples. On average 1.5% of the apples are bad, and these bad apples are randomly distributed between the crates. The restaurant owner opens a random crate and inspects each apple.

- If there are no bad apples in this crate, then the restaurant owner will buy the apples she needs from this farm.
- If more than 2 apples in this first crate are bad, then the restaurant owner will not buy from this farm.
- If only 1 or 2 apples in the first crate are bad, then a second crate is opened.
The restaurant owner will then only buy from this farm if the second crate contains at most 1 bad apple.

a) Find the probability that none of the apples in the first crate are bad.
b) Find the probability that more than 2 apples in the first crate are bad.
c) Find the probability that a second crate is opened.
d) What is the probability of the restaurant owner buying the apples she needs from this farm?

a) The average number of bad apples in each crate is $150 \times 0.015 = 2.25$.
So if X represents the number of bad apples in each crate, then $X \sim \text{Po}(2.25)$.
$P(X = 0) = \frac{e^{-2.25} \times 2.25^0}{0!} = e^{-2.25} = 0.1054$ (to 4 d.p.).

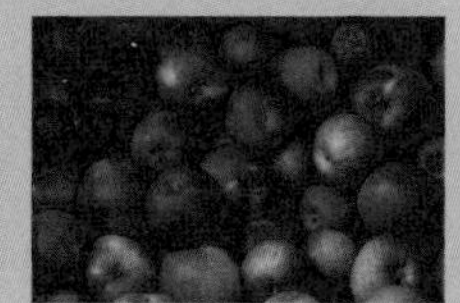
Definitely Poisson.

b) $P(X = 1) = \frac{e^{-2.25} \times 2.25^1}{1!} = e^{-2.25} \times 2.25 = 0.2371$ (to 4 d.p.).
$P(X = 2) = \frac{e^{-2.25} \times 2.25^2}{2!} = \frac{e^{-2.25} \times 2.25^2}{2} = 0.2668$ (to 4 d.p.).
So $P(X > 2) = 1 - P(X = 0) - P(X = 1) - P(X = 2) = 1 - 0.1054 - 0.2371 - 0.2668 = 0.3907$

c) A second crate is opened if $X = 1$ or $X = 2$. $P(X = 1 \text{ OR } X = 2) = 0.2371 + 0.2668 = 0.5039$

d) There are two ways the owner will buy apples from this farm:
- Either the first crate will contain no bad apples (probability = 0.1054),
- Or the first crate will contain 1 or 2 bad apples AND the second crate will contain 0 or 1 bad apples.

P(1st crate has 1 or 2 bad AND 2nd crate has 0 or 1 bad) = $0.5039 \times (0.1054 + 0.2371) = 0.1726$
So P(restaurant owner buys from this farm) = $0.1054 + 0.1726 = 0.278$

All it takes is one bad apple question and everything starts to go wrong...

I admit that apple question looks a nightmare at first... but just hold your nerve and take things nice and slowly. For example, in that last part, ask yourself: "What individual things need to happen before the restaurant owner will buy from this farm?" Work out the individual probabilities, add or multiply them as necessary, and Bob's your uncle.

S2 Section 3 — Practice Questions

Well, that's the section completed, which is as good a reason as most to celebrate. But wait... put that celebratory cup of tea on ice for a few minutes more, because you've still got some questions to answer to prove that you really do know everything. So try the questions... and if you get any wrong, do some more revision and try them again.

Warm-up Questions

Questions 1-9 are for AQA S2, Edexcel S2, OCR S2 and OCR MEI S2

1) If $X \sim \text{Po}(3.1)$, find (correct to 4 decimal places):
   a) $P(X = 2)$, b) $P(X = 1)$, c) $P(X = 0)$, d) $P(X < 3)$, e) $P(X \geq 3)$

2) If $X \sim \text{Po}(8.7)$, find (correct to 4 decimal places):
   a) $P(X = 2)$, b) $P(X = 1)$, c) $P(X = 0)$, d) $P(X < 3)$, e) $P(X \geq 3)$

3) For the following distributions, find: (i) $E(X)$, (ii) $\text{Var}(X)$, and (iii) the standard deviation of X.
   a) Po(8), b) Po(12.11) c) Po(84.2227)

4) For the following distributions, find: (i) $P(X \leq \mu)$, (ii) $P(X \leq \mu - \sigma)$
   a) Po(9), b) Po(4)

5) Which of the following would follow a Poisson distribution? Explain your answers.
   a) The number of defective products coming off a factory's production line in one day if defective products occur at random at an average of 25 per week.
   b) The number of heads thrown using a coin in 25 tosses if the probability of getting a head is always 0.5.
   c) The number of people joining a post-office queue each minute during lunchtime if people arrive at an average rate of 3 every five minutes.
   d) The total number of spelling mistakes in a document if mistakes are randomly made at an average rate of 3 per page.

6) In a radioactive sample, atoms decay at an average rate of 2000 per hour.
   State how the following quantities are distributed, giving as much detail as possible.
   a) The number of atoms decaying per minute.
   b) The number of atoms decaying per day.

7) Atoms in one radioactive sample decay at an average rate of 60 per minute, while in another they decay at an average rate of 90 per minute.
   a) How would the total number of atoms decaying each minute be distributed?
   b) How would the total number of atoms decaying each hour be distributed?

8) If $X \sim \text{Po}(8)$, use Poisson tables to find:
   a) $P(X \leq 2)$, b) $P(X \leq 7)$, c) $P(X \leq 5)$, d) $P(X < 9)$, e) $P(X \geq 8)$
   f) $P(X > 1)$, g) $P(X > 7)$, h) $P(X = 6)$, i) $P(X = 4)$, j) $P(X = 3)$

9) A gaggle of 100 geese is randomly scattered throughout a field measuring 10 m × 10m. What is the probability that in a randomly selected square metre of field, I find:
   a) no geese? b) 1 goose? c) 2 geese? d) more than 2 geese?

Question 10 is for Edexcel S2, OCR S2 and OCR MEI S2

10) Which of the following random variables could be approximated by a Poisson distribution? Where it is possible, state the Poisson distribution that could be used.
    a) $X \sim B(4, 0.4)$, b) $Y \sim B(700, 0.01)$, c) $W \sim B(850, 0.34)$
    d) $X \sim B(8, 0.1)$, e) $W \sim B(10\,000, 0.00001)$, f) $Y \sim B(80, 0.9)$ *(harder)*

S2 Section 3 — Practice Questions

Nearly there — just... one... more... page...

Exam Questions

Questions 1-2 are for AQA S2, Edexcel S2, OCR S2 and OCR MEI S2

1 a) State two conditions needed for a Poisson distribution to be a suitable model for a quantity.

(2 marks)

b) A birdwatcher knows that the number of chaffinches visiting a particular observation spot per hour follows a Poisson distribution with mean 7.

Find the probability that in a randomly chosen hour during the day:

(i) fewer than 4 chaffinches visit the observation spot,

(2 marks)

(ii) at least 7 chaffinches visit the observation spot,

(2 marks)

(iii) exactly 9 chaffinches visit the observation spot.

(2 marks)

c) The number of birds <u>other than</u> chaffinches visiting the same observation spot per hour can be modelled by the Poisson distribution Po(22).

Find the probability that exactly 3 birds (of any species) visit the observation spot in a random 15-minute period.

(4 marks)

2 The number of calls received at a call centre each hour can be modelled by a Poisson distribution with mean 20.

a) Find the probability that in a random 30-minute period:

(i) exactly 8 calls are received,

(3 marks)

(ii) more than 8 calls are received.

(2 marks)

b) For a Poisson distribution to be a suitable model, events have to occur independently. What is meant by "independently" in this context?

(1 mark)

Question 3 is for Edexcel S2, OCR S2 and OCR MEI S2

3 When a particular engineer is called out to fix a fault, the probability of him being unable to fix the fault is always 0.02.

a) The engineer's work is assessed after every 400 call-outs. The random variable X represents the number of faults the engineer is unable to fix over those 400 call-outs. Specify the statistical distribution that X will follow, stating the values of any parameters.

(2 marks)

b) (i) Under what conditions can a binomial distribution be approximated by a Poisson distribution?

(2 marks)

(ii) Write down a Poisson distribution that could be used to approximate X.

(1 mark)

(iii) Write down the mean and variance of your Poisson distribution.

(1 mark)

(iv) Using your Poisson approximation, calculate the probability that the engineer will be unable to fix fewer than 10 faults over a period of 400 call-outs.

(2 marks)

Probability Density Functions

This page is for AQA S2, Edexcel S2 and OCR S2

This section covers the same sorts of things as you've seen before with discrete random variables, only now the variables are continuous.

Continuous Random Variables take Any Value in a Range

1) With discrete random variables, there are 'gaps' between the possible values the random variable can take. The random variable's probability function tells you the probability of each of these values occurring.

   For example, if $X \sim B(3, 0.4)$, then you know that X can only take the values 0, 1, 2 or 3, and you could work out the probability of each of these values using the binomial probability function. You could even draw a graph of what this probability function looks like.

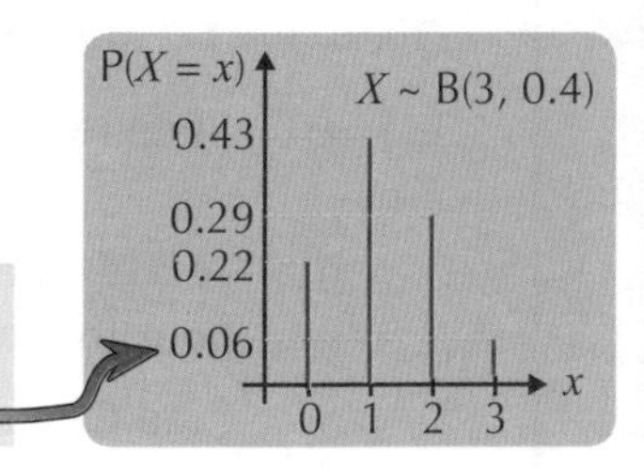

2) Continuous random variables are similar, but they can take any value within a certain range (e.g. they represent things like length, height, weight, etc.).

   So a continuous random variable X might be able to take any value between 0 and 4, for example. You can still draw a graph showing how likely X is to take values within this range. But instead of a series of bars, it would be a continuous line.

3) These graphs that show how likely continuous random variables are to take various values are called probability density functions (or p.d.f.s). Here, f(x) is a p.d.f.

4) It's actually the area under a p.d.f. that shows probability. For example, the shaded area shows the probability that this continuous random variable will take a value between 1 and 2.

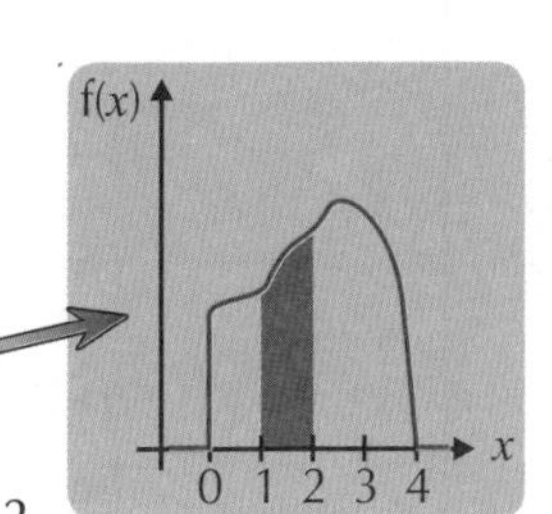

The Total Area under a p.d.f. is 1

Remember... it's the area under a p.d.f. that shows probability, and you find the area under a curve by integrating.

EXAMPLE a) Explain why a p.d.f. can never take negative values.
b) Explain why the total area under any p.d.f. must equal 1.

a) A p.d.f. can never be negative, since probabilities can never be negative.

b) The total area under a p.d.f. must always equal 1 since that's just the total probability of the random variable taking one of its possible values.

In maths-speak, this means $f(x) \geq 0$ for all x, and $\int_{-\infty}^{\infty} f(x)dx = 1$.

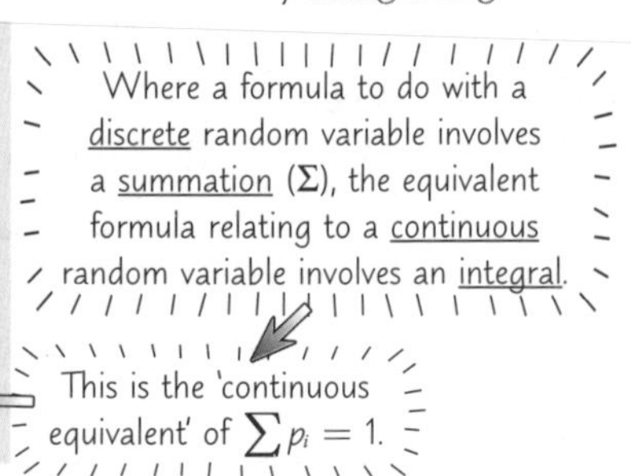

Find Probabilities by Calculating Areas

Some of the p.d.f.s you'll come across are defined "piecewise" (bit by bit). Don't let that faze you.

EXAMPLE The continuous random variable X has the probability density function below.

$$f(x) = \begin{cases} kx & \text{for } 0 < x < 4 \\ 0 & \text{otherwise} \end{cases}$$

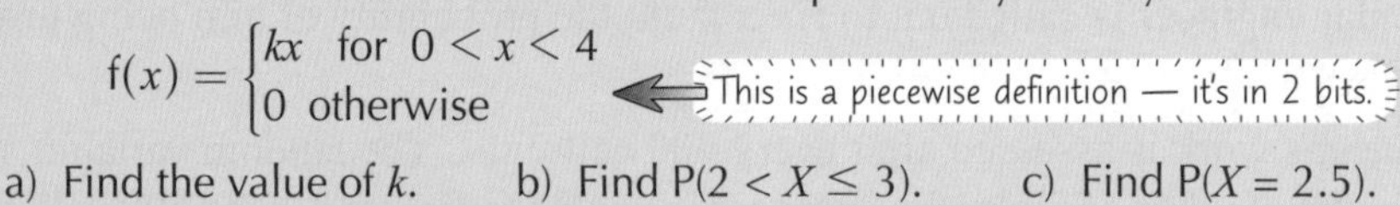

a) Find the value of k. b) Find $P(2 < X \leq 3)$. c) Find $P(X = 2.5)$.

a) The total area under the p.d.f. must equal 1.
Using a sketch of f(x), you can tell that $8k = 1$, or $k = 0.125$.

b) You need to find the area under the graph between $x = 2$ and $x = 3$.
Using the formula for the area of a trapezium, $P(2 < X \leq 3) = 0.3125$

c) The area under a graph at a single point is zero (since it would be the area of a trapezium with zero width). So $P(X = 2.5) = 0$.

The probability of a continuous random variable equalling any single value is always zero — it only makes sense to find the probability of it taking a value within a particular range. It also means that for a continuous random variable, $P(X < k) = P(X \leq k)$, for any k.

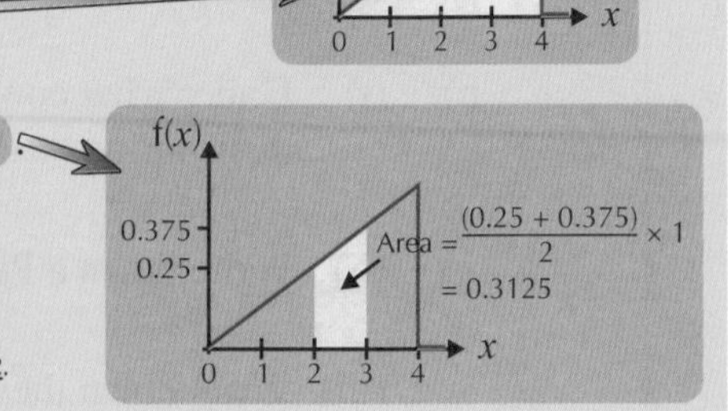

Sometimes, statistics all seems a little bit odd...

That thing about $P(X = x) = 0$ always seems weird to me. I mean... X has to take some value, so it seems peculiar that the probability of it taking any particular value equals zero. But that's the way it is. It makes a bit more sense if you remember that probabilities are represented by areas under a graph. Not many calculations here, but learn the ideas carefully.

Probability Density Functions

This page is for AQA S2, Edexcel S2 and OCR S2

It's time to put on your best integrating trousers, because you'll be finding more "areas under curves" on this page.

Some Probabilities Need to be Found by *Integrating*

Remember — probabilities are represented by areas, so if X has p.d.f. f(x):

$$P(a < X \leq b) = \int_a^b f(x)dx$$

EXAMPLE The continuous random variable X has the probability density function below.

$$f(x) = \begin{cases} x^2 + a & \text{for } 0 \leq x \leq 1 \\ 0 & \text{otherwise} \end{cases}$$

a) Sketch f(x), and find the value of a.

b) Find $P(X > \frac{1}{2})$.

a) The non-zero bit of the p.d.f. is a quadratic function, and so f(x) looks like this:
The area under the graph must equal 1, so integrate.

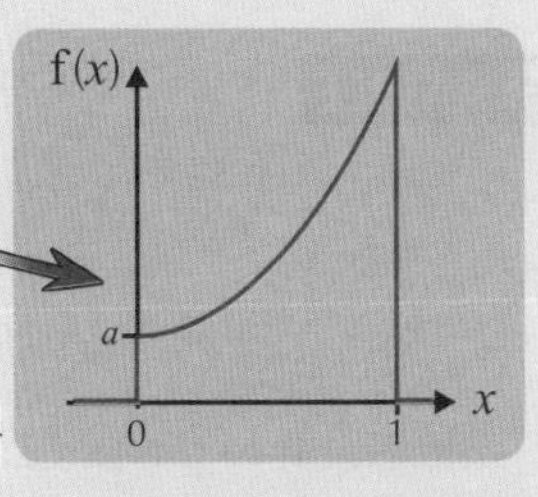

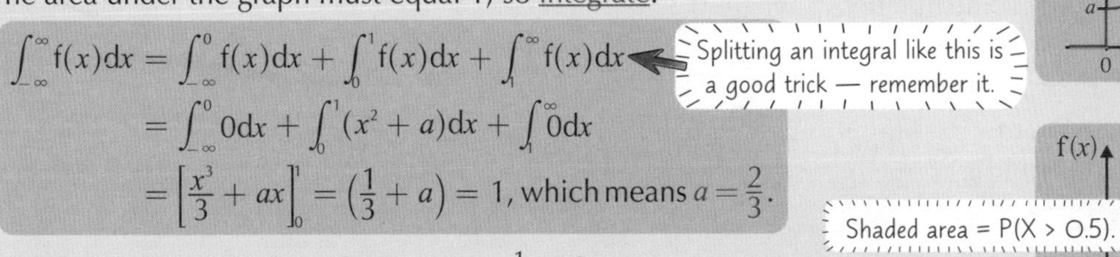

$$\int_{-\infty}^{\infty} f(x)dx = \int_{-\infty}^{0} f(x)dx + \int_{0}^{1} f(x)dx + \int_{1}^{\infty} f(x)dx$$

$$= \int_{-\infty}^{0} 0dx + \int_{0}^{1} (x^2 + a)dx + \int_{1}^{\infty} 0dx$$

$$= \left[\frac{x^3}{3} + ax\right]_0^1 = \left(\frac{1}{3} + a\right) = 1, \text{ which means } a = \frac{2}{3}.$$

Splitting an integral like this is a good trick — remember it.

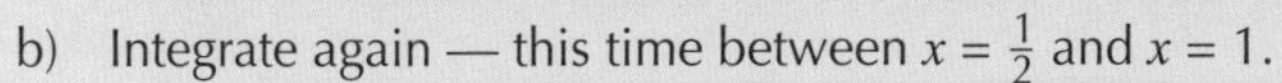

b) Integrate again — this time between $x = \frac{1}{2}$ and $x = 1$.

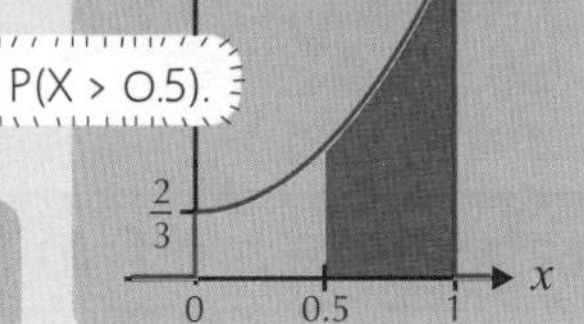

Shaded area = P(X > 0.5).

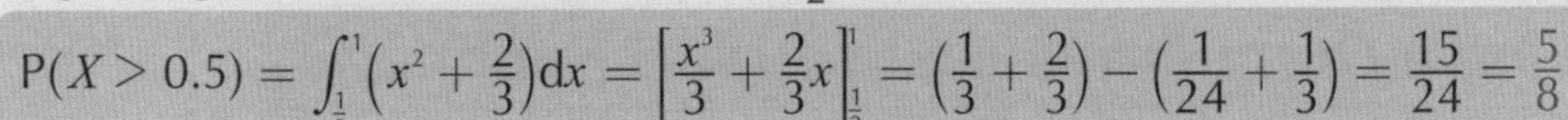

$$P(X > 0.5) = \int_{\frac{1}{2}}^{1} \left(x^2 + \frac{2}{3}\right)dx = \left[\frac{x^3}{3} + \frac{2}{3}x\right]_{\frac{1}{2}}^{1} = \left(\frac{1}{3} + \frac{2}{3}\right) - \left(\frac{1}{24} + \frac{1}{3}\right) = \frac{15}{24} = \frac{5}{8}$$

You Might Need to Spot a Function that's *NOT* a p.d.f.

EXAMPLE Which of the following could be probability density functions?

a) $f(x) = \begin{cases} 3x & \text{for } -1 \leq x \leq 1 \\ 0 & \text{otherwise} \end{cases}$ b) $g(x) = \begin{cases} kx & \text{for } 2 \leq x \leq 4 \\ 0 & \text{otherwise} \end{cases}$ c) $h(x) = \begin{cases} kx & \text{for } -2 \leq x \leq 2 \\ 0 & \text{otherwise} \end{cases}$

a) The graph of f(x) looks like this:
But a p.d.f. can never take negative values, so this cannot be a probability density function.

f(x), 3, −1, 1, x, −3

b) Since a p.d.f. can never take a negative value, k cannot be negative.

If k is positive, then the graph of g(x) looks like this,

and the total area under the graph is $\frac{2k + 4k}{2} \times 2 = 6k$.

So g(x) could be a p.d.f. as long as $k = \frac{1}{6}$.

g(x), 4k, 2k, 2, 4, x

The yellow area must equal 1 for g(x) to be a p.d.f.

c) If k is positive, then h(x) is negative for $-2 \leq x < 0$, so k cannot be positive.
If k is negative, then h(x) is negative for $0 < x \leq 2$, so k cannot be negative.

If $k = 0$, then $\int_{-\infty}^{\infty} h(x)dx = \int_{-\infty}^{\infty} 0dx = 0$, so h($x$) cannot be a p.d.f.

The integral $\int_{-\infty}^{\infty} h(x)dx$ must equal 1 for h(x) to be a p.d.f.

Three things you should definitely know about a p.d.f...

There's not really heaps to say about probability density functions. They can't be negative, the total area under a p.d.f. must equal 1, and you can find probabilities by finding areas under the p.d.f. between different limits. If you remember just those facts and can do a bit of integration, then you'll be well on the way to earning a few easy marks come exam time.

Cumulative Distribution Functions

This page is for AQA S2 and Edexcel S2

Now it's time for cumulative distribution functions (c.d.f.s). "Cumulative distribution function" sounds pretty complicated, but all it means is "the area under a p.d.f. up to a certain point".

A Cumulative Distribution Function shows $P(X \leq x)$

'Cumulative distribution functions' are sometimes called 'distribution functions'. These are exactly the same thing.

Cumulative Distribution Functions

If X is a continuous random variable with p.d.f. f(x), then its cumulative distribution function (c.d.f.) F(x) is given by:
$$F(x_0) = P(X \leq x_0) = \int_{-\infty}^{x_0} f(x)dx$$

Cumulative distribution functions are usually labelled with capital letters, e.g. **F(*x*)** — unlike p.d.f.s, which are usually labelled with lower case letters, e.g. **f(*x*)**.

EXAMPLE A continuous random variable X has probability density function f(x), where

$$f(x) = \begin{cases} 3x^2 & \text{for } 0 \leq x \leq 1 \\ 0 & \text{otherwise} \end{cases}$$

Find F(x_0) for some number x_0, where $0 \leq x_0 \leq 1$.

The yellow area shows the value of the cumulative distribution function at x_0.

f(x)
3
x_0 1

The graph of the p.d.f. is shown on the right.

To find F(x_0), you need to integrate between $-\infty$ and x_0:

$$F(x_0) = \int_{-\infty}^{x_0} f(x)dx = \int_{-\infty}^{0} f(x)dx + \int_{0}^{x_0} f(x)dx$$
$$= \int_{-\infty}^{0} 0dx + \int_{0}^{x_0} 3x^2dx = 0 + [x^3]_0^{x_0} = x_0^3$$

Integrate a p.d.f. to Find a Cumulative Distribution Function

The example above involved finding F(x) at a particular point. Now it's time to define F(x) everywhere.

EXAMPLE A continuous random variable X has probability density function f(x), where

$$f(x) = \begin{cases} 2x - 2 & \text{for } 1 \leq x \leq 2 \\ 0 & \text{otherwise} \end{cases}$$

Find the cumulative distribution function of X.

To find F(x), you need to integrate between $-\infty$ and x (so x actually needs to be the upper limit of your integral). To avoid having x as the limit of the integral and the variable you're integrating with respect to, it helps to use a different variable (e.g. 't') inside the integral. The t then disappears when you put in the limits.

$$F(x) = \int_{-\infty}^{1} f(t)dt + \int_{1}^{x} f(t)dt = \int_{-\infty}^{1} 0dt + \int_{1}^{x} (2t - 2)dt$$

For $1 \leq x \leq 2$:
$$= [t^2 - 2t]_1^x = x^2 - 2x - (1 - 2) = x^2 - 2x + 1$$

So
$$F(x) = \begin{cases} 0 & \text{for } x < 1 \\ x^2 - 2x + 1 & \text{for } 1 \leq x \leq 2 \\ 1 & \text{for } x > 2 \end{cases}$$

You must define F(x) for all possible values of x. And the 'pieces' should join together with 'no jumps' — the value of the c.d.f. at the end of one 'piece' must equal the value of the c.d.f. at the start of the next 'piece'. Here, F(1) = 0 using both the first and second 'pieces', and F(2) = 1 using both the second and third.

If you already know the c.d.f., then differentiate to find the p.d.f.

EXAMPLE Find the p.d.f. of the continuous random variable X if its cumulative distribution function F(x) is

$$F(x) = \begin{cases} 0 & \text{for } x < 0 \\ \frac{1}{2}(3x - x^3) & \text{for } 0 \leq x \leq 1 \\ 1 & \text{for } x > 1 \end{cases}$$

Differentiate the c.d.f. F(x) to find the p.d.f. f(x):
$$f(x) = \frac{dF(x)}{dx} = \begin{cases} \frac{3}{2} - \frac{3}{2}x^2 & \text{for } 0 \leq x \leq 1 \\ 0 & \text{otherwise} \end{cases}$$

OMG — too many three-letter abbreviations...

Well that wasn't too bad. That bit about using t instead of x looks odd at first but notice how the t disappears once you've done the integration and put in the limits. And don't forget to define F(x) and f(x) for all x between $-\infty$ and ∞.

Cumulative Distribution Functions

This page is for AQA S2 and Edexcel S2

Be Extra Careful if the p.d.f. is in '*Pieces*'

If a p.d.f. is defined piecewise, then you have to be careful with the c.d.f. where the 'pieces' join.

EXAMPLE Find the cumulative distribution function of X, whose probability density function is

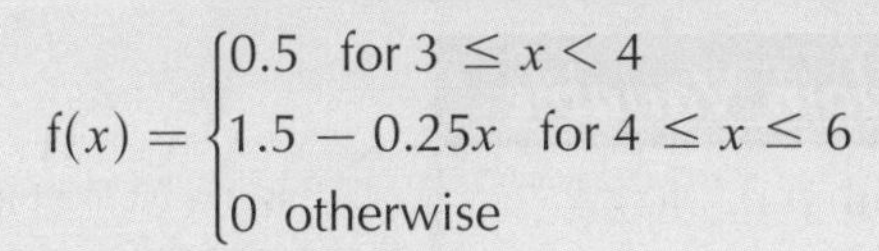

$$f(x) = \begin{cases} 0.5 & \text{for } 3 \le x < 4 \\ 1.5 - 0.25x & \text{for } 4 \le x \le 6 \\ 0 & \text{otherwise} \end{cases}$$

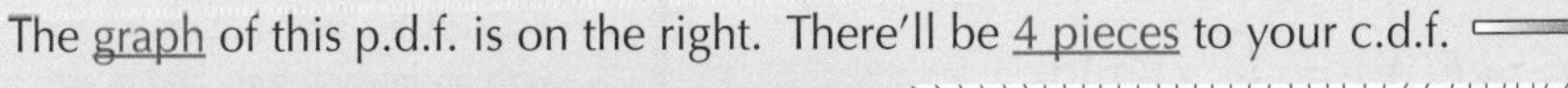

The graph of this p.d.f. is on the right. There'll be 4 pieces to your c.d.f.

Graph labels: f(x); Piece 1, Piece 2, Piece 3, Piece 4; 0.5, 0.25; x axis 0 1 2 3 4 5 6 7

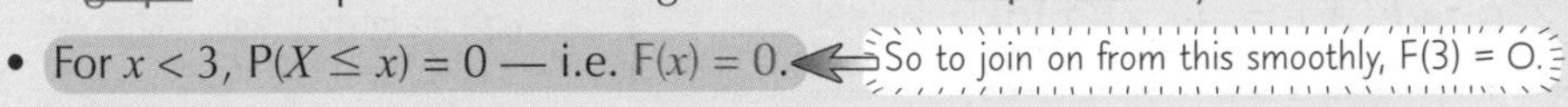

- For $x < 3$, $P(X \le x) = 0$ — i.e. $F(x) = 0$. ← So to join on from this smoothly, F(3) = 0.

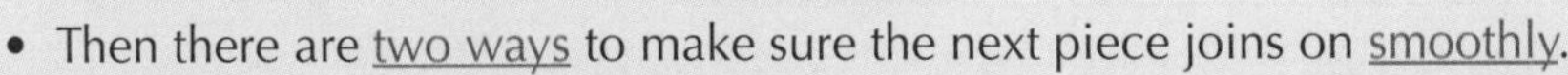

- Then there are two ways to make sure the next piece joins on smoothly.

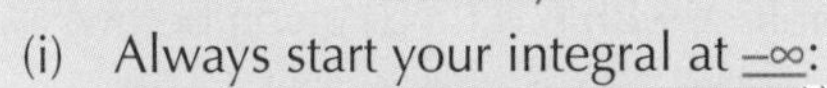

(i) Always start your integral at $-\infty$:

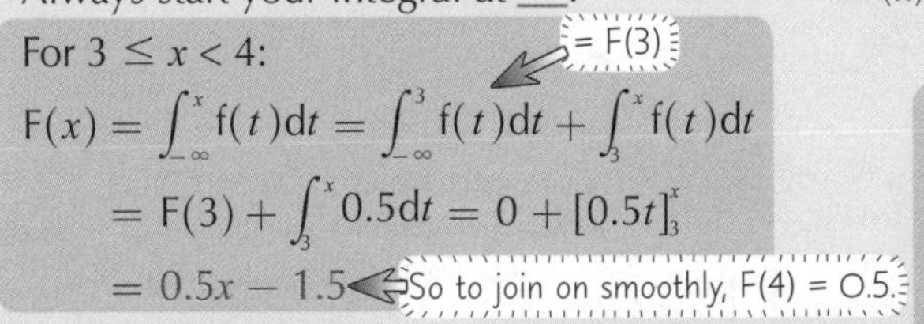

For $3 \le x < 4$:

$$F(x) = \int_{-\infty}^{x} f(t)dt = \int_{-\infty}^{3} f(t)dt + \int_{3}^{x} f(t)dt$$

(the first integral $= F(3)$)

$$= F(3) + \int_{3}^{x} 0.5dt = 0 + [0.5t]_3^x$$

$$= 0.5x - 1.5$$ ← So to join on smoothly, F(4) = 0.5.

For $4 \le x \le 6$:

$$F(x) = \int_{-\infty}^{4} f(t)dt + \int_{4}^{x} f(t)dt = F(4) + \int_{4}^{x} f(t)dt$$

(the first integral $= F(4)$)

$$= 0.5 + \int_{4}^{x} (1.5 - 0.25t)dt$$

$$= 0.5 + [1.5t - 0.125t^2]_4^x$$

$$= 0.5 + (1.5x - 0.125x^2) - (6 - 2)$$

$$= 1.5x - 0.125x^2 - 3.5$$

(ii) Use an indefinite integral and choose the constant of integration so that the join is 'smooth'.

For $3 \le x < 4$: ← Use x rather than t in this integral — since there are no limits, the t wouldn't disappear (see p118).

$$F(x) = \int f(x)dx = \int 0.5dx = 0.5x + k_1$$

But F(3) = 0 (to join the first piece of c.d.f. smoothly). So $k_1 = -1.5$, which gives $F(x) = 0.5x - 1.5$.

For $4 \le x \le 6$:

$$F(x) = \int f(x)dx = \int (1.5 - 0.25x)dx$$

$$= 1.5x - 0.125x^2 + k_2$$

But F(4) = 0.5 (using the previous part of the c.d.f.). So $k_2 = -3.5$, giving $F(x) = 1.5x - 0.125x^2 - 3.5$.

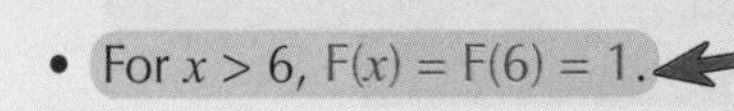

- For $x > 6$, $F(x) = F(6) = 1$.

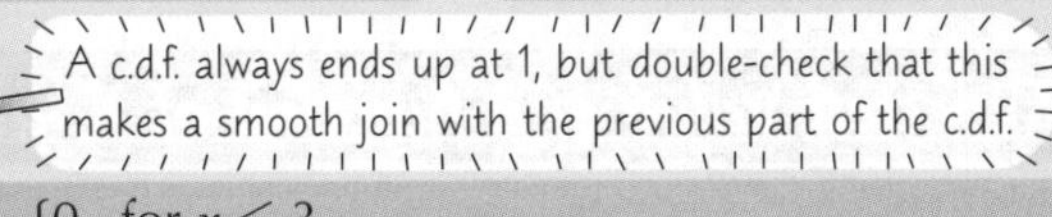

A c.d.f. always ends up at 1, but double-check that this makes a smooth join with the previous part of the c.d.f.

Put the bits together to get:

$$F(x) = \begin{cases} 0 & \text{for } x < 3 \\ 0.5x - 1.5 & \text{for } 3 \le x < 4 \\ 1.5x - 0.125x^2 - 3.5 & \text{for } 4 \le x \le 6 \\ 1 & \text{for } x > 6 \end{cases}$$

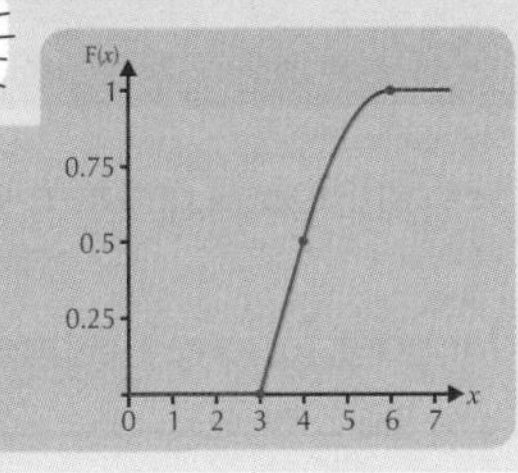

Find *All Sorts* of Probabilities Using a c.d.f.

EXAMPLE The cumulative distribution function F(x) of the continuous random variable X is given below.

$$F(x) = \begin{cases} 0 & \text{for } x < 0 \\ 0.5(3x - x^3) & \text{for } 0 \le x \le 1 \\ 1 & \text{for } x > 1 \end{cases}$$

Find: a) $P(X \le 0.5)$, b) $P(X > 0.25)$, c) $P(0.1 \le X \le 0.2)$, d) $P(X < 0.5)$

a) $P(X \le 0.5) = F(0.5) = 0.5 \times (3 \times 0.5 - 0.5^3) = 0.6875$.

b) $P(X > 0.25) = 1 - P(X \le 0.25) = 1 - F(0.25) = 1 - 0.5 \times (3 \times 0.25 - 0.25^3) = 1 - 0.3672 = 0.633$ (to 3 d.p.).

c) $P(0.1 \le X \le 0.2) = P(X \le 0.2) - P(X \le 0.1) = F(0.2) - F(0.1) = 0.296 - 0.1495 = 0.1465$.

d) $P(X < 0.5) = P(X \le 0.5) = 0.6875$. ← For a continuous random variable, $P(X \le k) = P(X < k)$ — since $P(X = k) = 0$.

Don't fall to pieces now — you've done the hardest bit...

Using a c.d.f. to find the probability of X falling within a particular range is similar to what you've seen before with the binomial and Poisson tables — you can work out anything you need, as long as you're prepared to think a bit.

Mean of a Continuous Random Variable

This page is for AQA S2, Edexcel S2 and OCR S2

You'll have seen something a bit like this before — except previously, the random variables were discrete and the formulas involved a summation (Σ). Here, the random variables are continuous, and you're going to need to integrate.

Integrate to Find the ***Mean*** of a Continuous Random Variable

Mean of a Continuous Random Variable

If X is a continuous random variable with p.d.f. f(x), then its mean (μ) or expected value (E(X)) is given by:

$$\mu = \mathrm{E}(X) = \int_{-\infty}^{\infty} x\mathrm{f}(x)\mathrm{d}x$$

This is a bit like the formula for the mean (expected value) of a discrete random variable — except the sigma (Σ) has been replaced with an integral sign, and p_i with f(x)dx.

EXAMPLE Find the expected value of the continuous random variable X with p.d.f. f(x) given below.

$$\mathrm{f}(x) = \begin{cases} \frac{3}{32}(4 - x^2) & \text{for } -2 \le x \le 2 \\ 0 & \text{otherwise} \end{cases}$$

$$\mathrm{E}(X) = \int_{-\infty}^{\infty} x\mathrm{f}(x)\mathrm{d}x = \int_{-2}^{2} x \cdot \frac{3}{32}(4 - x^2)\mathrm{d}x$$

$$= \int_{-2}^{2}\left(\frac{3}{8}x - \frac{3x^3}{32}\right)\mathrm{d}x = \left[\frac{3x^2}{16} - \frac{3x^4}{128}\right]_{-2}^{2}$$

$$= \left(\frac{3 \times 2^2}{16} - \frac{3 \times 2^4}{128}\right) - \left(\frac{3 \times (-2)^2}{16} - \frac{3 \times (-2)^4}{128}\right) = 0$$

You'd expect a mean of 0 here, since f(x) is symmetrical about the y-axis. So you didn't actually need to integrate.

y = f(x)
x
−2
2

You can find the ***Expected Value*** for a ***Function of X***

A function of X, g(X), is just an expression involving X, like X^2 or $3X + 5$. Finding E(g(X)) is simple — you do the same as if you were finding E(X), but replace x with g(x).

Expected Value of a function of X, g(X)

For a continuous random variable X with p.d.f. f(x), and a function of X, g(X): $\mathrm{E}(g(X)) = \int_{-\infty}^{\infty} g(x)\mathrm{f}(x)\mathrm{d}x$

See — x replaced with g(x).

And if your function is linear (of the form $aX + b$), it's easier still: $\mathrm{E}(aX + b) = a\mathrm{E}(X) + b$

EXAMPLE The continuous random variable X has p.d.f. f(x), where $\mathrm{f}(x) = \begin{cases} \frac{3}{37}x^2 & \text{for } 3 \le x \le 4 \\ 0 & \text{otherwise} \end{cases}$

Find: a) the expected value of X, μ,
b) E(X^2),
c) E($3X + 2$).

This p.d.f. isn't symmetrical in the interval [3, 4], so you have to integrate this time.

a) $\mu = \mathrm{E}(X) = \int_{-\infty}^{\infty} x\mathrm{f}(x)\mathrm{d}x = \int_{3}^{4} x \cdot \frac{3}{37}x^2\mathrm{d}x$

$$= \int_{3}^{4}\frac{3}{37}x^3\mathrm{d}x = \frac{3}{37}\left[\frac{x^4}{4}\right]_3^4 = \frac{3}{37 \times 4}(4^4 - 3^4) = \frac{3 \times (256 - 81)}{148} = \frac{525}{148}$$

Always check your mean looks sensible. Here, you'd expect the mean to be somewhere between 3 and 4 (so 525 ÷ 148 = 3.547... seems 'about right').

g(X) = X^2, so g(x) = x^2.

b) $\mathrm{E}(X^2) = \int_{-\infty}^{\infty} x^2\mathrm{f}(x)\mathrm{d}x = \int_{3}^{4} x^2 \cdot \frac{3}{37}x^2\mathrm{d}x = \int_{3}^{4}\frac{3}{37}x^4\mathrm{d}x$

$$= \frac{3}{37}\left[\frac{x^5}{5}\right]_3^4 = \frac{3}{37 \times 5}(4^5 - 3^5) = \frac{2343}{185}$$

c) $\mathrm{E}(3X + 2) = 3\mathrm{E}(X) + 2 = 3 \times \frac{525}{148} + 2 = \frac{1575 + 296}{148} = \frac{1871}{148}$

Variance of a Continuous Random Variable

This page is for AQA S2, Edexcel S2 and OCR S2

More integrating, I'm afraid. But stick with it — it'll soon be over.

Integrate to Find the *Variance* of a Continuous Random Variable

Variance of a Continuous Random Variable

If X is a continuous random variable with p.d.f. f(x), then its variance is given by:

$$\text{Var}(X) = \text{E}(X^2) - [\text{E}(X)]^2 = \text{E}(X^2) - \mu^2$$

$$= \int_{-\infty}^{\infty} x^2 \text{f}(x)\text{d}x - \mu^2$$

This is exactly the same formula that you've used before for discrete random variables.

EXAMPLE The continuous random variable X has p.d.f. f(x) given below, and a mean of 0. Find the variance of X.

$$\text{f}(x) = \begin{cases} \frac{3}{32}(4 - x^2) & \text{for } -2 \leq x \leq 2 \\ 0 & \text{otherwise} \end{cases}$$

You saw on page 120 that the mean of this p.d.f. is 0.

$$\text{Var}(X) = \text{E}(X^2) - \mu^2 = \int_{-\infty}^{\infty} x^2\text{f}(x)\text{d}x - \mu^2 = \int_{-2}^{2} x^2 \cdot \frac{3}{32}(4 - x^2)\text{d}x - 0^2$$

$$= \int_{-2}^{2}\left(\frac{3x^2}{8} - \frac{3x^4}{32}\right)\text{d}x = \left[\frac{x^3}{8} - \frac{3x^5}{160}\right]_{-2}^{2}$$

$$= \left(\frac{2^3}{8} - \frac{3 \times 2^5}{160}\right) - \left(\frac{(-2)^3}{8} - \frac{3 \times (-2)^5}{160}\right) = 0.8.$$

Find *Var(aX + b)* by *Squaring a* and *Getting Rid of b*

Again, this formula should bring back fond memories of the time you spent with discrete random variables...

Variance of (aX + b)

For a continuous random variable X with p.d.f. f(x): $\text{Var}(aX + b) = a^2\text{Var}(X)$

EXAMPLE The continuous random variable X has p.d.f. f(x), where $\text{f}(x) = \begin{cases} \frac{3}{37}x^2 & \text{for } 3 \leq x \leq 4 \\ 0 & \text{otherwise} \end{cases}$

If $\text{E}(X) = \frac{525}{148}$, find: a) the variance,

b) Var($3X + 2$).

See page 120 for the calculation of the mean of this p.d.f.

a) $\text{Var}(X) = \text{E}(X^2) - \mu^2 = \int_{-\infty}^{\infty} x^2\text{f}(x)\text{d}x - \left(\frac{525}{148}\right)^2 = \int_{3}^{4} x^2 \cdot \frac{3}{37}x^2\text{d}x - \left(\frac{525}{148}\right)^2$

$$= \frac{3}{37}\left[\frac{x^5}{5}\right]_3^4 - \left(\frac{525}{148}\right)^2 = \frac{3}{185}(4^5 - 3^5) - \left(\frac{525}{148}\right)^2 = 0.0815467... = 0.0815 \text{ (to 4 d.p.)}$$

b) $\text{Var}(3X + 2) = 3^2 \times \text{Var}(X) = 9 \times 0.0815467... = 0.734$ (to 3 d.p.).

Don't be fooled by the easy-looking formulas above...

...there are a couple of traps lurking here for the unwary:

1) Remember what goes inside the integral when you're calculating the variance... it's $x^2\text{f}(x)$.

2) Don't obsess so much about getting that integral right that you forget to subtract the square of the mean.

Mode, Median and Quartiles

This page is for AQA S2 and Edexcel S2. There's also a comment halfway down for OCR S2.

Mode = **Maximum** Value of a p.d.f.

This section is for Edexcel only

Mode of a Continuous Random Variable

If X is a continuous random variable with p.d.f. f(x), then its mode is the value of x where f(x) reaches its maximum.

EXAMPLE Find the modes of the continuous random variables whose p.d.f.s are given below.

a) X with p.d.f. $f(x) = \begin{cases} \frac{3}{16}(x^3 - 7x^2 + 10x) & \text{for } 0 \le x \le 2 \\ 0 & \text{otherwise} \end{cases}$ b) Y with p.d.f. $g(y) = \begin{cases} 2y - 2 & \text{for } 1 \le y \le 2 \\ 0 & \text{otherwise} \end{cases}$

a) As always, it's best to start with a sketch of the p.d.f.:

Since $x^3 - 7x^2 + 10x = x(x^2 - 7x + 10) = x(x - 5)(x - 2)$, f($x$) looks like this:

Now differentiate to find the maximum of f(x) in the range $0 \le x \le 2$.

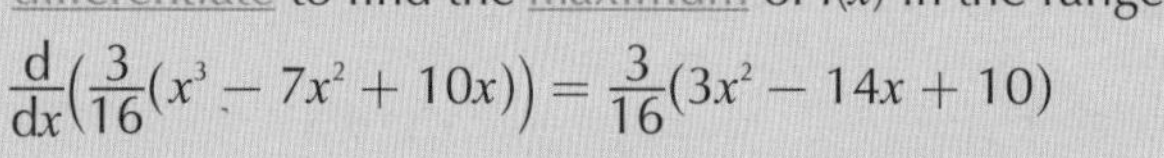

$$\frac{d}{dx}\left(\frac{3}{16}(x^3 - 7x^2 + 10x)\right) = \frac{3}{16}(3x^2 - 14x + 10)$$

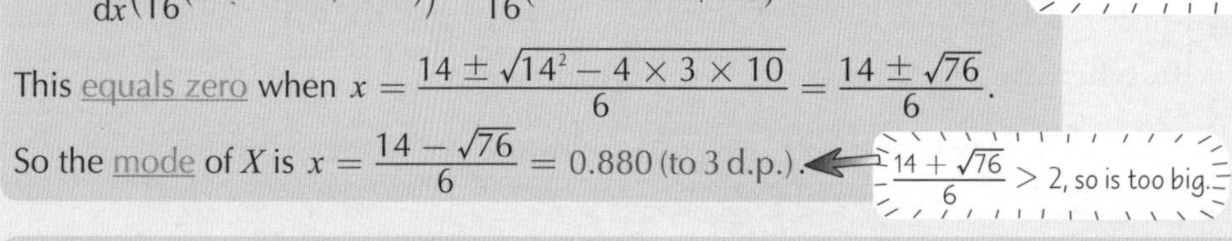

This equals zero when $x = \frac{14 \pm \sqrt{14^2 - 4 \times 3 \times 10}}{6} = \frac{14 \pm \sqrt{76}}{6}$.

So the mode of X is $x = \frac{14 - \sqrt{76}}{6} = 0.880$ (to 3 d.p.).

$\frac{14 + \sqrt{76}}{6} > 2$, so is too big.

$y = x^3 - 7x^2 + 10x$ is in grey. The p.d.f. f(x) is in red.

b) The maximum value of g(y) in the range $1 \le y \le 2$ is at $y = 2$. So the mode of Y is 2.

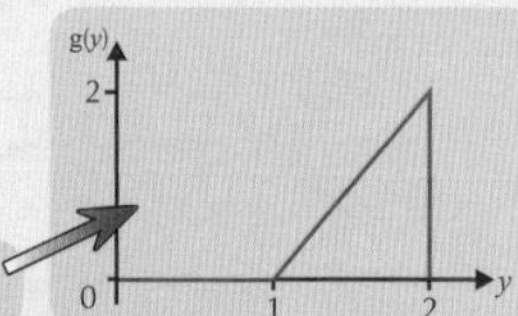

Find the **Median** and **Quartiles** using the c.d.f.

If you're doing **OCR**, you need to know how to **find the median** using the area under the p.d.f. The median, M, is the value which splits this area in half — i.e. P(X ≤ M) = 0.5. So, to find M, solve:

$$\int_{-\infty}^{M} f(x)\,dx = 0.5$$

Median and Quartiles of a Continuous Random Variable

If X is a continuous random variable with cumulative distribution function F(x), then:

- The median (Q_2) of X is given by $F(Q_2) = 0.5$.
- The lower quartile (Q_1) of X is given by $F(Q_1) = 0.25$.
- The upper quartile (Q_3) of X is given by $F(Q_3) = 0.75$.

If you're doing AQA, you might be asked to find percentiles. These are defined in a similar way. E.g. to find the 10th percentile, P_{10}, you'd solve $F(P_{10}) = 0.1$.

EXAMPLE The continuous random variable X has p.d.f. f(x), where $f(x) = \begin{cases} 0.4 & \text{for } 1 \le x < 2 \\ 0.4(x - 1) & \text{for } 2 \le x \le 3 \\ 0 & \text{otherwise} \end{cases}$

Find the interquartile range, $Q_3 - Q_1$.

Integrate to find the c.d.f., making sure the joins are smooth — I used the 'constants of integration' method (p119).

$$F(x) = \begin{cases} 0 & \text{for } x < 1 \\ 0.4x + k_1 & \text{for } 1 \le x < 2 \\ 0.2x^2 - 0.4x + k_2 & \text{for } 2 \le x \le 3 \\ 1 & \text{for } x > 3 \end{cases}$$

Choose k_1 and k_2 to give 'smooth joins'.

$$F(x) = \begin{cases} 0 & \text{for } x < 1 \\ 0.4x - 0.4 & \text{for } 1 \le x < 2 \\ 0.2x^2 - 0.4x + 0.4 & \text{for } 2 \le x \le 3 \\ 1 & \text{for } x > 3 \end{cases}$$

- The lower quartile (Q_1) is given by $F(Q_1) = 0.25$ — and you know Q_1 must be less than 2 (since F(2) = 0.4). So solve $0.4Q_1 - 0.4 = 0.25$, which gives $Q_1 = 1.625$.

Find the value of F(x) where the second and third pieces join to work out which piece contains each quartile.

- The upper quartile (Q_3) is given by $F(Q_3) = 0.75$ — and you know x must be greater than 2 (again, as F(2) = 0.4). So solve $0.2Q_3^2 - 0.4Q_3 + 0.4 = 0.75$, which gives $Q_3 = \frac{0.4 + \sqrt{0.44}}{0.4} = 1 + \frac{\sqrt{0.44}}{0.4}$ (using the quadratic formula).
- This means the interquartile range $= Q_3 - Q_1 = 1 + \frac{\sqrt{0.44}}{0.4} - 1.625 = 1.033$ (to 3 d.p.).

Mode, median — hmmm, that rings a vague bell from S1...

All sorts of mathematical bits and bobs are being used here — the quadratic formula, differentiation to find the maximum of a curve, and so on. That could happen in the exam, so if necessary, refresh your memory from earlier units.

S2 Section 4 — Practice Questions

It's the end of Section 4, and I'm going to assume that you know what you're supposed to do by now. And if you get any wrong, well... I'll say no more.

Warm-up Questions

Questions 1-5 are for AQA S2, Edexcel S2 and OCR S2

1) Find the value of k for each of the probability density functions below.

a) $f(x) = \begin{cases} kx \text{ for } 1 \leq x \leq 10 \\ 0 \text{ otherwise} \end{cases}$ b) $g(x) = \begin{cases} 0.2x + k \text{ for } 0 \leq x \leq 1 \\ 0 \text{ otherwise} \end{cases}$

2) For each of the probability density functions below, find: (i) $P(X < 1)$, (ii) $P(2 \leq X \leq 5)$, (iii) $P(X = 4)$.

a) $f(x) = \begin{cases} 0.08x \text{ for } 0 \leq x \leq 5 \\ 0 \text{ otherwise} \end{cases}$ b) $g(x) = \begin{cases} 0.02(10 - x) \text{ for } 0 \leq x \leq 10 \\ 0 \text{ otherwise} \end{cases}$

3) Find the exact value of k for each of the probability density functions below. Then for each p.d.f., find $P(X < 1)$.

a) $f(x) = \begin{cases} kx^2 \text{ for } 0 \leq x \leq 5 \\ 0 \text{ otherwise} \end{cases}$ b) $g(x) = \begin{cases} 0.1x^2 + kx \text{ for } 0 \leq x \leq 2 \\ 0 \text{ otherwise} \end{cases}$

4) Say whether the following are probability density functions. Explain your answers.

a) $f(x) = \begin{cases} 0.1x^2 + 0.2 \text{ for } 0 \leq x \leq 2 \\ 0 \text{ otherwise} \end{cases}$ b) $g(x) = \begin{cases} x \text{ for } -1 \leq x \leq 1 \\ 0 \text{ otherwise} \end{cases}$

5) The random variables X and Y have p.d.f.s. f(x) and g(y) respectively, where

$f(x) = \begin{cases} 0.08x \text{ for } 0 \leq x \leq 5 \\ 0 \text{ otherwise} \end{cases}$ and $g(y) = \begin{cases} 0.02(10 - y) \text{ for } 0 \leq y \leq 10 \\ 0 \text{ otherwise} \end{cases}$

a) Find the mean and variance of X and Y.

b) Find the mean and variance of $4X + 2$ and $3Y - 4$.

c) Find the median and interquartile range of X.

d) ***[Edexcel only]*** Find the mode of X.

Questions 6-7 are for AQA S2 and Edexcel S2

6) Find the cumulative distribution function (c.d.f.) for each of the following p.d.f.s.

a) $f(x) = \begin{cases} 0.08x \text{ for } 0 \leq x \leq 5 \\ 0 \text{ otherwise} \end{cases}$ b) $g(x) = \begin{cases} 0.02(10 - x) \text{ for } 0 \leq x \leq 10 \\ 0 \text{ otherwise} \end{cases}$

c) $h(x) = \begin{cases} 2x \text{ for } 0 \leq x \leq 0.5 \\ 1 \text{ for } 0.5 \leq x \leq 1 \\ 3 - 2x \text{ for } 1 \leq x \leq 1.5 \\ 0 \text{ otherwise} \end{cases}$ d) $m(x) = \begin{cases} 0.5 - 0.1x \text{ for } 2 \leq x \leq 4 \\ 0.1 \text{ for } 4 \leq x \leq 10 \\ 0 \text{ otherwise} \end{cases}$

7) Find the probability density function (p.d.f.) for each of the following c.d.f.s.

a) $F(x) = \begin{cases} 0 \text{ for } x < 0 \\ x^4 \text{ for } 0 \leq x \leq 1 \\ 1 \text{ for } x > 1 \end{cases}$ b) $G(x) = \begin{cases} 0 \text{ for } x < 1 \\ \frac{1}{100}(x - 1)^2 \text{ for } 1 \leq x < 6 \\ \frac{3}{8}x - 2 \text{ for } 6 \leq x \leq 8 \\ 1 \text{ for } x > 8 \end{cases}$

S2 Section 4 — Practice Questions

Think of this page as a game. To win the game, you have to get all the answers to the questions right. I know what you're thinking... it's a terrible game. In fact, it's not a game at all, but a shameless fib.

Exam Questions

Question 1 is for AQA S2, Edexcel S2 and OCR S2

1 The continuous random variable X has probability density function f(x), as defined below.

$$f(x) = \begin{cases} \frac{1}{k}(x+4) & \text{for } 0 \leq x \leq 2 \\ 0 & \text{otherwise} \end{cases}$$

a) Find the value of k. *(3 marks)*

b) Calculate E(X). *(3 marks)*

c) Calculate the variance of:

(i) X *(3 marks)*

(ii) $4X - 2$ *(2 marks)*

d) Find $P(0 < X < 1.5)$. *(2 marks)*

Parts e)-f) are for AQA S2 and Edexcel S2

e) Find the cumulative distribution function of X, F(x). *(5 marks)*

f) Find the median of X. *(4 marks)*

Part g) is for Edexcel S2

g) Write down the mode of X and describe the skew of X. *(2 marks)*

Question 2 is for AQA S2 and Edexcel S2

2 The continuous random variable X has cumulative distribution function F(x), as defined below.

$$F(x) = \begin{cases} 0 & \text{for } x < 1 \\ k(x-1) & \text{for } 1 \leq x < 3 \\ 0.5(x-2) & \text{for } 3 \leq x \leq 4 \\ 1 & \text{for } x > 4 \end{cases}$$

a) Calculate the value of k. *(2 marks)*

b) Calculate the interquartile range of X. *(5 marks)*

c) (i) Specify the probability density function of X, f(x). *(3 marks)*

(ii) Sketch the graph of f(x). *(1 mark)*

d) (i) Find the mean (μ) of X. *(3 marks)*

(ii) Find the variance (σ^2) of X. *(3 marks)*

(iii) Find $P(X < \mu - \sigma)$. *(2 marks)*

Continuous Uniform Distributions

This page is for AQA S2 and Edexcel S2

Continuous distributions don't have 'gaps' between possible values of the random variable.
But they do have probability density functions (p.d.f.s). See Section 4 if either of those facts took you by surprise.

A Continuous *Uniform* Distribution is *'Rectangular'*

A random variable with a continuous uniform distribution can take any value in a particular range, and its value is equally likely to be anywhere in the range.

EXAMPLE The continuous random variable X has a uniform distribution and can take any value from 1 to 5.
a) Sketch the graph of f(x), the probability density function (p.d.f.) of X.
b) Define f(x).

a) It's a uniform distribution, so within its range of possible values, f(x) is constant. Since the total area under the p.d.f. must equal 1 and the width of the rectangle is 5 – 1 = 4, its height must be 1 ÷ 4 = 0.25.

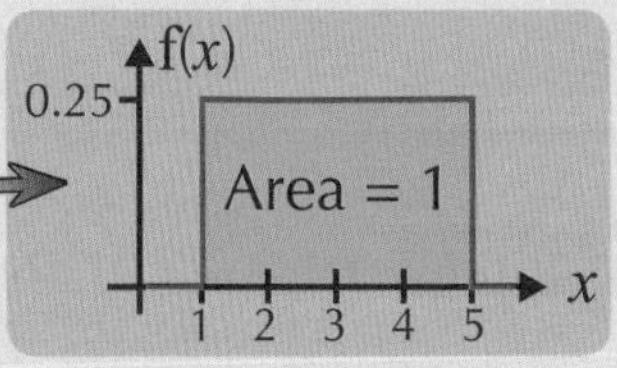

b) So f(x) is given by: $f(x) = \begin{cases} 0.25 & \text{for } 1 \le x \le 5 \\ 0 & \text{otherwise} \end{cases}$

You can go through the same process to work out the p.d.f. of any continuous uniform distribution.
But here's the general formula in a big box.

Continuous Uniform Distribution

If X is a random variable with a continuous uniform distribution, then the p.d.f. of X is:

$$f(x) = \begin{cases} \dfrac{1}{b-a} & \text{for } a \le x \le b \\ 0 & \text{otherwise} \end{cases} \qquad \text{for constants } a \text{ and } b.$$

You can write: $X \sim U[a, b]$.

These distributions can also be called 'rectangular'.

Remember... *Probability = Area Under a p.d.f.*

EXAMPLE If $X \sim U[8, 18]$, find:
a) $P(10 < X < 14.1)$, b) $P(X \le 14)$, c) $P(X < 14)$, d) $P(X \ge 10.5)$.

It's best to start off by drawing a sketch of the p.d.f.
You know the area of the rectangle (= 1) and its width (= 18 – 8 = 10), so its height must be 1 ÷ 10 = 0.1.

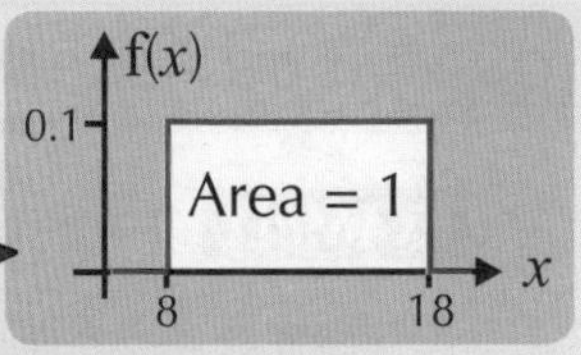

a) $P(10 < X < 14.1)$ = the area under the p.d.f. between x = 10 and x = 14.1.
This is (14.1 – 10) × 0.1 = 4.1 × 0.1 = 0.41

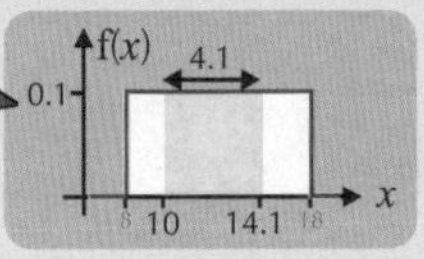

b) $P(X \le 14)$ = the area under the p.d.f. for $x \le 14$.
This is (14 – 8) × 0.1 = 6 × 0.1 = 0.6

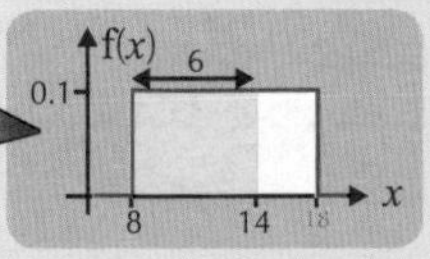

c) For a continuous distribution, $P(X \le k) = P(X < k)$ — so $P(X < 14) = P(X \le 14) = 0.6$

d) $P(X \ge 10.5)$ = the area under the p.d.f. for $x \ge 10.5$.
This is (18 – 10.5) × 0.1 = 7.5 × 0.1 = 0.75

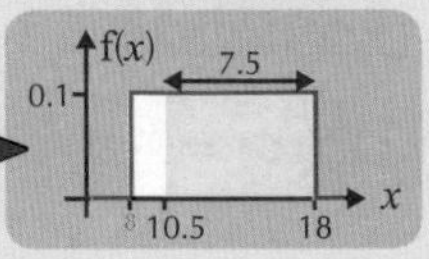

EXAMPLE If $X \sim U[4, 7]$ and $Y = 8X – 3$, write down the distribution of Y.

$Y \sim U[(8 \times 4 – 3), (8 \times 7 – 3)] = U[29, 53]$

Y will also follow a uniform distribution. Just put the limits of X in the formula for Y.

Working out the areas of rectangles — this page really is the limit...

Well, who'd have thought... a page on working out the areas of rectangles in S2. (Best enjoy it while the going's easy, mind.) There's not much more to say about this page really. So I'll keep my mouth shut and let you get on with the next page.

Continuous Uniform Distributions

This page is for AQA S2 and Edexcel S2

You'll be working out expected values and variances of continuous random variables on this page — so you can expect a bit of integration. See pages 120-121 for a bit more info if you don't remember why.

You Might be Asked to Prove these Formulas for E(X), Var(X) and F(x)

You could be asked to show why any of these formulas are true, so read this page very carefully.

Continuous Uniform Distributions

If $X \sim U[a, b]$, then:

(i) $E(X) = \frac{a+b}{2}$ (ii) $Var(X) = \frac{(b-a)^2}{12}$ (iii) $F(x) = \begin{cases} 0 \text{ for } x < a \\ \frac{x-a}{b-a} \text{ for } a \leq x \leq b \\ 1 \text{ for } x > b \end{cases}$

The cumulative distribution function (c.d.f.) — see p118.

EXAMPLE If $X \sim U[a, b]$, show: (i) $E(X) = \frac{a+b}{2}$, (ii) $Var(X) = \frac{(b-a)^2}{12}$, (iii) $F(x) = \frac{x-a}{b-a}$ for $a < x < b$

(i) From the symmetry of the p.d.f., the expected value of X must be $\frac{a+b}{2}$.

Or you can show this by integrating — see page 136, Q3.

(ii) $Var(X) = \int_{-\infty}^{\infty} x^2 f(x)dx - \mu^2 = \int_{-\infty}^{\infty} x^2 f(x)dx - \left(\frac{a+b}{2}\right)^2$ See p121.

But $\int_{-\infty}^{\infty} x^2 f(x)dx = \int_a^b x^2 \left(\frac{1}{b-a}\right)dx = \frac{1}{b-a}\int_a^b x^2 dx = \frac{1}{b-a}\left[\frac{x^3}{3}\right]_a^b = \frac{b^3-a^3}{3(b-a)} = \frac{b^2+ab+a^2}{3}$

$b^3 - a^3 = (b-a)(b^2+ab+a^2)$

So $Var(X) = \frac{b^2+ab+a^2}{3} - \left(\frac{a+b}{2}\right)^2 = \frac{4b^2+4ab+4a^2-3a^2-6ab-3b^2}{12}$

$= \frac{b^2-2ab+a^2}{12} = \frac{(b-a)^2}{12}$

As usual, the standard deviation of X equals the square root of the variance. So here it's $\frac{b-a}{2\sqrt{3}}$.

(iii) $F(x) = \int_{-\infty}^{x} f(t)dt = \int_a^x \frac{1}{b-a}dt = \frac{1}{b-a}\int_a^x 1dt = \frac{1}{b-a}[t]_a^x = \frac{x-a}{b-a}$

You Will be Asked to Use the Formulas for E(X), Var(X) and F(x)

EXAMPLE If $X \sim U[-7, 22]$, find: (i) $E(X)$, (ii) $Var(X)$, (iii) $F(x)$, (iv) $F(4)$.

(i) $E(X) = \frac{a+b}{2} = \frac{-7+22}{2} = 7.5$ (ii) $Var(X) = \frac{(b-a)^2}{12} = \frac{(22-(-7))^2}{12} = \frac{29^2}{12} = 70.083$ (to 3 d.p.).

(iii) $F(x) = \begin{cases} 0 \text{ for } x < -7 \\ \frac{x+7}{29} \text{ for } -7 \leq x \leq 22 \\ 1 \text{ for } x > 22 \end{cases}$ You need to write down all of this.

(iv) $F(4) = \frac{4+7}{29} = \frac{11}{29}$

EXAMPLE If $X \sim U[a, b]$, and $E(X) = Var(X) = 1$, find a and b.

$E(X) = \frac{a+b}{2} = 1$, so $a = 2 - b$

$Var(X) = \frac{(b-a)^2}{12} = 1$, so $b^2 - 2ab + a^2 = 12$, or $b^2 - 2b(2-b) + (2-b)^2 = 12$.

This gives $4b^2 - 8b + 4 = 12$, or $b^2 - 2b - 2 = 0$.

Solving with the quadratic formula gives $b = 1 + \sqrt{3}$, and then $a = 1 - \sqrt{3}$.

You need $b > a$, so use $b = 1 + \sqrt{3}$, not $b = 1 - \sqrt{3}$.

Continuous, Uniform — not the most exciting sounding topic, is it...

The formula for $F(x)$ won't be in your formula booklet, so you'll need to learn it. And you could be asked to show why any of the formulas on this page are true. That means you'll need to be able to write down all the working in the purple box.

Applications of U[a, b]

This page is for AQA S2 and Edexcel S2

Continuous uniform distributions describe things that are equally likely to take any value within an interval. So they're good for describing things that are completely random.

Use a **Uniform** Distribution When All Values are **Equally Likely**

EXAMPLE A runner's time over 100 m is measured as 12.3 seconds, to the nearest 0.1 second. Describe the distribution of the errors in the timing.

If the time was measured to the nearest 0.1 s, then the error could be anything up to 0.05 s above or below the recorded time. These errors are random — there's no reason to think they're likely to be high, low or in the middle. So if the random variable X represents the errors in the timing, then $X \sim U[-0.05, 0.05]$.

All the Usual **Probability Rules** Still Apply with Uniform Distributions

EXAMPLE If $X \sim U[4, 7]$ and $Y = 8X - 3$, find E(Y) and Var(Y).

$$E(X) = \frac{a+b}{2} = \frac{4+7}{2} = 5.5$$

$$E(Y) = E(8X - 3) = 8E(X) - 3 = 8 \times 5.5 - 3 = 41$$

See pages 120-121 for more info.

$$Var(X) = \frac{(b-a)^2}{12} = \frac{(7-4)^2}{12} = \frac{9}{12} = 0.75$$

$$Var(Y) = Var(8X - 3) = 8^2 \times Var(X) = 64 \times 0.75 = 48$$

EXAMPLE X and Y are independent random variables, with $X \sim U[1, 3]$ and $Y \sim U[0, 5]$. Find the probability that both X and Y take values greater than 1.2.

X and Y are independent, so $P(X > 1.2 \text{ and } Y > 1.2) = P(X > 1.2) \times P(Y > 1.2)$.

You could find each of these probabilities by integrating — but it's easier to sketch the p.d.f.s of X and Y, and take it from there.

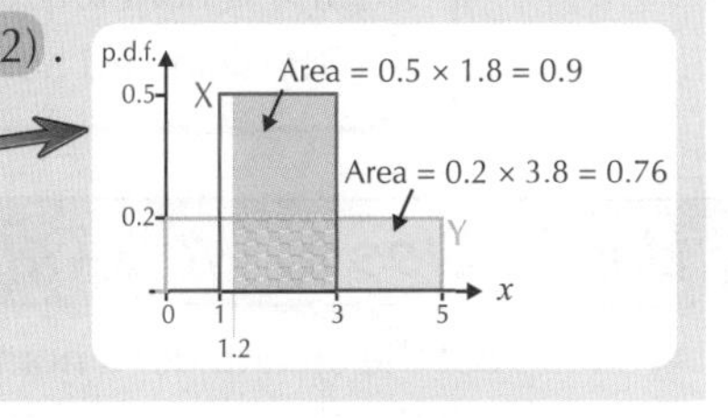

$P(X > 1.2) = 0.5 \times 1.8 = 0.9$ and $P(Y > 1.2) = 0.2 \times 3.8 = 0.76$

So $P(X > 1.2 \text{ and } Y > 1.2) = 0.9 \times 0.76 = 0.684$.

EXAMPLE $X \sim U[3, 12]$.

(a) Sketch the probability density function of X.

(b) Find E(X^2).

(a) It's a continuous uniform distribution, so the p.d.f. is a rectangle with an area of 1.

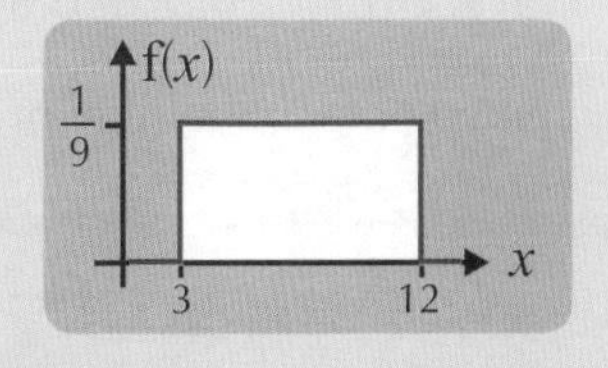

(b) You need to integrate to find E(X^2).

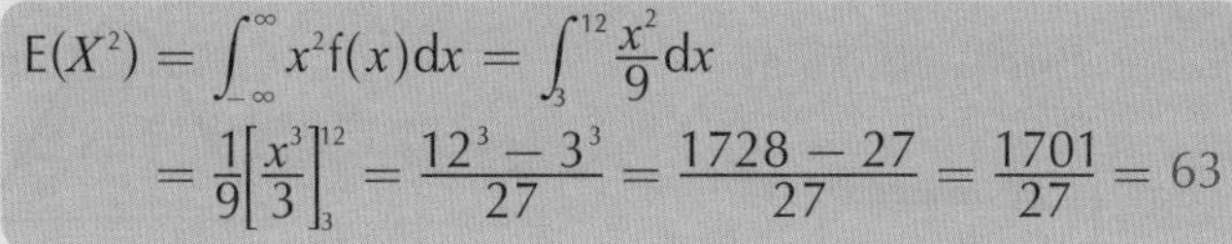

$$E(X^2) = \int_{-\infty}^{\infty} x^2 f(x)\,dx = \int_3^{12} \frac{x^2}{9}\,dx$$

$$= \frac{1}{9}\left[\frac{x^3}{3}\right]_3^{12} = \frac{12^3 - 3^3}{27} = \frac{1728 - 27}{27} = \frac{1701}{27} = 63$$

When I just can't do a question, I use X — where X ~ U[–500, 500]...

Questions on uniform distributions probably won't involve any really tricky maths, and they won't involve you using any of those awkward, fiddly tables. So they're potentially slightly easier marks... as long as you know what you're doing. So learn the stuff above, think carefully before you start pressing loads of buttons on your calculator, and Bob should be your uncle.

Normal Distributions

This page is for OCR S2 and OCR MEI S2

The normal distribution is everywhere in statistics. Everywhere, I tell you. So learn this well...

The Normal Distribution is 'Bell-Shaped'

1) Loads of things in real life are most likely to fall 'somewhere in the middle', and are much less likely to take extremely high or extremely low values. In this kind of situation, you often get a normal distribution.
2) If you were to draw a graph showing how likely different values are, you'd end up with a graph that looks a bit like a bell. There's a peak in the middle at the mean (or expected value). And the graph is symmetrical — so values the same distance above and below the mean are equally likely.

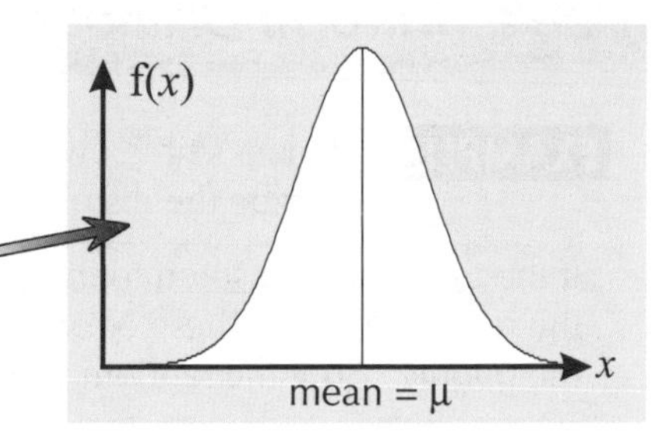

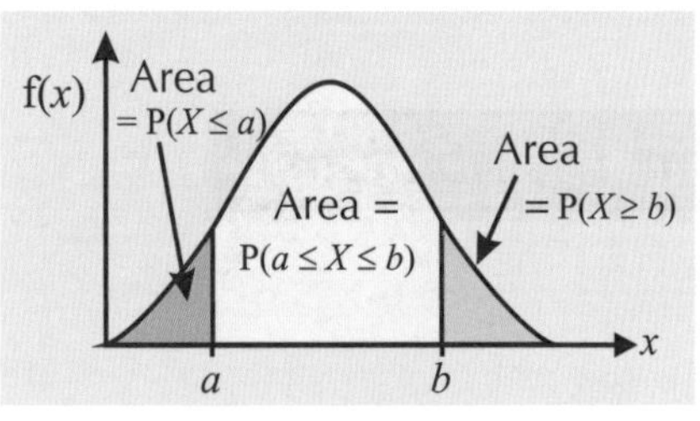

3) With a graph of a normal distribution, the probability of the random variable taking a value between two limits is the area under the graph between those limits. And since the total probability is 1, the total area under the graph must also be 1.

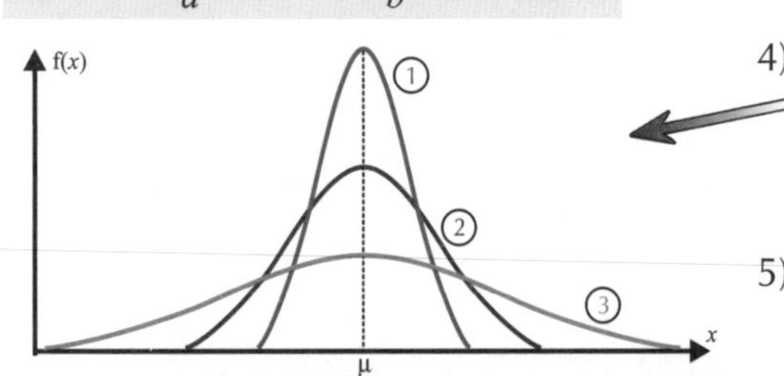

4) These three graphs all show normal distributions with the same mean (μ), but different variances (σ^2). Graph 1 has a small variance, and graph 3 has a larger variance — but the total area under all three curves is the same (= 1).
5) The most important normal distribution is the standard normal distribution, or Z — this has a mean of zero and a variance of 1.

Normal Distribution $N(\mu, \sigma^2)$

- If X is normally distributed with **mean** μ and **variance** σ^2, it's written $X \sim N(\mu, \sigma^2)$.
- The standard normal distribution Z has **mean** 0 and **variance** 1, i.e. $Z \sim N(0, 1)$.

Use Tables to Work Out Probabilities of Z

Working out the area under a normal distribution curve is usually hard. But for Z, there are tables you can use (see p160). You look up a value of z and these tables (labelled $\Phi(z)$) tell you the probability that $Z \le z$.

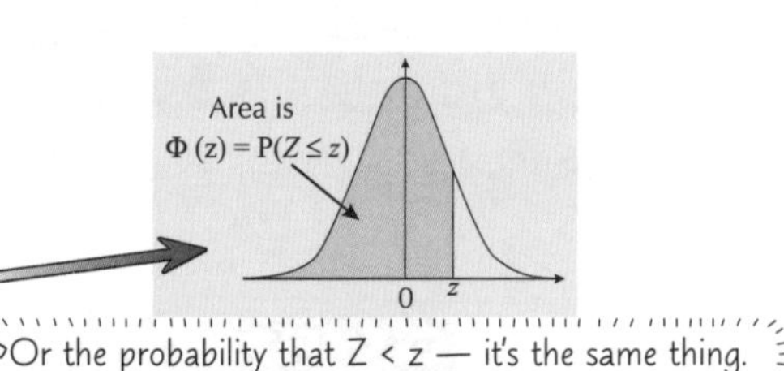

Or the probability that $Z < z$ — it's the same thing.

EXAMPLE Find the probability that:

a) $Z < 0.1$, b) $Z \le 0.64$, c) $Z > 0.23$, d) $Z \ge -0.42$, e) $Z \le -1.942$, f) $0.123 < Z \le 0.824$

Tables only tell you the probability of Z being less than a particular value — use a sketch to work out anything else.

a) $P(Z < 0.1) = 0.5398$ ← Just look up z = 0.1 in the tables.

b) $P(Z \le 0.64) = 0.7389$ ← For continuous distributions like Z (i.e. with 'no gaps' between its possible values): $P(Z \le 0.64) = P(Z < 0.64)$.

c) $P(Z > 0.23) = 1 - P(Z \le 0.23) = 1 - 0.5910 = 0.4090$

d) $P(Z \ge -0.42) = P(Z \le 0.42) = 0.6628$ ← Use the symmetry of the graph:

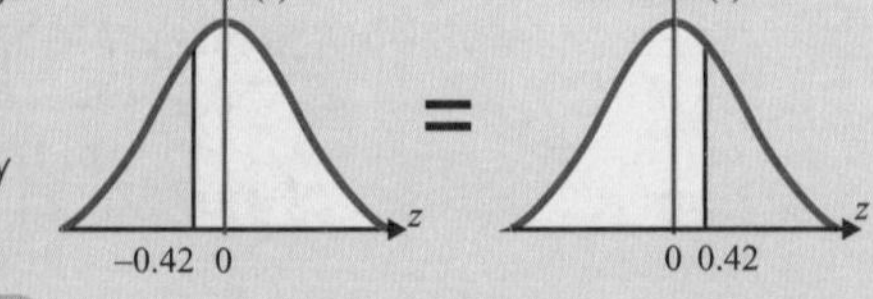

e) $P(Z \le -1.942) = P(Z \ge 1.942) = 1 - P(Z < 1.942) = 1 - 0.9739 = 0.0261$

f) $P(0.123 < Z \le 0.824) = P(Z \le 0.824) - P(Z \le 0.123) = 0.7950 - 0.5490 = 0.2460$

Again, draw a graph and use the symmetry:

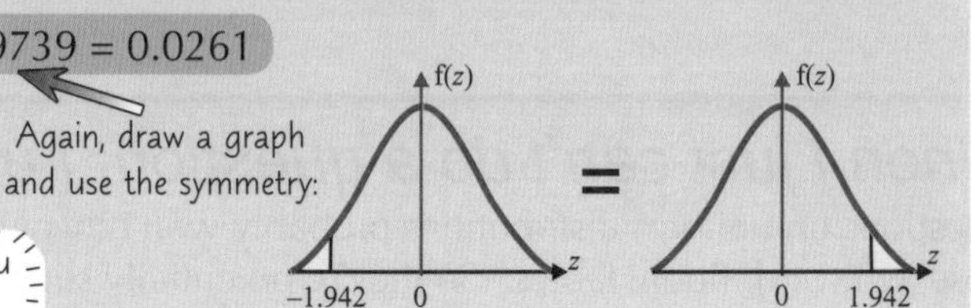

If you're doing OCR MEI, the real tables you'll get in the exam might give you slightly different values, but the method for reading them will be the same.

The Standard Normal Distribution, Z

This page is for OCR S2 and OCR MEI S2

You can use that big table of the normal distribution function ($\Phi(z)$) 'the other way round' as well — by starting with a probability and finding a value for z.

*Use **Tables** to Find a **Value** of z if You're Given a **Probability***

EXAMPLE If $P(Z < z) = 0.9554$, then what is the value of z?

Using the table for $\Phi(z)$ (the normal cumulative distribution function), $z = 1.700$.

All the probabilities in the table of $\Phi(z)$ are greater than 0.5, but you can still use the tables with values less than this. You'll most likely need to subtract the probability from 1, and then use a sketch.

EXAMPLE If $P(Z < z) = 0.2611$, then what is the value of z?

1. Subtract from 1 to get a probability greater than 0.5: $1 - 0.2611 = 0.7389$
2. If $P(Z < z) = 0.7389$, then from the table, $z = 0.640$.
3. So if $P(Z < z) = 0.2611$, then, $z = -0.640$.

If P(Z < z) = 0.2611, then z must be negative.

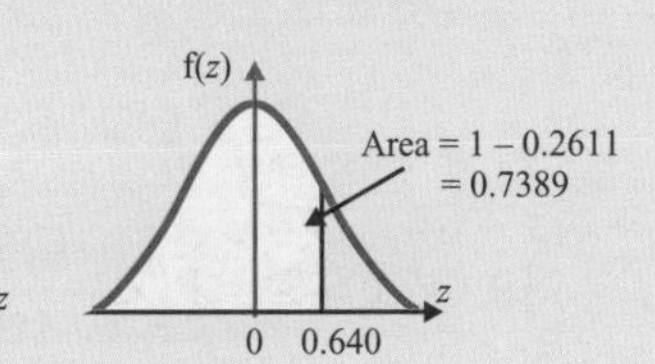

*The **Critical Values** (or **'Inverse'**) Table Tells You z if You're Given a **Probability***

You use the critical values table in a similar way — you start with a probability, and look up a value for z. Again, the probability you look up is the probability that z is less than a certain number.

EXAMPLE If $P(Z \leq z) = 0.9$, then what is the value of z?

Using the critical values table, $z = 1.282$.

Instead of a table of critical values, OCR MEI give you a table of the Inverse Normal function — it's a similar kind of thing though.

You might need to use a bit of imagination and a sketch to get the most out of the critical values table.

EXAMPLE Find z if: a) $P(Z < z) = 0.99$, b) $P(Z > z) = 0.1$, c) $P(Z < z) = 0.05$

a) If $P(Z < z) = 0.99$, then $z = 2.326$.

b) If $P(Z > z) = 0.1$, then $P(Z \leq z) = 0.9$. Using the critical values table, $z = 1.282$.

c) $p = 0.05$ isn't in the critical values table, so you have to look up $p = 1 - 0.05 = 0.95$ instead.

If $P(Z \leq z) = 0.95$, then $z = 1.645$, which means $P(Z > 1.645) = 0.05$.

Then $P(Z < -1.645) = 0.05$, so z must equal -1.645.

The medium of a random variable follows a paranormal distribution...

It's definitely worth sketching the graph when you're finding probabilities using a normal distribution — you're much less likely to make a daft mistake. So even if the question looks a simple one, draw a quick sketch — it's probably worth it.

Normal Distributions and Z-Tables

This page is for OCR S2 and OCR MEI S2

All normally-distributed variables can be transformed to Z — which is a marvellous thing.

Transform to Z by **Subtracting** μ, then **dividing by** σ

1) You can convert any normally-distributed variable to Z by:
   i) subtracting the mean, and then
   ii) dividing by the standard deviation.

This means that if you subtract μ from any numbers in the question and then divide by σ — you can use your tables for Z.

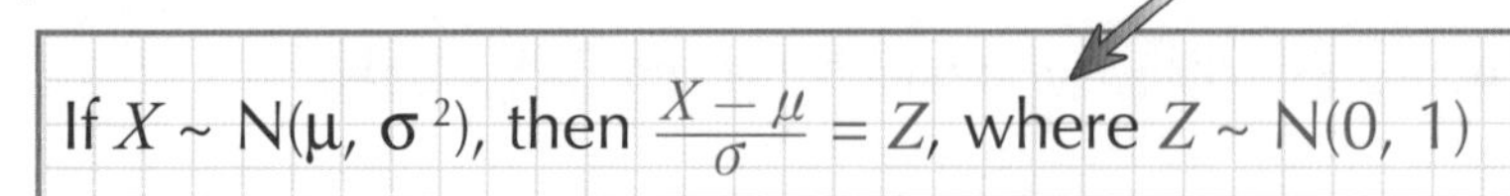

2) Once you've transformed a variable like this, you can use the Z-tables.

EXAMPLE If $X \sim N(5, 16)$ find: a) $P(X < 7)$, b) $P(X > 9)$, c) $P(5 < X < 11)$

Subtract μ (= 5) from any numbers and divide by σ (= $\sqrt{16}$ = 4) — then you'll have a value for $Z \sim N(0, 1)$.

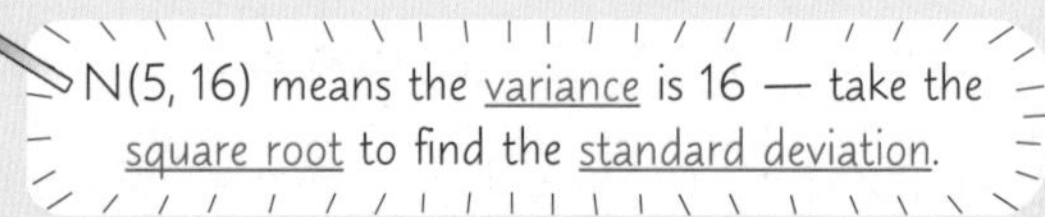

a) $P(X < 7) = P\left(Z < \frac{7-5}{4}\right) = P(Z < 0.5) = 0.6915$

Look up $P(Z < 0.5)$ in the big table on p.160.

b) $P(X > 9) = P\left(Z > \frac{9-5}{4}\right) = P(Z > 1) = 1 - P(Z < 1) = 1 - 0.8413 = 0.1587$

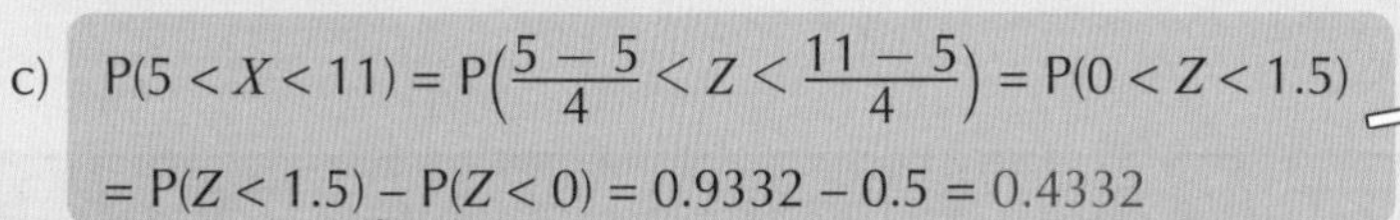

$= P(Z < 1.5) - P(Z < 0) = 0.9332 - 0.5 = 0.4332$

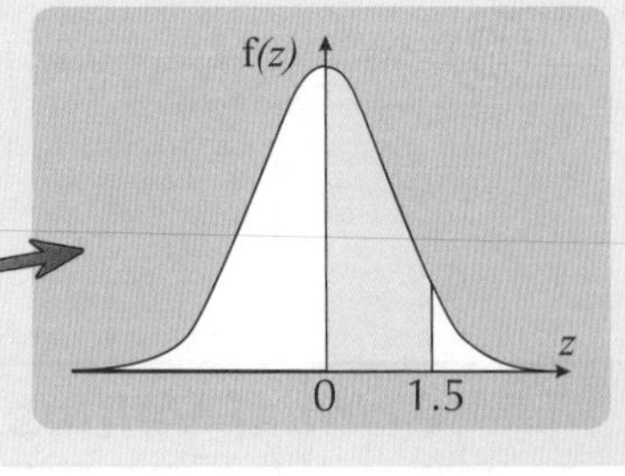

Find the area to the left of 1.5 and subtract the area to the left of 0.

The **Z-Distribution** Can be Used in **Real-Life** Situations

EXAMPLE The times taken by a group of people to complete an assault course are normally distributed with a mean of 600 seconds and a variance of 105 seconds. Find the probability that a randomly selected person took:
a) less than 575 seconds, b) more than 620 seconds.

If X represents the time taken in seconds, then $X \sim N(600, 105)$.
It's a normal distribution — so your first thought should be to try and 'standardise' it by converting it to Z.

a) Subtract the mean and divide by the standard deviation: $P(X < 575) = P\left(\frac{X-600}{\sqrt{105}} < \frac{575-600}{\sqrt{105}}\right) = P(Z < -2.440)$

So $P(Z < -2.440) = 1 - P(Z \leq 2.440) = 1 - 0.9927 = 0.0073$.

b) Again, subtract the mean and divide by the standard deviation:
$$P(X > 620) = P\left(\frac{X-600}{\sqrt{105}} > \frac{620-600}{\sqrt{105}}\right) = P(Z > 1.952) = 1 - P(Z \leq 1.952) = 1 - 0.9745 = 0.0255$$

Transform to Z and use Z-tables — I repeat: transform to Z and use Z-tables...

The basic idea is always the same — transform your normally-distributed variable to Z, and then use the Z-tables. Statisticians call this the "normal two-step". Well... some of them probably do, anyway. I admit, this stuff is all a bit weird and confusing at first. But as always, work through a few examples and it'll start to click. So get some practice.

Normal Distributions and Z-Tables

This page is for OCR S2 and OCR MEI S2

You might be given some probabilities and asked to find μ and σ. Just use the same old ideas...

Find μ and σ by First **Transforming** to Z

EXAMPLE $X \sim N(\mu, 2^2)$ and $P(X < 23) = 0.9015$. Find μ.

This is a normal distribution — so your first thought should be to convert it to Z.

① $P(X < 23) = P[\frac{X-\mu}{2} < \frac{23-\mu}{2}] = P[Z < \frac{23-\mu}{2}] = 0.9015$

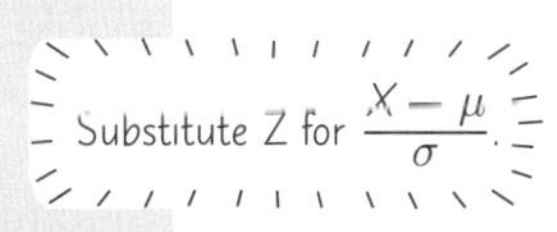

$p = 0.9015$ is one of the values in the table for $\Phi(z)$.

② If $P(Z < z) = 0.9015$, then $z = 1.290$

③ So $\frac{23-\mu}{2} = 1.290$ — now solve this to find $\mu = 23 - (2 \times 1.290) = 20.42$

EXAMPLE $X \sim N(53, \sigma^2)$ and $P(X < 50) = 0.1$. Find σ.

Again, this is a normal distribution — so you need to use that lovely standardising equation again.

① $P(X < 50) = P[Z < \frac{50-53}{\sigma}] = P[Z < -\frac{3}{\sigma}] = 0.1$.

Ideally, you'd look up 0.1 in the critical values table to find $-\frac{3}{\sigma}$.
Unfortunately, it isn't there, so you have to think a bit...

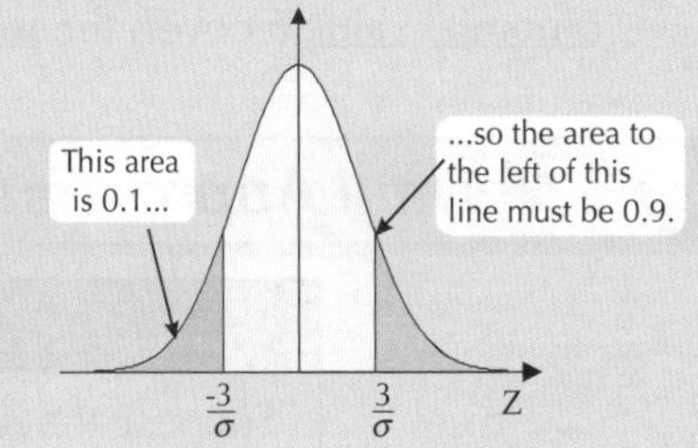

② $P[Z < -\frac{3}{\sigma}]$ is 0.1, so from the symmetry of the graph, $P[Z < \frac{3}{\sigma}]$ must be 0.9.

So look up 0.9 in the critical values table to find that:

$\frac{3}{\sigma} = 1.282$, or $\sigma = 2.34$ (to 3 sig. fig.)

If You Have to Find μ **and** σ, You'll Need to Solve **Simultaneous Equations**

EXAMPLE The random variable $X \sim N(\mu, \sigma^2)$. If $P(X < 9) = 0.5596$ and $P(X > 14) = 0.0322$, then find μ and σ.

① $P(X < 9) = P[Z < \frac{9-\mu}{\sigma}] = 0.5596$.

Using the table for $\Phi(z)$, this tells you that $\frac{9-\mu}{\sigma} = 0.150$, or $9 - \mu = 0.15\sigma$.

② $P(X > 14) = P[Z > \frac{14-\mu}{\sigma}] = 0.0322$, which means that $P[Z < \frac{14-\mu}{\sigma}] = 1 - 0.0322 = 0.9678$.

Using the table for $\Phi(z)$, this tells you that $\frac{14-\mu}{\sigma} = 1.850$, or $14 - \mu = 1.85\sigma$.

③ Subtract the equations: $(14 - \mu) - (9 - \mu) = 1.85\sigma - 0.15\sigma$, or $5 = 1.7\sigma$. This gives $\sigma = 5 \div 1.7 = 2.94$ (to 3 sig. fig.).
Now use one of the other equations to find μ: $\mu = 9 - (0.15 \times 2.94) = 8.56$ (to 3 sig. fig.).

The Norman distribution — came to England in 1066...

It's always the same — you always need to do the normal two-step of subtracting the mean and dividing by the standard deviation. Just make sure you don't use the variance by mistake — remember, in $N(\mu, \sigma^2)$, the second number always shows the variance. I'm sure you wouldn't make a mistake... it's just that I'm such a worrier and I do so want you to do well.

Normal Approximation to B(n, p)

This page is for Edexcel S2, OCR S2 and OCR MEI S2

If n is big, a binomial distribution (B(n, p)) can be tricky to work with. You saw the Poisson approximation on p112, but sometimes a normal distribution (Edexcel people, you met those in S1) is a better approximation. But there's a snag.

Use a **Continuity Correction** to Approximate a Binomial with a Normal

The binomial distribution is discrete but the normal distribution is continuous. To allow for this you need to use a continuity correction. Like a lot of this stuff, it sounds more complicated than it is.

- A binomially-distributed variable X is discrete, so you can work out P(X = 0), P(X = 1), etc.
- A normally-distributed variable is continuous, and so P(X = 0) = P(X = 1) = 0, etc.

So what you do is assume that the 'binomial 1' is spread out over the interval 0.5 - 1.5.

0 ⇐1⇒⇐2⇒⇐3⇒⇐4⇒
0.5 1.5 2.5 3.5 4.5

Then to approximate the binomial P(X = 1), you find the normal P(0.5 < X < 1.5).

Similarly, the 'binomial 2' is spread out over the interval 1.5 - 2.5, and so on.

Learn these **Continuity Corrections**

The interval you need to use with your normal distribution depends on the binomial probability you're trying to find out.

The general principle is the same, though — each binomial value b covers the interval from $b - \frac{1}{2}$ up to $b + \frac{1}{2}$.

Binomial	Normal	
P($X = b$)	P($b - \frac{1}{2} < X < b + \frac{1}{2}$)	
P($X \leq b$)	P($X < b + \frac{1}{2}$)	...to include b
P($X < b$)	P($X < b - \frac{1}{2}$)	...to exclude b
P($X \geq b$)	P($X > b - \frac{1}{2}$)	...to include b
P($X > b$)	P($X > b + \frac{1}{2}$)	...to exclude b

The **Normal Approximation** Only Works Well under **Certain Conditions**

Normal Approximation to the Binomial

Suppose the random variable X follows a binomial distribution, i.e. $X \sim$ **B(n, p)**.

If (i) $p \approx \frac{1}{2}$,

and (ii) n is large,

then $X \sim$ **N(np, npq)** (approximately), where $q = 1 - p$.

Since for a binomial distribution, $\mu = np$ and $\sigma^2 = npq$.

Even if p isn't all that close to 0.5, this approximation usually works fine as long as np and nq are both bigger than about 5.

EXAMPLE If $X \sim$ B(80, 0.4), use a suitable approximation to find: (i) P(X < 45) and (ii) P($X \geq 40$).

You need to make sure first that the normal approximation is suitable...

n is fairly large, and p is not far from $\frac{1}{2}$, so the normal approximation is valid.

Next, work out np and npq: $np = 80 \times 0.4 = 32$ and $npq = 80 \times 0.4 \times (1 - 0.4) = 19.2$ ← $q = 1 - p$

So the approximation you need is: $X \sim$ N(32, 19.2) ← So the standard deviation is $\sqrt{19.2}$.

Now apply a continuity correction, transform the variable to the standard normal distribution (Z), and use tables.

See page 160 for the normal distribution tables.

(i) You need P(X < 45) — so with the continuity correction this is P(X < 44.5).

$$P(X < 44.5) = P\left(\frac{X - 32}{\sqrt{19.2}} < \frac{44.5 - 32}{\sqrt{19.2}}\right) = P(Z < 2.853) = 0.9978$$

(ii) Now you need P($X \geq 40$) — with the continuity correction this is P(X > 39.5).

$$P(X > 39.5) = P\left(\frac{X - 32}{\sqrt{19.2}} > \frac{39.5 - 32}{\sqrt{19.2}}\right) = P(Z > 1.712) = 1 - P(Z \leq 1.712) = 1 - 0.9566 = 0.0434$$

Normal Approximation to B(n, p)

This page is for Edexcel S2, OCR S2 and OCR MEI S2

I know what you're thinking — you want to know just how good a normal approximation actually is. Well, let's see...

EXAMPLE: Newborn babies

The average number of births per year in a hospital is 228. If each baby is equally likely to be a boy or a girl, then use a suitable approximation to find the probability that next year:

(i) there will be more boys born than girls,

(ii) exactly 100 boys will be born.

Mean = np = 228 × 0.5 = 114
Variance = npq = 228 × 0.5 × 0.5 = 57

If X represents the number of boys born next year, then you can assume that $X \sim B(228, 0.5)$.
Since n is large, and p is 0.5, then you can use a normal approximation: $X \sim N(114, 57)$.

(i) You need to find $P(X > 114)$. With a continuity correction, this is $P(X > 114.5)$.
It's a normal distribution, so transform this to Z, the standard normal distribution.

$$P(X > 114.5) = P\left(\frac{X-114}{\sqrt{57}} > \frac{114.5-114}{\sqrt{57}}\right) = P(Z > 0.066)$$
$$= 1 - P(Z \leq 0.066) = 1 - 0.5263 = 0.4737$$

Using B(228, 0.5) instead of the normal approximation, you get 0.4736 — so this is a really good approximation.

(ii) With a continuity correction,
you need to find $P(99.5 < X < 100.5)$.

$$P(99.5 < X < 100.5) = P(X < 100.5) - P(X < 99.5)$$
$$= P\left(\frac{X-114}{\sqrt{57}} < \frac{100.5-114}{\sqrt{57}}\right) - P\left(\frac{X-114}{\sqrt{57}} < \frac{99.5-114}{\sqrt{57}}\right)$$
$$= P(Z < -1.788) - P(Z < -1.921)$$
$$= (1 - P(Z < 1.788)) - (1 - P(Z < 1.921))$$
$$= 1 - 0.9632 - 1 + 0.9727 = 0.0095$$

Using B(228, 0.5) instead of the normal approximation, you also get 0.0095.

Edexcel people: note that the tables you'll get in the exam are slightly different — you'll only be asked to look up values of z to 2 d.p., rather than 3 d.p.

EXAMPLE: Survival rates

a) On average, only 23% of the young of a particular species of bird survive to adulthood. If 80 chicks of this species are randomly selected, use a suitable approximation to find the probability that at least 30% of them survive.

b) If the survival rate were instead 18%, find the probability that more than three-quarters of the 80 chicks would die.

If X represents the number of survivors, then $X \sim B(80, 0.23)$.
Here, p isn't particularly close to 0.5, but n is quite large, so calculate np and nq:
$np = 80 \times 0.23 = 18.4$ and $nq = 80 \times (1 - 0.23) = 61.6$.

Both np and nq are much greater than 5, so a normal approximation should be okay to use — N(18.4, 14.168).

30% of 80 = 24, so with a continuity correction, you need to find $P(X > 23.5)$.

Mean = np = 80 × 0.23 = 18.4
Variance = npq = 80 × 0.23 × 0.77 = 14.168

$$P(X > 23.5) = P\left(Z > \frac{23.5-18.4}{\sqrt{14.168}}\right) = P(Z > 1.355) = 1 - P(Z \leq 1.355)$$
$$= 1 - 0.9123 = 0.0877$$

Using the original binomial distribution gives an answer of 0.0904, so this is a fairly good approximation.

b) This time, $X \sim B(80, 0.18)$, which means $np = 80 \times 0.18 = 14.4$ and $nq = 80 \times (1 - 0.18) = 65.6$.
So even though p is now quite far from 0.5, try the normal approximation — N(14.4, 11.808).
If more than three-quarters do not survive, that means $X < 20$, so you need to find $P(X < 19.5)$.

$$P(X < 19.5) = P\left(Z < \frac{19.5-14.4}{\sqrt{11.808}}\right) = P(Z < 1.484) = 0.9312$$

Using the original binomial distribution gives an answer of 0.9270, so this is another pretty good approximation.

Admit it — the normal distribution is the most amazing thing ever...

So the normal approximation works pretty well, even when p isn't really all that close to 0.5. But even so, you should always show that your approximation is 'suitable'. In fact, the question will usually tell you to use a 'suitable approximation', so part of your answer should be to show that you've made sure that it is actually okay. Remember that.

Normal Approximation to Po(λ)

This page is for Edexcel S2, OCR S2 and OCR MEI S2

More approximations, I'm afraid. But on the bright side, Mr Poisson is back. Good old Mr Poisson.

The Normal Approximation to **Po(λ)** Works Best if λ is **Big**

Normal Approximation to the Poisson Distribution

Suppose the random variable X follows a Poisson distribution, i.e. $X \sim \text{Po}(\lambda)$.
If λ is large, then (approximately) $X \sim N(\lambda, \lambda)$.

Since for a Poisson distribution, mean = variance = λ (see page 109).

Ideally, you want λ 'as large as possible' — but in practice as long as $\lambda > 10$, then you're fine.

Use a **Continuity Correction** to Approximate a Poisson with a Normal

Since a Poisson distribution is discrete (it can only take values 0, 1, 2...) but a normal distribution is continuous, you need to use a continuity correction. (See p132 for more about continuity corrections.)

EXAMPLE If $X \sim \text{Po}(49)$, find: a) $P(X < 50)$, b) $P(X \geq 45)$, c) $P(X = 60)$.

Since λ is large (it's greater than 10), you can use a normal approximation — $X \sim N(49, 49)$.

a) The continuity correction means you need to find $P(X < 49.5)$.

Transform to Z and use tables: $P(X < 49.5) = P\left(\frac{X-49}{7} < \frac{49.5-49}{7}\right) = P(Z < 0.071) = 0.5283$

b) This time you need to find $P(X > 44.5)$: $P(X > 44.5) = P\left(Z > \frac{44.5-49}{7}\right) = P(Z > -0.643)$
$= 1 - P(Z \leq -0.643) = P(Z \leq 0.643) = 0.7399$

c) $P(X = 60) = P(59.5 < X < 60.5) = P(X < 60.5) - P(X < 59.5)$
$= P\left(Z < \frac{60.5-49}{7}\right) - P\left(Z < \frac{59.5-49}{7}\right) = P(Z < 1.643) - P(Z < 1.5)$
$= 0.9498 - 0.9332 = 0.0166$

EXAMPLE A sloppy publishing company produces books containing an average of 25 random errors per page.
Use a suitable approximation to find the probability of: a) fewer than 20 errors on a particular page,
b) exactly 25 errors on a particular page.

The errors happen randomly, singly and (on average) at a constant rate, and so the number of errors that occur on a single page (X) will follow a Poisson distribution. Since there's an average of 25 errors per page, $X \sim \text{Po}(25)$.

a) Since λ is large (greater than 10), you can use a normal approximation — i.e. $X \sim N(25, 25)$.
You need to use a continuity correction here, so P(fewer than 20 errors) ≈ $P(X < 19.5)$.

Transform to Z and use tables: $P(X < 19.5) = P\left(\frac{X-25}{5} < \frac{19.5-25}{5}\right) = P(Z < -1.1)$
$= 1 - P(Z < 1.1)$
$= 1 - 0.8643 = 0.1357$

b) P(exactly 25 errors on a page) ≈ $P(24.5 < X < 25.5) = P(X < 25.5) - P(X < 24.5)$.

Transform to Z and use tables: $P(X < 25.5) - P(X < 24.5) = P\left(Z < \frac{25.5-25}{5}\right) - P\left(Z < \frac{24.5-25}{5}\right)$
$= P(Z < 0.1) - P(Z < -0.1)$
$= P(Z < 0.1) - (1 - P(Z < 0.1)) = 0.5398 - (1 - 0.5398) = 0.0796$

The abnormal approximation — guess wildly then say 'Close enough'...

Lordy, lordy... the number of times you've had to read 'transform something to Z and use tables'. But that's the thing... if you can find probabilities from a normal distribution, then it's bound to be worth marks in the exam. On a different note, continuity corrections are fairly easy to use — it's remembering to use one in the first place that can be a bit tricky.

More About Approximations

This page is for Edexcel S2, OCR S2 and OCR MEI S2

You need to know a few different approximations for S2 — and to be honest, it can all get a bit confusing. But although the picture below looks like a complex wiring diagram, it's actually an easy-to-use flowchart to sum up your options.

Approximate if You're **Told to**, or if Your **Tables** 'Don't Go High Enough'

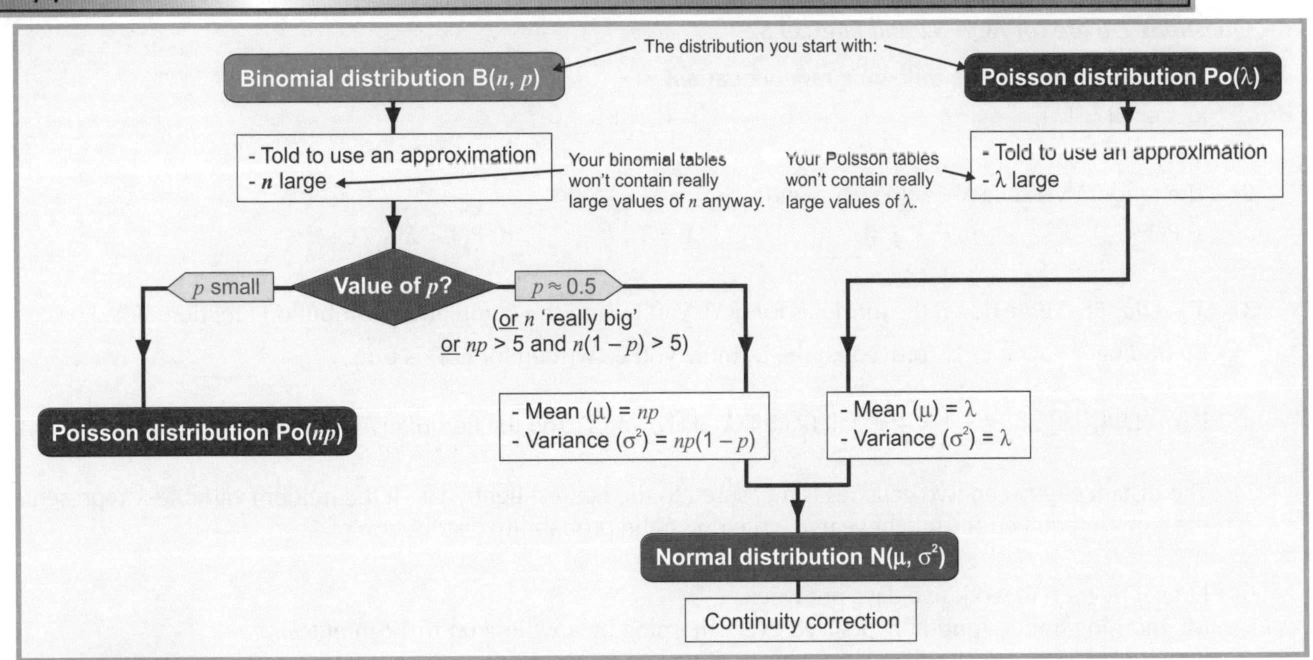

EXAMPLE:

A supermarket gives a customer a free bag when that customer can fit no more items into their existing bags. The number of bags given away was counted. It was found that 62% of customers needed at least one free bag, 2% of customers needed more than 5 free bags, and an average of 8 bags were given away each minute.

If the supermarket has 400 customers one particular morning, use suitable approximations to find:

a) the probability that more than 250 customers take at least one bag,

b) the probability that fewer than 10 customers take more than 5 bags,

c) the probability that more than 500 bags are given away in the first hour after the store opens.

a) Let X represent how many of the 400 customers take at least one bag.
Then $X \sim B(400, 0.62)$, and you need to find $P(X > 250)$.
Here, n is large and p is not too far from 0.5 — so approximate X with $N(248, 94.24)$.

$$P(X > 250) \approx P(X > 250.5) = P\left(\frac{X-248}{\sqrt{94.24}} > \frac{250.5-248}{\sqrt{94.24}}\right) = P(Z > 0.258) = 1 - P(Z \leq 0.258) = 0.3982$$

b) Let M represent the total number of customers taking more than 5 bags.
Then $M \sim B(400, 0.02)$, and you need to find $P(M < 10)$.
Here, n is large and p is very small — so use a Poisson approximation: $Po(8)$

$$P(M < 10) = P(M \leq 9) = 0.7166$$

Bags given away per minute ~ Po(8).
So bags per hour ~ Po(8 × 60).

c) Let R represent the number of bags given away in the first hour. This is not a fixed number of trials, so it's not a binomial distribution. But the period is fixed (1 hour), so it's Poisson — in fact, $R \sim Po(480)$.
You need to find $P(R > 500)$. Approximate R with $N(480, 480)$.

$$P(R > 500) \approx P(R > 500.5) = P\left(\frac{R-480}{\sqrt{480}} > \frac{500.5-480}{\sqrt{480}}\right) = P(Z > 0.936) = 1 - 0.8253 = 0.1747$$

Check your p — then decide what to do...

The trickiest decision you face when approximating is whether to approximate $B(n, p)$ with a Poisson or a normal distribution — and it all depends on p really. So once you've made that decision, you should be off and running. But don't forget that continuity correction if you're using the normal approximation. You have been warned.

S2 Section 5 — Practice Questions

You've come this far... don't give up now... only two more pages to go. And they're only questions — so it's not like there's loads more you're going to need to cram into your already crowded head. Loins girded? Good, here we go...

Warm-up Questions

Questions 1-6 are for AQA S2 and Edexcel S2

1) Sketch the p.d.f.s of the following random variables:
   a) $X \sim U[7, 11]$, b) $Y \sim U[-4, 18]$.

2) The random variable $X \sim U[0, 10]$. Find:
   a) $P(X < 4)$, b) $P(X \geq 8)$, c) $P(X = 5)$, d) $P(3 < X \leq 7)$.

3) $X \sim U[a, b]$. Write down the formulas for E(X), Var(X), and the cumulative distribution function of X, F(x).
   By finding $\int_{-\infty}^{\infty} x \cdot f(x)dx$, prove that the formula you've written for E(X) is true.

4) If $X \sim U[4, 19]$ and $Y = 6X - 3$, calculate E(X), E(Y), Var(X) and the cumulative distribution function of X, F(x).

5) The distance between two galaxies is measured to the nearest light year. If the random variable X represents the experimental error (in light years), write down the probability distribution of X.

6) I travel by train to work five days per week.
   My morning train is randomly delayed every morning by anything up to 12 minutes.
   If the delay is any greater than 8 minutes, I arrive late for work.
   a) Find the probability that I am late for work on a randomly chosen day during a particular week.
   b) Find the probability that I am late for work more than once during a single week.

Questions 7-10 are for OCR S2 and OCR MEI S2

7) Find the probability that: a) $Z < 0.84$, b) $Z \geq 1.55$, c) $Z < -2.10$, d) $0.102 < Z \leq 0.507$.

8) Find the value of z if: a) $P(Z < z) = 0.9131$, b) $P(Z > z) = 0.0359$.

9) If $X \sim N(50, 16)$ find: a) $P(X < 55)$, b) $P(X < 42)$, c) $P(X > 56)$, d) $P(47 < X < 57)$.

10) The random variable $X \sim N(\mu, \sigma^2)$. If $P(X < 15.2) = 0.9783$ and $P(X > 14.8) = 0.1056$, then find μ and σ.

Questions 11-13 are for Edexcel S2, OCR S2 and OCR MEI S2

11) The random variable X follows a binomial distribution: $X \sim B(100, 0.45)$.
    Using a normal approximation and continuity corrections, find:
    a) $P(X > 50)$, b) $P(X \leq 45)$, c) $P(40 < X \leq 47)$.

12) The random variable X follows a Poisson distribution: $X \sim Po(25)$.
    Using a normal approximation and continuity corrections, find:
    a) $P(X \leq 20)$, b) $P(X > 15)$, c) $P(20 \leq X < 30)$.

13) Seven people on average join the queue in the local post office every 15 minutes during the 7 hours it is open. The number of people working in the post office is constantly adjusted depending on how busy it is, with the result that there is a constant probability of 0.7 of any person being served within 1 minute.
    a) Find the probability of more than 200 people joining the queue in the post office on a particular day.
    b) If exactly 200 people come to the post office on a particular day, what is the probability that less than 70% of them are seen within a minute?

S2 Section 5 — Practice Questions

One last hurdle before you can consider yourself fully up to speed with continuous distributions...

Exam Questions

Question 1 is for AQA S2 and Edexcel S2

1 A 40 cm length of ribbon is cut in two at a random point.
The random variable X represents the length of the shorter piece of ribbon in cm.

a) Specify the probability distribution of X. *(2 marks)*

b) Sketch the probability density function of X. *(1 mark)*

c) Calculate $E(X)$ and $Var(X)$. *(3 marks)*

d) Find:
(i) $P(X > 5)$, *(1 mark)*

(ii) $P(X = 2)$. *(1 mark)*

Questions 2-4 are for Edexcel S2, OCR S2 and OCR MEI S2

2 The random variable X is binomially distributed with $X \sim B(100, 0.6)$.

a) (i) State the conditions needed for X to be well approximated by a normal distribution. *(2 marks)*

(ii) Explain why a continuity correction is necessary in these circumstances. *(2 marks)*

b) Using a suitable approximation, find:
(i) $P(X \geq 65)$ *(4 marks)*

(ii) $P(50 < X < 62)$ *(3 marks)*

3 The random variable X follows a binomial distribution: $X \sim B(n, p)$.
X is approximated by the normally distributed random variable Y.
Using this normal approximation, $P(X \leq 151) = 0.8944$ and $P(X > 127) = 0.9970$.

a) Find the mean and standard deviation of the normal approximation. *(8 marks)*

b) Use your results from a) to find n and p. *(4 marks)*

4 A factory has the capacity to increase its output by 50 items per week. A potential new customer has said it could sign a contract to order an average of 40 items per week, although the exact number of items needed each week will vary according to a Poisson distribution.

a) Specify a distribution that could be used to model the number of extra items that will be ordered per week if the new contract is signed. *(1 mark)*

b) Using a suitable approximation, find the probability that the number of items the potential new customer will order in a given week will exceed the factory's spare capacity. *(3 marks)*

c) The contract says that if the factory does not meet the new customer's order in two consecutive weeks, the factory must pay compensation. The factory's manager decides that he will only sign the contract if the probability of having to pay compensation in a given week is less than 0.01. Should the factory's manager sign the contract? Explain your answer. *(1 mark)*

Question 5 is for OCR S2 and OCR MEI S2

5 The lifetimes of a particular type of battery are normally distributed with mean μ and standard deviation σ. A student using these batteries finds that 25% last less than 20 hours and 90% last less than 30 hours. Find μ and σ. *(7 marks)*

Populations and Samples

This page is for Edexcel S2 and OCR S2

No time for any small talk, I'm afraid. It's straight on with the business of populations and how to find out about them.

A Population is a Group of people or items

For any statistical investigation, there will be a group of people or items that you want to find out about. This group is called the population.

This could be:

- All the students in a maths class
- All the penguins in Antarctica
- All the chocolate puddings produced by a company in a year

1) Populations are said to be finite if it's possible for someone to count how many members there are. E.g. all the students in a maths class would be a finite population.
2) If it's impossible to know exactly how many members there are, the population is said to be infinite.

A population might have a finite number of members in theory, but if it's impossible to count them all in practice, the population is infinite.

To collect information about your population, you can carry out a survey. This means questioning the people or examining the items.

E.g. all the blades of grass in a field would be an infinite population (even though you could count them all in theory).

A Census surveys the Whole Population

When you collect information from every member of a population, it's called a census.

To do this, your population needs to be finite. It also helps if it's fairly small and easily accessible — so that getting information from every member is a straightforward task.

You need to know the advantages and disadvantages of doing a census, so here they are:

Census — Advantages

1) You get accurate information about the population, because every member has been surveyed.
2) It's a true representation of the population — it's unbiased.

See the next page for more on bias.

Census — Disadvantages

1) For large populations, it takes a lot of time and effort to carry out.
2) This makes it expensive to do.
3) It can be difficult to make sure all members are surveyed. If some are missed, bias can creep in.
4) If the tested items are used up or damaged in some way, a census is impractical.

Watch out for anything that might make doing a census a silly idea.

A Sample is Part of a Population

If doing a census is impossible or impractical, you can find out about a population by questioning or examining just a selection of the people or items. This selected group is called a sample.

Before selecting your sample, you need to identify the sampling units — these are the individual members of the population. A full list of all the sampling units is called a sampling frame. This list must give a unique name or number to each sampling unit, and is used to represent the population when selecting a random sample (see the next page).

EXAMPLE A company produces 100 chocolate puddings every day, and each pudding is labelled with a unique product number. Every day, a sample of 5 puddings is eaten as part of a quality control test.

a) Why is it necessary for the company to take a sample rather than carry out a census?

If they ate all the puddings, there would be none left to sell.

b) Identify the sampling units.

The individual puddings

c) Suggest a sampling frame.

A list of all one hundred unique product numbers.

A sampling frame can only be produced when you know exactly who or what makes up the population.

I wouldn't mind sampling some chocolate puddings...

A slow start to the section, I'll grant you that. But just because there aren't any sums to do doesn't mean you can skim over this stuff. It's the basis of what's to come. And they like to ask you to identify sampling frames and units in exam questions.

Sampling

This page is for Edexcel S2 and OCR S2

This page is about the three Rs — Representative samples, Random sampling, and Raccoons...

A **Sample** needs to be **Representative** of its Population

Data collected from a sample is used to draw conclusions about the whole population (see p.140). So, it's vital that the sample is as much like the population as possible. A biased sample is one which doesn't fairly represent the population.

To Avoid Sampling Bias:

1. Select from the correct population and make sure none of the population is excluded — that means drawing up an accurate sampling frame and sticking to it. E.g. if you want to find out the views of residents from a particular street, your sample should only include residents from that street and should be chosen from a full list of the residents.
2. Select your sample at random — see below. Non-random sampling methods include things like the sampler just asking their friends — who might all give similar answers, or asking for volunteers — meaning they only get people with strong views on a subject.
3. Make sure all your sample members respond — otherwise the results could be biased. E.g. if some of your sampled residents are out when you go to interview them, it's important that you go back and get their views another time.

In **Random Sampling**, all Sampling Units are **Equally Likely** to be selected

In a Simple Random Sample...

- Every person or item in the population has an equal chance of being in the sample.
- Each selection is independent of every other selection.

Every single possible sample is equally likely.

1) To get a truly random sample, you need a complete list of the population — an accurate sampling frame.
2) To choose the sample, give every sampling unit a number. Then generate a list of random numbers and match them to the sampling units to select your sample.

Use a computer, calculator, dice or random number tables...

EXAMPLE A zoo has 80 raccoons. Describe how the random number table opposite could be used to select a sample of three of them, for a study on tail lengths.

8330	3992	1840
0330	1290	3237
9165	4815	0766

A list of all 80 raccoons.

1) First, draw up a sampling frame, giving each raccoon a two-digit number between 01 and 80.
2) Then, find the first three numbers between 01 and 80 from the table, (30, 39 and 18), and select the raccoons with the matching numbers.

You need to be able to **Justify choosing a Sample** over a Census

In most situations, it's more practical to survey a sample rather than carry out a census — make sure you can explain why. But remember, the downside is that your results might not be as reliable, either due to sampling bias, or just the natural variability between samples (see below).

Sampling — Advantages

1) Quicker and cheaper than a census, and easier to get hold of all the required information.
2) It's the only option when surveyed items are used up or damaged.

Sampling — Disadvantages

1) Variability between samples — each possible sample will give different results, so you could just happen to select one which doesn't accurately reflect the population. E.g. you could randomly pick the 3 raccoons with the longest tails in one sample and the 3 with the shortest tails in another. One way to reduce the likelihood of large variability is by using a large sample size. The more sampling units that are surveyed, the more reliable the information should be.
2) Samples can easily be affected by sampling bias.

A sample should be a miniature version of the population...

Using simple random sampling should reduce the risk of sampling bias, and so increase the likelihood of reliable results. Random sampling also means that the selected observations are independent random variables with the same distribution as the population. This is an important assumption for the theory in this section, and you'll see more about it on the next page.

Statistics and Sampling Distributions

This page is for Edexcel S2 and OCR S2

You're nearly at the end of A2 Statistics, so it's time to say what a statistic actually is.

Statistics are Calculated Using Only the Values from your Sample

Statistics

A statistic is a quantity calculated only from the known observations in a sample.
A statistic is a random variable — it takes different values for different samples.

It's important you can recognise what's a statistic and what's not...

EXAMPLE A random sample $X_1, ..., X_{10}$ is taken from a population with unknown mean, μ. State whether or not the following are statistics:

a) $X_{10} - X_1$ b) $\sum X_i$ c) $\sum X_i^2 - \mu$

a) Yes, b) Yes, c) No

c) depends on the unknown value of μ.

Back to what I was saying at the bottom of the last page... The 10 observations are independent random variables, defined as X_1, X_2, etc., and each takes a value from the population. A statistic calculated from these observations is also a random variable — its value depends on the values in the sample.

Parameters are quantities that describe the characteristics of a population — e.g. the mean, variance, or proportion that satisfy certain criteria. Certain statistics turn out to be very useful for testing theories about population parameters, or for estimating their values.

Every Statistic has a Sampling Distribution

Edexcel only

1) If you took a sample of observations and calculated a particular statistic, then took another sample and calculated the same statistic, then took another sample, etc., you'd end up with lots of values of the same statistic.
2) The probability distribution of a statistic is called the sampling distribution — it gives all the possible values of the statistic, along with the corresponding probabilities.

EXAMPLES

① A pirate's treasure chest contains a large number of coins. Unfortunately, the pirate has been diddled and the chest only contains 5p and 10p coins. The ratio of 5p to 10p coins is 4:1.

a) A random sample of 2 coins is taken from the chest. List all the possible samples.
b) Let X_i represent the ith coin in the sample. Find the probability distribution of X_i.
c) Find the sampling distribution of the sample mean $\bar{X} = \frac{\sum X_i}{n}$.

You can use the same general method for other statistics, e.g. the median.

a) The possible samples are: (5, 5), (5, 10), (10, 5), (10, 10)

b) The probability distribution specifies all the possible values X_i could take, and the probabilities of each value:

x_i	5	10
$P(X_i = x_i)$	0.8	0.2

The ratio 4:1 gives P(5) = 4/5 and P(10) = 1/5 — that's 0.8 and 0.2 as decimals.

c) Calculate the mean for each sample — e.g. (5, 5) gives (5 + 5)/2 = 5. Then work out the probabilities:
$P(\bar{X} = 5) = P(5, 5) = 0.8 \times 0.8 = 0.64$ ← You can multiply the probabilities because X_1 and X_2 are independent.
$P(\bar{X} = 7.5) = P(5, 10) \text{ or } P(10, 5) = (0.8 \times 0.2) + (0.2 \times 0.8) = 0.32$
$P(\bar{X} = 10) = P(10, 10) = 0.2 \times 0.2 = 0.04$
So the sampling distribution of $\bar{X}$ is

$\bar{x}$	5	7.5	10
$P(\bar{X} = \bar{x})$	0.64	0.32	0.04

Check that these probabilities add up to 1 — if not, you've made a mistake.

② A company makes celebrity-themed coat hangers. A constant proportion of the coat hangers are rejected for being too unrealistic. A random sample of 20 coat hangers is inspected. The random variable X_i is defined as: $X_i = 0$ if coat hanger i is acceptable, and $X_i = 1$ if it's faulty. Write down the sampling distribution of the statistic $Y = \sum X_i$.

Each coat hanger is an independent trial with constant probability, p, of being faulty, and Y is the number of faulty items. So, Y follows a binomial distribution — $Y \sim B(20, p)$ ← The proportion is unknown, so just call it p.

Knowing the sampling distribution of a statistic means you can use that statistic's value to say something about the population the sample was taken from (see page 144).

Statistics and Sampling Distributions

This page is for OCR S2 and OCR MEI S2

Now it's time to meet the stars of our statistics show — the sample mean $\overline{X}$, and partners μ and $\frac{\sigma^2}{n}$.

If X follows a Normal Distribution, the Sampling Distribution of $\overline{X}$ is Normal too

1) Suppose you've got a random variable X, with mean μ and variance σ^2.
2) When you take a sample of n readings from the distribution of X, you can work out the sample mean $\overline{X}$. If you now keep taking samples of size n from that distribution and working out the sample means, then you get a collection of sample means, drawn from the sampling distribution of $\overline{X}$. ($\overline{X}$ is a statistic.)
3) This sampling distribution also has mean μ, but the variance is $\frac{\sigma^2}{n}$.
4) And if X follows a normal distribution, then the sampling distribution of $\overline{X}$ will also be normal.

If $X \sim N(\mu, \sigma^2)$, then $\overline{X} \sim N\left(\mu, \frac{\sigma^2}{n}\right)$ ← This is the sampling distribution of $\overline{X}$.

EXAMPLE A continuous random variable X has the distribution $X \sim N(20, 16)$. The mean of a sample of 10 observations of X is defined as $\overline{X}$. Write down the sampling distribution of $\overline{X}$.

$X \sim N(20, 16)$, so $\overline{X} \sim N\left(20, \frac{16}{10}\right) \Rightarrow \overline{X} \sim N(20, 1.6)$

EXAMPLE A continuous random variable X has the distribution $N(45, 25)$. If $\overline{X}$ is the mean of n observations of X and $P(\overline{X} < 45.5) = 0.8413$, find the value of n.

① $X \sim N(45, 25)$, so $\overline{X} \sim N\left(45, \frac{25}{n}\right)$. Since $\overline{X}$ has a normal distribution, you can standardise it and use tables for Z (see p.130). Subtract the mean and divide by the standard deviation to get:

$$P(\overline{X} < 45.5) = P\left(Z < \frac{45.5 - 45}{\sqrt{25/n}}\right) = 0.8413$$

$\frac{\overline{X} - 45}{\sqrt{25/n}} \sim N(0, 1)$

(Here you're transforming $\overline{X}$ rather than an individual observation.)

② Looking up $P(Z < z) = 0.8413$ in the tables, you find that $z = 1$ — and so:

$$\frac{0.5}{5/\sqrt{n}} = 1 \Rightarrow 0.5 = \frac{5}{\sqrt{n}} \Rightarrow \sqrt{n} = 10 \Rightarrow n = 100$$

If you're doing OCR MEI, advance to p.142. If you pass 'Go', collect £200. OCR people need to know the following...

Even when X Isn't Normally Distributed, $\overline{X} \sim N(\mu, \sigma^2/n)$ (approx.) for Large n

The Central Limit Theorem tells you something about $\overline{X}$ — even if you know nothing at all about the distribution of X.

The Central Limit Theorem

Suppose you take a sample of n readings from any distribution with mean μ and variance σ^2.

- For large n, the distribution of the sample mean, $\overline{X}$, is approximately normal: $\overline{X} \sim N\left(\mu, \frac{\sigma^2}{n}\right)$.
- The bigger n is, the better the approximation will be. (For $n > 30$ it's pretty good.)

EXAMPLE A sample of size 50 is taken from a population with mean 20 and variance 10. Find the probability that the sample mean is less than 19.

① Since n (= 50) is quite large, you can use the Central Limit Theorem. Here $\overline{X} \sim N\left(20, \frac{10}{50}\right) = N(20, 0.2)$ (approx).

② You need $P(\overline{X} < 19)$. Standardising gives: $P(\overline{X} < 19) = P\left(Z < \frac{19 - 20}{\sqrt{0.2}}\right) = P(Z < -2.236)$

③ Now you can use your normal-distribution tables (see p160).

$P(Z < -2.236) = P(Z > 2.236) = 1 - P(Z \leq 2.236) = 1 - 0.9873 = 0.0127$

As always... if you need to, draw a sketch.

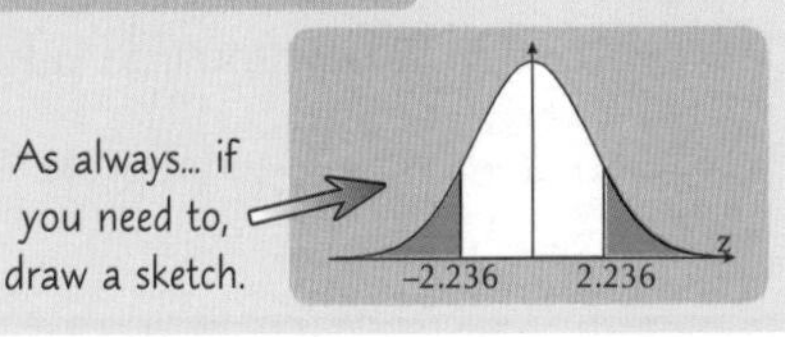

So if in doubt, it's probably normal then...

Now, you can't say the results on this page aren't useful. Especially that Central Limit Theorem. But remember, you can only apply the theorem for big n. Another vital thing to remember is that the variance of the sample mean is σ^2/n not σ^2.

Statistics and Sampling Distributions

This page is for OCR S2 and OCR MEI S2

Statistics are used to Estimate Population Parameters

1) Parameters are quantities that describe the characteristics of a population, such as the mean or variance.
   Statistics are used to estimate these parameters. For example, you can use the sample mean, $\overline{X} = \frac{\sum X_i}{n}$, to estimate the population mean, μ (and the sample mean is called an estimator of the population mean).
2) The sample mean is actually an unbiased estimator of the population mean — this means the expected value of the sample mean is the same as the population mean (a good thing).
3) However, if you have a sample $X_1, ..., X_n$ taken from a population, then $\frac{\sum(X_i - \overline{X})^2}{n} = \frac{\sum X_i^2}{n} - \left(\frac{\sum X_i}{n}\right)^2$ is not an unbiased estimator of the population variance, σ^2.
   For an unbiased estimate of the population variance, there's a slightly different formula.

This is the formula for variance you've seen previously.

Unbiased Estimators of Population Variance and Standard Deviation

These are unbiased estimators of the population variance (σ^2) and population standard deviation (σ):

Variance: $S^2 = \frac{n}{n-1}\left[\frac{\sum X_i^2}{n} - \left(\frac{\sum X_i}{n}\right)^2\right] = \frac{\sum(X_i - \overline{X})^2}{n-1}$

Standard deviation: $S = \sqrt{S^2}$

This one's in your formula booklet, but the one in red is more useful: "the mean of the squares minus the square of the mean, multiplied by $\frac{n}{n-1}$".

Greek letters, like σ, are often used for parameters, and Latin letters (e.g. S) are used for statistics.

EXAMPLE A random sample of 5 observations of X was taken from a population whose mean (μ) and variance (σ^2) are unknown. The data is summarised by $\sum x = 28.4$ and $\sum x^2 = 161.38$.

Find unbiased estimates of μ and σ^2.

$\bar{x} = \frac{\sum x}{n} = \frac{28.4}{5} = 5.68$, and $s^2 = \frac{5}{4}\left[\frac{161.38}{5} - \left(\frac{28.4}{5}\right)^2\right] = 0.017$

You should remember from S1 that if you've got lots of raw data it'll be in a table...

EXAMPLE The table shows how many sisters a random sample of 95 students from a school have.

a) Calculate unbiased estimates of the mean and variance for the whole school.

Number of sisters, x	0	1	2	3	4
Frequency, f	30	45	11	8	1

By adding extra rows to the table you can calculate:
$\sum f = 95, \sum fx = 95, \sum x^2 = 30$ and $\sum fx^2 = 177$

Number of sisters, x	0	1	2	3	4
Frequency, f	30	45	11	8	1
fx	0	45	22	24	4
x^2	0	1	4	9	16
fx^2	0	45	44	72	16

Then, $\bar{x} = \frac{\sum fx}{\sum f} = \frac{95}{95} = 1$

And $s^2 = \frac{\sum(x_i - \bar{x})^2 f_i}{n-1}$, where f_i is the frequency of the data value x_i.

Rearranging to $s^2 = \frac{n}{n-1}\left[\frac{\sum fx^2}{\sum f} - \left(\frac{\sum fx}{\sum f}\right)^2\right]$ gives you: $s^2 = \frac{95}{94}\left[\frac{177}{95} - 1^2\right] = \frac{95}{94} \times \frac{82}{95} = \frac{41}{47} = 0.87$ (2 d.p.)

For OCR only...

b) Another random sample of 30 is selected and the number of sisters the students have is recorded. If $\overline{X}$ is the mean of these 30 observations, calculate an estimate of $P(\overline{X} < 1.5)$.

Since n (= 30) is quite large, you can use the Central Limit Theorem to say that $\overline{X} \sim N\left(\mu, \frac{\sigma^2}{n}\right)$ (approx.).

You don't know the true values of μ and σ^2, but you can estimate them using the values you calculated in part a).

So $\overline{X} \sim N\left(1, \frac{41}{1410}\right)$ (approx.) and $P(\overline{X} < 1.5) = P\left(Z < \frac{1.5 - 1}{\sqrt{41/1410}}\right) = P(Z < 2.932) = 0.9983$

I rather like this page — but then I am biased...

... unlike $\overline{X}$ and S^2. You don't need to worry about understanding why these estimators are unbiased — just know that they are and that it's a good thing. So now you know two formulas for variance, which could be confusing, but at least they give you this one on the formula sheet. Remember, you only use this one when you're estimating population variance.

Confidence Intervals

This page is for AQA S2

You've met confidence intervals for the mean of a normal distribution before. Well now they're back. But there's a twist...

Remember, a **Confidence Interval** for the **Mean** should contain the true value of μ

A confidence interval for the mean of a distribution is $\left(\overline{X} - z\frac{\sigma}{\sqrt{n}}, \overline{X} + z\frac{\sigma}{\sqrt{n}}\right)$.

The z value to use depends on the level of confidence required. E.g. for a 95% C.I., use $z = 1.96$, since $P(-1.96 < Z < 1.96) = 0.95$.

This works fine as long as: 1) $\overline{X}$ follows a normal distribution,
and 2) You know σ (or can estimate it pretty well).

- If $\overline{X}$ follows a normal distribution, 1) is satisfied. Or if n is large, the Central Limit Theorem tells you that 1) is satisfied.
- And for large n, s^2 (see p.142) is a pretty good estimate of σ^2. But for small n, this method breaks down.

For **Unknown Variance** and **Small n**, you need the **t-Distribution**

1) If the population variance is unknown and the sample size is small ($n < 30$), you can still estimate σ^2 using s^2, but you can't assume that the standardised sample mean follows an N(0, 1) distribution. Instead, it has a *t*-distribution.
2) *t*-distributions are very similar to the standard normal distribution — their graphs are shaped like a bell and are symmetrical about a mean of zero. A *t*-distribution has one parameter, ν, called the degrees of freedom, and a *t*-distribution with ν degrees of freedom is written $t_{(\nu)}$.
3) So, going back to the standardised sample mean — this follows a *t*-distribution with $\nu = (n - 1)$ degrees of freedom. In other words: $\frac{\overline{X} - \mu}{S/\sqrt{n}} = T$ and $T \sim t_{(n-1)}$

ν is the Greek letter 'nu'.

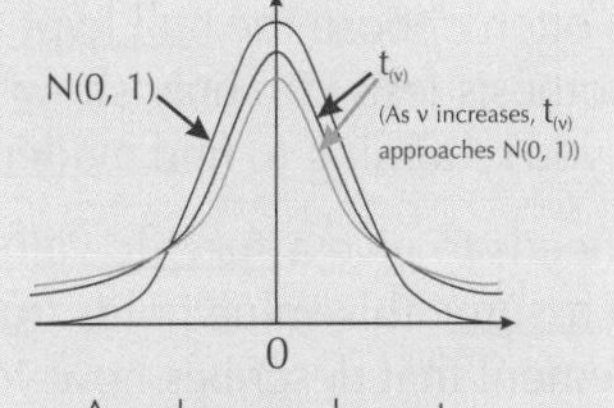

As n becomes large, $t_{(n-1)}$ becomes very close to N(0, 1).

So now you're ready for another big box and a second formula for finding confidence intervals:

Confidence Intervals — Normal Distribution with Unknown Variance

A confidence interval for the population mean when the sample size n is small is: $\left(\overline{x} - \frac{s}{\sqrt{n}}t_{(n-1)}\ ,\ \overline{x} + \frac{s}{\sqrt{n}}t_{(n-1)}\right)$

The *t*-tables give values of *t* at different percentage points (p) of the distribution.
The value you choose depends on the 'level of confidence'.
E.g., for a 95% confidence interval, choose $t_{(n-1)}$ so that: $P(-[t_{(n-1)}] < T < [t_{(n-1)}]) = 0.95$.

But hold your horses... The tables give values of t such that $P(T \le t) = p$. For a 95% C.I. you want $P(-t < T < t) = 0.95$, which means $P(T \le t) = 0.975$. So choose the value of $t_{(n-1)}$ for $p = 0.975$.

EXAMPLE A sample of 10 potato plants is selected from a field of potato plants, and the heights, h cm, are recorded. The heights are assumed to be normally distributed, and for the 10 plants: $\sum h = 350$ and $\sum h^2 = 12\,528$.
a) Calculate a 99% confidence interval for the mean height of all the potato plants in the field.
b) Comment on the claim that the mean height of all the potato plants is 42 cm.

a)
- Start by finding the sample mean, $\overline{x}$, and an estimate of the population variance, s^2.

  Using the formulas on the previous page: $\overline{x} = \frac{350}{10} = 35$ cm and $s^2 = \frac{10}{9}\left[\frac{12\,528}{10} - 35^2\right] = \frac{278}{9} = 30.889$ (3 d.p.)
- $n = 10$, so you need to look up $t_{(9)}$ in the t-distribution tables (p.167).
  For a 99% C.I., you need $P(T \le t) = 1 - 0.005 = 0.995$, so read across to $p = 0.995$ to get $t_{(9)} = 3.250$.
- So your confidence interval is: $\left(35 - \frac{\sqrt{278/9}}{\sqrt{10}} \times 3.25, 35 + \frac{\sqrt{278/9}}{\sqrt{10}} \times 3.25\right) = (29.29, 40.71)$ (2 d.p.)

b) The mean height claimed lies outside the confidence interval, so there is evidence to doubt this claim.

Useful Quote: Tea is liquid wisdom (Anonymous)...

You'll also see t-distributions called Student's t-distributions. William Sealy Gosset, the chap who first studied them, worked for Guinness at the time and wasn't allowed to publish research papers. So he used the pen name 'Student'. Interesting stuff.

Null and Alternative Hypotheses

This page is for AQA S2, Edexcel S2 and OCR S2

Hypothesis testing means checking if your theories about a population are consistent with the observations from your sample. The technical stuff on the next two pages might seem hard-going, but stick with it — it'll help later on.

*A **Hypothesis** is a **Statement** you want to **Test***

Hypothesis testing is about using sample data to test statements about population parameters.
Unfortunately, it comes with a fleet of terms you need to know.

- **Null Hypothesis (H_0)** — a statement about the value of a population parameter.
  Your data may allow you to reject this hypothesis.
- **Alternative Hypothesis (H_1)** — a statement that describes the value of the population parameter if H_0 is rejected.
- **Hypothesis test** — a statistical test that tests the claim made about a parameter by H_0 against that made by H_1.
  It tests whether H_0 should be rejected or not, using evidence from sample data.
- **Test Statistic** — a statistic calculated from sample data which is used to decide whether or not to reject H_0.

1) For any hypothesis test, you need to write two hypotheses — a null hypothesis and an alternative hypothesis.
2) You often choose the null hypothesis to be something you actually think is false. This is because hypothesis tests can only show that statements are false — they can't prove that things are true.
   So, you're aiming to find evidence for what you think is true, by disproving what you think is false.
3) H_0 needs to give a specific value to the parameter, since all your calculations will be based on this value.
   You assume this value holds true for the test, then see if your data allows you to reject it. H_1 is then a statement that describes how you think the value of the parameter differs from the value given by H_0.
4) The test statistic you choose depends on the parameter you're interested in. It should be a 'summary' of the sample data, and should have a sampling distribution that can be calculated using the parameter value specified by H_0.

EXAMPLE A 4-sided spinner has sides labelled A–D. Jemma thinks that the spinner is biased towards side A. She spins it 20 times and counts the number of times, Y, that she gets side A.

a) Write down a suitable null hypothesis to test Jemma's theory.
b) Write down a suitable alternative hypothesis.
c) Describe the test statistic Jemma should use.

a) If you assume the spinner is unbiased, each side has a probability of 0.25 of being spun.
   Let p = the probability of spinning side A. Then:
   $H_0: p = 0.25$ ← By assuming the spinner is unbiased, the parameter, p, can be given the specific value 0.25. Jemma is then interested in disproving this hypothesis.
b) If the spinner is biased towards side A,
   then the probability will be greater than 0.25. So: $H_1: p > 0.25$ ← This is what Jemma actually thinks.
c) The test statistic is Y, the number of times she gets side A. ← Assuming H_0 is true, the sampling distribution of Y is B(20, 0.25).

*Hypothesis Tests can be **One-Tailed** or **Two-Tailed***

The 'tailed' business is to do with the critical region used by the test — see next page.

For H_0: $\theta = a$, where θ is a parameter and a is a number:

1) The test is one-tailed if H_1 is specific about the value of θ compared to a, i.e. $H_1: \theta > a$, or $H_1: \theta < a$.
2) The test is two-tailed if H_1 specifies only that θ doesn't equal a, i.e. $H_1: \theta \neq a$.

Whether you use a one-tailed or a two-tailed test depends on how you define H_1. And that depends on what you want to find out about the parameter and any suspicions you might have about it.

E.g. in the example above, Jemma suspects that the probability of getting side A is greater than 0.25.
This is what she wants to test, so it is sensible to define $H_1: p > 0.25$.

If she wants to test for bias, but is unsure if it's towards or against side A, she could define $H_1: p \neq 0.25$.

A statistician's party game — pin two tails on the donkey...

Or should it be one? Anyway, a very important thing to remember is that the results of a hypothesis test are either 'reject H_0', or 'do not reject H_0' — which means you haven't found enough evidence to disprove H_0, and not that you've proved it.

Significance Levels and Critical Regions

This page is for AQA S2, Edexcel S2 and OCR S2

You use the value of your test statistic to decide whether or not to reject your null hypothesis. Poor little unloved H_0.

If your Data is Significant, Reject H_0

1) You would reject H_0 if the observed value of the test statistic is unlikely under the null hypothesis.
2) The significance level of a test (α) determines how unlikely the value needs to be before H_0 is rejected. It also determines the strength of the evidence that the test has provided — the lower the value of α, the stronger the evidence you have for saying H_0 is false. You'll usually be told what level to use — e.g. 1% ($\alpha = 0.01$), 5% ($\alpha = 0.05$), or 10% ($\alpha = 0.1$). α is also the probability of incorrectly rejecting H_0 — i.e. of getting extreme data by chance.
3) To decide whether your result is significant:
   - Define the sampling distribution of the test statistic under the null hypothesis.
   - Calculate the probability of getting a value that's at least as extreme as the observed value from this distribution.
   - If the probability is less than or equal to the significance level, reject H_0 in favour of H_1.

EXAMPLE Javed wants to test at the 5% level whether or not a coin is biased towards tails. He tosses the coin 10 times and records the number of tails, X. He gets 9 tails.

a) Define suitable hypotheses for p, the probability of getting tails.

b) State the condition under which Javed would reject H_0.

P(at least as extreme as 9) means 9 or more.

a) $H_0: p = 0.5$ and $H_1: p > 0.5$. b) Under H_0, $X \sim B(10, 0.5)$. If $P(X \geq 9) \leq 0.05$, Javed would reject H_0.

Significance level

The Critical Region is the Set of Significant Values

1) The critical region (CR) is the set of all values of the test statistic that would cause you to reject H_0. And the acceptance region is the set of values that would mean you do not reject H_0. The critical region is chosen so that P(test statistic is in CR, assuming H_0 is true) = α, or is as close to α as possible.

   The value on the boundary of the CR is called the critical value.
2) One-tailed tests have a single CR, containing the highest or lowest values. For two-tailed tests, the region is split into two — half at the lower end and half at the upper end. Each half has a probability of ½α.
3) To test whether your result is significant, find the critical region and if it contains the observed value, reject H_0.

Returning to the example above... Find the critical region for the test, at the 5% level.

- This is a one-tailed test with $H_1: p > 0.5$, so you're only interested in the upper end of the distribution.
- So, the critical region is the biggest possible set of 'high' values of X with a total probability of ≤ 0.05.
- Using the cumulative distribution tables for B(10, 0.5) you find that:
  $P(X \geq 9) = 0.011 < 0.05$, but $P(X \geq 8) = 0.055 > 0.05 \Rightarrow$ critical region is $X \geq 9$

So the critical value is 9 — values of 9 or 10 would cause you to reject H_0: $p = 0.5$.

In Hypothesis Testing, you can Work Out How Likely Things are to go Wrong

1) The actual significance level of a test is the probability of rejecting H_0 (assuming H_0 is true) when the test is carried out. This is often quite different from the level of significance originally specified in the test. E.g. for the above test, the actual significance level is $P(X \geq 9) = 0.011$, which is much lower than 0.05.

   Another way of describing the actual significance level is that it's the probability of incorrectly rejecting H_0 — i.e. rejecting H_0 when it's true. And another incorrect decision you can make is to not reject H_0 when it's not true.

   TYPE I ERROR — reject H_0 when H_0 is true   TYPE II ERROR — do not reject H_0 when H_0 is not true

2) You can calculate the probability of making each of these errors when doing a hypothesis test:

   P(Type I error) = P(value of test statistic is significant, assuming H_0 is true) — the actual significance level
   P(Type II error) = P(value of test statistic isn't significant, assuming H_1 is true) — where a value for H_1 is given

I repeat, X has entered the critical region — we have a significant situation...

Hope you've been following the last two pages closely. Basically, you need two hypotheses and the value of a test statistic calculated from sample data. By assuming H_0 is true, you can find the probabilities of the different values this statistic can take — if the observed value is unlikely enough, reject H_0. But bear in mind there's a chance you might have got it wrong...

Hypothesis Tests and Binomial Distributions

This page is for Edexcel S2 and OCR S2

OK, it's time to pick your best 'hypothesis testing' foot and put it firmly forward. It's also a good time to reacquaint yourself with binomial distributions before you go any further.

Use a *Hypothesis Test* to *Find Out* about the *Population Parameter p*

The first step in exam questions is to work out which distribution to use to model the situation. Words like 'proportion', 'percentage' or 'probability' are clues that it's binomial. Hypothesis tests for the binomial parameter p all follow the same general method — shown in the example below.

EXAMPLE:

In a past census of employees, 20% were in favour of a change to working hours. A later survey is carried out on a random sample of 30 employees, and 2 vote for a change to hours. The manager claims that there has been a decrease in the proportion of employees in favour of a change to working hours.

Stating your hypotheses clearly, test the manager's claim at the 5% level of significance.

1) Start by identifying the population parameter that you're going to test:

   Let p = proportion of employees in favour of change to hours.

2) Write null and alternative hypotheses for p.
   If you assume there's been no change in the proportion: $H_0: p = 0.2$

   The manager's interested in whether the proportion has decreased, so: $H_1: p < 0.2$

   You assume there's been no change in the value of the parameter, so you can give it a value of 0.2. The alternative hypothesis states what the manager actually thinks.

3) State the test statistic X — the number of 'successes', and its sampling distribution under H_0.
   $X \sim B(n, p)$ where n is the number in the sample and p is the probability of 'success' under H_0.

   Let X = number of employees in sample who are in favour of change. Under H_0, $X \sim B(30, 0.2)$.

   The sampling distribution of the test statistic uses the value $p = 0.2$.

4) State the significance level of the test. Here it's 5%, so $\alpha = 0.05$.

5) Test for significance by finding the probability of a value for your test statistic at least as extreme as the observed value. This is a one-tailed test and you're interested in the lower end of the distribution.
   So you want to find the probability of X taking a value less than or equal to 2.

   Using the binomial tables (see p.161): $P(X \leq 2) = 0.0442$, and since $0.0442 < 0.05$, the result is significant.

6) Now write your conclusion. Remember, hypothesis testing is about disproving the null hypothesis, or not disproving it — i.e. rejecting H_0, or not rejecting it. So that's how you need to word your conclusion.
   And don't forget to answer the original question:

   There is evidence at the 5% level of significance to reject H_0 and to support the manager's claim that the proportion in favour of change has decreased.

   Always say "there is evidence to reject H_0", or "there is insufficient evidence to reject H_0". Never talk about "accepting H_0" or "rejecting H_1".

And if you're asked (or prefer) to find a critical region, your test would look the same except for step 5...

5) Test for significance by finding the critical region for a test at this level of significance.
   This is a one-tailed test and you're interested in the lower end of the distribution.
   The critical region is the biggest possible set of 'low' values of X with a total probability of ≤ 0.05.

   Using the binomial tables: Try $X \leq 2$: $P(X \leq 2) = 0.0442 < 0.05$. Now try $X \leq 3$: $P(X \leq 3) = 0.1227 > 0.05$.
   So, CR is $X \leq 2$. These results fall in the CR, so the result is significant.

And finally — if you're doing OCR...

EXAMPLE Find: a) P(Type I error) and b) P(Type II error), given that the proportion in favour of change is actually 0.15.

a) P(test stat is significant, assuming H_0 true) = $P(X \leq 2 \mid X \sim B(30, 0.2)) = 0.0442$

b) P(test stat not significant, assuming H_1 true) = $P(X > 2 \mid X \sim B(30, 0.15)) = 1 - P(X \leq 2 \mid p = 0.15) = 1 - 0.1514 = 0.8486$

My hypothesis is — this is very likely to come up in the exam...

Make sure you learn this method inside out. Cover the page and outline the 6 steps. Then work through the example again.

Hypothesis Tests and Binomial Distributions

This page is for Edexcel S2 and OCR S2

A couple more examples here of the sorts of questions that might come up in the exam. Aren't I kind.

EXAMPLES:

Exam questions can look quite 'involved' and be a bit long-winded. But at the heart of them all is just a hypothesis test.

EXAMPLE 1 — USING CRITICAL REGIONS

Records show that the proportion of trees in a wood that suffer from a particular leaf disease is 15%. Chloe thinks that recent weather conditions might have affected this proportion. She examines a random sample of 20 of the trees.

a) Using a 10% level of significance, find the critical region for a two-tailed test of Chloe's theory. The probability of rejection in each tail should be less than 0.05.

b) Find the actual significance level of a test based on your critical region from part a).

Chloe finds that 8 of the sampled trees have the leaf disease.

c) Comment on this finding in relation to your answer to part a) and Chloe's theory.

Read part a) carefully — you could be asked to make the probability in each tail as close as possible to $\frac{1}{2}\alpha$, rather than less than.

a) Let p = proportion of trees with the leaf disease.

$H_0: p = 0.15$ $\quad H_1: p \neq 0.15$

Let X = number of sampled trees with the disease. Under H_0, $X \sim B(20, 0.15)$.

$\alpha = 0.1$, and since the test is two-tailed, the probability of X falling in each tail should be 0.05, at most.

This is a two-tailed test, so you're interested in both ends of the sampling distribution.
The lower tail is the biggest possible set of 'low' values of X with a total probability of ≤ 0.05.
The upper tail is the biggest possible set of 'high' values of X with a total probability of ≤ 0.05.

Using the tables:

Lower tail:
$P(X \leq 0) = 0.0388 < 0.05$
$P(X \leq 1) = 0.1756 > 0.05$

Upper tail: ← Look up $P(X \leq x - 1)$ and subtract from 1.
$P(X \geq 6) = 1 - P(X \leq 5) = 1 - 0.9327 = 0.0673 > 0.05$
$P(X \geq 7) = 1 - P(X \leq 6) = 1 - 0.9781 = 0.0219 < 0.05$

So, CR is $X = 0$ or $X \geq 7$.

You need to justify your CR by writing down the probabilities.

If you wanted the probability to be as close as possible to 0.05 instead of less than, the upper tail would be $X \geq 6$.

b) The actual significance level is $P(X = 0) + P(X \geq 7) = 0.0388 + 0.0219 = 0.0607$. ← You could also give this as 6.07%.

c) The observed value of 8 is in the critical region. So there is evidence at the 10% level of significance to reject H_0 and to support Chloe's theory that there has been a change in the proportion of affected trees.

And if you can't use the tables for your value of p, you have to use the binomial formula to work things out.

EXAMPLE 2 — WITHOUT USING TABLES

The proportion of pupils at a school who support the local football team is found to be $\frac{1}{8}$. Nigel attends a school nearby and claims that there is less support for the same local team at his school. In a random sample of 20 pupils from Nigel's school, 1 supports the local team. Use a 5% level of significance to test Nigel's claim.

Let p = proportion of pupils who support the local team.

$H_0: p = \frac{1}{8}$ $\quad H_1: p < \frac{1}{8}$

Let X = number of sampled pupils supporting the team. Under H_0, $X \sim B(20, \frac{1}{8})$. $\alpha = 0.05$.

Now you need to find the probability of getting a value less than or equal to 1. The tables don't have values for $p = \frac{1}{8}$, so you need to work out the probabilities individually and add them up:

See p.102 for the binomial formula. Here you need to use: $\binom{20}{x}(\frac{1}{8})^x(\frac{7}{8})^{20-x}$

$P(X \leq 1) = P(X = 0) + P(X = 1)$
$= (\frac{7}{8})^{20} + 20(\frac{1}{8})(\frac{7}{8})^{19} = 0.267$

$0.267 > 0.05$, so the result is not significant. There is insufficient evidence at the 5% level of significance to reject H_0 and to support Nigel's claim that there is less support for the team.

If your value of p isn't in the binomial tables, don't panic...

You won't find $p = 0.2438$ in the tables, or even $p = 0.24$, for that matter. But this isn't a problem as long as you know how to use the binomial probability function. And remember, the actual significance level isn't usually the same as α.

Hypothesis Tests and Poisson Distributions

This page is for Edexcel S2 and OCR S2

You can do exactly the same type of test, but for Poisson distributions. The Poisson parameter is λ.

Use a Hypothesis Test to Find Out about the Population Parameter λ

If λ is the rate at which an event occurs in a population and X is the number of those events that occur in a random interval, then X can be used as the test statistic for testing theories on λ.

EXAMPLE A bookshop sells copies of the book *'All you've never wanted to know about the Poisson distribution'* at a (surprisingly high) rate of 10 a week. The shop's manager decides to reduce the price of the book. In one randomly selected week after the price change, 16 copies are sold. Use a 10% level of significance to test whether there is evidence to suggest that sales of the book have increased.

Let λ = the rate at which copies of the book are sold per week.

$H_0: \lambda = 10$ $\quad H_1: \lambda > 10$

Let X = number of copies sold in a random week. Under H_0, $X \sim Po(10)$. $\alpha = 0.1$.

The mention of 'rate' tells you that the situation can be modelled by a Poisson distribution.

To find the probability of X taking a value greater than or equal to 16, use the Poisson tables (see p.166):

$P(X \geq 16) = 1 - P(X \leq 15) = 1 - 0.9513 = 0.0487$, and since $0.0487 < 0.1$, the result is significant.

Don't forget the conclusion:

There is evidence at the 10% level of significance to reject H_0 and to suggest that sales have increased.

If the Numbers are Awkward, you can use Approximations

If $X \sim B(n, p)$ for large n and small p, then X can be approximated by $Po(np)$.

See p.112 for more on the Poisson approximation.

EXAMPLE May has a crazy, 20-sided dice, which she thinks might be biased towards the number 1. She rolls the dice 100 times and gets 12 ones. Test May's theory at the 1% level of significance.

If the dice is unbiased, $P(1) = 1 \div 20 = 0.05$.

So, $H_0: p = 0.05$ and $H_1: p > 0.05$, where p is the probability of rolling a 1.
Let X = number of ones in 100 rolls. Under H_0, $X \sim B(100, 0.05)$. $\alpha = 0.01$.

You can't look up $n = 100$ in the tables, but n is big and p is quite small, so X can be approximated by $Po(5)$.
So using the Poisson tables:

$P(X \geq 12) = 1 - P(X \leq 11) = 1 - 0.9945 = 0.0055$. $0.0055 < 0.01$, so it's significant at the 1% level. There is very strong evidence to reject H_0 and to support May's claim that the dice is biased towards the number 1.

The evidence provided by this test is much stronger than the evidence provided by the test in the example above, because the value of α is much lower. So you can be more confident that you've correctly rejected H_0.

If $X \sim Po(\lambda)$ and λ is large, then the approximation $X \sim N(\lambda, \lambda)$ can be used. See p.132-135 for a reminder of normal approximations.

EXAMPLE An automated sewing machine produces faults randomly, at an average of 5 per day. The machine is serviced, and in a random five-day period, 20 faults are found. Using a suitable approximation, test at the 5% significance level whether the number of faults has decreased.

$H_0: \lambda = 5$ and $H_1: \lambda < 5$.

Let X = number of faults in 5 days. Under H_0, $X \sim Po(25) \Rightarrow X \sim N(25, 25)$.

λ is large, so you can approximate by using a normal distribution.

$P(X \leq 20) = P(X < 20.5) = P\left(Z < \frac{20.5 - 25}{\sqrt{25}}\right)$

Don't forget to include the continuity correction — see p.132. Here ≤ 20 becomes < 20.5.

$= P(Z < -0.9) = 1 - P(Z < 0.9) = 1 - 0.8159 = 0.1841$. $0.1841 > 0.05$, so there's insufficient evidence at the 5% level of significance to reject H_0. There is insufficient evidence to say that the number of faults has decreased.

Questions on hypothesis testing in exam papers ~ Po(2)...

The different types of questions on the last page can come up for Poisson too — e.g. working out critical regions, or using the Poisson probability formula (p.109) to calculate probabilities. Go through the first two examples on this page again, but this time carry out the tests by finding the critical region each time. And make sure you know the rules for approximating.

Hypothesis Tests and Normal Distributions

This page is for AQA S2, OCR S2 and OCR MEI S2

If you like normal distributions and you like hypothesis testing, you're going to *love* this page.

For a Normal Population with Known Variance — use a Z-Test

1) Suppose $X \sim N(\mu, \sigma^2)$, where you know the value of σ^2 but μ is unknown. If you take a random sample of n observations from the distribution of X, and calculate the sample mean $\overline{X}$, you can use your observed value $\bar{x}$ to test theories about the population mean μ.
2) You want to test whether your value of $\overline{X}$ is likely enough, under the hypothesised value of μ (remember that $\overline{X} \sim N(\mu, \frac{\sigma^2}{n})$).
3) To do this you need to use the classic 'normal trick' — standardise your value and compare it to N(0, 1). So...

The test statistic for a test of the mean of a normal distribution with known variance is: $Z = \frac{\overline{X} - \mu}{\sigma/\sqrt{n}} \sim N(0,1)$

And the critical region for Z depends on the significance level α.

E.g. for a two-tailed test at the 5% level of significance, the CR is $Z > 1.96$ or $Z < -1.96$, since $P(Z > 1.96) = 0.025$ and $P(Z < -1.96) = 0.025$.

EXAMPLE The times, in minutes, taken by the athletes in a running club to complete a certain run have been found to follow an N(12, 4) distribution. The coach increases the number of training sessions per week, and a random sample of 20 times run since the increase gives a mean time of 11.2 minutes. Assuming that the variance has remained unchanged, test at the 5% significance level whether there is evidence that the average time has decreased.

Let μ = mean time since increase in training sessions. Then $H_0: \mu = 12$, $H_1: \mu < 12$, $\alpha = 0.05$.

Under H_0, $\overline{X} \sim N(12, 4/20) \Rightarrow \overline{X} \sim N(12, 0.2)$ and $Z = \frac{\overline{X} - 12}{\sqrt{0.2}} \sim N(0,1)$. Since $\bar{x} = 11.2$, $z = \frac{11.2 - 12}{\sqrt{0.2}} = -1.789$

This is a one-tailed test and you're interested in the lower end of the distribution. So the critical value is z such that $P(Z < z) = 0.05$. Using the 'normal' tables you find that $P(Z > 1.645) = 0.05$, and so by symmetry, $P(Z < -1.645) = 0.05$. So CR is $Z < -1.645$.

CR = critical region

If you want, you can instead do the test by working out P(value at least as extreme as observed sample mean) and comparing it to α. So here you'd do: $P(\overline{X} \le 11.2) = P\left(Z \le \frac{11.2 - 12}{\sqrt{0.2}}\right) = 0.0367 < 0.05$, so reject H_0.

Since $z = -1.789 < -1.645$, the result is significant and there is evidence at the 5% level of significance to reject H_0 and to suggest that the average time has decreased.

The Z-test can also be used in the following situations:

- The population variance is unknown, but the sample size is large ($n > 30$) — just estimate σ with s (see p.142).
- *[AQA S2 and OCR S2]* As long as n is large, you can use the Central Limit Theorem (p.141) to approximate the sampling distribution of the mean of any distribution by a normal distribution, and so use the test for any population.

If the Variance is Unknown and the Sample Size is Small — use a T-Test

AQA only

The test statistic for a test of the mean of a normal distribution with unknown variance, given a small sample size, n, is: $T = \frac{\overline{X} - \mu}{S/\sqrt{n}} \sim t_{(n-1)}$

See p.143 for more on the t-distribution.

EXAMPLE A random sample of 8 observations from a normal distribution with mean μ and variance σ^2 produces the following results: $\bar{x} = 9.5$ and $s^2 = 2.25$. Test at the 1% level of significance the claim that $\mu > 9$.

$H_0: \mu = 9$, $H_1: \mu > 9$, $\alpha = 0.01$. σ^2 is unknown and $n = 8$, so you need to use a t-test with 7 degrees of freedom:

Under H_0, $T = \frac{\overline{X} - 9}{1.5/\sqrt{8}} \sim t_{(7)}$. Since $\bar{x} = 9.5$, $t = \frac{9.5 - 9}{1.5/\sqrt{8}} = 0.94$

The critical value is t such that $P(T > t) = 0.01$, which means looking up $P(T < t) = 0.99$ in the t-distribution tables. You find that $P(T < 2.998) = 0.99$, and so CR is $T > 2.998$.

Since $t = 0.94 < 2.998$, the result is not significant — there is no evidence at this level to support the claim that $\mu > 9$.

Chi-Squared (χ^2) Contingency Table Tests

This page is for AQA S2 and OCR MEI S2

Another page, another distribution, and another statistical table. But look — questions ahoy, so you're nearly there...

Contingency Tables show Observed Frequencies for Two Variables

1) Suppose you've got a sample of size n, and you're interested in two different variables for each of the n members — where each variable can be classified into different categories. E.g. eye colour (blue, brown, green, ...), favourite way of cooking potatoes (boiled, roast, baked, ...), etc. You can show this data in a contingency table.
2) You use the columns to show the categories for one of the variables, and the rows to show the other. You then fill in each cell in the table with the number of sample members that fit that particular combination of classes — e.g. 'blue eyes and boiled potatoes'.
3) With the data in this format, you can do a hypothesis test of whether the two variables are independent or linked.

Use a χ^2 Test to test whether Two Variables are Independent

χ is the Greek letter 'chi'.

EXAMPLE This table shows soil pH and plant growth for a sample of 100 plants. Test at the 5% level whether there is a link between soil pH and growth for these plants.

	Poor growth	Average growth	Good growth	Total
Acidic soil	12	16	4	32
Neutral soil	3	13	14	30
Alkaline soil	3	10	25	38
Total	18	39	43	100

Observed frequencies

① As always, the first step is to define the hypotheses:

H_0: the variables are independent, H_1: they're not independent

② Next you need to make a new table — adding the expected frequencies (E), assuming H_0 is true, for each cell in the original table.

Using the formula: $E = \frac{\text{(row total)} \times \text{(column total)}}{\text{Overall total } (n)}$, gives you this column:

E.g. for the first observation (12): (32 × 18) / 100 = 5.76. You get this formula for E by saying that under H_0 the ratio of the growth classes should be the same for each soil category. E.g. (18/100) of each row should have poor growth.

Observed frequency (O)	Expected Frequency (E)	$\frac{(O-E)^2}{E}$
12	5.76	6.76
3	5.4	1.066...
3	6.84	2.155...
16	12.48	0.992...
13	11.7	0.144...
10	14.82	1.567...
4	13.76	6.922...
14	12.9	0.093...
25	16.34	4.589...
100	100	**24.3**

③ Now you can calculate the test statistic: $X^2 = \sum \frac{(O_i - E_i)^2}{E_i}$,

where O_i and E_i are the observed and expected frequencies for cell i.

Add the third column shown to your table and sum the values. $X^2 = 24.3$

④ Under H_0, $X^2 \sim \chi^2_{(\nu)}$ (approx.) — it follows the chi-squared distribution with ν degrees of freedom.

For this approximation to be valid, all E_i must be greater than 5.

To get your value of ν: ν = [(no. of rows) – 1] × [(no. of columns) – 1] = 2 × 2 = 4

in original table

⑤ The significance level is 5%, so look up the 5% point of $\chi^2_{(4)}$ in the table. The critical value is x where $P(X > x) = 0.05$, which means looking up $P(X \le x) = 0.95$. Reject H_0 if $X^2 >$ this critical value.

This is just the top right corner of the χ^2 table. The table shows values of x satisfying $P(X \le x) = p$, for $X \sim \chi^2_{(\nu)}$.

0.1	0.9	0.95	0.975	0.99	0.995	p / ν
0.016	2.706	3.841	5.024	6.635	7.879	1
0.211	4.605	5.991	7.378	9.210	10.597	2
0.584	6.251	7.815	9.348	11.345	12.838	3
1.064	7.779	9.488	11.143	13.277	14.860	4

$P(X \le x) = 0.95$ for $x = 9.488$. $X^2 = 24.3 > 9.488$, so it's significant at this level. There is evidence to reject H_0 and to suggest that there is a link between soil pH and growth for these plants.

2 × 2 Contingency Tables are a Special Case

If your table has 2 columns and 2 rows, you can improve the X^2 approximation by using Yates' continuity correction:

The test statistic for a 2 × 2 table is: $X^2 = \sum \frac{(|O_i - E_i| - 0.5)^2}{E_i} \sim \chi^2_{(1)}$ (approx.)

Now you can prove that 'reading this book' and 'exam grade' are linked...

Well, that was a bit heavy going for the last page in the section — sorry about that. Still, it's really just another hypothesis test with the usual gubbins, plus some tables. The main formula will be given, but the Yates' correction needs learning.

S2 Section 6 — Practice Questions

Phew, that section really did have a bit of everything — you could say it was a smorgasbord of tasty S2 treats. To make sure you've fully digested everything on offer, finish off with these delicious practice questions.

Warm-up Questions

Questions 1-6 are for Edexcel S2 and OCR S2

1) The manager of a tennis club wants to know if members are happy with the facilities provided.
   a) Identify the population the manager is interested in.
   b) Identify the sampling units.
   c) Suggest a suitable sampling frame.

2) For each of the following situations, explain whether it would be more sensible to carry out a census or a sample survey:
   a) Marcel is in charge of a packaging department of 8 people. He wants to know the average number of items a person packs per day.
   b) A toy manufacturer produces batches of 500 toys. As part of a safety check, they want to test the toys to work out the strength needed to pull them apart.

3) Why is it a good idea to use simple random sampling to select a sample?

4) The weights of a population of jars of pickled onions have unknown mean μ and standard deviation σ. A random sample of 50 weights (X_1, ..., X_{50}) are recorded. Say whether each of these is a statistic or not:
   a) $\frac{X_{25}+X_{26}}{2}$ b) $\sum X_i - \sigma$ c) $\sum X_i^2 + \mu$ d) $\frac{\sum X_i}{50}$

5) a) Carry out the following test of the binomial parameter p. Let X represent the number of successes in a random sample of size 20: Test H_0: $p = 0.2$ against H_1: $p < 0.2$, at the 5% significance level, using $x = 2$.
   b) Carry out the following test of the Poisson parameter λ. Let X represent the number of events in a given interval and λ be the average rate at which they are assumed to occur in intervals of identical size: Test H_0: $\lambda = 2.5$ against H_1: $\lambda > 2.5$, at the 10% significance level, using $x = 4$.

6) a) Find the critical region for the following test where $X \sim B(10, p)$: Test H_0: $p = 0.3$ against H_1: $p < 0.3$, at the 5% significance level.
   b) Find the critical region for the following test where $X \sim Po(\lambda)$: Test H_0: $\lambda = 6$ against H_1: $\lambda < 6$, at the 10% significance level

Questions 7-8 are for OCR S2 and OCR MEI S2

7) If $X \sim N(8, 2)$, find $P(\overline{X} < 7)$ where $\overline{X}$ is the mean of a random sample of 10 observations of X.

8) A random sample was taken from a population whose mean (μ) and variance (σ^2) are unknown. Find unbiased estimates of μ and σ^2 if the sample values were: 8.4, 8.6, 7.2, 6.5, 9.1, 7.7, 8.1, 8.4, 8.5, 8.0.

Question 9 is for AQA S2, OCR S2 and OCR MEI S2

9) Carry out the following test of the mean, μ, of a normal distribution with variance $\sigma^2 = 9$. A random sample of 16 observations from the distribution was taken and the sample mean ($\overline{x}$) calculated. Test H_0: $\mu = 45$ against H_1: $\mu < 45$, at the 5% significance level, using $\overline{x} = 42$.

Question 10 is for AQA S2 and OCR MEI S2

10) A χ^2 contingency table test produces the test statistic $\chi^2 = 8.3$. By comparing this statistic to the $\chi^2_{(4)}$ distribution, test at the 1% level whether there is evidence of an association between the variables.

Questions 11-12 are for AQA S2

11) A random sample of size 12 is taken from a normally-distributed population with unknown mean, μ, and unknown variance, σ^2. Find a 95% confidence interval for μ if the sample mean is 50 and $s^2 = 0.7$.

12) Explain when you would use a t-test to test the mean of a population.

S2 Section 6 — Practice Questions

Aha, some exam-style practice questions to test whether you've got this section sussed. Wasn't expecting that.

Exam Questions

Question 1 is for Edexcel S2

1 Prize draw tickets are drawn from a container, inside which are large quantities of tickets numbered with '1' or '2'. The tickets correspond to 1 or 2 prizes. 70% have the number 1 and 30% have the number 2.

A random sample of 3 tickets is drawn from the container. Find the sampling distribution for the median, M, of the numbers on the tickets.

(7 marks)

Questions 2-3 are for Edexcel S2 and OCR S2

2 A tennis player serves a fault on her first serve at an average rate of 4 per service game. The player receives some extra coaching. In a randomly selected set of tennis, she serves 12 first-serve faults in 5 service games. She wants to test whether her average rate of first-serve faults has decreased.

a) Write down the conditions needed for the number of first-serve faults per service game to be modelled by a Poisson distribution.

(2 marks)

Assume the conditions you stated in part a) hold.

b) Using a suitable approximation, carry out the test at the 5% level of significance.

(6 marks)

3 The residents of a town are being asked their views on a plan to build a wind farm in the area. Environmental campaigners claim that 10% of the residents are against the plan. A random sample of 50 residents is surveyed.

a) State two reasons why surveying a random sample of 50 residents will allow reliable conclusions to be drawn.

(2 marks)

b) Using a 10% significance level, find the critical region for a two-tailed test of this claim. The probability of rejecting each tail should be as close as possible to 5%.

(5 marks)

c) State the probability of incorrectly rejecting H_0 using your critical region from part a).

(2 marks)

It's found that 4 of the sampled residents say they are against the plan.

d) Comment on this finding in relation to the environmental campaigners' claim.

(2 marks)

Question 4 is for AQA S2, OCR S2 and OCR MEI S2

4 The heights of trees in an area of woodland are known to be normally distributed with a mean of 5.1 m. A random sample of 100 trees from a second area of woodland is selected and the heights, X, of the trees are measured giving the following results:

$$\sum x = 490 \text{ and } \sum x^2 = 2421$$

a) Calculate unbiased estimates of the population mean, μ, and variance, σ^2, for this area.

(3 marks)

b) Test at the 1% level of significance whether the trees in the second area of woodland have a different mean height from the trees in the first area.

(6 marks)

Question 5 is for AQA S2 and OCR MEI S2

5 A random sample of 100 shoppers is surveyed to determine whether age is associated with favourite flavour of ice cream. The results are shown in the table.

		Age in full years		
		0 – 40	41 +	Total
Flavour of ice cream	Vanilla	10	14	24
	Chocolate	24	26	50
	Strawberry	7	4	11
	Mint choc chip	7	8	15
	Total	48	52	100

Use a χ^2 test at the 5% level of significance to examine whether there is any association between age and favourite flavour of ice cream.

(9 marks)

Correlation

This page is for OCR MEI S2. Note — the whole of Section 7 is for OCR MEI S2 only.

Correlation is all about how closely two quantities are linked. And it can involve a fairly hefty formula.

Draw a Scatter Diagram to see Patterns in Data

Sometimes variables are measured in pairs — maybe because you want to find out how closely they're linked.
These pairs of variables might be things like: — 'my age' and 'length of my feet', or
— 'temperature' and 'number of accidents on a stretch of road'.

You can plot readings from a pair of variables on a scatter diagram — this'll tell you something about the data.

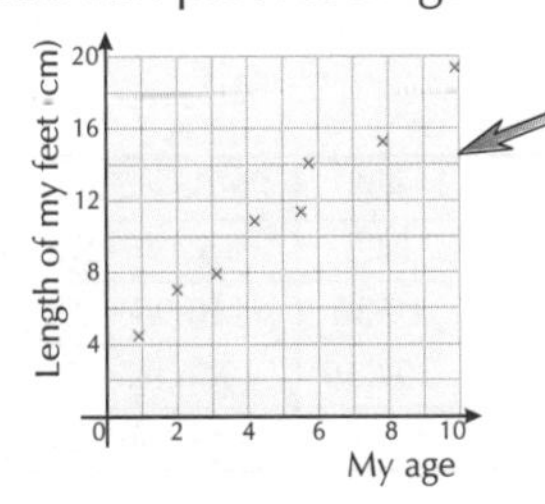

The variables 'my age' and 'length of my feet' seem linked — all the points lie close to a line. As I got older, my feet got bigger and bigger (though I stopped measuring when I was 10).

It's a lot harder to see any connection between the variables 'temperature' and 'number of accidents' — the data seems scattered pretty much everywhere.

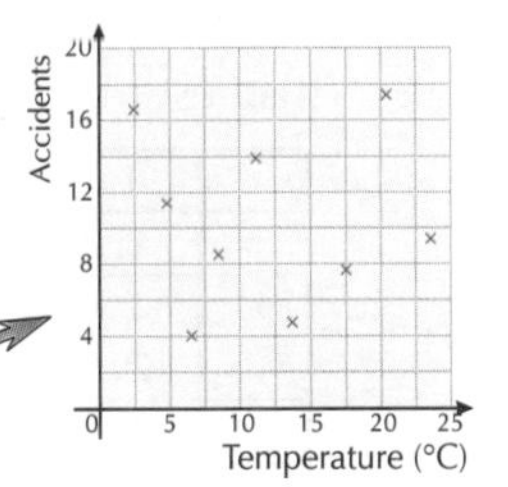

Correlation is a measure of How Closely variables are Linked

1) Sometimes, as one variable gets bigger, the other one also gets bigger — then the scatter diagram might look like the one on the right. Here, a line of best fit would have a positive gradient. The two variables are positively correlated (or there's a positive correlation between them).

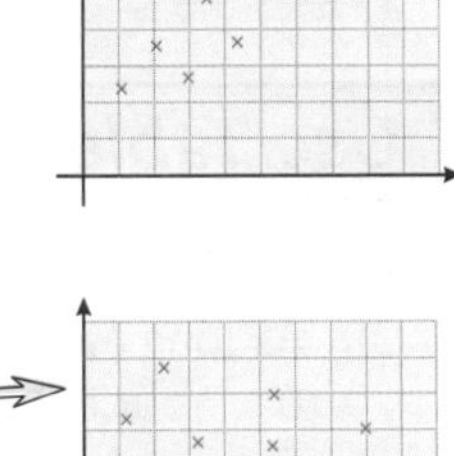

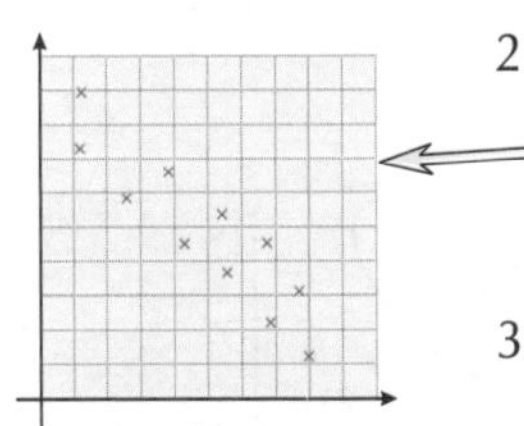

2) But if one variable gets smaller as the other one gets bigger, then the scatter diagram might look like this one — and the line of best fit would have a negative gradient. The two variables are negatively correlated (or there's a negative correlation between them).

3) And if the two variables aren't linked at all, you'd expect a random scattering of points — it's hard to say where the line of best fit would be. The variables aren't correlated (or there's no correlation).

WARNING

A strong correlation doesn't necessarily mean that one factor causes the other.
The number of televisions sold in Japan and the number of cars sold in America may well be correlated, but that doesn't mean that high TV sales in Japan cause high car sales in the US.

The Product-Moment Correlation Coefficient (r) measures Correlation

1) The Product-Moment Correlation Coefficient (PMCC, or r, for short) measures how close to a straight line the points on a scatter graph lie.

2) The PMCC is always between +1 and –1.
If all your points lie exactly on a straight line with a positive gradient (perfect positive correlation), $r = +1$.
If all your points lie exactly on a straight line with a negative gradient (perfect negative correlation), $r = -1$.

3) If $r = 0$ (or more likely, pretty close to 0), that would mean the variables aren't correlated.

4) The formula for the PMCC is a real stinker. But some calculators can work it out if you type in the pairs of readings, which makes life easier. Otherwise, just take it nice and slow.

In reality, you're unlikely to get a PMCC of +1 or –1 — your data points might lie close to a straight line, but it's unlikely they'd all be on it.

$$r = \frac{S_{xy}}{\sqrt{S_{xx}S_{yy}}} = \frac{\sum[x-\bar{x}][y-\bar{y}]}{\sqrt{(\sum[x-\bar{x}]^2)(\sum[y-\bar{y}]^2)}} = \frac{\sum xy - \frac{[\sum x][\sum y]}{n}}{\sqrt{\left(\sum x^2 - \frac{[\sum x]^2}{n}\right)\left(\sum y^2 - \frac{[\sum y]^2}{n}\right)}}$$

This is the easiest one to use, but it's still a bit hefty.

See page 156 for more about S_{xy} and S_{xx}.

WARNING

The PMCC is only a measure of a linear relationship between two variables (i.e. how close they'd be to a straight line if you plotted a scatter diagram). In this diagram, the PMCC would be low, but the two variables definitely look linked ('associated'). It looks like the points lie on a parabola (the shape of an x^2 curve) — not a straight line.

Correlation and Hypothesis Tests

This page is for OCR MEI S2

Don't rush questions on correlation. In fact, take your time and draw yourself a nice table.

EXAMPLE A researcher selects a random sample of fleas to investigate whether longer fleas tend to jump further. Her data recording the length of fleas (x mm) and the average distance they jump (y cm) is below.

Find the product-moment correlation coefficient (r) between the variables x and y.

x	1.6	2.0	2.1	2.1	2.5	2.8	2.9	3.3	3.4	3.8	4.1	4.4
y	11.4	11.8	11.5	12.2	12.5	12.0	12.9	13.4	12.8	13.4	14.2	14.3

1) There are 12 pairs of readings, so $n = 12$.
2) It's best to add a few extra rows to your table to work out the sums you need...

x	1.6	2	2.1	2.1	2.5	2.8	2.9	3.3	3.4	3.8	4.1	4.4	$35 = \Sigma x$
y	11.4	11.8	11.5	12.2	12.5	12	12.9	13.4	12.8	13.4	14.2	14.3	$152.4 = \Sigma y$
x^2	2.56	4	4.41	4.41	6.25	7.84	8.41	10.89	11.56	14.44	16.81	19.36	$110.94 = \Sigma x^2$
y^2	129.96	139.24	132.25	148.84	156.25	144	166.41	179.56	163.84	179.56	201.64	204.49	$1946.04 = \Sigma y^2$
xy	18.24	23.6	24.15	25.62	31.25	33.6	37.41	44.22	43.52	50.92	58.22	62.92	$453.67 = \Sigma xy$

Stick all these in the formula to get:
$$r = \frac{\left[453.67 - \frac{35 \times 152.4}{12}\right]}{\sqrt{\left[110.94 - \frac{35^2}{12}\right] \times \left[1946.04 - \frac{152.4^2}{12}\right]}} = \frac{9.17}{\sqrt{8.857 \times 10.56}} = 0.948$$
(to 3 s.f.)

This value of r looks pretty high. But before you can conclude whether this is evidence of x and y being correlated in the whole population, you need to test whether r differs from 0 in a statistically significant way.

*Use **Tables** to find **Critical Values** of r*

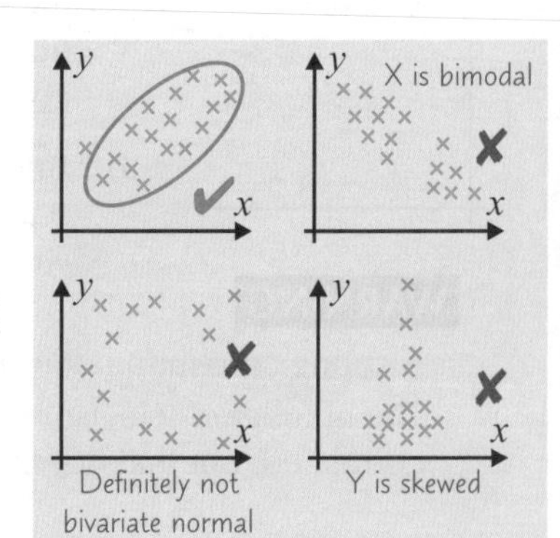

1) To test your value of r, you need the population to be 'bivariate normal'.
2) As long as the following two conditions are satisfied, everything is probably fine:
   i) Your data should come from a random sample.
   ii) The bulk of the data should form an ellipse (roughly) on a scatter diagram.
3) However, if one or both of your variables is skewed (i.e. asymmetrical) or bimodal (i.e. data points occur in two groups), then this test could give a false result.
4) The null hypothesis and the alternative hypothesis for the test are as follows:

   For a 1-tailed test: $H_0: r = 0$ and $H_1: r < 0$ (or $H_1: r > 0$)
   For a 2-tailed test: $H_0: r = 0$ and $H_1: r \neq 0$
5) Then as long as you know n and the significance level of your test, you can look up the critical value for r in tables.

EXAMPLE For the data in the above example, carry out a hypothesis test at the 5% significance level to investigate whether x and y are positively correlated in the flea population.

- Before carrying out the test, check you can draw a fairly neat ellipse around your points. Here, it looks safe to assume that the data is from a bivariate normal population.
- This is a 1-tailed test, because you want to test whether you have evidence for r being positive in the population. Use hypotheses — $H_0: r = 0$ and $H_1: r > 0$

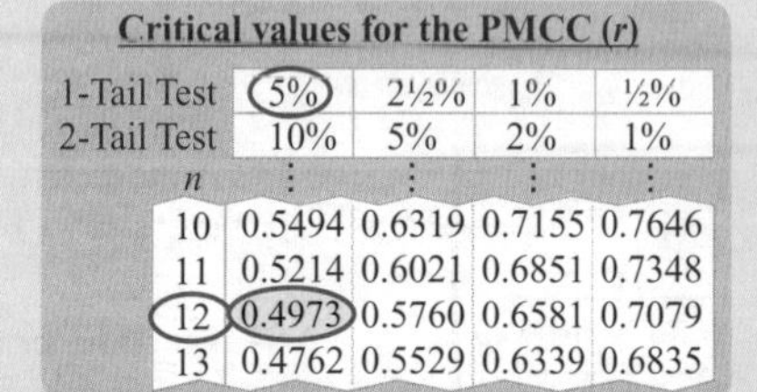

Critical values for the PMCC (r)

1-Tail Test	5%	2½%	1%	½%
2-Tail Test	10%	5%	2%	1%
n	⋮	⋮	⋮	⋮
10	0.5494	0.6319	0.7155	0.7646
11	0.5214	0.6021	0.6851	0.7348
12	0.4973	0.5760	0.6581	0.7079
13	0.4762	0.5529	0.6339	0.6835

Your formula booklet will contain a table showing critical values of r for n between 3 and 60.

- From tables, the critical value of r for $n = 12$ at a significance level of 5% is 0.4973. This means that under H_0, there is a probability of 5% that r will lie outside the range –0.4973 to 0.4973. So you can reject H_0 if r is outside this range.
- In this example, the value of r was 0.948, so you can reject H_0. So there is evidence at the 5% significance level that x and y are positively correlated.

What's a statistician's favourite soap — Correlation Street... (Boom boom)

Let me make one thing perfectly clear... this hypothesis test involving the PMCC will only be reliable if your data has come from a bivariate normal population, and so you must do that check of the scatter diagram. I hope that's perfectly clear.

Rank Correlation

This page is for OCR MEI S2

Spearman's Rank Correlation Coefficient (SRCC or r_s) works with Ranks

You can use the SRCC (or r_s, for short) when your data is a set of ranks. Ranks are the positions of the values when you put them in order — e.g. from biggest to smallest, or from best to worst, etc.

EXAMPLE At a dog show, two judges ranked 8 of the labradors (A-H) in the following order. Calculate the SRCC (r_s) between the sets of ranks, and comment on your result.

Position	1st	2nd	3rd	4th	5th	6th	7th	8th
Judge 1:	B	C	E	A	D	F	G	H
Judge 2:	C	B	E	D	F	A	G	H

First, make a table of the ranks of the 8 labradors — i.e. for each dog, write down where it came in the show.

Dog	A	B	C	D	E	F	G	H
Rank from Judge 1:	4	1	2	5	3	6	7	8
Rank from Judge 2:	6	2	1	4	3	5	7	8

Now for each dog, work out the difference (d) between the ranks from the two judges — you can ignore minus signs.

Dog	A	B	C	D	E	F	G	H
d	2	1	1	1	0	1	0	0

Take a deep breath, and add another row to your table — this time for d^2:

Dog	A	B	C	D	E	F	G	H	Total = $\sum d^2$
d^2	4	1	1	1	0	1	0	0	8

Then the SRCC is:

$$r_s = 1 - \frac{6\sum d^2}{n(n^2-1)}$$

You can ignore minus signs when you work out d, since only d² is used to work out the SRCC.

So here, $r_s = 1 - \frac{6 \times 8}{8 \times (8^2 - 1)} = 1 - \frac{48}{504} = 0.905$ (to 3 sig. fig.).

Interpret r_s in the same way as you'd interpret the PMCC (see p153).

This value for r_s is close to +1, so it appears the judges ranked the dogs in a pretty similar way.

To check this is evidence that the judges mark in a similar way more generally, perform a hypothesis test.

Critical Values of r_s are in Tables

There's no need to check anything about the distributions with this test.

1) You can use a hypothesis test to check whether a value of r_s is different from 0 in a statistically significant way. The process is very similar to the test on the previous page, but you need to use a different table of critical values.
2) The null and alternative hypotheses are:
   For a 1-tailed test — H_0: No association and H_1: Positive association (or H_1: Negative association)
   For a 2-tailed test — H_0: No association and H_1: Some association
3) As with the test described on p154, if your value of r_s is greater than the critical value (or less than the negative of the critical value), then you can reject H_0 and say that you have statistically significant evidence of an association.

EXAMPLE Test whether your value of r_s for the dog-show judges shows an association at the 1% significance level.

- Use a 2-tailed test — you're testing for a positive or negative association.
- Your hypotheses are — H_0: No association and H_1: Some association.
- The critical value for $n = 8$ at the 1% significance level is 0.8810. Since r_s is greater than the critical value, you can reject H_0 and conclude that you have evidence at the 1% significance level that there is some association between the judges' ranks.

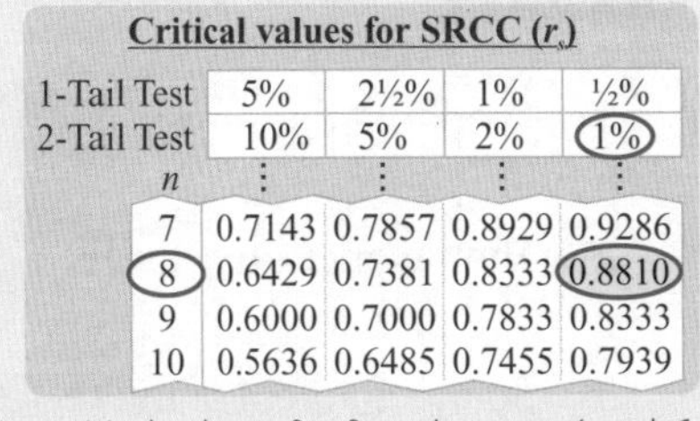
Critical values for SRCC (r_s)

1-Tail Test	5%	2½%	1%	½%
2-Tail Test	10%	5%	2%	1%
n	⋮	⋮	⋮	⋮
7	0.7143	0.7857	0.8929	0.9286
8	0.6429	0.7381	0.8333	0.8810
9	0.6000	0.7000	0.7833	0.8333
10	0.5636	0.6485	0.7455	0.7939

Your formula booklet will contain a table showing critical values of r_s for n between 4 and 60.

Spearman's correlation coefficient isn't rank — in fact, it's pretty cool...

You can also use Spearman's rank CC with 'normal' numerical data (like the flea data on p154), since it can sometimes recognise a 'non-linear association' between the variables much better than the PMCC. E.g. with the data on the right, Spearman would give a clearer result of an association than the PMCC.

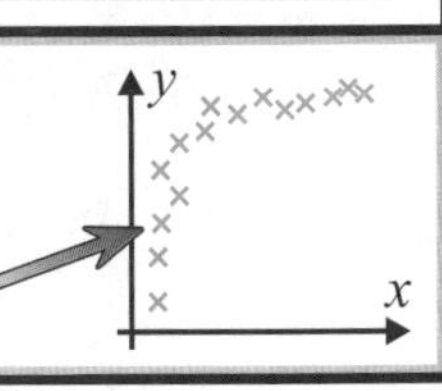

Linear Regression

This page is for OCR MEI S2

Linear regression is just fancy stats-speak for 'finding lines of best fit'. Not so scary now, eh...

Decide which is the **Independent Variable** and which is the **Dependent**

EXAMPLE The data below shows the load on a lorry, x (in tonnes), and the fuel efficiency, y (in km per litre).

x	5.1	5.6	5.9	6.3	6.8	7.4	7.8	8.5	9.1	9.8
y	9.6	9.5	8.6	8.0	7.8	6.8	6.7	6	5.4	5.4

1) The variable along the x-axis is the explanatory or independent variable — it's the variable you can control, or the one that you think is affecting the other. The variable 'load' goes along the x-axis here.
2) The variable up the y-axis is the response or dependent variable — it's the variable you think is being affected. In this example, this is the fuel efficiency.

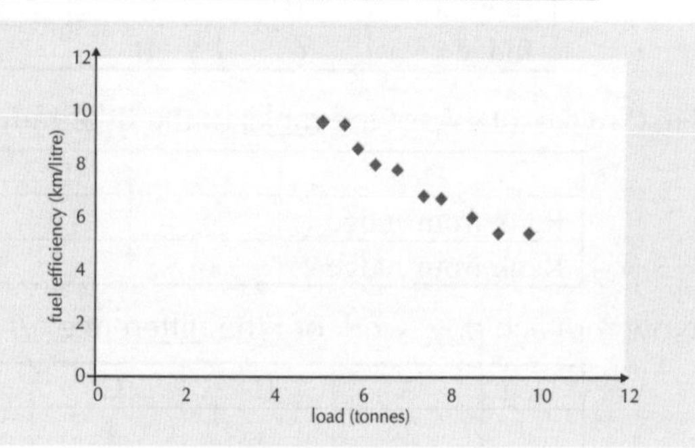

The **Regression Line** (Line of Best Fit) is in the form **y = a + bx**

To find the line of best fit for the above data you need to work out some sums.
Then it's quite easy to work out the equation of the line. If your line of best fit is $y = a + bx$, this is what you do...

① First work out these four sums — a table is probably the best way: $\sum x$, $\sum y$, $\sum x^2$, $\sum xy$.

x	5.1	5.6	5.9	6.3	6.8	7.4	7.8	8.5	9.1	9.8	$72.3 = \sum x$
y	9.6	9.5	8.6	8	7.8	6.8	6.7	6	5.4	5.4	$73.8 = \sum y$
x^2	26.01	31.36	34.81	39.69	46.24	54.76	60.84	72.25	82.81	96.04	$544.81 = \sum x^2$
xy	48.96	53.2	50.74	50.4	53.04	50.32	52.26	51	49.14	52.92	$511.98 = \sum xy$

② Then work out S_{xy}, given by: $S_{xy} = \sum(x-\bar{x})(y-\bar{y}) = \sum xy - \frac{(\sum x)(\sum y)}{n}$

and S_{xx}, given by: $S_{xx} = \sum(x-\bar{x})^2 = \sum x^2 - \frac{(\sum x)^2}{n}$

These are the same as the terms used to work out the PMCC (see p153).

③ The gradient (b) of your regression line is given by: $b = \frac{S_{xy}}{S_{xx}}$

④ And the intercept (a) is given by: $a = \bar{y} - b\bar{x}$.

⑤ Then the regression line is just: $y = a + bx$.

Loads of calculators will work out regression lines for you — but you still need to know this method, since they might give you just the sums from Step 1.

EXAMPLE Find the equation of the regression line of y on x for the data above.

The 'regression line of y on x' means that x is the independent variable, and y is the dependent variable.

1) Work out the sums: $\sum x = 72.3$, $\sum y = 73.8$, $\sum x^2 = 544.81$, $\sum xy = 511.98$.
2) Then work out S_{xy} and S_{xx}: $S_{xy} = 511.98 - \frac{72.3 \times 73.8}{10} = -21.594$, $S_{xx} = 544.81 - \frac{72.3^2}{10} = 22.081$
3) So the gradient of the regression line is: $b = \frac{-21.594}{22.081} = -0.978$ (to 3 sig. fig.)
4) And the intercept is: $a = \frac{\sum y}{n} - b\frac{\sum x}{n} = \frac{73.8}{10} - (-0.978) \times \frac{72.3}{10} = 14.451 = 14.5$ (to 3 sig. fig.)
5) This all means that your regression line is: $y = 14.5 - 0.978x$

Remember: $\bar{x} = \frac{\sum x}{n}$

The regression line always goes through the point $(\bar{x}, \bar{y})$.

Linear Regression

This page is for OCR MEI S2

So you've worked through the formulas and found a regression line. Now you need to know:
(i) what it means, (ii) if it's any good, and (iii) the dangers of using regression lines without due care and attention.

You should be able to Interpret your Regression Coefficients

On the previous page, you found the regression line of y (fuel efficiency in km per litre) on x (load in tonnes) was:

$$y = 14.5 - 0.978x$$

This tells you:

(i) for every extra tonne carried (i.e. as x increases by 1 unit), you'd expect the lorry's fuel efficiency (the corresponding value of y) to fall by 0.978 km per litre,

(ii) with no load ($x = 0$), you'd expect the lorry to do 14.5 km per litre of fuel.

Assuming the trend continues down to $x = 0$.

Residuals — the difference between Practice and Theory

A residual is the difference between an observed y-value and the y-value predicted by the regression line.

Residual = Observed y-value – Estimated y-value

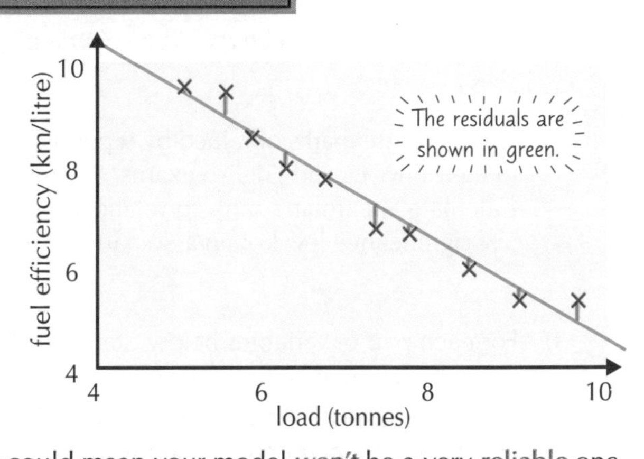

1) Residuals show the experimental error between the y-value that's observed and the y-value your regression line says it should be.
2) Residuals are shown by a vertical line from the actual point to the regression line.
3) Ideally, you'd like your residuals to be small — this would show your regression line fits the data well. If they're large (i.e. a high percentage of the dependent variable), then that could mean your model won't be a very reliable one.

EXAMPLE For the fuel efficiency example on the last page, calculate the residuals for: (i) $x = 5.6$, (ii) $x = 7.4$.

(i) When $x = 5.6$, the residual $= 9.5 - (-0.978 \times 5.6 + 14.5) = 0.477$ (to 3 sig. fig.)

(ii) When $x = 7.4$, the residual $= 6.8 - (-0.978 \times 7.4 + 14.5) = -0.463$ (to 3 sig. fig.)

A positive residual means the regression line is too low for that value of x.
A negative residual means the regression line is too high.

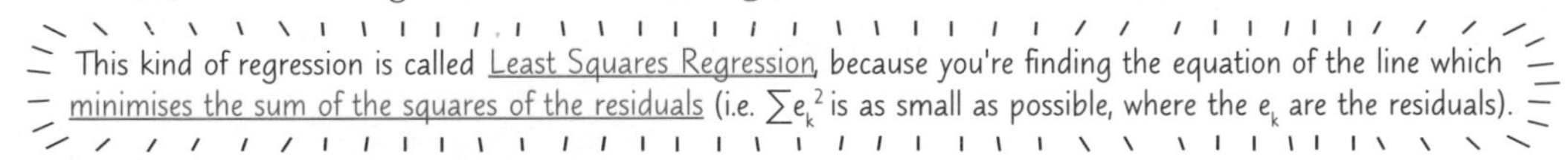
This kind of regression is called Least Squares Regression, because you're finding the equation of the line which minimises the sum of the squares of the residuals (i.e. $\sum e_k^2$ is as small as possible, where the e_k are the residuals).

Use Regression Lines With Care

You can use your regression line to predict values of y.
But it's best not to do this for x-values outside the range of your original table of values.

EXAMPLE Use your regression equation to estimate the value of y when: (i) $x = 7.6$, (ii) $x = 12.6$

(i) When $x = 7.6$, $y = -0.978 \times 7.6 + 14.5 = 7.07$ (to 3 sig. fig.). This should be a pretty reliable guess, since $x = 7.6$ falls in the range of x we already have readings for — this is called interpolation.

(ii) When $x = 12.6$, $y = -0.978 \times 12.6 + 14.5 = 2.18$ (to 3 sig. fig.). This may well be unreliable since $x = 12.6$ is bigger than the biggest x-value we already have — this is called extrapolation.

99% of all statisticians make sweeping statements...

Be careful with that extrapolation business — it's like me saying that because I grew at an average rate of 10 cm a year for the first few years of my life, by the time I'm 50 I should be 5 metres tall. Residuals are always errors in the values of y — these equations for working out the regression line all assume that you can measure x perfectly all the time.

S2 Section 7 — Practice Questions

That was a short section, but chock-full of fiddly terms and hefty equations. The only way to learn all those details is by using them — so stretch your maths muscles and take a jog around this obstacle course of practice questions.

Warm-up Questions

All questions are for OCR MEI S2

1) The table below shows the results of some measurements of randomly selected alcoholic cocktails. Here, x = total volume in ml, and y = percentage alcohol concentration by volume.

x	90	100	100	150	160	180	200	240	250	290	300
y	40	35	25	30	25	30	25	20	25	15	7

For the hypothesis tests on these pages, use critical values from either a formula booklet or the bits of tables on pages 154-5.

a) Draw a scatter diagram representing this information.

b) Calculate the product-moment correlation coefficient (PMCC) of these values.

c) Carry out a hypothesis test at a significance level of 5% to determine whether this is evidence of a correlation between the volume of alcoholic drinks and their alcohol concentration.

2) These are the marks obtained by 9 pupils in their Physics and English exams.

Physics	54	34	23	57	56	58	13	65	69
English	16	73	89	83	23	81	56	62	61

Calculate Spearman's rank correlation coefficient, and determine if this is evidence at a 5% significance level of an association between the marks in Physics and English exams.

3) For each pair of variables below, state which would be the dependent variable and which would be the independent variable.

a) • the annual number of volleyball-related injuries
 • the annual number of sunny days

b) • the annual number of rainy days
 • the annual number of Monopoly-related injuries

c) • a person's disposable income
 • a person's spending on luxuries

d) • the number of trips to the loo per day
 • the number of cups of tea drunk per day

e) • the number of festival tickets sold
 • the number of pairs of Wellington boots bought

4) The radius in mm, r, and the weight in grams, w, of 10 randomly selected blueberry pancakes are given in the table below.

r	48.0	51.0	52.0	54.5	55.1	53.6	50.0	52.6	49.4	51.2
w	100	105	108	120	125	118	100	115	98	110

a) Find: (i) $S_{rr} = \sum r^2 - \frac{(\sum r)^2}{n}$, (ii) $S_{rw} = \sum rw - \frac{(\sum r)(\sum w)}{n}$

The regression line of w on r has equation $w = a + br$.

b) Find b, the gradient of the regression line.

c) Find a, the intercept of the regression line on the w-axis.

d) Write down the equation of the regression line of w on r.

e) Use your regression line to estimate the weight of a blueberry pancake of radius 60 mm.

f) Comment on the reliability of your estimate, giving a reason for your answer.

S2 Section 7 — Practice Questions

Land ahoy, ye lily-livered yellow-bellies.
Just a quick heave-ho through these exam questions, then drop anchor and row ashore. Yaarrr, freedom...

Exam Questions

All questions are for OCR MEI S2

1 Thirteen Year-7 pupils in a school were randomly selected. For each pupil, two variables were recorded:
- the number of after-school catch-up sessions attended during a school year (x),
- their score in a maths test (y%).

The results are shown in the table below.

x	1	2	3	3	4	5	5	5	6	6	6	7	8
y	33	30	42	52	24	35	50	56	37	56	62	68	61

a) Represent this data on a scatter diagram. *(2 marks)*

b) Calculate the product-moment correlation coefficient (PMCC) between the two variables. *(4 marks)*

c) Carry out a hypothesis test at the 5% significance level to determine whether there is a correlation between the variables. *(6 marks)*

2 The following times (in seconds) were taken by eight different runners to complete distances of 20 metres and 60 metres.

Runner	A	B	C	D	E	F	G	H
20-metre time (x)	3.39	3.20	3.09	3.32	3.33	3.27	3.44	3.08
60-metre time (y)	8.78	7.73	8.28	8.25	8.91	8.59	8.90	8.05

a) Plot a scatter diagram to represent the data. *(2 marks)*

b) Find the equation of the regression line of y on x, and plot it on your scatter diagram. *(8 marks)*

c) Use the equation of the regression line to estimate the value of y when:
(i) $x = 3.15$, (ii) $x = 3.88$.
Comment on the reliability of your estimates. *(4 marks)*

d) Find the residuals for:
(i) $x = 3.32$ (ii) $x = 3.27$.
Illustrate them on your scatter diagram. *(4 marks)*

3 A journalist believes there is a positive correlation between the distance in miles, x, cycled during a training ride by cyclists from a particular cycling club and the number of calories, y, eaten at lunch. He gathers data from 10 randomly selected members of the club, and calculates the following statistics:

$$S_{xx} = 155\,440 \qquad S_{yy} = 395.5 \qquad S_{xy} = 6333$$

a) Use these values to calculate the product-moment correlation coefficient. *(2 marks)*

b) Carry out a hypothesis test at the 1% significance level to investigate the journalist's belief. State your hypotheses clearly. *(5 marks)*

c) State the assumption necessary for this test to be valid.
Explain how you might check whether this assumption is likely to be justified. *(2 marks)*

4 The equation of the line of regression for a set of data is $y = 211.599 + 9.602x$.

a) Use the equation of the regression line to estimate the value of y when:
(i) $x = 12.5$ (ii) $x = 14.7$. *(2 marks)*

b) Calculate the residuals if the respective observed y-values were $y = 332.5$ and $y = 352.1$. *(2 marks)*

S2 — STATISTICAL TABLES

Your tables may look slightly different to these — it depends on which exam board you're following. Make sure you understand how to use the tables you'll be using in your exam.

The normal distribution function

The table below shows $\Phi(z) = P(Z \leq z)$, where $Z \sim N(0, 1)$.

For negative z, use: $\Phi(-z) = 1 - \Phi(z)$.

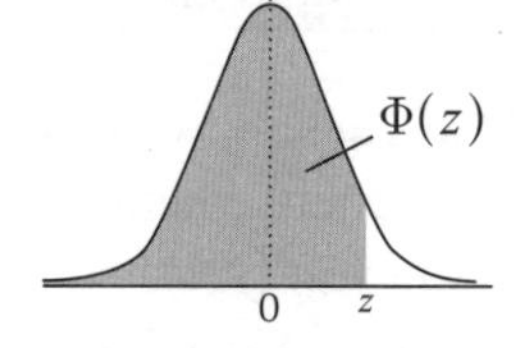

z	0	1	2	3	4	5	6	7	8	9	ADD 1	2	3	4	5	6	7	8	9
0.0	0.5000	0.5040	0.5080	0.5120	0.5160	0.5199	0.5239	0.5279	0.5319	0.5359	4	8	12	16	20	24	28	32	36
0.1	0.5398	0.5438	0.5478	0.5517	0.5557	0.5596	0.5636	0.5675	0.5714	0.5753	4	8	12	16	20	24	28	32	36
0.2	0.5793	0.5832	0.5871	0.5910	0.5948	0.5987	0.6026	0.6064	0.6103	0.6141	4	8	12	15	19	23	27	31	35
0.3	0.6179	0.6217	0.6255	0.6293	0.6331	0.6368	0.6406	0.6443	0.6480	0.6517	4	7	11	15	19	22	26	30	34
0.4	0.6554	0.6591	0.6628	0.6664	0.6700	0.6736	0.6772	0.6808	0.6844	0.6879	4	7	11	14	18	22	25	29	32
0.5	0.6915	0.6950	0.6985	0.7019	0.7054	0.7088	0.7123	0.7157	0.7190	0.7224	3	7	10	14	17	20	24	27	31
0.6	0.7257	0.7291	0.7324	0.7357	0.7389	0.7422	0.7454	0.7486	0.7517	0.7549	3	7	10	13	16	19	23	26	29
0.7	0.7580	0.7611	0.7642	0.7673	0.7704	0.7734	0.7764	0.7794	0.7823	0.7852	3	6	9	12	15	18	21	24	27
0.8	0.7881	0.7910	0.7939	0.7967	0.7995	0.8023	0.8051	0.8078	0.8106	0.8133	3	5	8	11	14	16	19	22	25
0.9	0.8159	0.8186	0.8212	0.8238	0.8264	0.8289	0.8315	0.8340	0.8365	0.8389	3	5	8	10	13	15	18	20	23
1.0	0.8413	0.8438	0.8461	0.8485	0.8508	0.8531	0.8554	0.8577	0.8599	0.8621	2	5	7	9	12	14	16	19	21
1.1	0.8643	0.8665	0.8686	0.8708	0.8729	0.8749	0.8770	0.8790	0.8810	0.8830	2	4	6	8	10	12	14	16	18
1.2	0.8849	0.8869	0.8888	0.8907	0.8925	0.8944	0.8962	0.8980	0.8997	0.9015	2	4	6	7	9	11	13	15	17
1.3	0.9032	0.9049	0.9066	0.9082	0.9099	0.9115	0.9131	0.9147	0.9162	0.9177	2	3	5	6	8	10	11	13	14
1.4	0.9192	0.9207	0.9222	0.9236	0.9251	0.9265	0.9279	0.9292	0.9306	0.9319	1	3	4	6	7	8	10	11	13
1.5	0.9332	0.9345	0.9357	0.9370	0.9382	0.9394	0.9406	0.9418	0.9429	0.9441	1	2	4	5	6	7	8	10	11
1.6	0.9452	0.9463	0.9474	0.9484	0.9495	0.9505	0.9515	0.9525	0.9535	0.9545	1	2	3	4	5	6	7	8	9
1.7	0.9554	0.9564	0.9573	0.9582	0.9591	0.9599	0.9608	0.9616	0.9625	0.9633	1	2	3	4	4	5	6	7	8
1.8	0.9641	0.9649	0.9656	0.9664	0.9671	0.9678	0.9686	0.9693	0.9699	0.9706	1	1	2	3	4	4	5	6	6
1.9	0.9713	0.9719	0.9726	0.9732	0.9738	0.9744	0.9750	0.9756	0.9761	0.9767	1	1	2	2	3	4	4	5	5
2.0	0.9772	0.9778	0.9783	0.9788	0.9793	0.9798	0.9803	0.9808	0.9812	0.9817	0	1	1	2	2	3	3	4	4
2.1	0.9821	0.9826	0.9830	0.9834	0.9838	0.9842	0.9846	0.9850	0.9854	0.9857	0	1	1	2	2	2	3	3	4
2.2	0.9861	0.9864	0.9868	0.9871	0.9875	0.9878	0.9881	0.9884	0.9887	0.9890	0	1	1	1	2	2	2	3	3
2.3	0.9893	0.9896	0.9898	0.9901	0.9904	0.9906	0.9909	0.9911	0.9913	0.9916	0	1	1	1	1	2	2	2	2
2.4	0.9918	0.9920	0.9922	0.9925	0.9927	0.9929	0.9931	0.9932	0.9934	0.9936	0	0	1	1	1	1	1	2	2
2.5	0.9938	0.9940	0.9941	0.9943	0.9945	0.9946	0.9948	0.9949	0.9951	0.9952	0	0	0	1	1	1	1	1	1
2.6	0.9953	0.9955	0.9956	0.9957	0.9959	0.9960	0.9961	0.9962	0.9963	0.9964	0	0	0	0	1	1	1	1	1
2.7	0.9965	0.9966	0.9967	0.9968	0.9969	0.9970	0.9971	0.9972	0.9973	0.9974	0	0	0	0	0	1	1	1	1
2.8	0.9974	0.9975	0.9976	0.9977	0.9977	0.9978	0.9979	0.9979	0.9980	0.9981	0	0	0	0	0	0	0	1	1
2.9	0.9981	0.9982	0.9982	0.9983	0.9984	0.9984	0.9985	0.9985	0.9986	0.9986	0	0	0	0	0	0	0	0	0

Critical values for the normal distribution

For $Z \sim N(0, 1)$, this table gives the value of z for which $P(Z \leq z) = p$.

p	0.75	0.90	0.95	0.975	0.99	0.995	0.9975	0.999
z	0.674	1.282	1.645	1.960	2.326	2.576	2.807	3.090

S2 — STATISTICAL TABLES

The binomial cumulative distribution function

The values below show $P(X \leq x)$, where $X \sim B(n, p)$.

	$p =$	0.05	0.10	0.15	0.20	0.25	0.30	0.35	0.40	0.45	0.50
$n = 5$	$x =$ 0	0.7738	0.5905	0.4437	0.3277	0.2373	0.1681	0.1160	0.0778	0.0503	0.0313
	1	0.9774	0.9185	0.8352	0.7373	0.6328	0.5282	0.4284	0.3370	0.2562	0.1875
	2	0.9988	0.9914	0.9734	0.9421	0.8965	0.8369	0.7648	0.6826	0.5931	0.5000
	3	1.0000	0.9995	0.9978	0.9933	0.9844	0.9692	0.9460	0.9130	0.8688	0.8125
	4	1.0000	1.0000	0.9999	0.9997	0.9990	0.9976	0.9947	0.9898	0.9815	0.9688
$n = 6$	$x =$ 0	0.7351	0.5314	0.3771	0.2621	0.1780	0.1176	0.0754	0.0467	0.0277	0.0156
	1	0.9672	0.8857	0.7765	0.6554	0.5339	0.4202	0.3191	0.2333	0.1636	0.1094
	2	0.9978	0.9842	0.9527	0.9011	0.8306	0.7443	0.6471	0.5443	0.4415	0.3438
	3	0.9999	0.9987	0.9941	0.9830	0.9624	0.9295	0.8826	0.8208	0.7447	0.6563
	4	1.0000	0.9999	0.9996	0.9984	0.9954	0.9891	0.9777	0.9590	0.9308	0.8906
	5	1.0000	1.0000	1.0000	0.9999	0.9998	0.9993	0.9982	0.9959	0.9917	0.9844
$n = 7$	$x =$ 0	0.6983	0.4783	0.3206	0.2097	0.1335	0.0824	0.0490	0.0280	0.0152	0.0078
	1	0.9556	0.8503	0.7166	0.5767	0.4449	0.3294	0.2338	0.1586	0.1024	0.0625
	2	0.9962	0.9743	0.9262	0.8520	0.7564	0.6471	0.5323	0.4199	0.3164	0.2266
	3	0.9998	0.9973	0.9879	0.9667	0.9294	0.8740	0.8002	0.7102	0.6083	0.5000
	4	1.0000	0.9998	0.9988	0.9953	0.9871	0.9712	0.9444	0.9037	0.8471	0.7734
	5	1.0000	1.0000	0.9999	0.9996	0.9987	0.9962	0.9910	0.9812	0.9643	0.9375
	6	1.0000	1.0000	1.0000	1.0000	0.9999	0.9998	0.9994	0.9984	0.9963	0.9922
$n = 8$	$x =$ 0	0.6634	0.4305	0.2725	0.1678	0.1001	0.0576	0.0319	0.0168	0.0084	0.0039
	1	0.9428	0.8131	0.6572	0.5033	0.3671	0.2553	0.1691	0.1064	0.0632	0.0352
	2	0.9942	0.9619	0.8948	0.7969	0.6785	0.5518	0.4278	0.3154	0.2201	0.1445
	3	0.9996	0.9950	0.9786	0.9437	0.8862	0.8059	0.7064	0.5941	0.4770	0.3633
	4	1.0000	0.9996	0.9971	0.9896	0.9727	0.9420	0.8939	0.8263	0.7396	0.6367
	5	1.0000	1.0000	0.9998	0.9988	0.9958	0.9887	0.9747	0.9502	0.9115	0.8555
	6	1.0000	1.0000	1.0000	0.9999	0.9996	0.9987	0.9964	0.9915	0.9819	0.9648
	7	1.0000	1.0000	1.0000	1.0000	1.0000	0.9999	0.9998	0.9993	0.9983	0.9961
$n = 9$	$x =$ 0	0.6302	0.3874	0.2316	0.1342	0.0751	0.0404	0.0207	0.0101	0.0046	0.0020
	1	0.9288	0.7748	0.5995	0.4362	0.3003	0.1960	0.1211	0.0705	0.0385	0.0195
	2	0.9916	0.9470	0.8591	0.7382	0.6007	0.4628	0.3373	0.2318	0.1495	0.0898
	3	0.9994	0.9917	0.9661	0.9144	0.8343	0.7297	0.6089	0.4826	0.3614	0.2539
	4	1.0000	0.9991	0.9944	0.9804	0.9511	0.9012	0.8283	0.7334	0.6214	0.5000
	5	1.0000	0.9999	0.9994	0.9969	0.9900	0.9747	0.9464	0.9006	0.8342	0.7461
	6	1.0000	1.0000	1.0000	0.9997	0.9987	0.9957	0.9888	0.9750	0.9502	0.9102
	7	1.0000	1.0000	1.0000	1.0000	0.9999	0.9996	0.9986	0.9962	0.9909	0.9805
	8	1.0000	1.0000	1.0000	1.0000	1.0000	1.0000	0.9999	0.9997	0.9992	0.9980
$n = 10$	$x =$ 0	0.5987	0.3487	0.1969	0.1074	0.0563	0.0282	0.0135	0.0060	0.0025	0.0010
	1	0.9139	0.7361	0.5443	0.3758	0.2440	0.1493	0.0860	0.0464	0.0233	0.0107
	2	0.9885	0.9298	0.8202	0.6778	0.5256	0.3828	0.2616	0.1673	0.0996	0.0547
	3	0.9990	0.9872	0.9500	0.8791	0.7759	0.6496	0.5138	0.3823	0.2660	0.1719
	4	0.9999	0.9984	0.9901	0.9672	0.9219	0.8497	0.7515	0.6331	0.5044	0.3770
	5	1.0000	0.9999	0.9986	0.9936	0.9803	0.9527	0.9051	0.8338	0.7384	0.6230
	6	1.0000	1.0000	0.9999	0.9991	0.9965	0.9894	0.9740	0.9452	0.8980	0.8281
	7	1.0000	1.0000	1.0000	0.9999	0.9996	0.9984	0.9952	0.9877	0.9726	0.9453
	8	1.0000	1.0000	1.0000	1.0000	1.0000	0.9999	0.9995	0.9983	0.9955	0.9893
	9	1.0000	1.0000	1.0000	1.0000	1.0000	1.0000	1.0000	0.9999	0.9997	0.9990

S2 — STATISTICAL TABLES

The binomial cumulative distribution function (continued)

p =	0.05	0.10	0.15	0.20	0.25	0.30	0.35	0.40	0.45	0.50
n = 12 x = 0	0.5404	0.2824	0.1422	0.0687	0.0317	0.0138	0.0057	0.0022	0.0008	0.0002
1	0.8816	0.6590	0.4435	0.2749	0.1584	0.0850	0.0424	0.0196	0.0083	0.0032
2	0.9804	0.8891	0.7358	0.5583	0.3907	0.2528	0.1513	0.0834	0.0421	0.0193
3	0.9978	0.9744	0.9078	0.7946	0.6488	0.4925	0.3467	0.2253	0.1345	0.0730
4	0.9998	0.9957	0.9761	0.9274	0.8424	0.7237	0.5833	0.4382	0.3044	0.1938
5	1.0000	0.9995	0.9954	0.9806	0.9456	0.8822	0.7873	0.6652	0.5269	0.3872
6	1.0000	0.9999	0.9993	0.9961	0.9857	0.9614	0.9154	0.8418	0.7393	0.6128
7	1.0000	1.0000	0.9999	0.9994	0.9972	0.9905	0.9745	0.9427	0.8883	0.8062
8	1.0000	1.0000	1.0000	0.9999	0.9996	0.9983	0.9944	0.9847	0.9644	0.9270
9	1.0000	1.0000	1.0000	1.0000	1.0000	0.9998	0.9992	0.9972	0.9921	0.9807
10	1.0000	1.0000	1.0000	1.0000	1.0000	1.0000	0.9999	0.9997	0.9989	0.9968
11	1.0000	1.0000	1.0000	1.0000	1.0000	1.0000	1.0000	1.0000	0.9999	0.9998
n = 15 x = 0	0.4633	0.2059	0.0874	0.0352	0.0134	0.0047	0.0016	0.0005	0.0001	0.0000
1	0.8290	0.5490	0.3186	0.1671	0.0802	0.0353	0.0142	0.0052	0.0017	0.0005
2	0.9638	0.8159	0.6042	0.3980	0.2361	0.1268	0.0617	0.0271	0.0107	0.0037
3	0.9945	0.9444	0.8227	0.6482	0.4613	0.2969	0.1727	0.0905	0.0424	0.0176
4	0.9994	0.9873	0.9383	0.8358	0.6865	0.5155	0.3519	0.2173	0.1204	0.0592
5	0.9999	0.9978	0.9832	0.9389	0.8516	0.7216	0.5643	0.4032	0.2608	0.1509
6	1.0000	0.9997	0.9964	0.9819	0.9434	0.8689	0.7548	0.6098	0.4522	0.3036
7	1.0000	1.0000	0.9994	0.9958	0.9827	0.9500	0.8868	0.7869	0.6535	0.5000
8	1.0000	1.0000	0.9999	0.9992	0.9958	0.9848	0.9578	0.9050	0.8182	0.6964
9	1.0000	1.0000	1.0000	0.9999	0.9992	0.9963	0.9876	0.9662	0.9231	0.8491
10	1.0000	1.0000	1.0000	1.0000	0.9999	0.9993	0.9972	0.9907	0.9745	0.9408
11	1.0000	1.0000	1.0000	1.0000	1.0000	0.9999	0.9995	0.9981	0.9937	0.9824
12	1.0000	1.0000	1.0000	1.0000	1.0000	1.0000	0.9999	0.9997	0.9989	0.9963
13	1.0000	1.0000	1.0000	1.0000	1.0000	1.0000	1.0000	1.0000	0.9999	0.9995
14	1.0000	1.0000	1.0000	1.0000	1.0000	1.0000	1.0000	1.0000	1.0000	1.0000
n = 20 x = 0	0.3585	0.1216	0.0388	0.0115	0.0032	0.0008	0.0002	0.0000	0.0000	0.0000
1	0.7358	0.3917	0.1756	0.0692	0.0243	0.0076	0.0021	0.0005	0.0001	0.0000
2	0.9245	0.6769	0.4049	0.2061	0.0913	0.0355	0.0121	0.0036	0.0009	0.0002
3	0.9841	0.8670	0.6477	0.4114	0.2252	0.1071	0.0444	0.0160	0.0049	0.0013
4	0.9974	0.9568	0.8298	0.6296	0.4148	0.2375	0.1182	0.0510	0.0189	0.0059
5	0.9997	0.9887	0.9327	0.8042	0.6172	0.4164	0.2454	0.1256	0.0553	0.0207
6	1.0000	0.9976	0.9781	0.9133	0.7858	0.6080	0.4166	0.2500	0.1299	0.0577
7	1.0000	0.9996	0.9941	0.9679	0.8982	0.7723	0.6010	0.4159	0.2520	0.1316
8	1.0000	0.9999	0.9987	0.9900	0.9591	0.8867	0.7624	0.5956	0.4143	0.2517
9	1.0000	1.0000	0.9998	0.9974	0.9861	0.9520	0.8782	0.7553	0.5914	0.4119
10	1.0000	1.0000	1.0000	0.9994	0.9961	0.9829	0.9468	0.8725	0.7507	0.5881
11	1.0000	1.0000	1.0000	0.9999	0.9991	0.9949	0.9804	0.9435	0.8692	0.7483
12	1.0000	1.0000	1.0000	1.0000	0.9998	0.9987	0.9940	0.9790	0.9420	0.8684
13	1.0000	1.0000	1.0000	1.0000	1.0000	0.9997	0.9985	0.9935	0.9786	0.9423
14	1.0000	1.0000	1.0000	1.0000	1.0000	1.0000	0.9997	0.9984	0.9936	0.9793
15	1.0000	1.0000	1.0000	1.0000	1.0000	1.0000	1.0000	0.9997	0.9985	0.9941
16	1.0000	1.0000	1.0000	1.0000	1.0000	1.0000	1.0000	1.0000	0.9997	0.9987
17	1.0000	1.0000	1.0000	1.0000	1.0000	1.0000	1.0000	1.0000	1.0000	0.9998
18	1.0000	1.0000	1.0000	1.0000	1.0000	1.0000	1.0000	1.0000	1.0000	1.0000

The binomial cumulative distribution function (continued)

	p =	0.05	0.10	0.15	0.20	0.25	0.30	0.35	0.40	0.45	0.50
n = 25	x = 0	0.2774	0.0718	0.0172	0.0038	0.0008	0.0001	0.0000	0.0000	0.0000	0.0000
	1	0.6424	0.2712	0.0931	0.0274	0.0070	0.0016	0.0003	0.0001	0.0000	0.0000
	2	0.8729	0.5371	0.2537	0.0982	0.0321	0.0090	0.0021	0.0004	0.0001	0.0000
	3	0.9659	0.7636	0.4711	0.2340	0.0962	0.0332	0.0097	0.0024	0.0005	0.0001
	4	0.9928	0.9020	0.6821	0.4207	0.2137	0.0905	0.0320	0.0095	0.0023	0.0005
	5	0.9988	0.9666	0.8385	0.6167	0.3783	0.1935	0.0826	0.0294	0.0086	0.0020
	6	0.9998	0.9905	0.9305	0.7800	0.5611	0.3407	0.1734	0.0736	0.0258	0.0073
	7	1.0000	0.9977	0.9745	0.8909	0.7265	0.5118	0.3061	0.1536	0.0639	0.0216
	8	1.0000	0.9995	0.9920	0.9532	0.8506	0.6769	0.4668	0.2735	0.1340	0.0539
	9	1.0000	0.9999	0.9979	0.9827	0.9287	0.8106	0.6303	0.4246	0.2424	0.1148
	10	1.0000	1.0000	0.9995	0.9944	0.9703	0.9022	0.7712	0.5858	0.3843	0.2122
	11	1.0000	1.0000	0.9999	0.9985	0.9893	0.9558	0.8746	0.7323	0.5426	0.3450
	12	1.0000	1.0000	1.0000	0.9996	0.9966	0.9825	0.9396	0.8462	0.6937	0.5000
	13	1.0000	1.0000	1.0000	0.9999	0.9991	0.9940	0.9745	0.9222	0.8173	0.6550
	14	1.0000	1.0000	1.0000	1.0000	0.9998	0.9982	0.9907	0.9656	0.9040	0.7878
	15	1.0000	1.0000	1.0000	1.0000	1.0000	0.9995	0.9971	0.9868	0.9560	0.8852
	16	1.0000	1.0000	1.0000	1.0000	1.0000	0.9999	0.9992	0.9957	0.9826	0.9461
	17	1.0000	1.0000	1.0000	1.0000	1.0000	1.0000	0.9998	0.9988	0.9942	0.9784
	18	1.0000	1.0000	1.0000	1.0000	1.0000	1.0000	1.0000	0.9997	0.9984	0.9927
	19	1.0000	1.0000	1.0000	1.0000	1.0000	1.0000	1.0000	0.9999	0.9996	0.9980
	20	1.0000	1.0000	1.0000	1.0000	1.0000	1.0000	1.0000	1.0000	0.9999	0.9995
	21	1.0000	1.0000	1.0000	1.0000	1.0000	1.0000	1.0000	1.0000	1.0000	0.9999
	22	1.0000	1.0000	1.0000	1.0000	1.0000	1.0000	1.0000	1.0000	1.0000	1.0000
n = 30	x = 0	0.2146	0.0424	0.0076	0.0012	0.0002	0.0000	0.0000	0.0000	0.0000	0.0000
	1	0.5535	0.1837	0.0480	0.0105	0.0020	0.0003	0.0000	0.0000	0.0000	0.0000
	2	0.8122	0.4114	0.1514	0.0442	0.0106	0.0021	0.0003	0.0000	0.0000	0.0000
	3	0.9392	0.6474	0.3217	0.1227	0.0374	0.0093	0.0019	0.0003	0.0000	0.0000
	4	0.9844	0.8245	0.5245	0.2552	0.0979	0.0302	0.0075	0.0015	0.0002	0.0000
	5	0.9967	0.9268	0.7106	0.4275	0.2026	0.0766	0.0233	0.0057	0.0011	0.0002
	6	0.9994	0.9742	0.8474	0.6070	0.3481	0.1595	0.0586	0.0172	0.0040	0.0007
	7	0.9999	0.9922	0.9302	0.7608	0.5143	0.2814	0.1238	0.0435	0.0121	0.0026
	8	1.0000	0.9980	0.9722	0.8713	0.6736	0.4315	0.2247	0.0940	0.0312	0.0081
	9	1.0000	0.9995	0.9903	0.9389	0.8034	0.5888	0.3575	0.1763	0.0694	0.0214
	10	1.0000	0.9999	0.9971	0.9744	0.8943	0.7304	0.5078	0.2915	0.1350	0.0494
	11	1.0000	1.0000	0.9992	0.9905	0.9493	0.8407	0.6548	0.4311	0.2327	0.1002
	12	1.0000	1.0000	0.9998	0.9969	0.9784	0.9155	0.7802	0.5785	0.3592	0.1808
	13	1.0000	1.0000	1.0000	0.9991	0.9918	0.9599	0.8737	0.7145	0.5025	0.2923
	14	1.0000	1.0000	1.0000	0.9998	0.9973	0.9831	0.9348	0.8246	0.6448	0.4278
	15	1.0000	1.0000	1.0000	0.9999	0.9992	0.9936	0.9699	0.9029	0.7691	0.5722
	16	1.0000	1.0000	1.0000	1.0000	0.9998	0.9979	0.9876	0.9519	0.8644	0.7077
	17	1.0000	1.0000	1.0000	1.0000	0.9999	0.9994	0.9955	0.9788	0.9286	0.8192
	18	1.0000	1.0000	1.0000	1.0000	1.0000	0.9998	0.9986	0.9917	0.9666	0.8998
	19	1.0000	1.0000	1.0000	1.0000	1.0000	1.0000	0.9996	0.9971	0.9862	0.9506
	20	1.0000	1.0000	1.0000	1.0000	1.0000	1.0000	0.9999	0.9991	0.9950	0.9786
	21	1.0000	1.0000	1.0000	1.0000	1.0000	1.0000	1.0000	0.9998	0.9984	0.9919
	22	1.0000	1.0000	1.0000	1.0000	1.0000	1.0000	1.0000	1.0000	0.9996	0.9974
	23	1.0000	1.0000	1.0000	1.0000	1.0000	1.0000	1.0000	1.0000	0.9999	0.9993
	24	1.0000	1.0000	1.0000	1.0000	1.0000	1.0000	1.0000	1.0000	1.0000	0.9998
	25	1.0000	1.0000	1.0000	1.0000	1.0000	1.0000	1.0000	1.0000	1.0000	1.0000

The binomial cumulative distribution function (continued)

	p =	0.05	0.10	0.15	0.20	0.25	0.30	0.35	0.40	0.45	0.50
n = 40	x = 0	0.1285	0.0148	0.0015	0.0001	0.0000	0.0000	0.0000	0.0000	0.0000	0.0000
	1	0.3991	0.0805	0.0121	0.0015	0.0001	0.0000	0.0000	0.0000	0.0000	0.0000
	2	0.6767	0.2228	0.0486	0.0079	0.0010	0.0001	0.0000	0.0000	0.0000	0.0000
	3	0.8619	0.4231	0.1302	0.0285	0.0047	0.0006	0.0001	0.0000	0.0000	0.0000
	4	0.9520	0.6290	0.2633	0.0759	0.0160	0.0026	0.0003	0.0000	0.0000	0.0000
	5	0.9861	0.7937	0.4325	0.1613	0.0433	0.0086	0.0013	0.0001	0.0000	0.0000
	6	0.9966	0.9005	0.6067	0.2859	0.0962	0.0238	0.0044	0.0006	0.0001	0.0000
	7	0.9993	0.9581	0.7559	0.4371	0.1820	0.0553	0.0124	0.0021	0.0002	0.0000
	8	0.9999	0.9845	0.8646	0.5931	0.2998	0.1110	0.0303	0.0061	0.0009	0.0001
	9	1.0000	0.9949	0.9328	0.7318	0.4395	0.1959	0.0644	0.0156	0.0027	0.0003
	10	1.0000	0.9985	0.9701	0.8392	0.5839	0.3087	0.1215	0.0352	0.0074	0.0011
	11	1.0000	0.9996	0.9880	0.9125	0.7151	0.4406	0.2053	0.0709	0.0179	0.0032
	12	1.0000	0.9999	0.9957	0.9568	0.8209	0.5772	0.3143	0.1285	0.0386	0.0083
	13	1.0000	1.0000	0.9986	0.9806	0.8968	0.7032	0.4408	0.2112	0.0751	0.0192
	14	1.0000	1.0000	0.9996	0.9921	0.9456	0.8074	0.5721	0.3174	0.1326	0.0403
	15	1.0000	1.0000	0.9999	0.9971	0.9738	0.8849	0.6946	0.4402	0.2142	0.0769
	16	1.0000	1.0000	1.0000	0.9990	0.9884	0.9367	0.7978	0.5681	0.3185	0.1341
	17	1.0000	1.0000	1.0000	0.9997	0.9953	0.9680	0.8761	0.6885	0.4391	0.2148
	18	1.0000	1.0000	1.0000	0.9999	0.9983	0.9852	0.9301	0.7911	0.5651	0.3179
	19	1.0000	1.0000	1.0000	1.0000	0.9994	0.9937	0.9637	0.8702	0.6844	0.4373
	20	1.0000	1.0000	1.0000	1.0000	0.9998	0.9976	0.9827	0.9256	0.7870	0.5627
	21	1.0000	1.0000	1.0000	1.0000	1.0000	0.9991	0.9925	0.9608	0.8669	0.6821
	22	1.0000	1.0000	1.0000	1.0000	1.0000	0.9997	0.9970	0.9811	0.9233	0.7852
	23	1.0000	1.0000	1.0000	1.0000	1.0000	0.9999	0.9989	0.9917	0.9595	0.8659
	24	1.0000	1.0000	1.0000	1.0000	1.0000	1.0000	0.9996	0.9966	0.9804	0.9231
	25	1.0000	1.0000	1.0000	1.0000	1.0000	1.0000	0.9999	0.9988	0.9914	0.9597
	26	1.0000	1.0000	1.0000	1.0000	1.0000	1.0000	1.0000	0.9996	0.9966	0.9808
	27	1.0000	1.0000	1.0000	1.0000	1.0000	1.0000	1.0000	0.9999	0.9988	0.9917
	28	1.0000	1.0000	1.0000	1.0000	1.0000	1.0000	1.0000	1.0000	0.9996	0.9968
	29	1.0000	1.0000	1.0000	1.0000	1.0000	1.0000	1.0000	1.0000	0.9999	0.9989
	30	1.0000	1.0000	1.0000	1.0000	1.0000	1.0000	1.0000	1.0000	1.0000	0.9997
	31	1.0000	1.0000	1.0000	1.0000	1.0000	1.0000	1.0000	1.0000	1.0000	0.9999
	32	1.0000	1.0000	1.0000	1.0000	1.0000	1.0000	1.0000	1.0000	1.0000	1.0000

The binomial cumulative distribution function (continued)

	$p =$	0.05	0.10	0.15	0.20	0.25	0.30	0.35	0.40	0.45	0.50
$n = 50$	$x =$ 0	0.0769	0.0052	0.0003	0.0000	0.0000	0.0000	0.0000	0.0000	0.0000	0.0000
	1	0.2794	0.0338	0.0029	0.0002	0.0000	0.0000	0.0000	0.0000	0.0000	0.0000
	2	0.5405	0.1117	0.0142	0.0013	0.0001	0.0000	0.0000	0.0000	0.0000	0.0000
	3	0.7604	0.2503	0.0460	0.0057	0.0005	0.0000	0.0000	0.0000	0.0000	0.0000
	4	0.8964	0.4312	0.1121	0.0185	0.0021	0.0002	0.0000	0.0000	0.0000	0.0000
	5	0.9622	0.6161	0.2194	0.0480	0.0070	0.0007	0.0001	0.0000	0.0000	0.0000
	6	0.9882	0.7702	0.3613	0.1034	0.0194	0.0025	0.0002	0.0000	0.0000	0.0000
	7	0.9968	0.8779	0.5188	0.1904	0.0453	0.0073	0.0008	0.0001	0.0000	0.0000
	8	0.9992	0.9421	0.6681	0.3073	0.0916	0.0183	0.0025	0.0002	0.0000	0.0000
	9	0.9998	0.9755	0.7911	0.4437	0.1637	0.0402	0.0067	0.0008	0.0001	0.0000
	10	1.0000	0.9906	0.8801	0.5836	0.2622	0.0789	0.0160	0.0022	0.0002	0.0000
	11	1.0000	0.9968	0.9372	0.7107	0.3816	0.1390	0.0342	0.0057	0.0006	0.0000
	12	1.0000	0.9990	0.9699	0.8139	0.5110	0.2229	0.0661	0.0133	0.0018	0.0002
	13	1.0000	0.9997	0.9868	0.8894	0.6370	0.3279	0.1163	0.0280	0.0045	0.0005
	14	1.0000	0.9999	0.9947	0.9393	0.7481	0.4468	0.1878	0.0540	0.0104	0.0013
	15	1.0000	1.0000	0.9981	0.9692	0.8369	0.5692	0.2801	0.0955	0.0220	0.0033
	16	1.0000	1.0000	0.9993	0.9856	0.9017	0.6839	0.3889	0.1561	0.0427	0.0077
	17	1.0000	1.0000	0.9998	0.9937	0.9449	0.7822	0.5060	0.2369	0.0765	0.0164
	18	1.0000	1.0000	0.9999	0.9975	0.9713	0.8594	0.6216	0.3356	0.1273	0.0325
	19	1.0000	1.0000	1.0000	0.9991	0.9861	0.9152	0.7264	0.4465	0.1974	0.0595
	20	1.0000	1.0000	1.0000	0.9997	0.9937	0.9522	0.8139	0.5610	0.2862	0.1013
	21	1.0000	1.0000	1.0000	0.9999	0.9974	0.9749	0.8813	0.6701	0.3900	0.1611
	22	1.0000	1.0000	1.0000	1.0000	0.9990	0.9877	0.9290	0.7660	0.5019	0.2399
	23	1.0000	1.0000	1.0000	1.0000	0.9996	0.9944	0.9604	0.8438	0.6134	0.3359
	24	1.0000	1.0000	1.0000	1.0000	0.9999	0.9976	0.9793	0.9022	0.7160	0.4439
	25	1.0000	1.0000	1.0000	1.0000	1.0000	0.9991	0.9900	0.9427	0.8034	0.5561
	26	1.0000	1.0000	1.0000	1.0000	1.0000	0.9997	0.9955	0.9686	0.8721	0.6641
	27	1.0000	1.0000	1.0000	1.0000	1.0000	0.9999	0.9981	0.9840	0.9220	0.7601
	28	1.0000	1.0000	1.0000	1.0000	1.0000	1.0000	0.9993	0.9924	0.9556	0.8389
	29	1.0000	1.0000	1.0000	1.0000	1.0000	1.0000	0.9997	0.9966	0.9765	0.8987
	30	1.0000	1.0000	1.0000	1.0000	1.0000	1.0000	0.9999	0.9986	0.9884	0.9405
	31	1.0000	1.0000	1.0000	1.0000	1.0000	1.0000	1.0000	0.9995	0.9947	0.9675
	32	1.0000	1.0000	1.0000	1.0000	1.0000	1.0000	1.0000	0.9998	0.9978	0.9836
	33	1.0000	1.0000	1.0000	1.0000	1.0000	1.0000	1.0000	0.9999	0.9991	0.9923
	34	1.0000	1.0000	1.0000	1.0000	1.0000	1.0000	1.0000	1.0000	0.9997	0.9967
	35	1.0000	1.0000	1.0000	1.0000	1.0000	1.0000	1.0000	1.0000	0.9999	0.9987
	36	1.0000	1.0000	1.0000	1.0000	1.0000	1.0000	1.0000	1.0000	1.0000	0.9995
	37	1.0000	1.0000	1.0000	1.0000	1.0000	1.0000	1.0000	1.0000	1.0000	0.9998
	38	1.0000	1.0000	1.0000	1.0000	1.0000	1.0000	1.0000	1.0000	1.0000	1.0000

S2 — STATISTICAL TABLES

The Poisson cumulative distribution function

The values below show $P(X \leq x)$, where $X \sim Po(\lambda)$.

$\lambda =$	0.5	1.0	1.5	2.0	2.5	3.0	3.5	4.0	4.5	5.0
$x =$ 0	0.6065	0.3679	0.2231	0.1353	0.0821	0.0498	0.0302	0.0183	0.0111	0.0067
1	0.9098	0.7358	0.5578	0.4060	0.2873	0.1991	0.1359	0.0916	0.0611	0.0404
2	0.9856	0.9197	0.8088	0.6767	0.5438	0.4232	0.3208	0.2381	0.1736	0.1247
3	0.9982	0.9810	0.9344	0.8571	0.7576	0.6472	0.5366	0.4335	0.3423	0.2650
4	0.9998	0.9963	0.9814	0.9473	0.8912	0.8153	0.7254	0.6288	0.5321	0.4405
5	1.0000	0.9994	0.9955	0.9834	0.9580	0.9161	0.8576	0.7851	0.7029	0.6160
6	1.0000	0.9999	0.9991	0.9955	0.9858	0.9665	0.9347	0.8893	0.8311	0.7622
7	1.0000	1.0000	0.9998	0.9989	0.9958	0.9881	0.9733	0.9489	0.9134	0.8666
8	1.0000	1.0000	1.0000	0.9998	0.9989	0.9962	0.9901	0.9786	0.9597	0.9319
9	1.0000	1.0000	1.0000	1.0000	0.9997	0.9989	0.9967	0.9919	0.9829	0.9682
10	1.0000	1.0000	1.0000	1.0000	0.9999	0.9997	0.9990	0.9972	0.9933	0.9863
11	1.0000	1.0000	1.0000	1.0000	1.0000	0.9999	0.9997	0.9991	0.9976	0.9945
12	1.0000	1.0000	1.0000	1.0000	1.0000	1.0000	0.9999	0.9997	0.9992	0.9980
13	1.0000	1.0000	1.0000	1.0000	1.0000	1.0000	1.0000	0.9999	0.9997	0.9993
14	1.0000	1.0000	1.0000	1.0000	1.0000	1.0000	1.0000	1.0000	0.9999	0.9998
15	1.0000	1.0000	1.0000	1.0000	1.0000	1.0000	1.0000	1.0000	1.0000	0.9999
16	1.0000	1.0000	1.0000	1.0000	1.0000	1.0000	1.0000	1.0000	1.0000	1.0000
17	1.0000	1.0000	1.0000	1.0000	1.0000	1.0000	1.0000	1.0000	1.0000	1.0000
18	1.0000	1.0000	1.0000	1.0000	1.0000	1.0000	1.0000	1.0000	1.0000	1.0000
19	1.0000	1.0000	1.0000	1.0000	1.0000	1.0000	1.0000	1.0000	1.0000	1.0000

$\lambda =$	5.5	6.0	6.5	7.0	7.5	8.0	8.5	9.0	9.5	10.0
$x =$ 0	0.0041	0.0025	0.0015	0.0009	0.0006	0.0003	0.0002	0.0001	0.0001	0.0000
1	0.0266	0.0174	0.0113	0.0073	0.0047	0.0030	0.0019	0.0012	0.0008	0.0005
2	0.0884	0.0620	0.0430	0.0296	0.0203	0.0138	0.0093	0.0062	0.0042	0.0028
3	0.2017	0.1512	0.1118	0.0818	0.0591	0.0424	0.0301	0.0212	0.0149	0.0103
4	0.3575	0.2851	0.2237	0.1730	0.1321	0.0996	0.0744	0.0550	0.0403	0.0293
5	0.5289	0.4457	0.3690	0.3007	0.2414	0.1912	0.1496	0.1157	0.0885	0.0671
6	0.6860	0.6063	0.5265	0.4497	0.3782	0.3134	0.2562	0.2068	0.1649	0.1301
7	0.8095	0.7440	0.6728	0.5987	0.5246	0.4530	0.3856	0.3239	0.2687	0.2202
8	0.8944	0.8472	0.7916	0.7291	0.6620	0.5925	0.5231	0.4557	0.3918	0.3328
9	0.9462	0.9161	0.8774	0.8305	0.7764	0.7166	0.6530	0.5874	0.5218	0.4579
10	0.9747	0.9574	0.9332	0.9015	0.8622	0.8159	0.7634	0.7060	0.6453	0.5830
11	0.9890	0.9799	0.9661	0.9467	0.9208	0.8881	0.8487	0.8030	0.7520	0.6968
12	0.9955	0.9912	0.9840	0.9730	0.9573	0.9362	0.9091	0.8758	0.8364	0.7916
13	0.9983	0.9964	0.9929	0.9872	0.9784	0.9658	0.9486	0.9261	0.8981	0.8645
14	0.9994	0.9986	0.9970	0.9943	0.9897	0.9827	0.9726	0.9585	0.9400	0.9165
15	0.9998	0.9995	0.9988	0.9976	0.9954	0.9918	0.9862	0.9780	0.9665	0.9513
16	0.9999	0.9998	0.9996	0.9990	0.9980	0.9963	0.9934	0.9889	0.9823	0.9730
17	1.0000	0.9999	0.9998	0.9996	0.9992	0.9984	0.9970	0.9947	0.9911	0.9857
18	1.0000	1.0000	0.9999	0.9999	0.9997	0.9993	0.9987	0.9976	0.9957	0.9928
19	1.0000	1.0000	1.0000	1.0000	0.9999	0.9997	0.9995	0.9989	0.9980	0.9965
20	1.0000	1.0000	1.0000	1.0000	1.0000	0.9999	0.9998	0.9996	0.9991	0.9984
21	1.0000	1.0000	1.0000	1.0000	1.0000	1.0000	0.9999	0.9998	0.9996	0.9993
22	1.0000	1.0000	1.0000	1.0000	1.0000	1.0000	1.0000	0.9999	0.9999	0.9997

S2 — STATISTICAL TABLES

Percentage points of the χ^2 distribution

The table shows values of x satisfying $P(X \leq x) = p$,
where X has the χ^2 distribution with v degrees of freedom.

p / v	0.005	0.01	0.025	0.05	0.1	0.9	0.95	0.975	0.99	0.995	p / v
1	0.00004	0.0002	0.001	0.004	0.016	2.706	3.841	5.024	6.635	7.879	1
2	0.010	0.020	0.051	0.103	0.211	4.605	5.991	7.378	9.210	10.597	2
3	0.072	0.115	0.216	0.352	0.584	6.251	7.815	9.348	11.345	12.838	3
4	0.207	0.297	0.484	0.711	1.064	7.779	9.488	11.143	13.277	14.860	4

Percentage points of the Student's t-distribution

The table shows values of x satisfying $P(X \leq x) = p$,
where X has the Student's t-distribution with v degrees of freedom.

p / v	0.9	0.95	0.975	0.99	0.995
1	**3.078**	**6.314**	**12.706**	**31.821**	**63.657**
2	**1.886**	**2.920**	**4.303**	**6.965**	**9.925**
3	**1.638**	**2.353**	**3.182**	**4.541**	**5.841**
4	**1.533**	**2.132**	**2.776**	**3.747**	**4.604**
5	**1.476**	**2.015**	**2.571**	**3.365**	**4.032**
6	**1.440**	**1.943**	**2.447**	**3.143**	**3.707**
7	**1.415**	**1.895**	**2.365**	**2.998**	**3.499**
8	**1.397**	**1.860**	**2.306**	**2.896**	**3.355**
9	**1.383**	**1.833**	**2.262**	**2.821**	**3.250**
10	**1.372**	**1.812**	**2.228**	**2.764**	**3.169**
11	**1.363**	**1.796**	**2.201**	**2.718**	**3.106**
12	**1.356**	**1.782**	**2.179**	**2.681**	**3.055**
13	**1.350**	**1.771**	**2.160**	**2.650**	**3.012**
14	**1.345**	**1.761**	**2.145**	**2.624**	**2.977**
15	**1.341**	**1.753**	**2.131**	**2.602**	**2.947**
16	**1.337**	**1.746**	**2.120**	**2.583**	**2.921**
17	**1.333**	**1.740**	**2.110**	**2.567**	**2.898**
18	**1.330**	**1.734**	**2.101**	**2.552**	**2.878**
19	**1.328**	**1.729**	**2.093**	**2.539**	**2.861**
20	**1.325**	**1.725**	**2.086**	**2.528**	**2.845**
21	**1.323**	**1.721**	**2.080**	**2.518**	**2.831**
22	**1.321**	**1.717**	**2.074**	**2.508**	**2.819**
23	**1.319**	**1.714**	**2.069**	**2.500**	**2.807**
24	**1.318**	**1.711**	**2.064**	**2.492**	**2.797**
25	**1.316**	**1.708**	**2.060**	**2.485**	**2.787**
26	**1.315**	**1.706**	**2.056**	**2.479**	**2.779**
27	**1.314**	**1.703**	**2.052**	**2.473**	**2.771**
28	**1.313**	**1.701**	**2.048**	**2.467**	**2.763**

Projectiles

This page is for Edexcel M2 and OCR M2

A 'projectile' is just any old object that's been lobbed through the air. When you're doing projectile questions you'll have to model the motion of particles in two dimensions whilst ignoring air resistance.

Split Velocity of Projection into Two Components

A particle projected with a speed u at an angle α to the horizontal has two components of initial velocity — one horizontal (parallel to the x-axis) and one vertical (parallel to the y-axis).
These are called x and y components, and they make projectile questions dead easy to deal with:

Here's the same information in a diagram:

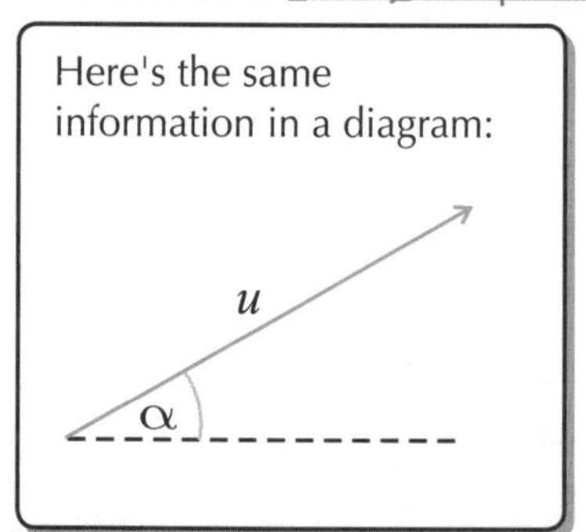

Split the velocity into its x and y components:

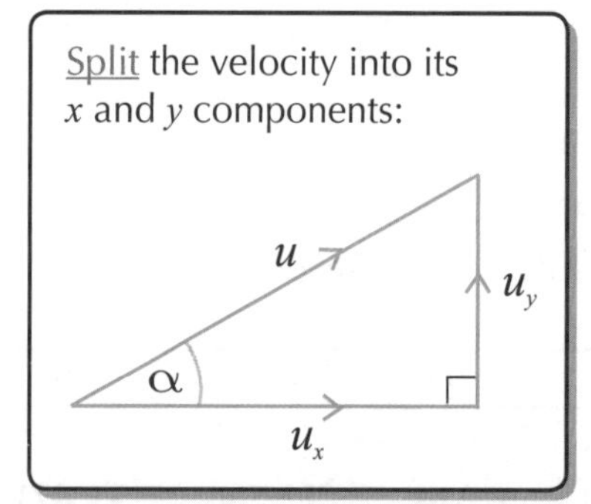

Finally, work out the values of the components using trigonometry:

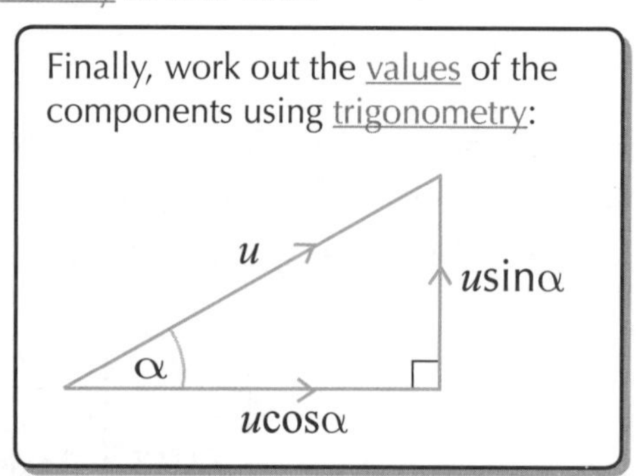

Split the Motion into Horizontal and Vertical Components too

Split everything you know about the motion into horizontal and vertical components too. Then you can deal with them separately using the 'uvast' equations from M1. The only thing that's the same in both directions is time — so this connects the two directions. Remember that the only acceleration is due to gravity — so horizontal acceleration is zero.

EXAMPLE

A stone is thrown horizontally with speed 10 ms^{-1} from a height of 2 m above the horizontal ground. Find the time taken for the stone to hit the ground and the horizontal distance travelled before impact. Find also the speed and direction of the stone after 0.5 s.

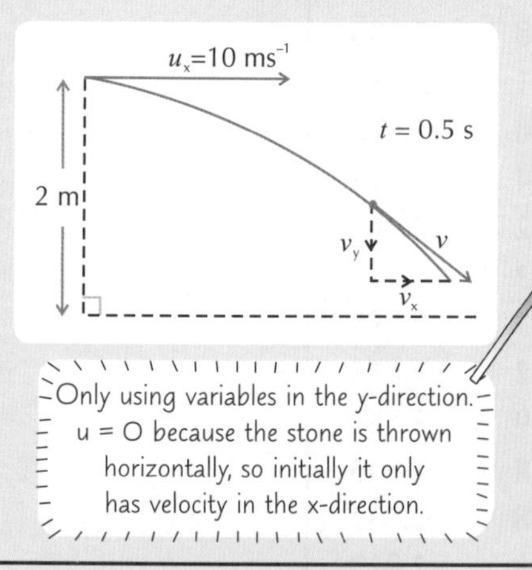

Only using variables in the y-direction. u = 0 because the stone is thrown horizontally, so initially it only has velocity in the x-direction.

Resolving vertically (take down as +ve):

$u = u_y = 0 \quad s = 2$
$a = 9.8 \quad t = ?$

$s = ut + \frac{1}{2}at^2$

$2 = 0 \times t + \frac{1}{2} \times 9.8 \times t^2$

$t = 0.639$ s (to 3 s.f.)
i.e. the stone lands after 0.639 seconds

Resolving horizontally (take right as +ve):

The same as for the vertical motion.

$u = u_x = 10 \quad s = ?$
$a = 0 \quad t = 0.6389$

$s = ut + \frac{1}{2}at^2$

$= 10 \times 0.6389 + \frac{1}{2} \times 0 \times 0.6389^2$

$= 6.39$ m

i.e. the stone has gone 6.39 m horizontally when it lands.

Now find the velocity after 0.5 s — again, keep the vertical and horizontal bits separate.

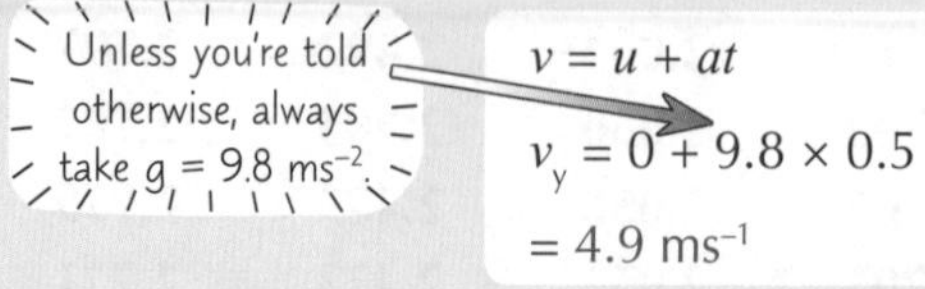

$v = u + at$

$v_x = 10 + 0 \times \frac{1}{2}$

$= 10$ ms^{-1}

v_x is always equal to u_x when there's no horizontal acceleration.

Now you can find the speed and direction...

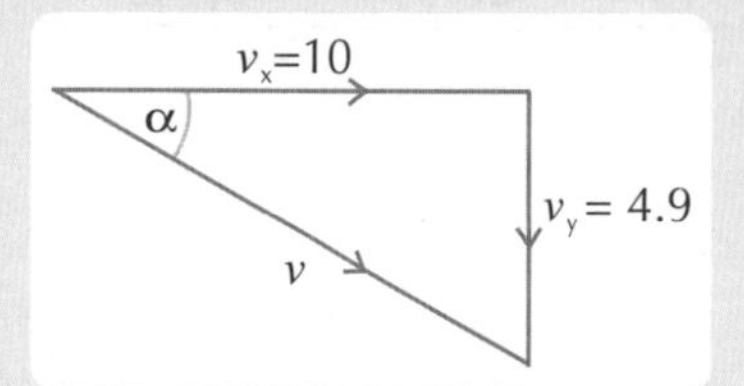

$v = \sqrt{4.9^2 + 10^2} = 11.1$ ms^{-1}

$\tan\theta = \frac{4.9}{10}$

So $\theta = 26.1°$ below horizontal

Projectiles

This page is for Edexcel M2 and OCR M2

EXAMPLE A cricket ball is projected with a speed of 30 ms^{-1} at an angle of 25° to the horizontal. Assume the ground is horizontal and the ball is struck from a point 1.5 m above the ground. Find:

a) the maximum height the ball reaches (h),

b) the horizontal distance travelled by the ball before it hits the ground (r),

c) the length of time the ball is at least 5 m above the ground.

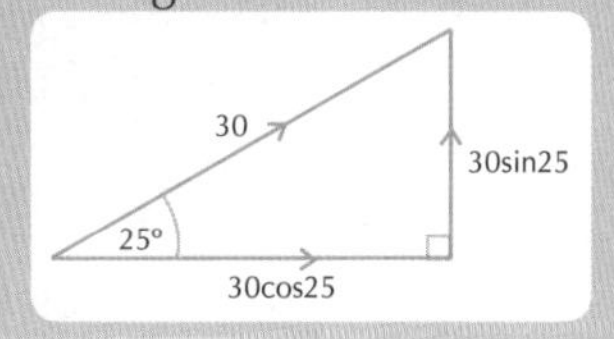

a) **Resolving vertically** (take up as +ve):

$u = 30\sin 25°$ $v = 0$
$a = -9.8$ $s = ?$

The ball will momentarily stop moving vertically when it reaches its maximum height.

$$v^2 = u^2 + 2as$$
$$0 = (30\sin 25°)^2 + 2(-9.8 \times s)$$
$$s = 8.201\text{m}$$
$$\text{h} = 8.201\text{m} + 1.5\text{m} = 9.70\text{m}$$

Don't forget to add the height from which the ball is hit.

b) **Resolving vertically** (take up as +ve):

The ground is 1.5 m <u>below</u> the ball's initial position.

$s = -1.5$
$a = -9.8$
$u = 30\sin 25°$
$t = ?$

$$s = ut + \frac{1}{2}at^2$$
$$-1.5 = (30\sin 25°)t - \frac{1}{2}(9.8)t^2$$
$$t^2 - 2.587t - 0.306 = 0$$
$$t = -0.11 \text{ or } \mathbf{t = 2.70\text{ s}}$$

Using the quadratic formula you get two answers, but time can't be negative, so forget about this answer.

Resolving horizontally (take right as +ve)

$s = \text{r}$ $u = 30\cos 25°$
$t = 2.70$ $a = 0$

r represents the ball's <u>range</u>.

$$s = ut + \frac{1}{2}at^2$$
$$\text{r} = 30\cos 25° \times 2.70 + \frac{1}{2} \times 0 \times 2.70^2$$
$$= 73.4\text{m}$$

c) **Resolving vertically** (take up as +ve):

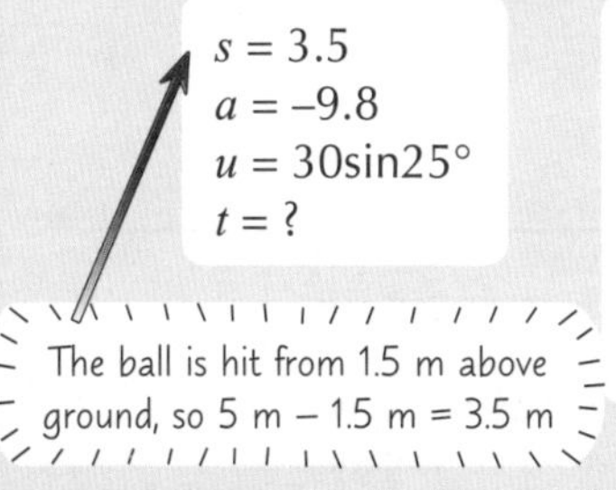

$$s = ut + \frac{1}{2}at^2$$
$$3.5 = (30\sin 25°)t - \frac{1}{2}(9.8)t^2$$
$$t^2 - 2.587t + 0.714 = 0$$
$$t = 0.31 \text{ or } t = 2.27\text{ s}$$

These are the two times when the ball is 5 m above the ground.

So, length of time at least 5 m above the ground:

2.27 s – 0.31 s = 1.96 s

*The **Components of Velocity** can be described using **i and j Vectors***

This bit is for Edexcel M2 only

Ho ho — **i** and **j** vectors. You'll remember those jokers from M1, no doubt. Well they're pretty useful in M2 as well:

EXAMPLE A stone is thrown from a point 1.2 metres above the horizontal ground. It travels for 4 seconds before landing on the ground. The stone is thrown with velocity ($2q\mathbf{i} + q\mathbf{j}$) ms^{-1}, where **i** and **j** are the horizontal and vertical unit vectors respectively. Find the value of q and the initial speed of the stone.

Resolving vertically (take up as +ve):

$u = q$ $t = 4$
$a = -9.8$ $s = -1.2$

The vertical component of velocity is q, the horizontal component is $2q$. Simples.

$$s = ut + \frac{1}{2}at^2$$
$$-1.2 = 4q - \frac{1}{2}(9.8)4^2$$
$$4q = 77.2$$
$$q = 19.3$$

Now find the initial speed of the stone:

$$u = \sqrt{(2q)^2 + q^2}$$
$$= \sqrt{38.6^2 + 19.3^2} = 43.2\text{ms}^{-1}$$

Projectiles

This page is for Edexcel M2 and OCR M2

Just one last example of projectile motion. But boy is it a beauty...

EXAMPLE

A golf ball is struck from a point A on a horizontal plane. When the ball has moved a horizontal distance x, its height above the plane is y. The ball is modelled as a particle projected with initial speed u ms^{-1} at an angle α.

a) Show that $y = x\tan\alpha - \dfrac{gx^2}{2u^2\cos^2\alpha}$.

This is called the 'Cartesian equation of the trajectory of a projectile'. Fancy.

The ball just passes over the top of a 10 m tall tree, which is 45 m away. Given that $\alpha = 45°$,

b) find the speed of the ball as it passes over the tree.

a) Displacement, acceleration and initial velocity are the only variables in the formula, so use these. Also use time, because that's the variable which connects the two components of motion. The formula includes motion in both directions (x and y), so form two equations and substitute one into the other:

Resolving horizontally (taking right as +ve):

$u_x = u\cos\alpha$ $\quad a = 0$

$s = x$ $\quad t = t$

Using $s = ut + \frac{1}{2}at^2$:

When you're using these variables, this is the obvious equation to use.

$x = u\cos\alpha \times t$

Rearrange to make t the subject:

$t = \dfrac{x}{u\cos\alpha}$ — call this **equation 1**

t doesn't appear in the final formula, so by making it the subject you can eliminate it.

Resolving vertically (taking up as +ve):

$u_y = u\sin\alpha$ $\quad a = -g$

$s = y$ $\quad t = t$

Using $s = ut + \frac{1}{2}at^2$:

It would be a massive pain to make t the subject here, so do it with the other equation.

$y = (u\sin\alpha \times t) - \frac{1}{2}gt^2$ — call this **equation 2**

t is the same horizontally and vertically, so you can substitute **equation 1** into **equation 2** and eliminate t:

$$y = u\sin\alpha \times \frac{x}{u\cos\alpha} - \frac{1}{2}g\left(\frac{x}{u\cos\alpha}\right)^2 = x\frac{\sin\alpha}{\cos\alpha} - \frac{1}{2}g\left(\frac{x^2}{u^2\cos^2\alpha}\right)$$

$\dfrac{\sin\theta}{\cos\theta} = \tan\theta$

$$= x\tan\alpha - \frac{gx^2}{2u^2\cos^2\alpha}$$ — as required.

b) Using the result from a), and substituting $x = 45$, $y = 10$ and $\alpha = 45°$:

$$10 = 45\tan 45° - \frac{9.8 \times 45^2}{2u^2 \times \cos^2 45°} = 45 - \frac{19\,845}{u^2}$$

Rearrange to find the speed of projection, u:

$35u^2 = 19\,845 \Rightarrow \boldsymbol{u = 23.81\text{ ms}^{-1}}$

If you need to round part way through a calculation, then round to more s.f. than your final answer will be rounded to. A better idea is not to round at all and use your calculator's memory.

Now resolve to find the components of the ball's velocity as it passes over the tree:

Resolving horizontally (taking right as +ve):

$v_x = u_x = 23.81\cos45 = \mathbf{16.84\text{ ms}^{-1}}$

Remember — with projectiles there's no horizontal acceleration, so v_x always equals u_x.

Resolving vertically (taking up as +ve):

$u_y = 23.81\sin45$ $\quad a = -g$

$s = 10$ $\quad v_y = ?$

Using $v^2 = u^2 + 2as$:

$v_y^2 = 283.46 - 2 \times 9.8 \times 10 = \mathbf{87.46}$

Don't bother finding the square root, as you need v_y^2 in the next step. Sneaky.

Now you can find the speed: $\quad v = \sqrt{v_x^2 + v_y^2} = 19.3\text{ ms}^{-1}$ (3 s.f.)

Projectiles — they're all about throwing up. Or across. Or slightly down...

You've used the equations of motion before, in M1, and there isn't much different here. The main thing to remember is that horizontal acceleration is zero — great news because it makes half the calculations as easy as a log-falling beginner's class.

Displacement, Velocity and Acceleration

This page is for AQA M2 and Edexcel M2

The "uvast" equations you saw back in M1 are all well and good when you've got a particle with constant acceleration. But when the acceleration of a particle varies with time, you need a few new tricks up your sleeve...

Differentiate to find Velocity and Acceleration from Displacement...

If you've got a particle moving in a straight line with acceleration that varies with time, you need to use calculus to find equations to describe the motion. (Look back at your C1 & C2 notes for a reminder about calculus.)

1) To find an equation for velocity, differentiate the equation for displacement with respect to time.
2) To find an equation for acceleration, differentiate the equation for velocity with respect to time. (Or differentiate the equation for displacement with respect to time twice.)

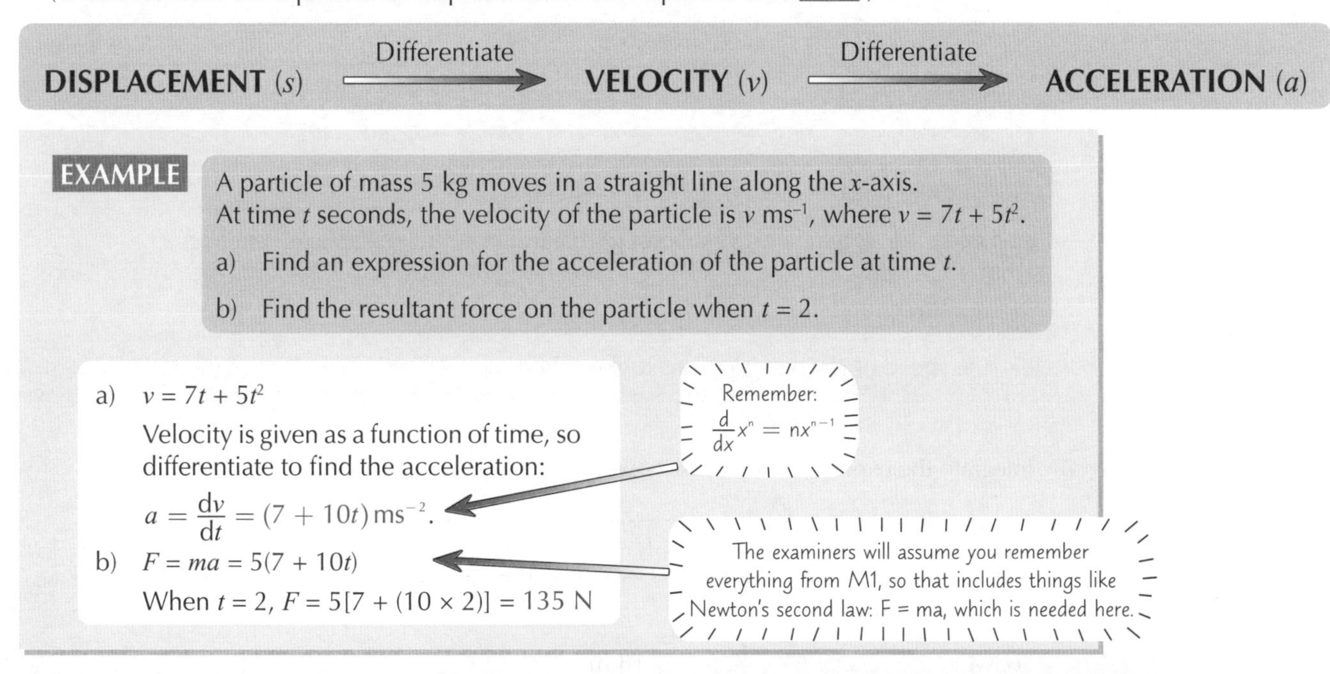

DISPLACEMENT (s) —Differentiate→ **VELOCITY** (v) —Differentiate→ **ACCELERATION** (a)

EXAMPLE

A particle of mass 5 kg moves in a straight line along the x-axis. At time t seconds, the velocity of the particle is v ms^{-1}, where $v = 7t + 5t^2$.

a) Find an expression for the acceleration of the particle at time t.

b) Find the resultant force on the particle when $t = 2$.

a) $v = 7t + 5t^2$

Velocity is given as a function of time, so differentiate to find the acceleration:

$a = \frac{dv}{dt} = (7 + 10t)\,\text{ms}^{-2}$.

b) $F = ma = 5(7 + 10t)$

When $t = 2$, $F = 5[7 + (10 \times 2)] = 135$ N

Remember: $\frac{d}{dx}x^n = nx^{n-1}$

The examiners will assume you remember everything from M1, so that includes things like Newton's second law: F = ma, which is needed here.

...and Integrate to find Velocity and Displacement from Acceleration

It's pretty similar if you're trying to go "back the other way", except you integrate rather than differentiate:

1) To find an equation for velocity, integrate the equation for acceleration with respect to time.
2) To find an equation for displacement, integrate the equation for velocity with respect to time.

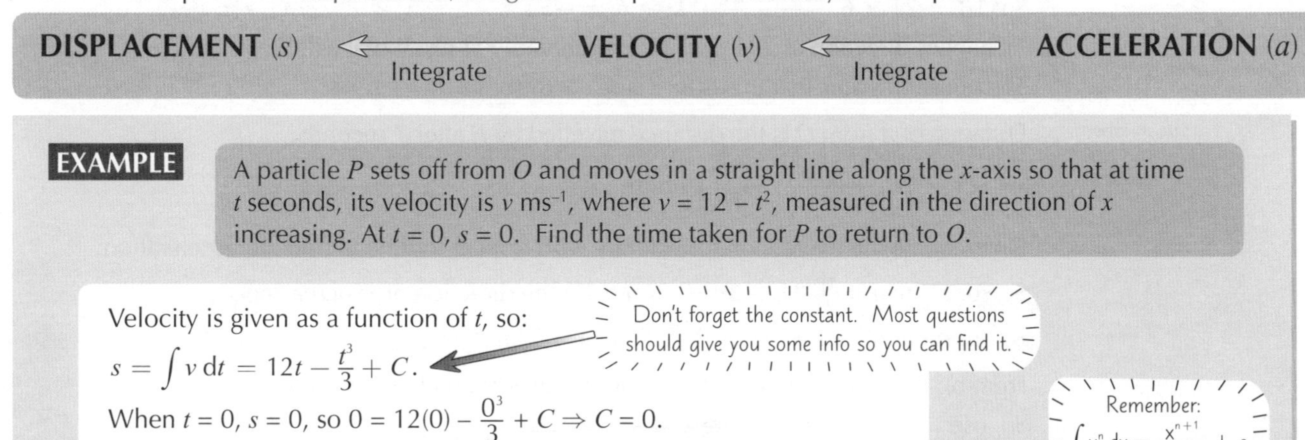

DISPLACEMENT (s) ←Integrate— **VELOCITY** (v) ←Integrate— **ACCELERATION** (a)

EXAMPLE

A particle P sets off from O and moves in a straight line along the x-axis so that at time t seconds, its velocity is v ms^{-1}, where $v = 12 - t^2$, measured in the direction of x increasing. At $t = 0$, $s = 0$. Find the time taken for P to return to O.

Velocity is given as a function of t, so:

$s = \int v\,dt = 12t - \frac{t^3}{3} + C$.

When $t = 0$, $s = 0$, so $0 = 12(0) - \frac{0^3}{3} + C \Rightarrow C = 0$.

P is at O when $s = 0$, i.e. when: $12t - \frac{t^3}{3} = 0 \Rightarrow t(36 - t^2) = 0$

i.e. when $t = 0$, 6 or -6. So time taken for P to return to O is 6 seconds.

Don't forget the constant. Most questions should give you some info so you can find it.

Remember: $\int x^n\,dx = \frac{x^{n+1}}{n+1} + c$

This can't be an answer, as you can't have a negative time.

Displacement, Velocity and Acceleration

This page is for Edexcel M2 only

Sometimes the **Velocity** is Defined by **More Than One Expression**

The velocity of a particle can sometimes be defined by different expressions for different values of t.
It just means that you have to deal with each time interval separately when you're differentiating and integrating.

These questions are a favourite with examiners — so make sure you understand what's going on in this example:

EXAMPLE

A particle P sets off from the origin at $t = 0$ and moves in a straight line along the x-axis in the direction of x increasing. The velocity of P after t seconds is v ms^{-1}, where v is given by:

$$v = \begin{cases} 2t - \frac{t^2}{4} & 0 \leqslant t \leqslant 6 \\ 21 - 3t & t > 6 \end{cases}$$

Make sure you read the question carefully, so you know which expression for v to use.

a) Find the displacement of P from O when $t = 6$.

At some $t > 6$, P reaches A, the point of maximum positive displacement from O.
From A, P begins to move in a straight line along the x-axis back towards the origin. Find:

b) the distance of A from O,

c) the speed of P when it returns to the origin.

a) Integrate the expression for velocity with respect to time to find displacement:

$s = \int v\,dt = \int \left(2t - \frac{t^2}{4}\right)dt = t^2 - \frac{t^3}{12} + C$ for $0 \leqslant t \leqslant 6$.

When $t = 0$, P is at the origin (i.e. $s = 0$) Use this to find C:

$0 = 0^2 - \frac{0^3}{12} + C \Rightarrow C = 0$.

So, when $t = 6$, $s = 6^2 - \frac{6^3}{12} + 0 = 18$ m

b) Again, integrate the expression for velocity with respect to time to find displacement:

$s = \int v\,dt = \int (21 - 3t)\,dt = 21t - \frac{3t^2}{2} + K$ for $t > 6$

From part a), when $t = 6$, $x = \mathbf{18}$.

Use this as your initial condition to find K, because even though t = 6 isn't in the interval t > 6, it is the lower limit of the interval.

So: $18 = 21 \times 6 - \frac{3 \times 36}{2} + K \Rightarrow \boldsymbol{K = -54}$

P changes direction at A, so will be momentarily at rest (i.e. v will be 0).

You're told this in the question.

The question states that this is for some $t > 6$, so: $v = 0 = 21 - 3t \Rightarrow \boldsymbol{t = 7}$.

Distance of A from O is the distance travelled by P after 7 seconds:

$s = (21 \times 7) - \frac{3(7^2)}{2} - 54 = 19.5$ m

c) Differentiate the expression for velocity with respect to time to find the acceleration:

So, for $t > 6$: $a = \frac{d}{dt}(21 - 3t) = -3$ ms^{-2} (in the direction of x increasing).

So, for motion back *towards* the origin, $a = 3$ ms^{-2}.

From b), the velocity of P at A is 0 and distance from O to A is 19.5 m

So, use $v^2 = u^2 + 2as$:

The acceleration here is constant, so you can use one of the uvast equations.

$v^2 = 0^2 + (2 \times 3 \times 19.5) \Rightarrow v = 10.8$ ms^{-1} (3 s.f.)

Displacement, Velocity and Acceleration

This page is for AQA M2 only

You might get given an equation of motion containing trig or exponential functions. If you've forgotten how to do calculus with these types of function, you'd better get those C3 and C4 notes out and brush up.

Use the *Chain Rule* for Functions of Functions

Once you've worked through a couple of examples, the Core stuff should all come flooding back to you...

EXAMPLE A particle sets off from the origin at $t = 0$ and moves along the x-axis. At time t seconds, the velocity of the particle is v ms^{-1}, where $v = 5t - 4\cos 4t + 8$. Find:

a) an expression for the acceleration of the particle at time t,

b) the displacement of the particle from the origin when $t = \pi$.

a) Split the equation for v into two parts:

$v = -4\cos 4t$ and $v = 5t + 8$

If you're struggling with the chain rule have a look at p. 45

Use the chain rule to differentiate the tricky bit — '$v = -4\cos 4t$':

let $u = 4t$, so $v = -4\cos u$

$\frac{du}{dt} = 4$ and $\frac{dv}{du} = 4\sin u$

Remember: $\frac{d}{dx}\cos x = -\sin x$

$\frac{dv}{dt} = \frac{dv}{du} \times \frac{du}{dt} = 16\sin 4t$

This bit is a doddle to differentiate:

$\frac{dv}{dt} = 5$

Add the two parts together to get the final expression for $a = \frac{dv}{dt}$:

$a = 16\sin 4t + 5$

b) Integrate v to find the displacement:

$$s = \int v\,dt = \int_0^{\pi} (5t - 4\cos 4t + 8)\,dt = \left[\frac{5}{2}t^2 - \sin 4t + 8t\right]_0^{\pi}$$

$$= \frac{5}{2}\pi^2 + 8\pi \text{ m } (= 49.8 \text{ m to 3 s.f.})$$

$\int \cos nx\,dx = \frac{1}{n}\sin nx$ (see page 68).

Doing a definite integral means you don't have to find a constant.

EXAMPLE At time t seconds, a particle moving in a straight line along the x-axis has displacement s m, where $s = 3e^{-2t} + 4t$. Find:

a) an expression for the velocity of the particle at time t seconds,

b) the range of values for the particle's velocity.

a) Differentiate s to get an expression for v, using the chain rule to differentiate '$s = 3e^{-2t}$':

Let $u = -2t$, so $s = 3e^u$,

$\Rightarrow \frac{ds}{du} = 3e^u$ and $\frac{du}{dt} = -2$

So, $\frac{ds}{dt} = \frac{ds}{du} \times \frac{du}{dt} = 3e^{-2t} \times -2 = -6e^{-2t}$

Differentiating the whole equation for s gives:

$\frac{ds}{dt} = v = -6e^{-2t} + 4$

b) t must be ≥ 0.
When $t = 0$, $e^{-2t} = 1$, so $v = -6 + 4 = -2$ ms^{-1}.
As $t \to \infty$, $e^{-2t} \to 0$, so $v \to 4$.
So, $-2 \leq v < 4$ ms^{-1}.

Remember — '$\to$' means 'tends to'.

You could also answer this question by sketching the graph of the function (see p. 34):

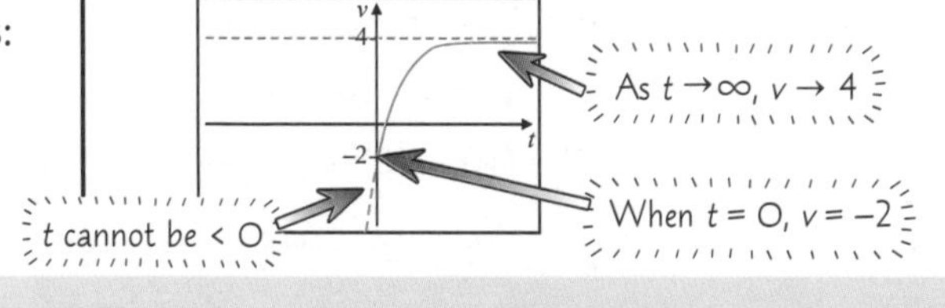

CGP driving tips #1 — differentiate velocity from displacement...

Calculus? In Mechanics? What fresh horror is this? Actually, it's really not that bad at all. Just make sure you know when to differentiate and when to integrate and then bang in the numbers you're given in the question to get the answer. Sorted.

Describing Motion Using Vectors

This page is for AQA M2 and Edexcel M2

I can tell that you loved the last few pages, but I know what you're thinking: "That's all fair enough mate, but what about when a particle is moving in two dimensions?" Well, you know I can't ignore a question like that, so here you go...

Differentiate and *Integrate* with *Vector Notation* for Motion on a *Plane*

1) When you've got a particle moving in two dimensions (i.e. on a plane), you can describe its position, velocity and acceleration using the unit vectors **i** and **j** (which you should remember from M1). This "**i** and **j**" notation shows the horizontal and vertical components of displacement, velocity or acceleration separately.
2) The relationship between displacement (position), velocity and acceleration from page 171 still applies to particles moving on a plane:

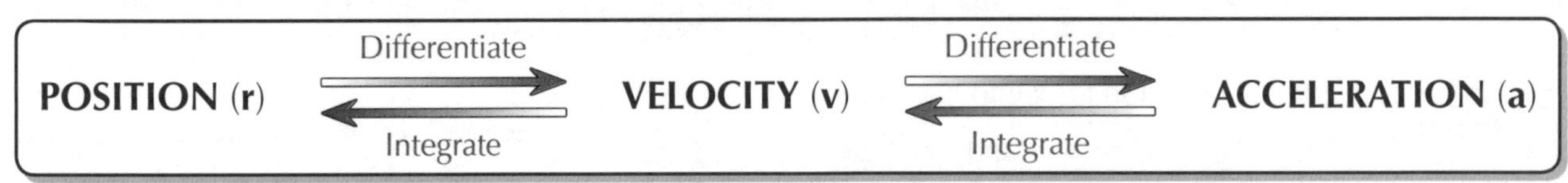

3) This means that you'll have to differentiate and integrate vectors written in **i and j** notation. Luckily, doing this is as easy as squeezing lemons — all you have to do is differentiate/integrate each component of the vector separately:

So, if $\mathbf{r} = x\mathbf{i} + y\mathbf{j}$ is a position vector, then:

velocity, $\mathbf{v} = \frac{d\mathbf{r}}{dt} = \frac{dx}{dt}\mathbf{i} + \frac{dy}{dt}\mathbf{j}$

The shorthand for $\frac{dr}{dt}$ is $\dot{r}$ (the single dot means differentiate r once with respect to time)...

and acceleration, $\mathbf{a} = \frac{d\mathbf{v}}{dt} = \frac{d^2\mathbf{r}}{dt^2} = \frac{d^2x}{dt^2}\mathbf{i} + \frac{d^2y}{dt^2}\mathbf{j}$.

...and the shorthand for $\frac{d^2r}{dt^2}$ is $\ddot{r}$ (the double dots mean differentiate r twice with respect to time).

It's a similar thing for integration:

If $\mathbf{v} = w\mathbf{i} + z\mathbf{j}$ is a velocity vector, then position, $\mathbf{r} = \int \mathbf{v}\,dt = \int (w\mathbf{i} + z\mathbf{j})\,dt = \left[\int w\,dt\right]\mathbf{i} + \left[\int z\,dt\right]\mathbf{j}$

Unfortunately, there's no snazzy shorthand for integration. Ahh well, easy come, easy go.

EXAMPLE:

A particle is moving on a horizontal plane so that at time t it has velocity v ms^{-1}, where

$$\mathbf{v} = (8 + 2t)\mathbf{i} + (t^3 - 6t)\mathbf{j}$$

At $t = 2$, the particle has a position vector of $(10\mathbf{i} + 3\mathbf{j})$ m with respect to a fixed origin O.

a) Find the acceleration of the particle at time t.

b) Show that the position of the particle relative to O when $t = 4$ is $\mathbf{r} = 38\mathbf{i} + 27\mathbf{j}$.

a) $\mathbf{a} = \dot{\mathbf{v}} = \frac{d\mathbf{v}}{dt}$

$= 2\mathbf{i} + (3t^2 - 6)\mathbf{j}$

Yep, that really is all there is to it.

b) $\mathbf{r} = \int \mathbf{v}\,dt$

$= (8t + t^2)\mathbf{i} + \left(\frac{t^4}{4} - 3t^2\right)\mathbf{j} + \mathbf{C}$

You still need a constant of integration, but it will be a vector with **i** and **j** components.

When $t = 2$, $\mathbf{r} = (10\mathbf{i} + 3\mathbf{j})$, so use this info to find the vector **C**:

$10\mathbf{i} + 3\mathbf{j} = 20\mathbf{i} - 8\mathbf{j} + \mathbf{C}$

$\Rightarrow \mathbf{C} = (10 - 20)\mathbf{i} + (3 - -8)\mathbf{j} = -10\mathbf{i} + 11\mathbf{j}$

Collect **i** and **j** terms and add/subtract to simplify.

So, $\mathbf{r} = (8t + t^2 - 10)\mathbf{i} + \left(\frac{t^4}{4} - 3t^2 + 11\right)\mathbf{j}$.

When $t = 4$, $\mathbf{r} = (32 + 16 - 10)\mathbf{i} + (64 - 48 + 11)\mathbf{j} = 38\mathbf{i} + 27\mathbf{j}$ — as required.

It's easy enough to extend all this to three dimensions. The position vector will be of the form $\mathbf{r} = x\mathbf{i} + y\mathbf{j} + z\mathbf{k}$, so you'll have three components to differentiate or integrate.

Describing Motion Using Vectors

This page is for AQA M2 and Edexcel M2

Watch out for Questions that include **Forces**

When you see "the action of a single force, **F** newtons" in one of these vector questions, you should immediately think **F** = m**a**, because you're almost certainly going to need it. Here are a couple of examples showing the examiners' faves:

EXAMPLE

A particle P is moving under the action of a single force, **F** newtons. The position vector, **r** m, of P after t seconds is given by

$$\mathbf{r} = (2t^3 - 3)\mathbf{i} + \frac{t^4}{2}\mathbf{j}$$

a) Find an expression for the acceleration of P at time t seconds.

b) P has mass 6 kg. Find the magnitude of **F** when $t = 3$.

a) $\mathbf{v} = \dot{\mathbf{r}} = 6t^2\mathbf{i} + 2t^3\mathbf{j}$
$\mathbf{a} = \dot{\mathbf{v}} = 12t\mathbf{i} + 6t^2\mathbf{j}$

b) At $t = 3$, $\mathbf{a} = 36\mathbf{i} + 54\mathbf{j}$

Using $\mathbf{F} = m\mathbf{a}$, substitute $m = 6$:

$\mathbf{F} = (6 \times 36)\mathbf{i} + (6 \times 54)\mathbf{j} = 216\mathbf{i} + 324\mathbf{j}$

$|\mathbf{F}| = \sqrt{216^2 + 324^2} = 389$ N (3 s.f.)

You could find the magnitude of **a** first instead if you wanted, then just multiply by m.

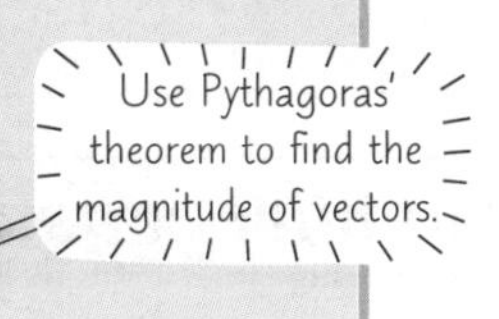

EXAMPLE

A particle of mass 4 kg moves in a plane under the action of a single force, **F** newtons.

At time t seconds, $\mathbf{F} = (24t\mathbf{i} - 8\mathbf{j})$ N

At time $t = 0$, the velocity of the particle is $(7\mathbf{i} + 22\mathbf{j})$ ms^{-1}.

a) The velocity of the particle at time t is **v** ms^{-1}. Show that

$$\mathbf{v} = (3t^2 + 7)\mathbf{i} + (22 - 2t)\mathbf{j}$$

b) Find the value of t when the particle is moving parallel to the vector **i**.

Here you have to work out the acceleration vector before you can do any integrating.

a) Use $\mathbf{F} = m\mathbf{a}$ to find an expression for the acceleration of the particle at time t:

$(24t\mathbf{i} - 8\mathbf{j}) = 4\mathbf{a} \Rightarrow \mathbf{a} = 6t\mathbf{i} - 2\mathbf{j}$

$\mathbf{v} = \int \mathbf{a}\,dt = \int (6t\mathbf{i} - 2\mathbf{j})\,dt = 3t^2\mathbf{i} - 2t\mathbf{j} + \mathbf{C}$

When $t = 0$, $\mathbf{v} = (7\mathbf{i} + 22\mathbf{j})$ ms^{-1}. Use this information to find **C**:

$7\mathbf{i} + 22\mathbf{j} = 0\mathbf{i} + 0\mathbf{j} + \mathbf{C} \Rightarrow \mathbf{C} = 7\mathbf{i} + 22\mathbf{j}$. So at time t,

$\mathbf{v} = 3t^2\mathbf{i} - 2t\mathbf{j} + 7\mathbf{i} + 22\mathbf{j}$

$= (3t^2 + 7)\mathbf{i} + (22 - 2t)\mathbf{j}$ — as required.

b) When the particle is moving parallel to the vector **i**, the **j** component of **v** is 0, and the **i** component is non-zero, so:

$22 - 2t = 0 \Rightarrow t = 11$

At $t = 11$, the **i** component of velocity is $(3 \times 11^2) + 7 = 370$, i.e. not zero.

So the particle is moving parallel to the vector **i** at 11 s.

Motion in two dimensions — it's plane simple...

Just remember to differentiate and integrate by treating each component separately, and pretty soon you'll be able to differentiate velocity vectors in 11-dimensional hyperspace. Just think how cool that'll look at the next sci-fi convention.

Using Differential Equations

This page is for AQA M2 only

Often, the hardest thing about M2 is working out which bits of maths you need to dredge up from the darkest recesses of your brain to help you answer a question. Although some people claim that's half the fun of it. Hmmm...

Form Differential Equations using Force = Mass × dv/dt

You should be pretty familiar with $F = ma$ by now, and on page 171, you were introduced to $a = \frac{dv}{dt}$.

Well guess what — you can stick these two equations together to form some super-useful differential equations:

EXAMPLE The combined mass of a cyclist and her bike is 75 kg. She is cycling along a horizontal road and stops pedalling when her velocity is 10 ms^{-1}. The only horizontal force acting on the cyclist and her bike is a resistive force of $15v$ N, where v is the velocity of the bike. Show that $\frac{dv}{dt} = -\frac{v}{5}$.

The resistive force is the only horizontal force acting on the cyclist, so the resultant force, F, is negative.

Resolving forces horizontally, using $F = ma$:

$$-15v = 75 \times \frac{dv}{dt} \quad \Rightarrow \quad \frac{dv}{dt} = -\frac{v}{5}$$

Solve Differential Equations by Separating the Variables

To make the most of these differential equations, you need to solve them. And that means bringing in another C4 skill — separating variables in differential equations. Head back to page 77 if you've forgotten all about it.

EXAMPLE a) In the example above, how long does it take the speed of the bike to fall to 3 ms^{-1}?

First rearrange the differential equation to get all the v terms on one side, and all the t terms on the other:

$$\frac{dv}{dt} = -\frac{v}{5} \Rightarrow \frac{1}{v}\,dv = -\frac{1}{5}\,dt$$

Next integrate both sides:

$$\int \frac{1}{v}\,dv = \int -\frac{1}{5}\,dt \Rightarrow [\ln|v|]_{10}^{3} = \left[-\frac{t}{5}\right]_0^T$$

Wondering where these limits appeared from? Velocity is 10 ms^{-1} at $t = 0$, when the cyclist stops pedalling. And velocity is 3 ms^{-1} at some unknown time, $t = T$. Easy.

So $\ln 3 - \ln 10 = -\frac{T}{5} \Rightarrow T = 6.02$ s (3 s.f.)

b) Show that $v = 10e^{-0.2t}$ for the cyclist.

As in part a), separate the variables and integrate both sides:

$$\int \frac{1}{v}\,dv = \int -\frac{1}{5}\,dt \Rightarrow \ln|v| = -\frac{t}{5} + C$$

An indefinite integral this time.

Take exponentials of both sides to find an expression for v:

$e^{\ln|v|} = e^{-0.2t + C} \Rightarrow |v| = e^{-0.2t} \times e^{C}$. So $v = ke^{-0.2t}$, for some constant k.

To do this step, just rewrite the constant e^c as another constant, k — see page 69.

Now find k. At $t = 0$, $v = 10$ ms^{-1}, so:

$10 = ke^{-(0.2 \times 0)} \Rightarrow k = 10$

So, $v = 10e^{-0.2t}$ as required.

If you have a suggestion about what we could use to fill this annoying bit of space, we'd like to hear it. Send us a postcard. Or a carrier pigeon.

dv/dt — I thought you got that from long-haul flights...

The questions you get on this stuff tend to be the same every time — form the differential equation, integrate it to get an equation connecting v and t, then use this equation to find t for a certain value of v. You just need to be able to manipulate yukky equations.

M2 Section 1 — Practice Questions

Well that wasn't such a bad intro to the world of M2. Before you crack on with more mechanical delights, I reckon it's time for some practice questions to make sure you've made sense of everything in this section. And because I'm nice, I'll start you off with some nice easy warm-up questions...

Warm-up Questions

Q1-3 are for Edexcel M2 and OCR M2

1) A particle is projected with initial velocity u ms^{-1} at an angle α to the horizontal. What is the initial velocity of the particle in the direction parallel to the horizontal in terms of u and α?

2) A rifle fires a bullet horizontally at 120 ms^{-1}. The target is hit at a horizontal distance of 60 m from the end of the rifle. Find how far the target is vertically below the end of the rifle. Take $g = 9.8$ ms^{-2}.

3) A golf ball takes 4 seconds to land after being hit with a golf club from a point on the horizontal ground. If it leaves the club with a speed of 22 ms^{-1}, at an angle of α to the horizontal, find α. Take $g = 9.8$ ms^{-2}.

Q4-6 are for AQA M2 and Edexcel M2

4) A particle sets off from the origin at $t = 0$ and moves along the x-axis with velocity $v = 8t^2 - 2t$. Find expressions for:
   a) the acceleration of the particle at time t, and b) the displacement of the particle at time t

5) A particle moving in a plane has position vector $\mathbf{r}$, where $\mathbf{r} = x\mathbf{i} + y\mathbf{j}$. What quantities are represented by the vectors $\dot{\mathbf{r}}$ and $\ddot{\mathbf{r}}$?

6) A particle sets off from the origin at $t = 0$ and moves in a plane with velocity $\mathbf{v} = 4t\mathbf{i} + t^2\mathbf{j}$. Find the position vector $\mathbf{r}$ and the acceleration vector $\mathbf{a}$ for the particle at time t.

Right, now you're warmed up and there's absolutely no danger of you pulling a maths muscle, it's time to get down to the serious business of practice exam questions.

Exam Questions

Whenever a numerical value of g is required in the questions below, take $g = 9.8$ ms^{-2}.

Q1-2 are for Edexcel M2 and OCR M2

1

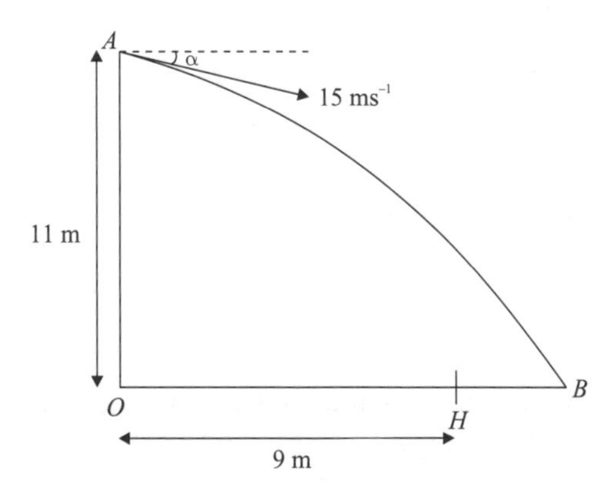

A stone is thrown from point A on the edge of a cliff, towards a point H, which is on horizontal ground. The point O is on the ground, 11 m vertically below the point of projection. The stone is thrown with speed 15 ms^{-1} at an angle α below the horizontal, where $\tan\alpha = \frac{3}{4}$.

The horizontal distance from O to H is 9 m.

The stone misses the point H and hits the ground at point B, as shown above. Find:

a) the time taken by the stone to reach the ground. *(5 marks)*

b) the horizontal distance the stone misses H by. *(3 marks)*

c) the speed of projection which would have ensured that the stone landed at H. *(5 marks)*

M2 Section 1 — Practice Questions

There's more where that came from. Oh yes indeed...

2 A stationary football is kicked with a speed of 20 ms^{-1}, at an angle of 30° to the horizontal, towards a goal 30 m away. The crossbar is 2.5 m above the level ground. Assuming the path of the ball is not impeded, determine whether the ball passes above or below the crossbar. What assumptions does your model make? *(6 marks)*

Q3 is for OCR M2 only

3 A stone is projected from horizontal ground with velocity U ms^{-1} at an angle θ° above the horizontal.

a) Given that the time it takes for the stone to land is $t = \sin\theta$ s, find U. *(2 marks)*

b) Show that the horizontal range, x, of the stone is $x = 2.45\sin 2\theta$ m. *(3 marks)*

c) Given that the range of the stone is 2 m, find the two possible values of t. *(5 marks)*

Q4 is for AQA M2 and Edexcel M2

4 A particle P is moving in a horizontal plane under the action of a single force **F** newtons. After t seconds, P has position vector:

$$\mathbf{r} = (2t^3 - 7t^2 + 12)\mathbf{i} + (3t^2 - 4t^3 - 7)\mathbf{j}\text{ m}$$

where the unit vectors **i** and **j** are in the directions of east and north respectively. Find:

a) an expression for the velocity of P after t seconds. *(2 marks)*

b) the speed of P when $t = \frac{1}{2}$, and the direction of motion of P at this time. *(3 marks)*

At $t = 2$, the magnitude of **F** is 170 N. Find:

c) the acceleration of P at $t = 2$, *(3 marks)*

d) the mass of the particle, *(3 marks)*

e) the value of t when **F** is acting parallel to **j**. *(3 marks)*

Q5-6 are for Edexcel M2 only

5 A particle sets off from the origin O at $t = 0$ and moves in a straight line along the x-axis. At time t seconds, the velocity of the particle is v ms^{-1} where

$$v = \begin{cases} 9t - 3t^2 & 0 \leqslant t \leqslant 4 \\ \dfrac{-192}{t^2} & t > 4 \end{cases}$$

Find:

a) the maximum speed of the particle in the interval $0 \leqslant t \leqslant 4$. *(4 marks)*

b) the displacement of the particle from O at

(i) $t = 4$ *(3 marks)*

(ii) $t = 6$ *(4 marks)*

M2 Section 1 — Practice Questions

6 A golf ball is hit from a tee at point O on the edge of a vertical cliff. Point O is 30 m vertically above A, the base of the cliff. The ball is hit with velocity $(14\mathbf{i} + 35\mathbf{j})$ ms^{-1} towards a hole, H, which lies on the horizontal ground. At time t seconds, the position of the ball is $(x\mathbf{i} + y\mathbf{j})$ m relative to O.
$\mathbf{i}$ and $\mathbf{j}$ are the horizontal and vertical unit vectors respectively.

a) By writing down expressions for x and y in terms of t, show that $y = \frac{5x}{2} - \frac{x^2}{40}$ *(4 marks)*

The ball lands on the ground at point B, 7 m beyond H, where AHB is a straight horizontal line.

b) Find the horizontal distance AB. *(3 marks)*

c) Find the speed of the ball as it passes through a point vertically above H. *(4 marks)*

Q7-10 are for AQA M2 only

7 A plane of mass m kg is taxiing along a runway. Its engines exert a horizontal driving force of $1000mv^{-1}$ N, and it experiences a horizontal force resistant to motion of $0.1mv^2$, where v is the velocity of the plane in ms^{-1}. Assume that these are the only two forces acting horizontally on the plane.

a) Show that the equation of motion of the plane is:

$$\frac{dv}{dt} = \frac{1000}{v} - 0.1v^2$$

(2 marks)

b) The pilot switches the engines off. Write down the new equation of motion for the plane. *(1 mark)*

c) If the pilot switched off the engines when the plane was travelling at 50 ms^{-1}, find the time it would take for the plane to slow down to 25 ms^{-1}. *(5 marks)*

8 At time t, the position vector, $\mathbf{r}$ m, of a particle relative to a fixed origin is given by:

$$\mathbf{r} = 3(\cos t)\mathbf{i} + 3(\sin t)\mathbf{j}$$

a) Show that the speed of the particle is constant, *(3 marks)*

b) Show that the particle is moving in a circle, and find the radius of this circle. *(3 marks)*

9 A particle moves in a straight line along the x-axis. At time t, the particle has acceleration a ms^{-2}, where:

$$a = 8t^2 + 6\sin 2t$$

When $t = 0$, the particle is stationary.

a) Find an expression for the particle's velocity at time t. *(4 marks)*

b) Show that when $t = \frac{\pi}{2}$, the particle's velocity is $\frac{\pi^3}{3} + 6$. *(2 marks)*

10 A 2 kg bowling ball moves along a smooth horizontal surface with initial velocity 12 ms^{-1}. The ball experiences a horizontal resistive force of kv N, where v is the velocity of the ball in ms^{-1} and k is a constant. Assuming that this is the only horizontal force experienced by the ball:

a) Show that: $\frac{dv}{dt} = -\frac{kv}{2}$ *(2 marks)*

b) Show that: $v = 12\sqrt{e^{-kt}}$ *(4 marks)*

Discrete Groups of Particles in 1 Dimension

This page is for AQA M2, Edexcel M2, OCR M2 and OCR MEI M2

Welcome to the Centre of Mass. No, not your local Catholic church...

For **Particles in a Line** — Combine **Moments** about the **Origin**

1) The weight of an object is considered to act at its centre of mass.
A group of objects also has a centre of mass, which isn't necessarily in the same position as any one of the objects.
2) It's often convenient to model these objects as particles (point masses) since the position of a particle is the position of its centre of mass. If a group of particles all lie in a horizontal line, then the centre of mass of the group will lie somewhere on the same line.

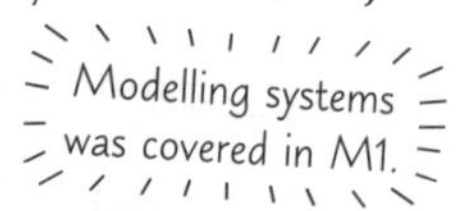

3) The moment (turning effect) of a particle from a fixed point is: weight (mass × gravity) × perpendicular distance from point
This is mgx if the fixed point and the particle are horizontally aligned (see p.193 for more on moments).
4) The moment of a group of particles in a horizontal line about a point in the horizontal line can be found by adding together all the individual moments about the point: Σmgx.
5) This has the same effect as the combined weight (Σmg) acting at the centre of mass of the whole group ($\bar{x}$).

Writing this as a formula:

$\Sigma mgx = \bar{x}\Sigma mg$

e.g. for 3 particles in a horizontal line:

$m_1gx_1 + m_2gx_2 + m_3gx_3 = \bar{x}(m_1g + m_2g + m_3g)$

$\Rightarrow m_1x_1 + m_2x_2 + m_3x_3 = \bar{x}(m_1 + m_2 + m_3)$

$\Rightarrow \Sigma mx = \bar{x}\Sigma m$

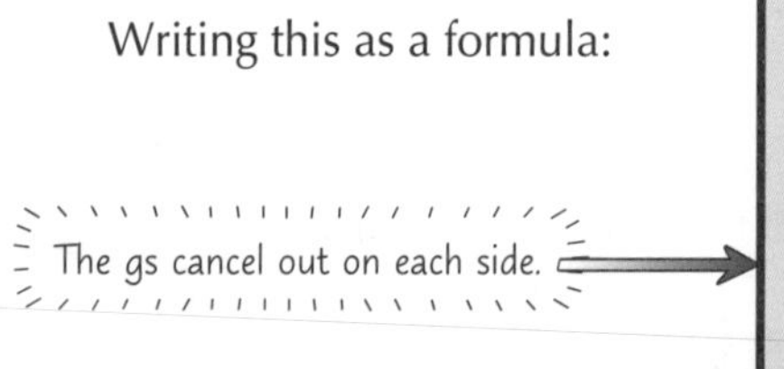

Use this simplified formula to find the centre of mass, $\bar{x}$, of a group of objects in a horizontal line.

EXAMPLE Three particles are placed at positions along the x-axis, as shown. Find the coordinates of the centre of mass of the group of particles, with respect to the origin, 0.

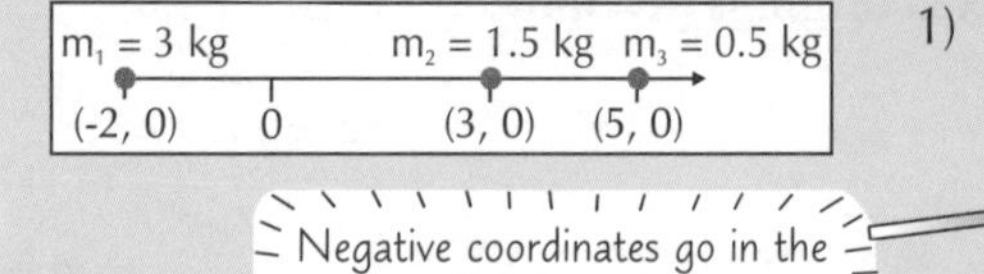

1) Use the formula $\Sigma mx = \bar{x}\Sigma m$ and put in what you know:
$m_1x_1 + m_2x_2 + m_3x_3 = \bar{x}(m_1 + m_2 + m_3)$
$\Rightarrow (3 \times -2) + (1.5 \times 3) + (0.5 \times 5) = \bar{x}(3 + 1.5 + 0.5)$
$\Rightarrow 1 = 5\bar{x} \quad \Rightarrow \quad \bar{x} = 0.2$
2) So the centre of mass of the group has the coordinates (0.2, 0)

Negative coordinates go in the formula just as they are.

Use $\bar{y}$ for Particles in a Vertical Line

It's the same for particles arranged in a vertical line. The centre of mass has the coordinate (0, $\bar{y}$).

$\Sigma my = \bar{y}\Sigma m$

EXAMPLE A light vertical rod AB has particles attached at various positions, as shown. At what height is the centre of mass of the rod?

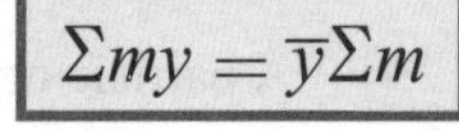

B
m_4 = 2 kg — 1 m
m_3 = 1 kg — 2 m
m_2 = 4 kg — 1 m
m_1 = 3 kg — 1 m
A

1) First, work out the positions of all the particles relative to a single point or 'origin'.
Since you're asked for the vertical height, pick point A at the bottom of the rod:
$y_1 = 1, y_2 = 2, y_3 = 4, y_4 = 5$.
2) Plug the numbers into the formula:
$\Sigma my = \bar{y}\Sigma m \quad \Rightarrow \quad m_1y_1 + m_2y_2 + m_3y_3 + m_4y_4 = \bar{y}(m_1 + m_2 + m_3 + m_4)$
$\Rightarrow (3 \times 1) + (4 \times 2) + (1 \times 4) + (2 \times 5) = \bar{y} \times (3 + 4 + 1 + 2)$
$\Rightarrow 25 = \bar{y} \times 10 \quad \Rightarrow \quad \bar{y} = 25 \div 10 = 2.5$.
3) Make sure you've answered the question — $\bar{y}$ is the vertical coordinate from the 'origin' which we took as the bottom of the rod. So the vertical height of the centre of mass is 2.5 m.

Take a moment to understand the basics...

Once you've got your head around what's going on with a system of particles, the number crunching is the easy part. You'll often have to tackle wordy problems where you first have to model a situation using rods and particles and things — you should be more than familiar with doing this from M1, and there's more practice to come later in the section.

Discrete Groups of Particles in 2 Dimensions

This page is for AQA M2, Edexcel M2, OCR M2 and OCR MEI M2

Let's face it, in the 'real world', you'll rarely come across a group in a perfectly orderly line (think of queuing up in the sales — madness). Luckily, the same principles apply in two dimensions — it's no harder than the stuff on the last page.

*Use the **Position Vector** $\bar{\mathbf{r}}$ for **Centre of Mass** of a **Group** on a **Plane***

There are two ways to find the centre of mass of a group of particles on a plane (i.e. in 2 dimensions, x and y, rather than just in a line). The quickest way uses position vectors, but I'll show you both methods and you can choose.

EXAMPLE Find the coordinates of the centre of mass of the system of particles shown in the diagram.

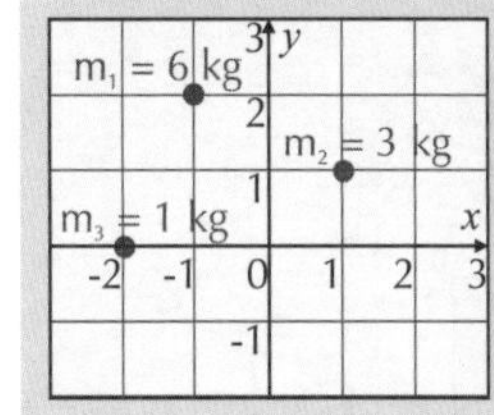

The Long Way — find the x and y coordinates separately:

1) Find the x coordinate of the centre of mass first (pretend they're in a horizontal line...)
$x_1 = -1$, $x_2 = 1$, $x_3 = -2$, so:
$m_1x_1 + m_2x_2 + m_3x_3 = \bar{x}(m_1 + m_2 + m_3) \Rightarrow (6 \times -1) + (3 \times 1) + (1 \times -2) = \bar{x}(6 + 3 + 1)$
$\Rightarrow \bar{x} = -\frac{5}{10} = \underline{-0.5}$.

2) Now find the y coordinate in the same way: $y_1 = 2$, $y_2 = 1$, $y_3 = 0$, so:
$m_1y_1 + m_2y_2 + m_3y_3 = \bar{y}(m_1 + m_2 + m_3) \Rightarrow (6 \times 2) + (3 \times 1) + (1 \times 0) = \bar{y}(6 + 3 + 1)$
$\Rightarrow \bar{y} = \frac{15}{10} = \underline{1.5}$.

3) So the centre of mass has the coordinates $\underline{(-0.5, 1.5)}$.

Column position vectors like these are just like coordinates standing upright: $\mathbf{r} = \binom{x}{y}$.

The Short Way — use position vectors:

1) Write out the position vector ($\mathbf{r}$) for each particle: $\mathbf{r}_1 = \binom{-1}{2}$, $\mathbf{r}_2 = \binom{1}{1}$, $\mathbf{r}_3 = \binom{-2}{0}$.

2) Use the formula, but replace the xs and ys with $\mathbf{r}$s: $\Sigma m\mathbf{r} = \bar{\mathbf{r}}\Sigma m \Rightarrow m_1\mathbf{r}_1 + m_2\mathbf{r}_2 + m_3\mathbf{r}_3 = \bar{\mathbf{r}}(m_1 + m_2 + m_3)$

$\Rightarrow 6\binom{-1}{2} + 3\binom{1}{1} + 1\binom{-2}{0} = \bar{\mathbf{r}}(6 + 3 + 1) \Rightarrow \binom{-6}{12} + \binom{3}{3} + \binom{-2}{0} = 10\bar{\mathbf{r}}$

$\Rightarrow \binom{-5}{15} = 10\bar{\mathbf{r}} \Rightarrow \bar{\mathbf{r}} = \binom{-0.5}{1.5}$. So the centre of mass has position vector $\binom{-0.5}{1.5}$, and coordinates $\underline{(-0.5, 1.5)}$.

Using this method the formula becomes: $\Sigma m\mathbf{r} = \bar{\mathbf{r}}\Sigma m$

*The **Formula** works for finding **Unknown Masses** and **Locations***

You won't always be asked to find the centre of mass of a system. You could be given the position of the centre of mass and asked to work out something else, like the mass or coordinates of a particle in the system. Use the same formula:

EXAMPLE The diagram shows the position of the centre of mass (COM) of a system of three particles attached to the corners of a light rectangular lamina. Find m_2.

A lamina is just a flat (2D) shape.

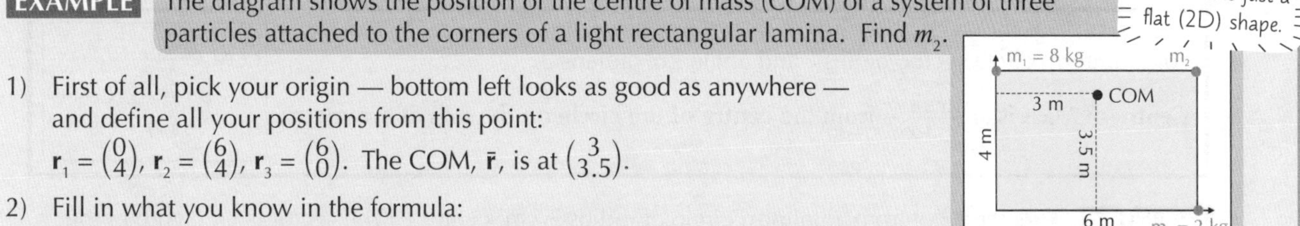

1) First of all, pick your origin — bottom left looks as good as anywhere — and define all your positions from this point:
$\mathbf{r}_1 = \binom{0}{4}$, $\mathbf{r}_2 = \binom{6}{4}$, $\mathbf{r}_3 = \binom{6}{0}$. The COM, $\bar{\mathbf{r}}$, is at $\binom{3}{3.5}$.

2) Fill in what you know in the formula:
$\Sigma m\mathbf{r} = \bar{\mathbf{r}}\Sigma m \Rightarrow m_1\mathbf{r}_1 + m_2\mathbf{r}_2 + m_3\mathbf{r}_3 = \bar{\mathbf{r}}(m_1 + m_2 + m_3)$
$\Rightarrow 8\binom{0}{4} + m_2\binom{6}{4} + 2\binom{6}{0} = \binom{3}{3.5} \times (8 + m_2 + 2) \Rightarrow \binom{0}{32} + \binom{6m_2}{4m_2} + \binom{12}{0} = \binom{3}{3.5} \times (m_2 + 10)$
$\Rightarrow \binom{6m_2 + 12}{4m_2 + 32} = \binom{3m_2 + 30}{3.5m_2 + 35}$

3) Pick either the top row or bottom row to solve the equation for m_2 (it should be the same in both), e.g.: Top — $6m_2 + 12 = 3m_2 + 30 \Rightarrow m_2 = \underline{6\text{ kg}}$. Bottom — $4m_2 + 32 = 3.5m_2 + 35 \Rightarrow m_2 = \underline{6\text{ kg}}$.

2D or not 2D — that is the question...

Well actually the question's more likely to be 'Find the centre of mass of the following system of particles...', but then I doubt that would have made Hamlet quite such a gripping tale. Make sure you can do this stuff with your eyes shut because you'll need it again later on, and there's also a new compulsory blindfolded section to the M2 exam this year...

Standard Uniform Laminas

This page is for AQA M2, Edexcel M2, OCR M2 and OCR MEI M2

A page full of shapes for you to learn, just like in little school. However, you need to be able to find the centres of mass of these uniform plane laminas, not just colour them in. Even if you can do it neatly inside the lines.

Use Lines of Symmetry with Regular and Standard Shapes

Uniform laminas have evenly spread mass, so the centre of mass is in the centre of the shape, on all the lines of symmetry. So for shapes with more than one line of symmetry, the centre of mass is where the lines of symmetry intersect.

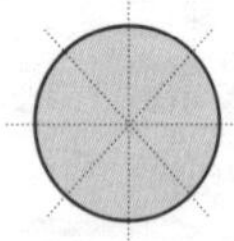
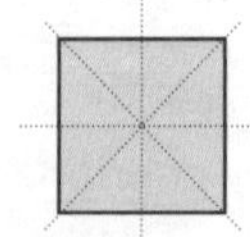
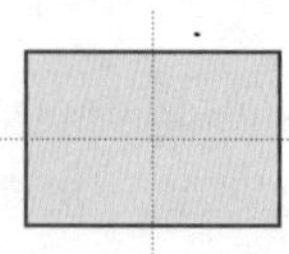

EXAMPLE Find the coordinates of the centre of mass of a uniform rectangular lamina with vertices A(–4, 7), B (2, 7), C(–4, –3) and D(2, –3).

1) A little sketch never goes amiss...
2) $\bar{x}$ is the midpoint of AB (or CD), i.e. $(-4 + 2) \div 2 = -1$.
3) $\bar{y}$ is the midpoint of AC (or BD), i.e. $(7 + -3) \div 2 = 2$.
4) So the centre of mass is at (–1, 2). Easy peasy lemon squeezy*.

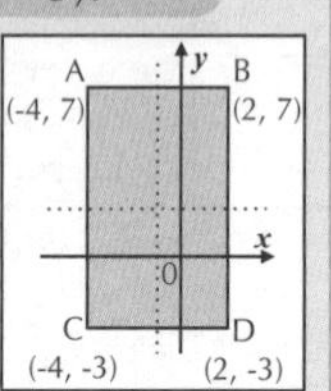

The Centre of Mass of a Triangle is the Centroid

1) In any triangle, the lines from each vertex to the midpoint of the opposite side are called medians.

In an equilateral triangle, the medians are lines of symmetry.

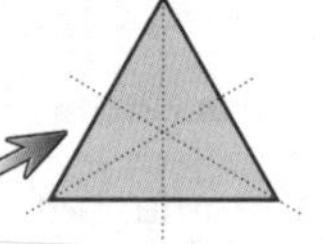

2) If you draw in the medians on any triangle, the point where they meet will be two thirds of the way up each median from each vertex. This point is the centroid, and it's the centre of mass in a uniform triangle.

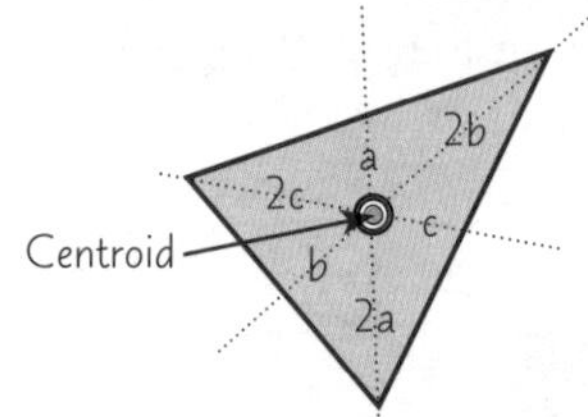

3) There's a formula for finding the coordinates of the centroid:

For a triangle with vertices at (x_1, y_1), (x_2, y_2) and (x_3, y_3):

Centre of Mass $(\bar{x}, \bar{y})$ is at $\left(\frac{x_1 + x_2 + x_3}{3}, \frac{y_1 + y_2 + y_3}{3}\right)$

(i.e. the mean x coordinate and mean y coordinate)

The 'two thirds' fact will be on the formula sheet, but you need to know how to use it.

Use the Formula to find the COM of a Sector of a Circle

This bit is for Edexcel M2, OCR M2 and OCR MEI M2 only

Finding the centre of mass of a sector of a circle is a bit harder, so the formula is given to you in the exam:

For a uniform circle sector, radius r and angle 2α radians:

Centre of Mass is at $\frac{2r\sin\alpha}{3\alpha}$ from the centre of the circle on the axis of symmetry.

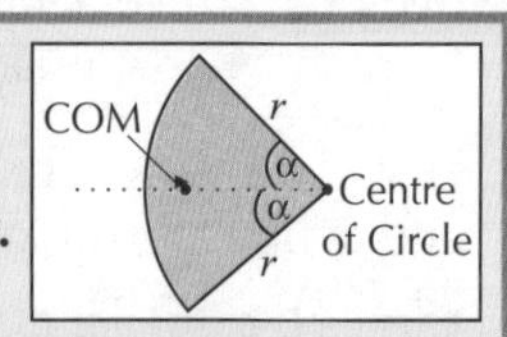

EXAMPLE A sector is cut from a uniform circle of radius 3 cm, centre P. The sector is an eighth of the whole circle. How far along the axis of symmetry is the centre of mass of the sector from P?

1) The angle of the sector is an eighth of the whole circle, so $2\alpha = \frac{2\pi}{8} \Rightarrow \alpha = \frac{\pi}{8}$.
2) Using the formula $\frac{2r\sin\alpha}{3\alpha}$, with $r = 3$ cm:

Centre of Mass = $\frac{2 \times 3 \times \sin\frac{\pi}{8}}{\frac{3\pi}{8}}$ = 1.95 cm from P (to 3 s.f.)

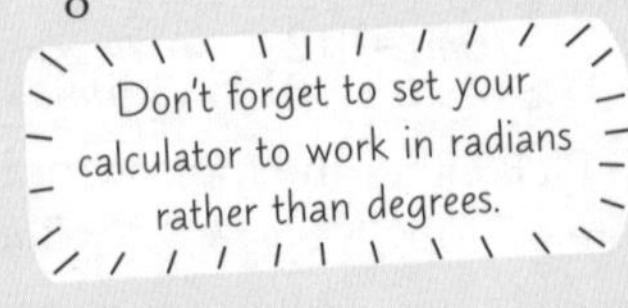

I love a lamina in uniform...

A nice easy page with lots of pretty shapes and colours. Before you start unleashing your inner toddler and demanding sweets and afternoon naps, make sure you fully understand what's been said on this page, because the tough stuff is coming right up. There's plenty of time for sweets and afternoon naps when the exams are over. Trust me...

*Squeezing lemons is actually quite tricky so I'm not sure where this saying comes from.

Composite Shapes

It's time to combine all the things covered so far in the section into one lamina lump. Yay.

For a Composite Shape — Find each COM Individually then Combine

This bit is for AQA M2, Edexcel M2, OCR M2 and OCR MEI M2

A composite shape is one that can be broken up into standard parts such as triangles, rectangles and circles.
Once you've found the COM of a part, imagine replacing it with a particle of the same mass in the position of the COM.
Do this for each part, then find the COM of the group of 'particles' — this is the COM of the composite shape.

EXAMPLE A house-shaped lamina is cut from a single piece of card, with dimensions as shown.
Find the location of the centre of mass of the shape in relation to the point O.

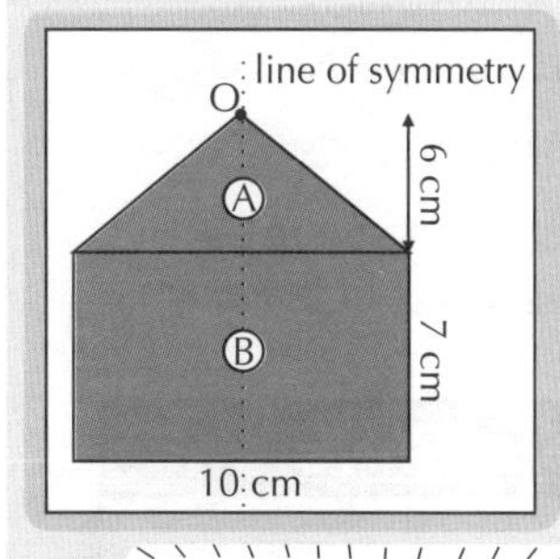

1) First, split up the shape into a triangle (A) and rectangle (B). As both bits are made of the same material, the masses of A and B are in proportion to their areas, so we can say $m_A = \frac{1}{2} \times 10 \times 6 = 30$, and $m_B = 10 \times 7 = 70$.
2) The shape has a line of symmetry, so the centre of mass must be on that line, directly below the point O.
3) Next find the vertical position of the centres of mass of both A and B individually:
   $y_A = \frac{2}{3}$ of the distance down from O $= \frac{2}{3} \times 6$ cm $= 4$ cm from O
   (since A is a triangle and the vertical line of symmetry from O is a median of the triangle — p.182)
   $y_B = 6$ cm $+ (7$ cm $\div 2) = 9.5$ cm from O
   (since B has a horizontal line of symmetry halfway down, but is 6 cm below O to start with)

Use symmetry where you can — but make sure you explain what you've done.

4) Treat the shapes as two particles positioned at the centres of mass of each shape, and use the formula from p.180:
   $\Sigma my = \bar{y}\Sigma m \Rightarrow m_A y_A + m_B y_B = \bar{y}(m_A + m_B)$
   $\Rightarrow (30 \times 4) + (70 \times 9.5) = \bar{y}(30 + 70) \Rightarrow 785 = 100\bar{y} \Rightarrow \bar{y} = 785 \div 100 = 7.85$ cm.
5) Make sure you've answered the question —
   The centre of mass of the whole shape is 7.85 cm vertically below O on the line of symmetry. Job done.

You can use the Removal Method for Some Shapes

This bit is for Edexcel M2, OCR M2 and OCR MEI M2

You may have a shape that looks like a 'standard' shape with other standard shapes 'removed' rather than stuck together.
The removal method is like the one above, except the individual centres of mass are subtracted rather than added.

EXAMPLE Find the coordinates of the centre of mass of the lamina shown — a circle of radius 3 with a quarter sector removed.

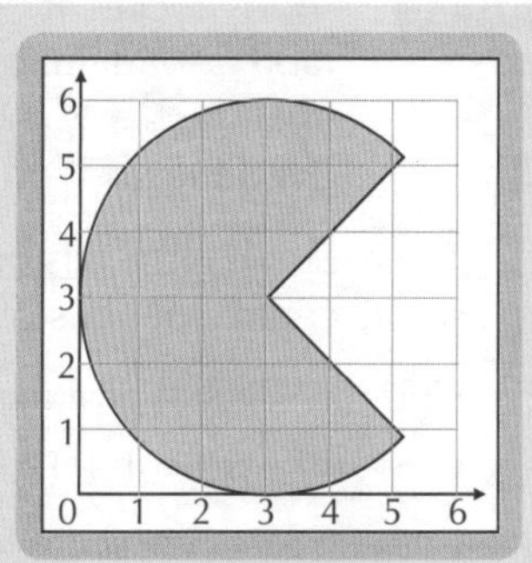

1) Let's call the 'whole' circle A and the sector that's been removed B.
   Since B is a quarter of A, we can say that the masses are $m_A = 4$ and $m_B = 1$.
2) We're working in 2D, so use position vectors $\mathbf{r}_A$ and $\mathbf{r}_B$ to describe the centres of mass. From the symmetry of the circle, $\mathbf{r}_A = \binom{3}{3}$.
   $\mathbf{r}_B$ can be worked out using the formula on p. 182:
   $y_B = 3$ (from the symmetry of the sector) and $x_B = 3 + \frac{2r\sin\alpha}{3\alpha}$.
   The angle of the sector, $2\alpha = \frac{\pi}{2} \Rightarrow \alpha = \frac{\pi}{4}$, and r = 3,
   so: $x_B = 3 + \frac{2 \times 3 \times \sin\frac{\pi}{4}}{\frac{3\pi}{4}} = 4.8006...$ So $\mathbf{r}_B = \binom{4.8006}{3}$.
3) Using the removal method, our formula becomes: $m_A\mathbf{r}_A - m_B\mathbf{r}_B = \bar{\mathbf{r}}(m_A - m_B)$
   $4\binom{3}{3} - 1\binom{4.8006}{3} = \bar{\mathbf{r}}(4 - 1) \Rightarrow \binom{12 - 4.8006}{12 - 3} = 3\bar{\mathbf{r}} \Rightarrow \bar{\mathbf{r}} = \binom{7.1994}{9} \div 3 = \binom{2.3998}{3}$.
4) So the coordinates of the centre of mass of the shape are (2.40, 3).

Waxing is another effective removal method...

Now you've got all you need to find the centre of mass of any lamina shape, so long as you can spot how it breaks up into circles, triangles, etc. Quite arty-farty this. Set out your working neatly though, especially for the more complicated shapes.

Centres of Mass in 3 Dimensions

This page is for OCR M2 and OCR MEI M2

I know how much you must have enjoyed this centre of mass stuff so far. But if you're anything like me, you'll be hungry for more. MORE. So, I present to you today's special — Centres of Mass in 3 Dimensions. Delicious.

Use **Symmetry** to find the COM of a **Uniform 3D Shape**

Remember — uniform means that the mass is evenly distributed.

1) As you've seen on the previous few pages, using the symmetry of a shape makes finding the COM tonnes easier.
2) This is just as true for uniform 3D shapes as it is for laminas.
3) For example, spheres, cubes, cylinders and cuboids all have their centre of mass right slap bang in the centre.
4) In fact, for any uniform 3D shape, the COM will be at the point where all the planes of symmetry intersect.
5) So, for a shape with an axis of rotational symmetry, the centre of mass will be somewhere on that axis.
6) Luckily, there's only a few shapes you're likely to get asked about on your exam, and for each one there's a formula telling you where the centre of mass is. Even better — they'll all be on the formula sheet in the exam.

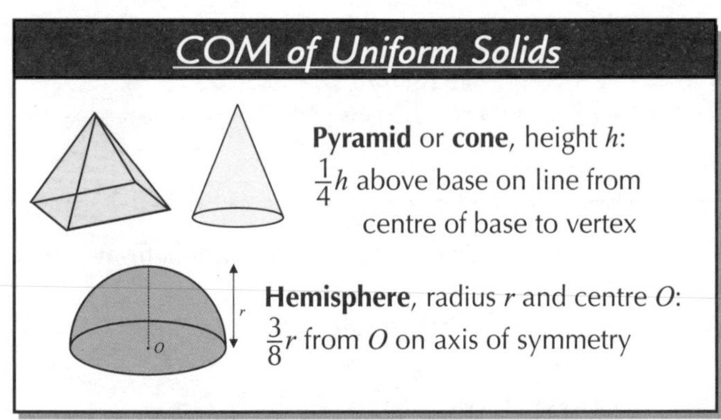

COM of Uniform Shells

Conical Shell, height h:
$\frac{1}{3}h$ above 'base' on line from centre of 'base' to vertex

Hemispherical shell, radius r and centre O:
$\frac{1}{2}r$ from O on axis of symmetry

A shell is basically a solid with all the insides scooped out. And with no base.

You also need to know how to find the centre of mass for a sphere, cube, cylinder and cuboid. But they is like well easy.

Use the same method for **Composite Shapes** in **3D** as in 2D

Just find the centre of mass for each shape individually, then combine.

EXAMPLE A toy rocket, R, is made up of a solid uniform cone, A, of height 4 cm and mass 0.5 kg and a solid uniform cylinder, B, of height 12 cm and mass 1 kg. A and B are joined so that the base of A coincides with one of the plane faces of B, as shown below. Find the position of the rocket's centre of mass in relation to O, the centre of its base.

1) The rocket has an axis of symmetry running through O and the vertex of A, so the centres of mass of A, B and R will all lie somewhere on this line.

   This simplifies the question to particles in a vertical line. Just like p. 180

2) Find the vertical position of centres of mass of A and B individually:

   $y_A = \frac{1}{4}(4) + 12 = 1 + 12 = 13$ cm above O

   (using the formula for the centre of mass of a solid uniform cone and the fact that the base of A is 12 cm above O to start with)

   $y_B = \frac{1}{2}h_B = \frac{1}{2}(12) = 6$ cm from O

   (since the cylinder has a horizontal plane of symmetry halfway up)

3) Now, just like you did in 2D, treat the shapes as two particles positioned at the centre of mass of each shape and use the formula from p. 180 to find the centre of mass of R:

   $\Sigma my = \bar{y}\Sigma m \Rightarrow m_A y_A + m_B y_B = m_R y_R$

   $\Rightarrow 0.5(13) + 1(6) = 1.5y_R \Rightarrow y_R = 12.5 \div 1.5 = 8\frac{1}{3}$ cm.

4 cm
A
12 cm
B
O

4) So the centre of mass of the rocket is $8\frac{1}{3}$ cm vertically above O on the axis of symmetry. Simple as that.

Centres of Mass in 3 Dimensions

This page is for OCR M2 and OCR MEI M2

As well as questions on 3D solids and shells, you might be asked about 3D shapes made from laminas.

Again, find each COM **Individually** then **Combine**

Isn't it great when you get to use the same method over and over again?

EXAMPLE A tray is made from 4 uniform rectangular laminas of the same material, A, B, C and D, arranged as shown below. Find the coordinates of the centre of mass of the tray referred to the axes shown and the distance, d, of this point from the origin, O.

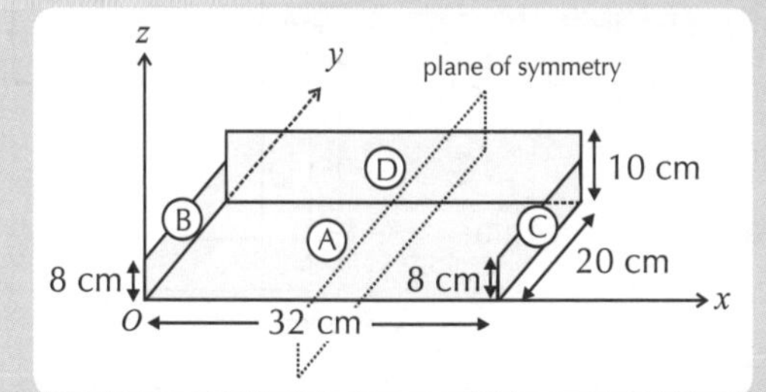

1) As A, B, C and D are made from the same material, their masses are in proportion to their areas. So:
$m_A = 32 \times 20 = 640$
$m_B = m_C = 20 \times 8 = 160$
$m_D = 32 \times 10 = 320$

2) The tray has a plane of symmetry at $x = 16$, so the x-coordinate of the tray's centre of mass must lie on this plane. So $\bar{x} = 16$.

This has simplified the problem to just finding the COM in two dimensions — y and z.

3) Now find the y-coordinate of the centre of mass of the tray:
The centres of mass of the individual laminas can be found using symmetry.
They are at the centre of each lamina — so $y_A = 10$, $y_B = 10$, $y_C = 10$ and $y_D = 20$.
Use the formula to find the y-coordinate of the centre of mass of the tray:

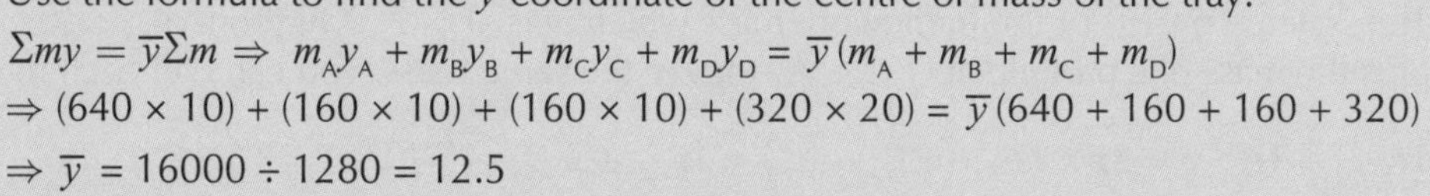

$$\Sigma my = \bar{y}\Sigma m \Rightarrow m_A y_A + m_B y_B + m_C y_C + m_D y_D = \bar{y}(m_A + m_B + m_C + m_D)$$
$$\Rightarrow (640 \times 10) + (160 \times 10) + (160 \times 10) + (320 \times 20) = \bar{y}(640 + 160 + 160 + 320)$$
$$\Rightarrow \bar{y} = 16000 \div 1280 = 12.5$$

4) And do the same thing for the z-coordinate:
$$\Sigma mz = \bar{z}\Sigma m \Rightarrow m_A z_A + m_B z_B + m_C z_C + m_D z_D = \bar{z}(m_A + m_B + m_C + m_D)$$
$$\Rightarrow (640 \times 0) + (160 \times 4) + (160 \times 4) + (320 \times 5) = \bar{z}(640 + 160 + 160 + 320)$$
$$\Rightarrow \bar{z} = 2880 \div 1280 = 2.25$$

Again, the z-coordinates of the COMs of A, B, C and D are found using symmetry.

So the coordinates of the centre of mass of the tray are $(\bar{x}, \bar{y}, \bar{z}) = (16, 12.5, 2.25)$.

5) Finally, use Pythagoras to find the distance, d, from O:
$$d = \sqrt{16^2 + 12.5^2 + 2.25^2} = 20.4 \text{ cm (3 s.f.)}$$

You can also use the **Formula** in **3 dimensions**

Sometimes you won't be able to use symmetry to simplify a 3D centre of mass question.
If you don't want to have to find each of the three coordinates individually, then you'll need to break out the vector equation from page 181 and use it with some 3D column vectors. Go on, it'll be fun.

EXAMPLE A system consists of 3 particles, A, B and C. A, B and C have coordinates (1, 2, 3), (1, –1, 0) and (–5, 4, 3) and masses 1 kg, 2 kg and 3 kg respectively. Find the centre of mass of the system.

Use the equation $\Sigma m\mathbf{r} = \bar{\mathbf{r}}\Sigma m$, where $\mathbf{r}_A = \begin{pmatrix}1\\2\\3\end{pmatrix}$, $\mathbf{r}_B = \begin{pmatrix}1\\-1\\0\end{pmatrix}$, $\mathbf{r}_C = \begin{pmatrix}-5\\4\\3\end{pmatrix}$ and $m_A = 1$ kg, $m_B = 2$ kg, $m_C = 3$ kg:

$$m_A\mathbf{r}_A + m_B\mathbf{r}_B + m_C\mathbf{r}_C = (m_A + m_B + m_C)\bar{\mathbf{r}} \Rightarrow 1\begin{pmatrix}1\\2\\3\end{pmatrix} + 2\begin{pmatrix}1\\-1\\0\end{pmatrix} + 3\begin{pmatrix}-5\\4\\3\end{pmatrix} = (1+2+3)\bar{\mathbf{r}} \Rightarrow \bar{\mathbf{r}} = \frac{1}{6}\begin{pmatrix}1+2-15\\2-2+12\\3+0+9\end{pmatrix} = \begin{pmatrix}-2\\2\\2\end{pmatrix}$$

So the coordinates of the centre of mass of the system are (–2, 2, 2).

'Centres of Mass 3D' — the must-miss movie of the year...

...contact your local cinema for details. Or, if that's not your thing, then spend some time practising centre of mass questions.

Frameworks

More pretty shapes, this time made from rods rather than laminas — imagine bending a wire coathanger into something shapely (and infinitely more useful since wire hangers are rubbish). These shapes are called frameworks.

Treat Each Side as a Rod with its own Centre of Mass

This bit is for AQA M2, Edexcel M2 and OCR MEI M2

In a framework, there's nothing in the middle, so all the mass is within the rods that make up the shape's edges. If the rods are straight and uniform, the centre of mass of each one is at the midpoint of the rod. Try to imagine each side of the shape as a separate rod, even if it's a single wire bent round into a shape.

EXAMPLES a) Find the coordinates of the centre of mass of the framework shown.

1) The black dots are the centres of mass of each of the rods that make up the frame — so the position vectors can simply be written down for each one, e.g. $\mathbf{r}_{AB} = \binom{1}{3.5}$.
2) The mass of each rod is proportional to its length, so $m_{AB} = 5$ etc.
3) You've now got the equivalent of a group of 6 particles, so put it all in the formula:

$m_{AB}\mathbf{r}_{AB} + m_{BC}\mathbf{r}_{BC} + m_{CD}\mathbf{r}_{CD} + m_{DE}\mathbf{r}_{DE} + m_{EF}\mathbf{r}_{EF} + m_{FA}\mathbf{r}_{FA} = \bar{\mathbf{r}}(m_{AB} + m_{BC} + m_{CD} + m_{DE} + m_{EF} + m_{FA})$

$5\binom{1}{3.5} + 5\binom{3.5}{6} + 2\binom{6}{5} + 2\binom{5}{4} + 3\binom{4}{2.5} + 3\binom{2.5}{1} = (5 + 5 + 2 + 2 + 3 + 3)\bar{\mathbf{r}}$

$\binom{5 + 17.5 + 12 + 10 + 12 + 7.5}{17.5 + 30 + 10 + 8 + 7.5 + 3} = 20\bar{\mathbf{r}} \Rightarrow \bar{\mathbf{r}} = \frac{1}{20}\binom{64}{76} = \binom{3.2}{3.8}$. So the coordinates are (3.2, 3.8).

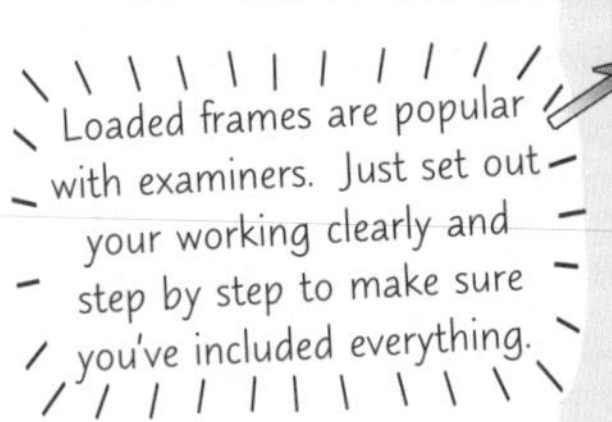

b) A particle with the same mass as the whole framework is attached to the frame at A. Find the new centre of mass of the system.

1) The system consists of the framework mass which acts at (3.2, 3.8) (from a), plus the mass of a particle at (1, 1). As they're the same mass, you can call each mass '1'.
2) $1\mathbf{r}_{Frame} + 1\mathbf{r}_{Particle} = 2\bar{\mathbf{r}} \Rightarrow \binom{3.2}{3.8} + \binom{1}{1} = 2\bar{\mathbf{r}} \Rightarrow \bar{\mathbf{r}} = \binom{4.2}{4.8} \div 2 = \binom{2.1}{2.4}$.

So the coordinates of the new COM are (2.1, 2.4).

Arcs have their own Formula

This bit is for Edexcel M2 and OCR MEI M2

Arcs are parts of the edge of a circle, and just like sectors have their own formula to work out their centre of mass:

For a uniform arc of a circle of radius r and angle 2α radians:

Centre of Mass is at $\frac{r\sin\alpha}{\alpha}$ from the centre of the circle on the axis of symmetry.

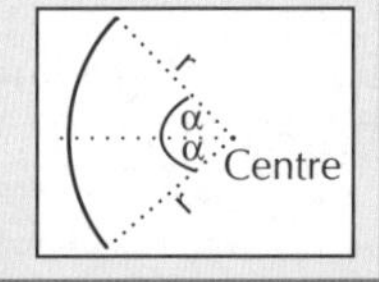

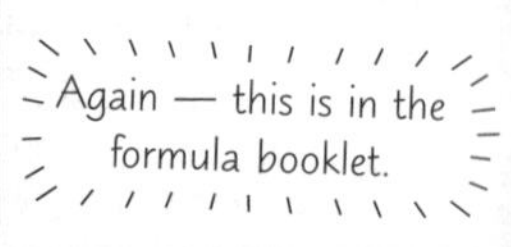

EXAMPLE A wire is bent to form a sector of a circle with a radius of 4 cm and angle $\phi = \frac{\pi}{3}$, as shown. Find, to 3 s.f., the horizontal distance of the centre of mass of the framework from the point C.

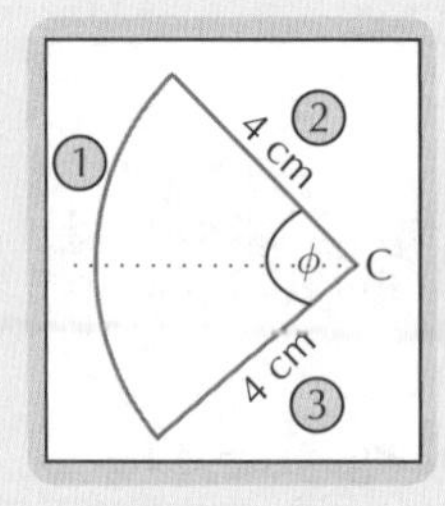

1) The mass of each rod that forms the 3 edges of the shape is in proportion to its length. The length of the arc is $\frac{4\pi}{3}$ (from arc length = $r\theta$ — see C2), and the other two sides are each 4.
2) To find the COM of the arc, use the formula $\frac{r\sin\alpha}{\alpha}$ (where $2\alpha = \frac{\pi}{3}$, so $\alpha = \frac{\pi}{6}$):

$x_1 = \frac{4 \times \sin\frac{\pi}{6}}{\frac{\pi}{6}} = \frac{12}{\pi}$ cm from C.

3) Then for the straight rods: each has their centre of mass 2 cm along their length, and the horizontal distance from C can be found using basic trig:

$x_2 = x_3 = 2 \text{ cm} \times \cos\frac{\pi}{6} = \sqrt{3}$ cm. (They're the same because of the symmetry.)

4) Treat the 3 rods like particles on a horizontal line: $m_1x_1 + m_2x_2 + m_3x_3 = \bar{x}(m_1 + m_2 + m_3)$

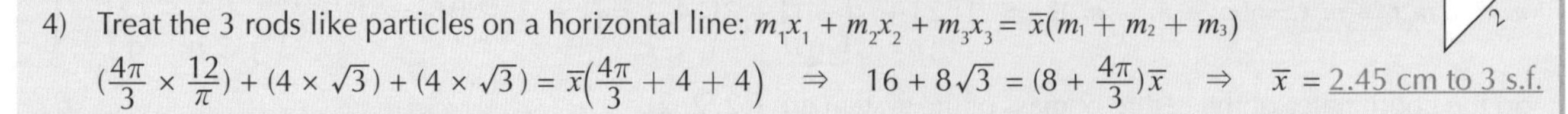

$\left(\frac{4\pi}{3} \times \frac{12}{\pi}\right) + (4 \times \sqrt{3}) + (4 \times \sqrt{3}) = \bar{x}\left(\frac{4\pi}{3} + 4 + 4\right) \Rightarrow 16 + 8\sqrt{3} = \left(8 + \frac{4\pi}{3}\right)\bar{x} \Rightarrow \bar{x} = 2.45$ cm to 3 s.f.

Arc-asm is the lowest form of wit — yet still funnier than this gag...

There's been a lot to take in on the last few pages. You'll notice every 'new' bit needs you to do all the 'old' bits too, and more besides. Maths is kinda like that. Make sure you're astoundingly marvellous at the section so far before you move on.

Laminas in Equilibrium

This is what the whole section's been working up to. The raison d'être for centres of mass, if you'll pardon my French. The position of the centre of mass will tell you what happens when you hang it up or tilt it. Très intéressant, non?

Laminas **Hang** with the Centre of Mass **Directly Below** the **Pivot**

This bit is for AQA M2, Edexcel M2, OCR M2 and OCR MEI M2

When you suspend a shape, either from a point on its edge or from a pivot point within the shape, it will hang in equilibrium so that the centre of mass is vertically below the suspension point.

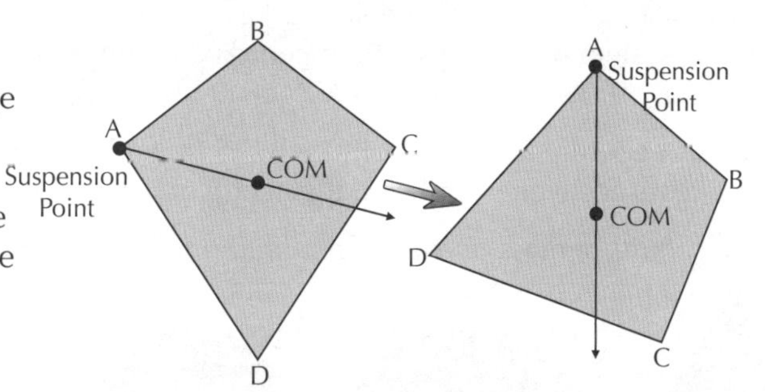

Knowing where the centre of mass lies will let you work out the angle that the shape hangs at.

This also applies to composite shapes — the centre of mass of the composite shape hangs directly below the pivot.

EXAMPLE In the shape above, A is at (0, 6), C is at (8, 6), and the COM is at (4, 5). Find, in radians to 3 s.f., the angle AC makes with the vertical when the shape is suspended from A.

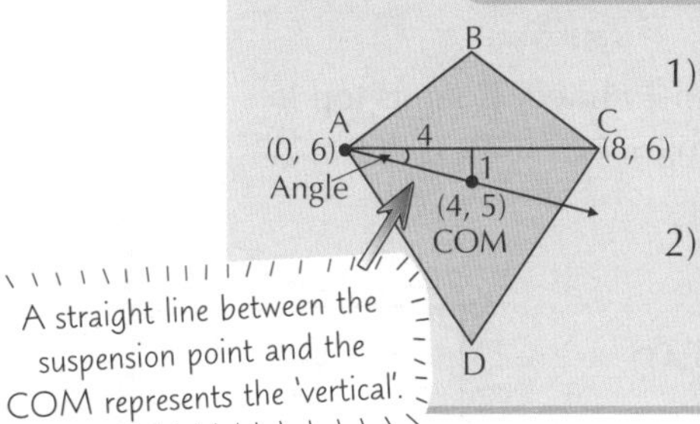

1) Do a little sketch of the shape showing the lengths you know. Draw in the line representing the vertical from the suspension point to the COM and label the angle you need to find.

2) The angle should now be an easy piece of trig away:
Angle = $\tan^{-1}\frac{1}{4}$ = 0.245 radians to 3 s.f.

In an exam question, you'll usually have to find the position of the centre of mass first and THEN do this bit to finish.

Shapes **Topple** if the COM is not **Directly Above** the **Bottom Edge**

This bit is for Edexcel M2, OCR M2 and OCR MEI M2

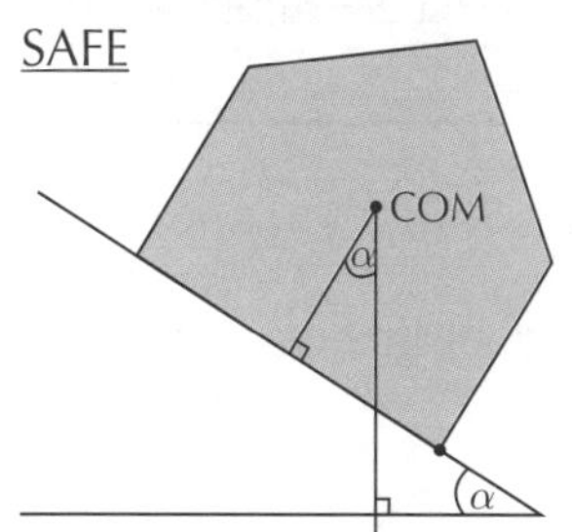

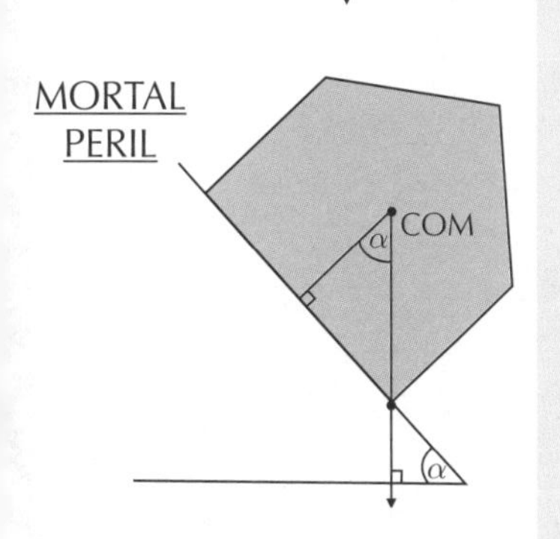

Tilting a shape on an inclined plane will make it topple over eventually (assuming there's enough friction to stop it sliding).

To make it fall over, you need to incline the plane above an angle, α, where the centre of mass is vertically above the bottom corner or edge of the shape, as shown in the pictures on the left.

EXAMPLE The house-shaped lamina from p. 183 is in equilibrium on a plane inclined at an angle α. Find the value of α at the point where the shape is about to topple (in rads to 3 s.f.).

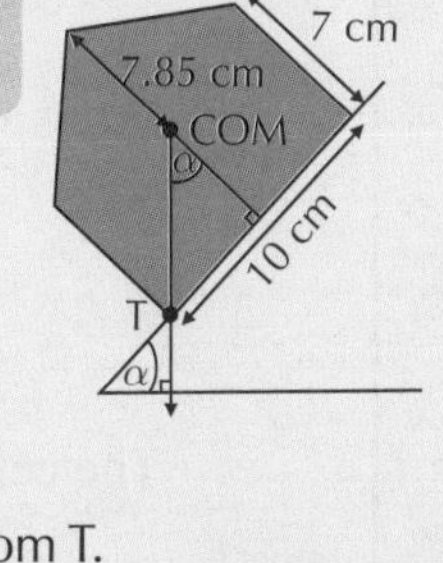

1) Draw in a line between the COM and the corner point (T) of the shape — this line will be vertical at the tipping point.

Use what you know about the position of the COM to draw a right-angled triangle containing α.
Height of COM from bottom edge = 7 + 6 − 7.85 = 5.15 cm.
COM is also halfway along the bottom edge, i.e. 10 ÷ 2 = 5 cm from T.

2) Use basic trig to work out the size of the angle:
$\alpha = \tan^{-1}\left(\frac{5}{5.15}\right)$ = 0.771 rads to 3 s.f.

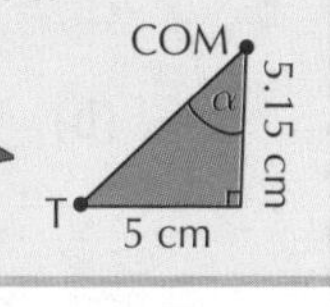

Don't hang around — get practising or you're heading for a fall...

This page may seem deceptively easy because in both examples it's assumed you've already found the centre of mass (that's what all the other pages were about in case you'd forgotten). In the exam you'll more than likely have to find the COM first and then work out the hanging or toppling angles. Luckily, there's plenty of practice at doing this on the next few pages...

M2 Section 2 — Practice Questions

Hurrah and huzzah — it's time to put your slick Section 2 skills (try saying that in a hurry) to the test.
As with all strenuous exercise, you need to warm up properly — and as if by chance, look what we have here...

Warm-up Questions

Q1-4 are for AQA M2 [except 3b), 3c) and 4], Edexcel M2, OCR M2 and OCR MEI M2

1) Three particles have mass $m_1 = 1$ kg, $m_2 = 2$ kg, and $m_3 = 3$ kg.
Find the centre of mass of the system of particles if their coordinates are, respectively:
a) (1, 0), (2, 0), (3, 0) b) (0, 3), (0, 2), (0, 1) c) (3, 4), (3, 1), (1, 0)

2) A system of particles located at coordinates A(0, 0), B(0, 4), C(5, 4) and D(5, 0) have masses m kg, $2m$ kg, $3m$ kg and 12 kg respectively. Find m, if the centre of mass of the system is at (3.5, 2).

3) Find the coordinates of the centres of mass of each of the uniform laminas shown below.

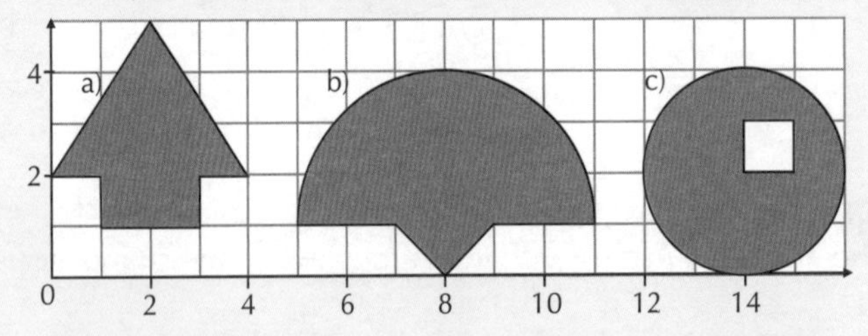

4) A square uniform lamina of width 10 cm has a smaller square of width 2 cm cut from its top left corner. Find the vertical distance of the centre of mass of the remaining shape from its top edge.

Q5 is for AQA M2, Edexcel M2 and OCR MEI M2

5) a) A light square framework has side lengths of 5 cm.
Masses are attached to the corners as shown on the right.
Find the distance of the centre of mass from: i) side AB ii) side AD.

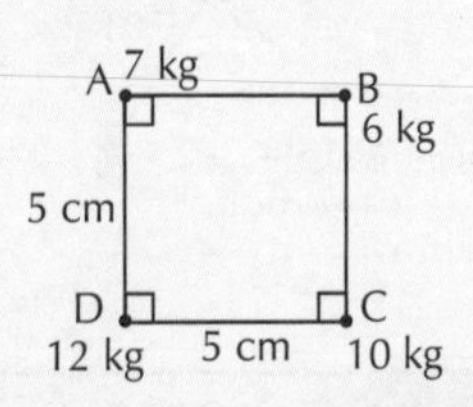

b) The frame is suspended from corner A and hangs in equilibrium.
Find the angle AB makes with the vertical. Give your answer in degrees to the nearest degree.

The universe is full of seemingly unanswerable questions to ponder.
Fortunately for those not particularly inclined towards philosophy, there are some perfectly good answerable ones here.

Exam Questions

Q1-2 are for AQA M2, Edexcel M2, OCR M2 and OCR MEI M2

1 The diagram below shows three particles attached to a light rectangular lamina at coordinates A(1, 3), B(5, 1) and C(4, y).

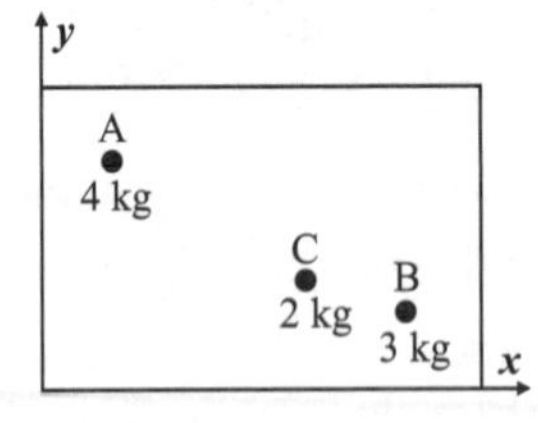

The centre of mass of the system is at $(\bar{x}, 2)$.

(a) Show that $y = 1.5$. *(3 marks)*

(b) Show that $\bar{x} = 3$. *(3 marks)*

The light lamina is replaced with a uniform rectangle PQRS, having a mass of 6 kg and vertices at P(0, 0), Q(0, 5), R(7, 5) and S(7, 0). Particles A, B and C remain at their existing coordinates.

(c) Find the coordinates of the new centre of mass of the whole system. *(6 marks)*

M2 Section 2 — Practice Questions

2 A cardboard 'For Sale' sign is modelled as a uniform lamina consisting of two squares and an isosceles triangle. The line of symmetry through the triangle coincides with that of the larger square, as shown.

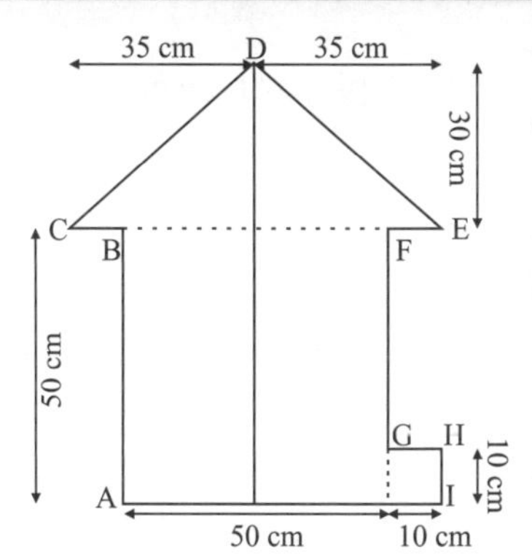

(a) Show that the centre of mass of the sign, to 3 s.f., is 25.8 cm from AB and 34.5 cm from AI. *(6 marks)*

The sign, with a mass of 1 kg, is suspended from the point D, and hangs in equilibrium, at an angle. A small weight, modelled as a particle, is attached at A, so that the sign hangs with AI horizontal.

(b) Find the mass of the particle needed to make the sign hang in this way. Give your answer in kg to 3 s.f. *(3 marks)*

Q3-5 are for Edexcel M2, OCR M2 and OCR MEI M2

3 A stencil is made from a uniform sheet of metal by removing a quarter of a circle with centre at the point O.

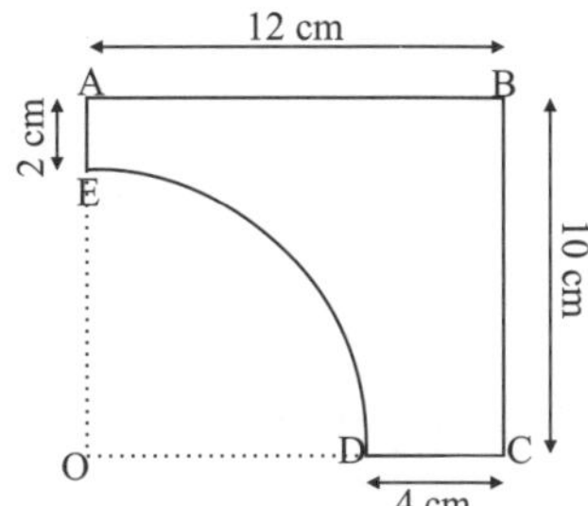

(a) Taking the point O as the origin, find the coordinates of the centre of mass of the stencil, to 3 s.f. Assume the stencil can be modelled as a lamina. *(7 marks)*

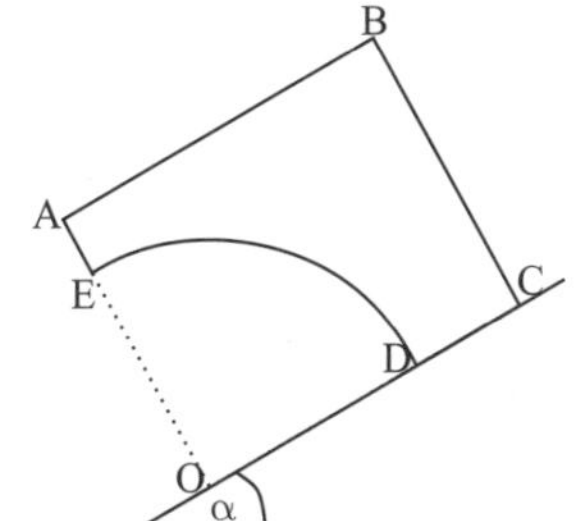

Particles are added to the stencil to adjust the centre of mass so that it now acts at coordinates (9, 6) from O. The stencil rests in equilibrium on a rough inclined plane, as shown. The angle of incline is increased until the shape is just about to fall over, balanced on the point D.

(b) Find the angle of incline above which the shape will topple. Give your answer in radians to 3 s.f. *(3 marks)*

4 A wire sculpture is modelled as a frame made from two uniform rods, one straight with mass $2m$, and one a semicircular arc with mass πm, with 2 particles of mass $3m$ and $4m$ attached to each corner, as shown:

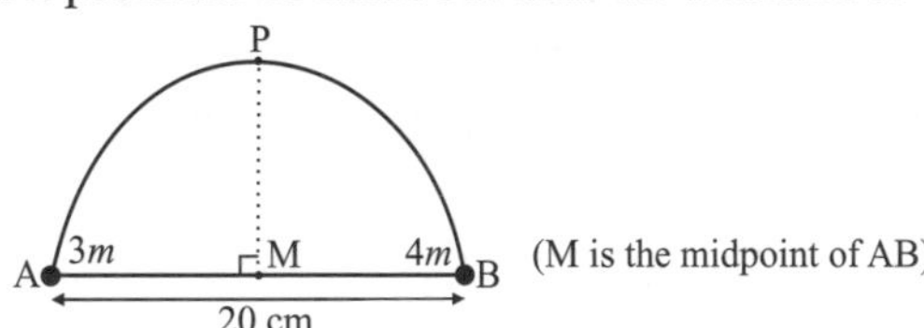

(a) Find, in cm to 4 decimal places, the distance of the centre of mass of the loaded framework from:

(i) MP. *(3 marks)*

(ii) AB. *(3 marks)*

The sculpture is suspended from the point P, and hangs in equilibrium.

(b) Find the angle MP makes with the vertical. Give your answer in radians to 3 s.f. *(3 marks)*

M2 Section 2 — Practice Questions

5 A piece of jewellery is made by cutting a triangle from a thin circle of metal, as shown below.

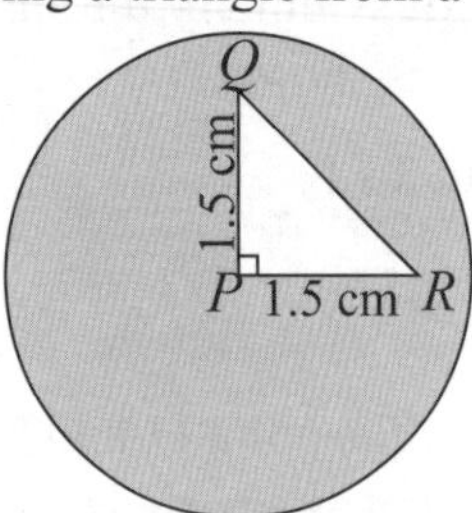

P is the centre of the circle, which has a radius of 2 cm. Triangle PQR is right-angled and isosceles. The shape can be modelled as a uniform lamina.

a) Show that the centre of mass of the shape is 0.070 cm from P, to 3 decimal places. *(5 marks)*

The shape hangs in equilibrium from a pin at point Q, about which it is able to freely rotate. The pin can be modelled as a smooth peg.

b) Find the angle that PQ makes with the vertical.
Give your answer in degrees, to 1 decimal place. *(3 marks)*

Q6-7 are for OCR M2 and OCR MEI M2 only

6 The shape shown below is formed by folding a uniform lamina through 90° about the z-axis so the edge AO coincides with the y-axis and the edge OB with the x-axis.

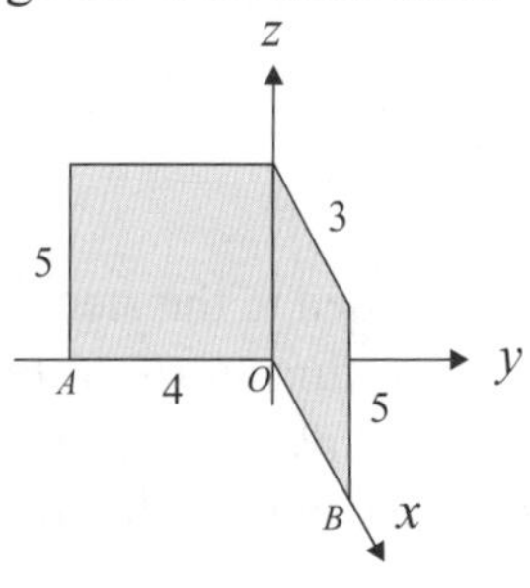

Find the coordinates of the centre of mass of the shape, referred to the axes shown. *(6 marks)*

7 A traffic cone, C, is modelled as an open conical shell, A, of height 1.2 m and base radius 0.15 m attached to a square lamina, B, of side 0.5 m. A and B are combined so that the vertex of A is vertically above the centre of B, the point O. It is assumed that A and B are made from the same uniform material.

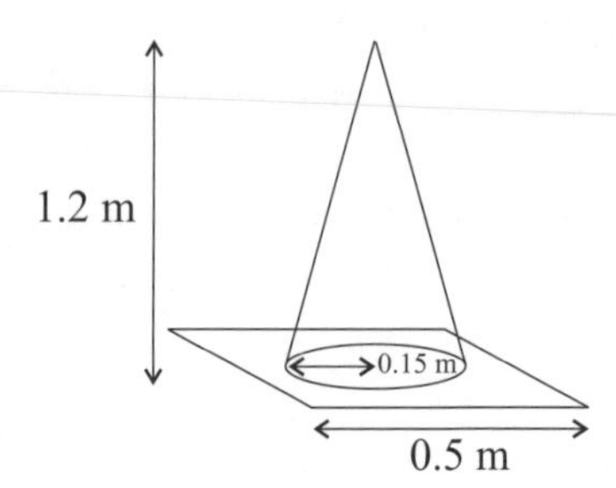

Find the position of the centre of mass of the cone in relation to O.
Use the fact that the curved surface of a cone has area πrl, where l is the length of the sloped edge. *(6 marks)*

Friction

This page is for OCR MEI M2 only

Friction tries to prevent motion, but don't let it prevent you getting marks in the exam — revise this page and it won't.

Friction Tries to **Prevent Motion**

Push hard enough and a particle will move, even though there's friction opposing the motion. A friction force, F, has a maximum value. This depends on the roughness of the surface and the value of the normal reaction from the surface.

$F \leq \mu R$ OR $F \leq \mu N$ (where R and N both stand for normal reaction)

μ has no units. μ is pronounced as 'mu'.

μ is called the "coefficient of friction". The rougher the surface, the bigger μ gets.

EXAMPLE

What range of values can a friction force take in resisting a horizontal force P acting on a particle Q, of mass 12 kg, resting on a rough horizontal plane which has a coefficient of friction of 0.4? (Take $g = 9.8$ ms^{-2}.)

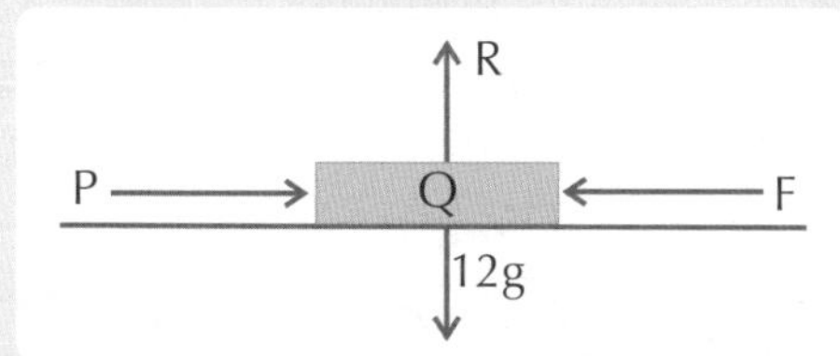

Resolving vertically: $R = 12g$

Use formula from above:

$F \leq \mu R$

$F \leq 0.4(12g)$

$F \leq 47.04$ N

So friction can take any value between 0 and 47.04 N, depending on how large P is.

If $P \leq 47.04$ N then Q remains in equilibrium. If P = 47.04 N then Q is on the point of sliding — i.e. friction is at its limit. If P > 47.04 then Q will start to move.

Limiting Friction is When Friction is at **Maximum ($F = \mu R$)**

EXAMPLE

A particle of mass 6 kg is placed on a rough horizontal plane which has a coefficient of friction of 0.3. A horizontal force Q is applied to the particle. Describe what happens if Q is: a) 16 N b) 20 N
Take $g = 9.8$ ms^{-2}

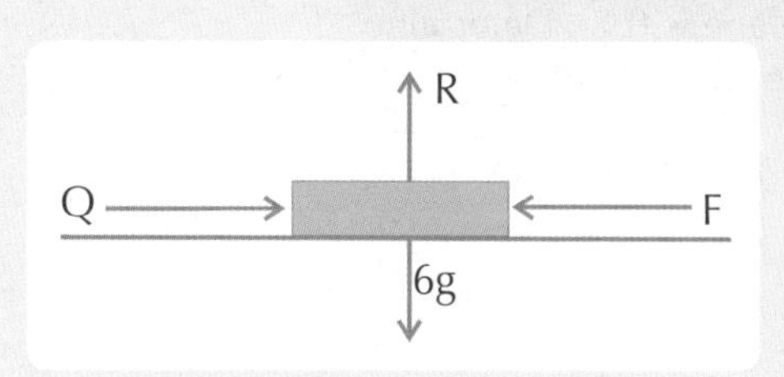

Resolving vertically: $R = 6g$

Using formula above:

$F \leq \mu R$

$F \leq 0.3(6g)$

$F \leq 17.64$ N

a) Since Q < 17.64 it won't move.

b) Since Q > 17.64 it'll start moving. No probs.

EXAMPLE

A particle of mass 4 kg at rest on a rough horizontal plane is being pushed by a horizontal force of 30 N. Given that the particle is on the point of moving, find the coefficient of friction.

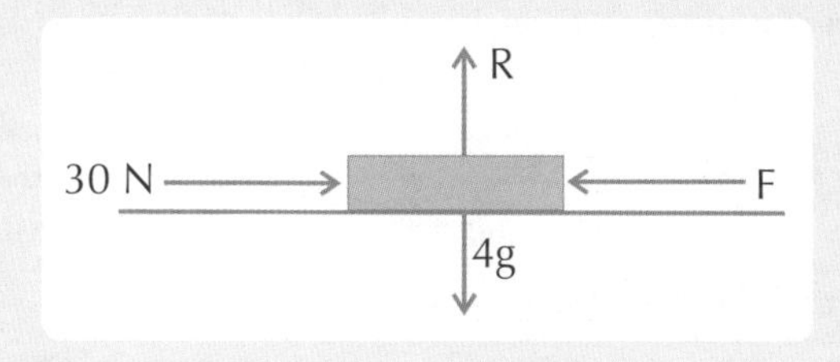

Resolving horizontally: F = 30

Resolving vertically: $R = 4g$

The particle's about to move, so friction is at its limit:

$F = \mu R$

$30 = \mu(4g)$

$\mu = \frac{30}{4g} = 0.77$

Sometimes friction really rubs me up the wrong way...

Friction can be a right nuisance, but without it we'd just slide all over the place, which would be worse (I imagine).

Friction and Inclined Planes

Solving these problems involves careful use of $F_{net} = ma$, $F = \mu R$ and the equations of motion.

Use F = ma in Two Directions for Inclined Plane questions

For inclined slope questions, it's much easier to resolve forces parallel and perpendicular to the plane's surface. Remember that friction always acts in the opposite direction to the motion.

This example is for OCR MEI M2 only

EXAMPLE A small body of weight 20 N accelerates from rest and moves a distance of 5 m down a rough plane angled at 15° to the horizontal. Draw a force diagram and find the coefficient of friction between the body and the plane given that the motion takes 6 seconds. Take g = 9.8 ms^{-2}.

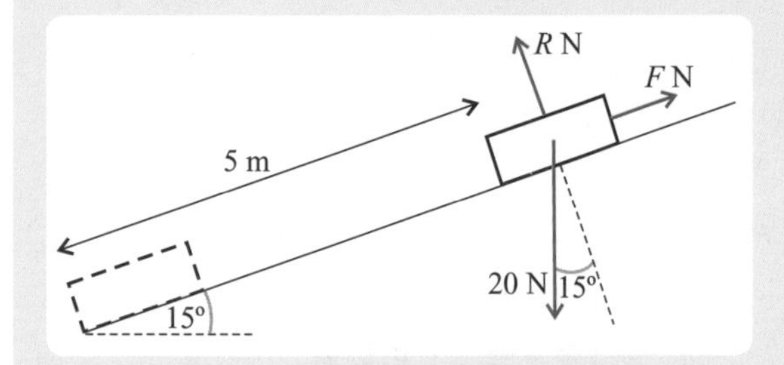

Taking down the slope as positive:

$u = 0, \quad s = 5, \quad t = 6, \quad a = ?$

Use one of the equations of motion: $s = ut + \frac{1}{2}at^2$

$5 = (0 \times 6) + (\frac{1}{2}a \times 6^2)$ so $\mathbf{a = 0.2778\ ms^{-2}}$

Resolving in ↖ direction: $F_{net} = ma$

$R - 20\cos15° = \frac{20}{g} \times 0$

So: $R = 20\cos15° = \mathbf{19.32\ N}$

There is no acceleration perpendicular to the slope.

Resolving in ↙ direction: $F_{net} = ma$

$20\sin15° - F = \frac{20}{g} \times 0.2778$

$\mathbf{F = 4.609\ N}$

It's sliding, so $F = \mu R$

$4.609 = \mu \times 19.32$

$\mu = 0.24$ (to 2 d.p.)

This example is for OCR M2 and OCR MEI M2 only

EXAMPLE An object rests on a rough plane inclined at an angle of θ° to the horizontal. The coefficient of friction between the object and the plane is 0.7. Determine the smallest value of θ for which the object will begin to slide down the plane.

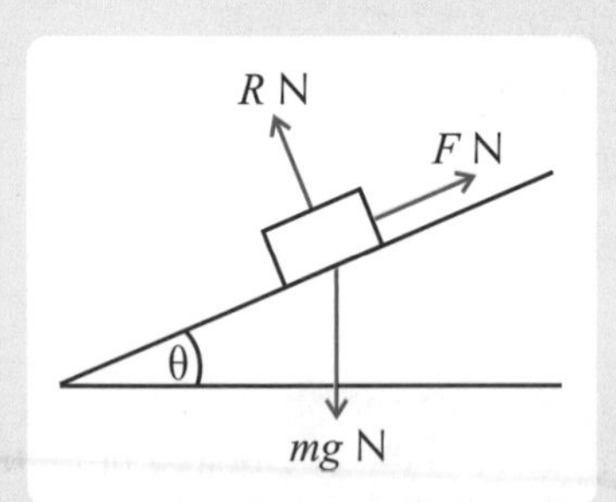

Resolving in ↙ direction: $F_{net} = ma$

$mg\sin\theta - F = 0$

$\Rightarrow F = mg\sin\theta$, call this equation ①.

Resolving in ↖ direction: $F_{net} = ma$

$R - mg\cos\theta = 0$

$\Rightarrow R = mg\cos\theta$, call this equation ②.

When the object is about to slide, friction is limiting, so $F = \mu R$.

So, using equations ① and ②:

$F = \mu R \Rightarrow mg\sin\theta = \mu mg\cos\theta$

Cancel mg: $\sin\theta = \mu\cos\theta \Rightarrow \mu = \tan\theta$

So, $\theta = \tan^{-1}\mu = \tan^{-1}(0.7) = 35.0°$ (3 s.f.)

In general, a particle on an inclined plane will slide if $\tan\theta \geq \mu$.

Inclined planes — nothing to do with suggestible Boeing 737s...

The main thing to remember about inclined planes is that you can choose to resolve in any two directions as long as they're perpendicular. It makes sense to choose the directions that involve doing as little work as possible. Obviously.

Moments

In this lifetime there are moments — moments of joy and of sorrow, and those moments where you have to answer questions on moments in exams.

Moments are **Clockwise** or **Anticlockwise**

This bit is for AQA M2, OCR M2 and OCR MEI M2

A 'moment' is the turning effect a force has around a point.
The larger the force, and the greater the distance from the point, then the larger the moment.

Moment = Force × Perpendicular Distance

EXAMPLE A plank 2 m long is attached horizontally to a ship at one end, O, as shown. A bird lands on the other end of the plank, applying a force of 15 N. Model the plank as a light rod and find the moment applied by the bird.

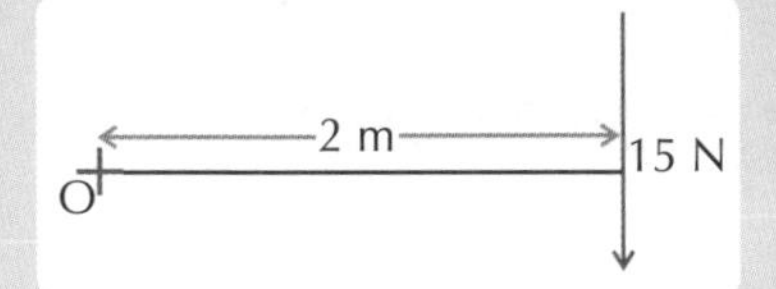

Moment $= F \times d$
$= 15 \times 2$
$= 30$ Nm

The units are just newtons × metres = Nm. Couldn't they have thought of a cleverer name?

Actually, Moment = Force × Perpendicular Distance from **Force's Line of Action**

This bit is for Edexcel M2, OCR M2 and OCR MEI M2

Often, you'll be given a distance between the point and the force, but this distance won't be perpendicular to the force's 'line of action'. You'll need to resolve to find the perpendicular distance.

EXAMPLE Find the sum of the moments of the forces shown about the point A.

2 N, 5 N, 1 m, 2 m, 60°, A

Calculating the clockwise moment is simple as the line of action is perpendicular to A:

$2 \times 1 = 2$ Nm

The anticlockwise moment is trickier as the line of action of the force isn't perpendicular to A. There are two ways to go about finding the moment — by finding the perpendicular distance or finding the perpendicular component of the force.

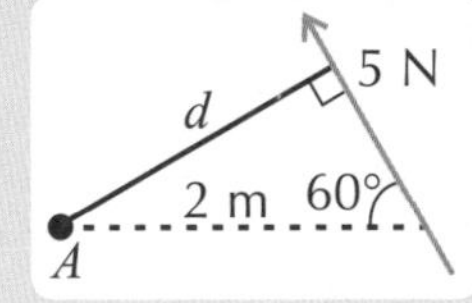

Finding the perpendicular distance:

$d = 2\sin 60°$

So, moment $= 5 \times 2\sin 60° = 10\sin 60° = 5\sqrt{3}$ Nm

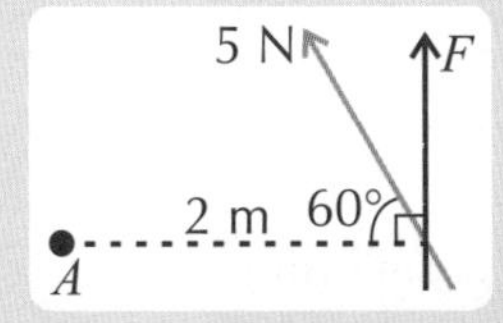

Finding the perpendicular component of the force:

$F = 5\sin 60°$

So, moment $= 5\sin 60° \times 2 = 10\sin 60° = 5\sqrt{3}$ Nm

Both methods give the same moment. Just choose whichever you find simplest — and be sure to show your workings.

We can now find the sum of the moments (in this case, taking anticlockwise as negative):

Clockwise + anticlockwise moments $= 2 + (-5\sqrt{3}) = -6.66$ Nm $= 6.66$ Nm anticlockwise

Resolve the force, Luke — use the perpendicular distance...

Why do I want to write a musical every time I read a page about moments? Clearly a sci-fi epic would be more appropriate.

Moments

This page is for AQA M2, Edexcel M2, OCR M2 and OCR MEI M2

Some of the stuff on the next two pages might look very familiar to you from M1.
Watch out though — whichever exam board you're doing, there'll be some new stuff in there.

In Equilibrium Moments Total Zero

In equilibrium, the moments about any point total zero — so anticlockwise moments = clockwise moments.

EXAMPLE Two weights of 30 N and 45 N are placed on a light 8 m beam. The 30 N weight is at one end of the beam, as shown, whilst the other weight is a distance d from the midpoint M. The beam is horizontal and held in equilibrium by a single wire with tension T attached at M. Find T and the distance d.

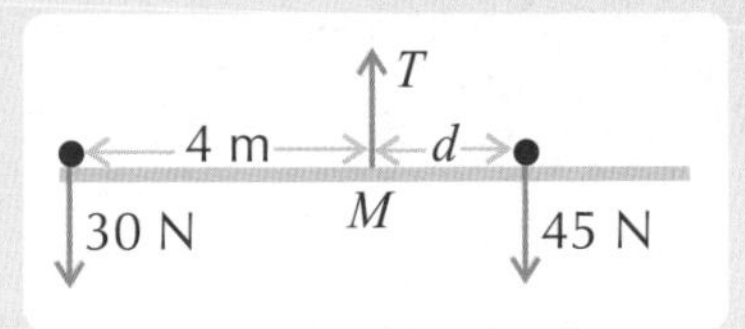

Resolving vertically: $30 + 45 = T = 75$ N

Take moments about M:
Clockwise Moment = Anticlockwise Moment
$45 \times d = 30 \times 4$
$d = \frac{120}{45} = 2\frac{2}{3} = 2.67$ m

This example is for Edexcel M2, OCR M2 and OCR MEI M2 only

EXAMPLE A rod, AB, of length 6 m is held in equilibrium in a horizontal position by two strings, as shown. By taking moments, find the mass, m, of the rod. Take $g = 9.8$ ms^{-2}.

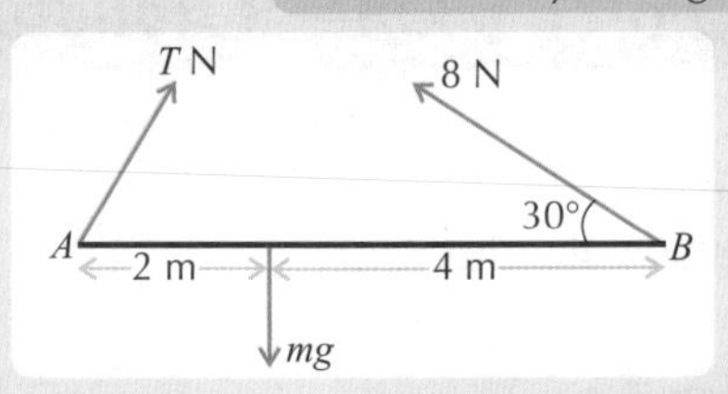

By taking moments about A:
clockwise moments = anticlockwise moments
$2mg = 6\sin30° \times 8$
$mg = 12 \Rightarrow m = 12 \div 9.8 = 1.22$ kg

Although you can take moments about any point (even one not on the rod), it's always easier to take moments about a point that has an unknown force going through it.

The Weight acts at the Centre of a Uniform rod

Mostly, you'll be dealing with rods. A model rod has negligible thickness, so you only need to consider where along its length the centre of mass lies. If the rod is uniform then the weight acts at the centre of the rod.

EXAMPLE A 6 m long uniform beam AB of weight 40 N is supported at A by a vertical reaction R. AB is held horizontal by a vertical wire attached 1 m from the other end. A particle of weight 30 N is placed 2 m from the support R. Find the tension T in the wire and the force R.

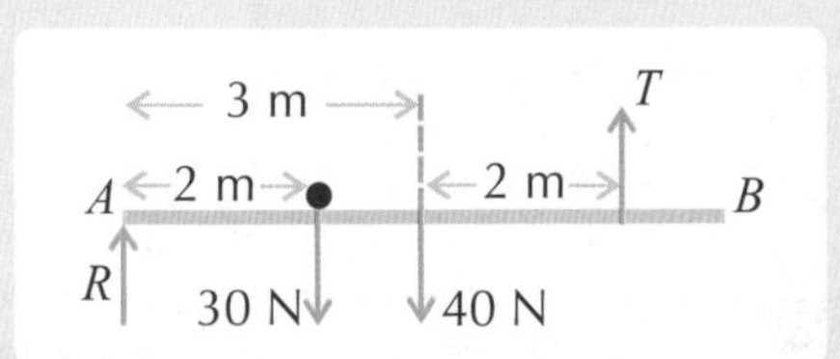

Take moments about A.
Clockwise Moment = Anticlockwise Moment
$(30 \times 2) + (40 \times 3) = T \times 5$
$T = 36$ N

Resolve vertically: $T + R = 30 + 40$
So: $R = 34$ N

This example is for Edexcel M2, OCR M2 and OCR MEI M2 only

EXAMPLE A uniform rod, AB, of length l m and mass m kg is suspended in horizontal equilibrium by two inextensible wires, with tensions as shown. Find m. Take $g = 9.8$ ms^{-2}.

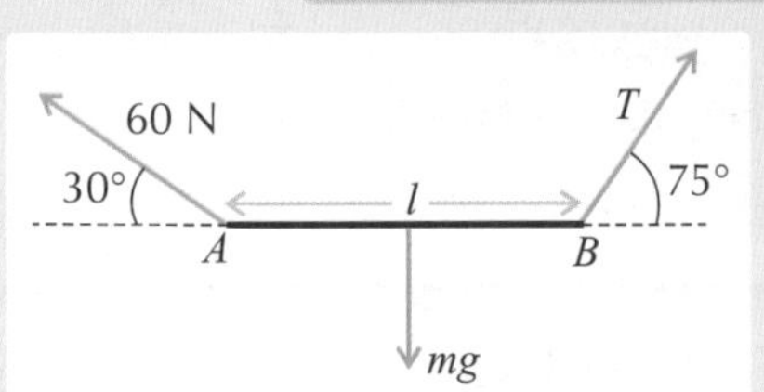

Take moments about B:
$60\sin30° \times l = mg \times \frac{1}{2}l$
so $m = \frac{60l}{gl} = \frac{60}{g} = 6.12$ kg

Again, it makes sense to take moments about B so you don't have to worry about the unknown force T.

Moments

This page is for AQA M2, Edexcel M2, OCR M2 and OCR MEI M2

You can **Calculate** the Centre of Mass for **Non-Uniform** rods

If the weight acts at an unknown point along a rod, the point can be found in the usual way — by taking moments. You might also have to resolve the forces horizontally or vertically to find some missing information.

EXAMPLE A plank of mass 5 kg is supported horizontally by a vertical string attached at a point B, as shown. One end of the plank, A, rests upon a pole. The tension in the string is T and the normal reaction at the pole is 70 N. A particle, P, of mass 9 kg rests on the plank 2 m from A, as shown. The system is in equilibrium. Find T and the distance, x, between A and the centre of mass of the plank.

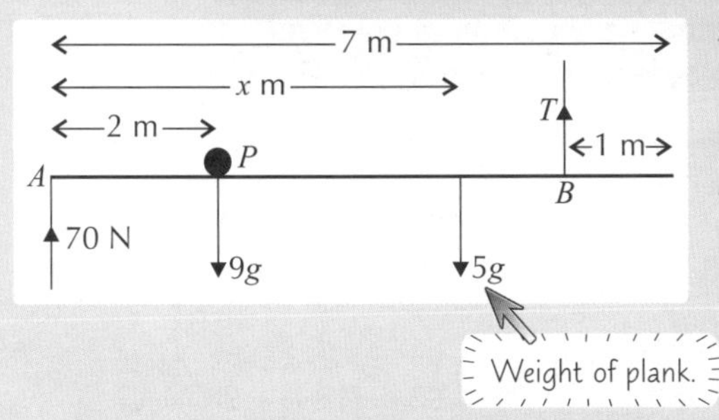

Weight of plank.

Resolve vertically: upward forces = downward forces:

$T + 70 = 9g + 5g$

so, $T = 14g - 70 = 67.2$ N

Moments about A: clockwise moments = anticlockwise moments:

$(9g \times 2) + (5g \times x) = 67.2 \times 6$

$49x = 403.2 - 176.4$

so $x = \frac{226.8}{49} = 4.63$ m (3 s.f.)

EXAMPLE A Christmas banner, AB, is attached to a ceiling by two pieces of tinsel at A and C, where $BC = 0.6$ m. The banner has mass 8 kg and is held in a horizontal position. The tensions in the pieces of tinsel are equal. The banner can be modelled as a non-uniform rod held in equilibrium and the tinsel as light strings. Find the tension in the tinsel and the distance, x, between A and the centre of mass of the rod. Take $g = 9.8$ ms^{-2}.

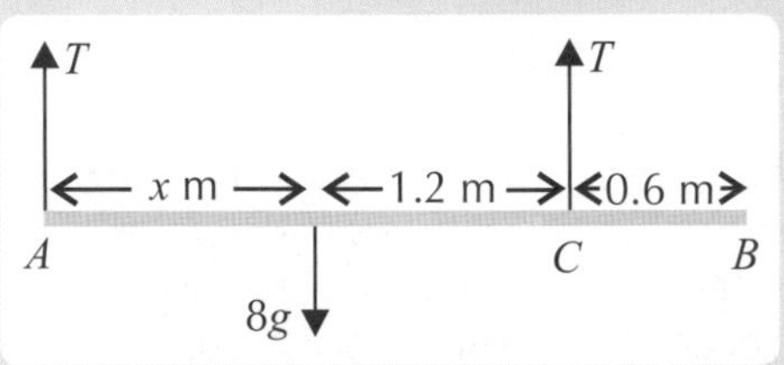

Resolve vertically:

$2T = 8g \Rightarrow T = 4g = 39.2$ N

Take moments about A:

$8g \times x = T \times (x + 1.2)$

$\Rightarrow 8gx = 4g(x + 1.2)$

$\Rightarrow 2x = x + 1.2 \Rightarrow x = 1.2$ m

The tinsel at A snaps and a downward force is applied at B to keep the banner horizontal. Find the magnitude of the force applied at B and the tension in the tinsel attached at C.

Take moments about C:

$8g \times 1.2 = F_B \times 0.6$

So, $F_B = \frac{94.1}{0.6} = 157$ N (3 s.f.)

Resolve vertically:

$T_C = 8g + 157$

So, $T_C = 235$ N (3 s.f.)

This example is for Edexcel M2, OCR M2 and OCR MEI M2 only

EXAMPLE A non-uniform rod, AB, of mass 2 kg and length 1 m, is suspended in equilibrium at an angle of θ to the vertical by two vertical strings, as shown. The tensions in the strings are T N and 12 N respectively. Find the distance, x, from A to the rod's centre of mass. Take $g = 9.8$ ms^{-2}.

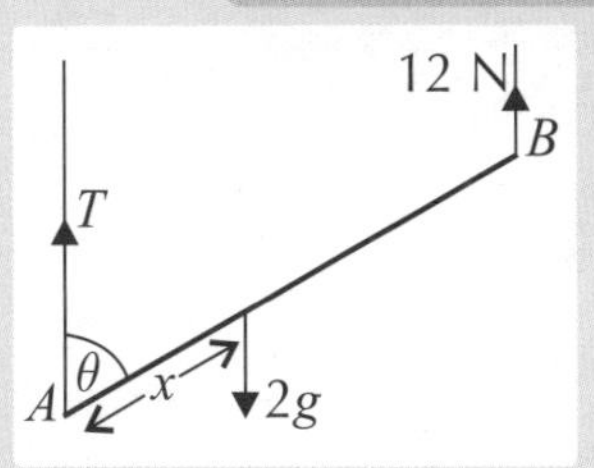

Taking moments about A:

$2g\sin\theta \times x = 12\sin\theta \times 1$

$2\sin\theta$ cancels, so:

$gx = 6$

$x = 0.612$ m (3 s.f.)

Significant moments in life — birthdays, exams, exam results...

Don't worry that the models used in Mechanics aren't that realistic — it's the ability to do the maths (and pass those exams) that counts. Anyway, you can worry about real life when you're done with school (when there'll be fewer exams).

Rigid Bodies

This page is for Edexcel M2, OCR M2 and OCR MEI M2

If reactions are at a weird angle, rather than horizontal, vertical or perpendicular to something else, then it's easier to think of them as two components — in two nice, convenient perpendicular directions.

Reactions can have Horizontal and Vertical Components

If a rod is connected to a plane (such as a wall) by a hinge or pivot and the forces holding it in equilibrium aren't parallel, then the reaction at the wall won't be perpendicular to the wall. Don't panic though, components are super-helpful here.

EXAMPLE A uniform rod, AB, is freely hinged on a vertical wall. The rod is held in horizontal equilibrium by a light inextensible string attached at a point C, 0.4 m from the end B at an angle of 45° to the rod, as shown. Given that the rod is 1.2 m long and has mass 2 kg, find the tension in the string and the magnitude of the reaction at the wall.

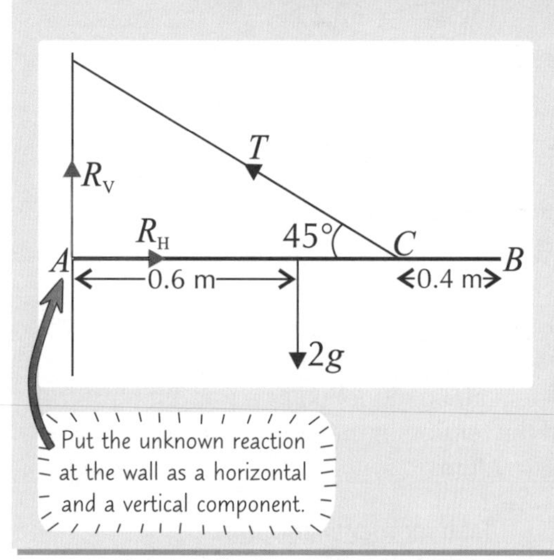

Put the unknown reaction at the wall as a horizontal and a vertical component.

Moments about A:
$T\sin45° \times (1.2 - 0.4) = 2g \times 0.6$
so, $T\sin45° = 14.7$
and $T = 20.8$ N (3 s.f.)

Choose A so that you can ignore the unknown reaction components while finding T.

Resolving horizontally:
$R_H = T\cos45° = 20.79\cos45°$
so, $R_H = 14.7$ N

Resolving vertically:
$R_V + T\sin45° = 2g$
so $R_V = 4.9$ N

The rod is in equilibrium, so the resultant force in any direction is zero.

Magnitude of reaction:
$|R| = \sqrt{R_H^2 + R_V^2} = \sqrt{14.7^2 + 4.9^2}$
so $|R| = 15.5$ N (3 s.f.)

EXAMPLE A non-uniform rod, AB, of length $6a$ and mass 4 kg is supported by a light strut at an angle of 70° to a vertical wall, as shown. The distance from A to the centre of mass, X, of the rod is xa m. The strut exerts a thrust of 16 N at the centre of the rod. A particle of weight 2 N is placed at B. Find x, and the magnitude and direction of the reaction at A.

Moments about A:
$16\cos70° \times 3a = (4g \times xa) + (2 \times 6a)$
so $4gxa = 4.417a$
and $x = 0.113$

Don't worry if you're not sure which directions the reaction components act in. You'll just get negative numbers if you're wrong (the magnitude will be the same).

Resolving horizontally:
$R_H = 16\sin70°$
$\Rightarrow R_H = 15.04$ N

Resolving vertically:
$R_V + 4g + 2 = 16\cos70°$
so $R_V = 16\cos70° - 39.2 - 2 \Rightarrow R_V = -35.73$ N

R_V is negative — so it must go upwards instead of downwards.

Magnitude of reaction:
$|R| = \sqrt{R_H^2 + R_V^2} = \sqrt{15.04^2 + 35.73^2}$
so $|R| = 38.8$ N

Direction of reaction:

$$\tan\theta = \frac{15.04}{35.73}$$

so $\theta = 22.8°$ to the wall

I'm trying to resist making a pun about rigor mortis...

...so I'll just tell you that it's due to irreversible muscular contraction caused by a shortage of adenosine triphosphate. Nice.

Rigid Bodies and Friction

Where would we be without friction? Well, using a ladder would certainly be trickier. Before getting too distracted by that thought you should really revise this page instead — ladders are featured, I promise.

Friction lets you assume the Reaction is Perpendicular

This bit is for Edexcel M2, OCR M2 and OCR MEI M2

From the previous page, you know that a rod attached to a wall has a reaction at the wall with a horizontal and vertical component. If the rod is held by friction instead, then the frictional force 'replaces' the vertical component.

EXAMPLE A rod, AB, rests against a rough vertical wall and is held in limiting equilibrium perpendicular to the wall by a light inextensible string attached at B at an angle of θ, as shown, where $\tan\theta = \frac{7}{17}$. The tension in the string is 42 N. The length AB is 5.5 m and the centre of mass is located 3.8 m from B. Find the mass of the rod, m, and the coefficient of friction, μ, between the wall and the rod.

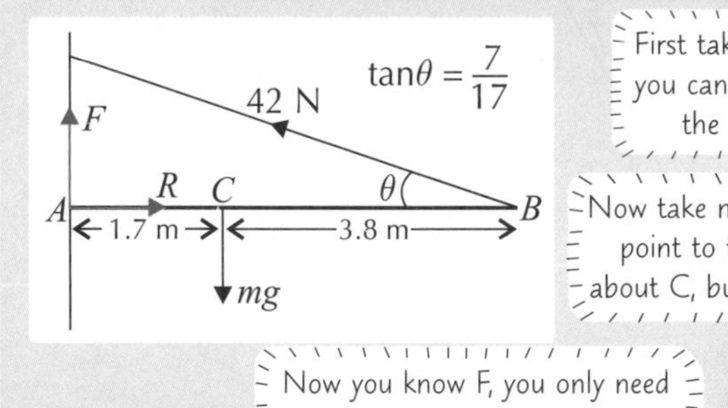

First take moments about A so you can find mg while ignoring the unknowns F and R.

Moments about A: $mg \times 1.7 = 42\sin\theta \times 5.5$
so $mg = 51.7$ N
and $m = 5.3$ kg

Now take moments about a different point to find F. I've taken them about C, but you could have used B.

Moments about C: $1.7 \times F = 3.8 \times 42\sin\theta$
so $F = 35.7$ N

Now you know F, you only need to find R before you can find μ.

Resolving horizontally:
$R = 42\cos\theta$
$R = 38.8$ N

Limiting equilibrium, so $F = \mu R$:
$35.7 = 38.8\mu$
so $\mu = 0.92$

Limiting equilibrium has shown up before — it means that the body is on the point of moving.

Multiple Surfaces can exert a Frictional Force

This bit is for AQA M2, Edexcel M2, OCR M2 and OCR MEI M2

'Ladder' questions, where a rod rests at an angle against the ground and a wall, are common in M2 exams. Often, the ground is modelled as rough and the wall as smooth. Can't take these things for granted though...

The 4 possible combinations of surfaces for 'ladder' questions:

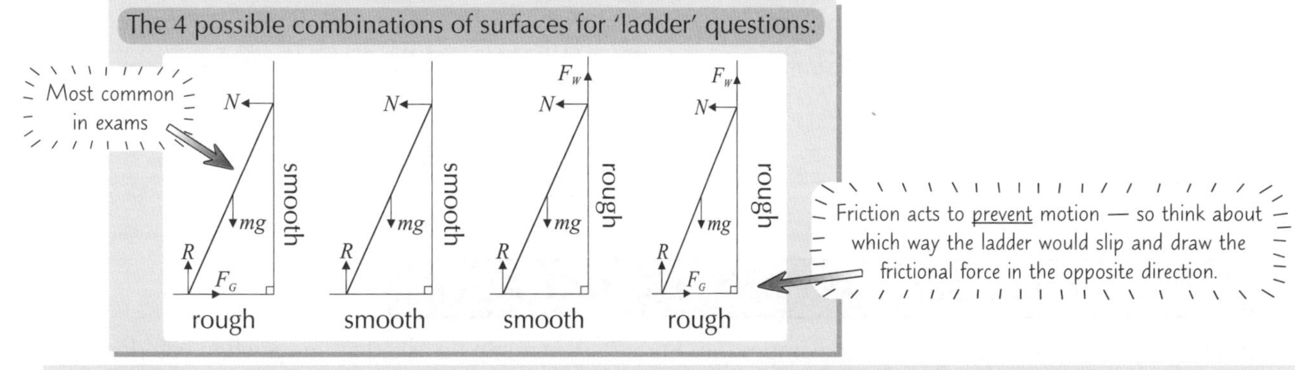

EXAMPLE A ladder rests against a smooth wall at an angle of 65° to the rough ground, as shown. The ladder has mass 1.3 kg and length $5x$ m. A cat of mass 4.5 kg sits on the ladder at C, $4x$ m from the base. The ladder is in limiting equilibrium. Model the ladder as a uniform rod and the cat as a particle. Find the coefficient of friction between the ground and the ladder.

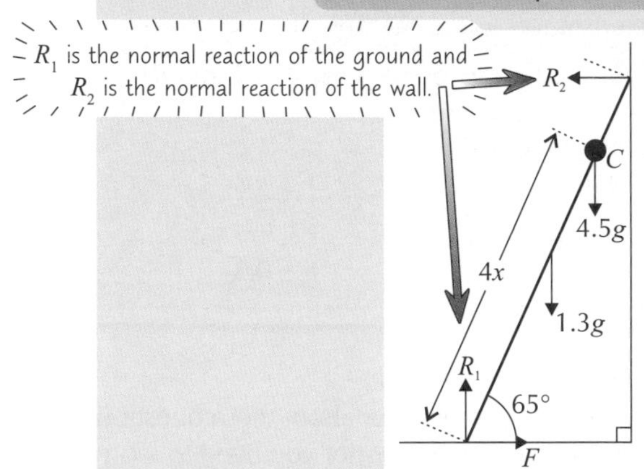

Resolving horizontally: $F = R_2$

Take moments about the base of the ladder to find R_2:
$R_2\sin65° \times 5x = (1.3g\cos65° \times 2.5x) + (4.5g\cos65° \times 4x)$
$4.532xR_2 = 13.46x + 74.55x$
so, $R_2 = \frac{88.01x}{4.532x} = 19.42$ N

Resolve vertically to find R_1:
$R_1 = 1.3g + 4.5g$
$\Rightarrow R_1 = 56.84$ N

The ladder is in limiting equilibrium, so $F = \mu R_1$:
Resolving horizontally shows $F = R_2$, so, $19.42 = 56.84\mu$
and $\mu = 0.342$ (to 3 s.f.)

Rigid Bodies and Friction

This page is for Edexcel M2, OCR M2 and OCR MEI M2

A Reaction is always Perpendicular to the Surface

Sometimes a body may be leaning against a surface that isn't vertical. This blows my mind.

EXAMPLE A uniform ladder of length 3 m rests against a smooth wall slanted at 10° to the vertical, as shown. The ladder is at an angle of 60° to the ground. The magnitude of the normal reaction of the wall is 18 N. Find the mass of the ladder.

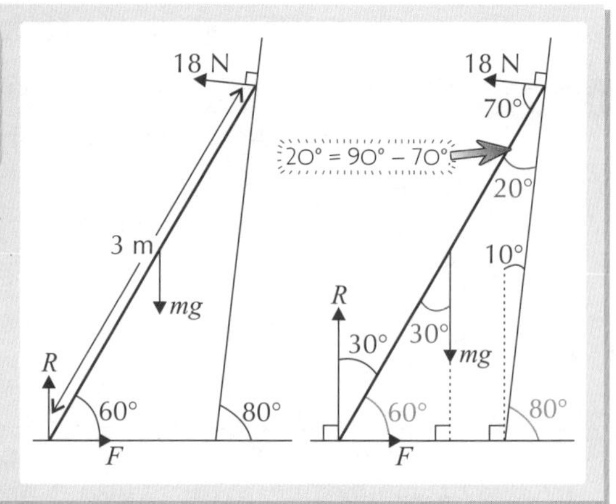

If the wall was vertical then the angles shown in red would be identical as both the weight and the wall would be perpendicular to the ground. Simple. However, the 18 N reaction at the wall is perpendicular to the wall, so be careful when resolving forces relative to the ladder.

Moments about the base of the ladder:
$mg\sin 30° \times 1.5 = 18\sin 70° \times 3$
so $m = 6.9$ kg

Bodies can be Supported Along Their Lengths

If a rod is resting on something along its length then the reaction is perpendicular to the rod.

EXAMPLE A uniform rod, AB, rests with end A on rough ground and upon a smooth peg at C, 0.9 m from B. The rod has length 3.3 m and weight 10 N. A particle, P, with weight 25 N is placed at B. Given that the rod is held in limiting equilibrium at an angle of 28° above the horizontal, find the magnitude of the normal reaction, R_2, at the peg and the friction, F, between the rod and the ground.

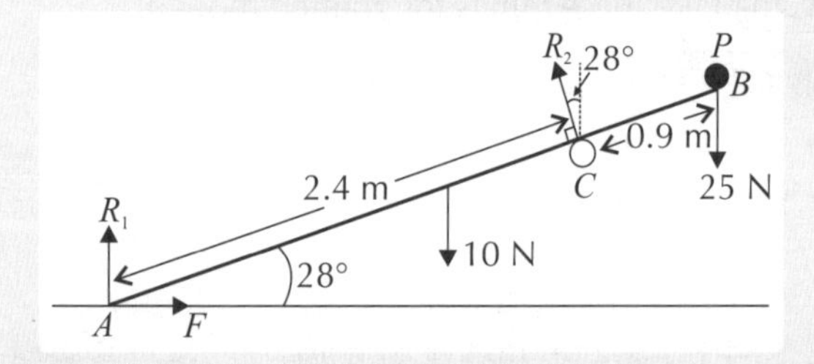

Moments about A:
$2.4R_2 = (10\cos 28° \times 1.65) + (25\cos 28° \times 3.3)$
so $R_2 = 36.4$ N

It's a uniform rod, so its weight acts in the middle.

Resolving horizontally:
$F = R_2\sin 28° = 36.4\sin 28°$
$\Rightarrow F = 17.1$ N

If you Don't Know that equilibrium is Limiting, $F \leq \mu R$

In limiting equilibrium, friction is at its maximum (i.e. $F = \mu R$). You might be asked to find μ when you don't know if equilibrium is limiting. Just find it in the same way as if equilibrium was limiting, but replace $F = \mu R$ with $F \leq \mu R$.

EXAMPLE A rough peg supports a rod at a point B, 0.2 m from one end of the rod, as shown. The other end of the rod, A, rests on a smooth horizontal plane. The rod is 1.5 m long, with its centre of mass located 1.2 m from A at point C. Given that the rod is in equilibrium at an angle of 15° to the horizontal plane and that the friction at the peg exerts a force of 8 N, show that $\mu \geq 0.3$.

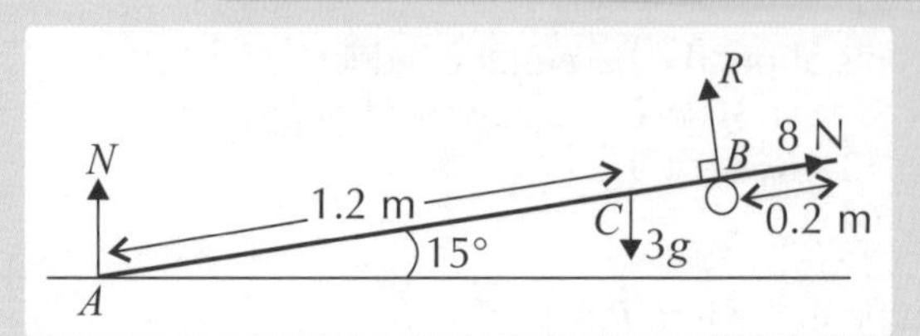

You know F, but to find μ you also need to know R.
Take moments about A to find R:
$R \times (1.5 - 0.2) = 3g\cos 15° \times 1.2$
so $R = 26.2$ N

$F \leq \mu R$
so $\mu \geq \frac{8}{26.2}$
$\mu \geq 0.3$

A body can also be supported along its length by a bed...

... but rough ground and a smooth peg sound much more comfortable. The thing to remember about these questions is to keep an eye on whether the points of contact are rough or smooth — that tells you whether or not you need to worry about friction. It's a bit of a pain, but you've just got to take the rough with the smooth I guess...

Laminas and Moments

This page is for OCR M2 and OCR MEI M2

I'm getting a bit fed up with thin rods and beams — I'm ready to take it to another dimension. Pay close attention.

*Remember to measure to the **Line of Action** of a Force*

You probably don't need me to tell you again that to find the moment of a force about a point, you need to use the perpendicular distance from the point to the line of action of the force. Well, it's that 'line of action' bit that starts to become more important when you're dealing with 2D shapes:

EXAMPLE A uniform lamina, $ABCD$, of weight 8 N is freely hinged to the corner of a wall at A. The lamina is held in equilibrium by a vertical force F acting at point C as shown. Find the value of F by taking moments about A.

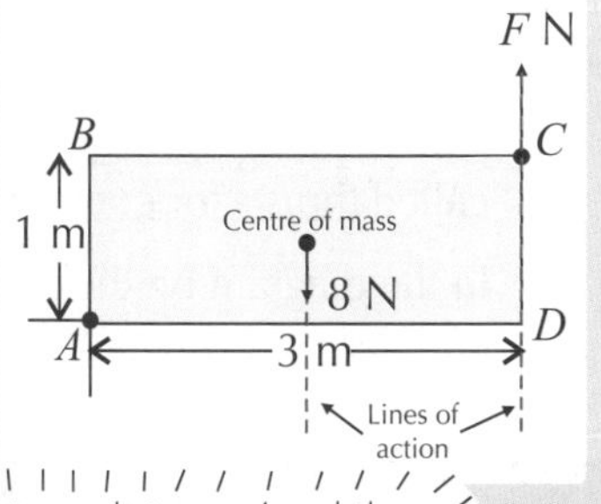

1) The lamina is uniform, so the mass acts at the centre. The perpendicular distance from A to the line of action of the lamina's weight is the horizontal distance from AB to the centre of mass (1.5 m).
2) The force F also acts vertically and the perpendicular distance from A to the line of action of the force is 3 m.
3) So, taking moments about A: $8 \times 1.5 = F \times 3 \Rightarrow F = 4$ N

The vertical distances between A and the forces don't matter. Because both forces are only acting vertically, you only need to use the horizontal distance to work out the moments.

*Split **Angled forces** into **Perpendicular components**, then take Moments*

1) When you've got a lamina held in equilibrium at a funny angle, the easiest thing to do is just split each force acting on the lamina into two perpendicular components.
2) Then, when you take moments, just use the perpendicular distance to the line of action of each component.
3) And remember — when you take moments about a point on a 2D shape, you need to look for forces acting in all directions. (With 1D shapes like rods, you can ignore any forces acting parallel to the rod when taking moments.)

EXAMPLE A house-shaped lamina of mass 8 kg is smoothly pivoted at point P and is supported in equilibrium by a light, inextensible vertical wire at point Q, as shown. The base of the lamina makes an angle of 15° with the horizontal. Find the tension, T, in the wire.

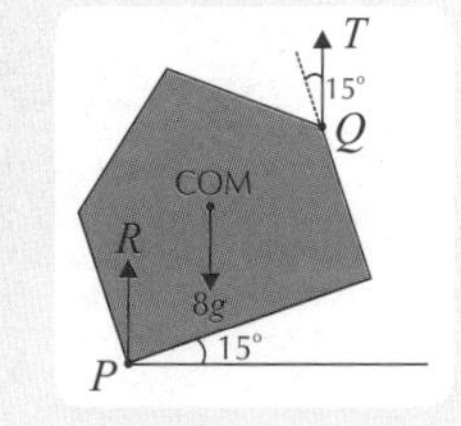

This is the house-shaped lamina from page 183. On page 183, the centre of mass of the shape was found to be 7.85 cm below the vertex of the triangle, on the line of symmetry.

Split the forces acting on the shape into components acting parallel and perpendicular to the base of the lamina:

You have to calculate this length because it's measured from a different point than on p. 183 — it's 13 – 7.85 = 5.15 cm from P.

Now take moments about P (so R can be ignored):

- Total clockwise moment = $(8g\cos15 \times 5) + (T\sin15 \times 7)$
- Total anticlockwise moment = $(8g\sin15 \times 5.15) + (T\cos15 \times 10)$

The lamina is in equilibrium, so:

$(8g\cos15 \times 5) + (T\sin15 \times 7) = (8g\sin15 \times 5.15) + (T\cos15 \times 10)$

$\Rightarrow 7T\sin15 - 10T\cos15 = (8g\sin15 \times 5.15) - (8g\cos15 \times 5)$

$$\Rightarrow T = \frac{(8g\sin15 \times 5.15) - (8g\cos15 \times 5)}{7\sin15 - 10\cos15} = 34.9 \text{ N (3 s.f.)}$$

Let's just take a moment to resolve the issue...

You don't have to split the forces into perpendicular components first — you could just take moments straight away. But more often than not, that leads to tricky Pythagoras and trig to get the distances right. I reckon this way makes it a bit easier.

Forces in Frameworks

This page is for OCR MEI M2 only

Remember frameworks from page 186? Well they're about to thrust themselves back into your life. Awesome.

Internal Forces in rods can be *Thrusts* or *Tensions*

In a framework of light, pin-jointed rods, each rod experiences a single internal force, which is constant throughout the rod. There are two types of internal force:

1) Forces acting 'inwards', towards the centres of the rods, are known as tensions.
2) Forces acting 'outwards', towards the ends of the rods, are called thrusts (or compressions).
3) In the diagram on the right, rods AB and BC are in tension, and rod AC is in compression.

Internal Forces in a Framework

ABC is a framework of light, rigid rods freely pin-jointed together at A, B and C.

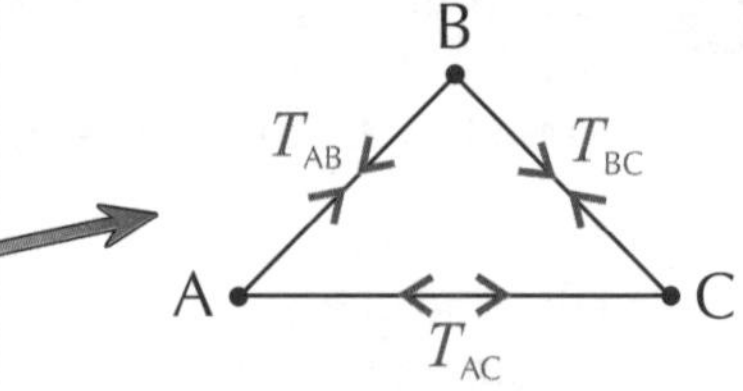

T_{AB}**,** T_{BC} **and** T_{AC} **are the internal forces acting in the rods.**

Resolve at Each Joint to find *Unknown Forces*

1) The rods in a framework are in equilibrium, so the resultant force at each joint will be zero.
2) This means you can resolve forces at each joint to find any unknown forces. ← This is likely to require some deft use of trig. Again.
3) To resolve at a joint, you only consider the forces acting at that joint. Don't forget that there may also be external forces acting at the joint, such as a weight or reaction forces.
4) Working out whether each internal force is a tension or thrust is easy. Just assume that all the internal forces are one type (e.g. tensions), and then if any forces turn out negative, then they must be the other (i.e. thrusts). Cool, eh?

EXAMPLE The diagram shows a framework of light rigid rods AB and AC in a vertical plane. The rods are freely pin-jointed to each other at A and to fixed points on a vertical wall at B and C. A load of weight 4 N is attached at point A.

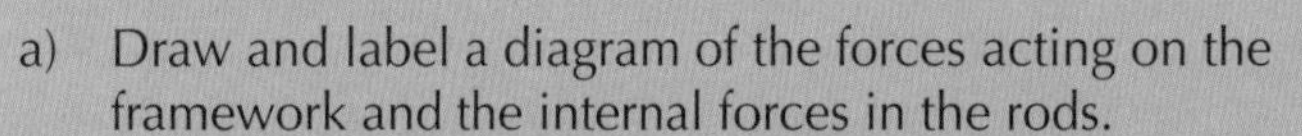

a) Draw and label a diagram of the forces acting on the framework and the internal forces in the rods.

b) Calculate the internal forces in the two rods, and state whether they are tensions or thrusts.

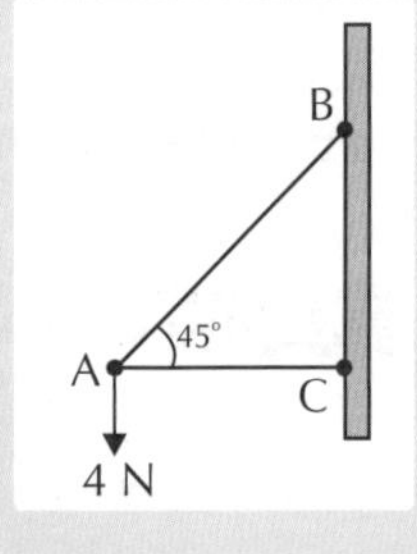

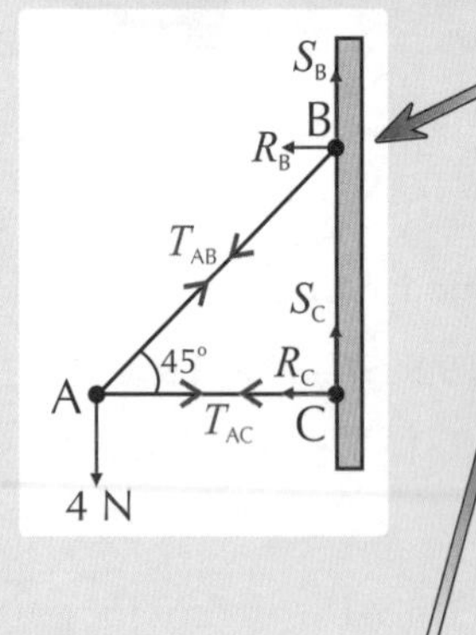

a) The diagram of the forces acting on the framework can be seen here. The forces S_B, R_B, S_C and R_C are the reaction forces of the wall on the framework, and T_{AB} and T_{AC} are the internal forces acting on the rods. Both internal forces have been drawn as tensions.

b) Resolve at A to find the internal forces:

Vertically:

$4 = T_{AB}\sin 45$

$\Rightarrow\ T_{AB} = 4 \div \sin 45 = 4\sqrt{2}$ N

Horizontally:

$T_{AB}\cos 45 + T_{AC} = 0$

$\Rightarrow\ T_{AC} = -T_{AB}\cos 45 = \frac{-4\sqrt{2}}{\sqrt{2}} = -4$ N

As the internal forces were all assumed to be tensions, T_{AB} is a tension of $4\sqrt{2}$ N and T_{AC} is a thrust of 4 N.

Resolve at A rather than B or C so you don't have to deal with the unknown reactions on the wall.

Remember that internal forces are the same throughout a rod — so the magnitude of T_{AC} is 4 N at both A and C. That would be handy to know if you were trying to find R_C.

I'm sensing a lot of tension in the room — maybe it's all this talk of rigid rods...

There's nothing too tricky on this page. Once you get your head around resolving at a point, it's all stuff you've done before. Make sure you understand that 'assuming all to be tensions' business, and then crack on with some questions...

M2 Section 3 — Practice Questions

Time to make like a tree and ~~leaf leave sway gently in the breeze~~. Darn it, that analogy wasn't really working... Anyway, time to be a dedicated student and practise your statics know-how. Ace.

Warm-up Questions

Take $g = 9.8 \text{ ms}^{-2}$ in each of these questions.

Q1 is for OCR MEI M2 only

1) A brick of mass 1.2 kg is sliding down a rough plane which is inclined at 25° to the horizontal. Given that its acceleration is 0.3 ms^{-2} down the plane, find the coefficient of friction between the brick and the plane.

Q2 is for AQA M2, OCR M2 and OCR MEI M2

2) A 60 kg uniform beam AE of length 14 m is in equilibrium, supported by two vertical ropes attached to B and D as shown.

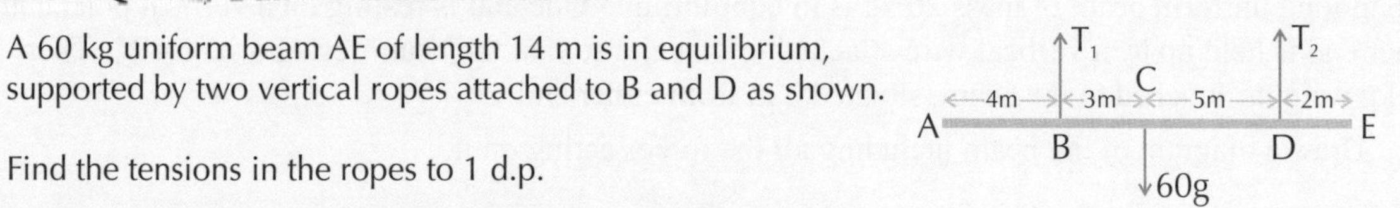

Find the tensions in the ropes to 1 d.p.

Q3-4 are for AQA M2, Edexcel M2, OCR M2 and OCR MEI M2

3) What is meant by a 'non-uniform rod'?

4) A uniform ladder, of length l m, is placed on rough horizontal ground and rests against a smooth vertical wall at an angle of 20° to the wall. Draw a diagram illustrating this system with forces labelled. State what assumptions you would make.

Q5-6 are for Edexcel M2, OCR M2 and OCR MEI M2

5) Calculate the perpendicular distance from the particle, P, to the forces shown in the diagrams below:

a) A, P, 39.6°, 0.8 m

b)

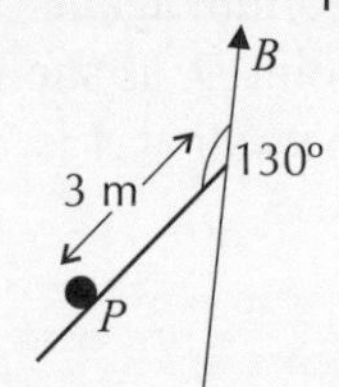

c)

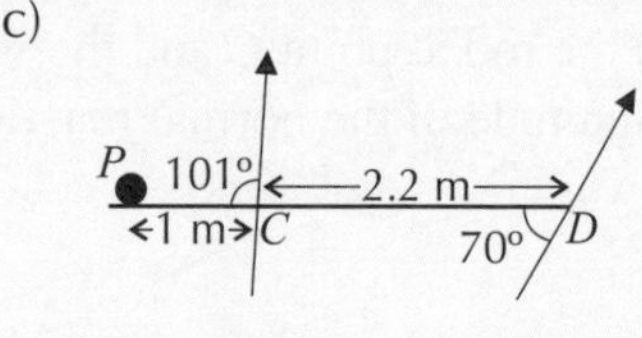

6) a) Calculate the magnitude of T in this diagram.

b) Find the value of x.

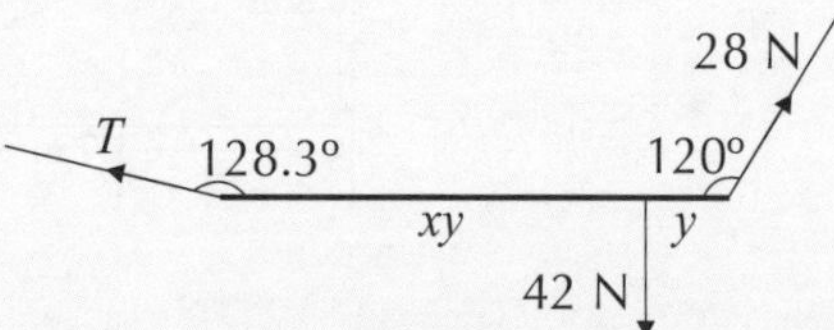

Those practice questions should've been a doddle. Time to step it up a notch with some questions more like those you'll get in the exam. In the words of a fictional dance-squad commander, "don't let me down".

Exam Questions

Whenever a numerical value of g is required in the following questions, take $g = 9.8 \text{ ms}^{-2}$.

Q1 is for Edexcel M2, OCR M2 and OCR MEI M2

1 A non-uniform rod, AB, is freely hinged at a vertical wall. It is held in horizontal equilibrium by a beam attached to the wall at C at an angle of 55°. The tension in the beam is 30 N, as shown. The rod has mass 2 kg, centred 0.4 m from A. The total length of the rod is x m.

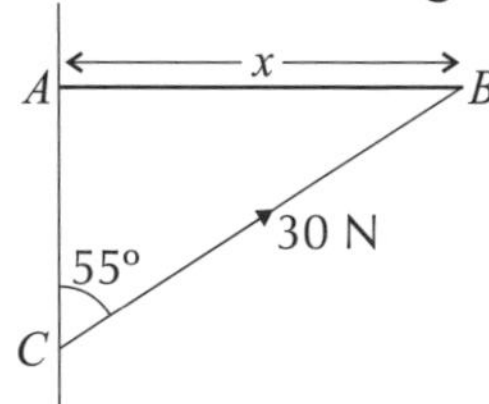

a) Find the length of the rod, x.

(3 marks)

b) Find the magnitude and direction of the reaction at A.

(5 marks)

M2 Section 3 — Practice Questions

If the previous question was a struggle, then you know what to do — go back a few pages and have another look. These questions will still be here while you're gone. Lurking.

Q2-3 are for AQA M2, OCR M2 and OCR MEI M2

2 A horizontal uniform beam with length x and weight 18 N is held in equilibrium by 2 strings. One string is attached to one end of the beam and the other at point A, 3 m from the first string. The tension in the string at point A is 12 N.
Show that $x = 4$ m.

(3 marks)

3 A 6 m long uniform beam of mass 20 kg is in equilibrium. One end is resting on a vertical pole, and the other end is held up by a vertical wire attached to that end so that the beam rests horizontally. There are two 10 kg weights attached to the beam, situated 2 m from either end.

a) Draw a diagram of the beam including all the forces acting on it.

(2 marks)

Find, in terms of g:

b) T, the tension in the wire.

(3 marks)

c) R, the normal reaction at the pole.

(2 marks)

Q4-5 are for Edexcel M2, OCR M2 and OCR MEI M2

4 A uniform rod, AB, is held in limiting horizontal equilibrium against a rough wall by an inextensible string connected to the rod at point C and the wall at point D, as shown. A particle of mass m kg rests at point B. The magnitude of the normal reaction of the wall at A is 72.5 N. The mass of the rod is 3 kg.

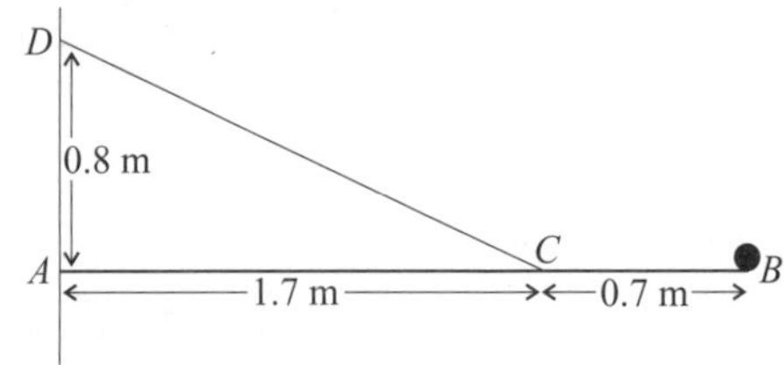

a) Find the tension, T, in the string.

(4 marks)

b) Find m.

(3 marks)

c) Find the magnitude of the frictional force, F, between the wall and the rod.

(3 marks)

5 A uniform rod of mass m kg rests in equilibrium against rough horizontal ground at point A and a smooth peg at point B, making an angle of θ with the ground, as shown. The rod is l m long and B is $\frac{3}{4}l$ from A.

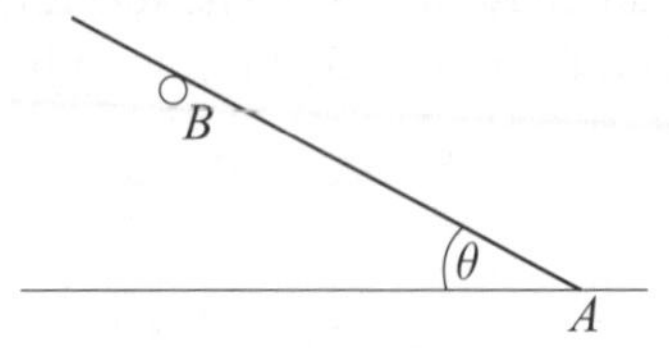

a) Show that the perpendicular reaction at the peg, $P = \frac{2}{3}mg\cos\theta$.

(3 marks)

b) Given that $\sin\theta = \frac{3}{5}$, find the range of values which the coefficient of friction between the rod and the ground could take.

(6 marks)

M2 Section 3 — Practice Questions

Q6 is for AQA M2, Edexcel M2, OCR M2 and OCR MEI M2

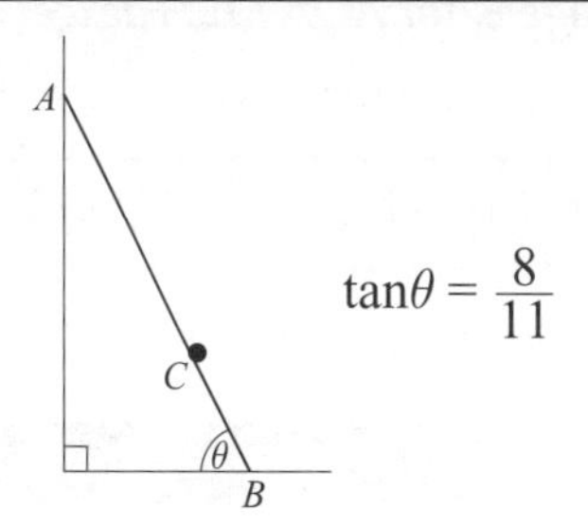

6 A uniform ladder, AB, is positioned against a smooth vertical wall and rests upon rough horizontal ground at an angle of θ, as shown. Clive stands on the ladder at point C, two-thirds of the way along the ladder's length from A. The ladder is 4.2 m long and weighs 180 N. The normal reaction at A is 490 N.

The ladder rests in limiting equilibrium. Model Clive as a particle and find:

a) the mass of Clive, m, to the nearest kg.

(4 marks)

b) the coefficient of friction, μ, between the ground and the ladder.

(5 marks)

Q7-8 are for OCR MEI M2 only

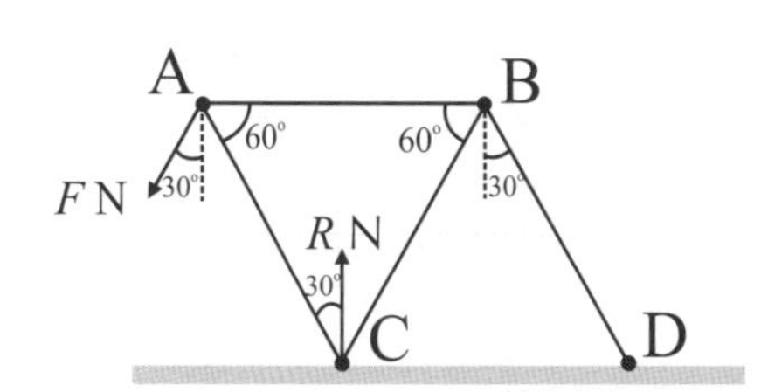

7 The diagram shows a framework of light, rigid rods freely pin-jointed to each other at A, B and C and to a fixed horizontal surface at D. The framework rests on the horizontal surface at C and experiences a reaction force of R N. There is also a force of F N acting at A at an angle of 30° to the vertical, as shown.

a) Draw a diagram showing the internal forces acting on the rods. Also include R and F in your diagram.

(1 mark)

b) Find the magnitude of each of the internal forces in terms of F and state whether each rod is under tension or compression. Also find R in terms of F.

(7 marks)

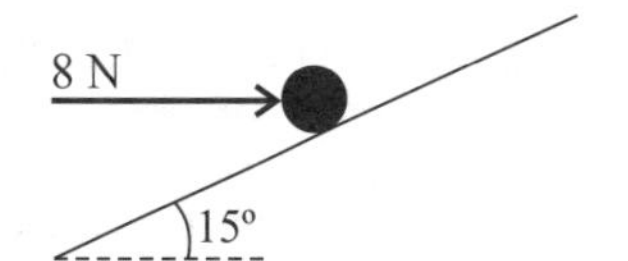

8 A horizontal force of 8 N just stops a mass of 7 kg from sliding down a plane inclined at 15° to the horizontal, as shown.

a) Calculate the coefficient of friction between the mass and the plane to 2 d.p.

(5 marks)

b) The 8 N force is now removed. Find how long the mass takes to slide a distance of 3 m down the line of greatest slope.

(7 marks)

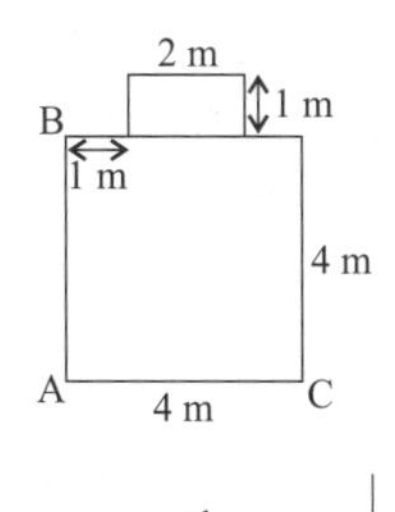

Q9 is for OCR M2 and OCR MEI M2

9 A shape is made up of a uniform square lamina and a uniform rectangular lamina as shown. The centre of mass of the shape is located 2 m from AB and 2.28 m from AC.

The shape is smoothly pivoted at A and rests in equilibrium against a smooth vertical wall. The angle between AC and the horizontal ground is 30°. Given that the weight of the shape is 4 N, find the magnitude of the reaction force, R, of the wall on the shape.

(4 marks)

Work Done

This page is for AQA M2, Edexcel M2, OCR M2 and OCR MEI M2

Hello, good evening, welcome to Section 4 — where work and energy are tonight's chef's specials...

You Can Find the **Work Done** by a Force Over a Certain **Distance**

When a force is acting on a particle, you can work out the work done by the force using the formula:

Work done = force (F) × distance moved in the direction of the force (s)

For F in newtons, and s in metres, the unit of work done is joules (J).

E.g. if an object is pushed 4 m across a horizontal floor by a force of magnitude 12 N acting horizontally, the work done by the force will be 12 × 4 = 48 J

EXAMPLE A rock is dragged across horizontal ground by a rope attached to the rock at an angle of 25° to the horizontal. Given that the work done by the force is 470 J and the tension in the rope is 120 N, find the distance the rock is moved.

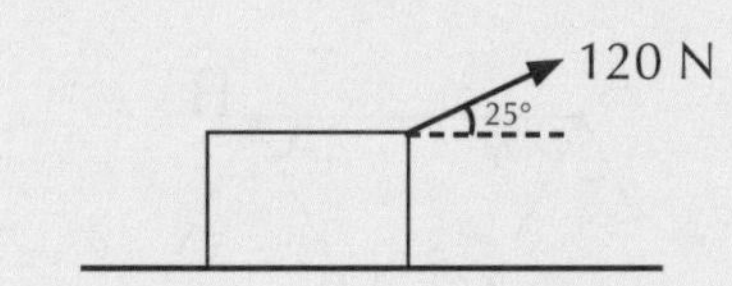

Work = horizontal component of force × s

$470 = 120\cos25 \times s$

$s = 4.32$ m

Because the force and the distance moved have to be in the same direction.

EXAMPLE A sack of flour of mass m kg is attached to a vertical rope and raised h m at a constant speed. Show that the work done against gravity by the tension in the rope, T, can be expressed as mgh.

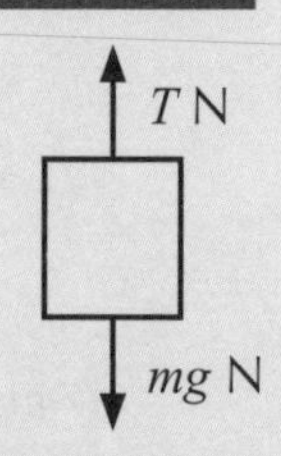

Resolve vertically:

$F = ma$

$T - mg = m \times 0$

$\Rightarrow T = mg$

Work done $= Fs$

$= T \times h$

$= mgh$

Work and Gravity

You can always use the formula mgh to find the work done by a force against gravity.

A Particle Moving **Up a Rough Slope** does Work against **Friction and Gravity**

EXAMPLE A block of mass 3 kg is pulled 9 m up a rough plane inclined at an angle of 20° to the horizontal by a force, T. The block moves at a constant speed. The work done by T against friction is 154 J.

Find:
a) the work done by T against gravity
b) the coefficient of friction between the block and the plane.

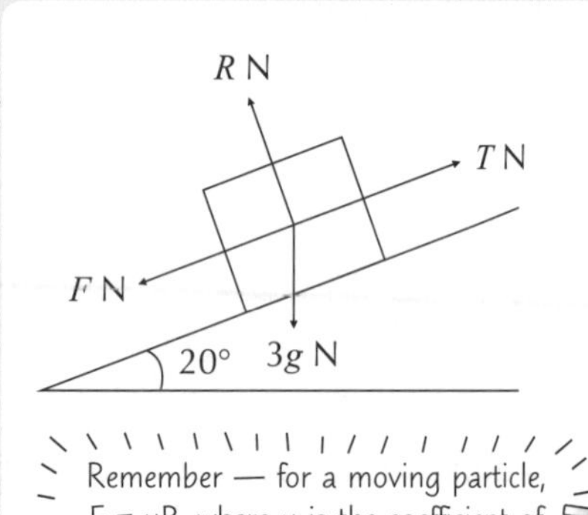

a) Work done against gravity $= mgh$

$= 3g \times 9\sin20$

$=$ **90.5 J** (3 s.f.)

You need to use the vertical height, because it's only vertically that T does work against gravity.

b) Resolve perpendicular to the slope to find R:

$R - 3g\cos20 = m \times 0$

$\Rightarrow R = 3g\cos20$

Particle is moving, so:

$F = \mu R$

$= \mu \times 3g\cos20$

Remember — for a moving particle, F = μR, where μ is the coefficient of friction and R is the normal reaction.

Work done by T against friction

$= F \times s = 154$

So, $\mu \times 3g\cos20 \times 9 = 154$

$\mu =$ **0.619** (3 s.f.)

The bit of T that is working against friction must be equal to F as the block is moving with constant speed (i.e. a = 0), so you can just use F here.

My work done = coffee × flapjack...

The really important thing to remember from this page is that the distance moved must be in the same direction as the force. Also, don't forget that a particle moving at constant velocity has no resultant force acting on it — this makes resolving forces easy.

Kinetic and Potential Energy

This page is for AQA M2, Edexcel M2, OCR M2 and OCR MEI M2

Here are a couple of jokers you might remember from GCSE Science. I know, I know — Science. This means we're skirting dangerously close to the real world here. :| Don't be too afraid though — it's not as scary as you might think...

A *Moving Particle* Possesses *Kinetic Energy*

Any particle that is moving has kinetic energy (K.E.). You can find the kinetic energy of a particle using the formula:

$$\text{K.E.} = \frac{1}{2}mv^2$$

You need to learn this formula — you won't be given it in the exam.

If mass, m, is measured in kg and velocity, v, in ms^{-1}, then kinetic energy is measured in joules.

EXAMPLE An ice skater of mass 60 kg is moving at a constant velocity of 8 ms^{-1}. Find the ice skater's kinetic energy.

Kinetic energy $= \frac{1}{2}mv^2$

$= \frac{1}{2} \times 60 \times 8^2 = 1920 \text{ J}$

Work Done is the same as the *Change* in a Particle's *Kinetic Energy*

The work done by a force to change the velocity of a particle moving horizontally is equal to the change in that particle's kinetic energy:

Work done = change in kinetic energy

$$\text{Work done} = \frac{1}{2}mv^2 - \frac{1}{2}mu^2 = \frac{1}{2}m(v^2 - u^2)$$

EXAMPLE A particle P of mass 6 kg is pulled along a rough horizontal plane by a force of 40 N, acting parallel to the plane. The particle travels 4 m in a straight line between two points on the plane, A and B. The coefficient of friction between P and the plane is 0.35.

a) Find the work done against friction in moving P from A to B.

At B, P has a speed of 8 ms^{-1}.

b) Calculate the speed of P at A.

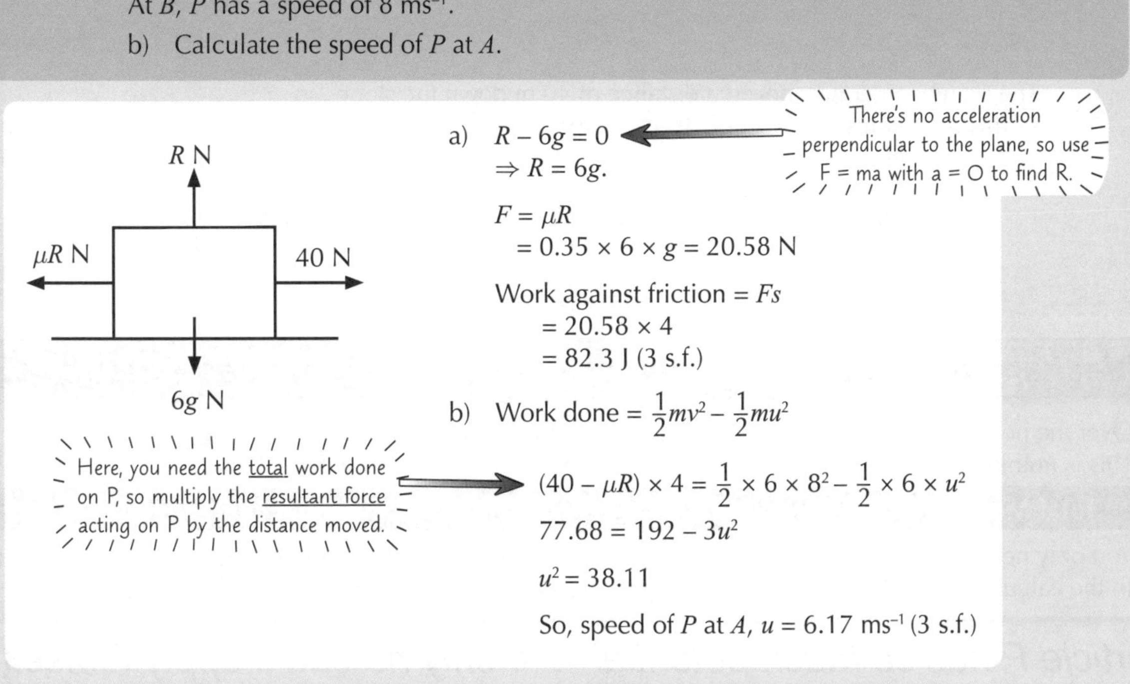

a) $R - 6g = 0$
$\Rightarrow R = 6g$.

There's no acceleration perpendicular to the plane, so use F = ma with a = 0 to find R.

$F = \mu R$
$= 0.35 \times 6 \times g = 20.58 \text{ N}$

Work against friction $= Fs$
$= 20.58 \times 4$
$= 82.3 \text{ J (3 s.f.)}$

b) Work done $= \frac{1}{2}mv^2 - \frac{1}{2}mu^2$

Here, you need the total work done on P, so multiply the resultant force acting on P by the distance moved.

$(40 - \mu R) \times 4 = \frac{1}{2} \times 6 \times 8^2 - \frac{1}{2} \times 6 \times u^2$

$77.68 = 192 - 3u^2$

$u^2 = 38.11$

So, speed of P at A, $u = 6.17 \text{ ms}^{-1}$ (3 s.f.)

Kinetic and Potential Energy

This page is for AQA M2, Edexcel M2, OCR M2 and OCR MEI M2

Gravitational Potential Energy is all about a Particle's **Height**

The gravitational potential energy (G.P.E.) of a particle can be found using the formula:

$$\text{G.P.E.} = mgh$$

You need to learn this formula as well.

If mass (m) is measured in kg, acceleration due to gravity (g) in ms^{-2} and the vertical height (h) above some base level in m, then G.P.E. is measured in joules, J.

The greater the height of a particle above the 'base level', the greater that particle's gravitational potential energy.

EXAMPLE

A lift and its occupants have a combined mass of 750 kg. The lift moves vertically from the ground to the first floor of a building, 6.1 m above the ground. After pausing, it moves vertically to the 17th floor, 64.9 m above the ground. Find the gravitational potential energy gained by the lift and its occupants in moving:

a) from the ground floor to the first floor,

b) from the first floor to the 17th floor.

a) G.P.E. gained $= mg \times$ increase in height
$= 750 \times 9.8 \times 6.1$
$= 44\,800$ J (3 s.f.)

b) G.P.E. gained $= mg \times$ increase in height
$= 750 \times 9.8 \times (64.9 - 6.1)$
$= 432\,000$ J (3 s.f.)

Gravitational Potential Energy Always uses the **Vertical Height**

When you're working out the gravitational potential energy of a particle, the value of h you use should always, always, always be the vertical height above the 'base level'. This means that for a particle moving on a slope, it's only the vertical component of the distance you're interested in:

EXAMPLE A skateboarder and her board have a combined mass of 65 kg. The skateboarder starts from rest at a point X and freewheels down a slope inclined at 15° to the horizontal. She travels 40 m down the line of greatest slope. Find the gravitational potential energy lost by the skateboarder.

The skateboarder has moved a distance of 40 m down the slope, so this is a vertical distance of: 40sin15° m.

G.P.E. $= mgh$
$= 65 \times 9.8 \times 40\sin 15°$
$= 6590$ J $= 6.59$ kJ (both to 3 s.f.)

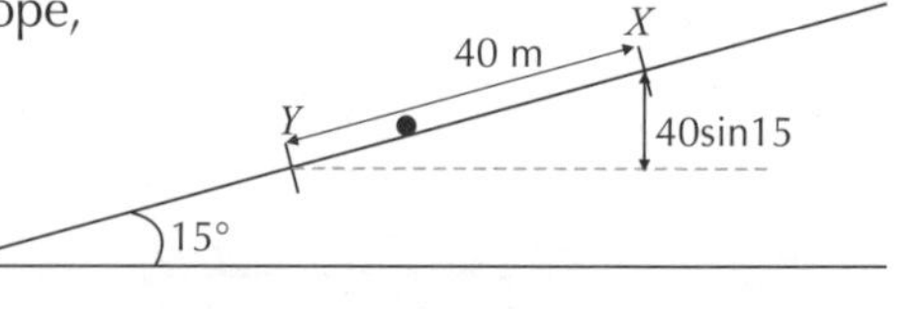

Mechanical Energy is the **Sum** of a Particle's **Kinetic and Potential Energies**

Over the next couple of pages, you're going to see a fair bit about 'mechanical energy'.
This is nothing to freak out about — it's just the sum of the kinetic and potential energies of a particle:

Total Mechanical Energy = Kinetic Energy + Gravitational Potential Energy (+ Elastic Potential Energy)

You only need to know about elastic potential energy if you're doing AQA M2. You can find out all about it on p. 209. In the calculations on the next couple of pages, you can ignore elasticity, and just use K.E. and G.P.E.

Particle P has so much potential — if only he could apply himself...

There shouldn't be anything earth-shattering on these two pages — I'd bet my completed 1994-95 Premier League sticker album that you've seen both of these types of energy before*. Still, it's worth refreshing yourself for what comes next.

*I won't though, it's too dear to me. Ahh, shinies...

The Work-Energy Principle

This page is for AQA M2, Edexcel M2, OCR M2 and OCR MEI M2

Those pages refreshing your memory on potential and kinetic energy weren't just for fun and giggles. Behold...

Learn the **Principle of Conservation of Mechanical Energy...**

The principle of conservation of mechanical energy says that:

If there are no external forces doing work on an object, the total mechanical energy of the object will remain constant.

An external force is any force other than the weight of the object, e.g. friction, air resistance, tension in a rope, etc. This means that the sum of potential and kinetic energies remains the same throughout an object's motion. This is a pretty useful bit of knowledge:

EXAMPLE

A BASE jumper with mass 88 kg jumps from a ledge on a building, 150 m above the ground. He falls with an initial velocity of 6 ms^{-1} towards the ground. He releases his parachute at a point 60 m above the ground.

a) Find the initial kinetic energy of the jumper.

b) Use the principle of conservation of mechanical energy to find the jumper's kinetic energy and speed at the point where he releases his parachute.

c) State one assumption you have made in modelling this situation.

a) Initial K.E. = $\frac{1}{2}mu^2 = \frac{1}{2} \times 88 \times (6)^2$
$= 1584 = 1580$ J (3 s.f.)

b) Decrease in G.P.E. as he falls:
$mgh = 88 \times 9.8 \times (150 - 60)$
$= 77\,616$ J

You can just use the change in height here, as it's the change in P.E. that you're interested in.

Using conservation of mechanical energy: Increase in K.E. = Decrease in G.P.E.

So, K.E. when parachute released: Initial K.E. = Decrease in G.P.E.

$\frac{1}{2}mv^2$ = Decrease in G.P.E. + Initial K.E.
$= 77616 + 1584 = 79\,200$ J

Rearrange $\frac{1}{2}mv^2 = 79\,200$ to find the speed of the jumper when parachute is released:

$$v = \sqrt{\frac{79\,200}{\frac{1}{2} \times 88}} = 42.4 \text{ ms}^{-1} \text{ (3 s.f.)}$$

If you don't assume this, then you can't use the principle of conservation of energy.

c) That the only force acting on the jumper is his weight.

...and the **Work-Energy Principle**

1) As you saw above, if there are no external forces doing work on an object, then the total mechanical energy of the object remains constant.
2) So, if there *is* an external force doing work on an object, then the total mechanical energy of the object must change.
3) This leads to the work-energy principle:

The work done on an object by external forces is equal to the change in the total mechanical energy of that object.

4) The work-energy principle is pretty similar to the result on page 205. It's generally more useful though, because you can use it for objects moving in any direction — not just horizontally.

Turn the page for a **HOT** and **SEXY** example...

The Work-Energy Principle

This page is for AQA M2, Edexcel M2, OCR M2 and OCR MEI M2

As promised, a lovely example of the work-energy principle...

Example

A particle of mass 3 kg is projected up a rough plane inclined at an angle θ to the horizontal, where $\tan\theta = \frac{5}{12}$. The particle moves through a point A at a speed of 11 ms^{-1}.
The particle continues to move up the line of greatest slope and comes to rest at a point B before sliding back down the plane. The coefficient of friction between the particle and the slope is $\frac{1}{3}$.

a) Use the work-energy principle to find the distance AB.

b) Find the speed of the particle when it returns to A.

a) Let x = distance from A to B.

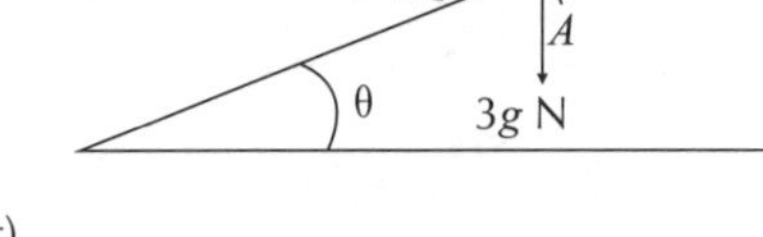

You're told to use the work-energy principle, so first find the change in total mechanical energy:

Change in K.E. of the particle = Final K.E. – Initial K.E.

$$= \frac{1}{2}mv^2 - \frac{1}{2}mu^2 = 0 - \left(\frac{1}{2} \times 3 \times 11^2\right) = \mathbf{-181.5\ J}$$

Change in G.P.E. of the particle = mg × (change in height)

$$= 3gx\sin\theta = 3gx \times \frac{5}{13} = \mathbf{\frac{15gx}{13}\ J}$$

$\tan\theta = \frac{5}{12} \Rightarrow \sin\theta = \frac{5}{13}$

So, change in total mechanical energy

$$= \mathbf{-181.5 + \frac{15gx}{13}}$$

The only external force doing work on the particle is the frictional force, F.
So you need to find the work done by F.

First, resolve perpendicular to slope:

$$R - 3g\cos\theta = 0 \Rightarrow R = 3g \times \frac{12}{13} = \frac{36g}{13}$$

$\tan\theta = \frac{5}{12} \Rightarrow \cos\theta = \frac{12}{13}$

$$F = \mu R = \frac{1}{3} \times \frac{36g}{13} = \frac{12g}{13}$$

Displacement is negative because the particle is moving in the opposite direction to F.

$$\text{Work done by } F = Fs = \frac{12g}{13} \times -x = \mathbf{-\frac{12gx}{13}}$$

Using the work-energy principle:
Change in total mechanical energy = Work done by F

$$\text{So: } -181.5 + \frac{15gx}{13} = -\frac{12gx}{13}$$

$$\frac{27gx}{13} = 181.5$$

$$x = \frac{181.5 \times 13}{27g} = 8.92 \text{ m (3 s.f.)}$$

b) The particle moves from A, up to B and back down to A, so overall change in G.P.E. = 0

So, the change in total mechanical energy between the first and second time the particle is at A is just the change in Kinetic Energy, i.e. Final K.E. – Initial K.E. = $\frac{1}{2}mv^2 - \frac{1}{2}mu^2$

Work done on the particle = $Fs = F \times -2x$

$$= \frac{12g}{13} \times -2(8.917) = \mathbf{-161.3}$$

The particle has travelled the distance AB twice and is always moving in the opposite direction to the frictional force.

Using the work-energy principle:

$$\frac{1}{2} \times 3 \times v^2 - \frac{1}{2} \times 3 \times 11^2 = -161.3$$

u = 11 ms^{-1}, as this is the speed of the particle when it's first at A.

$$\frac{3}{2}v^2 = 181.5 - 161.3 \quad \Rightarrow \quad v = 3.67 \text{ ms}^{-1} \text{ (3 s.f.)}$$

Does this mean we can save energy by doing less work...

There are a few different ways you could tackle part b). You could just look at the motion back down the slope and look at the K.E. gained and the G.P.E. lost. Or you could resolve parallel to the slope, work out the acceleration and use $v^2 = u^2 + 2as$.
If the question doesn't tell you what method to use, you'll get marks for using any correct method. Correct being the key word.

Elastic Potential Energy

This page is for AQA M2 only

This page is all about stretching stuff. Exciting or what?

Use Hooke's Law to find Tension in a String or Spring

1) Hooke's law is a formula for finding the tension (T) in a string or spring that has been stretched a distance e. You need to learn this formula — you won't be given it in the exam.

$$T = \frac{\lambda}{l}e$$

You can also use this formula to find the force in a spring which has been compressed a distance e.

2) l is the natural length of a spring or elastic string — i.e. its length when it is not being stretched.

3) The modulus of elasticity (λ) is a measure of how easily something can be stretched in a way that it will return to its original shape afterwards. If T has units of N, then the units of λ will also be N.

EXAMPLE A wooden block is suspended in equilibrium from an elastic string. The string is extended from its natural length of 5 m to a length of 8 m. Given that the modulus of elasticity of the string is 30 N, find the mass of the block.

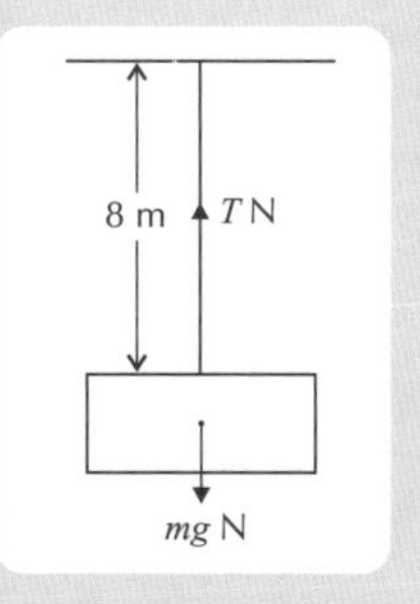

Use Hooke's law to find the tension in the string: $T = \frac{\lambda}{l}e = \frac{30}{5} \times (8 - 5) = 18\text{N}.$

The block is in equilibrium, so resolving forces vertically: $T = mg \Rightarrow 18 = mg$

$\Rightarrow m = 18 \div 9.8 = 1.84$ kg (3 s.f.).

Elastic Potential Energy is energy stored in a Stretched String or Spring

1) When an elastic string (or spring) is stretched, the work done in stretching it is converted into Elastic Potential Energy (E.P.E.), which is stored in the string.

2) You've already seen that W, the work done by a constant force over a distance x m, is given by $W = Fx$. But tension in an elastic string is not constant — it's a variable force.

The E.P.E. stored in a compressed spring can be found exactly the same way.

3) To find the work done by a variable force, you need to integrate the force with respect to distance, i.e. $W = \int F\,dx$.

4) This can be used to find a formula for E.P.E. It's another formula you need to know I'm afraid. But even worse, you could be asked to derive it in the exam. I know — absolutely shocking.

The formula for Elastic Potential Energy

From Hooke's law, the tension in an elastic string extended a distance of x m is: $T = \frac{\lambda}{l}x$

So, work done against tension:

$$W = \int_0^e T\,dx = \int_0^e \frac{\lambda}{l}x\,dx = \left[\frac{\lambda}{2l}x^2\right]_0^e = \frac{\lambda}{2l}e^2.$$

It might seem odd setting the limits of the integral as [0, e] rather than [0, x] for an extension by x m, but it's 'bad maths' to use the variable you are integrating as a limit, so another variable is picked to represent the length of extension.

The potential energy stored in the string is equal to the work done, so $\text{E.P.E.} = \frac{\lambda}{2l}e^2$

This is the baby you've got to learn — there's an example using it on the next page.

Mechanical Energy also includes Elastic Potential Energy

We can no longer ignore the E.P.E. term from the equation on page 206. So:

Total Mechanical Energy = Kinetic Energy + Gravitational Potential Energy + Elastic Potential Energy

1) Including the extra term in the equation opens up a whole new array of exciting questions which you could be asked about the Principle of Conservation of Mechanical Energy and the Work-Energy Principle from page 207.

2) It's nothing scary — just remember that if the only forces acting on a particle are its weight and tension in an elastic string or spring, then the total mechanical energy will be constant.

3) And if any other forces are acting on a particle, the work done by them is equal to the change in total mechanical energy. Simples.

Turn the page for a **COLD** and **UGLY** example...

Elastic Potential Energy

This page is for AQA M2 only

Well, that's pretty much all the theory on E.P.E. Now it's time to hone your skills with this beast of an example.

EXAMPLE

One end of a light elastic string is attached at point O to a smooth plane inclined at an angle of 30° to the horizontal, as shown.
The other end of the string is attached to a particle of mass 8 kg.
The string has a natural length of 1 m and modulus of elasticity 40 N.

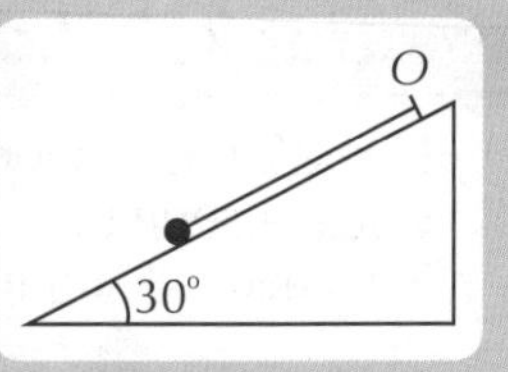

The particle is released from rest at O and slides down the slope.

a) Find the length of the string when the particle's acceleration is zero.

b) The string extends to a total length of x m before the particle first comes to rest. Show that $x^2 - 3.96x + 1 = 0$.

c) Hence find the distance from O at which the particle first comes to rest.

The particle is held at point A, 3 m down the slope from O, where it is released from rest and moves up the slope with speed v ms^{-1}.

'Taut' means that the string is stretched to some length greater than its natural length.

d) Show that, while the string is taut, the particle's motion satisfies the equation $v^2 = -5y^2 + 19.8y - 14.4$, where y is the particle's distance from O down the slope.

e) Find the speed of the particle at the point where the string becomes slack.

a) Resolve parallel to the slope to find the tension in the string, T:

$T - mg\sin 30 = 8 \times 0 \Rightarrow T = 8 \times 9.8 \times \sin 30 = \mathbf{39.2\ N}$.

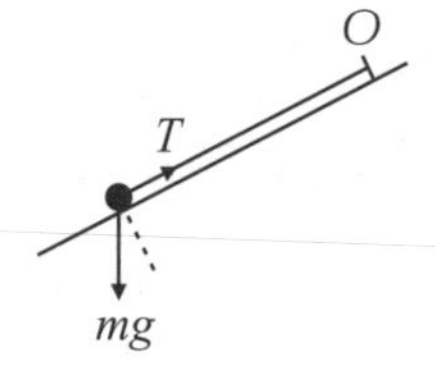

Now use Hooke's law to find the length of extension at this point:

$T = \frac{\lambda}{l}e \Rightarrow e = \frac{Tl}{\lambda} = \frac{39.2 \times 1}{40} = \mathbf{0.98\ m}$.

So the length of the string at this point is $l + e = 1 + 0.98 = 1.98$ m.

b) The particle starts and ends at rest, so the change in the particle's kinetic energy is zero.
So, by the conservation of mechanical energy: Change in G.P.E. = Change in E.P.E.

G.P.E. lost $= mgh = 8 \times 9.8 \times x\sin 30 = 39.2x$.

E.P.E. gained $= \frac{\lambda}{2l}e^2 = \frac{40}{2 \times 1} \times (x - 1)^2 = 20(x - 1)^2$

So $20(x - 1)^2 = 39.2x \Rightarrow x^2 - 2x + 1 = 1.96x \Rightarrow x^2 - 3.96x + 1 = 0$.

The total length of the extended string is x m, and its natural length is 1 m, so the length of extension is $(x - 1)$ m.

c) Solving $x^2 - 3.96x + 1 = 0$ using the quadratic formula gives

$x = \mathbf{3.69}$ or $\mathbf{0.271}$.

0.271 can be ignored as we're told that the string is extended, so x must be greater than 1 (the natural length of the string). So the particle is first stationary when it is 3.69 m from O (3 s.f.).

d) At point A, the particle has no K.E. and no G.P.E. (if the level of A is taken as the 'base level').

So, total mechanical energy $= 0 + 0 +$ E.P.E. $= \frac{\lambda}{2l}e^2 = \frac{40}{2 \times 1} \times (3 - 1)^2 = \mathbf{80\ J}$

By the conservation of mechanical energy, during the motion of the particle after release from A:

$\frac{\lambda}{2l}e^2 + mgh + \frac{1}{2}mv^2 = 80 \quad \Rightarrow \quad 20(y - 1)^2 + 78.4(3 - y)\sin 30 + 4v^2 = 80,$

which rearranges to $v^2 = -5y^2 + 19.8y - 14.4$ — as required.

The vertical distance between A (3 m from O) and the particle's current position (y m from O).

e) The string becomes slack when it is no longer stretched, i.e. when it is at its natural length.
So, substitute $y = 1$ into the equation from part d) and solve for v:

$v^2 = -5 + 19.8 - 14.4 \Rightarrow v = 0.63$ ms^{-1} (2 s.f.)

I'm just Hooked on elastic potential energy...

All in all quite a nice little topic, if you ask me. You need to know the formulas, and the Principle of Conservation of Mechanical Energy always crops up somewhere along the way. As long as you're down with that, then it's all good.

Power

This page is for AQA M2, Edexcel M2, OCR M2 and OCR MEI M2

Power is the Rate at which Work is done on an Object

Power is a measure of the rate at which a force does work on an object.
The unit for power is the watt, where 1 watt (1 W) = 1 joule per second.

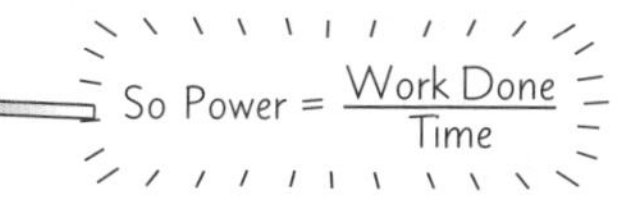

For an engine producing a driving force of F Newtons, and moving a vehicle at a speed of v ms^{-1}, the power in watts can be found using the formula:

Power = $F \times v$

$\text{Power} = \frac{\text{Work Done}}{\text{Time}} = \frac{\text{Force} \times \text{Distance}}{\text{Time}} = \text{Force} \times \text{Velocity}$

This is the formula you'll end up using most of the time — those examiners can't resist a question about engines.
But don't forget what power means, just in case they throw you a wild one — it's the rate of doing work.

EXAMPLE

A train of mass 500 000 kg is travelling along a straight horizontal track with a constant speed of 20 ms^{-1}. The train experiences a constant resistance to motion of magnitude 275 000 N.

a) Find the rate at which the train's engine is working. Give your answer in kW.

b) The train now moves up a hill inclined at 2° to the horizontal. If the engine continues to work at the same rate and the magnitude of the non-gravitational resistance to motion remains the same, find the new constant speed of the train.

a) Call the driving force of the train T N and the speed of the train u ms^{-1}.
Resolve parallel to the slope to find T:
$T - 275\,000 = m \times 0$
So $T = 275\,000$ N
Power = $T \times u = 275\,000 \times 20 = 5500$ kW.

b) Call the new driving force T' and resolve parallel to the slope:
$T' - 275\,000 - 500\,000g\sin 2° = m \times 0$
$\Rightarrow T' = 275\,000 + 500\,000g\sin 2°$ N $= 446\,008$ N
Power = $T' \times v$
$5\,500\,000 = 446\,008 \times v \quad \Rightarrow \quad v = \frac{5\,500\,000}{446\,008} = 12.3$ ms^{-1} (3 s.f.)

EXAMPLE

A tractor of mass 3000 kg is moving down a hill inclined at an angle of θ to the horizontal, where $\sin\theta = \frac{1}{24}$. The acceleration of the tractor is 1.5 ms^{-2} and its engine is working at a constant rate of 30 kW. Find the magnitude of the non-gravitational resistance to motion at the instant when the tractor is travelling at a speed of 8 ms^{-1}.

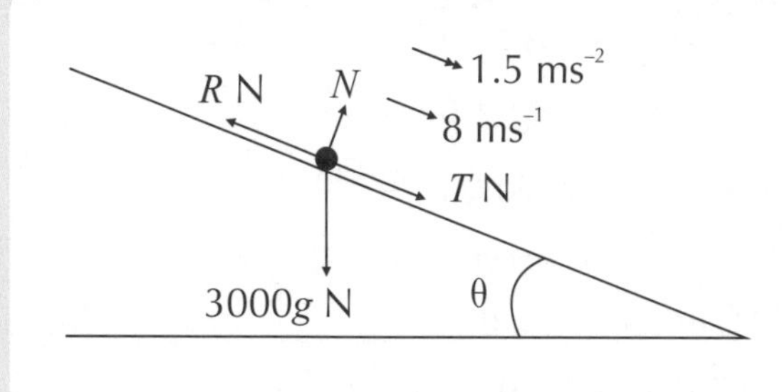

Use Power = $F \times v$ to find T:
$30\,000 = T \times 8 \Rightarrow T = 3750$ N
Resolve parallel to the slope: $T + mg\sin\theta - R = ma$
$3750 + (3000 \times 9.8 \times \frac{1}{24}) - R = 3000 \times 1.5$
$R = 3750 + 1225 - 4500$
$R = 475$ N

Add the component of weight, as the tractor is moving down the slope.

There is acceleration here, so this term doesn't disappear for once.

All together now — Watt's the unit for power...

These power questions all revolve around the use of $F = ma$, $P = Fv$ and maybe the occasional Power = Work ÷ Time.
Learn these and you're set for life. Well, maybe not life, but at least your exams. And what is life without exams?

Power

This page is for AQA M2 and OCR M2

Be prepared for a **Variable Resistive Force**

There's a good chance you'll get a power question where the resistive force isn't constant — it'll be dependent on the velocity of whatever's moving. Like the examples on the previous page, these questions require resolving of forces and the careful use of $F = ma$.

EXAMPLE A car of mass 1200 kg travels on a straight horizontal road. It experiences a resistive force of magnitude $30v$ N, where v is the car's speed in ms^{-1}. The maximum speed of the car on this road is 70 ms^{-1}. Find:

a) the car's maximum power,

b) the car's maximum possible acceleration when its speed is 40 ms^{-1}.

a) When the car is travelling at its maximum speed, acceleration is zero, and so the driving force of the car, T, must be equal to the resistive force, i.e. $T = 30v$.

Now use Power = Force × Velocity to give $P = 30v^2 = 30 \times 70^2 = 147$ kW.

b) Call the new driving force of the car F.

Power = Force × Velocity $\Rightarrow F = \frac{147\,000}{40} = 3675$ N

Resolve forces horizontally:

$3675 - 30v = ma$

$3675 - (30 \times 40) = 1200a \Rightarrow a = 2.06 \text{ ms}^{-2}$ (3 s.f.).

Maximum acceleration will only be possible when engine is working at maximum power.

EXAMPLE A van of mass 1500 kg moves up a road inclined at an angle of 5° to the horizontal. The van's engine works at a constant rate of 25 kW and the van experiences a resistive force of magnitude kv N, where k is a constant and v is the van's speed in ms^{-1}. At the point where the van has speed 8 ms^{-1}, its acceleration is 0.5 ms^{-2}.

a) Show that $k = 137$ to 3 s.f.

b) Using $k = 137$, show that U, the maximum speed of the van up this road, satisfies the equation: $U^2 + 9.35U - 182 = 0$, where coefficients are given to 3 s.f.

a) Use Power = $T \times v$, to find T, the driving force of the van's engine:

$T = \frac{\text{Power}}{v} = \frac{25\,000}{8} = 3125$ N

Resolve forces parallel to the slope: $T - kv - mg\sin 5 = ma$

So: $3125 - 8k - (1500 \times 9.8 \times \sin 5) = 1500 \times 0.5$

And rearrange: $k = \frac{1}{8}(3125 - (1500 \times 9.8 \times \sin 5) - (1500 \times 0.5)) = 137$ (3 s.f.)

b) From Power = Fv, the driving force of the van's engine at speed U is $\frac{25\,000}{U}$ N.

Again, resolve forces parallel to the slope: $\frac{25\,000}{U} - 137U - mg\sin 5 = 0$

Multiply throughout by U: $25\,000 - 137U^2 - (mg\sin 5 \times U) = 0$

Rearrange and simplify (to 3 s.f.) to give: $U^2 + 9.35U - 182 = 0$ — as required.

The van is travelling at maximum speed, so acceleration is zero.

The Power of Love — a Variable Resistive Force from Above...

Well that pretty much wraps up this section on work and energy. Plenty of formulas to learn and plenty of fun force diagrams to draw. If you're itching for some practice at all this then turn over and crack on. Even if you're not, do it anyway.

M2 Section 4 — Practice Questions

I don't know about you, but I enjoyed that section. Lots of engines and energy and blocks moving on slopes and GRRRRR look how manly I am as I do work against friction. *Ahem* sorry about that. Right-oh — practice questions...

Warm-up Questions

These questions are for AQA M2, Edexcel M2, OCR M2 and OCR MEI M2

1) A crate is pushed across a smooth horizontal floor by a force of 250 N, acting in the direction of motion. Find the work done in pushing the crate 3 m.
2) A crane lifts a concrete block 12 m vertically at constant speed. If the crane does 34 kJ of work against gravity, find the mass of the concrete block. Take $g = 9.8$ ms^{-2}.
3) A horse of mass 450 kg is cantering at a speed of 13 ms^{-1}. Find the horse's kinetic energy.
4) An ice skater of mass 65 kg sets off from rest. After travelling 40 m in a straight line across horizontal ice, she has done 800 J of work. Find the speed of the ice skater at this point.
5) A particle of mass 0.5 kg is projected upwards from ground level and reaches a maximum height of 150 m above the ground. Find the increase in the particle's gravitational potential energy. Take $g = 9.8$ ms^{-2}.
6) State the principle of conservation of mechanical energy.
   Explain why you usually need to model an object as a particle if you are using this principle.
7) A jubilant cowboy throws his hat vertically upwards with a velocity of 5 ms^{-1}. Use conservation of energy to find the maximum height the hat reaches above the point of release. Take $g = 9.8$ ms^{-2}.
8) State the work-energy principle. Explain what is meant by an 'external force'.
9) A car's engine is working at a rate of 350 kW. If the car is moving with speed 22 ms^{-1}, find the driving force of the engine.

Well those warm-up questions should have got your maths juices flowing, and you should now be eager to move on to something a bit more exam-like. It's probably best not to ask what maths juice is.

Exam Questions

Whenever a numerical value of g is required in the questions below, take $g = 9.8$ ms^{-2}.

Q1-5 are for AQA M2, Edexcel M2, OCR M2 and OCR MEI M2

1

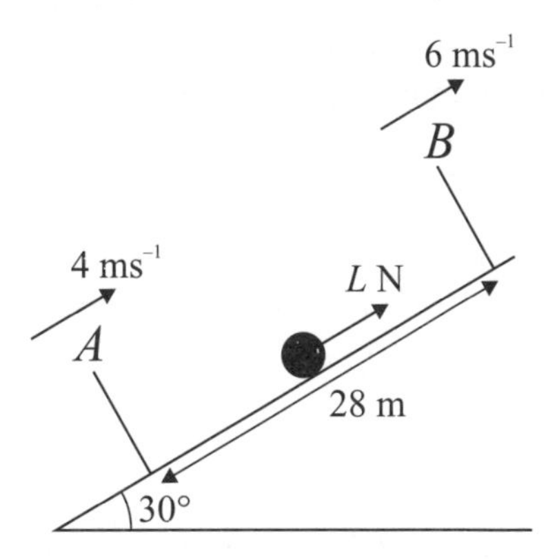

A skier is pulled up a sloping plane by a force, L, acting parallel to the plane which is inclined at an angle of 30° to the horizontal. The skier and his skis have a combined mass of 90 kg and he experiences a constant frictional force of 66 N as he moves up the slope. The skier passes through two gates, A and B, which are 28 m apart. His speed at gate A is 4 ms^{-1}. At gate B, his speed has increased to 6 ms^{-1}. Find:

a) the increase in the skier's total mechanical energy as he moves from gate A to gate B, *(5 marks)*

b) the magnitude of the force, L, pulling the skier up the slope. *(3 marks)*

M2 Section 4 — Practice Questions

I hope you've still got the energy left to power through this last bit of work. I don't want to have to force you...

2 A stone of mass 0.3 kg is dropped down a well. The stone hits the surface of the water in the well with a speed of 20 ms^{-1}.

a) Calculate the kinetic energy of the stone as it hits the water. *(2 marks)*

b) By modelling the stone as a particle and using conservation of energy, find the height above the surface of the water from which the stone was dropped. *(3 marks)*

c) When the stone hits the water, it begins to sink vertically and experiences a constant resistive force of 23 N. Use the work-energy principle to find the depth the stone has sunk to when the speed of the stone is reduced to 1 ms^{-1}. *(5 marks)*

3 A van of mass 2700 kg is travelling at a constant speed of 16 ms^{-1} up a road inclined at an angle of 12° to the horizontal. The non-gravitational resistance to motion is modelled as a single force of magnitude of 800 N.

a) Find the rate of work of the engine. *(4 marks)*

When the van passes a point A, still travelling at 16 ms^{-1}, the engine is switched off and the van comes to rest without braking, a distance x m from A. If all resistance to motion remains constant, find:

b) the distance x, *(4 marks)*

c) the time taken for the van to come to rest. *(4 marks)*

4

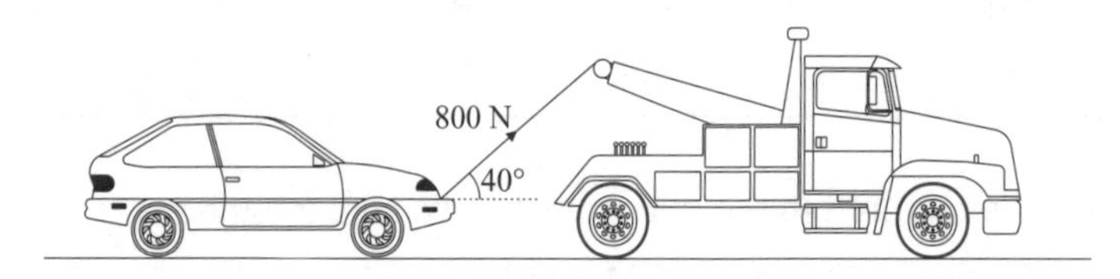

A car of mass 1500 kg is towed 320 m along a straight horizontal road by a rope attached to a pick-up truck. The rope is attached to the car at an angle of 40° to the horizontal and the tension in the rope is 800 N. The car experiences a constant resistance to motion from friction.

a) Find the work done by the towing force. *(3 marks)*

b) Over the 320 m, the car increases in speed from 11 ms^{-1} to 16 ms^{-1}. Assuming that the magnitude of the towing force remains constant at 800 N, find the coefficient of friction between the car and the road. *(4 marks)*

5 A cyclist is riding up a road at a constant speed of 4 ms^{-1}. The road is inclined at an angle α to the horizontal. The cyclist is working at a rate of 250 W and experiences a constant non-gravitational resistance to motion of magnitude 35 N. The cyclist and his bike have a combined mass of 88 kg.

a) Find the angle of the slope, α. *(4 marks)*

b) The cyclist now increases his work rate to 370 W. If all resistances to motion remain unchanged, find the cyclist's acceleration when his speed is 4 ms^{-1}. *(4 marks)*

M2 Section 4 — Practice Questions

Q6-7 are for AQA M2 only

6 A block of mass 3 kg is attached to one end of a light elastic string of natural length 2 m. The other end of the string is attached to a fixed point A. The weight of the block extends the length of the string to 5 m. The system is in equilibrium, with the block hanging directly below A. Find:

a) the modulus of elasticity of the string, *(3 marks)*

b) the elastic potential energy in the string. *(2 marks)*

The block is pulled down to a distance of 8 m directly below A, where it is released from rest and begins to move upwards.

c) Find the speed of the block when it is a distance of 3 m below A. *(4 marks)*

7 A block of weight 10 N is attached to one end of a light elastic string, the other end of which is O, a point on a vertical wall.
The block is placed on a rough horizontal surface, as shown, where the coefficient of friction between the block and the surface is $\mu = 0.5$. The string has natural length 5 m and modulus of elasticity 50 N.

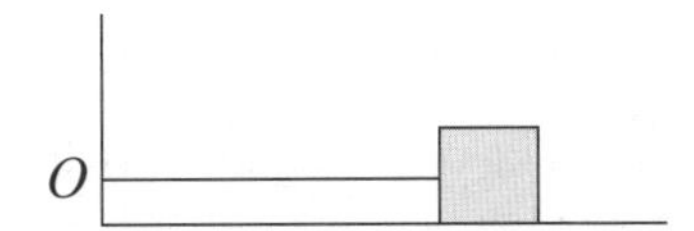

a) The block is held a horizontal distance d m from O, where $d > 5$.
Find an expression for the elastic potential energy of the system in terms of d. *(2 marks)*

The block is released from rest d m from O. The subsequent motion results in the block coming to rest just as it returns to O.

b) Find d. *(7 marks)*

Q8 is for AQA M2 and OCR M2

8 A car of mass 1000 kg experiences a resistive force of magnitude kv N, where k is a constant and v ms^{-1} is the car's speed. The car travels up a slope inclined at an angle of $\theta°$ to the horizontal, where $\sin\theta = 0.1$. The power generated by the car is 20 kW and its speed up the slope remains constant at 10 ms^{-1}.

a) Show that $k = 102$. *(3 marks)*

b) The car's maximum power output is 50 kW.

(i) Show that, up this slope, the car's maximum possible speed, u, satisfies the equation

$$102u^2 + 980u - 50\,000 = 0.$$

(4 marks)

(ii) Hence find the car's maximum possible speed on this slope. *(2 marks)*

The car reaches the top of the slope and begins travelling on a flat horizontal road.
The power increases to 21 kW and the resistive force remains at kv N.

b) Find the acceleration of the car when its speed is 12 ms^{-1}. *(3 marks)*

Linear Momentum

This page is for OCR MEI M2 only

Momentum has Magnitude and Direction

Momentum is a measure of how much "umph" a moving object has, due to its mass and velocity.
Total momentum before a collision equals total momentum after a collision.
This idea is called "Conservation of Momentum".
Because it's a vector, the sign of the velocity in momentum is important.

Momentum = Mass × Velocity

The unit of momentum is $kgms^{-1}$ or Ns

EXAMPLE Particles A and B, each of mass 5 kg, move in a straight line with velocities $6\ ms^{-1}$ and $2\ ms^{-1}$ respectively. After collision mass A continues in the same direction with velocity $4.2\ ms^{-1}$. Find the velocity of B after impact.

Before

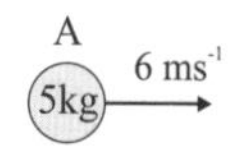

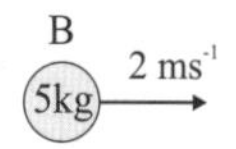

After

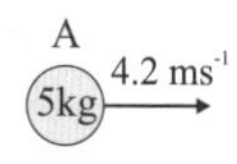

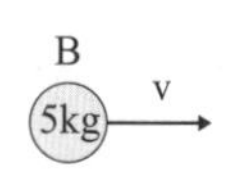

Taking 'right' as positive:

Momentum A + Momentum B = Momentum A + Momentum B

$$(5 \times 6) + (5 \times 2) = (5 \times 4.2) + (5 \times v)$$

$$40 = 21 + 5v$$

So $v = 3.8\ ms^{-1}$ in the same direction as before

Draw 'before' and 'after' diagrams to help you see what's going on.

Stick to saying 'same' or 'opposite' direction, rather than left or right — there's less chance of confusion.

EXAMPLE Particles A and B of mass 6 kg and 3 kg are moving towards each other at speeds of $2\ ms^{-1}$ and $1\ ms^{-1}$ respectively. Given that B rebounds with speed $3\ ms^{-1}$ in the opposite direction to its initial velocity, find the velocity of A after the collision.

Before

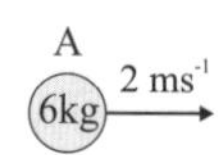

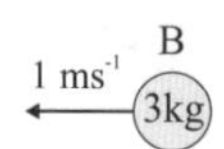

After

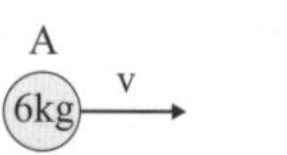

Taking 'right' as positive:

$$(6 \times 2) + (3 \times -1) = (6 \times v) + (3 \times 3)$$

$$9 = 6v + 9$$

$$v = 0$$

Masses Joined Together have the Same Velocity

Particles that stick together after impact are said to "coalesce". After that you can treat them as just one object.

EXAMPLE Two particles of mass 40 g and M kg move towards each other with speeds of $6\ ms^{-1}$ and $3\ ms^{-1}$ respectively. Given that the particles coalesce after impact and move with a speed of $2\ ms^{-1}$ in the same direction as that of the 40 g particle's initial velocity, find M.

Before

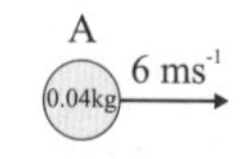

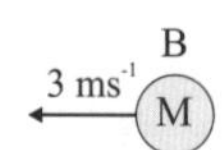

After

(M + 0.04) kg — 2 ms⁻¹

$$(0.04 \times 6) + (M \times -3) = (M + 0.04) \times 2$$

$$0.24 - 3M = 2M + 0.08$$

$$5M = 0.16$$

$$M = 0.032 \text{ kg}$$

Don't forget to convert all masses to the same units.

EXAMPLE A lump of ice of mass 0.1 kg is slid across the smooth surface of a frozen lake with speed $4\ ms^{-1}$. It collides with a stationary stone of mass 0.3 kg. The lump of ice and the stone then move in opposite directions to each other with the same speeds. Find their speed.

Before

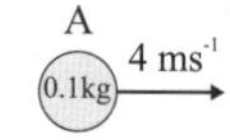

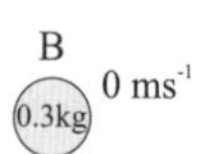

After

A: v ← 0.1kg   B: 0.3kg → v

$$(0.1 \times 4) + (0.3 \times 0) = (0.1 \times -v) + 0.3v$$

$$0.4 = -0.1v + 0.3v$$

$$v = 2\ ms^{-1}$$

Ever heard of Hercules?

Well, he carried out 12 tasks. Nothing to do with momentum, but if you're feeling sorry for yourself for doing M2, think on.

Impulse

This page is for OCR M2 and OCR MEI M2

An impulse changes the momentum of a particle in the direction of motion.

*Impulse is **Change in Momentum***

To work out the impulse that's acted on an object, just subtract the object's initial momentum from its final momentum. Impulse is measured in newton seconds (Ns).

Impulse = $mv - mu$

EXAMPLE A body of mass 500 g is travelling in a straight line. Find the magnitude of the impulse needed to increase its velocity from 2 ms^{-1} to 5 ms^{-1}.

Impulse = Change in momentum

$= mv - mu$

$= (0.5 \times 5) - (0.5 \times 2)$

$= 1.5$ Ns

This is called the impulse-momentum principle. Ooooh, aaaaaah.

Momentum = mass × velocity

EXAMPLE A 20 g ball is dropped 1 m onto horizontal ground. Immediately after rebounding the ball has a speed of 2 ms^{-1}. Find the impulse given to the ball by the ground. How high does the ball rebound? Take $g = 9.8$ ms^{-2}.

Take 'down' as positive.

First you need to work out the ball's speed as it reaches the ground:

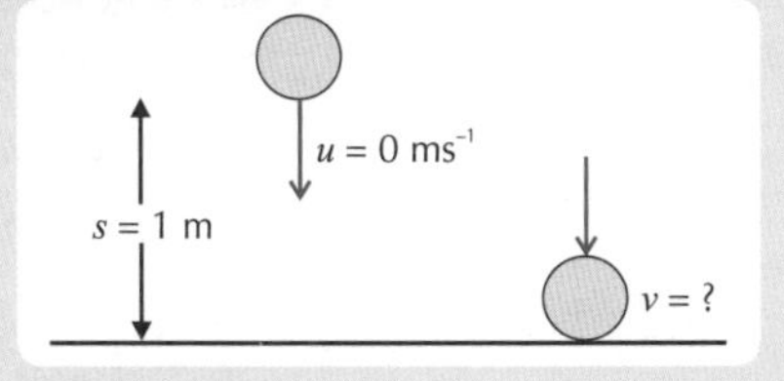

List the variables you're given:
$u = 0$
$s = 1$
$a = 9.8$
$v = ?$

The ball was dropped, so it started with u = 0.

Acceleration due to gravity.

Choose an equation containing u, s, a and v:

$v^2 = u^2 + 2as$

$v^2 = 0^2 + (2 \times 9.8 \times 1)$

$\mathbf{v = 4.43}$ **ms^{-1}**

The sign is really important. Make sure that down is positive throughout this part of the question.

Now work out the impulse as the ball hits the ground and rebounds:

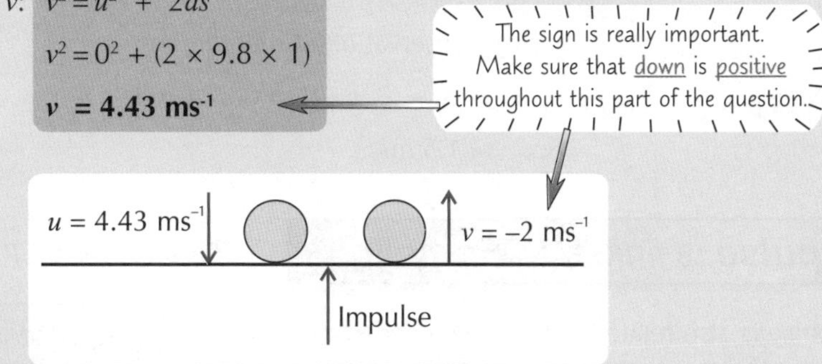

Impulse $= mv - mu$

$= (0.02 \times -2) - (0.02 \times 4.43)$

$= -0.129$ Ns (to 3 s.f.)

Finally you need to use a new equation of motion to find s (the greatest height the ball reaches after the bounce). This time take 'up' as positive:

List the variables:
$u = 2$
$v = 0$
$a = -9.8$
$s = ?$

$v = 0$ at the ball's greatest height.

a is negative because the ball is decelerating.

$v^2 = u^2 + 2as$

$0^2 = 2^2 + (2 \times -9.8 \times s)$

$s = 0.204$ m (to 3 s.f.)

Impulse

Impulses always ***Balance*** *in* ***Collisions***

This bit is for OCR M2 and OCR MEI M2

During impact between particles A and B, the impulse that A gives to B is the same as the impulse that B gives to A, but in the opposite direction.

EXAMPLE

A mass of 2 kg moving at 2 ms^{-1} collides with a mass of 3 kg which is moving in the same direction at 1 ms^{-1}. The 2 kg mass continues to move in the same direction at 1 ms^{-1} after impact. Find the impulse given by the 2 kg mass to the other mass.

Using "conservation of momentum":

$(2 \times 2) + (3 \times 1) = (2 \times 1) + 3v$

So $v = 1\frac{2}{3}$ ms^{-1}

Before

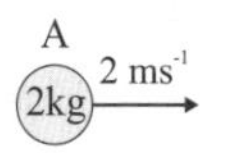

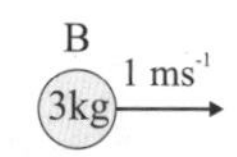

After

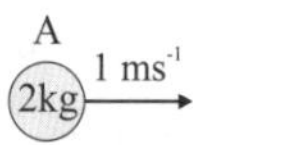

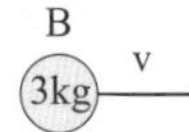

Impulse (on B) = $mv - mu$ (for B)

$= (3 \times 1\frac{2}{3}) - (3 \times 1)$

$= 2$ Ns

The impulse B gives to A is $(2 \times 1) - (2 \times 2) = -2$ **Ns**. Aside from the different direction, you can see it's the same — so you didn't actually need to find v for this question.

EXAMPLE

Two snooker balls A and B have a mass of 0.6 kg and 0.9 kg respectively. The balls are initially at rest on a snooker table. Ball A is given an impulse of magnitude 4.5 Ns towards ball B. Modelling the snooker table as a smooth horizontal plane, find the speed of ball A before it collides with B.

Impulse = $mv - mu$

$\Rightarrow 4.5 = (0.6 \times v) - (0.6 \times 0) = 0.6v$

$\Rightarrow v = \frac{4.5}{0.6} = 7.5$ ms^{-1}

The balls collide and move away in the direction A was travelling before the collision. Find the speed of ball A after the collision, given that the speed of ball B is 4 ms^{-1}.

Using conservation of momentum:

$(0.6 \times 7.5) + (0.9 \times 0) = 0.6v + (0.9 \times 4)$

So, $v = 1.5$ ms^{-1}

Impulse *is linked to* ***Force*** *too*

This bit is for OCR MEI M2 only

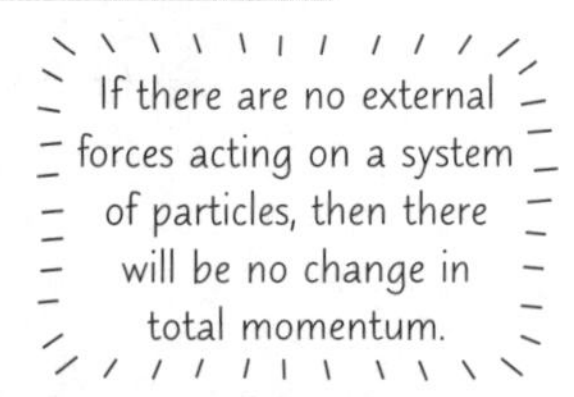

Impulse is also related to the force needed to change the momentum and the time it takes.

Impulse = Force × Time

EXAMPLE

A 0.9 tonne car increases its speed from 30 kmh^{-1} to 40 kmh^{-1}. Given that the maximum additional constant forward force the car's engine can produce is 1 kN, find the shortest time it will take to achieve this change in speed.

To change kmh^{-1} to ms^{-1}, multiply by 1000 (to change km to m), then divide by 3600 (to change h^{-1} to s^{-1}).

Impulse = $mv - mu$

$= (900 \times \frac{40\,000}{3600}) - (900 \times \frac{30\,000}{3600})$

$= 2500$ Ns

Now use Impulse = Force × Time:

$2500 = 1000 \times t$

$t = 2.5$ s

Doctor, this man is sick — 'im pulse is very weak...

...a little bit like that pun actually. Anyway, naff humour aside, make sure you've got your head round all of this straight line momentum and impulse stuff, 'cos we're about to throw a whole new dimension into the mix. I bet you can't wait.

Momentum and Impulse

This page is for Edexcel M2 and OCR MEI M2 only

It's time to move into the next dimension — look out for those **i** and **j** vectors, we're going 2D...

An *Impulse* causes a *Change* in *Momentum*

All moving objects have momentum (mass (kg) × velocity (ms^{-1})). If an object receives an impulse (I — measured in newton seconds, or Ns) its momentum will change. The size of the change is the size of the impulse.

Impulse = final momentum – initial momentum

$I = mv - mu$ or $\mathbf{I} = m\mathbf{v} - m\mathbf{u}$ ← This is the vector form.

Since velocity is a vector, momentum and impulse are vectors too.

EXAMPLE A ball (m = 0.1 kg) travels with a velocity of (5**i** + 12**j**) ms^{-1} before receiving an impulse of **I** Ns. If the ball's new velocity is (15**i** + 22**j**) ms^{-1}, find **I**.

1) Don't be put off by the vector notation. Plug the info in the formula as usual: $\mathbf{I} = m\mathbf{v} - m\mathbf{u}$, where $m = 0.1$, $\mathbf{v} = 15\mathbf{i} + 22\mathbf{j}$ and $\mathbf{u} = 5\mathbf{i} + 12\mathbf{j}$.
2) $\mathbf{I} = 0.1(15\mathbf{i} + 22\mathbf{j}) - 0.1(5\mathbf{i} + 12\mathbf{j})$
   $= 1.5\mathbf{i} + 2.2\mathbf{j} - 0.5\mathbf{i} - 1.2\mathbf{j}$
   $= 1\mathbf{i} + 1\mathbf{j} = \mathbf{i} + \mathbf{j}$.
3) So the ball received an impulse of **(i + j) Ns.**

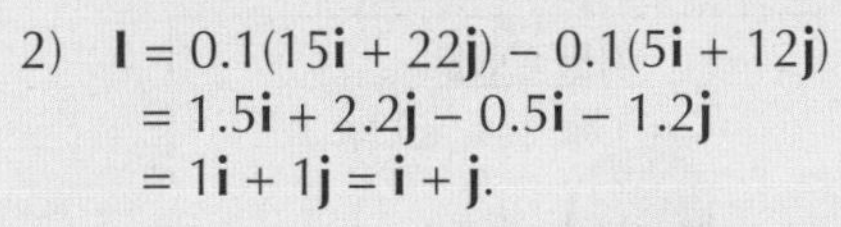

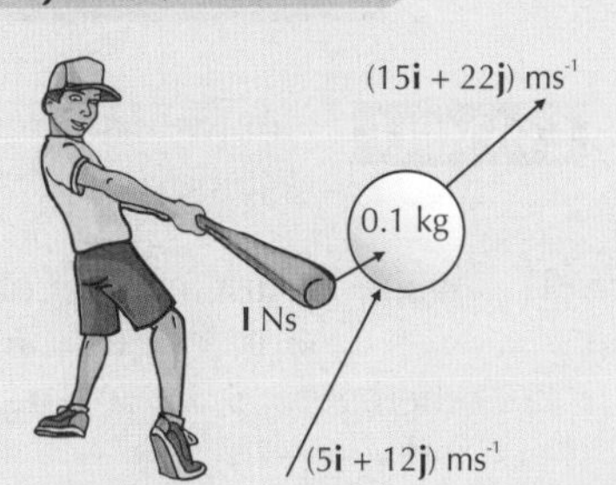

Use *Pythagoras* and *Trig* for the *Magnitude* and *Angle* of *Impulse*

With vectors, you can use the horizontal **i** component and the vertical **j** component to form a right-angled triangle. Then simply use basic trig and Pythagoras to find any angles, or the magnitude (scalar size) of impulses or velocities.

EXAMPLE A badminton player smashes a shuttlecock (m = 0.005 kg) with an impulse of (0.035**i** – 0.065**j**) Ns. If the shuttle was initially travelling at (–3**i** + **j**) ms^{-1}, find its speed after the smash, and the angle it makes with the horizontal.

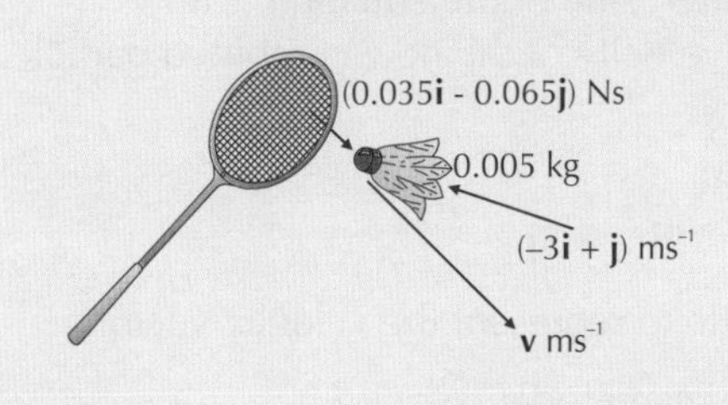

1) Find the final velocity as a vector first, so: $\mathbf{I} = m\mathbf{v} - m\mathbf{u}$, where $\mathbf{I} = 0.035\mathbf{i} - 0.065\mathbf{j}$, $m = 0.005$ and $\mathbf{u} = -3\mathbf{i} + \mathbf{j}$.

   $0.035\mathbf{i} - 0.065\mathbf{j} = 0.005\mathbf{v} - 0.005(-3\mathbf{i} + \mathbf{j})$
   $\Rightarrow 0.035\mathbf{i} - 0.065\mathbf{j} = 0.005\mathbf{v} + 0.015\mathbf{i} - 0.005\mathbf{j}$
   $\Rightarrow 0.005\mathbf{v} = 0.035\mathbf{i} - 0.065\mathbf{j} - 0.015\mathbf{i} + 0.005\mathbf{j}$
   $\Rightarrow 0.005\mathbf{v} = 0.02\mathbf{i} - 0.06\mathbf{j}$
   $\Rightarrow \mathbf{v} = (0.02\mathbf{i} - 0.06\mathbf{j}) \div 0.005 = 4\mathbf{i} - 12\mathbf{j}$.

2) Draw a right-angled triangle of the velocity vector:

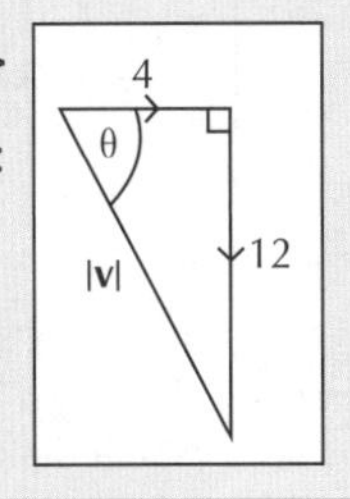

3) Use Pythagoras to find the speed (the magnitude of the velocity):
   $|\mathbf{v}| = \sqrt{4^2 + 12^2} = 12.6\ ms^{-1}$ to 3 s.f.
4) Use trig to find the angle of motion with the horizontal:
   $\theta = \tan^{-1}\left(\frac{12}{4}\right) = 71.6°$ (3 s.f.) below the horizontal.

Finding the magnitude of a vector this way should be familiar to you from M1.

The perils of internet shopping — the midnight impulse buy...

This page isn't too bad — it's just a case of combining things you already know. It often helps to write down all the bits of information you know before you start plugging numbers into formulas, especially when you've got quite a wordy question where some key info mightn't be immediately obvious. If this page just wasn't enough for you, then boy are you in for a treat.

Momentum and Impulse

This page is for Edexcel M2 and OCR MEI M2

The 'principle of conservation of linear momentum' is really just a lengthy title that boils down to 'momentum in = momentum out'.

Momentum is Conserved when things are Free to Move

When two things collide that are free to move around, they exert an equal and opposite impulse on each other — these impulses 'cancel out' so there is no impulse for the overall system. No impulse means no change in momentum:

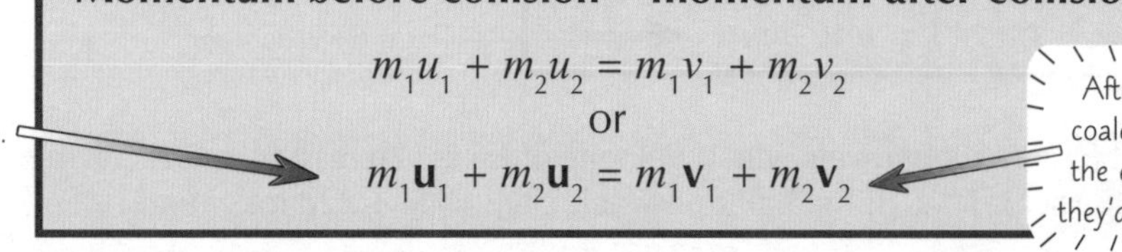

After the collision, the two things might coalesce (stick together) — so this side of the equation would just be $(m_1 + m_2)\mathbf{v}$, as they'd move together with the same velocity.

EXAMPLE a) Two particles A and B, shown below, collide. Following the collision they move separately at different velocities. Find B's velocity after the collision.

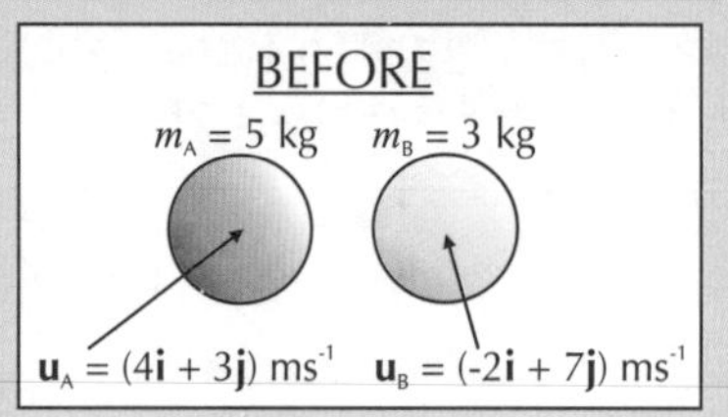

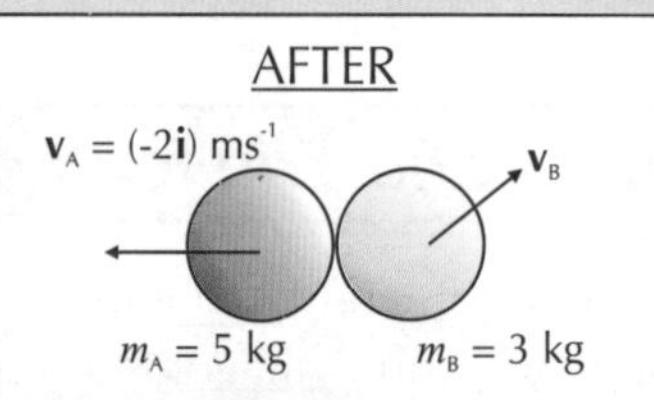

1) Again, it's just a matter of plugging the numbers in:
   $m_A\mathbf{u}_A + m_B\mathbf{u}_B = m_A\mathbf{v}_A + m_B\mathbf{v}_B$,
   where $m_A = 5$, $m_B = 3$, $\mathbf{u}_A = 4\mathbf{i} + 3\mathbf{j}$, $\mathbf{u}_B = -2\mathbf{i} + 7\mathbf{j}$ and $\mathbf{v}_A = -2\mathbf{i}$.
2) $5(4\mathbf{i} + 3\mathbf{j}) + 3(-2\mathbf{i} + 7\mathbf{j}) = 5(-2\mathbf{i}) + 3\mathbf{v}_B$
   $\Rightarrow 20\mathbf{i} + 15\mathbf{j} - 6\mathbf{i} + 21\mathbf{j} = -10\mathbf{i} + 3\mathbf{v}_B$
   $\Rightarrow 3\mathbf{v}_B = 24\mathbf{i} + 36\mathbf{j}$
   $\Rightarrow \mathbf{v}_B = 8\mathbf{i} + 12\mathbf{j}$

So B's new velocity is $(8\mathbf{i} + 12\mathbf{j})$ ms^{-1}.

EXAMPLE b) If the two particles coalesce instead, find their combined speed after the collision, and the direction they travel in.

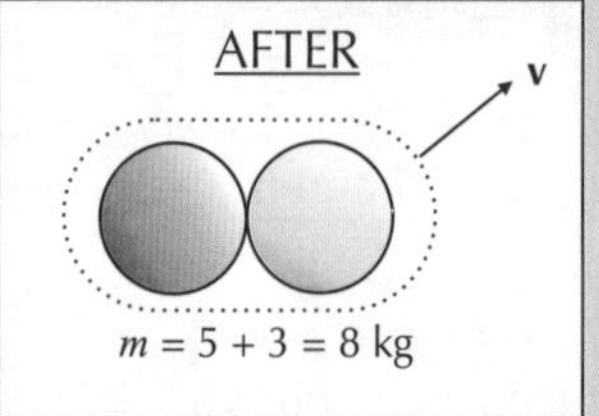

1) Start off as in a), but the 'after' side of the equation is now $(m_A + m_B)\mathbf{v}$, where $\mathbf{v}$ is the velocity of the new combined particle.
   $5(4\mathbf{i} + 3\mathbf{j}) + 3(-2\mathbf{i} + 7\mathbf{j}) = (5 + 3)\mathbf{v}$
   $\Rightarrow 20\mathbf{i} + 15\mathbf{j} - 6\mathbf{i} + 21\mathbf{j} = 8\mathbf{v}$
   $\Rightarrow 8\mathbf{v} = 14\mathbf{i} + 36\mathbf{j} \Rightarrow \mathbf{v} = 1.75\mathbf{i} + 4.5\mathbf{j}$
2) Draw a right-angled triangle to represent the velocity vector.

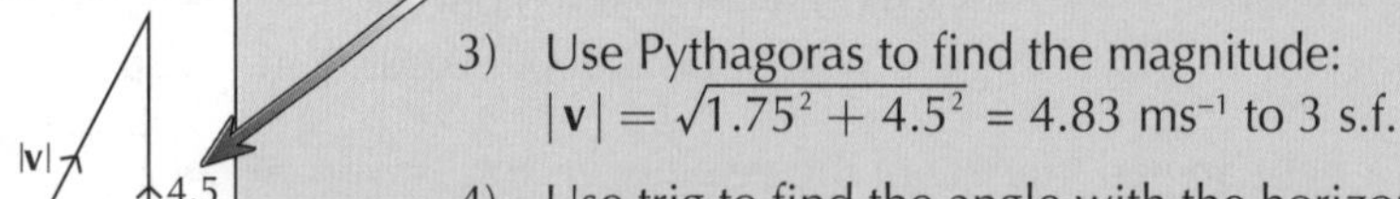

3) Use Pythagoras to find the magnitude:
   $|\mathbf{v}| = \sqrt{1.75^2 + 4.5^2} = 4.83$ ms^{-1} to 3 s.f.
4) Use trig to find the angle with the horizontal:
   $\theta = \tan^{-1}\left(\frac{4.5}{1.75}\right) = 68.7°$ to 3 s.f.

|v|, 4.5, θ, 1.75

So the combined particles move away from the collision at 4.83 ms^{-1}, at an angle of 68.7° to the horizontal.

'Connected' particles, joined by a 'light inelastic string' (see M1) can be dealt with in the same way as coalesced particles — when the string between a connected pair is taut, they act as one particle travelling at a common speed.

Mo' mentum mo' problems...

The theory here is the same as it was in 1 dimension — you've just got to mind your **i**'s and **j**'s. I find drawing a picture of the situation helps when you're trying to visualise the particles bouncing in all directions. Or, you could draw inspirational doodles and think about that emo type you fancy who sits at the back of class. Less productive though...

Collisions

This page is for Edexcel M2, OCR M2 and OCR MEI M2

Oh yes, you've not seen the last of those colliding particles. If you like things loud and dramatic, think demolition balls and high speed crashes. If you're anything like me though you'll be picturing a nice sedate game of snooker.

The *Coefficient of Restitution* is always between *0 and 1*

When two particles collide in a direct impact (i.e. they're moving on the same straight line), the speeds they bounce away at depend on the coefficient of restitution, e. This is known as Newton's Law of Restitution, and looks like this:

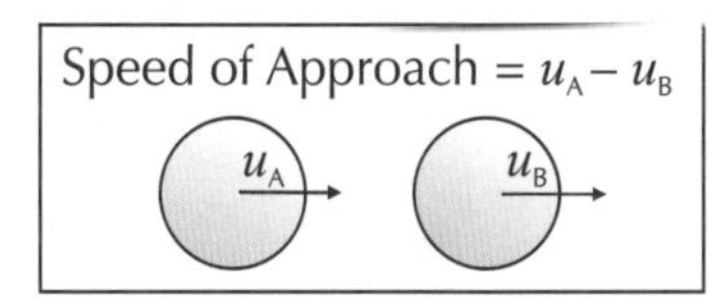

$$e = \frac{\text{speed of separation of particles}}{\text{speed of approach of particles}}$$

$$e = \frac{v_B - v_A}{u_A - u_B}$$

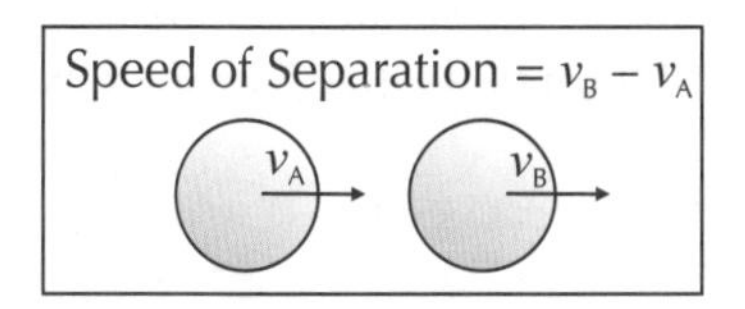

1) The value of e depends on the material that the particles are made of.
2) e always lies between 0 and 1.
3) When $e = 0$ the particles are called 'inelastic', and they'll coalesce.
4) When $e = 1$ the particles are 'perfectly elastic' and they'll bounce apart with no loss of speed.

Balls of modelling clay would be near the $e = 0$ end of the scale, while ping pong balls are nearer to $e = 1$.

EXAMPLE Two particles collide as shown. Find the coefficient of restitution.

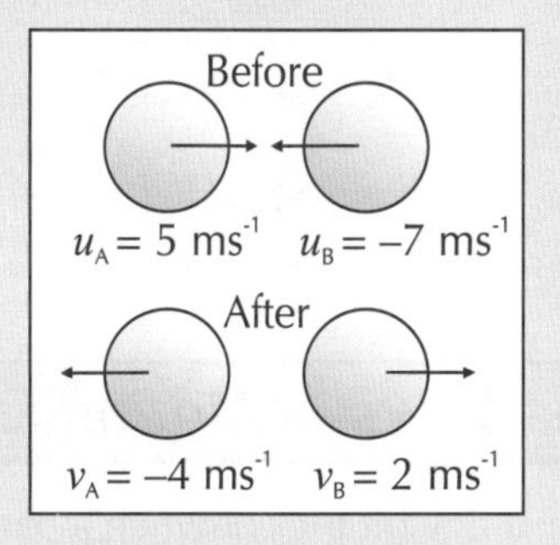

1) Firstly, work out the speeds of approach and separation, taking care with positives and negatives:
Speed of approach $= u_A - u_B = 5 - (-7) = 12$ ms^{-1}.
Speed of separation $= v_B - v_A = 2 - (-4) = 6$ ms^{-1}.

Think of 'left to right' as positive, and so particles travelling 'right to left' will have a negative velocity.

2) Use $e = \frac{\text{speed of separation of particles}}{\text{speed of approach of particles}}$:

$e = \frac{6}{12} = 0.5$. So the coefficient of restitution is 0.5.

For *Two Unknown Speeds* — use *Momentum Conservation* too

Often you'll be given the value of e and asked to find the velocities of both particles either before or after impact. As there are two unknowns, you'll need to use the formula for conservation of momentum (on p.216) along with the Law of Restitution to form simultaneous equations.

EXAMPLE Two particles, A and B, are moving in opposite directions in the same straight line, as shown. If $e = \frac{1}{3}$, find the velocities of both particles after impact.

$m_A = 4$ kg $\quad m_B = 12$ kg

$u_A = 10$ ms^{-1} $\quad u_B = -2$ ms^{-1}

1) Use $e = \frac{v_B - v_A}{u_A - u_B}$ to get the first equation:

$\frac{1}{3} = \frac{v_B - v_A}{10 - (-2)} \Rightarrow v_B - v_A = 4$. Call this **equation 1**.

2) Use $m_A u_A + m_B u_B = m_A v_A + m_B v_B$ to get the second equation:
$(4 \times 10) + (12 \times -2) = 4v_A + 12v_B$
$\Rightarrow 16 = 4v_A + 12v_B \Rightarrow v_A + 3v_B = 4$. Call this **equation 2**.

3) **Equation 1** + **equation 2** gives:
$4v_B = 8$, so $v_B = 2$ ms^{-1} (i.e. 2 ms^{-1} going left to right).

4) Substituting in **equation 1** gives:
$2 - v_A = 4$, so $v_A = -2$ ms^{-1} (i.e. 2 ms^{-1} going right to left).

Collisions

This page is for Edexcel M2, OCR M2 and OCR MEI M2

There's a saying in Stoke-on-Trent that goes: 'cost kick a bo againt a wo till it bosses?'*
Well, that's kinda what this next bit's about, a.k.a. 'the collision of a particle with a plane surface'.

The **Law of Restitution** also works with a **Smooth Plane Surface**

Particles don't just collide with each other. They can collide with a fixed flat surface — such as when a ball is kicked against a vertical wall, or dropped onto a horizontal floor.

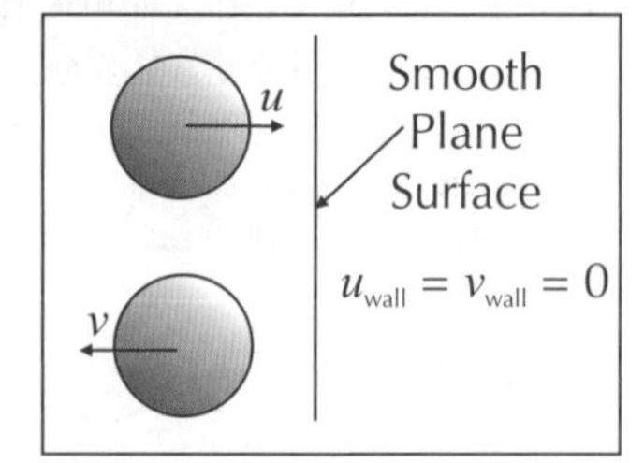

As long as the surface can be modelled as smooth (i.e. no friction) and perpendicular to the motion of the particle, the law can be simplified to:

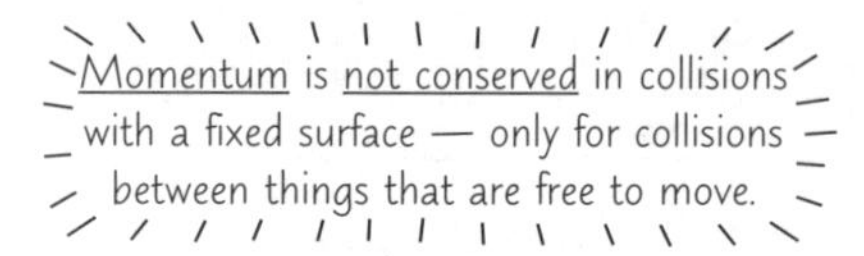

$$e = \frac{\text{speed of rebound of particle}}{\text{speed of approach of particle}} = \frac{v}{u}$$

EXAMPLE A ball rolling along a smooth horizontal floor at 6 ms^{-1} hits a smooth vertical wall, with a coefficient of restitution $e = 0.65$. Find the speed of the ball as it rebounds.

BEFORE

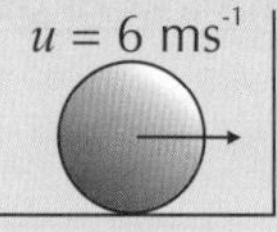

AFTER

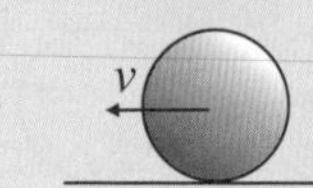

Using $e = \frac{v}{u}$:

$0.65 = \frac{v}{6} \Rightarrow v = 0.65 \times 6 = 3.9$ ms^{-1}.

So the ball rebounds at a speed of 3.9 ms^{-1}. Piece of cake.

Use the **Laws of Motion** for things being **Dropped**

Things get a tiny bit trickier when a particle is dropped onto a horizontal surface because acceleration under gravity comes into play. You should be pretty nifty with equations of motion now though — just remember to use them here.

EXAMPLE A basketball is dropped vertically from rest at a height of 1.4 m onto a horizontal floor. It rebounds to a height of 0.9 m. Find e for the impact with the floor.

1) Assuming the ball is a particle, and the floor is smooth, we can use $e = \frac{v}{u}$.
   For the diagram shown, this would be $e = \frac{u_2}{v_1}$, as we need the velocity just before the impact (v_1) and the velocity just after (u_2).

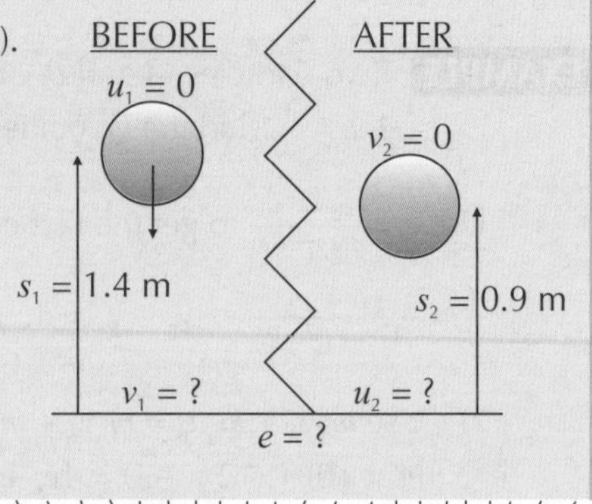

2) Using $v^2 = u^2 + 2as$ before the impact with the floor (where $a = g \approx 9.8$ ms^{-2}):
   $v_1^2 = 0 + 2 \times 9.8 \times 1.4 = 27.44$
   $\Rightarrow v_1 = 5.238$ ms^{-1} to 4 s.f.
3) Using $v^2 = u^2 + 2as$ after the impact with the floor (where $a = -g$ since the motion is against gravity):
   $0 = u_2^2 + 2 \times -9.8 \times 0.9$
   $\Rightarrow u_2 = 4.2$ ms^{-1}.
4) Finally, we can find e: $e = \frac{u_2}{v_1} = \frac{4.2}{5.238} = 0.802$ to 3 s.f.

There's loads about using the equations of motion in M1, as well as in Section 1 of M2 (see p 168-170).

Dating Tip #107 — Avoid them if they're on the rebound...

Just when you were thinking this section was a load of balls, along come walls and floors to shake things up a bit. The Law of Restitution is a pretty straightforward formula, but chances are there'll be added complications in the exam questions. Learn how to tackle the four types of question on these last two pages and you'll be laughing.

*For those unfamiliar with Potteries dialect, this means 'Can you kick a ball against a wall until it burst

Oblique Collisions

This page is for OCR MEI M2 only

Don't let the title worry you — there's nothing too difficult on this page. It's pretty similar to what was on the last page, but now the collisions are at funny angles. It's time to get your trig on. Again.

In an **Oblique Impact** with a Plane, Impulse acts **Perpendicular** to the Plane

1) When an object collides with a smooth plane at an oblique angle (i.e. not perpendicular to the surface), the impulse on the object acts perpendicular to the plane.
2) This means that only the component of the object's velocity in the direction perpendicular to the plane is changed by the collision. The component of velocity parallel to the surface remains unchanged.
3) The direction of the perpendicular component of velocity will be reversed by the collision (because the object is moving away from the surface after the collision instead of towards it).
4) The Law of Restitution still applies in oblique collisions. The magnitude of the perpendicular component of velocity after the collision is the original magnitude multiplied by e, the coefficient of restitution between the object and the surface.

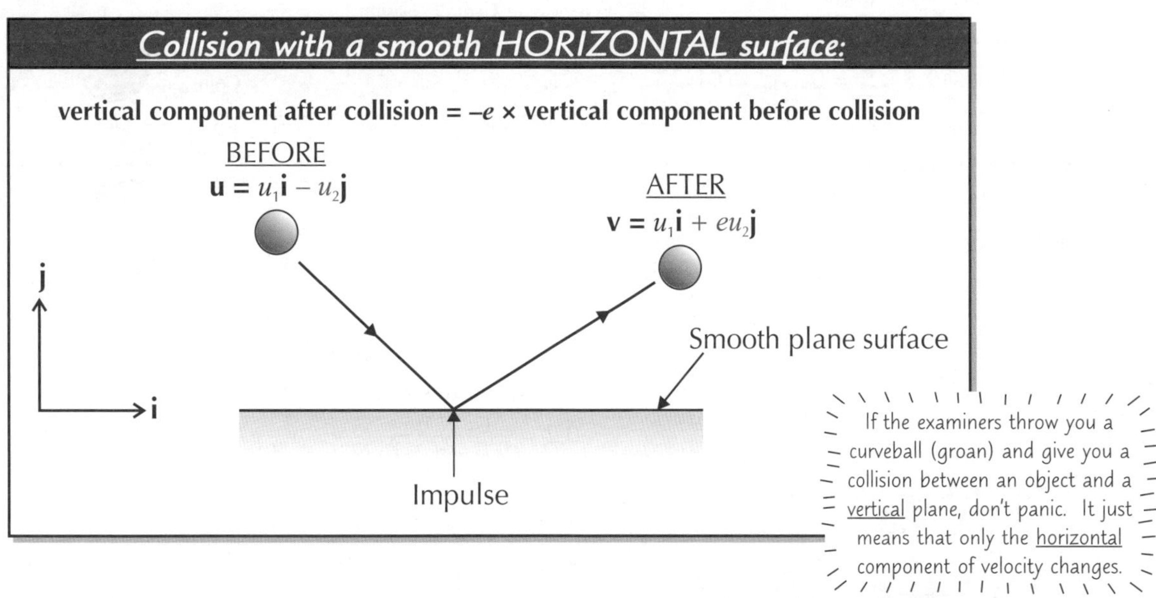

If the examiners throw you a curveball (groan) and give you a collision between an object and a vertical plane, don't panic. It just means that only the horizontal component of velocity changes.

EXAMPLE

A tennis ball hits the horizontal ground with velocity $12\mathbf{i} - 8\mathbf{j}$ ms^{-1} (where $\mathbf{i}$ and $\mathbf{j}$ are the horizontal and vertical unit vectors respectively), and rebounds at an angle of $\alpha°$ to $\mathbf{i}$. By modelling the ground as a smooth plane surface with coefficient of restitution $e = 0.5$, find v, the speed of the ball after the collision, and α.

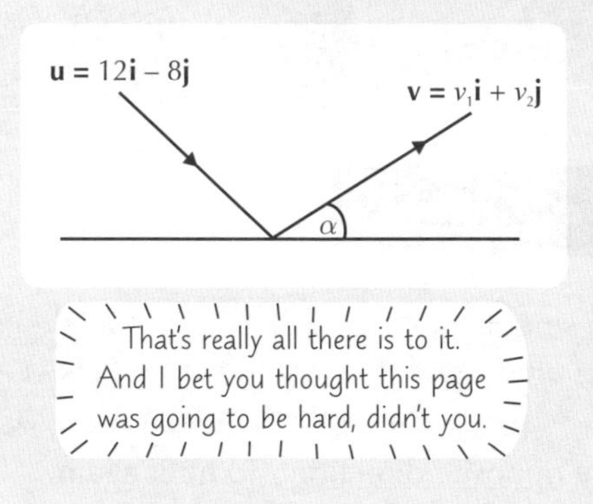

1) The ground is parallel to $\mathbf{i}$, so the component of velocity in this direction is unchanged, i.e. $v_1 = 12$.
2) Perpendicular to the wall, the component of the velocity is reversed and multiplied by e, so $v_2 = -0.5(-8) = 4$.
3) So the velocity of the ball after the collision is $\mathbf{v} = 12\mathbf{i} + 4\mathbf{j}$ ms^{-1}. You can now calculate α: $\tan\alpha = \frac{4}{12} \Rightarrow \alpha = 18.4°$ (3 s.f.)
4) Use Pythagoras to find the speed after the collision: $v = \sqrt{12^2 + 4^2} = 12.6$ ms^{-1} (3 s.f.)

That's really all there is to it. And I bet you thought this page was going to be hard, didn't you.

Feel like you're on a collision course with your exams?

You might not always be given an object's velocity in terms of $\mathbf{i}$ and $\mathbf{j}$ vectors. In that case, you just have to do a bit of trig to find the components parallel and perpendicular to the surface — then it's just the same as the method on this page. Lovely.

Complex Collisions

This page is for Edexcel M2 and OCR MEI M2

You've had an easy ride so far this section, but now it's time to fasten your seatbelt, don your crash helmet, and prepare for some pretty scary collisions. Don't say I didn't warn you...

Solve *Successive* Collisions *Step by Step...*

Think of this as a multi-particle pile-up. One particle collides with another, which then shoots off to collide with a third. No extra maths required, but quite a bit of extra thinking.

EXAMPLE Particles P, Q and R are travelling at different speeds along the same smooth straight line, as shown. Particles P and Q collide first ($e = 0.6$), then Q goes on to collide with R ($e = 0.2$). What are the velocities of P, Q and R after the second collision?

0.1 kg: P 20 ms⁻¹ → ; 0.4 kg: Q 5 ms⁻¹ → ; 2 kg: R ← 1 ms⁻¹

1) Take things step by step. Forget about R for the moment and concentrate on the first collision — the one between P and Q:

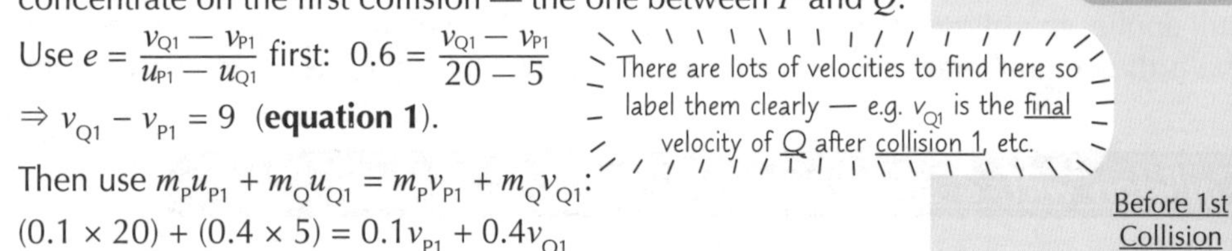

Use $e = \frac{v_{Q1} - v_{P1}}{u_{P1} - u_{Q1}}$ first: $0.6 = \frac{v_{Q1} - v_{P1}}{20 - 5}$

$\Rightarrow v_{Q1} - v_{P1} = 9$ (**equation 1**).

Then use $m_P u_{P1} + m_Q u_{Q1} = m_P v_{P1} + m_Q v_{Q1}$:

$(0.1 \times 20) + (0.4 \times 5) = 0.1v_{P1} + 0.4v_{Q1}$

$\Rightarrow 4 = 0.1v_{P1} + 0.4v_{Q1} \quad \Rightarrow \quad v_{P1} + 4v_{Q1} = 40$ (**equation 2**).

Equation 1 + **equation 2** gives:

$5v_{Q1} = 49 \quad \Rightarrow \quad v_{Q1} = 9.8 \text{ ms}^{-1}$.

Substituting in **equation 1** gives:

$9.8 - v_{P1} = 9 \quad \Rightarrow \quad v_{P1} = 9.8 - 9 = 0.8 \text{ ms}^{-1}$.

Before 1st Collision: 0.1 kg P 20 ms⁻¹ →, 0.4 kg Q 5 ms⁻¹ →; $e = 0.6$
After 1st Collision: P 0.8 ms⁻¹ →, Q 9.8 ms⁻¹ →

2) For the second collision, which is between Q and R: $e = \frac{v_{R2} - v_{Q2}}{u_{Q2} - u_{R2}}$.
u_{Q2} is the same as the velocity of Q after the first collision — you found this above (9.8 ms⁻¹), so:

$0.2 = \frac{v_{R2} - v_{Q2}}{9.8 - (-1)} \quad \Rightarrow \quad v_{R2} - v_{Q2} = 2.16$ (**equation 3**).

Then $m_Q u_{Q2} + m_R u_{R2} = m_Q v_{Q2} + m_R v_{R2}$:

$(0.4 \times 9.8) + (2 \times -1) = 0.4v_{Q2} + 2v_{R2}$

$\Rightarrow 1.92 = 0.4v_{Q2} + 2v_{R2} \quad \Rightarrow \quad 0.2v_{Q2} + v_{R2} = 0.96$ (**equation 4**).

Equation 4 – **equation 3** gives:

$1.2v_{Q2} = -1.2 \quad \Rightarrow \quad v_{Q2} = -1 \text{ ms}^{-1}$.

Substituting in **equation 3** gives:

$v_{R2} - (-1) = 2.16 \quad \Rightarrow \quad v_{R2} = 2.16 - 1 = 1.16 \text{ ms}^{-1}$.

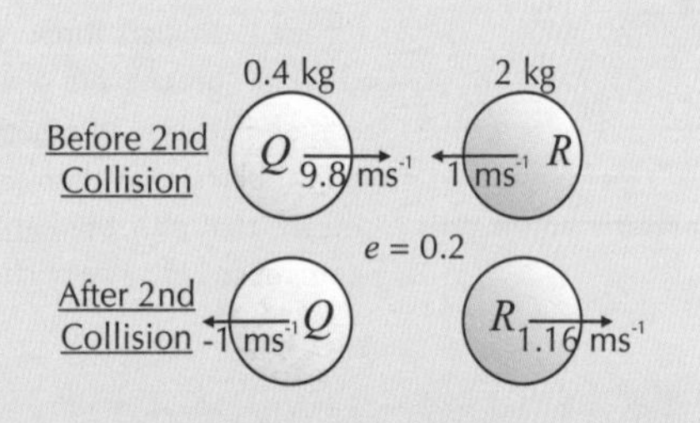

3) Velocities after both collisions are: $P = 0.8 \text{ ms}^{-1}$, $Q = -1 \text{ ms}^{-1}$ and $R = 1.16 \text{ ms}^{-1}$:

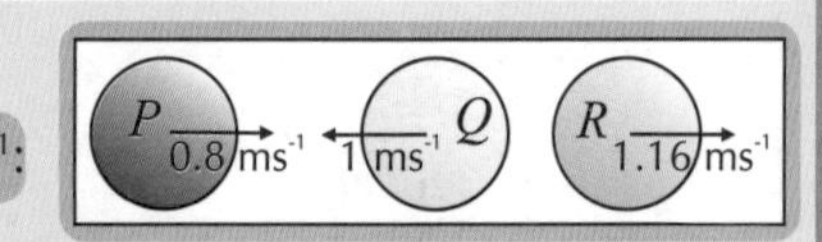

... as well as *Subsequent Collisions* with a *Plane Surface*

EXAMPLE Following the second collision, P is removed and R hits a smooth vertical wall at a right angle. How big would e have to be for this impact to allow R to collide again with Q, assuming Q is moving with velocity –1 ms⁻¹?

1) Think things through carefully. Q is currently going at 1 ms⁻¹ in the opposite direction. To hit it again, R needs to bounce off the wall with a rebound speed higher than 1 ms⁻¹, so it can 'catch up'. So $v_{R3} > 1$.
2) For the impact with the wall, $e = \frac{v_{R3}}{u_{R3}} \quad \Rightarrow \quad v_{R3} = eu_{R3}$, and so $eu_{R3} > 1$.
3) From the example above, $u_{R3} = v_{R2} = 1.16 \text{ ms}^{-1}$, so $1.16e > 1 \quad \Rightarrow \quad e > \frac{1}{1.16} \quad \Rightarrow \quad e > 0.8620...$
4) So, to 3 s.f., e must be higher than 0.862 for R to collide again with Q.

Complex Collisions

This page is for Edexcel M2 only

Well they do say 'what goes up must come down'. And up again. And down again. Just look at the economy.

*Particles may have **Successive Rebounds** before coming to **Rest***

EXAMPLE A ball falls from a height of 10 m and rebounds several times from the ground, where $e = 0.8$ for each impact. Find the height the ball reaches after each of the first three bounces, stating any assumptions.

1) Some assumptions — the ball is a particle, air resistance can be ignored, it falls vertically onto a horizontal, smooth, plane surface, under a constant acceleration downwards of $g = 9.8 \text{ ms}^{-2}$.

2) For each bounce use $v^2 = u^2 + 2as$ to find the approach speed to the ground and the Law of Restitution, $e = \frac{v}{u}$, to find the rebound speed (p. 222). Then use $v^2 = u^2 + 2as$ again to find the height the ball reaches (s) after the bounce.

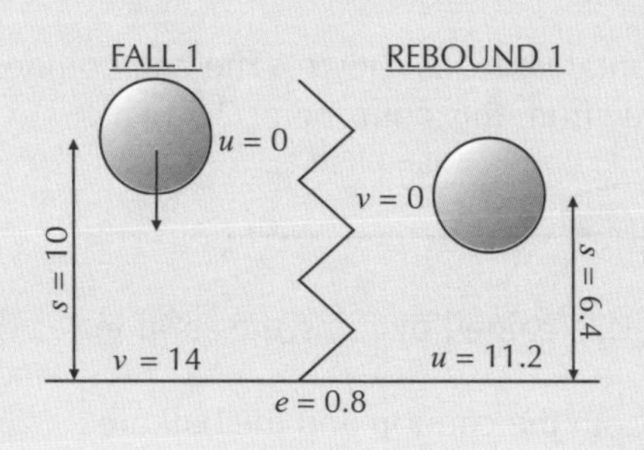

Set v to zero here because we want to know how far it will go upwards (with an acceleration of $-g$) before it stops and falls back down.

BOUNCE 1

Falling: (Taking 'down' as positive.)
$v^2 = u^2 + 2as$ where $u = 0$, $a = 9.8$ and $s = 10$:
$v^2 = 0 + (2 \times 9.8 \times 10) \quad \Rightarrow \quad v = \sqrt{2 \times 9.8 \times 10} = 14 \text{ ms}^{-1}$.

Colliding:
$e = \frac{v}{u} \quad \Rightarrow \quad v = eu$, where v is the velocity just after the impact, $e = 0.8$ and u is the velocity just before the impact (i.e. 14 ms^{-1} as found above).
$\Rightarrow v = 0.8 \times 14 = 11.2 \text{ ms}^{-1}$ (upwards).

Rebounding: (Taking 'up' as positive.)
$v^2 = u^2 + 2as$, where $v = 0$, $a = -9.8$, and u is the velocity just after the impact (i.e. 11.2 ms^{-1} as found above).
$0 = 11.2^2 + (2 \times -9.8)s \Rightarrow$ height of first rebound $= s = \frac{11.2^2}{2 \times 9.8} = \underline{6.4 \text{ m}}$.

BOUNCE 2

Falling: (Taking 'down' as positive.)
The motion as the ball rises then falls is symmetrical — it covers the same distance under the same acceleration on the 2nd fall as it did on the 1st rebound. So it hits the floor the second time with the same speed it left it at, 11.2 ms^{-1}.

If you're not convinced, put the numbers in the equation of motion again to see for yourself.

Colliding:
Again, $v = eu$, where v is the velocity after the 2nd impact, $e = 0.8$ and $u = 11.2 \text{ ms}^{-1}$ (velocity just before impact). $v = 0.8 \times 11.2 = 8.96 \text{ ms}^{-1}$ (upwards).

Rebounding: (Taking 'up' as positive.)
$v^2 = u^2 + 2as$, where $v = 0$, $u = 8.96$ (velocity after impact) and $a = -9.8$:
$0 = 8.96^2 + (2 \times -9.8)s \Rightarrow$ height of second rebound $= s = \frac{8.96^2}{2 \times 9.8} = \underline{4.10 \text{ m}}$ (3 s.f.)

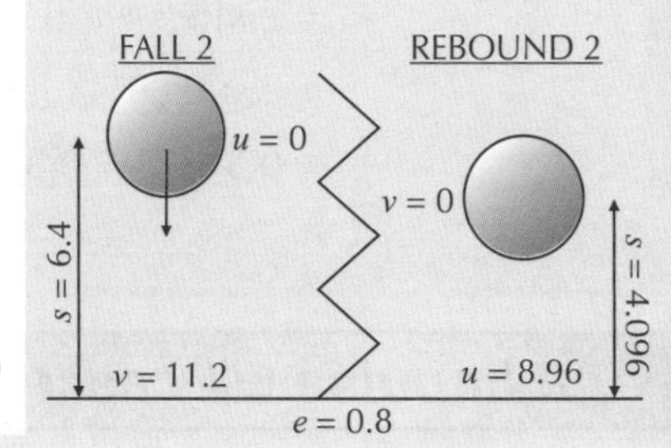

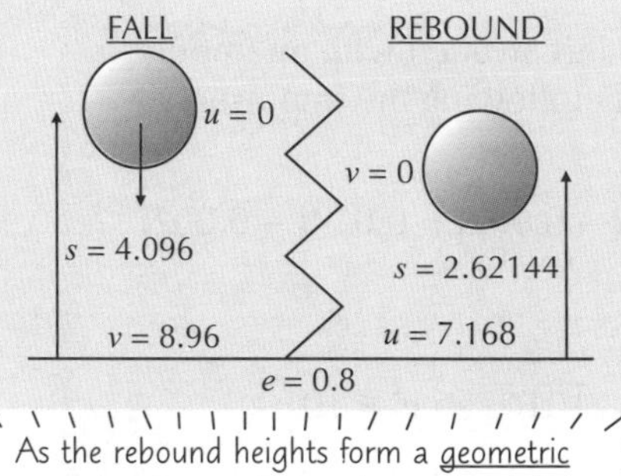

As the rebound heights form a geometric progression, you could use the sum to infinity formula (see C2) to work out the total distance the bouncing particle will travel before stopping.

BOUNCE 3

Falling: (Taking 'down' as positive.)
Using the symmetry of the vertical motion, velocity just before 3rd impact = velocity after 2nd impact = 8.96 ms^{-1}.

Colliding:
$v = eu$, where v is the velocity after the 3rd impact, $e = 0.8$ and $u = 8.96 \text{ ms}^{-1}$.
$v = 0.8 \times 8.96 = 7.168 \text{ ms}^{-1}$ (upwards).

Rebounding: (Taking 'up' as positive.)
$v^2 = u^2 + 2as$, where $v = 0$, $u = 7.168$ and $a = -9.8$:
$0 = 7.168^2 + (2 \times -9.8)s \Rightarrow$ height of third rebound $= s = \frac{7.168^2}{2 \times 9.8} = \underline{2.62 \text{ m}}$ (3 s.f.)

Bouncin's what particles do best...

That's as complex as it gets — just break it down into steps, bounce by bounce. It's a bit like those dance mat games, except less sweaty. And I doubt you'll find Madonna doing M2 maths in one of her videos, dressed in a neon pink leotard...

Collisions and Energy

This page is for Edexcel M2 and OCR MEI M2

Almost the end of the section, and I guess your energy might be waning. Most things lose kinetic energy when they collide — you need to know how to work out how much. It's enough to make you want a quiet lie down...

Kinetic Energy is only Conserved in Perfectly Elastic Collisions

For any collision where $e < 1$, some kinetic energy will be lost (it changes into things like heat and sound). The formula for working out how much has been lost is fairly straightforward:

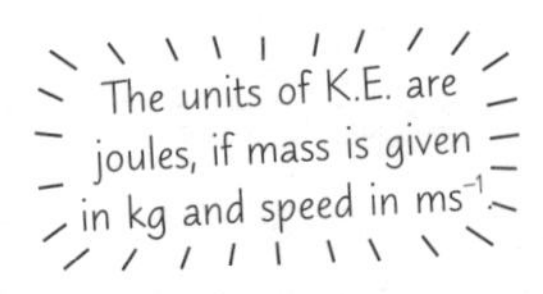

Loss of K.E. on Impact	$= \text{Total K.E. before} - \text{Total K.E. after}$ $= (\frac{1}{2}m_1u_1^2 + \frac{1}{2}m_2u_2^2) - (\frac{1}{2}m_1v_1^2 + \frac{1}{2}m_2v_2^2)$

For velocities given in vector (**i** and **j**) form, find their magnitude (speed) to put into the K.E. formula.

The tricky bit is finding the u's and v's to put in the formula...

EXAMPLE A tiny cannon fires a ball in a straight line across a smooth horizontal table, as shown. The ball collides directly with another, stationary, ball with $e = 0.7$, and moves away from this collision at 7.5 ms^{-1}.

$m_c = 0.05$ kg $\quad m_1 = m_2 = 0.001$ kg

C 1 2

a) Find the loss of K.E. when the balls collide.

1) We first need to find u_1 (the speed of the fired ball before it hits the other) and v_2 (the final speed of the other ball). Use the law of restitution and conservation of momentum (as on p. 221) where $e = 0.7$, $v_1 = 7.5$, and $u_2 = 0$.

2) $e = \frac{v_2 - v_1}{u_1 - u_2} \Rightarrow 0.7 = \frac{v_2 - 7.5}{u_1 - 0} \Rightarrow v_2 - 0.7u_1 = 7.5$ (**eqn 1**).

   $m_1u_1 + m_2u_2 = m_1v_1 + m_2v_2$ and since $m_1 = m_2$, they cancel:
   $u_1 + 0 = 7.5 + v_2 \Rightarrow u_1 - v_2 = 7.5$ (**eqn 2**).

   Eqn 1 + eqn 2: $0.3u_1 = 15 \Rightarrow u_1 = 50$ ms^{-1}.

   Sub in **eqn 2**: $50 - v_2 = 7.5 \Rightarrow v_2 = 50 - 7.5 = 42.5$ ms^{-1}.

3) Finally, putting all the values in the K.E. formula:

$$\begin{aligned}\text{Loss of K.E.} &= (\tfrac{1}{2}m_1u_1^2 + \tfrac{1}{2}m_2u_2^2) - (\tfrac{1}{2}m_1v_1^2 + \tfrac{1}{2}m_2v_2^2)\\ &= \tfrac{1}{2}m[(u_1^2 + u_2^2) - (v_1^2 + v_2^2)]\\ &= \tfrac{1}{2} \times 0.001 \times [(50^2 + 0^2) - (7.5^2 + 42.5^2)]\\ &= 0.31875 = 0.319 \text{ J to 3 s.f.}\end{aligned}$$

b) Find the K.E. gained by firing the cannon.

1) Since both the cannon and the ball are stationary before firing, there is no initial K.E. The gain in K.E. is simply $\frac{1}{2}m_cv_c^2 + \frac{1}{2}m_1v_1^2$, where v_1 is the speed of the ball after firing, i.e. 50 ms^{-1}, as calculated in part a). You need to work out the velocity of the cannon (v_c) though.

2) Momentum is conserved so:
   $m_cu_c + m_1u_1 = m_cv_c + m_1v_1$
   $\Rightarrow 0 + 0 = 0.05v_c + (0.001 \times 50)$
   $\Rightarrow v_c = -(0.001 \times 50) \div 0.05 = -1$ ms^{-1}.
   (i.e. the cannon moves backwards at 1 ms^{-1}).

3) Gain in K.E. $= \frac{1}{2}m_cv_c^2 + \frac{1}{2}m_1v_1^2$
   $= (\frac{1}{2} \times 0.05 \times (-1)^2) + (\frac{1}{2} \times 0.001 \times 50^2)$
   $= 1.275 = 1.28$ J to 3 s.f.

An Impulse will cause a Change in K.E.

EXAMPLE A fly of mass 0.002 kg is moving at a velocity of $(2\mathbf{i} - \mathbf{j})$ ms^{-1} when it is swatted with an impulse of $(0.01\mathbf{i} - 0.06\mathbf{j})$ Ns. How much kinetic energy is gained by the fly following the impulse?

1) Using the impulse formula from p. 219: $I = m\mathbf{v} - m\mathbf{u}$, so
   $0.01\mathbf{i} - 0.06\mathbf{j} = 0.002\mathbf{v} - 0.002(2\mathbf{i} - \mathbf{j}) \Rightarrow 0.002\mathbf{v} = 0.01\mathbf{i} - 0.06\mathbf{j} + 0.004\mathbf{i} - 0.002\mathbf{j} = 0.014\mathbf{i} - 0.062\mathbf{j}$
   $\Rightarrow \mathbf{v} = (0.014\mathbf{i} - 0.062\mathbf{j}) \div 0.002 = (7\mathbf{i} - 31\mathbf{j})$ ms^{-1}.

2) The initial speed of the fly $|\mathbf{u}| = \sqrt{2^2 + 1^2} = \sqrt{5}$, so $u^2 = 5$.
   After the impulse this becomes $|\mathbf{v}| = \sqrt{7^2 + 31^2} = \sqrt{1010}$, so $v^2 = 1010$.

3) Increase in K.E. $= \frac{1}{2}mv^2 - \frac{1}{2}mu^2 = (\frac{1}{2} \times 0.002 \times 1010) - (\frac{1}{2} \times 0.002 \times 5) = 1.005$ J.

The formula's been tweaked to suit the situation — there's only one 'particle', and there will be an increase rather than a loss in K.E.

I'm not lazy — I'm just conserving my kinetic energy...

There are plenty of different situations where you could be asked to find a change in kinetic energy — but they all use pretty much the same formula, and no doubt require you to calculate some speeds. Just think it through logically to decide whether K.E. will go up or down or whatever. Now make yourself a quick bevvy and a light snack — it's practice time...

M2 Section 5 — Practice Questions

Well that's been a crash course in collisions (ho ho). Don't just sit and hope that you've understood it all — come and have a go. Have a practice lap first...

Warm-up Questions

Q1 is for OCR MEI M2 only

1) Each diagram represents the motion of two particles moving in a straight line.
Find the missing mass or velocity (all masses are in kg and all velocities are in ms^{-1}).

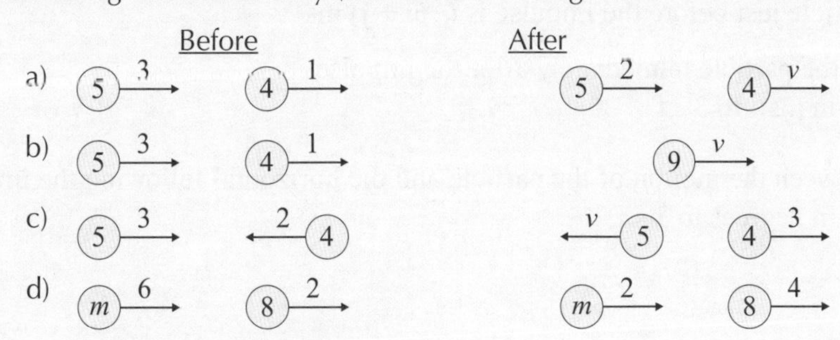

Q2-3 are for OCR M2 and OCR MEI M2 only

2) An impulse of 2 Ns acts against a ball of mass 300 g moving with a velocity of 5 ms^{-1}.
Find the ball's new velocity.

3) A particle of mass 450 g is dropped 2 m onto a floor. It rebounds to two thirds of its original height.
Find the impulse given to the particle by the ground.

Q4-7 are for Edexcel M2 and OCR MEI M2 only

4) Find the velocity of a particle of mass 0.1 kg, travelling at $(\mathbf{i} + \mathbf{j})$ ms^{-1}, after receiving an impulse of:
a) $2\mathbf{i} + 5\mathbf{j}$ Ns b) $-3\mathbf{i} + \mathbf{j}$ Ns c) $-\mathbf{i} - 6\mathbf{j}$ Ns d) $4\mathbf{i}$ Ns.

5) A 2 kg particle, travelling at $(4\mathbf{i} - \mathbf{j})$ ms^{-1}, receives an impulse, $\mathbf{Q}$, changing its velocity to $(-2\mathbf{i} + \mathbf{j})$ ms^{-1}.
Find:
a) $\mathbf{Q}$ b) $|\mathbf{Q}|$, in Ns to 3 s.f. c) the angle $\mathbf{Q}$ makes with $\mathbf{i}$, in degrees to 3 s.f.

6) Two particles *A* and *B* collide, where $m_A = 0.5$ kg and $m_B = 0.4$ kg.
Their initial velocities are $\mathbf{u}_A = (2\mathbf{i} + \mathbf{j})$ ms^{-1} and $\mathbf{u}_B = (-\mathbf{i} - 4\mathbf{j})$ ms^{-1}. Find, to 3 s.f.:
a) the speed of B after impact if A moves away from the collision at a velocity of $(-\mathbf{i} - 2\mathbf{j})$ ms^{-1},
b) their combined speed after the impact if they coalesce instead.

7) Find the loss in kinetic energy when a particle of mass 2 kg travelling at 3 ms^{-1} collides with a stationary particle of mass 3 kg on a smooth horizontal plane surface, where $e = 0.3$.

Q8-9 are for Edexcel M2, OCR M2 and OCR MEI M2 only

8) Two particles travelling directly towards each other at the same speed collide. The impact causes one particle to stop, and the other to go in the opposite direction at half its original speed. Find the value of *e*.

9) A particle of mass 1 kg travelling at 10 ms^{-1} on a horizontal plane has a collision, where $e = 0.4$.
Find the particle's rebound speed if it collides head-on with:
a) a smooth vertical wall, b) a particle of mass 2 kg travelling at 12 ms^{-1} towards it.

Q10 is for OCR MEI M2 only

10) A particle of mass 2 kg collides with a smooth horizontal surface at a velocity of $\mathbf{u} = 4\mathbf{i} - \mathbf{j}$ ms^{-1}, where $\mathbf{i}$ and $\mathbf{j}$ are the horizontal and vertical unit vectors respectively. The coefficient of restitution for the impact is $e = 0.5$. Find the velocity, $\mathbf{v}$, of the particle after the impact and the kinetic energy lost in the collision.

Q11-12 are for Edexcel M2 only

11) Particles *A* (mass 1 kg), *B* (4 kg) and *C* (5 kg) travel in the same line at speeds of $3u$, $2u$ and u, respectively.
If *A* collides with *B* first ($e = \frac{1}{4}$), then *B* with *C* ($e = \frac{1}{3}$), determine whether *A* and *B* will collide again.

12) A stationary particle drops vertically from a height of 1 m and rebounds from a smooth horizontal plane surface with $e = 0.5$. Find the height that it reaches after its first, second and third bounce.

M2 Section 5 — Practice Questions

Ready to notch it up a gear? Think you're the Stig of M2?
Well rev her up and let rip — just watch out for those hairpin bends.

Exam Questions

Q1 is for Edexcel M2 and OCR MEI M2

1 A particle of mass 0.4 kg receives an impulse of $(3\mathbf{i} - 8\mathbf{j})$ Ns.
The velocity of the particle just before the impulse is $(-6\mathbf{i} + \mathbf{j})$ ms^{-1}.

a) Find the speed of the particle immediately after the impulse.
Give your answer in ms^{-1} to 3 s.f.

(5 marks)

b) Find the angle between the motion of the particle and the horizontal following the impulse.
Give your answer in degrees to 3 s.f.

(2 marks)

Q2-4 are for Edexcel M2, OCR M2 [Except 4 b)] and OCR MEI M2

2 A marble of mass 0.02 kg, travelling at 2 ms^{-1}, collides with another, stationary, marble of mass 0.06 kg.
Both can be modelled as smooth spheres on a smooth horizontal plane.
If the collision is perfectly elastic, find the speed of each marble immediately after the collision.

(4 marks)

3 Particles P (of mass $2m$) and Q (of mass m), travelling in a straight line towards each other at the same speed (u) on a smooth horizontal plane surface, collide with a coefficient of restitution of $\frac{3}{4}$.

a) Show that the collision reverses the direction of both particles,
with Q having eight times the rebound speed of P.

(6 marks)

Following the collision, Q goes on to collide with a smooth vertical wall, perpendicular to its path.
The coefficient of restitution for the impact with the wall is e_{wall}.
Q goes on to collide with P again on the rebound from the wall.

b) Show that $e_{wall} > \frac{1}{8}$.

(3 marks)

c) Suppose that $e_{wall} = \frac{3}{5}$. If after the second collision with P, Q continues to move away from the wall, but with a speed of 0.22 ms^{-1}, find the value of u, the initial speed of both particles, in ms^{-1}.

(7 marks)

4 A particle of mass $2m$, travelling at a speed of $3u$ on a smooth horizontal plane, collides directly with a particle of mass $3m$ travelling at $2u$ in the same direction. The coefficient of restitution is $\frac{1}{4}$.

a) Find expressions for the speeds of both particles after the collision.
Give your answers in terms of u.

(4 marks)

b) Show that the amount of kinetic energy lost in the collision is $\frac{9mu^2}{16}$.

(4 marks)

M2 Section 5 — Practice Questions

Encore encore, more more more...

Q5 is for Edexcel M2 only

5 Particles A (mass m), B (mass $2m$) and C (mass $4m$) lie on a straight line, as shown:

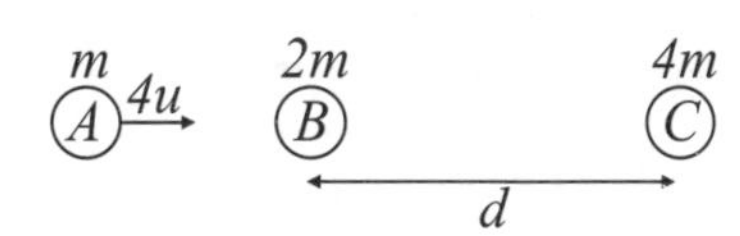

B and C are initially stationary when A collides with B at a speed of $4u$ ($u > 0$), causing B to collide with C. The coefficient of restitution between B and C is $2e$, where e is the coefficient of restitution between A and B.

a) Show that the collision between A and B does not reverse the direction of A. *(7 marks)*

By the time B and C collide, A has travelled a distance of $\frac{d}{4}$ since the first collision.

b) Show that $e = \frac{1}{3}$. *(3 marks)*

c) Hence find, in terms of u, the speed of C following its collision with B. *(5 marks)*

Q6-8 are for OCR MEI M2 only

6 Two particles of mass 0.8 kg and 1.2 kg are travelling in the same direction along a straight line with speeds of 4 ms^{-1} and 2 ms^{-1} respectively until they collide. After the collision the 0.8 kg mass has a velocity of 2.5 ms^{-1} in the same direction. The 1.2 kg mass then continues with its new velocity until it collides with a mass of m kg travelling with a speed of 4 ms^{-1} in the opposite direction to it.
Given that both particles are brought to rest by this collision, find the mass m. *(4 marks)*

7 A particle, A, of mass 7 kg is moving on a smooth horizontal plane with velocity $6\mathbf{i} - 4\mathbf{j}$ ms^{-1}.
Another particle, B, of mass 2 kg is at rest. B experiences a force of magnitude F N for 5 seconds, which causes it to begin moving with velocity $\mathbf{v}_B$ ms^{-1}.
B collides with A and the two particles coalesce to form a new particle, C, with velocity $3\mathbf{i} + 4\mathbf{j}$ ms^{-1}.
Find $\mathbf{v}_B$ and F. *(7 marks)*

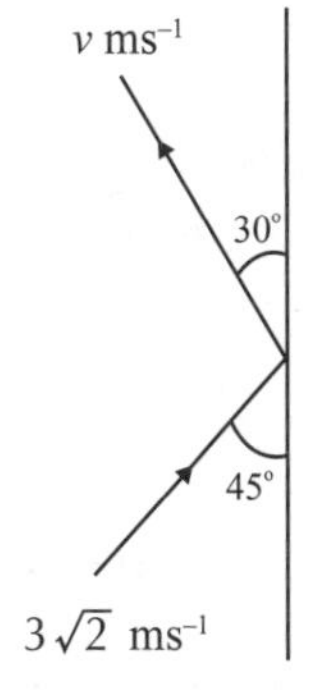

8 A particle of mass 1 kg bounces off a smooth vertical metal plate. It hits the plate at an angle of 45° to the vertical and with speed $u = 3\sqrt{2}$ ms^{-1}. It rebounds at an angle of 30°. By modelling the plate as a smooth vertical plane, find:

a) v, the speed of the particle after the collision, *(3 marks)*

b) the coefficient of restitution, e, for the collision, *(3 marks)*

c) the kinetic energy lost in the collision. *(2 marks)*

Circular Motion

This page is for AQA M2 and OCR M2

Things often move in circular paths — and it's a whole new world of motion to sink your teeth into.

Angular Speed is Measured in *Radians per Second*

1) You can measure the speed of a particle travelling in a circle in two different ways — linear speed and angular speed:

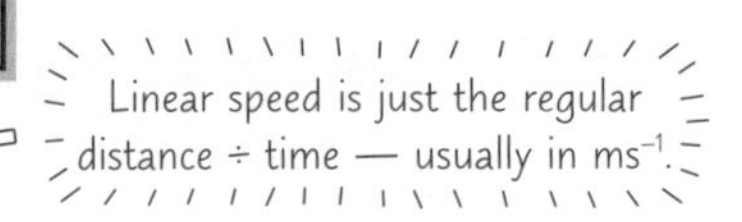

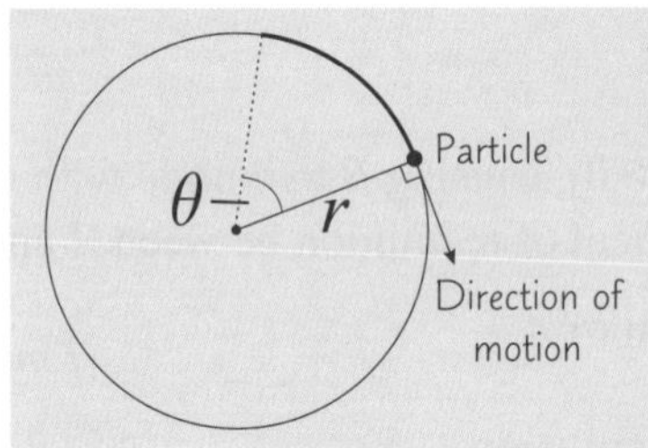

The angular speed, ω, is how quickly the radius, r, is turning — or the rate of change of θ. There's a formula for working it out:

$$\omega = \frac{\theta}{t}$$

Angular speed — radians per second.

Angle radius has moved through — in radians.

Time taken — in seconds.

You need to learn the angular speed formula. Make sure you do.

2) It's really important to measure the angle in RADIANS, or this lovely equation linking angular and linear speed won't work:

Linear speed = radius × angular speed
(ms^{-1}) (m) (radians s^{-1})

$$v = r\omega$$

3) You might have to convert from units such as revolutions per minute to radians per second, so learn these:

360° = 2π radians = 1 revolution

EXAMPLE A particle moves in a horizontal circle, completing 600 revolutions per minute. What is its angular speed?

Find θ: 600 revolutions = 600 × 2π radians = 1200π radians.

Now find ω: $\omega = \frac{\theta}{t} = \frac{1200\pi}{60} = 20\pi$ radians s^{-1}

The time must be in seconds.

Direction is Always *Changing,* so the *Velocity* is *Changing* too

1) The direction of something moving in a circle is always parallel to the tangent of the circle — so it's constantly changing. Velocity has magnitude and direction, so changing direction means changing velocity.

2) If something's velocity is changing, it must be accelerating. So even if a particle is moving in a circle with a constant speed, it will still be accelerating.

This is a strange one — take a moment to make sure you get it.

3) The acceleration is always directed towards the centre of the circle, perpendicular to the direction of motion. It's called radial acceleration and there's a couple of formulas to learn:

One using angular speed...

$$a = r\omega^2$$

...and one using linear speed.

$$a = \frac{v^2}{r}$$

4) There must be a force acting on the particle to produce the acceleration. And there is. It's called the centripetal force, it always acts towards the centre of the circle, and you just use the old $F = ma$ formula to find it.

EXAMPLE A particle moves with an angular speed of 20π rad s^{-1} around a horizontal circle of radius 0.25 m.
a) Find the magnitude of its acceleration.

$$a = r\omega^2 = 0.25 \times (20\pi)^2 = 100\pi^2 \text{ ms}^{-2}$$

In the case of a particle on a string, the centripetal force is provided by the tension in the string.

b) A light string connects the particle above to the centre of the circle. Find the tension in the string if the particle's mass is 3 kg.

Resolving horizontally:
$F = ma \Rightarrow T = mr\omega^2$
$= 3 \times 100\pi^2 = 300\pi^2$ N

I propose the motion that we stop going round in circles and move on...

You need to learn all the formulas on this page — you won't be given them in the exam. Remember that θ should be in radians when you're calculating ω and remember to use the radial acceleration in $F = ma$ to find the centripetal force.

Conical Pendulums

This page is for AQA M2 and OCR M2

Conical pendulum questions aren't as tricky as they look — they're usually just a case of resolving a force and then plonking things into the equations on the previous page.

Resolve Tension in a Conical Pendulum into Components

1) If you dangle an object at the end of a string, then twirl it round so the weight moves in a horizontal circle, you've made a conical pendulum.
2) There are only two forces acting on the object — its weight and the tension in the string.
3) The vertical component of the tension in the string supports the weight of the object, and the horizontal component is the centripetal force causing the radial acceleration.

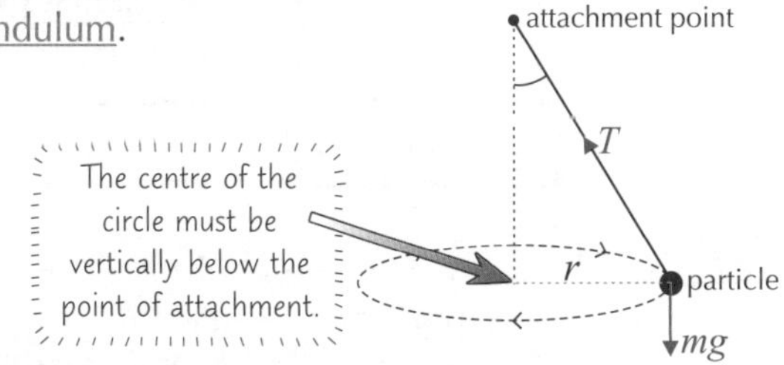

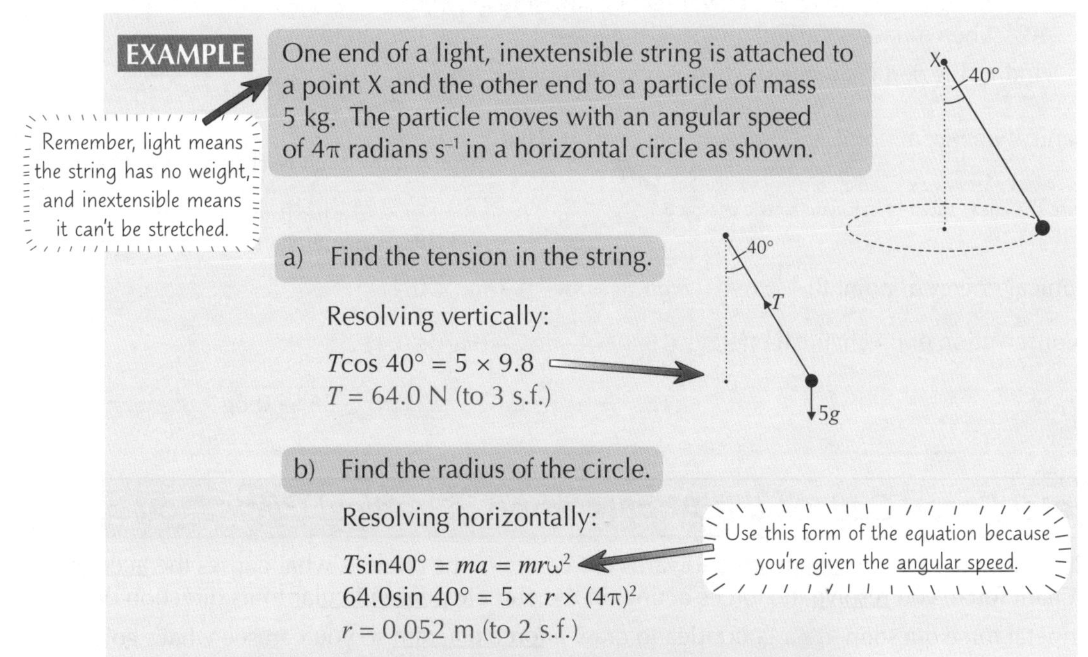

EXAMPLE One end of a light, inextensible string is attached to a point X and the other end to a particle of mass 5 kg. The particle moves with an angular speed of 4π radians s^{-1} in a horizontal circle as shown.

a) Find the tension in the string.

Resolving vertically:

$T\cos 40° = 5 \times 9.8$
$T = 64.0$ N (to 3 s.f.)

b) Find the radius of the circle.

Resolving horizontally:

$T\sin 40° = ma = mr\omega^2$
$64.0\sin 40° = 5 \times r \times (4\pi)^2$
$r = 0.052$ m (to 2 s.f.)

Don't be Confused by Slight Variations in Questions

The conical pendulums in most exam questions are just like those above.
However, watch out for special pendulums that they sometimes throw in to spice things up:

The single string through a ring:

The tension in both parts of the string is the same — and it's equal to the weight of the particle hanging vertically:

$T = m_1 g$

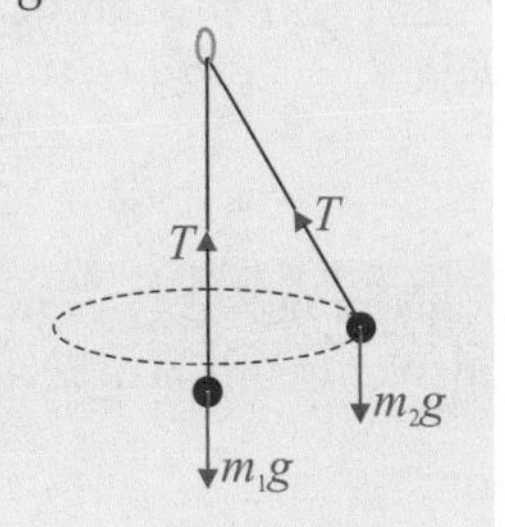

The two string pendulum:

Each string has a separate tension. You have to include the horizontal or vertical components of both strings when resolving forces,

e.g. Resolving vertically:
$T_1\cos\alpha + T_2\cos\beta = mg$

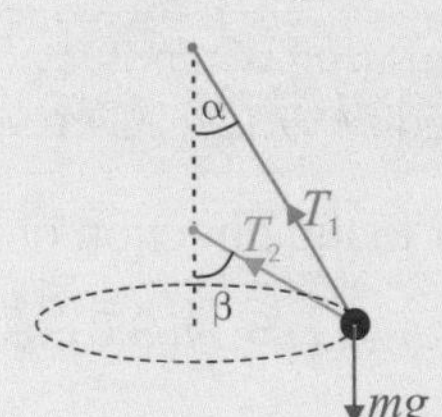

The second string might be horizontal — that makes life easier, cos there'll be no vertical component of tension. Brilliant.

A big deep breath... And... resolve...

You might have to use radians and degrees in the same question — so stay alert. You'll need angles in radians for finding the angular speed, but in degrees for finding the components of forces. Don't worry — just a few more pages to go and then you'll be entering the CGP holodeck. I've programmed it to simulate M2 questions so beautiful, you'll weep actual tears.

Vertical Circular Motion

This page is for AQA M2 only

So far you've only seen particles moving in horizontal circles — well, you need to know about vertical circles too. The thing with vertical circles is that the speed isn't the same all the way round. It increases on the way down and decreases on the way up. All this on top of the ever-changing acceleration due to the changing direction. AAAH.

Energy is Always Conserved in Vertical Circular Motion

The centripetal force always acts perpendicular to the direction of motion, so does no work in this direction. So, for something moving in a vertical circle, you can use conservation of mechanical energy — see page 207.

The sum of G.P.E. and Kinetic Energy is the same at any point on the circle.

EXAMPLE A particle of mass m kg is attached to a light, inextensible string of length 0.4 m. The other end of the string is attached to a fixed point, O, and the particle moves in a vertical circle. It has speed $v_A = 6$ ms^{-1} when it passes through point A, the lowest point on the circle. Find v_B, the particle's speed when it reaches point B, shown.

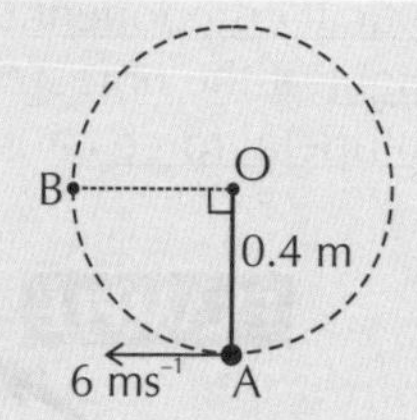

Total mechanical energy at point A: $\frac{1}{2}mv_A^2 = \frac{1}{2}m \times 6^2 = 18m$

A is the lowest point, so say that it has no GPE here.

Always use the linear speed to calculate kinetic energy.

Point B is higher, so it has both kinetic energy and GPE, relative to the 'base level'.

Total mechanical energy at point B: $\frac{1}{2}mv_B^2 + mgh = \frac{1}{2}mv_B^2 + (mg \times 0.4)$

So, by the conservation of mechanical energy: $18m = \frac{1}{2}mv_B^2 + (mg \times 0.4)$

$$18 = \frac{1}{2}v_B^2 + 0.4g \Rightarrow v_B = \sqrt{36 - 0.8g} \Rightarrow v_B = 5.31\text{ms}^{-1}$$

A Centripetal Force Causes the Acceleration in Vertical Circles too

As with any circular motion, the resultant force towards the centre of the circle is what causes the acceleration. To find the resultant force, you resolve the forces acting on the particle perpendicular to its direction of motion.

With any centripetal force question, it's a good idea to draw a force diagram so you can see what's going on.

EXAMPLE a) Find the tension in the string when the particle in the example above is at point A.

First draw a force diagram — there are only two forces acting on the particle, the tension in the string, T, and the particle's weight.

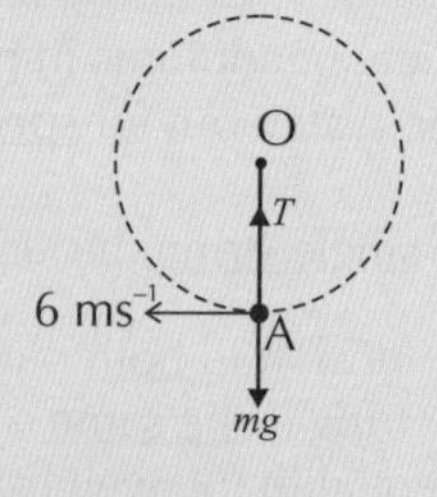

Now, resolving vertically:

$$T - mg = \frac{mv^2}{r} \Rightarrow T - mg = \frac{m \times 6^2}{0.4} \Rightarrow T = mg + \left(\frac{m \times 6^2}{0.4}\right) \Rightarrow T = 99.8m$$

$\frac{v^2}{r}$ is the formula for radial acceleration from page 230

b) Find the tension in the string when it is at an angle of 60° to OA, given that at this point, the particle has a speed of v ms^{-1}. Give your answer in terms of m and v.

Resolve perpendicular to the direction of motion:

$$T - mg\cos 60 = \frac{mv^2}{0.4} \Rightarrow T = m\left(9.8\cos 60 + \frac{v^2}{0.4}\right) \Rightarrow T = m(4.9 + 2.5v^2)$$

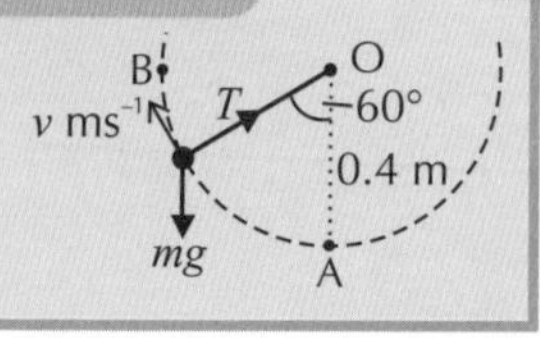

Centripetals — in the middle of the flower...

Remember that the speed of a particle moving in a vertical circle changes. Don't start assuming it's moving with constant speed. That would be BAD. Well, not as bad as a meteor crashing into Earth and causing a tsunami which wipes out the entire population, except for the US government, Paris Hilton and her pet chihuahua who all get to go in the secret bunker.

Vertical Circular Motion

This page is for AQA M2 only
Being attached to a string isn't the only way a particle can get to move in a circle...

A **Bead on a Wire** has a **Reaction Force**

1) Questions often involve a bead sliding around a smooth ring or circular wire.
2) There's no tension in strings here. Instead, the wire exerts a reaction force on the bead.

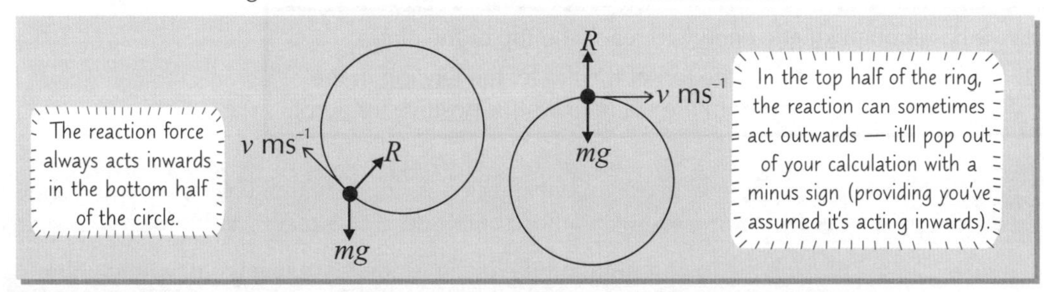

EXAMPLE A bead of mass m kg moves around a smooth ring of radius 3 m, as shown. If its speed at point A is $2v$ ms^{-1} and its speed at point B is v ms^{-1}, find the reaction of the ring on the bead at point B in terms of m and g.

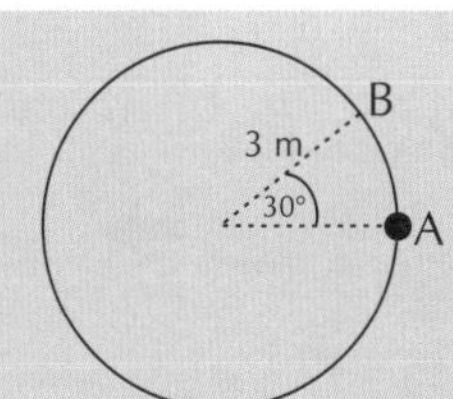

1) Draw a force diagram:

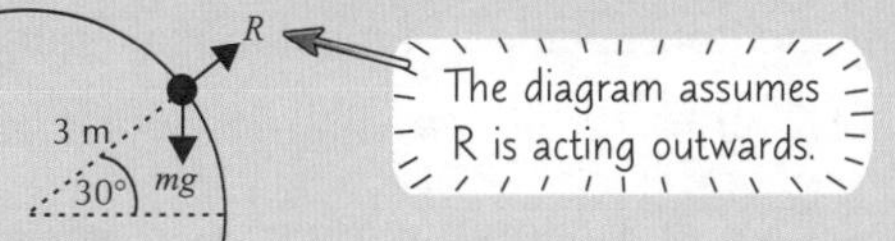

2) Resolve perpendicular to the direction of motion:

$$mg\sin 30° - R = \frac{mv^2}{r}$$

$$\Rightarrow R = \frac{mg}{2} - \frac{mv^2}{3}$$

You've got a v in your expression, but you need it in terms of m and g only.

3) Find v using conservation of mechanical energy:

$$\frac{1}{2}m(2v)^2 = \frac{1}{2}mv^2 + (mg \times 3\sin 30°)$$

$$4v^2 = v^2 + 3g \Rightarrow v^2 = g$$

Taking the level of A as the 'base level' for G.P.E.

4) You can now eliminate v from the equation for R:

$$R = \frac{mg}{2} - \frac{mg}{3} \Rightarrow R = \frac{mg}{6}$$

Particles can move on **Circular Surfaces**

1) A particle can move in a vertical circle on the inside of a horizontal cylinder.
2) You treat it just like the bead on a ring above — except that the reaction of the surface on the particle always acts inwards.
3) You can also have a particle rolling over the surface of a hemisphere (or over the top half of a horizontal cylinder).
4) Again, treat it just like the bead on a ring — but remember that the reaction of the surface on the particle always acts outwards.

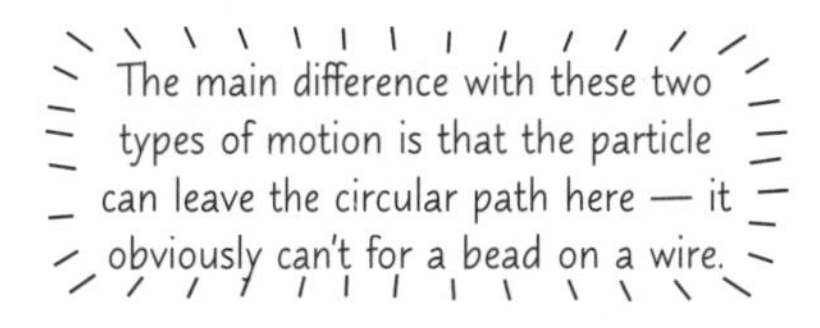

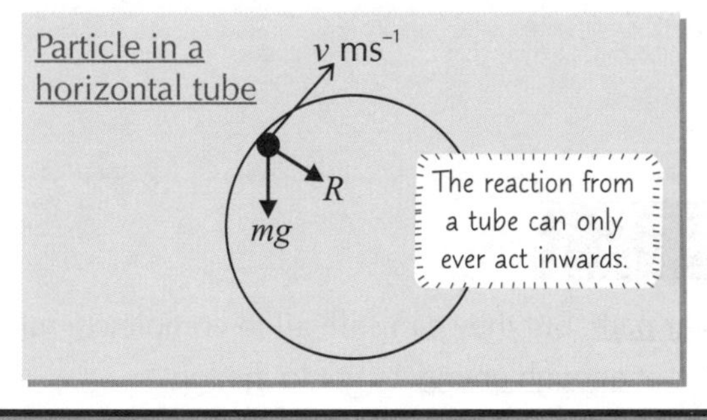

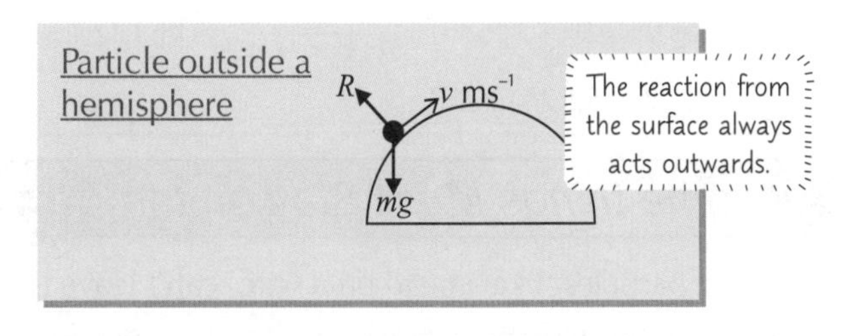

To fall or not to fall — Bard on a wire...

So particles inside cylinders are exactly like particles on strings, except the string tension is replaced by a reaction force. And they're in a cylinder. With no string in sight. Beads on wires are kind of like particles on strings. There's a reaction force instead of a tension, but this isn't always directed towards the centre of the circle. My mind is literally boggled.

More Vertical Circular Motion

This page is for AQA M2 only

As you'll know from swinging conkers round, they'll only do complete circles if you swing 'em hard enough.

Some things can Leave the Circular Path

Some things, like particles at the ends of strings, or particles on the surfaces of cylinders, will leave the circular path if their speed isn't sufficient. If such a particle is set into vertical circular motion, it'll only complete the circle if:

1) It has enough kinetic energy to reach the top of the circle.
2) When the particle is at the top of the circle, the tension in the string or the reaction from the cylinder wall is positive (or zero).

EXAMPLE A particle is attached to one end of a light, inextensible string of length 0.6 m, the other end of which is fixed. The particle is set into motion with a horizontal speed of 5 ms⁻¹. Will it form a complete circle?

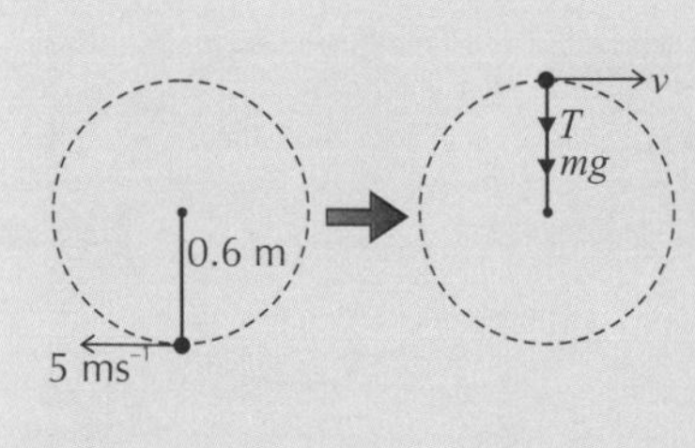

Taking the lowest point on the circle as the 'base level' (i.e. where G.P.E. = 0):

Energy of particle at lowest point: $\frac{1}{2}m \times 5^2 = 12.5m$

Total mechanical energy of particle at the top: $\frac{1}{2}mv^2 + mgh = \frac{1}{2}mv^2 + 1.2mg$

So, assuming that the particle completes a full circle, by the conservation of mechanical energy:

$\frac{1}{2}mv^2 + 1.2mg = 12.5m \Rightarrow v^2 = 25 - (2.4 \times 9.8)$

$\Rightarrow v = \sqrt{25 - (2.4 \times 9.8)} = \mathbf{1.217\ ms^{-1}}$

If something goes nutty crackers in here and you find yourself trying to take the square root of a negative number, then the particle doesn't have enough energy to complete the circle.

Resolve vertically: $T + mg = \frac{m(1.217^2)}{0.6} \Rightarrow T = m(2.47 - 9.8) \Rightarrow \mathbf{T = -7.33m}$

The tension is negative, so the particle doesn't make a complete circle.

You Can Find Where a Particle Leaves the Circle

At the point where a particle leaves the circular path, the tension in the string, or the reaction of the surface, is zero. So the component of the weight acting towards the centre of the circle is just enough to maintain the radial acceleration.

EXAMPLE A particle is set in motion at 3 ms⁻¹ vertically upwards from point A on the inside surface of a smooth pipe of radius 0.5 m. Find the value of θ (shown) when the particle leaves the pipe wall.

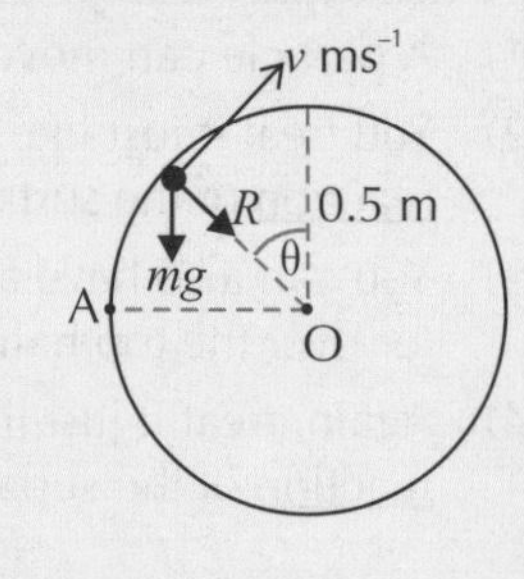

First find the speed of the particle when it leaves the wall. By the conservation of mechanical energy, taking the horizontal line through O as the 'base level':

$\frac{1}{2}m \times 3^2 = \frac{1}{2}mv^2 + (mg \times 0.5\cos\theta) \Rightarrow 4.5 = \frac{1}{2}v^2 + 0.5g\cos\theta \Rightarrow v^2 = 9 - g\cos\theta$

Leave it as v^2 — you need v^2 in the next step.

When the particle leaves the wall, $R = 0$. So, resolving perpendicular to the direction of motion, $mg\cos\theta = \frac{mv^2}{r}$:

$mg\cos\theta = \frac{m(9 - g\cos\theta)}{0.5} \Rightarrow 0.5g\cos\theta = 9 - g\cos\theta$

$1.5g\cos\theta = 9 \Rightarrow \cos\theta = \frac{9}{1.5 \times 9.8} = 0.6122...$

And so $\theta = 52.2°$ (3 s.f.)

But some things Must Stay on the Circular Path

1) Some particles, like a bead on a wire, can't leave the circular path, but they can still fail to complete a full circle.
2) For particles like these, you only need to check that they've got enough energy to get to the top. You don't need to worry whether any forces are positive or negative.

Me loop de loop didn't quite loop de loop...

So in a question about whether something will manage to complete the circle, always think about whether it can leave the circle or not. It makes a difference to what you have to show. Don't forget tension always acts towards the circle's centre.

M2 Section 6 — Practice Questions

When you design your own fairground, you'll need to know about circular motion — so it really is a vital life skill. It'll also be handy for the M2 exam, in which you're virtually guaranteed one, if not two, questions on it.

Warm-up Questions

Take $g = 9.8$ ms^{-2} in each of these questions.

Q1-3 are for AQA M2 and OCR M2

1) A particle attached to a light, inextensible string of length 3 m moves in a horizontal circle with constant speed. Leaving your answers in terms of π, find the particle's angular speed and its acceleration if:
   a) the particle takes 1.5 seconds to complete 1 revolution.
   b) the particle complete 15 revolutions in one minute.
   c) the string moves through 160° in one second.
   d) the linear speed of the particle is 10 ms^{-1}.

2) A particle with mass 2 kg moves in a horizontal circle of radius 0.4 m with constant speed. Find the centripetal force acting on the particle if:
   a) the particle's angular speed is 10π radians s^{-1}.
   b) the particle's linear speed is 4 ms^{-1}.

3) For the conical pendulum shown on the right, find:
   a) the tension in the string (which is light and inextensible).
   b) the radius of the circle described by the pendulum given that the particle has a speed of v ms^{-1}.

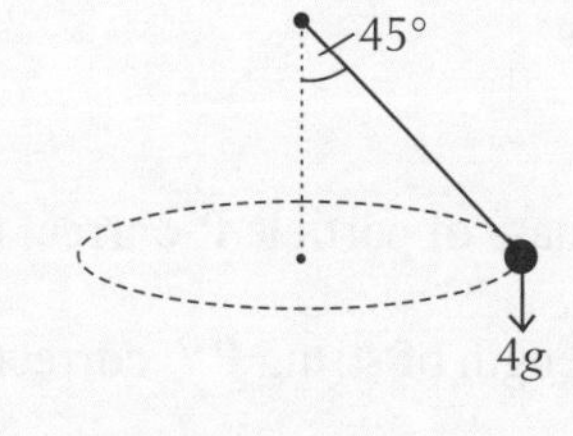

Q4-5 are for AQA M2 only

4) A particle of mass m kg is set into vertical circular motion inside a horizontal pipe from point A, vertically above the circle's centre. At point A it has a speed of 2.2 ms^{-1}.
   a) Draw a force diagram for the particle when it is at point B, shown in the diagram.
   b) Use conservation of mechanical energy to find v_B, the particle's speed at point B.
   c) Find, in terms of m, the normal reaction of the pipe on the particle when it is at point B.

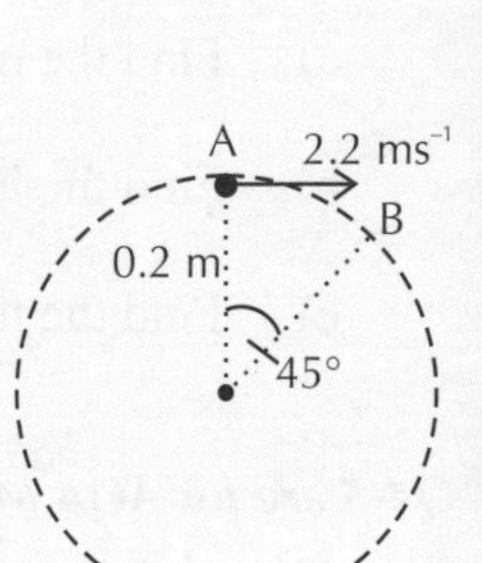

5) A particle is set into vertical circular motion with a horizontal speed of 9 ms^{-1} from the lowest point of a circle. The circular path has a 2 m radius.
   a) Show that the particle will complete the circle if it is a bead on a smooth wire but not if it's a bead at the end of a light inextensible string.
   b) Find the angle the string makes with the vertical when the bead leaves the circular path.

More on the merry-go-round of M2 maths magic...

Exam Questions

Whenever a numerical value of g is required in the following questions, take $g = 9.8$ ms^{-2}.

Q1 is for AQA M2 only

1 A 2 kg particle is attached to one end of a light, inextensible string of length 0.5 m. The other end of the string is attached to a fixed point O. The particle is released from rest at point A (shown in the diagram) and moves in a circular path in a vertical plane. Find, in terms of g:

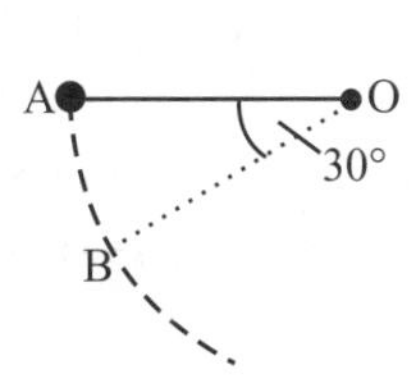

a) the particle's speed at point B, where $\angle OAB = 30°$,

(3 marks)

b) the tension in the string when the particle is at point B,

(3 marks)

c) the angular speed of the particle at point B.

(2 marks)

M2 Section 6 — Practice Questions

Q2-3 are for AQA M2 and OCR M2

2 A quad bike of mass 500 kg is travelling with constant speed around a horizontal circular track with a radius of 30 m. A frictional force acts towards the centre of the track, and there is no resistance to motion.
If the coefficient of friction between the quad bike and the track is 0.5, find the greatest speed the quad bike can go round the track without slipping. Give your answer to 3 significant figures.

(5 marks)

3 Particle P is attached to two light inextensible strings. The other end of each string is attached to a vertical rod XY, as shown in the diagram. The tension in string PX is 55 N, and the tension in string PY is 80 N. The particle moves in a horizontal circle about Y with a constant speed of 3 ms^{-1}.

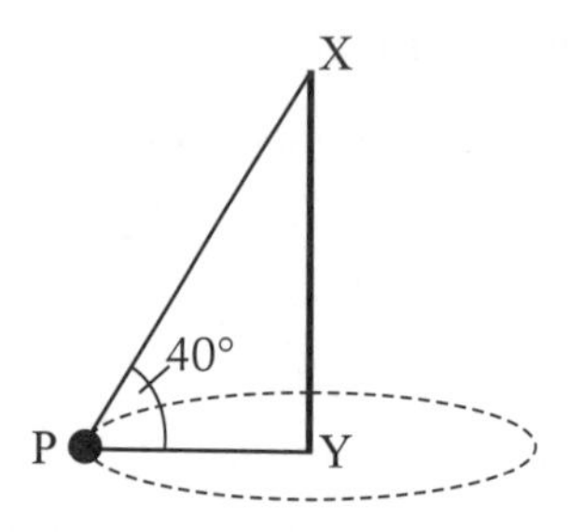

a) Find the mass of particle P, correct to three significant figures. *(2 marks)*

b) Find the length of string PY, correct to three significant figures. *(3 marks)*

c) Find the number of revolutions the particle will make in one minute. *(3 marks)*

Q4-5 are for AQA M2 only

4 A bead with mass m moves in vertical circular motion about O on a smooth ring with radius 1 m. At point A, shown in the diagram, the bead has a speed of $\sqrt{20}$ ms^{-1}.

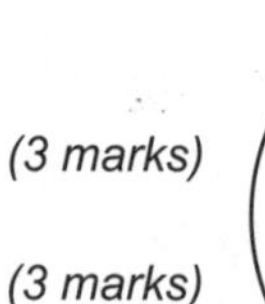

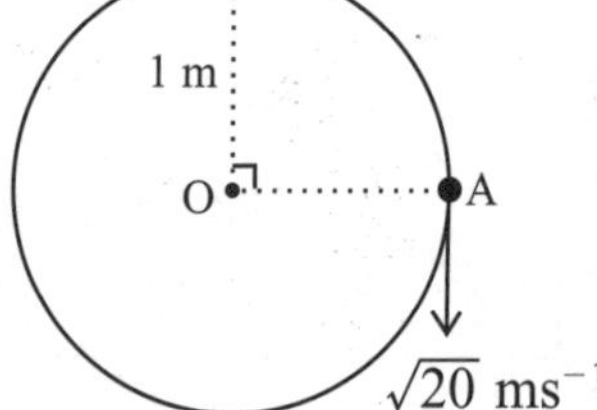

a) Show that the bead's speed at the highest point of the circle, B, is $\sqrt{0.4}$ ms^{-1}. *(3 marks)*

b) Find the reaction of the ring on the bead at point B in terms of m. *(3 marks)*

c) Find, to three significant figures, the minimum speed at point A that will allow a complete circle to be made. *(3 marks)*

5 A particle at the end of a light, inextensible string of length 5 m is set into vertical circular motion about a point O from position J, the lowest point on the circle. The particle's initial speed is 15 ms^{-1}.
When the particle reaches position K, it has speed v ms^{-1} and the string is at an angle of θ to the vertical.

K
θ
O
5 m
15 ms^{-1}
J

a) Show that $v^2 = 225 - 10g(1 + \cos\theta)$. *(4 marks)*

b) Show that the particle will not complete the circle. *(5 marks)*

c) Find the value of θ when the string first becomes slack. *(5 marks)*

Answers

Core Section 1 — Functions

Warm-up Questions

1) a) Range $f(x) \geq -16$. This is a function, and it's one-to-one (the domain is restricted so every x-value is mapped to only one value of $f(x)$).

b) To find the range of this function, you need to find the minimum point of $x^2 - 7x + 10$ — do this by completing the square: $x^2 - 7x + 10 = (x - 3.5)^2 - 12.25 + 10$
$= (x - 3.5)^2 - 2.25$.
As $(x - 3.5)^2 \geq 0$ the minimum value of $x^2 - 7x + 10$ is -2.25, so the range is $f(x) \geq -2.25$.
This is a function, and it's many-to-one (as more than one x-value is mapped to the same value of $f(x)$).

You could also have found the minimum point by differentiating, setting the derivative equal to 0 and solving for x.

c) Range $f(x) \geq 0$. It's not a function as $f(x)$ doesn't exist for $x < 0$.

d) Sketch the graph for this one:

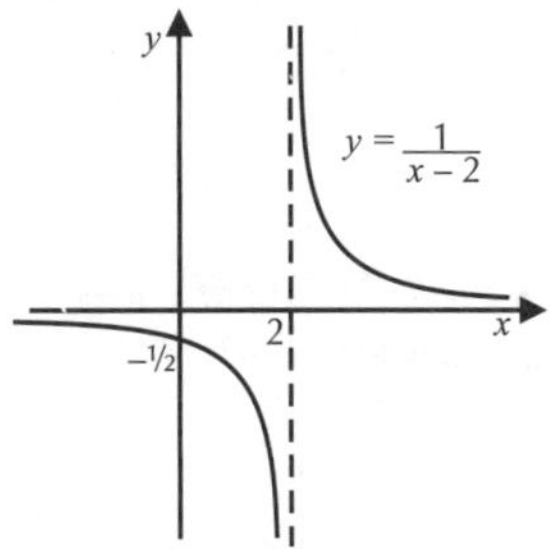

From the graph, the range is $f(x) \in \mathbb{R}$, $f(x) \neq 0$.
This is not a function as it's not defined for $x = 2$.

2) a) $fg(2) = f(2(2) + 3) = f(7) = 3/7$.
$gf(1) = g(3/1) = g(3) = 2(3) + 3 = 9$.
$fg(x) = f(2x + 3) = \frac{3}{2x + 3}$.

b) $fg(2) = f(2 + 4) = f(6) = 3(6^2) = 3 \times 36 = 108$.
$gf(1) = g(3(1^2)) = g(3) = 3 + 4 = 7$.
$fg(x) = f(x + 4) = 3(x + 4)^2$.

3) f is a one-to-one function so it has an inverse. The domain of the inverse is the range of the function and vice versa, so the domain of $f^{-1}(x)$ is $x \geq 3$ and the range is $f^{-1}(x) \in \mathbb{R}$.

4) Let $y = f(x)$. Then $y = \sqrt{2x - 4}$

$$y^2 = 2x - 4$$
$$y^2 + 4 = 2x$$
$$x = \frac{y^2 + 4}{2} = \frac{y^2}{2} + 2$$

Writing in terms of x and $f^{-1}(x)$ gives the inverse function as $f^{-1}(x) = \frac{x^2}{2} + 2$, which has domain $x \geq 0$ (as the range of f is $x \geq 0$) and range $f^{-1}(x) \geq 2$.

5) a)

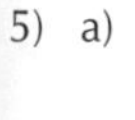

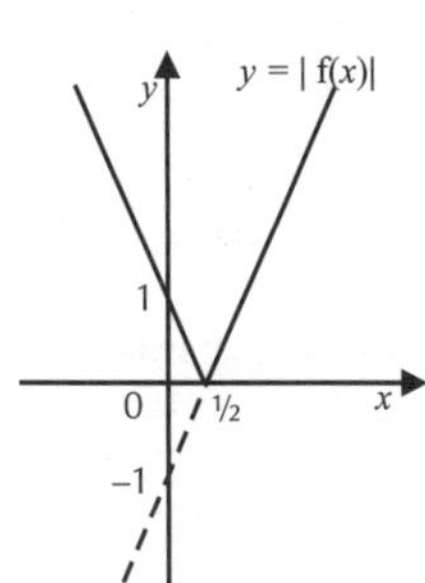

b)

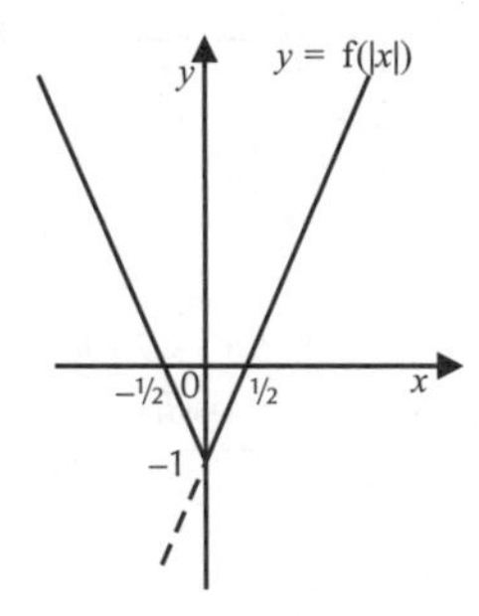

6)

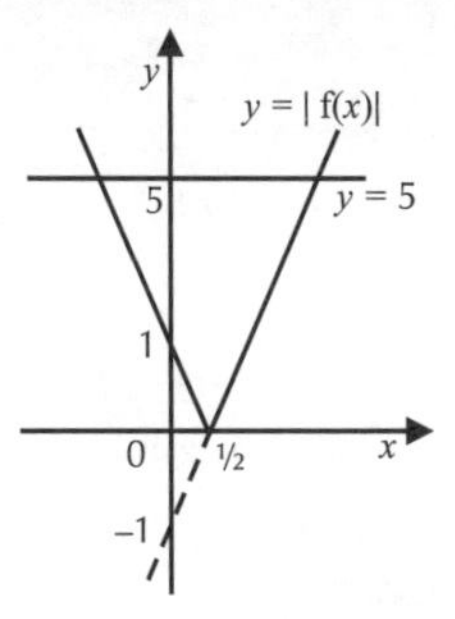

From the graph, $|2x - 1| = 5$ has 2 solutions, one where $2x - 1 = 5$ (so $x = 3$) and one where $-(2x - 1) = 5$ (so $x = -2$).

7) First, square both sides: $(2x + 1)^2 = (x + 4)^2$

$$4x^2 + 4x + 1 = x^2 + 8x + 16$$
$$3x^2 - 4x - 15 = 0$$
$$(3x + 5)(x - 3) = 0.$$

So $x = -\frac{5}{3}$ or $x = 3$.

You could have solved this one by sketching the graphs instead, then solving the equations 2x + 1 = x + 4 and −(2x + 1) = x + 4.

8)

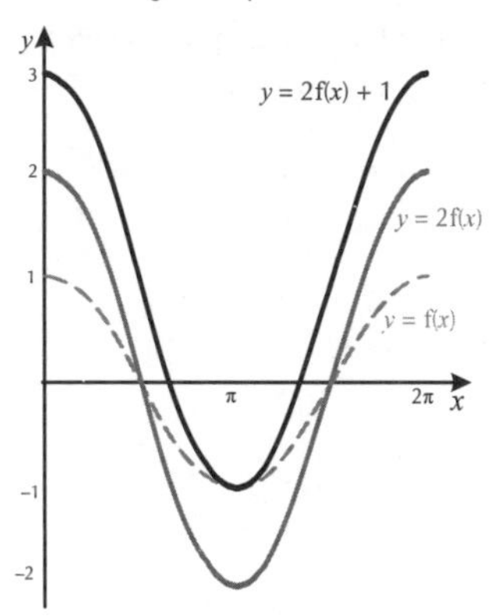

9) a) $y = x^2$ is even (but not periodic).

b) $y = \tan x$ is odd and periodic (with period π).

c) $y = x^3$ is odd (but not periodic).

Exam Questions

1 a)

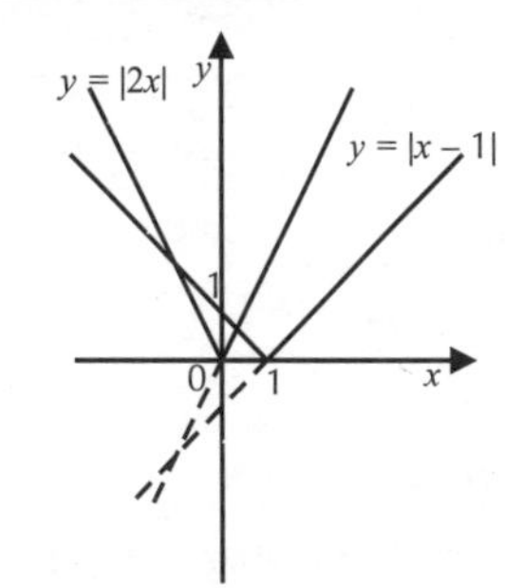

[3 marks available — 1 mark for y = |2x| (touching axes at (0, 0)), 2 marks for y = |x − 1| (1 mark for reflection in x-axis for x < 1, 1 mark for crossing y-axis at y = 1)]

Answers

b) $(2x)^2 = (x - 1)^2$
$4x^2 = x^2 - 2x + 1$ ***[1 mark]***
$3x^2 + 2x - 1 = 0$ ***[1 mark]***
$(3x - 1)(x + 1) = 0$
So $x = \frac{1}{3}$ or $x = -1$ ***[1 mark]***.
You could also have solved the equations $-(2x) = -(x - 1)$ and $2x = -(x - 1)$ — you'd have got the same solutions.

c) $|2x| \leq |x - 1|$ when $y = |2x|$ is equal to or underneath $y = |x - 1|$ ***[1 mark]***. From the graph, this occurs between the solutions in b), i.e. $-1 \leq x \leq \frac{1}{3}$ ***[1 mark]***.

2 To transform the curve $y = x^3$ into $y = (x - 1)^3$, move it 1 unit ***[1 mark]*** horizontally to the right ***[1 mark]***. To transform this into the curve $y = 2(x - 1)^3$, stretch it vertically ***[1 mark]*** by a scale factor of 2 ***[1 mark]***. Finally, to transform into the curve $y = 2(x - 1)^3 + 4$, the whole curve is moved 4 units ***[1 mark]*** upwards ***[1 mark]***.

3 a) $gf(x) = g(x^2 - 3)$ ***[1 mark]*** $= \frac{1}{x^2 - 3}$ ***[1 mark]***

b) $\frac{1}{x^2 - 3} = \frac{1}{6} \Rightarrow x^2 - 3 = 6 \Rightarrow x^2 = 9 \Rightarrow x = 3, x = -3$
[3 marks available — 1 mark for rearranging to solve equation, 1 mark for each correct solution]

4 a) $fg(6) = f(\sqrt{(3 \times 6) - 2}) = f(\sqrt{16})$ ***[1 mark]***
$= f(4) = 2^4 = 16$ ***[1 mark]***

b) $gf(2) = g(2^2) = g(4)$ ***[1 mark]***
$= \sqrt{(3 \times 4) - 2} = \sqrt{10}$ ***[1 mark]***

c) (i) First, write $y = g(x)$ and rearrange to make x the subject:
$y = \sqrt{3x - 2}$
$\Rightarrow y^2 = 3x - 2$
$\Rightarrow y^2 + 2 = 3x$
$\Rightarrow \frac{y^2 + 2}{3} = x$ ***[1 mark]***
Then replace x with $g^{-1}(x)$ and y with x: $g^{-1}(x) = \frac{x^2 + 2}{3}$ ***[1 mark]***.

(ii) $fg^{-1}(x) = f\left(\frac{x^2 + 2}{3}\right)$ ***[1 mark]***
$= 2^{\frac{x^2+2}{3}}$ ***[1 mark]***

5 a) The range of f is $f(x) > 0$ ***[1 mark]***.

b) (i) Let $y = f(x)$. Then $y = \frac{1}{x + 5}$.
Rearrange this to make x the subject:
$y(x + 5) = 1$
$\Rightarrow x + 5 = \frac{1}{y}$ ***[1 mark]***
$\Rightarrow x = \frac{1}{y} - 5$ ***[1 mark]***
Finally, write out in terms of x and $f^{-1}(x)$:
$f^{-1}(x) = \frac{1}{x} - 5$ ***[1 mark]***.

(ii) The domain of the inverse is the same as the range of the function, so $x > 0$ ***[1 mark]***. The range of the inverse is the same as the domain of the function, so $f^{-1}(x) > -5$ ***[1 mark]***.

c)
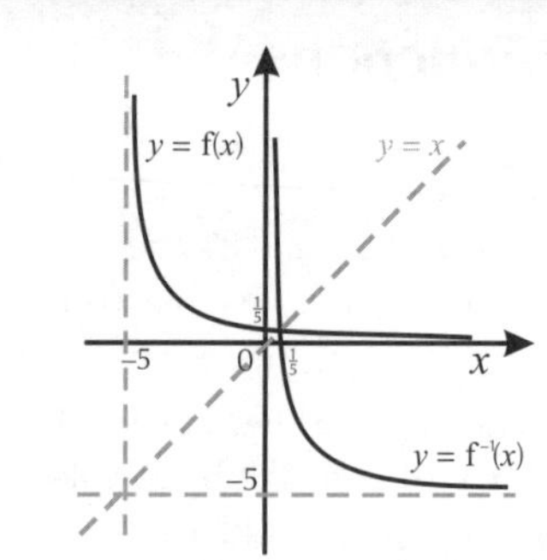

[2 marks available — 1 mark for each correct curve, each with correct intersections and asymptotes as shown]

6 a)
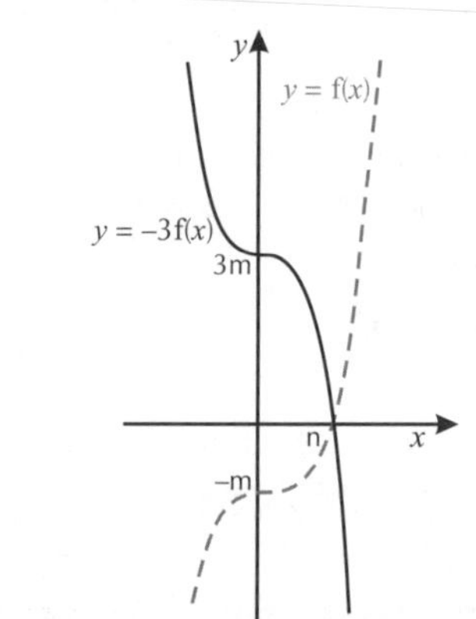

[2 marks available — 1 mark for reflecting in x-axis at x = n, 1 mark for crossing y-axis at y = m]

b)

[2 marks available — 1 mark for reflecting in y-axis and 1 mark for crossing y-axis at y = 3m (due to stretch by scale factor 3)]

c)
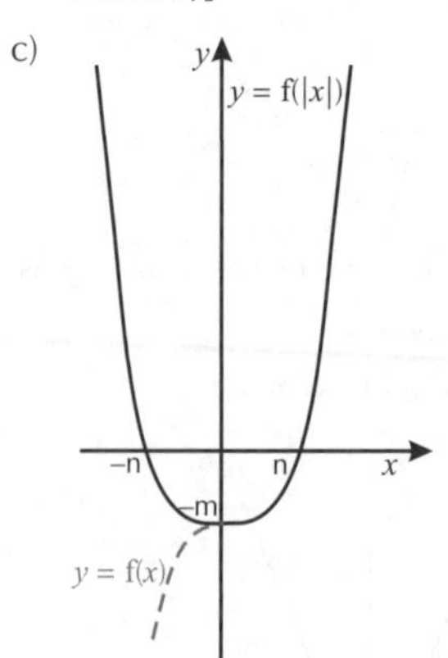

[2 marks available — 1 mark for reflecting in y-axis and 1 mark for crossing the x-axis at −n]

Answers

Core Section 2 — Algebra

Warm-up Questions

1) a) $\frac{4x^2-25}{6x-15}=\frac{(2x+5)(2x-5)}{3(2x-5)}=\frac{2x+5}{3}$

b) $\frac{2x+3}{x-2}\times\frac{4x-8}{2x^2-3x-9}$

$=\frac{2x+3}{x-2}\times\frac{4(x-2)}{(2x+3)(x-3)}$

$=\frac{4}{x-3}$

c) $\frac{x^2-3x}{x+1}\div\frac{x}{2}=\frac{x(x-3)}{x+1}\times\frac{2}{x}$

$=\frac{x-3}{x+1}\times 2=\frac{2(x-3)}{x+1}$

2) a) $\frac{x}{2x+1}+\frac{3}{x^2}+\frac{1}{x}=\frac{x\cdot x^2}{x^2(2x+1)}+\frac{3(2x+1)}{x^2(2x+1)}+\frac{x(2x+1)}{x^2(2x+1)}$

$=\frac{x^3+6x+3+2x^2+x}{x^2(2x+1)}=\frac{x^3+2x^2+7x+3}{x^2(2x+1)}$

b) $\frac{2}{x^2-1}-\frac{3x}{x-1}+\frac{x}{x+1}$

$=\frac{2}{(x+1)(x-1)}-\frac{3x(x+1)}{(x+1)(x-1)}+\frac{x(x-1)}{(x+1)(x-1)}$

$=\frac{2-3x^2-3x+x^2-x}{(x+1)(x-1)}=\frac{2-2x^2-4x}{(x+1)(x-1)}$

$=\frac{2(1-x^2-2x)}{(x+1)(x-1)}$

3)

$$\begin{array}{r l}
 & x^2-2x+7\ \text{r}-9 \\
x+4 & \overline{)\,x^3+2x^2-x+19} \\
- & \underline{x^3+4x^2} \\
 & -2x^2-x \\
- & \underline{-2x^2-8x} \\
 & 7x+19 \\
- & \underline{7x+28} \\
 & -9
\end{array}$$

so $(x^3+2x^2-x+19)\div(x+4)=x^2-2x+7$ remainder -9.

For Q4 and 5, you can use the substitution method or the equating coefficients method. I've just shown one method for each.

4) a) $\frac{4x+5}{(x+4)(2x-3)}\equiv\frac{A}{(x+4)}+\frac{B}{(2x-3)}$

$4x+5\equiv A(2x-3)+B(x+4)$

Using substitution method:

substitute $x=-4$: $-11=-11A \Rightarrow A=1$

substitute $x=1.5$: $11=5.5B \Rightarrow B=2$

$\frac{4x+5}{(x+4)(2x-3)}\equiv\frac{1}{(x+4)}+\frac{2}{(2x-3)}$

b) $\frac{-7x-7}{(3x+1)(x-2)}\equiv\frac{A}{(3x+1)}+\frac{B}{(x-2)}$

$-7x-7\equiv A(x-2)+B(3x+1)$

Using equating coefficients method:

coefficients of x: $-7=A+3B$

constants: $-7=-2A+B$

Solving simultaneously: $A=2$, $B=-3$

$\frac{-7x-7}{(3x+1)(x-2)}\equiv\frac{2}{(3x+1)}-\frac{3}{(x-2)}$

c) $\frac{x-18}{(x+4)(3x-4)}\equiv\frac{A}{(x+4)}+\frac{B}{(3x-4)}$

$x-18\equiv A(3x-4)+B(x+4)$

Using substitution method:

substitute $x=-4$: $-22=-16A \Rightarrow A=\frac{11}{8}$.

And using equating coefficients method:

coefficients of x: $1=3A+B$

Substituting $A=\frac{11}{8}$: $1=3A+B \Rightarrow B=1-\frac{33}{8}=-\frac{25}{8}$.

$\frac{x-18}{(x+2)(3x-4)}\equiv\frac{11}{8(x+2)}-\frac{25}{8(3x-4)}$

Don't worry if you get fractions for your coefficients — just put the numerator on the top and the denominator on the bottom.

d) Factorise the denominator:

$\frac{6+4y}{9-y^2}\equiv\frac{6+4y}{(3-y)(3+y)}\equiv\frac{A}{(3-y)}+\frac{B}{(3+y)}$

$6+4y\equiv A(3+y)+B(3-y)$

Using substitution method:

substitute $y=3$: $18=6A \Rightarrow A=3$

substitute $y=-3$: $-6=6B \Rightarrow B=-1$

$\frac{6+4y}{9-y^2}\equiv\frac{3}{(3-y)}-\frac{1}{(3+y)}$

e) $\frac{-11x^2+6x+11}{(2x+1)(3-x)(x+2)}\equiv\frac{A}{(2x+1)}+\frac{B}{(3-x)}+\frac{C}{(x+2)}$

$-11x^2+6x+11$

$\equiv A(3-x)(x+2)+B(2x+1)(x+2)+C(2x+1)(3-x)$

Using substitution method:

substitute $x=3$: $-70=35B \Rightarrow B=-2$

substitute $x=-2$: $-45=-15C \Rightarrow C=3$

substitute $x=-0.5$: $5.25=5.25A \Rightarrow A=1$

$\frac{-11x^2+6x+11}{(2x+1)(3-x)(x+2)}\equiv\frac{1}{(2x+1)}-\frac{2}{(3-x)}+\frac{3}{(x+2)}$

f) $\frac{6x^2+17x+5}{x(x+2)^2}\equiv\frac{A}{x}+\frac{B}{(x+2)}+\frac{C}{(x+2)^2}$

$6x^2+17x+5\equiv A(x+2)^2+Bx(x+2)+Cx$

substitute $x=-2$: $-5=-2C \Rightarrow C=\frac{5}{2}$

substitute $x=0$: $5=4A \Rightarrow A=\frac{5}{4}$

coefficients of x^2: $6=A+B$

substitute $A=\frac{5}{4}$: $6=\frac{5}{4}+B \Rightarrow B=\frac{19}{4}$

$\frac{6x^2+19x+8}{x(x+2)^2}\equiv\frac{5}{4x}+\frac{19}{4(x+2)}+\frac{5}{2(x+2)^2}$

g) $\frac{-18x+14}{(2x-1)^2(x+2)}\equiv\frac{A}{(2x-1)}+\frac{B}{(2x-1)^2}+\frac{C}{(x+2)}$

$-18x+14\equiv A(2x-1)(x+2)+B(x+2)+C(2x-1)^2$

substitute $x=-2$: $50=25C \Rightarrow C=2$

substitute $x=0.5$: $5=2.5B \Rightarrow B=2$

coefficients of x^2: $0=2A+4C$

substitute $C=2$: $0=2A+8 \Rightarrow A=-4$

$\frac{-18x+14}{(2x-1)^2(x+2)}\equiv\frac{-4}{(2x-1)}+\frac{2}{(2x-1)^2}+\frac{2}{(x+2)}$

h) Factorise the denominator:

$\frac{8x^2-x-5}{x^3-x^2}\equiv\frac{8x^2-x-5}{x^2(x-1)}\equiv\frac{A}{x}+\frac{B}{x^2}+\frac{C}{(x-1)}$

$8x^2-x-5\equiv Ax(x-1)+B(x-1)+Cx^2$

coefficients of x^2: $8=A+C$ (eq. 1)

coefficients of x: $-1=-A+B$ (eq. 2)

constants: $-5=-B \Rightarrow B=5$

Answers

substitute $B = 5$ in eq. 2: $-1 = -A + 5 \Rightarrow A = 6$

substitute $A = 6$ in eq. 1: $8 = 6 + C \Rightarrow C = 2$

$$\frac{8x^2 - x - 5}{x^3 - x^2} \equiv \frac{6}{x} + \frac{5}{x^2} + \frac{2}{(x-1)}$$

5) a) Expand the denominator:

$$\frac{2x^2 + 18x + 26}{(x+2)(x+4)} \equiv \frac{2x^2 + 18x + 26}{x^2 + 6x + 8}$$

Divide the fraction:

$$x^2 + 6x + 8 \overline{)2x^2 + 18x + 26} \quad \text{quotient } 2$$
$$2x^2 + 12x + 16$$
$$6x + 10$$

$$\frac{2x^2 + 18x + 26}{(x+2)(x+4)} \equiv 2 + \frac{6x + 10}{(x+2)(x+4)}$$

Now express $\frac{6x+10}{(x+2)(x+4)}$ as partial fractions:

$$\frac{6x+10}{(x+2)(x+4)} \equiv \frac{A}{(x+2)} + \frac{B}{(x+4)}$$

$6x + 10 \equiv A(x + 4) + B(x + 2)$

substitute $x = -4$: $-14 = -2B \Rightarrow B = 7$

substitute $x = -2$: $-2 = 2A \Rightarrow A = -1$

So overall $\frac{2x^2 + 18x + 26}{(x+2)(x+4)} \equiv 2 - \frac{1}{(x+2)} + \frac{7}{(x+4)}$

b) Expand the denominator:

$$\frac{3x^2 + 9x + 2}{x(x+1)} \equiv \frac{3x^2 + 9x + 2}{x^2 + x}$$

Divide the fraction:

$$x^2 + x \overline{)3x^2 + 9x + 2} \quad \text{quotient } 3$$
$$3x^2 + 3x$$
$$6x + 2$$

$$\frac{3x^2 + 9x + 2}{x(x+1)} \equiv 3 + \frac{6x+2}{x(x+1)}$$

Now express $\frac{6x+2}{x(x+1)}$ as partial fractions:

$$\frac{6x+2}{x(x+1)} \equiv \frac{A}{x} + \frac{B}{(x+1)}$$

$6x + 2 \equiv A(x + 1) + Bx$

substitute $x = -1$: $-4 = -B \Rightarrow B = 4$

substitute $x = 0$: $2 = A$

So overall $\frac{3x^2 + 9x + 2}{x(x+1)} \equiv 3 + \frac{2}{x} + \frac{4}{(x+1)}$

c) Expand the denominator:

$$\frac{24x^2 - 70x + 53}{(2x-3)^2} \equiv \frac{24x^2 - 70x + 53}{4x^2 - 12x + 9}$$

Divide the fraction:

$$4x^2 - 12x + 9 \overline{)24x^2 - 70x + 53} \quad \text{quotient } 6$$
$$24x^2 - 72x + 54$$
$$2x - 1$$

$$\frac{24x^2 - 70x + 53}{(2x-3)^2} \equiv 6 + \frac{2x-1}{(2x-3)^2}$$

Now express $\frac{2x-1}{(2x-3)^2}$ as partial fractions:

$$\frac{2x-1}{(2x-3)^2} \equiv \frac{A}{(2x-3)} + \frac{B}{(2x-3)^2}$$

$2x - 1 \equiv A(2x - 3) + B$

substitute $x = 1.5$: $2 = B$

substitute $x = 0$, and $B = 2$: $-1 = -3A + 2 \Rightarrow A = 1$

So overall $\frac{24x^2 - 70x + 53}{(2x-3)^2} \equiv 6 + \frac{1}{(2x-3)} + \frac{2}{(2x-3)^2}$

d) Expand the denominator:

$$\frac{3x^3 - 2x^2 - 2x - 3}{(x+1)(x-2)} \equiv \frac{3x^3 - 2x^2 - 2x - 3}{x^2 - x - 2}$$

Divide the fraction using $f(x) = q(x)d(x) + r(x)$:

$3x^3 - 2x^2 - 2x - 3 = (Ax + B)(x^2 - x - 2) + Cx + D$

coefficients of x^3: $3 = A$

coefficients of x^2: $-2 = -A + B$

substitute $A = 3$: $-2 = -3 + B \Rightarrow B = 1$

coefficients of x: $-2 = -2A - B + C$

substitute $A = 3$ and $B = 1$: $-2 = -6 - 1 + C \Rightarrow C = 5$

constants: $-3 = -2B + D$

substitute $B = 1$: $-3 = -2 + D \Rightarrow D = -1$

$$\frac{3x^3 - 2x^2 - 2x - 3}{(x+1)(x-2)} \equiv 3x + 1 + \frac{5x-1}{(x+1)(x-2)}$$

Now express $\frac{5x-1}{(x+1)(x-2)}$ as partial fractions:

$$\frac{5x-1}{(x+1)(x-2)} \equiv \frac{M}{(x+1)} + \frac{N}{(x-2)}$$

$5x - 1 \equiv M(x - 2) + N(x + 1)$

substitute $x = 2$: $9 = 3N \Rightarrow N = 3$

substitute $x = -1$: $-6 = -3M \Rightarrow M = 2$

So overall $\frac{3x^3 - 2x^2 - 2x - 3}{(x+1)(x-2)} \equiv 3x + 1 + \frac{2}{(x+1)} + \frac{3}{(x-2)}$

I used the remainder theorem to divide this fraction, because it's a bit trickier than the rest have been. You could have used the remainder theorem in 5)a)-c) too, but I reckon they were easier to do with long division.

6) $(1 + 2x)^3 = 1 + 3(2x) + \frac{3 \times 2}{1 \times 2}(2x)^2 + \frac{3 \times 2 \times 1}{1 \times 2 \times 3}(2x)^3$

$= 1 + 6x + 12x^2 + 8x^3$

You could have used Pascal's Triangle to get the coefficients here. I've done it the long way because I like to show off.

7) a) $(1 + x)^{-4}$

$\approx 1 + (-4)x + \frac{-4 \times -5}{1 \times 2}x^2 + \frac{-4 \times -5 \times -6}{1 \times 2 \times 3}x^3$

$= 1 - 4x + 10x^2 - 20x^3$

b) $(1 - 3x)^{-3} \approx 1 + (-3)(-3x) + \frac{-3 \times -4}{1 \times 2}(-3x)^2 + \frac{-3 \times -4 \times -5}{1 \times 2 \times 3}(-3x)^3$

$= 1 + 9x + 54x^2 + 270x^3$

Be extra careful with terms like $(-3x)^2$... remember to square everything in the brackets — the x, the 3 and the minus.

c) $(1 - 5x)^{\frac{1}{2}}$

$\approx 1 + \frac{1}{2}(-5x) + \frac{\frac{1}{2} \times -\frac{1}{2}}{1 \times 2}(-5x)^2 + \frac{\frac{1}{2} \times -\frac{1}{2} \times -\frac{3}{2}}{1 \times 2 \times 3}(-5x)^3$

$= 1 - \frac{5}{2}x - \frac{25}{8}x^2 - \frac{125}{16}x^3$

8) a) $\left|\frac{dx}{c}\right| < 1$ (or $|x| < \left|\frac{c}{d}\right|$)

b) 7) a): expansion valid for $|x| < 1$

7) b): expansion valid for $|-3x| < 1 \Rightarrow |-3||x| < 1 \Rightarrow |x| < \frac{1}{3}$

7) c): expansion valid for $|-5x| < 1 \Rightarrow |-5||x| < 1 \Rightarrow |x| < \frac{1}{5}$

Answers

9) a) $(3+2x)^{-2} = \left(3\left(1+\frac{2}{3}x\right)\right)^{-2} = \frac{1}{9}\left(1+\frac{2}{3}x\right)^{-2}$

$\approx \frac{1}{9}\left(1+(-2)\left(\frac{2}{3}x\right)+\frac{-2\times-3}{1\times 2}\left(\frac{2}{3}x\right)^2\right)$

$= \frac{1}{9}\left(1-\frac{4}{3}x+\frac{4}{3}x^2\right)$

$= \frac{1}{9}-\frac{4}{27}x+\frac{4}{27}x^2$

This expansion is valid for $\left|\frac{2x}{3}\right|<1 \Rightarrow \frac{2}{3}|x|<1 \Rightarrow |x|<\frac{3}{2}$.

b) $(8-x)^{\frac{1}{3}} = \left(8\left(1-\frac{1}{8}x\right)\right)^{\frac{1}{3}} = 2\left(1-\frac{1}{8}x\right)^{\frac{1}{3}}$

$\approx 2\left(1+\frac{1}{3}\left(-\frac{1}{8}x\right)+\frac{\frac{1}{3}\times-\frac{2}{3}}{1\times 2}\left(-\frac{1}{8}x\right)^2\right)$

$= 2\left(1-\frac{1}{24}x-\frac{1}{576}x^2\right)$

$= 2-\frac{1}{12}x-\frac{1}{288}x^2$

This expansion is valid for $\left|\frac{-x}{8}\right|<1 \Rightarrow \frac{|-1||x|}{8}<1$

$\Rightarrow |x|<8$.

10) The simplest way to disprove the statement is to find a counter-example. Try some values of n and see if the statement is true for them:

$n=3 \Rightarrow n^2-n-1 = 3^2-3-1 = 5$ — prime

$n=4 \Rightarrow n^2-n-1 = 4^2-4-1 = 11$ — prime

$n=5 \Rightarrow n^2-n-1 = 5^2-5-1 = 19$ — prime

$n=6 \Rightarrow n^2-n-1 = 6^2-6-1 = 29$ — prime

$n=7 \Rightarrow n^2-n-1 = 7^2-7-1 = 41$ — prime

$n=8 \Rightarrow n^2-n-1 = 8^2-8-1 = 55$ — not prime

n^2-n-1 is not prime when n = 8.

So the statement is false.

Sometimes good old trial and error is the easiest way to find a counter-example. Don't forget, if you've been told to disprove a statement like this, then a counter-example must exist.

Exam Questions

1 $\frac{2x^2-9x-35}{x^2-49} = \frac{(2x+5)(x-7)}{(x+7)(x-7)} = \frac{2x+5}{x+7}$

[3 marks available — 1 mark for factorising the numerator, 1 mark for factorising the denominator and 1 mark for correct answer (after cancelling)]

2 a) $f(x) = (9-4x)^{-\frac{1}{2}} = (9)^{-\frac{1}{2}}\left(1-\frac{4}{9}x\right)^{-\frac{1}{2}} = \frac{1}{3}\left(1-\frac{4}{9}x\right)^{-\frac{1}{2}}$

$= \frac{1}{3}\left(1+\left(-\frac{1}{2}\right)\left(-\frac{4}{9}x\right)+\frac{(-\frac{1}{2})\times(-\frac{3}{2})}{1\times 2}\left(-\frac{4}{9}x\right)^2 + \frac{(-\frac{1}{2})\times(-\frac{3}{2})\times(-\frac{5}{2})}{1\times 2\times 3}\left(-\frac{4}{9}x\right)^3+\ldots\right)$

$= \frac{1}{3}\left(1+\left(-\frac{1}{2}\right)\left(-\frac{4}{9}x\right)+\frac{(\frac{3}{4})}{2}\left(-\frac{4}{9}x\right)^2+\frac{(-\frac{15}{8})}{6}\left(-\frac{4}{9}x\right)^3+\ldots\right)$

$= \frac{1}{3}\left(1+\left(-\frac{1}{2}\right)\left(-\frac{4}{9}x\right)+\frac{3}{8}\left(-\frac{4}{9}x\right)^2+\left(-\frac{5}{16}\right)\left(-\frac{4}{9}x\right)^3+\ldots\right)$

$= \frac{1}{3}\left(1+\frac{2}{9}x+\frac{2}{27}x^2+\frac{20}{729}x^3+\ldots\right)$

$= \frac{1}{3}+\frac{2}{27}x+\frac{2}{81}x^2+\frac{20}{2187}x^3+\ldots$

[5 marks available in total:

- ***1 mark for factorising out $(9)^{-\frac{1}{2}}$ or $\frac{1}{3}$***
- ***1 mark for expansion of an expression of the form $(1+ax)^{-\frac{1}{2}}$***
- ***2 marks for the penultimate line of working — 1 for the first two terms in brackets correct, 1 for the 3rd and 4th terms in brackets correct.***
- ***1 mark for the final answer correct]***

Multiplying out those coefficients can be pretty tricky. Don't try to do things all in one go — you won't be penalised for writing an extra line of working, but you probably will lose marks if your final answer's wrong.

b) $(2-x)\left(\frac{1}{3}+\frac{2}{27}x+\frac{2}{81}x^2+\frac{20}{2187}x^3+\ldots\right)$

You only need the first three terms of the expansion, so just write the terms up to x^2 when you multiply out the brackets:

$= \frac{2}{3}+\frac{4}{27}x+\frac{4}{81}x^2+\ldots$

$\quad -\frac{1}{3}x-\frac{2}{27}x^2+\ldots$

$= \frac{2}{3}-\frac{5}{27}x-\frac{2}{81}x^2+\ldots$

[4 marks available in total:

- ***1 mark for multiplying your answer to part (a) by (2 – x)***
- ***1 mark for multiplying out brackets to find constant term, two x-terms and two x^2-terms.***
- ***1 mark for correct constant and x-terms in final answer***
- ***1 mark for correct x^2-term in final answer]***

3 Expand the denominator:

$\frac{18x^2-15x-62}{(3x+4)(x-2)} \equiv \frac{18x^2-15x-62}{3x^2-2x-8}$

Divide the fraction:

$$\begin{array}{r} 6 \\ 3x^2-2x-8\overline{)18x^2-15x-62} \\ -\underline{18x^2-12x-48} \\ -3x-14 \end{array}$$

Watch out for the negative signs here. You're subtracting the bottom line from the top, so be sure to get it right.

$\frac{18x^2-15x-62}{(3x+4)(x-2)} \equiv 6+\frac{-3x-14}{(3x+4)(x-2)}$

$A = 6$ ***[1 mark]***

$\frac{-3x-14}{(3x+4)(x-2)} \equiv \frac{B}{(3x+4)}+\frac{C}{(x-2)}$

$-3x-14 \equiv B(x-2)+C(3x+4)$ ***[1 mark]***

substitute $x=2$: $-20 = 10C \Rightarrow C = -2$ ***[1 mark]***

coefficients of x: $-3 = B+3C$

$-3 = B-6 \Rightarrow B = 3$ ***[1 mark]***

I used equating coefficients for the last bit, because I realised that I'd need to substitute $-\frac{4}{3}$ in for x, and I really couldn't be bothered.

4 a) $36x^2+3x-10 \equiv A(1-3x)^2+B(4+3x)(1-3x)+C(4+3x)$ ***[1 mark]***

Let $x=\frac{1}{3}$, then $4+1-10 = 5C \Rightarrow -5 = 5C \Rightarrow C = -1$ ***[1 mark]***

Let $x=-\frac{4}{3}$, then $64-4-10 = 25A \Rightarrow 50 = 25A \Rightarrow A = 2$ ***[1 mark]***

Equate the terms in x^2:

$36 = 9A-9B = 18-9B \Rightarrow -18 = 9B \Rightarrow B = -2$ ***[1 mark]***

Answers

b) $f(x) = \frac{2}{(4+3x)} - \frac{2}{(1-3x)} - \frac{1}{(1-3x)^2}$

Expand each term separately:

$2(4+3x)^{-1} = 2\left(4\left(1+\frac{3}{4}x\right)\right)^{-1} = \frac{1}{2}\left(1+\frac{3}{4}x\right)^{-1}$

$= \frac{1}{2}\left(1 + (-1)\left(\frac{3}{4}x\right) + \frac{(-1)\times(-2)}{1\times 2}\left(\frac{3}{4}x\right)^2 + ...\right)$

$= \frac{1}{2}\left(1 - \frac{3}{4}x + \frac{9}{16}x^2 + ...\right) = \frac{1}{2} - \frac{3}{8}x + \frac{9}{32}x^2 + ...$

Now the second term: $-2(1-3x)^{-1} =$

$-2\left(1 + (-1)(-3x) + \frac{(-1)\times(-2)}{1\times 2}(-3x)^2 + ...\right)$

$= -2(1 + 3x + 9x^2 + ...) = -2 - 6x - 18x^2 + ...$

And the final term: $-(1-3x)^{-2} =$

$-\left(1 + (-2)(-3x) + \frac{(-2)\times(-3)}{1\times 2}(-3x)^2 + ...\right)$

$= -(1 + 6x + 27x^2 + ...) = -1 - 6x - 27x^2 + ...$

Putting it all together gives

$\frac{1}{2} - \frac{3}{8}x + \frac{9}{32}x^2 - 2 - 6x - 18x^2 - 1 - 6x - 27x^2$

$= -\frac{5}{2} - \frac{99}{8}x - \frac{1431}{32}x^2 + ...$

[6 marks available in total:
- ***1 mark for rewriting f(x) in the form $A(4+3x)^{-1} + B(1-3x)^{-1} + C(1-3x)^{-2}$***
- ***1 mark for correct binomial expansion of $(4+3x)^{-1}$***
- ***1 mark for correct binomial expansion of $(1-3x)^{-1}$***
- ***1 mark for correct binomial expansion of $(1-3x)^{-2}$***
- ***1 mark for correct constant and x-terms in final answer***
- ***1 mark for correct x^2-term in final answer]***

You know what, I can't think of anything else remotely useful, witty or interesting to say about binomials... Seriously, I'm going to have to resort to slightly weird jokes in a minute... You've been warned...

c) Expansion of $(4+3x)^{-1}$ is valid for $\left|\frac{3x}{4}\right| < 1 \Rightarrow \frac{3|x|}{4} < 1$

$\Rightarrow |x| < \frac{4}{3}$

Expansions of $(1-3x)^{-1}$ and $(1-3x)^{-2}$ are valid for

$\left|\frac{-3x}{1}\right| < 1 \Rightarrow \frac{|-3||x|}{1} < 1 \Rightarrow |x| < \frac{1}{3}$

The combined expansion is valid for the narrower of these two ranges. So the expansion of f(x) is valid for $|x| < \frac{1}{3}$.

[2 marks available in total:
- ***1 mark for identifying the valid range of the expansion of f(x) as being the narrower of the two valid ranges shown***
- ***1 mark for correct answer]***

5 a) $(16+3x)^{\frac{1}{4}} = 16^{\frac{1}{4}}\left(1+\frac{3}{16}x\right)^{\frac{1}{4}} = 2\left(1+\frac{3}{16}x\right)^{\frac{1}{4}}$

$\approx 2\left(1 + \left(\frac{1}{4}\right)\left(\frac{3}{16}x\right) + \frac{\frac{1}{4}\times -\frac{3}{4}}{1\times 2}\left(\frac{3}{16}x\right)^2\right)$

$= 2\left(1 + \left(\frac{1}{4}\right)\left(\frac{3}{16}x\right) + \left(-\frac{3}{32}\right)\left(\frac{9}{256}x^2\right)\right)$

$= 2\left(1 + \frac{3}{64}x - \frac{27}{8192}x^2\right)$

$= 2 + \frac{3}{32}x - \frac{27}{4096}x^2$

[5 marks available in total:
- ***1 mark for factorising out $16^{\frac{1}{4}}$ or 2***
- ***1 mark for expansion of an expression of the form $(1+ax)^{\frac{1}{4}}$***
- ***2 marks for the penultimate line of working — 1 for the first two terms in brackets correct, 1 for the 3rd term in brackets.***
- ***1 mark for the final answer correct]***

b) (i) $16 + 3x = 12.4 \Rightarrow x = -1.2$

So $(12.4)^{\frac{1}{4}} \approx 2 + \frac{3}{32}(-1.2) - \frac{27}{4096}(-1.2)^2$

$= 2 - 0.1125 - 0.0094921875$

$= 1.878008$ (to 6 d.p.)

[2 marks available in total:
- ***1 mark for substituting x = –1.2 into the expansion from part (a)***
- ***1 mark for correct answer]***

(ii) Percentage error

$= \left|\frac{\text{real value} - \text{estimate}}{\text{real value}}\right| \times 100$ ***[1 mark]***

$= \left|\frac{\sqrt[4]{12.4} - 1.878008}{\sqrt[4]{12.4}}\right| \times 100$

$= \frac{|1.876529... - 1.878008|}{1.876529...} \times 100$

$= 0.0788\%$ (to 3 s.f.) ***[1 mark]***

Why did the binomial expansion cross the road?
Don't be silly, binomial expansions can't move independently...

6 a) $\left(1 - \frac{4}{3}x\right)^{-\frac{1}{2}}$

$\approx 1 + \left(-\frac{1}{2}\right)\left(-\frac{4}{3}x\right) + \frac{\left(-\frac{1}{2}\right)\times\left(-\frac{3}{2}\right)}{1\times 2}\left(-\frac{4}{3}x\right)^2 + \frac{\left(-\frac{1}{2}\right)\times\left(-\frac{3}{2}\right)\times\left(-\frac{5}{2}\right)}{1\times 2\times 3}\left(-\frac{4}{3}x\right)^3$

$= 1 + \left(-\frac{1}{2}\right)\left(-\frac{4}{3}x\right) + \frac{\left(\frac{3}{4}\right)}{2}\left(-\frac{4}{3}x\right)^2 + \frac{\left(-\frac{15}{8}\right)}{6}\left(-\frac{4}{3}x\right)^3$

$= 1 + \left(-\frac{1}{2}\right)\left(-\frac{4}{3}x\right) + \frac{3}{8}\left(\frac{16}{9}x^2\right) + \left(-\frac{15}{48}\right)\left(-\frac{64}{27}x^3\right)$

$= 1 + \frac{2}{3}x + \frac{2}{3}x^2 + \frac{20}{27}x^3$

[4 marks available in total:
- ***1 mark for writing out binomial expansion formula with $n = -\frac{1}{2}$***
- ***1 mark for writing out binomial expansion formula substituting $-\frac{4}{3}x$ for x***
- ***1 mark for correct constant and x-terms in final answer***
- ***1 mark for correct x^2- and x^3-terms in final answer]***

b) $\sqrt{\frac{27}{(3-4x)}} = \sqrt{\frac{27}{3\left(1-\frac{4}{3}x\right)}} = \sqrt{\frac{9}{\left(1-\frac{4}{3}x\right)}} = \frac{3}{\sqrt{\left(1-\frac{4}{3}x\right)}}$

$= 3\left(1-\frac{4}{3}x\right)^{-\frac{1}{2}}$

$\approx 3\left(1 + \frac{2}{3}x + \frac{2}{3}x^2\right)$

$= 3 + 2x + 2x^2$

So $a = 3$, $b = 2$, $c = 2$.

Expansion is valid for $\left|-\frac{4}{3}x\right| < 1 \Rightarrow \left|-\frac{4}{3}\right||x| < 1 \Rightarrow |x| < \frac{3}{4}$

[3 marks available in total:
- ***1 mark for showing expression is equal to $3\left(1-\frac{4}{3}x\right)^{-\frac{1}{2}}$***
- ***1 mark for using expansion from part (b) to find the correct values of a, b and c.***
- ***1 mark for correct valid range]***

Doctor, doctor, I keep thinking I'm a binomial expansion...
I'm sorry, I don't think I can help you, I'm a cardiologist.

Answers

7 First put $x = -6$ into both sides of the identity $x^3 + 15x^2 + 43x - 30 \equiv (Ax^2 + Bx + C)(x + 6) + D$:
$(-6)^3 + 15(-6)^2 + 43(-6) - 30 = D \Rightarrow 36 = D$ ***[1 mark]***. Now set $x = 0$ to get $-30 = 6C + D$, so $C = -11$ ***[1 mark]***. Equating the coefficients of x^3 gives $1 = A$. Equating the coefficients of x^2 gives $15 = 6A + B$, so $B = 9$ ***[1 mark]***.
So $x^3 + 15x^2 + 43x - 30 = (x^2 + 9x - 11)(x + 6) + 36$.

You could also do this question by algebraic long division — you just have to use your answer to work out A, B, C and D.

8 a) Proof by exhaustion: let n be even. $n^2 - n = n(n - 1)$.
If n is even, $n - 1$ is odd so $n(n - 1)$ is even (as even × odd = even). This means that $n(n - 1) - 1$ is odd ***[1 mark]***.
Let n be odd. If n is odd, $n - 1$ is even, so $n(n - 1)$ is even (as odd × even = even). This means that $n(n - 1) - 1$ is odd ***[1 mark]***. As any integer n has to be either odd or even, $n^2 - n - 1$ is odd for any value of n ***[1 mark]***.

b) As $n^2 - n - 1$ is odd, $n^2 - n - 2$ is even ***[1 mark]***. The product of even numbers is also even ***[1 mark]***, so as $(n^2 - n - 2)^3$ is the product of 3 even numbers, it will always be even ***[1 mark]***.

9 Expand the denominator:
$$\frac{-80x^2 + 49x - 9}{(5x-1)(2-4x)} \equiv \frac{-80x^2 + 49x - 9}{-20x^2 + 14x - 2}$$
Divide the fraction:
$$\begin{array}{r} 4 \\ -20x^2 + 14x - 2\,\overline{)\,-80x^2 + 49x - 9} \\ \underline{-80x^2 + 56x - 8} \\ -7x - 1 \end{array}$$
$$\frac{-80x^2 + 49x - 9}{(5x-1)(2-4x)} \equiv 4 + \frac{-7x - 1}{(5x-1)(2-4x)} \quad \textbf{[1 mark]}$$
$$\frac{-7x - 1}{(5x-1)(2-4x)} \equiv \frac{A}{(5x-1)} + \frac{B}{(2-4x)}$$
$-7x - 1 \equiv A(2 - 4x) + B(5x - 1)$ ***[1 mark]***
substitute $x = 0.5$: $-4.5 = 1.5B \Rightarrow B = -3$ ***[1 mark]***
coefficients of x: $-7 = -4A + 5B$
substitute $B = -3$: $-7 = -4A - 15$
$8 = -4A \Rightarrow A = -2$ ***[1 mark]***

10 a) (i) $\sqrt{\frac{1+2x}{1-3x}} = \frac{\sqrt{1+2x}}{\sqrt{1-3x}} = (1 + 2x)^{\frac{1}{2}}(1 - 3x)^{-\frac{1}{2}}$ ***[1 mark]***
$$(1+2x)^{\frac{1}{2}} \approx 1 + \tfrac{1}{2}(2x) + \frac{(\frac{1}{2}) \times (-\frac{1}{2})}{1 \times 2}(2x)^2$$
$= 1 + x - \frac{1}{2}x^2$ ***[1 mark]***
$$(1-3x)^{-\frac{1}{2}} \approx 1 + \left(-\tfrac{1}{2}\right)(-3x) + \frac{(-\frac{1}{2}) \times (-\frac{3}{2})}{1 \times 2}(-3x)^2$$
$= 1 + \frac{3}{2}x + \frac{27}{8}x^2$ ***[1 mark]***
$\sqrt{\frac{1+2x}{1-3x}} \approx \left(1 + x - \frac{1}{2}x^2\right)\left(1 + \frac{3}{2}x + \frac{27}{8}x^2\right)$ ***[1 mark]***
$\approx 1 + \frac{3}{2}x + \frac{27}{8}x^2 + x + \frac{3}{2}x^2 - \frac{1}{2}x^2$
(ignoring any terms in x^3 or above)
$= 1 + \frac{5}{2}x + \frac{35}{8}x^2$ ***[1 mark]***

(ii) Expansion of $(1 + 2x)^{\frac{1}{2}}$ is valid for $|2x| < 1 \Rightarrow |x| < \frac{1}{2}$
Expansion of $(1 - 3x)^{-\frac{1}{2}}$ is valid for $|-3x| < 1$
$\Rightarrow |-3||x| < 1 \Rightarrow |x| < \frac{1}{3}$
The combined expansion is valid for the narrower of these two ranges.
So the expansion of $\sqrt{\frac{1+2x}{1-3x}}$ is valid for $|x| < \frac{1}{3}$.
[2 marks available in total:
- ***1 mark for identifying the valid range of the expansion as being the narrower of the two valid ranges shown***
- ***1 mark for correct answer]***

b) $x = \frac{2}{15} \Rightarrow \sqrt{\frac{1+2x}{1-3x}} = \sqrt{\frac{1+\frac{4}{15}}{1-\frac{6}{15}}} = \sqrt{\frac{\frac{19}{15}}{\frac{9}{15}}} = \sqrt{\frac{19}{9}} = \frac{1}{3}\sqrt{19}$
[1 mark]
$$\sqrt{19} \approx 3\left(1 + \tfrac{5}{2}\left(\tfrac{2}{15}\right) + \tfrac{35}{8}\left(\tfrac{2}{15}\right)^2\right)$$
$$= 3\left(1 + \tfrac{1}{3} + \tfrac{7}{90}\right)$$
$$= 3\left(\tfrac{127}{90}\right)$$
$= \frac{127}{30}$ ***[1 mark]***

The binomial expansion walks into a bar and asks for a pint. The barman says, "I'm sorry, I can't serve alcohol in a joke that may be read by under-18s."

11 a) Expand the denominator:
$$\frac{3x^2 + 12x - 11}{(x+3)(x-1)} \equiv \frac{3x^2 + 12x - 11}{x^2 + 2x - 3}$$
Divide the fraction:
$$\begin{array}{r} 3 \\ x^2 + 2x - 3\,\overline{)\,3x^2 + 12x - 11} \\ \underline{-\ 3x^2 + \ 6x - \ 9} \\ 6x - \ 2 \end{array}$$
$\Rightarrow -2 + 6x$ ***[1 mark]***
$$\frac{3x^2 + 12x - 11}{(x+3)(x-1)} \equiv 3 + \frac{-2 + 6x}{(x+3)(x-1)}$$
$A = 3$ ***[1 mark]***, $B = -2$ ***[1 mark]***, $C = 6$ ***[1 mark]***

b) $\frac{3x^2 + 12x - 11}{(x+3)(x-1)} \equiv 3 + \frac{-2 + 6x}{(x+3)(x-1)} \equiv 3 + \frac{M}{(x+3)} + \frac{N}{(x-1)}$
$-2 + 6x \equiv M(x - 1) + N(x + 3)$
substitute $x = 1$: $4 = 4N \Rightarrow N = 1$ ***[1 mark]***
substitute $x = -3$: $-20 = -4M \Rightarrow M = 5$ ***[1 mark]***
So overall $\frac{3x^2 + 12x - 11}{(x+3)(x-1)} \equiv 3 + \frac{5}{(x+3)} + \frac{1}{(x-1)}$ ***[1 mark]***

12 a) $13x - 17 \equiv A(2x - 1) + B(5 - 3x)$ ***[1 mark]***
Let $x = \frac{1}{2}$, then $\frac{13}{2} - 17 = B(5 - \frac{3}{2}) \Rightarrow -\frac{21}{2} = \frac{7}{2}B \Rightarrow B = -3$
[1 mark]
Let $x = \frac{5}{3}$, then $\frac{65}{3} - 17 = A(\frac{10}{3} - 1) \Rightarrow \frac{14}{3} = \frac{7}{3}A \Rightarrow A = 2$
[1 mark]

b) (i) $(2x - 1)^{-1} = -(1 - 2x)^{-1}$ ***[1 mark]***
$$\approx -\left(1 + (-1)(-2x) + \frac{(-1) \times (-2)}{1 \times 2}(-2x)^2\right)$$
$$= -(1 + 2x + 4x^2)$$
$= -1 - 2x - 4x^2$ ***[1 mark]***

Answers

(ii) $(5-3x)^{-1} = 5^{-1}\left(1-\frac{3}{5}x\right)^{-1} = \frac{1}{5}\left(1-\frac{3}{5}x\right)^{-1}$

$\approx \frac{1}{5}\left(1+(-1)\left(-\frac{3}{5}x\right)+\frac{(-1)\times(-2)}{1\times 2}\left(-\frac{3}{5}x\right)^2\right)$

$= \frac{1}{5}\left(1+\frac{3}{5}x+\frac{9}{25}x^2\right)$

$= \frac{1}{5}+\frac{3}{25}x+\frac{9}{125}x^2$

[5 marks available in total:

- ***1 mark for factorising out 5^{-1} or $\frac{1}{5}$***
- ***1 mark for expansion of an expression of the form $(1 + ax)^{-1}$***
- ***2 marks for the penultimate line of working — 1 mark for the first two terms in brackets correct, 1 mark for the 3rd term in brackets correct.***
- ***1 mark for the final answer correct]***

c) $\frac{13x-17}{(5-3x)(2x-1)} = \frac{2}{(5-3x)} - \frac{3}{(2x-1)}$

$= 2(5-3x)^{-1} - 3(2x-1)^{-1}$ ***[1 mark]***

$\approx 2\left(\frac{1}{5}+\frac{3}{25}x+\frac{9}{125}x^2\right) - 3(-1-2x-4x^2)$

$= \frac{2}{5}+\frac{6}{25}x+\frac{18}{125}x^2+3+6x+12x^2$

$= \frac{17}{5}+\frac{156}{25}x+\frac{1518}{125}x^2$

[1 mark]

Core Section 3 — Trigonometry Warm-up Questions

1) a) $\sin^{-1}\frac{1}{\sqrt{2}} = \frac{\pi}{4}$

b) $\cos^{-1}0 = \frac{\pi}{2}$

c) $\tan^{-1}\sqrt{3} = \frac{\pi}{3}$

2) See p25.

3) a) cosec 30° = 2 (since sin 30° = 0.5)

b) sec 30° = $\frac{2}{\sqrt{3}}$ (since cos 30° = $\frac{\sqrt{3}}{2}$)

c) cot 30° = $\sqrt{3}$ (since tan 30° = $\frac{1}{\sqrt{3}}$)

4) See p26.

5) Divide the whole identity by $\cos^2\theta$ to get:

$\frac{\cos^2\theta}{\cos^2\theta} + \frac{\sin^2\theta}{\cos^2\theta} \equiv \frac{1}{\cos^2\theta}$

$\Rightarrow 1 + \tan^2\theta \equiv \sec^2\theta$

(as sin/cos ≡ tan and 1/cos ≡ sec)

6) Using the identities $\text{cosec}^2\theta \equiv 1 + \cot^2\theta$ and $\sin^2\theta + \cos^2\theta \equiv 1$, the LHS becomes:

$(\text{cosec}^2\theta - 1) + (1 - \cos^2\theta) \equiv \text{cosec}^2\theta - \cos^2\theta$,

which is the same as the RHS.

7) $\frac{\pi}{12} = \frac{\pi}{3} - \frac{\pi}{4}$, so use the addition formula for $\cos(A - B)$:

$\cos\frac{\pi}{12} = \cos\left(\frac{\pi}{3}-\frac{\pi}{4}\right) = \cos\frac{\pi}{3}\cos\frac{\pi}{4} + \sin\frac{\pi}{3}\sin\frac{\pi}{4}$

As $\cos\frac{\pi}{3} = \frac{1}{2}$, $\cos\frac{\pi}{4} = \frac{1}{\sqrt{2}}$, $\sin\frac{\pi}{3} = \frac{\sqrt{3}}{2}$ and $\sin\frac{\pi}{4} = \frac{1}{\sqrt{2}}$,

putting these values into the equation gives:

$\cos\frac{\pi}{3}\cos\frac{\pi}{4} + \sin\frac{\pi}{3}\sin\frac{\pi}{4} = \left(\frac{1}{2}\cdot\frac{1}{\sqrt{2}}\right) + \left(\frac{\sqrt{3}}{2}\cdot\frac{1}{\sqrt{2}}\right)$

$= \frac{1}{2\sqrt{2}} + \frac{\sqrt{3}}{2\sqrt{2}} = \frac{1+\sqrt{3}}{2\sqrt{2}} = \frac{\sqrt{2}(1+\sqrt{3})}{4} = \frac{\sqrt{2}+\sqrt{6}}{4}$

You could also have used $\frac{\pi}{12} = \frac{\pi}{4} - \frac{\pi}{6}$ in your answer.

8) $\sin(A + B) = \sin A \cos B + \cos A \sin B$.

As $\sin A = \frac{4}{5}$, $\cos A = \frac{3}{5}$ (from the right-angled triangle with sides of length 3, 4 and 5) and as $\sin B = \frac{7}{25}$, $\cos B = \frac{24}{25}$ (from the right-angled triangle with sides 7, 24 and 25).

Putting these values into the equation gives:

$\sin A\cos B + \cos A\sin B = \left(\frac{4}{5}\cdot\frac{24}{25}\right) + \left(\frac{3}{5}\cdot\frac{7}{25}\right)$

$= \frac{96}{125} + \frac{21}{125} = \frac{117}{125}$ (= 0.936)

9) $\cos 2\theta \equiv \cos^2\theta - \sin^2\theta$

$\cos 2\theta \equiv 2\cos^2\theta - 1$

$\cos 2\theta \equiv 1 - 2\sin^2\theta$

10) $\sin 2\theta = -\sqrt{3}\sin\theta \Rightarrow \sin 2\theta + \sqrt{3}\sin\theta = 0$

$2\sin\theta\cos\theta + \sqrt{3}\sin\theta = 0$

$\sin\theta(2\cos\theta + \sqrt{3}) = 0$

So either $\sin\theta = 0$, so θ = 0°, 180°, 360° or

$2\cos\theta + \sqrt{3} = 0 \Rightarrow \cos\theta = -\frac{\sqrt{3}}{2}$

so θ = 150° or 210°. The set of values for θ is 0°, 150°, 180°, 210°, 360°.

If you don't know where the 180°, 360°, 210° etc. came from, you need to go back over your C2 notes...

11) $a\cos\theta + b\sin\theta = R\cos(\theta - \alpha)$ or

$b\sin\theta + a\cos\theta = R\sin(\theta + \alpha)$

12) $5\sin\theta - 6\cos\theta = R\sin(\theta - \alpha)$

$= R\sin\theta\cos\alpha - R\cos\theta\sin\alpha$ (using the addition rule for sin).

Equating coefficients of $\sin\theta$ and $\cos\theta$ gives:

1. $R\cos\alpha = 5$ and 2. $R\sin\alpha = 6$.

Dividing 2. by 1. to find α: $\frac{R\sin\alpha}{R\cos\alpha} = \tan\alpha$, so $\frac{6}{5} = \tan\alpha$

Solving this gives α = 50.19°.

To find R, square equations 1. and 2., then square root:

$R = \sqrt{5^2+6^2} = \sqrt{25+36} = \sqrt{61}$, so

$5\sin\theta - 6\cos\theta = \sqrt{61}\sin(\theta - 50.19°)$.

13) Use the sin addition formulas:

$\sin(x + y) \equiv \sin x\cos y + \cos x\sin y$

$\sin(x - y) \equiv \sin x\cos y - \cos x\sin y$

Take the second away from the first:

$\sin(x + y) - \sin(x - y) \equiv 2\cos x\sin y$.

Let $A = x + y$ and $B = x - y$, so that $x = ½(A + B)$ and $y = ½(A - B)$. Then $\sin A - \sin B = 2\cos\left(\frac{A+B}{2}\right)\sin\left(\frac{A-B}{2}\right)$.

Hint: to get the formulas for x and y in terms of A and B, you need to treat A = x + y and B = x − y as a pair of simultaneous equations.

14) Start by putting the LHS over a common denominator:

$\frac{\cos\theta}{\sin\theta} + \frac{\sin\theta}{\cos\theta} \equiv \frac{\cos\theta\cos\theta}{\sin\theta\cos\theta} + \frac{\sin\theta\sin\theta}{\sin\theta\cos\theta}$

$\equiv \frac{\cos^2\theta + \sin^2\theta}{\sin\theta\cos\theta} \equiv \frac{1}{\sin\theta\cos\theta}$

Answers

(using the identity $\sin^2\theta + \cos^2\theta \equiv 1$).

Now, $\sin 2\theta \equiv 2\sin\theta\cos\theta$, so $\sin\theta\cos\theta = \frac{1}{2}\sin 2\theta$.

So $\frac{1}{\sin\theta\cos\theta} \equiv \frac{1}{\frac{1}{2}\sin 2\theta} \equiv 2\,\text{cosec}\,2\theta$, which is the same as the RHS.

Exam Questions

1 a)

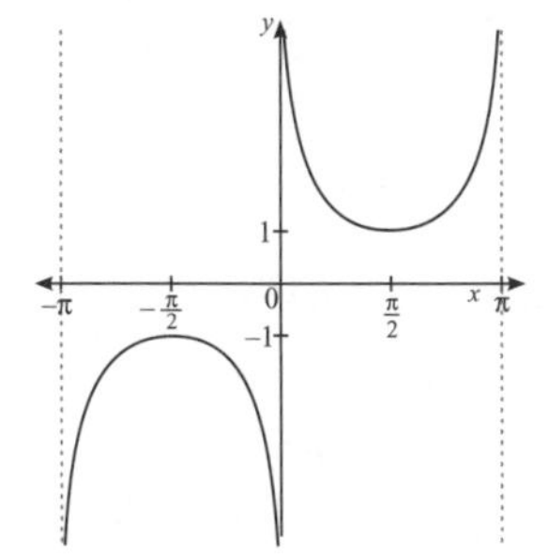

[3 marks available — 1 mark for n-shaped curve in third quadrant and u-shaped curve in first quadrant, 1 mark for asymptotes at 0 and ±π and 1 mark for max/min points of the curves at –1 and 1]

b) If $\text{cosec}\,x = \frac{5}{4} \Rightarrow \frac{1}{\sin x} = \frac{5}{4} \Rightarrow \sin x = \frac{4}{5}$ ***[1 mark]***.

Solving this for x gives $x = 0.927, 2.21$

[1 mark for each solution, lose a mark if answers aren't given to 3 s.f.].

The second solution can be found by sketching y = sin x:

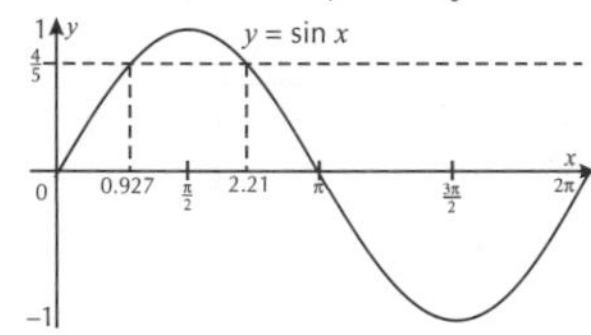

You can see that there are two solutions, one at 0.927, and the other at π – 0.927 = 2.21.

c) $\text{cosec}\,x = 3\sec x \Rightarrow \frac{1}{\sin x} = \frac{3}{\cos x} \Rightarrow \frac{\cos x}{\sin x} = 3$

$\Rightarrow \frac{1}{\tan x} = 3$ so $\tan x = \frac{1}{3}$

Solving for x gives $x = -2.82, 0.322$, ***[1 mark for appropriate rearranging, 1 mark for each solution.]***.

Again, you need to sketch a graph to find the second solution:

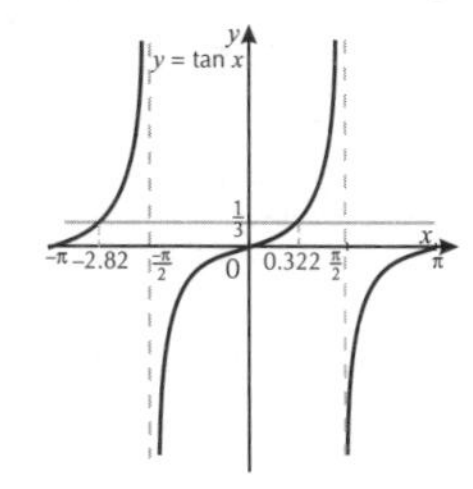

You can see that there are two solutions in the given range, one at 0.322 (this is the one you get from your calculator) and one at –π + 0.322 = –2.82.

2 a) $\frac{2\sin x}{1-\cos x} - \frac{2\cos x}{\sin x} \equiv \frac{2\sin^2 x - 2\cos x + 2\cos^2 x}{\sin x(1-\cos x)}$ ***[1 mark]***

$\equiv \frac{2 - 2\cos x}{\sin x(1-\cos x)}$ ***[1 mark]***

$\equiv \frac{2(1-\cos x)}{\sin x(1-\cos x)}$ ***[1 mark]***

$\equiv \frac{2}{\sin x} \equiv 2\,\text{cosec}\,x$ ***[1 mark]***

b) $2\,\text{cosec}\,x = 4$

$\text{cosec}\,x = 2$ <u>OR</u> $\sin x = \frac{1}{2}$ ***[1 mark]***

$x = \frac{\pi}{6}$ ***[1 mark]***, $x = \frac{5\pi}{6}$ ***[1 mark]***.

3 a) (i) Rearrange the identity $\sec^2\theta \equiv 1 + \tan^2\theta$ to get $\sec^2\theta - 1 \equiv \tan^2\theta$, then replace $\tan^2\theta$ in the equation:

$3\tan^2\theta - 2\sec\theta = 5$

$3(\sec^2\theta - 1) - 2\sec\theta - 5 = 0$ ***[1 mark]***

$3\sec^2\theta - 3 - 2\sec\theta - 5 = 0$

so $3\sec^2\theta - 2\sec\theta - 8 = 0$ ***[1 mark]***

(ii) To factorise this, let $y = \sec\theta$, so the equation becomes $3y^2 - 2y - 8 = 0$, so $(3y + 4)(y - 2) = 0$ ***[1 mark]***. Solving for y gives $y = -\frac{4}{3}$ or $y = 2$. As $y = \sec\theta$, this means that $\sec\theta = -\frac{4}{3}$ or $\sec\theta = 2$ ***[1 mark]***. $\sec\theta = \frac{1}{\cos\theta}$, so $\cos\theta = -\frac{3}{4}$ or $\cos\theta = \frac{1}{2}$ ***[1 mark]***.

b) Let $\theta = 2x$. From above, we know that the solutions to $3\tan^2\theta - 2\sec\theta = 5$ satisfy $\cos\theta = -\frac{3}{4}$ or $\cos\theta = \frac{1}{2}$.

The range for x is $0 \le x \le 180°$, so as $\theta = 2x$, the range for θ is $0 \le \theta \le 360°$ ***[1 mark]***. Solving these equations for θ gives $\theta = 138.59°, 221.41°$ and $\theta = 60°, 300°$ ***[1 mark]***. So, as $\theta = 2x$, $x = \frac{1}{2}\theta$, so $x = 69.30°, 110.70°, 30°, 150°$ ***[1 mark]***.

Once you have the values 60° and 138.59°, you can sketch the graph to find the other values:

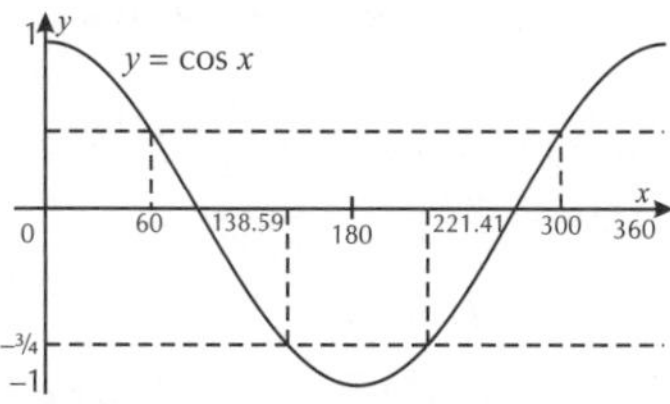

There is a solution at 360 – 60 = 300°, and another at 360 – 138.59 = 221.41°. Don't be fooled by the 2x in this question — you don't need to use the double angle formulas for this one.

4 a) The start and end points of the cos curve (with restricted domain) are (0, 1) and (π, –1), so the coordinates of the start point of arccos (point A) are (–1, π) ***[1 mark]*** and the coordinates of the end point (point B) are (1, 0) ***[1 mark]***.

b) $y = \arccos x$, that is, $y = \cos^{-1}x$, so $x = \cos y$ ***[1 mark]***.

c) $\arccos x = 2$, so $x = \cos 2$ ***[1 mark]*** $\Rightarrow x = -0.416$ ***[1 mark]***.

5 a) $9\sin\theta + 12\cos\theta \equiv R\sin(\theta + \alpha)$. Using the sin addition formula, $9\sin\theta + 12\cos\theta \equiv R\sin\theta\cos\alpha + R\cos\theta\sin\alpha$.

Equating coefficients of $\sin\theta$ and $\cos\theta$ gives:

$R\cos\alpha = 9$ and $R\sin\alpha = 12$ ***[1 mark]***.

$\frac{R\sin\alpha}{R\cos\alpha} = \tan\alpha$, so $\tan\alpha = \frac{12}{9} = \frac{4}{3}$

Solving this gives $\alpha = 0.927$

[1 mark — no other solutions in given range].

$R = \sqrt{9^2 + 12^2} = \sqrt{81 + 144} = \sqrt{225} = 15$ ***[1 mark]***, so $9\sin\theta + 12\cos\theta = 15\sin(\theta + 0.927)$.

Answers

b) If $9\sin\theta + 12\cos\theta = 3$, then from part a), $15\sin(\theta + 0.927) = 3$, so $\sin(\theta + 0.927) = 0.2$. The range for θ is $0 \le \theta \le 2\pi$, which becomes $0.927 \le \theta + 0.927 \le 7.210$. Solving the equation gives $(\theta + 0.927) = 0.201$ ***[1 mark]***. As this is outside the range, use a sketch to find values that are in the range:

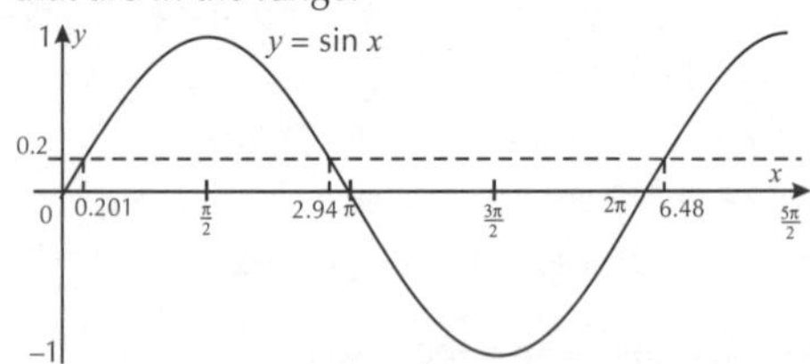

From the graph, it is clear that there are solutions at $\pi - 0.201 = 2.94$ and at $2\pi + 0.201 = 6.48$, so $(\theta + 0.927) = 2.940, 6.48$ ***[1 mark for each value]***, so $\theta = 2.01, 5.56$ ***[1 mark for each solution]***.

Be careful with the range — if you hadn't extended the range to $2\pi + 0.927$, you would have missed one of the solutions.

6 $\sin 3x \equiv \sin(2x + x) \equiv \sin 2x \cos x + \cos 2x \sin x$ ***[1 mark]***
$\equiv (2\sin x \cos x)\cos x + (1 - 2\sin^2 x)\sin x$ ***[1 mark]***
$\equiv 2\sin x \cos^2 x + \sin x - 2\sin^3 x$
$\equiv 2\sin x(1 - \sin^2 x) + \sin x - 2\sin^3 x$ ***[1 mark]***
$\equiv 2\sin x - 2\sin^3 x + \sin x - 2\sin^3 x$
$\equiv 3\sin x - 4\sin^3 x$ ***[1 mark]***

7 a) $5\cos\theta + 12\sin\theta \equiv R\cos(\theta - \alpha)$. Using the cos addition formula, $5\cos\theta + 12\sin\theta \equiv R\cos\theta\cos\alpha + R\sin\theta\sin\alpha$.
Equating coefficients gives:
$R\cos\alpha = 5$ and $R\sin\alpha = 12$ ***[1 mark]***.
$\frac{R\sin\alpha}{R\cos\alpha} = \tan\alpha$, so $\tan\alpha = \frac{12}{5}$ ***[1 mark]***.
Solving this gives $\alpha = 67.38°$ ***[1 mark]***.
$R = \sqrt{5^2 + 12^2} = \sqrt{25 + 144} = \sqrt{169} = 13$ ***[1 mark]***,
so $5\cos\theta + 12\sin\theta = 13\cos(\theta - 67.38°)$.

b) From part (a), if $5\cos\theta + 12\sin\theta = 2$, that means $13\cos(\theta - 67.38°) = 2$, so $\cos(\theta - 67.38°) = \frac{2}{13}$ ***[1 mark]***.
The range for θ is $0 \le \theta \le 360°$,
which becomes $-67.38° \le \theta - 67.38° \le 292.62°$ ***[1 mark]***.
Solving the equation gives $\theta - 67.38 = 81.15, 278.85$ ***[1 mark]***,
so $\theta = 148.53°, 346.23°$ ***[1 mark for each value]***.

Look at the cos graph to get the second solution of $\theta - 67.38°$:

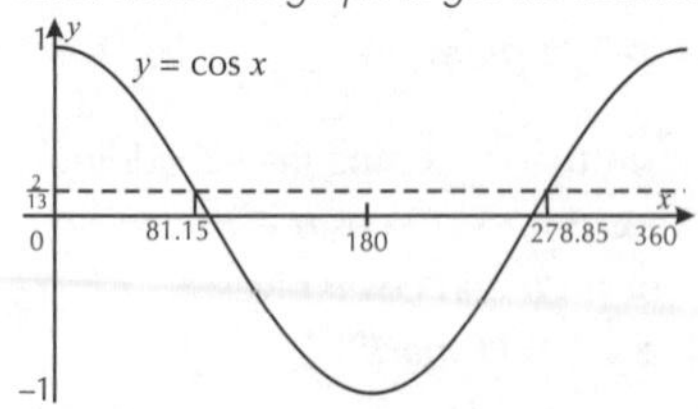

There are two solutions, one at 81.15°, and the other at 360 – 81.15 = 278.85°.

c) The minimum points of the cos curve have a value of –1, so as $5\cos\theta + 12\sin\theta = 13\cos(\theta - 67.38°)$, the minimum value of $5\cos\theta + 12\sin\theta$ is –13 ***[1 mark]***. Hence the minimum value of $(5\cos\theta + 12\sin\theta)^3$ is $(-13)^3 = -2197$ ***[1 mark]***.

Core Section 4 — Exponentials and Logs

Warm-up Questions

1) a)-d)

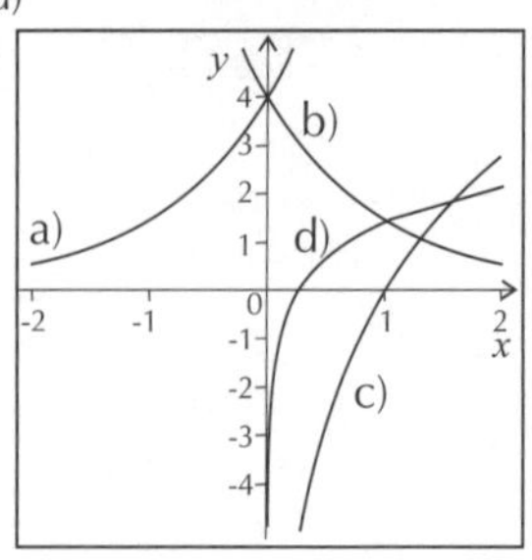

2) a) $e^{2x} = 6 \Rightarrow 2x = \ln 6 \Rightarrow x = \ln 6 \div 2 = 0.8959$ to 4 d.p.

b) $\ln(x + 3) = 0.75 \Rightarrow x + 3 = e^{0.75} \Rightarrow x = e^{0.75} - 3 = -0.8830$ to 4 d.p.

c) $3e^{-4x+1} = 5 \Rightarrow e^{-4x+1} = \frac{5}{3} \Rightarrow e^{4x-1} = \frac{3}{5} \Rightarrow 4x - 1 = \ln\frac{3}{5}$
$\Rightarrow x = (\ln\frac{3}{5} + 1) \div 4 = 0.1223$ to 4 d.p.

d) $\ln x + \ln 5 = \ln 4 \Rightarrow \ln(5x) = \ln 4 \Rightarrow 5x = 4$
$\Rightarrow x = 0.8000$ to 4 d.p.

3) a) $\ln(2x - 7) + \ln 4 = -3 \Rightarrow \ln(4(2x - 7)) = -3$
$\Rightarrow 8x - 28 = e^{-3} \Rightarrow x = \frac{e^{-3} + 28}{8}$ or $\frac{1}{8e^3} + \frac{7}{2}$.

b) $2e^{2x} + e^x = 3$, so if $y = e^x$, $2y^2 + y - 3 = 0$, which will factorise to: $(2y + 3)(y - 1) = 0$, so $e^x = -1.5$ (not possible), and $e^x = 1$, so $x = 0$ is the only solution.

4) a) $y = 2 - e^{x+1}$

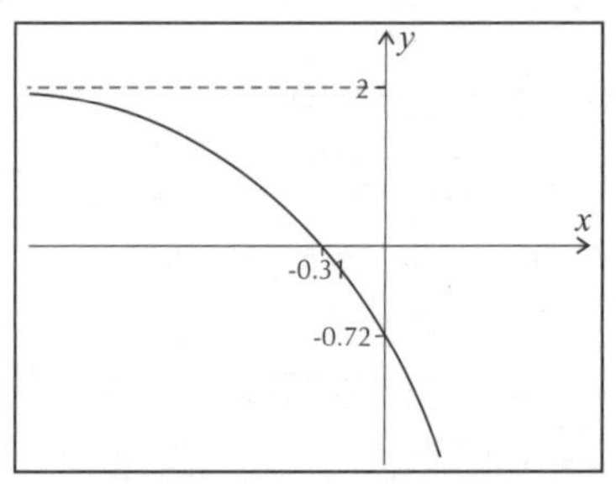

Goes through (0, –0.72) and (–0.31, 0), with asymptote at $y = 2$.

b) $y = 5e^{0.5x} + 5$

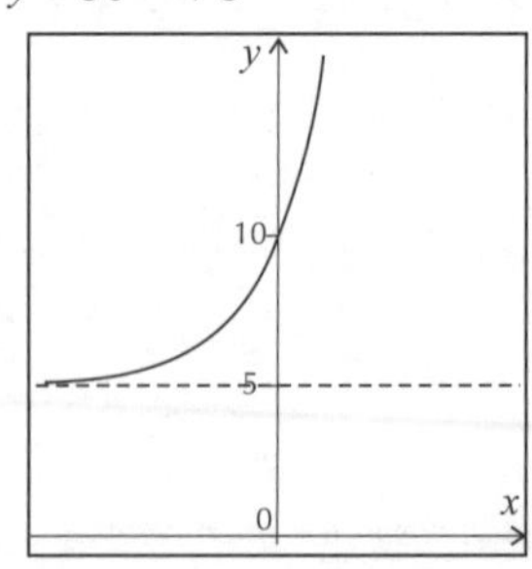

Goes through (0, 10), with asymptote at $y = 5$.

Answers

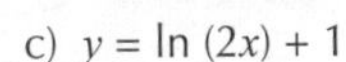

c) $y = \ln(2x) + 1$

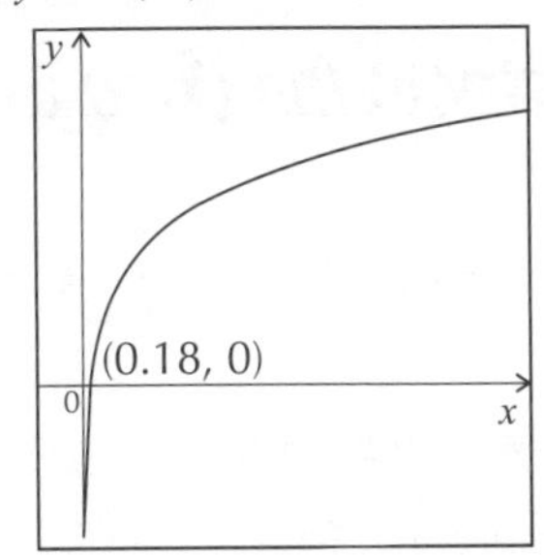

Goes through (0.18, 0), with asymptote at $x = 0$.

d) $y = \ln(x + 5)$

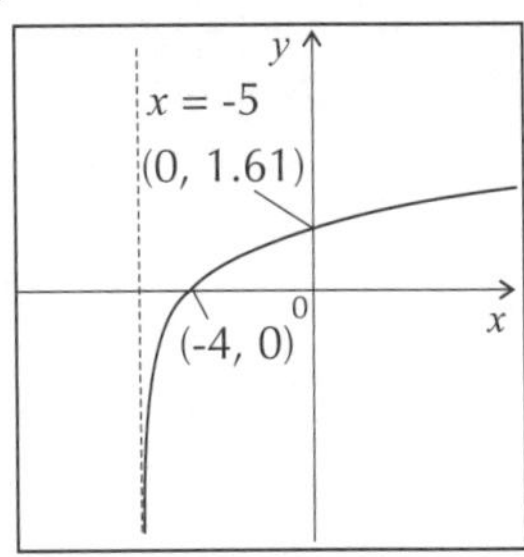

Goes through (0, 1.61) and (–4, 0), with asymptote at $x = -5$.

You can use your 'graph transformation' skills to work out what they'll look like, e.g. d) is just y = ln x shifted 5 to the left.

5) a) $V = 7500e^{-0.2t}$, so when $t = 0$, $V = 7500 \times e^0 = £7500$.

b) $V = 7500 \times e^{(-0.2 \times 10)} = £1015$ to the nearest £.

c) When $V = 500$, $500 = 7500e^{-0.2t}$

$\Rightarrow e^{-0.2t} = \frac{500}{7500} \Rightarrow e^{0.2t} = \frac{7500}{500} \Rightarrow 0.2t = \ln\frac{7500}{500} = 2.7080...$

$\Rightarrow t = 2.7080... \div 0.2 = 13.54$ years.

So it will be 13.54 years old before the value falls below £500.

d)

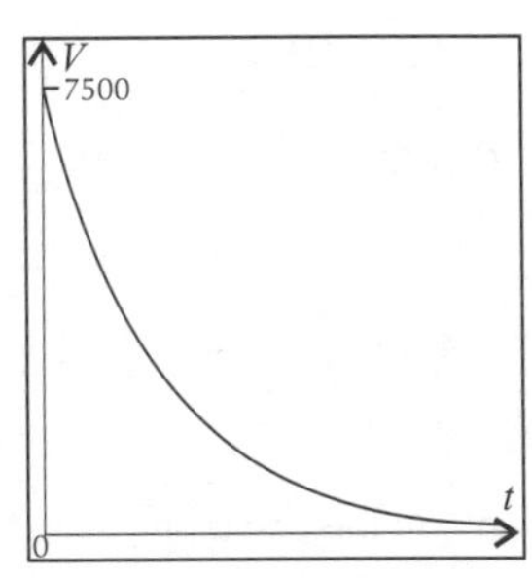

Goes through (0, 7500) with an asymptote at $y = 0$.

Exam Questions

1 a) $6e^x = 3 \Rightarrow e^x = 0.5$ ***[1 mark]*** $\Rightarrow x = \ln 0.5$ ***[1 mark]***.

b) $e^{2x} - 8e^x + 7 = 0$.

(This looks like a quadratic, so use $y = e^x$...)

If $y = e^x$, then $y^2 - 8y + 7 = 0$. This will factorise to give:

$(y - 7)(y - 1) = 0 \Rightarrow y = 7$ and $y = 1$.

So $e^x = 7 \Rightarrow x = \ln 7$, and $e^x = 1 \Rightarrow x = \ln 1 = 0$.

[4 marks available — 1 mark for factorisation of a quadratic, 1 mark for both solutions for e^x, and 1 mark for each correct solution for x.]

c) $4 \ln x = 3 \Rightarrow \ln x = 0.75$ ***[1 mark]*** $\Rightarrow x = e^{0.75}$ ***[1 mark]***.

d) $\ln x + \frac{24}{\ln x} = 10$

(You need to get rid of that fraction, so multiply through by ln x...)

$(\ln x)^2 + 24 = 10 \ln x$

$\Rightarrow (\ln x)^2 - 10 \ln x + 24 = 0$

(...which looks like a quadratic, so use y = ln x...)

$y^2 - 10y + 24 = 0 \Rightarrow (y - 6)(y - 4) = 0$

$\Rightarrow y = 6$ or $y = 4$.

So $\ln x = 6 \Rightarrow x = e^6$, or $\ln x = 4 \Rightarrow x = e^4$.

[4 marks available — 1 mark for factorisation of a quadratic, 1 mark for both solutions for ln x, and 1 mark for each correct solution for x.]

2 $y = e^{ax} + b$

The sketch shows that when $x = 0$, $y = -6$, so:

$-6 = e^0 + b$ ***[1 mark]***

$-6 = 1 + b \Rightarrow b = -7$ ***[1 mark]***.

The sketch also shows that when $y = 0$, $x = \frac{1}{4}\ln 7$, so:

$0 = e^{(\frac{a}{4}\ln 7)} - 7$ ***[1 mark]***

$\Rightarrow e^{(\frac{a}{4}\ln 7)} = 7$

$\Rightarrow \frac{a}{4}\ln 7 = \ln 7 \Rightarrow \frac{a}{4} = 1 \Rightarrow a = 4$ ***[1 mark]***.

The asymptote occurs as $x \to -\infty$, so $e^{4x} \to 0$,

and since $y = e^{4x} - 7$, $y \to -7$.

So the equation of the asymptote is $y = -7$ ***[1 mark]***.

3 a) $2e^x + 18e^{-x} = 20$

(Multiply through by e^x to remove the e^{-x}, since $e^x \times e^{-x} = 1$)

$2e^{2x} + 18 = 20e^x$

$\Rightarrow 2e^{2x} - 20e^x + 18 = 0 \Rightarrow e^{2x} - 10e^x + 9 = 0$

(This now looks like a quadratic equation, so use $y = e^x$ to simplify...)

$y^2 - 10y + 9 = 0$

$\Rightarrow (y - 1)(y - 9) = 0 \Rightarrow y = 1$ or $y = 9$.

So $e^x = 1 \Rightarrow x = 0$

or $e^x = 9 \Rightarrow x = \ln 9$.

[4 marks available — 1 mark for factorisation of a quadratic, 1 mark for both solutions for e^x, and 1 mark for each correct exact solution for x.]

b) $2 \ln x - \ln 3 = \ln 12$

$\Rightarrow 2 \ln x = \ln 12 + \ln 3$

(Use the log laws to simplify at this point...)

$\Rightarrow \ln x^2 = \ln 36$ ***[1 mark]***

$\Rightarrow x^2 = 36$ ***[1 mark]***

$\Rightarrow x = 6$ ***[1 mark]***

(x must be positive as ln (–6) does not exist.)

4 a) $y = \ln(4x - 3)$, and $x = a$ when $y = 1$.

$1 = \ln(4a - 3) \Rightarrow e^1 = 4a - 3$ ***[1 mark]***

$\Rightarrow a = (e^1 + 3) \div 4 = 1.43$ to 2 d.p. ***[1 mark]***.

b) The curve can only exist when $4x - 3 > 0$ ***[1 mark]***

so $x > 3 \div 4$, $x > 0.75$. If $x > b$, then $b = 0.75$ ***[1 mark]***.

Answers

c)

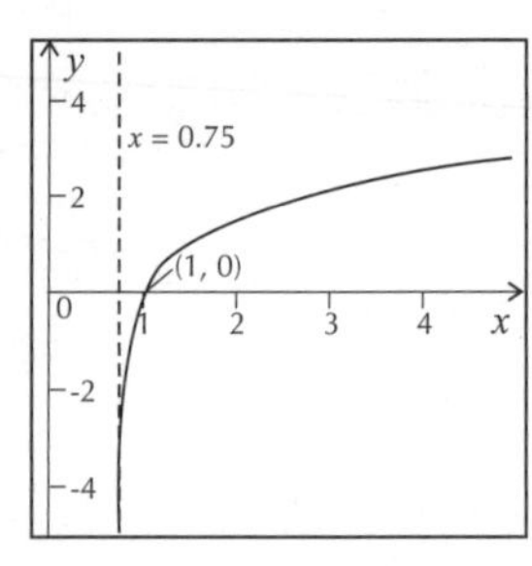

When $y = 0$, $4x - 3 = e^0 = 1$, so $x = 1$. As $x \to \infty$, $y \to \infty$ gradually. From (b), there will be an asymptote at $x = 0.75$.

[2 marks available — 1 mark for correct shape including asymptote at x = 0.75, 1 mark for (1, 0) as a point on the graph.]

5 a) When $t = 0$ (i.e. when the mink were introduced to the habitat) $M = 74 \times e^0 = 74$, so there were 74 mink originally ***[1 mark]***.

b) After 3 years, $M = 74 \times e^{0.6 \times 3}$ ***[1 mark]*** = 447 mink ***[1 mark]***.

You can't round up here as there are only 447 whole mink.

c) For $M = 10\,000$:
$10\,000 = 74e^{0.6t}$
$\Rightarrow e^{0.6t} = 10\,000 \div 74 = 135.1351$
$\Rightarrow 0.6t = \ln 135.1351 = 4.9063$ ***[1 mark]***
$\Rightarrow t = 4.9063 \div 0.6 = 8.2$ years to reach 10 000, so it would take 9 complete years for the population to exceed 10 000 ***[1 mark]***.

d)

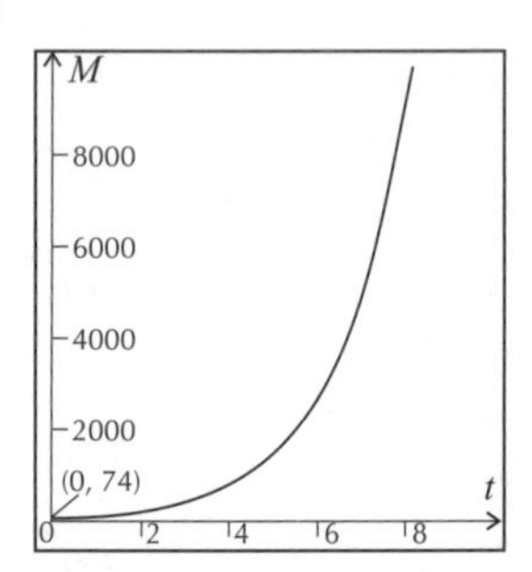

[2 marks available — 1 mark for correct shape of graph, 1 mark for (0, 74) as a point on the graph.]

6 a) B is the value of A when $t = 0$.
From the table, $B = 50$ ***[1 mark]***.

b) Substitute $t = 5$ and $A = 42$ into $A = 50e^{-kt}$:
$42 = 50e^{-5k} \Rightarrow e^{-5k} = \frac{42}{50} \Rightarrow e^{5k} = \frac{50}{42}$ ***[1 mark]***
$\Rightarrow 5k = \ln\left(\frac{50}{42}\right) = 0.17435$
$\Rightarrow k = 0.17435 \div 5 = 0.0349$ to 3 s.f. ***[1 mark]***.

c) $A = 50e^{-0.0349t}$ (using values from (a) and (b)), so when $t = 10$, $A = 50 \times e^{-0.0349 \times 10}$ ***[1 mark]***
= 35 to the nearest whole ***[1 mark]***.

d) The half-life will be the value of t when A reaches half of the original value of 50, i.e. when $A = 25$.
$25 = 50e^{-0.0349t}$
$\Rightarrow \frac{25}{50} = e^{-0.0349t} \Rightarrow \frac{50}{25} = e^{0.0349t} \Rightarrow e^{0.0349t} = 2$ ***[1 mark]***.
$0.0349t = \ln 2$ ***[1 mark]***
$\Rightarrow t = \ln 2 \div 0.0349 = 20$ days to the nearest day ***[1 mark]***.

Core Section 5 — Coordinate Geometry in the (x, y) Plane

Warm-up Questions

1) a) Substitute the values of t into the parametric equations to find the corresponding values of x and y:

$t = 0 \Rightarrow x = \frac{6-0}{2} = 3,\ y = 2(0)^2 + 0 + 4 = 4$

$t = 1 \Rightarrow x = \frac{6-1}{2} = 2.5,\ y = 2(1)^2 + 1 + 4 = 7$

$t = 2 \Rightarrow x = \frac{6-2}{2} = 2,\ y = 2(2)^2 + 2 + 4 = 14$

$t = 3 \Rightarrow x = \frac{6-3}{2} = 1.5,\ y = 2(3)^2 + 3 + 4 = 25$

b) Use the given values in the parametric equations and solve for t:
(i) $\frac{6-t}{2} = -7 \Rightarrow t = 20$
(ii) $2t^2 + t + 4 = 19$
$\Rightarrow 2t^2 + t - 15 = 0$
$\Rightarrow (2t - 5)(t + 3) = 0$
$\Rightarrow t = 2.5,\ t = -3$

c) Rearrange the parametric equation for x to make t the subject:
$x = \frac{6-t}{2} \Rightarrow 2x = 6 - t \Rightarrow t = 6 - 2x$
Now substitute this into the parametric equation for y:
$y = 2t^2 + t + 4$
$= 2(6 - 2x)^2 + (6 - 2x) + 4$
$= 2(36 - 24x + 4x^2) + 10 - 2x$
$y = 8x^2 - 50x + 82$.

2) a) Substitute the values of θ into the parametric equations to find the corresponding values of x and y:
(i) $x = 2\sin\frac{\pi}{4} = \frac{2}{\sqrt{2}} = \sqrt{2}$
$y = \cos^2\frac{\pi}{4} + 4 = \left(\cos\frac{\pi}{4}\right)^2 + 4 = \left(\frac{1}{\sqrt{2}}\right)^2 + 4 = \frac{1}{2} + 4 = \frac{9}{2}$
So the coordinates are $\left(\sqrt{2}, \frac{9}{2}\right)$.
(ii) $x = 2\sin\frac{\pi}{6} = 2 \times \frac{1}{2} = 1$
$y = \cos^2\frac{\pi}{6} + 4 = \left(\cos\frac{\pi}{6}\right)^2 + 4 = \left(\frac{\sqrt{3}}{2}\right)^2 + 4 = \frac{3}{4} + 4 = \frac{19}{4}$
So the coordinates are $\left(1, \frac{19}{4}\right)$.

b) Use the identity $\cos^2\theta = 1 - \sin^2\theta$ in the equation for y so both equations are in terms of $\sin\theta$:
$y = \cos^2\theta + 4$
$= 1 - \sin^2\theta + 4$
$= 5 - \sin^2\theta$
Rearrange the equation for x to get $\sin^2\theta$ in terms of x:
$x = 2\sin\theta \Rightarrow \frac{x}{2} = \sin\theta \Rightarrow \sin^2\theta = \frac{x^2}{4}$
So $y = 5 - \sin^2\theta \Rightarrow y = 5 - \frac{x^2}{4}$

c) $x = 2\sin\theta$, and $-1 \le \sin\theta \le 1$ so $-2 \le x \le 2$.

I know what you're thinking — this answer section would be brightened up immensely by a cheery song-and-dance number. Sorry, no such luck I'm afraid. Here's the next answer instead...

Answers

3) Use the identity $\cos 2\theta = 1 - 2\sin^2\theta$ in the equation for y:

$y = 3 + 2\cos 2\theta$
$= 3 + 2(1 - 2\sin^2\theta)$
$= 5 - 4\sin^2\theta$

Rearrange the equation for x to get $\sin^2\theta$ in terms of x:

$x = \frac{\sin\theta}{3} \Rightarrow 3x = \sin\theta \Rightarrow \sin^2\theta = 9x^2$

So $y = 5 - 4\sin^2\theta$
$\Rightarrow y = 5 - 4(9x^2)$
$\Rightarrow y = 5 - 36x^2$

4) a) (i) On the y-axis:

$x = 0 \Rightarrow t^2 - 1 = 0 \Rightarrow t = \pm 1$

If $t = 1$, $y = 4 + \frac{3}{1} = 7$

If $t = -1$, $y = 4 + \frac{3}{-1} = 1$

So the curve crosses the y-axis at (0, 1) and (0, 7).

(ii) Substitute the parametric equations into the equation of the line:

$x + 2y = 14$
$\Rightarrow (t^2 - 1) + 2(4 + \frac{3}{t}) = 14$
$\Rightarrow t^2 - 1 + 8 + \frac{6}{t} = 14$
$\Rightarrow t^2 - 7 + \frac{6}{t} = 0$
$\Rightarrow t^3 - 7t + 6 = 0$
$\Rightarrow (t - 1)(t^2 + t - 6) = 0$
$\Rightarrow (t - 1)(t - 2)(t + 3) = 0$
$\Rightarrow t = 1, t = 2, t = -3$

When $t = 1$, $x = 0$, $y = 7$ (from part (i))

When $t = 2$, $x = 2^2 - 1 = 3$, $y = 4 + \frac{3}{2} = 5.5$

When $t = -3$, $x = (-3)^2 - 1 = 8$, $y = 4 + \frac{3}{-3} = 3$

So the curve crosses the line $x + 2y = 14$ at (0, 7), (3, 5.5) and (8, 3).

b) Use $\int y\,dx = \int y \frac{dx}{dt}\,dt$

$\frac{dx}{dt} = 2t$, so $\int y\,dx = \int \left(4 + \frac{3}{t}\right)2t\,dt = \int 8t + 6\,dt$

5 a) $r = 7$, centre = (0, 0).

b) $r = 5$, centre = (2, –1).

Exam Questions

1 a) Substitute the given value of θ into the parametric equations:

$\theta = \frac{\pi}{3} \Rightarrow x = 1 - \tan\frac{\pi}{3} = 1 - \sqrt{3}$

$y = \frac{1}{2}\sin\left(\frac{2\pi}{3}\right) = \frac{1}{2}\left(\frac{\sqrt{3}}{2}\right) = \frac{\sqrt{3}}{4}$

So $P = \left(1 - \sqrt{3}, \frac{\sqrt{3}}{4}\right)$

[2 marks available — 1 mark for substituting $\theta = \frac{\pi}{3}$ into the parametric equations, 1 mark for both coordinates of P correct.]

b) Use $y = -\frac{1}{2}$ to find the value of θ:

$-\frac{1}{2} = \frac{1}{2}\sin 2\theta \Rightarrow \sin 2\theta = -1$
$\Rightarrow 2\theta = -\frac{\pi}{2}$
$\Rightarrow \theta = -\frac{\pi}{4}$

You can also find θ using the parametric equation for x, with x = 2.

[2 marks available — 1 mark for substituting given x- or y-value into the correct parametric equation, 1 mark for finding the correct value of θ.]

c) $x = 1 - \tan\theta \Rightarrow \tan\theta = 1 - x$

$y = \frac{1}{2}\sin 2\theta$
$= \frac{1}{2}\left(\frac{2\tan\theta}{1 + \tan^2\theta}\right)$
$= \frac{\tan\theta}{1 + \tan^2\theta}$
$= \frac{(1 - x)}{1 + (1 - x)^2}$
$= \frac{1 - x}{1 + 1 - 2x + x^2}$
$= \frac{1 - x}{x^2 - 2x + 2}$

[3 marks available — 1 mark for using the given identity to rearrange one of the parametric equations, 1 mark for eliminating θ from the parametric equation for y, 1 mark for correctly expanding to give the Cartesian equation given in the question.]

Just think, if you lived in the Bahamas, you could be doing this revision on the beach. (Please ignore that comment if you actually do live in the Bahamas. Or anywhere else where you can revise on the beach.)

2 a) Substitute $y = 1$ into the parametric equation for y:

$t^2 - 2t + 2 = 1$
$\Rightarrow t^2 - 2t + 1 = 0$
$\Rightarrow (t - 1)^2 = 0$
$\Rightarrow t = 1$ ***[1 mark]***

So a is the value of x when $t = 1$.

$a = t^3 + t = 1^3 + 1 = 2$ ***[1 mark]***

b) Substitute the parametric equations for x and y into the equation of the line:

$8y = x + 6$
$\Rightarrow 8(t^2 - 2t + 2) = (t^3 + t) + 6$ ***[1 mark]***
$\Rightarrow 8t^2 - 16t + 16 = t^3 + t + 6$
$\Rightarrow t^3 - 8t^2 + 17t - 10 = 0$

We know that this line passes through K, and from a) we know that $t = 1$ at K, so $t = 1$ is a solution of this equation, and $(t - 1)$ is a factor:

$\Rightarrow (t - 1)(t^2 - 7t + 10) = 0$ ***[1 mark]***
$\Rightarrow (t - 1)(t - 2)(t - 5) = 0$

So $t = 2$ at L and $t = 5$ at M. ***[1 mark]***

If you got stuck on this bit, go back and look up 'factorising cubics' in your AS notes.

Substitute $t = 2$ and $t = 5$ back into the parametric equations: ***[1 mark]***

If $t = 2$, then $x = 2^3 + 2 = 10$
and $y = 2^2 - 2(2) + 2 = 2$

If $t = 5$, then $x = 5^3 + 5 = 130$
and $y = 5^2 - 2(5) + 2 = 17$

So $L = (10, 2)$ ***[1 mark]***
and $M = (130, 17)$ ***[1 mark]***

3 a) Use the x- or y-coordinate of H in the relevant parametric equation to find θ:

At H, $3 + 4\sin\theta = 5$
$\Rightarrow 4\sin\theta = 2$
$\Rightarrow \sin\theta = \frac{1}{2}$
$\Rightarrow \theta = \frac{\pi}{6}$

OR

At H, $\frac{1 + \cos 2\theta}{3} = \frac{1}{2}$
$\Rightarrow 1 + \cos 2\theta = \frac{3}{2}$

Answers

$\Rightarrow \cos 2\theta = \frac{1}{2}$

$\Rightarrow 2\theta = \frac{\pi}{3}$

$\Rightarrow \theta = \frac{\pi}{6}$

[2 marks available — 1 mark for substituting one coordinate of H into the correct parametric equation, 1 mark finding the correct value of θ.]

b) $R = \int_{-1}^{5} y \, dx$

To get the integral with respect to θ, we need to use

$\int y \, dx = \int y \frac{dx}{d\theta} \, d\theta$ ***[1 mark]***

$\frac{dx}{d\theta} = 4\cos\theta$ ***[1 mark]***

Change the limits of the integral:

$x = 5 \Rightarrow \theta = \frac{\pi}{6}$, from part a)

$x = -1 \Rightarrow 3 + 4\sin\theta = -1$

$\Rightarrow 4\sin\theta = -4$

$\Rightarrow \sin\theta = -1$

$\Rightarrow \theta = -\frac{\pi}{2}$ ***[1 mark]***

So $R = \int_{-1}^{5} y \, dx = \int_{-\frac{\pi}{2}}^{\frac{\pi}{6}} y \frac{dx}{d\theta} \, d\theta$

$= \int_{-\frac{\pi}{2}}^{\frac{\pi}{6}} \left(\frac{1 + \cos 2\theta}{3}\right)(4\cos\theta) d\theta$ ***[1 mark]***

$= \int_{-\frac{\pi}{2}}^{\frac{\pi}{6}} \frac{4}{3}(1 + \cos 2\theta)(\cos\theta) d\theta$

$= \int_{-\frac{\pi}{2}}^{\frac{\pi}{6}} \frac{4}{3}(2\cos^2\theta)(\cos\theta) d\theta$

(Using $\cos 2\theta \equiv 2\cos^2\theta - 1$) ***[1 mark]***

$= \frac{8}{3}\int_{-\frac{\pi}{2}}^{\frac{\pi}{6}} \cos^3\theta \, d\theta$

c) Rearrange the parametric equation for x to make $\sin\theta$ the subject:

$x = 3 + 4\sin\theta \Rightarrow \sin\theta = \frac{x - 3}{4}$ ***[1 mark]***

Use the identity $\cos 2\theta = 1 - 2\sin^2\theta$ to rewrite the parametric equation for y in terms of $\sin\theta$:

$y = \frac{1 + \cos 2\theta}{3}$

$= \frac{1 + (1 - 2\sin^2\theta)}{3}$ ***[1 mark]***

$= \frac{2 - 2\sin^2\theta}{3}$

$= \frac{2}{3}(1 - \sin^2\theta)$

$= \frac{2}{3}\left(1 - \left(\frac{x-3}{4}\right)^2\right)$ ***[1 mark]***

$= \frac{2}{3}\left(1 - \frac{(x-3)^2}{16}\right)$

$= \frac{2}{3}\left(\frac{16 - (x^2 - 6x + 9)}{16}\right)$

$= \frac{2}{3}\left(\frac{-x^2 + 6x + 7}{16}\right)$

$= \frac{-x^2 + 6x + 7}{24}$ ***[1 mark]***

d) $-\frac{\pi}{2} \leq \theta \leq \frac{\pi}{2} \Rightarrow -1 \leq \sin\theta \leq 1$

$\Rightarrow -4 \leq 4\sin\theta \leq 4$

$\Rightarrow -1 \leq 3 + 4\sin\theta \leq 7$

$\Rightarrow -1 \leq x \leq 7$ ***[1 mark]***

As Shakespeare himself might have put it "That section was a ruddy pain in the backside, but at least it's finished."*

**Arnold Shakespeare (1948–)*

4 a) The area $R = \int_{4}^{18} y \, dx$ ***[1 mark]***

$\frac{dx}{dt} = 2t + 3$ ***[1 mark]***

Change the limits of the integral:

$x = 18 \Rightarrow t^2 + 3t - 18 = 0 \Rightarrow (t - 3)(t + 6) = 0 \Rightarrow t = 3, t = -6$

$x = 4 \Rightarrow t^2 + 3t - 4 = 0 \Rightarrow (t - 1)(t + 4) = 0 \Rightarrow t = 1, t = -4$

$t > 0$, so we can ignore the negative values of t, so the limits are $t = 3$ and $t = 1$. ***[1 mark]***

So $R = \int_{4}^{18} y \, dx = \int_{1}^{3} y \frac{dx}{dt} \, dt$

$= \int_{1}^{3} \left(t^2 + \frac{1}{t^3}\right)(2t + 3) \, dt$

$= \int_{1}^{3} \left(\frac{t^5 + 1}{t^3}\right)(2t + 3) \, dt$

$= \int_{1}^{3} \frac{(t^5 + 1)(2t + 3)}{t^3} \, dt$ ***[1 mark]***

b) $R = \int_{1}^{3} \frac{(t^5 + 1)(2t + 3)}{t^3} \, dt$

$= \int_{1}^{3} \frac{2t^6 + 3t^5 + 2t + 3}{t^3} \, dt$

$= \int_{1}^{3} \frac{2t^6}{t^3} + \frac{3t^5}{t^3} + \frac{2t}{t^3} + \frac{3}{t^3} \, dt$

$= \int_{1}^{3} 2t^3 + 3t^2 + 2t^{-2} + 3t^{-3} \, dt$ ***[1 mark]***

$= \left[\frac{t^4}{2} + t^3 - 2t^{-1} - \frac{3}{2}t^{-2}\right]_1^3$ ***[1 mark]***

$= \left(\frac{81}{2} + 27 - \frac{2}{3} - \frac{1}{6}\right) - \left(\frac{1}{2} + 1 - 2 - \frac{3}{2}\right)$ ***[1 mark]***

$= \frac{200}{3} - (-2)$

$= \frac{206}{3}$ ***[1 mark]***

5 The circle has a radius of 5, so the centre will have coordinates $(6 - 5, 4 - 5) = (1, -1)$.

$a = 1$ ***[1 mark]***, $b = -1$ ***[1 mark]***.

Answers

Core Section 6 — Differentiation 1

Warm-up Questions

1) a) $y = u^{\frac{1}{2}} \Rightarrow \frac{dy}{du} = \frac{1}{2}u^{-\frac{1}{2}} = \frac{1}{2\sqrt{u}} = \frac{1}{2\sqrt{x^3+2x^2}}$

$u = x^3 + 2x^2 \Rightarrow \frac{du}{dx} = 3x^2 + 4x$

$\Rightarrow \frac{dy}{dx} = \frac{3x^2+4x}{2\sqrt{x^3+2x^2}}$.

b) $y = u^{-\frac{1}{2}} \Rightarrow \frac{dy}{du} = -\frac{1}{2}u^{-\frac{3}{2}} = -\frac{1}{2(\sqrt{u})^3} = -\frac{1}{2(\sqrt{x^3+2x^2})^3}$

$u = x^3 + 2x^2 \Rightarrow \frac{du}{dx} = 3x^2 + 4x$

$\Rightarrow \frac{dy}{dx} = -\frac{3x^2+4x}{2(\sqrt{x^3+2x^2})^3}$.

c) $y = e^u \Rightarrow \frac{dy}{du} = e^u = e^{5x^2}$.

$u = 5x^2 \Rightarrow \frac{du}{dx} = 10x$

$\Rightarrow \frac{dy}{dx} = 10xe^{5x^2}$.

d) $y = \ln u \Rightarrow \frac{dy}{du} = \frac{1}{u} = \frac{1}{(6-x^2)}$

$u = 6 - x^2 \Rightarrow \frac{du}{dx} = -2x$

$\Rightarrow \frac{dy}{dx} = -\frac{2x}{(6-x^2)}$.

2) a) $x = 2e^y \Rightarrow \frac{dx}{dy} = 2e^y \Rightarrow \frac{dy}{dx} = \frac{1}{2e^y}$.

b) $x = \ln u$ where $u = 2y + 3$

$\frac{dx}{du} = \frac{1}{u} = \frac{1}{2y+3}$ and $\frac{du}{dy} = 2 \Rightarrow \frac{dx}{dy} = \frac{2}{2y+3}$

$\Rightarrow \frac{dy}{dx} = \frac{2y+3}{2} = y + 1.5$.

3) a) For $f(x) = y = \sin^2(x + 2)$, use the chain rule twice:

$y = u^2$, where $u = \sin(x + 2)$

$\frac{dy}{du} = 2u = 2\sin(x+2)$ and $\frac{du}{dx} = \cos(x+2)\cdot 1$ (by chain rule)

$\Rightarrow \frac{dy}{dx} = f'(x) = 2\sin(x+2)\cos(x+2)$ $[= \sin(2x+4)]$.

b) $f(x) = y = 2\cos 3x$:

$y = 2\cos u$, where $u = 3x$

$\frac{dy}{du} = -2\sin u = -2\sin 3x$ and $\frac{du}{dx} = 3$

$\Rightarrow \frac{dy}{dx} = f'(x) = -6\sin 3x$.

c) $f(x) = y = \sqrt{\tan x} = (\tan x)^{\frac{1}{2}}$:

$y = u^{\frac{1}{2}}$, where $u = \tan x$

$\frac{dy}{du} = \frac{1}{2}u^{-\frac{1}{2}} = \frac{1}{2\sqrt{u}} = \frac{1}{2\sqrt{\tan x}}$ and $\frac{du}{dx} = \sec^2 x$

$\Rightarrow \frac{dy}{dx} = f'(x) = \frac{\sec^2 x}{2\sqrt{\tan x}}$.

4) a) For $y = e^{2x}(x^2 - 3)$, use the product rule:

$u = e^{2x} \Rightarrow \frac{du}{dx} = 2e^{2x}$ (from the chain rule),

$v = x^2 - 3 \Rightarrow \frac{dv}{dx} = 2x$.

$\frac{dy}{dx} = u\frac{dv}{dx} + v\frac{du}{dx} = 2xe^{2x} + 2e^{2x}(x^2 - 3) = 2e^{2x}(x^2 + x - 3)$.

When $x = 0$, $\frac{dy}{dx} = 2e^0(0 + 0 - 3) = 2 \times 1 \times -3 = -6$.

b) For $y = \ln x \sin x$, use the product rule:

$u = \ln x \Rightarrow \frac{du}{dx} = \frac{1}{x}$,

$v = \sin x \Rightarrow \frac{dv}{dx} = \cos x$.

$\frac{dy}{dx} = u\frac{dv}{dx} + v\frac{du}{dx} = \ln x \cos x + \frac{\sin x}{x}$.

When $x = 1$, $\frac{dy}{dx} = \ln 1 \cos 1 + \frac{\sin 1}{1} = 0 + \sin 1$
$= 0.841$ (to 3 s.f.).

5) For $y = \frac{6x^2+3}{4x^2-1}$, use the quotient rule:

$u = 6x^2 + 3 \Rightarrow \frac{du}{dx} = 12x$,

$v = 4x^2 - 1 \Rightarrow \frac{dv}{dx} = 8x$.

$\frac{dy}{dx} = \frac{v\frac{du}{dx} - u\frac{dv}{dx}}{v^2} = \frac{12x(4x^2-1) - 8x(6x^2+3)}{(4x^2-1)^2}$.

At (1, 3), $x = 1$ and so gradient =

$\frac{dy}{dx} = \frac{12(4-1) - 8(6+3)}{(4-1)^2} = \frac{36-72}{9} = -4$.

Equation of a straight line is:

$y - y_1 = m(x - x_1)$, where m is the gradient.

So the equation of the tangent at (1, 3) is:

$y - 3 = -4(x - 1) \Rightarrow y = -4x + 7$ (or equivalent).

6) $y = \operatorname{cosec}(3x - 2)$, so use chain rule:

$y = \operatorname{cosec} u$ where $u = 3x - 2$

$\frac{dy}{du} = -\operatorname{cosec} u \cot u = -\operatorname{cosec}(3x-2)\cot(3x-2)$

and $\frac{du}{dx} = 3$,

$\Rightarrow \frac{dy}{dx} = -3\operatorname{cosec}(3x-2)\cot(3x-2)$

$= \frac{-3}{\sin(3x-2)\tan(3x-2)}$.

When $x = 0$, $\frac{dy}{dx} = \frac{-3}{\sin(-2)\tan(-2)} = 1.51$ (to 3 s.f.).

7) For $y = \frac{e^x}{\sqrt{x}}$, use the quotient rule:

$u = e^x \Rightarrow \frac{du}{dx} = e^x$,

$v = x^{\frac{1}{2}} \Rightarrow \frac{dv}{dx} = \frac{1}{2}x^{-\frac{1}{2}} = \frac{1}{2\sqrt{x}}$.

$\frac{dy}{dx} = \frac{v\frac{du}{dx} - u\frac{dv}{dx}}{v^2} = \frac{e^x\sqrt{x} - \frac{e^x}{2\sqrt{x}}}{(\sqrt{x})^2}$.

Multiplying top and bottom by $2\sqrt{x}$ gives:

$\frac{dy}{dx} = \frac{2xe^x - e^x}{2x\sqrt{x}} = \frac{e^x(2x-1)}{2x\sqrt{x}}$.

At the stationary point, $\frac{dy}{dx} = 0$,

$\Rightarrow \frac{e^x(2x-1)}{2x\sqrt{x}} = 0 \Rightarrow e^x(2x - 1) = 0$,

so either $e^x = 0$ or $2x - 1 = 0$. e^x does not exist at 0,

so the stationary point must be at $2x - 1 = 0$, $x = \frac{1}{2}$.

To find out the nature of the stationary point,

differentiate again: $\frac{dy}{dx} = \frac{e^x(2x-1)}{2x\sqrt{x}}$, so use quotient rule and product rule:

$u = e^x(2x - 1) \Rightarrow$ using product rule $\frac{du}{dx} = 2e^x + e^x(2x - 1)$
$= e^x(2x - 1 + 2) = e^x(2x + 1)$.

$v = 2x^{\frac{3}{2}} \Rightarrow \frac{dv}{dx} = 3x^{\frac{1}{2}} = 3\sqrt{x}$.

$\frac{d^2y}{dx^2} = \frac{v\frac{du}{dx} - u\frac{dv}{dx}}{v^2} = \frac{2x\sqrt{x}e^x(2x+1) - 3\sqrt{x}e^x(2x-1)}{(2x\sqrt{x})^2}$

$= \frac{\sqrt{x}e^x(4x^2 - 4x + 3)}{4x^3}$.

When $x = \frac{1}{2}$, $\frac{d^2y}{dx^2} > 0$, so it is a minimum point.

Give yourself a big pat on the back if you survived question 7.

Answers

Exam Questions

1 a) For $x = \sqrt{y^2 + 3y}$, find $\frac{dx}{dy}$ first (using the chain rule):

$x = u^{\frac{1}{2}}$ where $u = y^2 + 3y$.

$\frac{dx}{du} = \frac{1}{2}u^{-\frac{1}{2}} = \frac{1}{2\sqrt{u}} = \frac{1}{2\sqrt{y^2 + 3y}}$ ***[1 mark].***

$\frac{du}{dy} = 2y + 3$ ***[1 mark].***

So $\frac{dx}{dy} = \frac{2y + 3}{2\sqrt{y^2 + 3y}}$ ***[1 mark].***

(Now, flip the fraction upside down for dy/dx...)

$\frac{dy}{dx} = \frac{2\sqrt{y^2 + 3y}}{2y + 3}$ ***[1 mark].***

At the point (2, 1), $y = 1$, so:

$\frac{dy}{dx} = \frac{2\sqrt{1^2 + 3}}{2 + 3} = \frac{4}{5} = 0.8$ ***[1 mark].***

b) Equation of a straight line is:

$y - y_1 = m(x - x_1)$, where m is the gradient.

For the tangent at (2, 1), $y_1 = 1$, $x_1 = 2$, and $m = \frac{dy}{dx} = 0.8$.

So the equation is:

$y - 1 = 0.8(x - 2) \Rightarrow y = 0.8x - 0.6$ (or equivalent fractions)

[2 marks available — 1 mark for correct substitution of (2, 1) and gradient from (a), and 1 mark for final answer.]

2 a) For $y = \sqrt{e^x + e^{2x}}$, use the chain rule:

$y = u^{\frac{1}{2}}$ where $u = e^x + e^{2x}$.

$\frac{dy}{du} = \frac{1}{2}u^{-\frac{1}{2}} = \frac{1}{2\sqrt{u}} = \frac{1}{2\sqrt{e^x + e^{2x}}}$ ***[1 mark].***

$\frac{du}{dx} = e^x + 2e^{2x}$ ***[1 mark].***

So $\frac{dy}{dx} = \frac{e^x + 2e^{2x}}{2\sqrt{e^x + e^{2x}}}$ ***[1 mark].***

b) For $y = 3e^{2x+1} - \ln(1 - x^2) + 2x^3$, use the chain rule for the first 2 parts separately:

For $y = 3e^{2x+1}$, $y = 3e^u$ where $u = 2x + 1$, so $\frac{dy}{du} = 3e^u = 3e^{2x+1}$

and $\frac{du}{dx} = 2$, so $\frac{dy}{dx} = 6e^{2x+1}$ ***[1 mark].***

For $y = \ln(1 - x^2)$, $y = \ln u$ where $u = 1 - x^2$,

so $\frac{dy}{du} = \frac{1}{u} = \frac{1}{(1 - x^2)}$ and $\frac{du}{dx} = -2x$, so $\frac{dy}{dx} = -\frac{2x}{(1 - x^2)}$

[1 mark].

So overall: $\frac{dy}{dx} = 6e^{2x+1} + \frac{2x}{(1 - x^2)} + 6x^2$ ***[1 mark].***

3 For $y = \frac{e^x + x}{e^x - x}$, use the quotient rule:

$u = e^x + x \Rightarrow \frac{du}{dx} = e^x + 1$.

$v = e^x - x \Rightarrow \frac{dv}{dx} = e^x - 1$.

$\frac{dy}{dx} = \frac{v\frac{du}{dx} - u\frac{dv}{dx}}{v^2} = \frac{(e^x - x)(e^x + 1) - (e^x + x)(e^x - 1)}{(e^x - x)^2}$.

When $x = 0$, $e^x = 1$, and $\frac{dy}{dx} = \frac{(1 - 0)(1 + 1) - (1 + 0)(1 - 1)}{(1 - 0)^2}$

$= \frac{2 - 0}{1^2} = 2$.

[3 marks available — 1 mark for finding u, v and their derivatives, 1 mark for dy/dx (however rearranged), and 1 mark for dy/dx = 2 when x = 0.]

4 a) For $f(x) = 4 \ln 3x$, use the chain rule:

$y = 4 \ln u$ where $u = 3x$, so $\frac{dy}{du} = \frac{4}{u} = \frac{4}{3x}$, and $\frac{du}{dx} = 3$

[1 mark for both], so $f'(x) = \frac{dy}{dx} = \frac{12}{3x} = \frac{4}{x}$ ***[1 mark].***

So for $x = 1$, $f'(1) = 4$ ***[1 mark].***

b) Equation of a straight line is:

$y - y_1 = m(x - x_1)$, where m is the gradient.

For the tangent at $x_1 = 1$, $y_1 = 4\ln 3$, and $m = \frac{dy}{dx} = 4$.

So the equation is:

$y - 4\ln 3 = 4(x - 1) \Rightarrow y = 4(x - 1 + \ln 3)$ (or equivalent).

[3 marks available — 1 mark for finding y = 4ln 3, 1 mark for correct substitution of (1, 4ln3) and gradient from (a), and 1 mark for correct final answer.]

5 $f(x) = \sec x = \frac{1}{\cos x}$, so using the quotient rule:

$u = 1 \Rightarrow \frac{du}{dx} = 0$ and $v = \cos x \Rightarrow \frac{dv}{dx} = -\sin x$.

$\frac{dy}{dx} = \frac{v\frac{du}{dx} - u\frac{dv}{dx}}{v^2} = \frac{(\cos x \cdot 0) - (1 \cdot - \sin x)}{\cos^2 x} = \frac{\sin x}{\cos^2 x}$.

Since $\tan x = \frac{\sin x}{\cos x}$, and $\sec x = \frac{1}{\cos x}$,

$f'(x) = \frac{dy}{dx} = \frac{\sin x}{\cos x} \times \frac{1}{\cos x} = \sec x \tan x$.

[4 marks available — 1 mark for correct identity for sec x, 1 mark for correct entry into quotient rule, 1 mark for correct answer from quotient rule, and 1 mark for correct rearrangement to sec x tan x.]

6 a) For $y = \ln(3x + 1) \sin(3x + 1)$, use the product rule and the chain rule:

Product rule: $u = \ln(3x + 1)$ and $v = \sin(3x + 1)$.

Using the chain rule for $\frac{du}{dx} = \frac{3}{3x + 1}$ ***[1 mark].***

Using the chain rule for $\frac{dv}{dx} = 3\cos(3x + 1)$ ***[1 mark].***

So $\frac{dy}{dx} = u\frac{dv}{dx} + v\frac{du}{dx}$

$= [\ln(3x + 1) \cdot 3\cos(3x + 1)] + [\sin(3x + 1) \cdot \frac{3}{3x + 1}]$ ***[1 mark]***

$= 3\ln(3x + 1)\cos(3x + 1) + \frac{3\sin(3x + 1)}{3x + 1}$ ***[1 mark].***

b) For $y = \frac{\sqrt{x^2 + 3}}{\cos 3x}$, use the quotient rule and the chain rule:

Quotient rule: $u = \sqrt{x^2 + 3}$ and $v = \cos 3x$.

Using the chain rule for $\frac{du}{dx} = \frac{2x}{2\sqrt{x^2 + 3}} = \frac{x}{\sqrt{x^2 + 3}}$ ***[1 mark].***

Using the chain rule for $\frac{dv}{dx} = -3\sin 3x$ ***[1 mark].***

So $\frac{dy}{dx} = \frac{v\frac{du}{dx} - u\frac{dv}{dx}}{v^2} = \frac{\left[\cos 3x \cdot \frac{x}{\sqrt{x^2 + 3}}\right] - \left[\sqrt{x^2 + 3} \cdot - 3\sin 3x\right]}{\cos^2 3x}$

[1 mark]. Then multiply top and bottom by $\sqrt{x^2 + 3}$ to get:

$\frac{dy}{dx} = \frac{x\cos 3x + 3(x^2 + 3)\sin 3x}{(\sqrt{x^2 + 3})\cos^2 3x} = \frac{x + 3(x^2 + 3)\tan 3x}{(\sqrt{x^2 + 3})\cos 3x}$

[1 mark].

c) For $y = \sin^3(2x^2)$, use the chain rule twice:

$y = u^3$ where $u = \sin(2x^2)$.

$\frac{dy}{du} = 3u^2 = 3\sin^2(2x^2)$ ***[1 mark].***

$\frac{du}{dx} = 4x\cos(2x^2)$ (using chain rule again) ***[1 mark].***

So $\frac{dy}{dx} = 12x\sin^2(2x^2)\cos(2x^2)$ ***[1 mark].***

Answers

d) For $y = 2\text{cosec}\,(3x)$, use the chain rule:

$y = 2\text{ cosec } u$ where $u = 3x$.

$\frac{dy}{du} = -2\text{cosec } u \cot u = -2\text{cosec } 3x \cot 3x$.

$\frac{du}{dx} = 3$ ***[1 mark for both]***,

so $\frac{dy}{dx} = -6\text{cosec } 3x \cot 3x$ ***[1 mark]***.

7 For $y = \sin^2 x - 2\cos 2x$, use the chain rule on each part:

For $y = \sin^2 x$, $y = u^2$ where $u = \sin x$, so $\frac{dy}{du} = 2u = 2\sin x$ and $\frac{du}{dx} = \cos x$, so $\frac{dy}{dx} = 2\sin x\cos x$ ***[1 mark]***.

For $y = 2\cos 2x$, $y = 2\cos u$ where $u = 2x$, so $\frac{dy}{du} = -2\sin u = -2\sin 2x$ and $\frac{du}{dx} = 2$, so $\frac{dy}{dx} = -4\sin 2x$ ***[1 mark]***.

Overall $\frac{dy}{dx} = 2\sin x\cos x + 4\sin 2x$.

(Think 'double angle formula' for the sin x cos x...)

$\sin 2x \equiv 2\sin x\cos x$, so:

$\frac{dy}{dx} = \sin 2x + 4\sin 2x = 5\sin 2x$ ***[1 mark]***.

(For gradient of the tangent, put the x value into dy/dx...)

Gradient of the tangent when $x = \frac{\pi}{12}$ is:

$5 \times \sin\frac{\pi}{6} = 2.5$ ***[1 mark]***.

8 For $x = \sin 4y$, $\frac{dx}{dy} = 4\cos 4y$ ***[1 mark]*** (using chain rule),

and so $\frac{dy}{dx} = \frac{1}{4\cos 4y}$ ***[1 mark]***.

At $(0, \frac{\pi}{4})$, $y = \frac{\pi}{4}$ and so $\frac{dy}{dx} = \frac{1}{4\cos\pi} = -\frac{1}{4}$ ***[1 mark]***.

(This is the gradient of the tangent at that point, so to find the gradient of the normal do –1 ÷ gradient of tangent...)

Gradient of normal at $(0, \frac{\pi}{4}) = -1 \div -\frac{1}{4} = 4$ ***[1 mark]***.

Equation of a straight line is:

$y - y_1 = m(x - x_1)$, where m is the gradient.

For the normal at $(0, \frac{\pi}{4})$, $x_1 = 0$, $y_1 = \frac{\pi}{4}$, and $m = 4$.

So the equation is:

$y - \frac{\pi}{4} = 4(x - 0)$ ***[1 mark]*** $\Rightarrow y = 4x + \frac{\pi}{4}$ (or equivalent) ***[1 mark]***.

9 a) For $y = e^x \sin x$, use the product rule:

$u = e^x \Rightarrow \frac{du}{dx} = e^x$

$v = \sin x \Rightarrow \frac{dv}{dx} = \cos x$

So $\frac{dy}{dx} = u\frac{dv}{dx} + v\frac{du}{dx} = (e^x \cdot \cos x) + (\sin x \cdot e^x)$
$= e^x(\cos x + \sin x)$ ***[1 mark]***.

At the turning points, $\frac{dy}{dx} = 0$, so:

$e^x(\cos x + \sin x) = 0$ ***[1 mark]***

$\Rightarrow$ turning points are when $e^x = 0$ or $\cos x + \sin x = 0$.

e^x cannot be 0, so the turning points are when $\cos x + \sin x = 0$ ***[1 mark]***

$\Rightarrow \sin x = -\cos x \Rightarrow \frac{\sin x}{\cos x} = -1 \Rightarrow \tan x = -1$ ***[1 mark]***.

Look back at C2 for the graph of tan x to help you find all the solutions — it repeats itself every π radians...

There are two solutions for $\tan x = -1$ in the interval $-\pi \le x \le \pi$: $x = -\frac{\pi}{4}$ and $x = \pi - \frac{\pi}{4} = \frac{3\pi}{4}$,

so the values of x at each turning point are $-\frac{\pi}{4}$ ***[1 mark]*** and $\frac{3\pi}{4}$ ***[1 mark]***.

b) To determine the nature of the turning points, find $\frac{d^2y}{dx^2}$ at the points:

For $\frac{dy}{dx} = e^x(\cos x + \sin x)$, use the product rule:

$u = e^x \Rightarrow \frac{du}{dx} = e^x$

$v = \cos x + \sin x \Rightarrow \frac{dv}{dx} = \cos x - \sin x$, so:

$\frac{d^2y}{dx^2} = u\frac{dv}{dx} + v\frac{du}{dx} = [e^x \cdot (\cos x - \sin x)] + [(\cos x + \sin x) \cdot e^x]$
$= 2e^x \cos x$ ***[1 mark]***.

When $x = -\frac{\pi}{4}$, $\frac{d^2y}{dx^2} > 0$ ***[1 mark]***,
so this is a minimum point ***[1 mark]***.

When $x = \frac{3\pi}{4}$, $\frac{d^2y}{dx^2} < 0$ ***[1 mark]***,
so this is a maximum point ***[1 mark]***.

Core Section 7 — Differentiation 2

Warm-up Questions

1) a) $\frac{dx}{dt} = 2t$, $\frac{dy}{dt} = 9t^2 - 4$, so $\frac{dy}{dx} = \frac{dy}{dt} \div \frac{dx}{dt} = \frac{9t^2 - 4}{2t}$

b) The stationary points are when $\frac{9t^2 - 4}{2t} = 0$

$\Rightarrow 9t^2 = 4 \Rightarrow t = \pm\frac{2}{3}$

$t = \frac{2}{3} \Rightarrow x = \left(\frac{2}{3}\right)^2 = \frac{4}{9}$, $y = 3\left(\frac{2}{3}\right)^3 - 4\left(\frac{2}{3}\right) = \frac{8}{9} - \frac{8}{3} = -\frac{16}{9}$

$t = -\frac{2}{3} \Rightarrow x = \left(-\frac{2}{3}\right)^2 = \frac{4}{9}$,

$y = 3\left(-\frac{2}{3}\right)^3 - 4\left(-\frac{2}{3}\right) = -\frac{8}{9} + \frac{8}{3} = \frac{16}{9}$

So the stationary points are $\left(\frac{4}{9}, -\frac{16}{9}\right)$ and $\left(\frac{4}{9}, \frac{16}{9}\right)$.

2) a) Differentiate each term separately with respect to x:

$\frac{d}{dx}4x^2 - \frac{d}{dx}2y^2 = \frac{d}{dx}7x^2y$

Differentiate $4x^2$ first:

$\Rightarrow 8x - \frac{d}{dx}2y^2 = \frac{d}{dx}7x^2y$

Differentiate $2y^2$ using chain rule:

$\Rightarrow 8x - \frac{d}{dy}2y^2\frac{dy}{dx} = \frac{d}{dx}7x^2y$

$\Rightarrow 8x - 4y\frac{dy}{dx} = \frac{d}{dx}7x^2y$

Differentiate $7x^2y$ using product rule:

$\Rightarrow 8x - 4y\frac{dy}{dx} = 7x^2\frac{d}{dx}y + y\frac{d}{dx}7x^2$

$\Rightarrow 8x - 4y\frac{dy}{dx} = 7x^2\frac{dy}{dx} + 14xy$

Rearrange to make $\frac{dy}{dx}$ the subject:

$\Rightarrow (4y + 7x^2)\frac{dy}{dx} = 8x - 14xy$

$\Rightarrow \frac{dy}{dx} = \frac{8x - 14xy}{4y + 7x^2}$

For implicit differentiation questions, you need to know what you're doing with the chain rule and product rule. If you're struggling to keep up with what's going on here, go back and refresh your memory.

b) Differentiate each term separately with respect to x:

$\frac{d}{dx}3x^4 - \frac{d}{dx}2xy^2 = \frac{d}{dx}y$

Differentiate $3x^4$ first:

$\Rightarrow 12x^3 - \frac{d}{dx}2xy^2 = \frac{dy}{dx}$

Differentiate $2xy^2$ using product rule:

Answers

$\Rightarrow 12x^3 - \left(y^2\frac{d}{dx}2x + 2x\frac{d}{dy}y^2\frac{dy}{dx}\right) = \frac{dy}{dx}$

$\Rightarrow 12x^3 - 2y^2 - 4xy\frac{dy}{dx} = \frac{dy}{dx}$

Rearrange to make $\frac{dy}{dx}$ the subject:

$\Rightarrow (1 + 4xy)\frac{dy}{dx} = 12x^3 - 2y^2$

$\Rightarrow \frac{dy}{dx} = \frac{12x^3 - 2y^2}{1 + 4xy}$

c) Use the product rule to differentiate each term separately with respect to x:

$\frac{d}{dx}\cos x \sin y = \frac{d}{dx}xy$

$\Rightarrow \cos x\frac{d}{dx}(\sin y) + \sin y\frac{d}{dx}(\cos x) = x\frac{d}{dx}y + y\frac{d}{dx}x$

Use the chain rule on $\frac{d}{dx}(\sin y)$:

$\Rightarrow \cos x\frac{d}{dy}(\sin y)\frac{dy}{dx} + \sin y\frac{d}{dx}(\cos x) = x\frac{d}{dx}y + y\frac{d}{dx}x$

$\Rightarrow (\cos x \cos y)\frac{dy}{dx} - \sin y \sin x = x\frac{dy}{dx} + y$

Rearrange to make $\frac{dy}{dx}$ the subject:

$\Rightarrow (\cos x \cos y - x)\frac{dy}{dx} = y + \sin x \sin y$

$\Rightarrow \frac{dy}{dx} = \frac{\sin x \sin y + y}{\cos x \cos y - x}$

Make sure you learn how to differentiate trig functions. Chances are they'll come up in your A2 exam. And even if they don't, that sort of skill will make you a hit at parties. Trust me.

3) a) At $(1, -4)$, $\frac{dy}{dx} = \frac{8x - 14xy}{4y + 7x^2}$

$= \frac{8(1) - 14(1)(-4)}{4(-4) + 7(1)^2} = \frac{8 + 56}{-16 + 7} = -\frac{64}{9}$

b) At $(1, 1)$, $\frac{dy}{dx} = \frac{12x^3 - 2y^2}{1 + 4xy}$

$= \frac{12(1)^3 - 2(1)^2}{1 + 4(1)^2(1)} = \frac{12 - 2}{1 + 4} = \frac{10}{5} = 2$

So the gradient of the normal is $-\frac{1}{2}$.

4) Take the log of both sides of the equation:

$y = a^x \Rightarrow \ln y = \ln a^x$

$\Rightarrow \ln y = x \ln a$ (using log laws)

Now use implicit differentiation to find $\frac{dy}{dx}$:

$\ln y = x \ln a \Rightarrow \frac{d}{dx}(\ln y) = \frac{d}{dx}(x \ln a)$

$\Rightarrow \frac{d}{dy}(\ln y)\frac{dy}{dx} = \ln a$

$\Rightarrow \frac{1}{y}\frac{dy}{dx} = \ln a$

$\Rightarrow \frac{dy}{dx} = y \ln a$

So as $y = a^x$, $\frac{dy}{dx} = a^x \ln a$

5) $A = 2(x)(2x) + 2(x)(3x) + 2(2x)(3x)$

$= 4x^2 + 6x^2 + 12x^2$

$= 22x^2$

So $\frac{dA}{dx} = 44x$

$V = (x)(2x)(3x) = 6x^3$

So $\frac{dV}{dx} = 18x^2$

By the chain rule:

$\frac{dA}{dt} = \frac{dA}{dx} \times \frac{dx}{dt} = 44x \times \frac{dx}{dt}$

To find $\frac{dx}{dt}$, use the chain rule again:

$\frac{dx}{dt} = \frac{dx}{dV} \times \frac{dV}{dt} = \frac{1}{\left(\frac{dV}{dx}\right)} \times \frac{dV}{dt} = \frac{1}{18x^2} \times 3 = \frac{1}{6x^2}$

So $\frac{dA}{dt} = 44x \times \frac{1}{6x^2} = \frac{22}{3x}$

Exam Questions

1 a) c is the value of y when $x = 2$. If $x = 2$, then

$6x^2y - 7 = 5x - 4y^2 - x^2 \Rightarrow 6(2)^2y - 7 = 5(2) - 4y^2 - (2)^2$

$\Rightarrow 24y - 7 = 6 - 4y^2$

$\Rightarrow 4y^2 + 24y - 13 = 0$

$\Rightarrow (2y + 13)(2y - 1) = 0$

$\Rightarrow y = -6.5$ or $y = 0.5$ ***[1 mark]***

$c > 0$, so $c = 0.5$ ***[1 mark]***

b) (i) Q is another point on C where $y = 0.5$.

If $y = 0.5$, then $6x^2y - 7 = 5x - 4y^2 - x^2$

$\Rightarrow 6x^2(0.5) - 7 = 5x - 4(0.5)^2 - x^2$ ***[1 mark]***

$\Rightarrow 3x^2 - 7 = 5x - 1 - x^2$

$\Rightarrow 4x^2 - 5x - 6 = 0$

$\Rightarrow (x - 2)(4x + 3)$

$\Rightarrow x = 2$ or $x = -0.75$

$x \neq 2$, as $x = 2$ at the other point where T crosses C.

So the coordinates of Q are $(-0.75, 0.5)$. ***[1 mark]***

(ii) To find the gradient of C, use implicit differentiation.

Differentiate each term separately with respect to x:

$\frac{d}{dx}6x^2y - \frac{d}{dx}7 = \frac{d}{dx}5x - \frac{d}{dx}4y^2 - \frac{d}{dx}x^2$ ***[1 mark]***

Differentiate x-terms and constant terms:

$\Rightarrow \frac{d}{dx}6x^2y - 0 = 5 - \frac{d}{dx}4y^2 - 2x$ ***[1 mark]***

Differentiate y-terms using chain rule:

$\Rightarrow \frac{d}{dx}6x^2y = 5 - \frac{d}{dy}4y^2\frac{dy}{dx} - 2x$

$\Rightarrow \frac{d}{dx}6x^2y = 5 - 8y\frac{dy}{dx} - 2x$ ***[1 mark]***

Differentiate xy-terms using product rule:

$\Rightarrow 6x^2\frac{dy}{dx} + y\frac{d}{dx}6x^2 = 5 - 8y\frac{dy}{dx} - 2x$

$\Rightarrow 6x^2\frac{dy}{dx} + 12xy = 5 - 8y\frac{dy}{dx} - 2x$ ***[1 mark]***

Rearrange to make $\frac{dy}{dx}$ the subject:

$\Rightarrow 6x^2\frac{dy}{dx} + 8y\frac{dy}{dx} = 5 - 2x - 12xy$

$\Rightarrow \frac{dy}{dx} = \frac{5 - 2x - 12xy}{6x^2 + 8y}$ ***[1 mark]***

So at $Q = (-0.75, 0.5)$,

$\frac{dy}{dx} = \frac{5 - 2\left(-\frac{3}{4}\right) - 12\left(-\frac{3}{4}\right)\left(\frac{1}{2}\right)}{6\left(-\frac{3}{4}\right)^2 + 8\left(\frac{1}{2}\right)} = \frac{5 + \frac{3}{2} + \frac{9}{2}}{\frac{27}{8} + 4}$

$= \frac{11}{\left(\frac{59}{8}\right)} = 11 \times \frac{8}{59} = \frac{88}{59}$ ***[1 mark]***

2 a) (i) Using implicit differentiation:

$3e^x + 6y = 2x^2y \Rightarrow \frac{d}{dx}3e^x + \frac{d}{dx}6y = \frac{d}{dx}2x^2y$ ***[1 mark]***

$\Rightarrow 3e^x + 6\frac{dy}{dx} = 2x^2\frac{dy}{dx} + y\frac{d}{dx}2x^2$

$\Rightarrow 3e^x + 6\frac{dy}{dx} = 2x^2\frac{dy}{dx} + 4xy$ ***[1 mark]***

$\Rightarrow 2x^2\frac{dy}{dx} - 6\frac{dy}{dx} = 3e^x - 4xy$

$\Rightarrow \frac{dy}{dx} = \frac{3e^x - 4xy}{2x^2 - 6}$ ***[1 mark]***

Answers

(ii) At the stationary points of C, $\frac{dy}{dx} = 0$

$\Rightarrow \frac{3e^x - 4xy}{2x^2 - 6} = 0$ ***[1 mark]***

$\Rightarrow 3e^x - 4xy = 0$

$\Rightarrow y = \frac{3e^x}{4x}$ ***[1 mark]***

b) Substitute $y = \frac{3e^x}{4x}$ into the original equation of curve C:

$3e^x + 6y = 2x^2y \Rightarrow 3e^x + 6\frac{3e^x}{4x} = 2x^2\frac{3e^x}{4x}$ ***[1 mark]***

$\Rightarrow 1 + \frac{3}{2x} = \frac{x}{2}$

$\Rightarrow x^2 - 2x - 3 = 0$ ***[1 mark]***

$\Rightarrow (x + 1)(x - 3) = 0$

$\Rightarrow x = -1$ or $x = 3$

$x = -1 \Rightarrow y = \frac{3e^{-1}}{4(-1)} = -\frac{3}{4e}$

$x = 3 \Rightarrow y = \frac{3e^3}{4(3)} = \frac{1}{4}e^3$

So the stationary points of C are $(-1, -\frac{3}{4e})$ and $(3, \frac{1}{4}e^3)$

[2 marks — 1 mark for each correct pair of coordinates]

Don't forget — if the question asks you for an exact answer, that usually means leaving it in terms of something like π or ln or, in this case, e.

3 a) $y = 4^x \Rightarrow \frac{dy}{dx} = 4^x \ln 4$ ***[1 mark]***

So $\frac{dy}{dx} = \ln 4 \Rightarrow 4^x = 1 \Rightarrow x = 0 \Rightarrow y = 4^0 = 1$

$\frac{dy}{dx} = \ln 4$ at coordinates (0, 1) ***[1 mark]***

b) Use the chain rule to find $\frac{dy}{dx}$:

Let $u = (x - 4)^3$.

Then $\frac{dy}{dx} = \frac{dy}{du} \times \frac{du}{dx} = \frac{d}{du}(4^u) \times \frac{d}{dx}(x-4)^3$ ***[1 mark]***

$= (4^u \ln 4)(3(x - 4)^2 \times 1)$

(using chain rule again to find $\frac{du}{dx}$)

$= 4^{(x-4)^3}\, 3(x - 4)^2 \ln 4$ ***[1 mark]***

So when $x = 3$, $\frac{dy}{dx} = 4^{(3-4)^3}\, 3(3 - 4)^2 \ln 4$ ***[1 mark]***

$\frac{dy}{dx} = 4^{-1}\, 3 \ln 4 = \frac{3}{4} \ln 4 = 1.040$ (to 3 d.p.) ***[1 mark]***

4 a) Start by differentiating x and y with respect to θ:

$\frac{dy}{d\theta} = 2\cos\theta$ ***[1 mark]***

$\frac{dx}{d\theta} = 3 + 3\sin 3\theta$ ***[1 mark]***

$\frac{dy}{dx} = \frac{dy}{d\theta} \div \frac{dx}{d\theta} = \frac{2\cos\theta}{3 + 3\sin 3\theta}$ ***[1 mark]***

b) (i) We need the value of θ at $(\pi + 1, \sqrt{3})$:

$y = 2\sin\theta = \sqrt{3}$, for $-\pi \le \theta \le \pi \Rightarrow \theta = \frac{\pi}{3}$ or $\frac{2\pi}{3}$ ***[1 mark]***

If $\theta = \frac{\pi}{3}$, then $x = 3\theta - \cos 3\theta = \pi - \cos\pi = \pi + 1$.

If $\theta = \frac{2\pi}{3}$, then $x = 3\theta - \cos 3\theta = 2\pi - \cos 2\pi = 2\pi - 1$.

So at $(\pi + 1, \sqrt{3})$, $\theta = \frac{\pi}{3}$ ***[1 mark]***

$\theta = \frac{\pi}{3} \Rightarrow \frac{dy}{dx} = \frac{2\cos\frac{\pi}{3}}{3 + 3\sin\pi} = \frac{2(\frac{1}{2})}{3 + 0} = \frac{1}{3}$ ***[1 mark]***

(ii) $\theta = \frac{\pi}{6} \Rightarrow x = \frac{\pi}{2} - \cos\frac{\pi}{2} = \frac{\pi}{2} - 0 = \frac{\pi}{2}$

$\theta = \frac{\pi}{6} \Rightarrow y = 2\sin\frac{\pi}{6} = 2 \times \frac{1}{2} = 1$

So $\theta = \frac{\pi}{6}$ at the point $(\frac{\pi}{2}, 1)$ ***[1 mark]***

$\theta = \frac{\pi}{6} \Rightarrow \frac{dy}{dx} = \frac{2\cos\frac{\pi}{6}}{3 + 3\sin\frac{\pi}{2}} = \frac{2\left(\frac{\sqrt{3}}{2}\right)}{3 + 3(1)} = \frac{\sqrt{3}}{6}$ ***[1 mark]***

Gradient of normal $= -\frac{1}{\left(\frac{dy}{dx}\right)} = -\frac{6}{\sqrt{3}} = -\frac{6\sqrt{3}}{3}$

$= -2\sqrt{3}$ ***[1 mark]***

So the normal is $y = -2\sqrt{3}x + c$ for some c.

$\Rightarrow 1 = -2\sqrt{3} \times \frac{\pi}{2} + c = -\pi\sqrt{3} + c$

$\Rightarrow c = 1 + \pi\sqrt{3}$

The equation of the normal is $y = -2\sqrt{3}x + 1 + \pi\sqrt{3}$ ***[1 mark]***

5 a) First find the value of t when $y = -6$:

$y = 2 - t^3 = -6 \Rightarrow t^3 = 8 \Rightarrow t = 2$ ***[1 mark]***

$\Rightarrow x = 2^2 + 2(2) - 3 = 5$

Now find the gradient of the curve:

$\frac{dy}{dt} = -3t^2$, $\frac{dx}{dt} = 2t + 2$

So $\frac{dy}{dx} = \frac{dy}{dt} \div \frac{dx}{dt} = \frac{-3t^2}{2t + 2}$ ***[1 mark]***

So when $t = 2$, $\frac{dy}{dx} = \frac{-3(2)^2}{2(2) + 2} = \frac{-12}{6} = -2$ ***[1 mark]***

So the tangent at $y = -6$ is

$y = -2x + c \Rightarrow -6 = -2(5) + c \Rightarrow c = 4$

The equation of L is $y = -2x + 4$ ***[1 mark]***

b) (i) Sub $y = 2 - t^3$ and $x = t^2 + 2t - 3$ into the equation of L:

$y = -2x + 4$

$\Rightarrow 2 - t^3 = -2(t^2 + 2t - 3) + 4$ ***[1 mark]***

$\Rightarrow 2 - t^3 = -2t^2 - 4t + 10$

$\Rightarrow t^3 - 2t^2 - 4t + 8 = 0$

We know from part (a) that $t = 2$ is a root, so take out $(t - 2)$ as a factor:

$\Rightarrow (t - 2)(t^2 - 4) = 0$ ***[1 mark]***

$\Rightarrow (t - 2)(t + 2)(t - 2) = 0$

$\Rightarrow t = 2$ or $t = -2$ ***[1 mark]***

So t must be -2 at P.

$t = -2 \Rightarrow x = (-2)^2 + 2(-2) - 3 = -3$, $y = 2 - (-2)^3 = 10$.

The coordinates of P are $(-3, 10)$. ***[1 mark]***

(ii) At P, $t = -2$, so $\frac{dy}{dx} = \frac{-3(-2)^2}{2(-2) + 2} = \frac{-12}{-2} = 6$. ***[1 mark]***

So the gradient of the normal at P is

$-\frac{1}{\left(\frac{dy}{dx}\right)} = -\frac{1}{6}$ ***[1 mark]***

The equation of the normal at P is

$y = -\frac{1}{6}x + c \Rightarrow 10 = -\frac{(-3)}{6} + c \Rightarrow c = \frac{19}{2}$

So the normal to the curve at point P is

$y = -\frac{1}{6}x + \frac{19}{2}$ ***[1 mark]***

Answers

6 a) Start by finding the missing side length of the triangular faces. Call the missing length s:

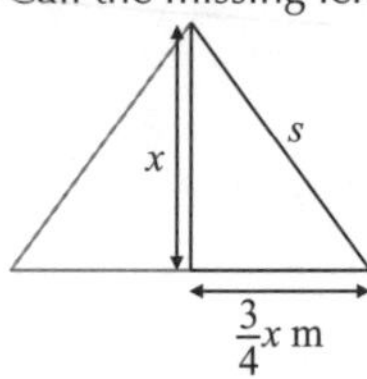

$s = \sqrt{x^2 + \left(\frac{3}{4}x\right)^2}$
$= \sqrt{x^2 + \frac{9}{16}x^2}$
$= \sqrt{\frac{25}{16}x^2}$
$= \frac{5}{4}x$ ***[1 mark]***

Now find A by adding up the area of each of the faces:
$A = 2(\frac{1}{2} \times \frac{3}{2}x \times x) + (\frac{3}{2}x \times 4x) + 2(\frac{5}{4}x \times 4x)$ ***[1 mark]***
$= \frac{3}{2}x^2 + 6x^2 + 10x^2$
$= \frac{35}{2}x^2$ ***[1 mark]***

b) $\frac{dA}{dt} = 0.07$
$A = \frac{35}{2}x^2 \Rightarrow \frac{dA}{dx} = 35x$ ***[1 mark]***
Using chain rule, $\frac{dx}{dt} = \frac{dx}{dA} \times \frac{dA}{dt}$ ***[1 mark]***
$= \frac{1}{\left(\frac{dA}{dx}\right)} \times \frac{dA}{dt} = \frac{1}{35x} \times 0.07$
$= \frac{0.07}{35 \times 0.5} = 0.004 \text{ m s}^{-1}$ ***[1 mark]***

c) First you need to figure out what the question is asking for. 'Find the rate of change of V' means we're looking for $\frac{dV}{dt}$.
Start by finding an expression for V:
$V = (\frac{1}{2} \times \frac{3}{2}x \times x) \times 4x = 3x^3$ ***[1 mark]***
So $\frac{dV}{dx} = 9x^2$ ***[1 mark]***
Using chain rule, $\frac{dV}{dt} = \frac{dV}{dx} \times \frac{dx}{dt}$ ***[1 mark]***
$= 9x^2 \times \frac{0.07}{35x} = \frac{9(1.2)^2 \times 0.07}{35 \times 1.2} = 0.0216 \text{ m s}^{-1}$ ***[1 mark]***

Core Section 8 — Numerical Methods

Warm-up Questions

1) There are 2 roots (graph crosses the x-axis twice in this interval).

2) a) Sin (2 × 3) = −0.2794... and sin (2 × 4) = 0.9893...
Since sin(2x) is a continuous function, the change of sign means there is a root between 3 and 4.

b) ln (2.1 − 2) + 2 = −0.3025...
and ln (2.2 − 2) + 2 = 0.3905...
Since the function is continuous for $x > 2$, the change of sign means there is a root between 2.1 and 2.2.

c) Rearrange first to give $x^3 - 4x^2 - 7 = 0$, then:
$4.3^3 - 4 \times (4.3^2) - 7 = -1.453$ and
$4.5^3 - 4 \times (4.5^2) - 7 = 3.125$.
The function is continuous, so the change of sign means there is a root between 4.3 and 4.5.

3) If 1.2 is a root to 1 d.p. then there should be a sign change for f(x) between the upper and lower bounds:
$f(1.15) = 1.15^3 + 1.15 - 3 = -0.3291...$
$f(1.25) = 1.25^3 + 1.25 - 3 = 0.2031...$
There is a change of sign, and the function is continuous, so the root must lie between 1.15 and 1.25, so to 1 d.p. the root is at $x = 1.2$.

4) $x_1 = -\frac{1}{2}\cos(-1) = -0.2701...$
$x_2 = -\frac{1}{2}\cos(-0.2701...) = -0.4818...$
$x_3 = -\frac{1}{2}\cos(-0.4818...) = -0.4430...$
$x_4 = -\frac{1}{2}\cos(-0.4430...) = -0.4517...$
$x_5 = -\frac{1}{2}\cos(-0.4517...) = -0.4498...$
$x_6 = -\frac{1}{2}\cos(-0.4498...) = -0.4502...$
x_4, x_5 and x_6 all round to −0.45, so to 2 d.p. $x = -0.45$.

5) $x_1 = \sqrt{\ln 2 + 4} = 2.1663...$
$x_2 = \sqrt{\ln 2.1663... + 4} = 2.1847...$
$x_3 = \sqrt{\ln 2.1847... + 4} = 2.1866...$
$x_4 = \sqrt{\ln 2.1866... + 4} = 2.1868...$
$x_5 = \sqrt{\ln 2.1868... + 4} = 2.1868...$
x_3, x_4 and x_5 all round to 2.187, so to 3 d.p. $x = 2.187$.

6) a) i) $2x^2 - x^3 + 1 = 0 \Rightarrow 2x^2 - x^3 = -1$
$\Rightarrow x^2(2 - x) = -1 \Rightarrow x^2 = \frac{-1}{2-x} \Rightarrow x = \sqrt{\frac{-1}{2-x}}$.
ii) $2x^2 - x^3 + 1 = 0 \Rightarrow x^3 = 2x^2 + 1$
$\Rightarrow x = \sqrt[3]{2x^2 + 1}$.
iii) $2x^2 - x^3 + 1 = 0 \Rightarrow 2x^2 = x^3 - 1$
$\Rightarrow x^2 = \frac{x^3 - 1}{2} \Rightarrow x = \sqrt{\frac{x^3 - 1}{2}}$.

b) Using $x_{n+1} = \sqrt{\frac{-1}{2 - x_n}}$ with $x_0 = 2.3$ gives:
$x_1 = \sqrt{\frac{-1}{2 - 2.3}} = 1.8257...$
$x_2 = \sqrt{\frac{-1}{2 - 1.8257...}}$ has no real solution
so this formula does not converge to a root.

Using $x_{n+1} = \sqrt[3]{2x_n^2 + 1}$ with $x_0 = 2.3$ gives:
$x_1 = \sqrt[3]{2 \times (2.3)^2 + 1} = 2.2624...$
$x_2 = \sqrt[3]{2 \times (2.2624...)^2 + 1} = 2.2398...$

Answers

$x_3 = \sqrt[3]{2 \times (2.2398...)^2 + 1} = 2.2262...$
$x_4 = \sqrt[3]{2 \times (2.2262...)^2 + 1} = 2.2180...$
$x_5 = \sqrt[3]{2 \times (2.2180...)^2 + 1} = 2.2131...$
$x_6 = \sqrt[3]{2 \times (2.2131...)^2 + 1} = 2.2101...$
$x_7 = \sqrt[3]{2 \times (2.2101...)^2 + 1} = 2.2083...$

x_5, x_6 and x_7 all round to 2.21, so to 2 d.p. $x = 2.21$ is a root.

Using $x_{n+1} = \sqrt{\frac{x_n^3 - 1}{2}}$ with $x_0 = 2.3$ gives:

$x_1 = \sqrt{\frac{2.3^3 - 1}{2}} = 2.3629...$

$x_2 = \sqrt{\frac{2.3629...^3 - 1}{2}} = 2.4691...$

$x_3 = \sqrt{\frac{2.4691...^3 - 1}{2}} = 2.6508...$

$x_4 = \sqrt{\frac{2.6508...^3 - 1}{2}} = 2.9687...$

This sequence is diverging so does not converge to a root. The only formula that converges to a root is $x_{n+1} = \sqrt[3]{2x_n^2 + 1}$.

7)

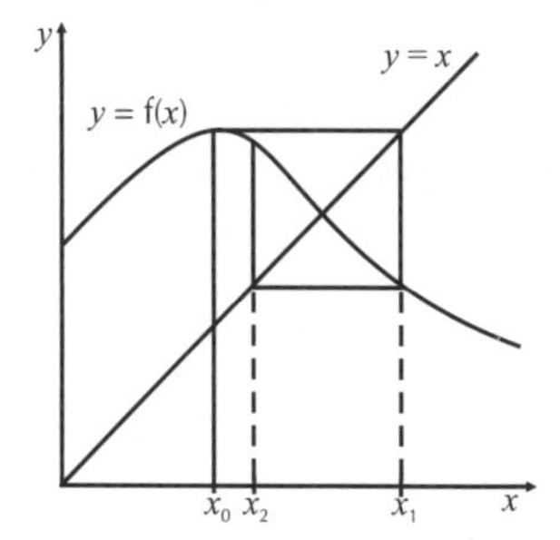

Exam Questions

1 a) There will be a change of sign between f(0.7) and f(0.8) if p lies between 0.7 and 0.8.
$f(0.7) = (2 \times 0.7 \times e^{0.7}) - 3 = -0.1807...$ ***[1 mark]***
$f(0.8) = (2 \times 0.8 \times e^{0.8}) - 3 = 0.5608...$ ***[1 mark]***
f(x) is continuous, and there is a change of sign, so $0.7 < p < 0.8$ ***[1 mark]***.

b) If $2xe^x - 3 = 0$, then $2xe^x = 3 \Rightarrow xe^x = \frac{3}{2}$
$\Rightarrow x = \frac{3}{2e^x} \Rightarrow x = \frac{3}{2}e^{-x}$.
[2 marks available — 1 mark for partial rearrangement, 1 mark for correct final answer.]

c) $x_{n+1} = \frac{3}{2}e^{-x_n}$ and $x_0 = 0.7$, so:

$x_1 = \frac{3}{2}e^{-0.7} = 0.74487... = 0.7449$ to 4 d.p.

$x_2 = \frac{3}{2}e^{-0.74487...} = 0.71218... = 0.7122$ to 4 d.p.

$x_3 = \frac{3}{2}e^{-0.71218...} = 0.73585... = 0.7359$ to 4 d.p.

$x_4 = \frac{3}{2}e^{-0.73585...} = 0.71864... = 0.7186$ to 4 d.p.

[3 marks available — 1 mark for x_1 correct, 1 mark for x_2 correct, 1 mark for all 4 correct.]

d) If the root of f(x) = 0, p, is 0.726 to 3 d.p. then there must be a change of sign in f(x) between the upper and lower bounds of p.
Lower bound = 0.7255.
$f(0.7255) = (2 \times 0.7255 \times e^{0.7255}) - 3 = -0.0025...$
Upper bound = 0.7265.
$f(0.7265) = (2 \times 0.7265 \times e^{0.7265}) - 3 = 0.0045...$
f(x) is continuous, and there's a change of sign, so $p = 0.726$ to 3 d.p.

[3 marks available — 1 mark for identifying upper and lower bounds, 1 mark for finding value of the function at both bounds, 1 mark for indicating that the change in sign and the fact that it's a continuous function shows the root is correct to the given accuracy.]

2 a) Where $y = \sin 3x + 3x$ and $y = 1$ meet,
$\sin 3x + 3x = 1 \Rightarrow \sin 3x + 3x - 1 = 0$ ***[1 mark]***.

$x = a$ is a root of this equation, so if $x = 0.1$ and $x = 0.2$ produce different signs, then a lies between them. So for the continuous function $f(x) = \sin 3x + 3x - 1$:
$f(0.1) = \sin(3 \times 0.1) + (3 \times 0.1) - 1 = -0.4044...$ ***[1 mark]***
$f(0.2) = \sin(3 \times 0.2) + (3 \times 0.2) - 1 = 0.1646...$ ***[1 mark]***
There is a change of sign, so $0.1 < a < 0.2$ ***[1 mark]***.

b) $\sin 3x + 3x = 1 \Rightarrow 3x = 1 - \sin 3x \Rightarrow x = \frac{1}{3}(1 - \sin 3x)$.
[2 marks available — 1 mark for partial rearrangement, 1 mark for correct final answer.]

c) $x_{n+1} = \frac{1}{3}(1 - \sin 3x_n)$ and $x_0 = 0.2$:
$x_1 = \frac{1}{3}(1 - \sin(3 \times 0.2)) = 0.1451...$ ***[1 mark]***
$x_2 = \frac{1}{3}(1 - \sin(3 \times 0.1451...)) = 0.1927...$
$x_3 = \frac{1}{3}(1 - \sin(3 \times 0.1927...)) = 0.1511...$
$x_4 = \frac{1}{3}(1 - \sin(3 \times 0.1511...)) = 0.1873...$
So $x_4 = 0.187$ to 3 d.p. ***[1 mark]***.

3 a) $x_{n+1} = \sqrt[3]{x_n^2 - 4}$, $x_0 = -1$:
$x_1 = \sqrt[3]{(-1)^2 - 4} = -1.44224... = -1.4422$ to 4 d.p.
$x_2 = \sqrt[3]{(-1.4422...)^2 - 4} = -1.24287... = -1.2429$ to 4 d.p.
$x_3 = \sqrt[3]{(-1.2428...)^2 - 4} = -1.34906... = -1.3491$ to 4 d.p.
$x_4 = \sqrt[3]{(-1.3490...)^2 - 4} = -1.29664... = -1.2966$ to 4 d.p.
[3 marks available — 1 mark for x_1 correct, 1 mark for x_2 correct, 1 mark for all 4 correct.]

b) If b is a root of $x^3 - x^2 + 4 = 0$, then $x^3 - x^2 + 4 = 0$ will rearrange to form $x = \sqrt[3]{x^2 - 4}$, the iteration formula used in (a).

(This is like finding the iteration formula in reverse...)

$x^3 - x^2 + 4 = 0 \Rightarrow x^3 = x^2 - 4 \Rightarrow x = \sqrt[3]{x^2 - 4}$, and so b must be a root of $x^3 - x^2 + 4 = 0$.

[2 marks available — 1 mark for stating that b is a root if one equation can be rearranged into the other, 1 mark for correct demonstration of rearrangement.]

c) If the root of $f(x) = x^3 - x^2 + 4 = 0$, b, is -1.315 to 3 d.p. then there must be a change of sign in f(x) between the upper and lower bounds of b, which are -1.3145 and -1.3155.
$f(-1.3145) = (-1.3145)^3 - (-1.3145)^2 + 4 = 0.00075...$
$f(-1.3155) = (-1.3155)^3 - (-1.3155)^2 + 4 = -0.00706...$
f(x) is continuous, and there's a change of sign, so $b = -1.315$ to 3 d.p.

[3 marks available — 1 mark for identifying upper and lower bounds, 1 mark for finding value of the function at both bounds, 1 mark for indicating that the change in sign and the fact that it's a continuous function shows the root is correct to the given accuracy.]

Answers

4 a) For $f(x) = \ln(x + 3) - x + 2$, there will be a change in sign between f(3) and f(4) if the root lies between those values.

$f(3) = \ln(3 + 3) - 3 + 2 = 0.7917...$ ***[1 mark]***
$f(4) = \ln(4 + 3) - 4 + 2 = -0.0540...$ ***[1 mark]***

There is a change of sign, and the function is continuous for $x > -3$, so the root, m, must lie between 3 and 4 ***[1 mark]***.

b) $x_{n+1} = \ln(x_n + 3) + 2$, and $x_0 = 3$, so:
$x_1 = \ln(3 + 3) + 2 = 3.7917...$
$x_2 = \ln(3.7917... + 3) + 2 = 3.9157...$
$x_3 = \ln(3.9157... + 3) + 2 = 3.9337...$
$x_4 = \ln(3.9337... + 3) + 2 = 3.9364...$
$x_5 = \ln(3.9364... + 3) + 2 = 3.9367...$

So $m = 3.94$ to 2 d.p.

[3 marks available — 1 mark for correct substitution of x_0 to find x_1, 1 mark for evidence of correct iterations up to x_5, 1 mark for correct final answer to correct accuracy.]

c) From b), $m = 3.94$ to 2 d.p. If this is correct then there will be a change of sign in f(x) between the upper and lower bounds of m, which are 3.935 and 3.945.
$f(3.935) = \ln(3.935 + 3) - 3.935 + 2 = 0.00158...$
$f(3.945) = \ln(3.945 + 3) - 3.945 + 2 = -0.00697...$

f(x) is continuous for $x > -3$, and there's a change of sign, so $m = 3.94$ is correct to 2 d.p.

[3 marks available — 1 mark for identifying upper and lower bounds, 1 mark for finding value of the function at both bounds, 1 mark for indicating that the change in sign and the fact that it's a continuous function shows the root is correct to the given accuracy.]

d)
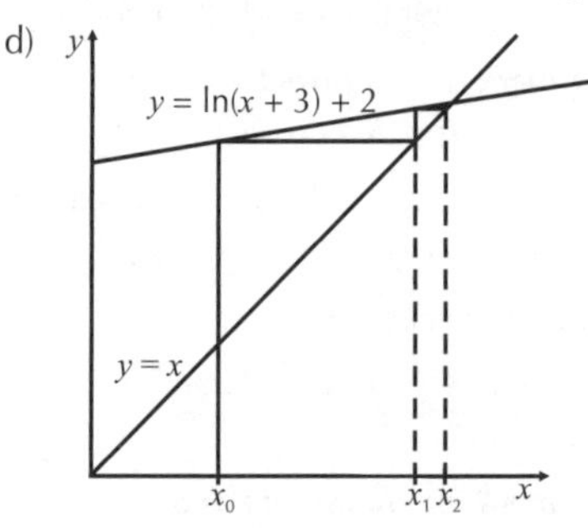

[2 marks available — 1 mark for converging staircase, 1 mark for correct positions of x_1 and x_2]

Core Section 9 — Integration 1

Warm-up Questions

1) $2e^{2x} + C$

2) $\frac{1}{3}e^{3x-5} + C$

3) $\frac{2}{3}\ln|x| + C$

4) $\ln|2x + 1| + C$

5) a) Just integrate each term separately: $\int \cos 4x \, dx = \frac{1}{4}\sin 4x$, $\int \sec^2 7x = \frac{1}{7}\tan 7x \, dx$. Putting these bits together gives: $\frac{1}{4}\sin 4x - \frac{1}{7}\tan 7x + C$.

b) Again, just integrate each term separately: $\int 6\sec 3x \tan 3x \, dx = 2\sec 3x$, $\int -\operatorname{cosec}^2\frac{x}{5} \, dx = 5\cot\frac{x}{5}$.
Putting these bits together gives: $2\sec 3x + 5\cot\frac{x}{5} + C$.

6) $\ln|\sin x| + C$

7) $e^{x^3} + C$

8) $4\ln|x^5 + x^3 - 3x| + C$

9) From the identity $\sec^2 x \equiv 1 + \tan^2 x$, write $2\tan^2 3x$ as $2\sec^2 3x - 2$. The integral becomes:
$\int 2\sec^2 3x - 2 + 2 \, dx$
$= \int 2\sec^2 3x \, dx = \frac{2}{3}\tan 3x + C$.

10) Use the double angle formula for tan $2x$ to write $\frac{2\tan 3x}{1 - \tan^2 3x}$ as $\tan 6x$. The integral becomes
$\int \tan 6x \, dx = \int \frac{\sin 6x}{\cos 6x} \, dx = -\frac{1}{6}\ln|\cos 6x| + C$.

Don't forget to double the coefficient of x when you use the double angle formula.

Exam Questions

1 a) $-\frac{1}{2}e^{(5-6x)} + C$

[1 mark for answer in the form $ke^{(5-6x)}$, 1 mark for the correct value of k]

b) $-\ln|\cot x + 2x| + C$

[1 mark for answer in the form $k\ln|f(x)|$, 1 mark for correct value of k and 1 mark for correct function f(x). Lose 1 mark if C is missed off both answers a) and b)]

2 Use the identity $\operatorname{cosec}^2 x \equiv 1 + \cot^2 x$ to write $2\cot^2 x$ as $2\operatorname{cosec}^2 x - 2$ ***[1 mark]***. The integral becomes:
$\int 2\operatorname{cosec}^2 x - 2 \, dx = -2\cot x - 2x + C$
[1 mark for $-2\cot x$, 1 mark for $-2x + C$].

Answers

Core Section 10 — Integration 2

Warm-up Questions

1) If $u = e^x - 1$, then $\frac{du}{dx} = e^x$ (so $\frac{du}{e^x} = dx$) and $e^x + 1 = u + 2$.
Substituting this into the integral gives:

$$\int e^x(u+2)u^2\frac{du}{e^x} = \int (u+2)u^2 du$$
$$= \int u^3 + 2u^2 du = \frac{1}{4}u^4 + \frac{2}{3}u^3 + C$$
$$= \frac{1}{4}(e^x - 1)^4 + \frac{2}{3}(e^x - 1)^3 + C$$

Make sure you put $u = e^x - 1$ back into your final answer.

2) If $u = \sec x$, then $\frac{du}{dx} = \sec x \tan x$ (so $\frac{du}{\sec x \tan x} = dx$).
Change the limits: when $x = \frac{\pi}{4}$, $u = \sec\frac{\pi}{4} = \sqrt{2}$ and when $x = \frac{\pi}{3}$, $u = \sec\frac{\pi}{3} = 2$. Substituting into the integral gives:

$$\int_{\sqrt{2}}^{2} \sec x \tan x\, u^3 \frac{du}{\sec x \tan x} = \int_{\sqrt{2}}^{2} u^3\, du = [\tfrac{1}{4}u^4]_{\sqrt{2}}^{2}$$
$$= [\tfrac{1}{4}(2)^4] - [\tfrac{1}{4}(\sqrt{2})^4]$$
$$= \tfrac{1}{4}(16) - \tfrac{1}{4}(4) = 4 - 1 = 3.$$

3) Let $u = \ln x$ and let $\frac{dv}{dx} = 3x^2$. So $\frac{du}{dx} = \frac{1}{x}$ and $v = x^3$.
Putting these into the formula gives:

$$\int 3x^2 \ln x\, dx = [x^3 \ln x] - \int \frac{x^3}{x} dx = [x^3 \ln x] - \int x^2 dx$$
$$= x^3 \ln x - \tfrac{1}{3}x^3 + C = x^3(\ln x - \tfrac{1}{3}) + C$$

4) Let $u = 4x$, and let $\frac{dv}{dx} = \cos 4x$. So $\frac{du}{dx} = 4$ and $v = \frac{1}{4}\sin 4x$.
Putting these into the formula gives:

$$\int 4x \cos 4x\, dx = 4x(\tfrac{1}{4}\sin 4x) - \int 4(\tfrac{1}{4}\sin 4x) dx$$
$$= x \sin 4x + \tfrac{1}{4}\cos 4x + C$$

5) If $y = \frac{1}{x}$ then $y^2 = \frac{1}{x^2}$. Putting this into the integral gives:

$$V = \pi\int_2^4 \frac{1}{x^2} dx = \pi\Big[-\frac{1}{x}\Big]_2^4 = \pi\Big[\Big(-\frac{1}{4}\Big) - \Big(-\frac{1}{2}\Big)\Big] = \frac{\pi}{4}.$$

6) First, rearrange the equation to get it in terms of x^2:
$y = x^2 + 1 \Rightarrow x^2 = y - 1$. Putting this into the formula:

$$V = \pi\int_1^3 y - 1\, dy = \pi\Big[\frac{1}{2}y^2 - y\Big]_1^3$$
$$= \pi\Big[\Big(\frac{1}{2}(9) - 3\Big) - \Big(\frac{1}{2}(1) - 1\Big)\Big] = 2\pi.$$

7) If $x = t^2$, then $\frac{dx}{dt} = 2t$. Find the limits in terms of t:
$4 = t^2 \Rightarrow t = 2$ (as $t > 0$) and $9 = t^2 \Rightarrow t = 3$ (as $t > 0$).
$y = \frac{1}{t}$, so $y^2 = \frac{1}{t^2}$. Putting all this into the formula gives:

$$\pi\int_{t=2}^{t=3} y^2 \frac{dx}{dt} dt = \pi\int_2^3 \frac{2t}{t^2} dt = \pi\int_2^3 \frac{2}{t} dt$$
$$= \pi[2\ln|t|]_2^3 = \pi(2\ln 3 - 2\ln 2)$$
$$= \pi(\ln 9 - \ln 4) = \pi\ln\frac{9}{4}$$

It's really easy to forget to change the limits from x to t — ignore this warning at your own risk.

8) First, find the partial fractions: If
$\frac{3x+10}{(2x+3)(x-4)} \equiv \frac{A}{2x+3} + \frac{B}{x-4}$,
then $3x + 10 \equiv A(x - 4) + B(2x + 3)$.
From this, you get the simultaneous equations $3 = A + 2B$ and $10 = -4A + 3B$. Solving these gives $A = -1$ and $B = 2$:
$\frac{3x+10}{(2x+3)(x-4)} = \frac{-1}{2x+3} + \frac{2}{x-4}$.
Now putting this into the integral gives:

$$\int \frac{-1}{2x+3} + \frac{2}{x-4} dx = -\frac{1}{2}\ln|2x+3| + 2\ln|x-4| + C$$

9) $\frac{dy}{dx} = \frac{1}{y}\cos x \Rightarrow y\, dy = \cos x\, dx$

so $\int y\, dy = \int \cos x\, dx \Rightarrow \frac{y^2}{2} = \sin x + C_0$

$\Rightarrow y^2 = 2\sin x + C_1$ (where $C_1 = 2C_0$)

10) a) $\frac{dS}{dt} = kS$

b) Solving the differential equation above gives:
$\frac{dS}{dt} = kS \Rightarrow \int \frac{1}{S} dS = \int k\, dt$
$\ln|S| = kt + C$
$S = Ae^{kt}$ where $A = e^C$
For the initial population, $t = 0$. Put $S = 30$ and $t = 0$ into the equation to find the value of A: $30 = Ae^0 \Rightarrow A = 30$.
Now, use $S = 150$, $k = 0.2$ and $A = 30$ to find t:
$150 = 30e^{0.2t} \Rightarrow 5 = e^{0.2t} \Rightarrow \ln 5 = 0.2t$, so $t = 8.047$.
It will take the squirrels 8 weeks before they can take over the forest.
Go squirrels go!

11) For the Trapezium Rule using 4 strips, $h = 1.5$ and you need to work out the values of y at $x = 0, 1.5, 3, 4.5, 6$:
$x_0 = 0, y_0 = -108, x_1 = 1.5, y_1 = -1.6875, x_2 = 3, y_2 = 0,$
$x_3 = 4.5, y_3 = 413.4375, x_4 = 6, y_4 = 5400.$
Putting these values into the formula gives:
$A \approx 0.75[-108 + 2(-1.6875 + 0 + 413.4375) + 5400]$
$= 4586.625.$
Using 6 strips, $h = 1$ and you need the y-values for $x = 0, 1, 2, 3, 4, 5, 6$. You already know the values for $x = 0, 3, 6$.
$(x_0 = 0, y_0 = -108), x_1 = 1, y_1 = 0, x_2 = 2, y_2 = 0, (x_3 = 3, y_3 = 0),$
$x_4 = 4, y_4 = 108, x_5 = 5, y_5 = 1152, (x_6 = 6, y_6 = 5400).$
Putting these values into the formula gives:
$A \approx \frac{1}{2}[-108 + 2(0 + 0 + 0 + 108 + 1152) + 5400] = 3906.$
To calculate the percentage error, you must first work out the exact value of the integral:

$$\int_0^6 (6x - 12)(x^2 - 4x + 3)^2 dx = [(x^2 - 4x + 3)^3]_0^6$$
$$= (6^2 - 4(6) + 3)^3 - (0 - 0 + 3)^3$$
$$= 3375 - 27 = 3348.$$

This uses the formula $\int (n+1)f'(x)[f(x)]^n dx = [f(x)]^{n+1} + C$.

Now work out the error for 4 strips:
$\frac{4586.625 - 3348}{3348} \times 100 = 37.00\%$ (4 s.f.)
and for 6 strips: $\frac{3906 - 3348}{3348} \times 100 = 16.67\%$ (4 s.f.).
Neither of these estimates was particularly accurate — but the one with more strips had a lower % error (as you would expect).

12) a) The width of each strip is $\frac{4-1}{6} = 0.5$, so you need y-values for $x = 1, 1.5, 2, 2.5, 3, 3.5$ and 4:
$x_0 = 1, y_0 = 1.0986, x_1 = 1.5, y_1 = 1.1709, x_2 = 2, y_2 = 1.2279,$
$x_3 = 2.5, y_3 = 1.2757, x_4 = 3, y_4 = 1.3170, x_5 = 3.5,$
$y_5 = 1.3535, x_6 = 4, y_6 = 1.3863.$ Putting these values into the formula gives:
$A \approx \frac{1}{3}0.5[(1.0986 + 1.3863) + 4(1.1709 + 1.2757 + 1.3535) + 2(1.2279 + 1.3170)] = 3.7959$ (4 d.p.).
If you'd had a go at this using the Mid-Ordinate Rule, you'd have ended up with $A \approx 3.797$ (to 3 d.p.).

b) Simpson's Rule only works with an even number of strips, so 5 strips won't work.

Answers

13) The width of each strip is $\frac{4-2}{4} = 0.5$, so $x_0 = 2$, $x_1 = 2.5$, $x_2 = 3$, $x_3 = 3.5$ and $x_4 = 4$. You need y-values for the midpoints of these strips, i.e. $x = 2.25$, 2.75, 3.25 and 3.75:
$x_{0.5} = 2.25$, $y_{0.5} = 6.5004$, $x_{1.5} = 2.75$, $y_{1.5} = 16.0098$,
$x_{2.5} = 3.25$, $y_{2.5} = 36.8667$, $x_{3.5} = 3.75$, $y_{3.5} = 80.9241$.
Putting these values into the formula gives:
$A \approx 0.5[6.5004 + 16.0098 + 36.8667 + 80.9241]$
$= 70.1505$ (4 d.p.).

Using Simpson's Rule, you'd have got $A \approx 71.996$ (3 d.p.).

Exam Questions

1 $V = \pi\int_{\frac{\pi}{4}}^{\frac{\pi}{3}} y^2\,dx = \pi\int_{\frac{\pi}{4}}^{\frac{\pi}{3}} \text{cosec}^2 x\,dx = \pi[-\cot x]_{\frac{\pi}{4}}^{\frac{\pi}{3}}$
$= \pi[(-\cot\frac{\pi}{3}) - (-\cot\frac{\pi}{4})]$
$= \pi[-\frac{1}{\sqrt{3}} + 1] = 1.328$ (3 d.p.).
[3 marks available — 1 mark for correct function for y^2, 1 mark for integrating, 1 mark for substituting in values of x to obtain correct answer]

Don't forget π here.

2 a) When $x = \frac{\pi}{2}$, $y = \frac{\pi}{2}\sin\frac{\pi}{2} = \frac{\pi}{2} = 1.5708$ ***[1 mark]***,
and when $x = \frac{3\pi}{4}$, $y = \frac{3\pi}{4}\sin\frac{3\pi}{4} = 1.6661$ ***[1 mark]***

b) The width of each strip (h) is $\frac{\pi}{4}$, so the Trapezium Rule is:
$A = \frac{1}{2}\frac{\pi}{4}[0 + 2(0.5554 + 1.5708 + 1.6661) + 0]$
$= \frac{\pi}{8}[2(3.7923)] = 2.978$ (3 d.p.).
[4 marks available — 1 mark for correct value of h, 2 marks for correct use of formula, 1 mark for correct answer]

c) Let $u = x$, so $\frac{du}{dx} = 1$. Let $\frac{dv}{dx} = \sin x$, so $v = -\cos x$
[1 mark for both parts correct]. Using integration by parts,
$\int_0^{\pi} x\sin x\,dx = [-x\cos x]_0^{\pi} - \int_0^{\pi} -\cos x\,dx$ ***[1 mark]***
$= [-x\cos x]_0^{\pi} + [\sin x]_0^{\pi}$ ***[1 mark]***
$= (\pi - 0) + (0) = \pi$ ***[1 mark]***

If you'd tried to use $u = \sin x$, you'd have ended up with a more complicated function to integrate ($x^2\cos x$).

d) To find the percentage error, divide the difference between the approximate answer and the exact answer by the exact answer and multiply by 100:
$\frac{\pi - 2.978}{\pi} \times 100 = 5.2\%$ (2 s.f.).
[2 marks available — 1 mark for appropriate method and 1 mark for correct answer]

3 If $u = \ln x$, then $\frac{du}{dx} = \frac{1}{x}$, so $x\,du = dx$. Changing the limits: when $x = 1$, $u = \ln 1 = 0$. When $x = 2$, $u = \ln 2$. Substituting all this into the integral gives:
$\int_1^2 \frac{8}{x}(\ln x + 2)^3\,dx = \int_0^{\ln 2}\frac{8}{x}(u + 2)^3 x\,du = \int_0^{\ln 2} 8(u + 2)^3\,du$
$= [2(u + 2)^4]_0^{\ln 2}$
$= [2(\ln 2 + 2)^4] - [2(0 + 2)^4]$
$= 105.21 - 32 = 73.21$ (4 s.f.).
[6 marks available — 1 mark for finding substitution for dx, 1 mark for finding correct limits, 1 mark for correct integral in terms of u, 2 marks for correct integration (1 for an answer in the form $k(u + 2)^n$, 1 mark for correct values of k and n), 1 mark for final answer (to 4 s.f.)]

4 The width of each strip is $\frac{3-1}{4} = 0.5$, so $x_0 = 1$, $x_1 = 1.5$, $x_2 = 2$, $x_3 = 2.5$ and $x_4 = 3$. You need y-values for the midpoints of these strips, i.e. $x = 1.25$, 1.75, 2.25 and 2.75:
$x_{0.5} = 1.25$, $y_{0.5} = 2.3784$, $x_{1.5} = 1.75$, $y_{1.5} = 3.3636$,
$x_{2.5} = 2.25$, $y_{2.5} = 4.7568$, $x_{3.5} = 2.75$, $y_{3.5} = 6.7272$.
Putting these values into the formula gives:
$A \approx 0.5[2.3784 + 3.3636 + 4.7568 + 6.7272] = 8.613$ (4 s.f.).
[3 marks available — 1 mark for calculating y-values for the midpoint of each strip, 1 mark for correct use of formula, 1 mark for correct answer]

5 a) First, rearrange the equation to get it in terms of x^2:
$y = \frac{1}{x^2} \Rightarrow x^2 = \frac{1}{y}$. Putting this into the formula:
$V = \pi\int_1^3 \frac{1}{y}\,dy = \pi[\ln y]_1^3$
$= \pi[(\ln 3) - (\ln 1)] = \pi\ln 3$.
[5 marks available — 1 mark for rearranging equation, 1 mark for correct formula for volume, 1 mark for correct integration, 1 mark for substituting in limits, 1 mark for final answer (in terms of π and ln)]

b) The width of each strip is $\frac{5-1}{4} = 1$, so you need y-values for $x = 1$, 2, 3, 4 and 5:
$x_0 = 1$, $y_0 = 0.25$, $x_1 = 2$, $y_1 = 0.1$, $x_2 = 3$, $y_2 = 0.0556$,
$x_3 = 4$, $y_3 = 0.0357$, $x_4 = 5$, $y_4 = 0.025$.
Putting these values into the formula gives:
$A \approx \frac{1}{3}1[(0.25 + 0.025) + 4(0.1 + 0.0357) + 2(0.0556)]$
$= 0.310$ (3 s.f.).
[4 marks available — 1 mark for correct x-values, 1 mark for correct y-values, 1 mark for correct use of formula, 1 mark for final answer]

6 To find the volume of the solid formed when R is rotated, find the volume for each curve separately then subtract. First, find the volume when the area under the curve $y = \sqrt{\sin x}$ is rotated about the x-axis:
As $y = \sqrt{\sin x}$, $y^2 = \sin x$.
$V = \pi\int_{\frac{\pi}{3}}^{\frac{2\pi}{3}} \sin x\,dx = \pi[-\cos x]_{\frac{\pi}{3}}^{\frac{2\pi}{3}}$ ***[1 mark]***
$= \pi\left(-\left(-\frac{1}{2}\right) - -\frac{1}{2}\right) = \pi$ ***[1 mark]***.
Now find the volume for $y = e^{-0.5x}$:
$y^2 = (e^{-0.5x})^2 = e^{-x}$ ***[1 mark]***.
$V = \pi\int_{\frac{\pi}{3}}^{\frac{2\pi}{3}} e^{-x}\,dx = \pi[-e^{-x}]_{\frac{\pi}{3}}^{\frac{2\pi}{3}}$ ***[1 mark]***
$= \pi(-e^{-\frac{2\pi}{3}} - (-e^{-\frac{\pi}{3}})) = 0.7156$ ***[1 mark]***.
[1 mark for using correct formula for the volume of revolution for both curves]
To find the volume you want, you need to subtract 0.7156 from π: $\pi - 0.7156 = 2.426$ (4 s.f.) ***[1 mark]***.

Answers

7 a) $\frac{dy}{dx} = \frac{\cos x \cos^2 y}{\sin x} \Rightarrow \frac{1}{\cos^2 y} dy = \frac{\cos x}{\sin x} dx$

$\Rightarrow \int \sec^2 y \, dy = \int \frac{\cos x}{\sin x} dx$

$\Rightarrow \tan y = \ln|\sin x| + C$

[4 marks available — 1 mark for separating the variables into functions of x and y, 1 mark for correct integration of RHS, 1 mark for correct integration of LHS, 1 mark for general solution]

b) If $y = \pi$ when $x = \frac{\pi}{6}$, that means that

$\tan\pi = \ln|\sin\frac{\pi}{6}| + C$

$0 = \ln\left|\frac{1}{2}\right| + C$ ***[1 mark]***

As $\ln \frac{1}{2} = \ln 1 - \ln 2 = -\ln 2$ (as $\ln 1 = 0$), it follows that $C = \ln 2$.

So $\tan y = \ln|\sin x| + \ln 2$ or $\tan y = \ln|2 \sin x|$ ***[1 mark]***.

This is the particular solution — you found the general solution in part a).

8 a) $\frac{dm}{dt} = k\sqrt{m}$, $k > 0$ ***[1 mark for RHS, 1 mark for LHS]***

b) First solve the differential equation to find m:

$\frac{dm}{dt} = k\sqrt{m} \Rightarrow \frac{1}{\sqrt{m}} dm = k \, dt$

$\Rightarrow \int m^{-\frac{1}{2}} dm = \int k \, dt$ ***[1 mark]***

$\Rightarrow 2m^{\frac{1}{2}} = kt + C$

$\Rightarrow m = \left(\frac{1}{2}(kt + C)\right)^2 = \frac{1}{4}(kt + C)^2$ ***[1 mark]***

At the start of the campaign, $t = 0$. Putting $t = 0$ and $m = 900$ into the equation gives: $900 = \frac{1}{4}(0 + C)^2 \Rightarrow 3600 = C^2 \Rightarrow C = 60$ (C must be positive, otherwise the sales would be decreasing). ***[1 mark]***.

This gives the equation $m = \frac{1}{4}(kt + 60)^2$ ***[1 mark]***.

c) Substituting $t = 5$ and $k = 2$ into the equation gives: $m = \frac{1}{4}((2 \times 5) + 60)^2 = 1225$ tubs sold.

[3 marks available — 2 marks for substituting correct values of t and k, 1 mark for answer]

Core Section 11 — Vectors

Warm-up Questions

1) Any multiples of the vectors will do:

a) e.g. **a** and 4**a**

b) e.g. 6**i** + 8**j** – 4**k** and 9**i** + 12**j** – 6**k**

c) e.g. $\begin{pmatrix} 2 \\ 4 \\ -2 \end{pmatrix}$ and $\begin{pmatrix} 4 \\ 8 \\ -4 \end{pmatrix}$

2) a) **b** – **a** b) **a** – **b** c) **b** – **c** d) **c** – **a**

3) 2**i** – 4**j** + 5**k**

4) a) $\sqrt{3^2 + 4^2 + (-2)^2} = \sqrt{29}$

b) $\sqrt{1^2 + 2^2 + (-1)^2} = \sqrt{6}$

5) a) $\sqrt{(3-1)^2 + (-1-2)^2 + (-2-3)^2} = \sqrt{38}$

b) $\sqrt{1^2 + 2^2 + 3^2} = \sqrt{14}$

c) $\sqrt{3^2 + (-1)^2 + (-2)^2} = \sqrt{14}$

6) a) $\mathbf{r} = (4\mathbf{i} + \mathbf{j} + 2\mathbf{k}) + t(3\mathbf{i} + \mathbf{j} - \mathbf{k})$ or $\mathbf{r} = \begin{pmatrix} 4 \\ 1 \\ 2 \end{pmatrix} + t\begin{pmatrix} 3 \\ 1 \\ -1 \end{pmatrix}$

b) $\mathbf{r} = (2\mathbf{i} - \mathbf{j} + \mathbf{k}) + t((2\mathbf{j} + 3\mathbf{k}) - (2\mathbf{i} - \mathbf{j} + \mathbf{k}))$

$\Rightarrow \mathbf{r} = (2\mathbf{i} - \mathbf{j} + \mathbf{k}) + t(-2\mathbf{i} + 3\mathbf{j} + 2\mathbf{k})$

or $\mathbf{r} = \begin{pmatrix} 2 \\ -1 \\ 1 \end{pmatrix} + t\begin{pmatrix} -2 \\ 3 \\ 2 \end{pmatrix}$

7) E.g. If $t = 1$, $((3 + 1(-1)), (2 + 1(3)), (4 + 1(0))) = (2, 5, 4)$

If $t = 2$, $((3 + 2(-1)), (2 + 2(3)), (4 + 2(0))) = (1, 8, 4)$

If $t = -1$, $((3 + -1(-1)), (2 + -1(3)), (4 + -1(0))) = (4, -1, 4)$

8) a) $(3\mathbf{i} + 4\mathbf{j}) \cdot (\mathbf{i} - 2\mathbf{j} + 3\mathbf{k}) = 3 - 8 + 0 = -5$

b) $\begin{pmatrix} 4 \\ 2 \\ 1 \end{pmatrix} \cdot \begin{pmatrix} 3 \\ -4 \\ -3 \end{pmatrix} = (4 \times 3) + (2 \times -4) + (1 \times -3) = 1$

9) a) $\begin{pmatrix} 2 \\ -1 \\ 2 \end{pmatrix} + t\begin{pmatrix} -4 \\ 6 \\ -2 \end{pmatrix} = \begin{pmatrix} 3 \\ 2 \\ 4 \end{pmatrix} + u\begin{pmatrix} -1 \\ 3 \\ 0 \end{pmatrix}$

Where the lines intersect, these 3 equations are true:

$2 - 4t = 3 - u$

$-1 + 6t = 2 + 3u$

$2 - 2t = 4$

Solve the third equation to give $t = -1$.
Substituting $t = -1$ in either of the other equations gives $u = -3$.
Substituting $t = -1$ and $u = -3$ in the remaining equation gives a true result, so the lines intersect.

Substituting $t = -1$ in the first vector equation gives the position vector of the intersection point:

$\begin{pmatrix} 6 \\ -7 \\ 4 \end{pmatrix}$

You'll often have to solve a pair of equations simultaneously (both variables will usually be in all three equations).

Answers

b) To find the angle between the lines, only consider the direction components of the vector equations:

$$\begin{pmatrix}-4\\6\\-2\end{pmatrix}\cdot\begin{pmatrix}-1\\3\\0\end{pmatrix} = 4 + 18 + 0 = 22$$

magnitude of 1st vector: $\sqrt{(-4)^2 + 6^2 + (-2)^2} = \sqrt{56}$
magnitude of 2nd vector: $\sqrt{(-1)^2 + 3^2 + (0)^2} = \sqrt{10}$

$$\cos\theta = \frac{22}{\sqrt{56}\sqrt{10}} = \Rightarrow \theta = 21.6°$$

10) Find values for a, b and c that give a scalar product of 0 when the two vectors are multiplied together.

$(3\mathbf{i} + 4\mathbf{j} - 2\mathbf{k}).(a\mathbf{i} + b\mathbf{j} + c\mathbf{k}) = 3a + 4b - 2c = 0$
E.g. $a = 2$, $b = 1$, $c = 5$
Perpendicular vector = $(2\mathbf{i} + \mathbf{j} + 5\mathbf{k})$

Just pick values for a and b, then see what value of c is needed to make the scalar product zero.

11) The normal vector $\mathbf{n} = \begin{pmatrix}1\\3\\-3\end{pmatrix}$.

Call the position vector of the given point **a**.
Then the coefficients of x, y and z in the Cartesian equation are the components of **n**, and the constant term is

$$d = -\mathbf{a}\cdot\mathbf{n} = -\begin{pmatrix}2\\2\\4\end{pmatrix}\cdot\begin{pmatrix}1\\3\\-3\end{pmatrix} = -[(2 \times 1) + (2 \times 3) + (4 \times -3)] = 4$$

So the Cartesian equation of the plane is
$n_1x + n_2y + n_3z + d = 0$
$\Rightarrow x + 3y - 3z + 4 = 0$

12) Label the points: A = (1, –2, 5), B = (6, 2, –3), C = (4, 0, 2)
Then a vector equation for the plane is $\mathbf{r} = \mathbf{a} + \lambda\mathbf{b} + \mu\mathbf{c}$, where

$\mathbf{a}$ = position vector of A = $\begin{pmatrix}1\\-2\\5\end{pmatrix}$,

$$\mathbf{b} = \overrightarrow{AB} = -\begin{pmatrix}1\\-2\\5\end{pmatrix} + \begin{pmatrix}6\\2\\-3\end{pmatrix} = \begin{pmatrix}5\\4\\-8\end{pmatrix},$$

$$\mathbf{c} = \overrightarrow{AC} = -\begin{pmatrix}1\\-2\\5\end{pmatrix} + \begin{pmatrix}4\\0\\2\end{pmatrix} = \begin{pmatrix}3\\2\\-3\end{pmatrix}.$$

So the vector equation is $\mathbf{r} = \begin{pmatrix}1\\-2\\5\end{pmatrix} + \lambda\begin{pmatrix}5\\4\\-8\end{pmatrix} + \mu\begin{pmatrix}3\\2\\-3\end{pmatrix}$

Exam Questions

1 a) $\overrightarrow{AB} = \mathbf{b} - \mathbf{a} = \begin{pmatrix}3\\2\\1\end{pmatrix} - \begin{pmatrix}1\\5\\9\end{pmatrix} = \begin{pmatrix}2\\-3\\-8\end{pmatrix}$

[2 marks available — 1 mark for attempting to subtract position vector a from position vector b, 1 mark for correct answer.]

b) l_1: $\mathbf{r} = \mathbf{c} + \mu(\mathbf{d} - \mathbf{c}) = \begin{pmatrix}-2\\4\\3\end{pmatrix} + \mu\left(\begin{pmatrix}5\\-1\\-7\end{pmatrix} - \begin{pmatrix}-2\\4\\3\end{pmatrix}\right)$ ***[1 mark]***

$\mathbf{r} = \begin{pmatrix}-2\\4\\3\end{pmatrix} + \mu\begin{pmatrix}7\\-5\\-10\end{pmatrix}$ ***[1 mark]***

c) Equation of line through AB:

$\overrightarrow{AB}$: $\mathbf{r} = \mathbf{a} + t(\mathbf{b} - \mathbf{a}) = \begin{pmatrix}1\\5\\9\end{pmatrix} + t\begin{pmatrix}2\\-3\\-8\end{pmatrix}$ ***[1 mark]***

At intersection of lines:

$\begin{pmatrix}1\\5\\9\end{pmatrix} + t\begin{pmatrix}2\\-3\\-8\end{pmatrix} = \begin{pmatrix}-2\\4\\3\end{pmatrix} + \mu\begin{pmatrix}7\\-5\\-10\end{pmatrix}$ ***[1 mark]***

Any two of: $1 + 2t = -2 + 7\mu$
$5 - 3t = 4 - 5\mu$
$9 - 8t = 3 - 10\mu$ ***[1 mark]***

Solving any two equations simultaneously gives
$t = 2$ or $\mu = 1$ ***[1 mark]***

Substituting $t = 2$ in the equation of the line through AB (or $\mu = 1$ in the equation for l_1) gives: (5, –1, –7) ***[1 mark]***

d) i) Vectors needed are $\begin{pmatrix}2\\-3\\-8\end{pmatrix}$ and $\begin{pmatrix}7\\-5\\-10\end{pmatrix}$ (direction vector of l_1).

$\begin{pmatrix}2\\-3\\-8\end{pmatrix}\cdot\begin{pmatrix}7\\-5\\-10\end{pmatrix} = 14 + 15 + 80 = 109$ ***[1 mark]***

magnitude of 1st vector: $\sqrt{2^2 + (-3)^2 + (-8)^2} = \sqrt{77}$
magnitude of 2nd vector:
$\sqrt{7^2 + (-5)^2 + (-10)^2} = \sqrt{174}$ ***[1 mark]***

$\cos\theta = \frac{109}{\sqrt{77}\sqrt{174}}$ ***[1 mark]***

$\Rightarrow \theta = 19.7°$ ***[1 mark]***

ii) Draw a diagram:

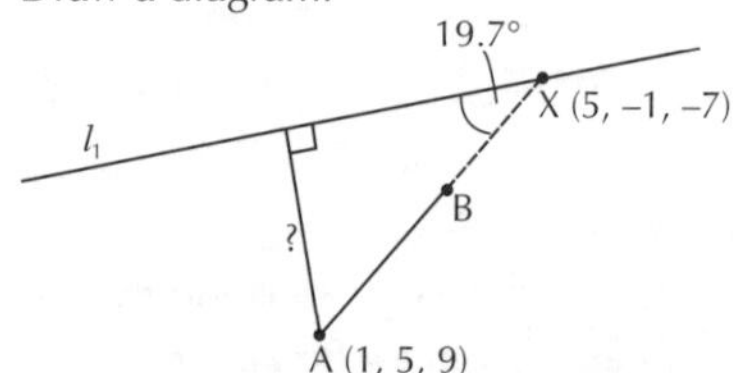

[1 mark for showing that the shortest distance is perpendicular to l_1]

X is the intersection point found in part c)
— (5, –1, –7)
Distance from A to X =

$\sqrt{(5-1)^2 + (-1-5)^2 + (-7-9)^2} = \sqrt{308}$
[1 mark]

Now you've got a right-angled triangle, so just use trig to find the side you want:
Shortest distance from A to l_1
$= \sqrt{308} \times \sin 19.7°$ ***[1 mark]*** = 5.9 units ***[1 mark]***

The tricky thing here is figuring out how to go about it. Drawing a diagram definitely helps you see what you know and what you need to work out. Often, you'll be meant to use something you worked out in a previous part of the question.

2 a) $-3(\mathbf{i} - 4\mathbf{j} + 2\mathbf{k}) = -3\mathbf{i} + 12\mathbf{j} - 6\mathbf{k}$ ***[1 mark]***

b) **i** component: $3 + (\mu \times 1) = 2$ gives $\mu = -1$ ***[1 mark]***
So $\mathbf{r} = (3\mathbf{i} - 3\mathbf{j} - 2\mathbf{k}) - 1(\mathbf{i} - 4\mathbf{j} + 2\mathbf{k}) = 2\mathbf{i} + \mathbf{j} - 4\mathbf{k}$ ***[1 mark]***
This is the position vector of the point A(2, 1, –4)

Answers

c) B lies on l_2 so it has position vector
$\mathbf{b} = (10\mathbf{i} - 21\mathbf{j} + 11\mathbf{k}) + \lambda(-3\mathbf{i} + 12\mathbf{j} - 6\mathbf{k})$ ***[1 mark]***

So $\overrightarrow{AB} = \mathbf{b} - \mathbf{a}$
$= ((10\mathbf{i} - 21\mathbf{j} + 11\mathbf{k}) + \lambda(-3\mathbf{i} + 12\mathbf{j} - 6\mathbf{k})) - (2\mathbf{i} + \mathbf{j} - 4\mathbf{k})$
$= (8 - 3\lambda)\mathbf{i} + (-22 + 12\lambda)\mathbf{j} + (15 - 6\lambda)\mathbf{k}$ ***[1 mark]***

You know the scalar product of the direction vector of l_1 and $\overrightarrow{AB}$ must equal zero as they're perpendicular:
$(\mathbf{i} - 4\mathbf{j} + 2\mathbf{k}).((8 - 3\lambda)\mathbf{i} + (-22 + 12\lambda)\mathbf{j} + (15 - 6\lambda)\mathbf{k})$ ***[1 mark]***
$= (8 - 3\lambda) + (88 - 48\lambda) + (30 - 12\lambda)$
$= 126 - 63\lambda = 0$
$\Rightarrow \lambda = 2$ ***[1 mark]***

Substitute in $\lambda = 2$ to find the position vector $\mathbf{b}$:
$\mathbf{b} = (10\mathbf{i} - 21\mathbf{j} + 11\mathbf{k}) + 2(-3\mathbf{i} + 12\mathbf{j} - 6\mathbf{k})$ ***[1 mark]***
$= 4\mathbf{i} + 3\mathbf{j} - \mathbf{k}$

Position vector of B $= 4\mathbf{i} + 3\mathbf{j} - \mathbf{k}$ ***[1 mark]***

You could have multiplied $\overrightarrow{AB}$ by the direction bit of the l_2 vector equation, as $\overrightarrow{AB}$ is perpendicular to both l_1 and l_2. But the numbers for the l_1 vector are smaller, making your calculations easier.

d) $\overrightarrow{AB} = \mathbf{b} - \mathbf{a}$
$= (4\mathbf{i} + 3\mathbf{j} - \mathbf{k}) - (2\mathbf{i} + \mathbf{j} - 4\mathbf{k}) = 2\mathbf{i} + 2\mathbf{j} + 3\mathbf{k}$ ***[1 mark]***
$|\overrightarrow{AB}| = \sqrt{2^2 + 2^2 + 3^2} = \sqrt{17} = 4.1$ ***[1 mark]***

3 a) At an intersection point: $\begin{pmatrix}3\\0\\-2\end{pmatrix} + \lambda\begin{pmatrix}1\\3\\-2\end{pmatrix} = \begin{pmatrix}0\\2\\1\end{pmatrix} + \mu\begin{pmatrix}2\\-5\\-3\end{pmatrix}$

[1 mark]

This gives equations: $3 + \lambda = 2\mu$
$3\lambda = 2 - 5\mu$
$-2 - 2\lambda = 1 - 3\mu$ ***[1 mark]***

Solving the first two equations simultaneously gives:
$\lambda = -1,\ \mu = 1$ ***[1 mark]***
Substituting these values in the third equation gives:
$-2 - 2(-1) = 1 - 3(1) \Rightarrow 0 \neq -2$ ***[1 mark]***

So the lines don't intersect.

You could have solved any two of the equations simultaneously, then substituted the results in the remaining equation to show that they don't work and there's no intersection point.

b) (i) At the intersection point of PQ and l_1:

$\begin{pmatrix}3\\0\\-2\end{pmatrix} + \lambda\begin{pmatrix}1\\3\\-2\end{pmatrix} = \begin{pmatrix}5\\4\\-9\end{pmatrix} + t\begin{pmatrix}0\\2\\3\end{pmatrix}$ ***[1 mark]***

This gives equations: $3 + \lambda = 5$
$3\lambda = 4 + 2t$
$-2 - 2\lambda = -9 + 3t$

[1 mark for any two equations]

Solving two of these equations gives: $\lambda = 2$, $t = 1$
[1 mark]

Intersection point $= \begin{pmatrix}5\\4\\-9\end{pmatrix} + 1\begin{pmatrix}0\\2\\3\end{pmatrix} = \begin{pmatrix}5\\6\\-6\end{pmatrix} = (5, 6, -6)$
[1 mark]

(ii) If perpendicular, the scalar product of direction vectors of lines will equal 0:

$\begin{pmatrix}0\\2\\3\end{pmatrix} \cdot \begin{pmatrix}1\\3\\-2\end{pmatrix}$ ***[1 mark]***

$= (0 \times 1) + (2 \times 3) + (3 \times -2) = 0$ ***[1 mark]***

(iii) Call intersection point X.

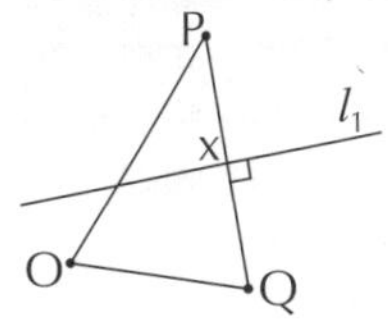

$\overrightarrow{PX} = \mathbf{x} - \mathbf{p}$

$= \begin{pmatrix}5\\6\\-6\end{pmatrix} - \begin{pmatrix}5\\8\\-3\end{pmatrix} = \begin{pmatrix}0\\-2\\-3\end{pmatrix}$ ***[1 mark]***

$\overrightarrow{OQ} = \overrightarrow{OP} + 2\overrightarrow{PX} = \begin{pmatrix}5\\8\\-3\end{pmatrix} + 2\begin{pmatrix}0\\-2\\-3\end{pmatrix}$ ***[1 mark]***

$= \begin{pmatrix}5\\4\\-9\end{pmatrix}$ ***[1 mark]***

The trick with this one is to realise that point Q lies the same distance from the intersection point as P does — drawing a quick sketch will definitely help.

4 a) $(\overrightarrow{OA}).(\overrightarrow{OB})$ ***[1 mark]***
$= (3\mathbf{i} + 2\mathbf{j} + \mathbf{k}).(3\mathbf{i} - 4\mathbf{j} - \mathbf{k}) = 9 - 8 - 1 = 0$ ***[1 mark]***
Therefore, side OA is perpendicular to side OB, and the triangle has a right angle. ***[1 mark]***

You could also have found the lengths |OA|, |OB| and |AB| and shown by Pythagoras that AOB is a right-angled triangle ($|AB|^2 = |OA|^2 + |OB|^2$).

b) $\overrightarrow{BA} = \mathbf{a} - \mathbf{b} = (3\mathbf{i} + 2\mathbf{j} + \mathbf{k}) - (3\mathbf{i} - 4\mathbf{j} - \mathbf{k}) = (6\mathbf{j} + 2\mathbf{k})$ ***[1 mark]***
$\overrightarrow{BO} = -3\mathbf{i} + 4\mathbf{j} + \mathbf{k}$
$\overrightarrow{BA} \cdot \overrightarrow{BO} = 24 + 2 = 26$ ***[1 mark]***
$|\overrightarrow{BA}| = \sqrt{6^2 + 2^2} = \sqrt{40}$
$|\overrightarrow{BO}| = \sqrt{(-3)^2 + 4^2 + 1^2} = \sqrt{26}$ ***[1 mark]***
$\cos\angle ABO = \dfrac{\overrightarrow{BA} \cdot \overrightarrow{BO}}{|\overrightarrow{BA}| \cdot |\overrightarrow{BO}|} = \dfrac{26}{\sqrt{40}\sqrt{26}}$ ***[1 mark]***
$\angle ABO = 36.3°$ ***[1 mark]***

c) (i) $\overrightarrow{AC} = \mathbf{c} - \mathbf{a} = (3\mathbf{i} - \mathbf{j}) - (3\mathbf{i} + 2\mathbf{j} + \mathbf{k}) = (-3\mathbf{j} - \mathbf{k})$ ***[1 mark]***
$|\overrightarrow{AC}| = \sqrt{(-3)^2 + (-1)^2} = \sqrt{10}$
$|\overrightarrow{OC}| = \sqrt{3^2 + (-1)^2} = \sqrt{10}$ ***[1 mark]***
Sides AC and OC are the same length, so the triangle is isosceles. ***[1 mark]***

(ii) You know side lengths AC and OC from part c)(i). Calculate length of OA:
$|\overrightarrow{OA}| = \sqrt{3^2 + 2^2 + 1^2} = \sqrt{14}$ ***[1 mark]***

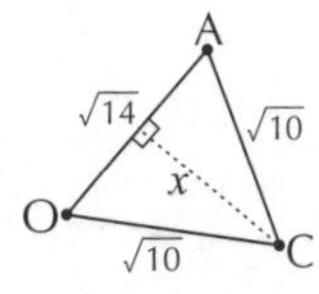

Now find the height of the triangle, x, using Pythagoras:

$x = \sqrt{(\sqrt{10})^2 - \left(\dfrac{\sqrt{14}}{2}\right)^2} = \sqrt{6.5}$ ***[1 mark]***

Area $= ½(\text{base} \times \text{height})$
$= ½(\sqrt{14} \times \sqrt{6.5})$ ***[1 mark]***
$= 4.77$ square units ***[1 mark]***

Answers

d) (i) $\mathbf{r} = \mathbf{a} + t(\mathbf{b} - \mathbf{a})$

$\mathbf{r} = (3\mathbf{i} + 2\mathbf{j} + \mathbf{k}) + t((3\mathbf{i} - 4\mathbf{j} - \mathbf{k}) - (3\mathbf{i} + 2\mathbf{j} + \mathbf{k}))$ ***[1 mark]***

$\mathbf{r} = (3\mathbf{i} + 2\mathbf{j} + \mathbf{k}) + t(-6\mathbf{j} - 2\mathbf{k})$ ***[1 mark]***

(ii) $\mathbf{k}$ component: $1 - 2t = 1$, $t = 0$ ***[1 mark]***

$\mathbf{r} = (3\mathbf{i} + 2\mathbf{j} + \mathbf{k}) + 0(6\mathbf{j} + 2\mathbf{k}) = 3\mathbf{i} + 2\mathbf{j} + \mathbf{k}$

$a = 3$ ***[1 mark]***, $b = 2$ ***[1 mark]***

5 a) Use the coefficients of x, y and z in the equation of each plane as the components of the normal vectors.

So a normal to A is $\mathbf{n}_1 = \begin{pmatrix} 3 \\ 4 \\ 2 \end{pmatrix}$,

and a normal to B is $\mathbf{n}_2 = \begin{pmatrix} 1 \\ -1 \\ 6 \end{pmatrix}$ ***[1 mark for both correct]***.

b) The angle between two planes is equal to the angle between their normal vectors ***[1 mark]***.

So $\cos\theta = \frac{\mathbf{n}_1 \cdot \mathbf{n}_2}{|\mathbf{n}_1||\mathbf{n}_2|} = \frac{(3 \times 1) + (4 \times -1) + (2 \times 6)}{\sqrt{3^2 + 4^2 + 2^2}\sqrt{1^2 + (-1)^2 + 6^2}}$

$= \frac{11}{\sqrt{29}\sqrt{38}}$

[1 mark for correct scalar product in numerator]

$= 0.3314$

So $\theta = \cos^{-1} 0.3314 = 70.6°$ ***[1 mark]***

c) $\mathbf{r} = \begin{pmatrix} 5 \\ 1 \\ 3 \end{pmatrix} + \lambda\begin{pmatrix} -3 \\ 2 \\ 0 \end{pmatrix} = \begin{pmatrix} 5 - 3\lambda \\ 1 + 2\lambda \\ 3 \end{pmatrix}$

So at the point of intersection,

$x = 5 - 3\lambda$, $y = 1 + 2\lambda$, $z = 3$ ***[1 mark]***.

Substitute these into the equation of plane A:

$3x + 4y + 2z = 1$

$\Rightarrow 3(5 - 3\lambda) + 4(1 + 2\lambda) + 2(3) = 1$ ***[1 mark]***

$\Rightarrow 15 - 9\lambda + 4 + 8\lambda + 6 = 1$

$\Rightarrow 25 - \lambda = 1$

$\Rightarrow \lambda = 24$ ***[1 mark]***

So at the point of intersection,

$x = 5 - 3(24) = -67$, $y = 1 + 2(24) = 49$, $z = 3$

The point of intersection is $(-67, 49, 3)$ ***[1 mark]***.

Answers

S2 Section 1 — Discrete Random Variables

Warm-up Questions

1) a) All the probabilities have to add up to 1.
So $0.5 + k + k + 3k = 0.5 + 5k = 1$, i.e. $5k = 0.5$, i.e. $k = 0.1$.

b) $P(Y < 2) = P(Y = 0) + P(Y = 1) = 0.5 + 0.1 = 0.6$.

2) There are 5 possible outcomes, and the probability of each of them is k. That means $5k = 1$, so $k = 1 \div 5 = 0.2$.
Mean of X = $(0 \times 0.2) + (1 \times 0.2) + (2 \times 0.2) + (3 \times 0.2) + (4 \times 0.2) = 2$.
$E(X^2) = (0 \times 0.2) + (1 \times 0.2) + (4 \times 0.2) + (9 \times 0.2) + (16 \times 0.2) = 6$, so variance of X = $E(X^2) - [E(X)]^2 = 6 - 4 = 2$.

3) a) As always, the probabilities have to add up to 1, so
$k = 1 - \left(\frac{1}{6} + \frac{1}{2} + \frac{5}{24}\right) = 1 - \frac{21}{24} = \frac{3}{24} = \frac{1}{8}$

b) $E(X) = \left(1 \times \frac{1}{6}\right) + \left(2 \times \frac{1}{2}\right) + \left(3 \times \frac{1}{8}\right) + \left(4 \times \frac{5}{24}\right)$
$= \frac{4 + 24 + 9 + 20}{24} = \frac{57}{24} = \frac{19}{8}$

$E(X^2) = \left(1^2 \times \frac{1}{6}\right) + \left(2^2 \times \frac{1}{2}\right) + \left(3^2 \times \frac{1}{8}\right) + \left(4^2 \times \frac{5}{24}\right)$
$= \frac{4 + 48 + 27 + 80}{24} = \frac{159}{24} = \frac{53}{8}$

$Var(X) = E(X^2) - [E(X)]^2 = \frac{53}{8} - \left(\frac{19}{8}\right)^2$
$= \frac{424 - 361}{64} = \frac{63}{64}$

c) $E(2X - 1) = 2E(X) - 1 = 2 \times \frac{19}{8} - 1 = \frac{30}{8} = \frac{15}{4}$

$Var(2X - 1) = 2^2 Var(X) = 4 \times \frac{63}{64} = \frac{63}{16}$

4) a) $E(X) = (1 \times 0.1) + (2 \times 0.2) + (3 \times 0.25) + (4 \times 0.2) + (5 \times 0.1) + (6 \times 0.15) = 3.45$

b) $Var(X) = E(X^2) - (E(X))^2$
$E(X^2) = (1 \times 0.1) + (4 \times 0.2) + (9 \times 0.25) + (16 \times 0.2) + (25 \times 0.1) + (36 \times 0.15) = 14.25$
So $Var(X) = 14.25 - 3.45^2 = 2.3475$

c) $E\left(\frac{2}{X}\right) = \sum\left(\frac{2}{x_i} \times p_i\right) = \left(\frac{2}{1} \times 0.1\right) + \left(\frac{2}{2} \times 0.2\right) + \left(\frac{2}{3} \times 0.25\right) + \left(\frac{2}{4} \times 0.2\right) + \left(\frac{2}{5} \times 0.1\right) + \left(\frac{2}{6} \times 0.15\right) = 0.757$ (3 d.p.)

You could have calculated part c) by doing 2[E(1/X)] instead — if you'd wanted to show off your ability to deal with linear functions too.

Exam Questions

1 a) The probability of getting 3 heads is: $\frac{1}{2} \times \frac{1}{2} \times \frac{1}{2} = \frac{1}{8}$
[1 mark]

The probability of getting 2 heads is: $3 \times \frac{1}{2} \times \frac{1}{2} \times \frac{1}{2} = \frac{3}{8}$
(multiply by 3 because any of the three coins could be the tail — the order in which the heads and the tail occur isn't important).
[1 mark]

Similarly the probability of getting 1 head is:
$3 \times \frac{1}{2} \times \frac{1}{2} \times \frac{1}{2} = \frac{3}{8}$
And the probability of getting no heads is $\frac{1}{2} \times \frac{1}{2} \times \frac{1}{2} = \frac{1}{8}$
So the probability of 1 or no heads = $\frac{3}{8} + \frac{1}{8} = \frac{1}{2}$ ***[1 mark]***

Hence the probability distribution of X is:

x	20p	10p	*nothing*
$P(X = x)$	$\frac{1}{8}$	$\frac{3}{8}$	$\frac{1}{2}$

[1 mark]

b) You need the probability that X >10p ***[1 mark]***
This is just P(X = 20p) = $\frac{1}{8}$ ***[1 mark]***
Easy peasy. The difficult question is — why would anyone play such a rubbish game?

2 a) All the probabilities must add up to 1, so
$2k + 3k + k + k = 1$, i.e. $7k = 1$, and so $k = \frac{1}{7}$. ***[1 mark]***

b) $P(X > 2) = P(X = 3) = \frac{1}{7}$, using part a) ***[1 mark]***
All probabilities add up to one! OK, that's the last time I'm going to say it. Scout's honour.

3 a) There are 10 possible values,
so $10k = 1$ and $k = 1 \div 10 = 0.1$.

x	*0*	*1*	*2*	*3*	*4*	*5*	*6*	*7*	*8*	*9*
P(X = x)	*0.1*	*0.1*	*0.1*	*0.1*	*0.1*	*0.1*	*0.1*	*0.1*	*0.1*	*0.1*

[1 mark]

b) Mean = 0 + 0.1 + 0.2 + 0.3 + 0.4 + 0.5 + 0.6 + 0.7 + 0.8 + 0.9 = 4.5 ***[1 mark]***
Variance = $E(X^2) - [E(X)]^2$
$E(X^2) = 0 + 0.1 + 0.4 + 0.9 + 1.6 + 2.5 + 3.6 + 4.9 + 6.4 + 8.1 = 28.5$ ***[1 mark]***
So variance = $28.5 - 4.5^2 = 8.25$ ***[1 mark]***

c) $P(X < 4.5) = P(X = 0) + P(X = 1) + P(X = 2) + P(X = 3) + P(X = 4)$ ***[1 mark]***
= 0.5 ***[1 mark]***

4 a) P(X = 1) = a, P(X = 2) = 2a, P(X = 3) = 3a.
Therefore the total probability is 3a + 2a + a = 6a.
This must equal 1, so $a = \frac{1}{6}$. ***[1 mark]***

b) $E(X) = \left(1 \times \frac{1}{6}\right) + \left(2 \times \frac{2}{6}\right) + \left(3 \times \frac{3}{6}\right) = \frac{1 + 4 + 9}{6}$ ***[1 mark]***
$= \frac{7}{3}$ ***[1 mark]***

c) $E(X^2) = Var(X) + [E(X)]^2 = \frac{5}{9} + \left(\frac{7}{3}\right)^2 = \frac{5 + 49}{9}$ ***[1 mark]***
$= \frac{54}{9} = 6$ ***[1 mark]***

d) $E(3X + 4) = 3E(X) + 4 = 3 \times \frac{7}{3} + 4 = 11$ ***[1 mark]***
$Var(3X + 4) = 3^2 Var(X) = 9 \times \frac{5}{9}$ ***[1 mark]***
$= 5$ ***[1 mark]***

Answers

5 a) $E(X) = (0 \times 0.4) + (1 \times 0.3) + (2 \times 0.2) + (3 \times 0.1)$
$= 0 + 0.3 + 0.4 + 0.3$ ***[1 mark]***
$= 1$ ***[1 mark]***

b) $E(6X + 8) = 6E(X) + 8 = 6 + 8$ ***[1 mark]***
$= 14$ ***[1 mark]***

c) The formula for variance is $Var(X) = E(X^2) - [E(X)^2]$
So first work out $E(X^2)$:
$E(X^2) = (0^2 \times 0.4) + (1^2 \times 0.3) + (2^2 \times 0.2) + (3^2 \times 0.1)$
$= 0.3 + 0.8 + 0.9$ ***[1 mark]***
$= 2$ ***[1 mark]***

Then complete the formula also using your answer to part a):
$E(X^2) - [E(X)^2] = 2 - (1^2)$ ***[1 mark]***
$= 1$ ***[1 mark]***

d) $Var(aX + b) = a^2Var(X)$
$Var(5 - 3X) = (-3)^2Var(X) = 9Var(X) = 9 \times 1$
[1 mark]
$= 9$ ***[1 mark]***

S2 Section 2 — The Binomial Distribution

Warm-up Questions

1) a) There are 21 objects altogether, so if all the balls were different colours, there would be 21! ways to arrange them. But since 15 of the objects are identical, you need to divide this figure by 15!. So there are 21! ÷ 15! = 39 070 080 possible arrangements.

b) There are $\frac{16!}{4!4!4!4!} = 63\,063\,000$ possible arrangements.

You'd be a while counting all these on your fingers.

2) a) $P(5\text{ heads}) = 0.5^5 \times 0.5^5 \times \binom{10}{5}$
$= 0.5^{10} \times \frac{10!}{5!5!} = 0.246$ (to 3 sig.fig.).

b) $P(9\text{ heads}) = 0.5^9 \times 0.5 \times \binom{10}{9}$
$= 0.5^{10} \times \frac{10!}{9!1!} = 0.00977$ (to 3 sig.fig.).

3) a) Binomial — there are a fixed number of independent trials (30) with two possible results ('prime' / 'not prime'), a constant probability of success, and the random variable is the total number of successes.

b) Binomial — there are a fixed number of independent trials (however many students are in the class) with two possible results ('heads' / 'tails'), a constant probability of success, and the random variable is the total number of successes.

c) Not binomial — the probability of being dealt an ace changes each time, since the total number of cards decreases as each card is dealt.

d) Not binomial — the number of trials is not fixed.

It's weird to have to write actual sentences in a maths exam, but be ready for it.

4) a) Use tables with $n = 10$ and $p = 0.5$.
If X represents the number of heads, then:
$P(X \geq 5) = 1 - P(X < 5) = 1 - P(X \leq 4)$
$= 1 - 0.3770 = 0.6230$

b) $P(X \geq 9) = 1 - P(X < 9) = 1 - P(X \leq 8)$
$= 1 - 0.9893 = 0.0107$

You do have to be prepared to monkey around with the numbers the tables give you.

5) a) You can't use tables here (because they don't include $p = 0.27$ or $n = 14$), so you have to use the probability function.
$P(X = 4) = \binom{14}{4} \times 0.27^4 \times (1 - 0.27)^{10}$
$= 0.229$ (to 3 sig.fig.)

b) $P(X < 2) = P(X = 0) + P(X = 1)$
$= \binom{14}{0} \times 0.27^0 \times (1 - 0.27)^{14}$
$+ \binom{14}{1} \times 0.27^1 \times (1 - 0.27)^{13}$
$= 0.012204... + 0.063195...$
$= 0.0754$ (to 3 sig.fig.)

c) $P(5 < X \leq 8) = P(X = 6) + P(X = 7) + P(X = 8)$
$= \binom{14}{6} \times 0.27^6 \times (1 - 0.27)^8$
$+ \binom{14}{7} \times 0.27^7 \times (1 - 0.27)^7$
$+ \binom{14}{8} \times 0.27^8 \times (1 - 0.27)^6$
$= 0.093825... + 0.039660... + 0.012835...$
$= 0.146$ (to 3 sig.fig.)

6) For parts a)-c), use tables with $n = 25$ and $p = 0.15$.

a) $P(X \leq 3) = 0.4711$

b) $P(X \leq 7) = 0.9745$

c) $P(X \leq 15) = 1.0000$

For parts d)-f), define a new random variable $T \sim B(15, 0.35)$. Then use tables with $n = 15$ and $p = 0.35$.

d) $P(Y \leq 3) = P(T \geq 12) = 1 - P(T < 12)$
$= 1 - P(T \leq 11) = 1 - 0.9995 = 0.0005$

e) $P(Y \leq 7) = P(T \geq 8) = 1 - P(T < 8)$
$= 1 - P(T \leq 7) = 1 - 0.8868 = 0.1132$

f) $P(Y \leq 15) = 1$ (since 15 is the maximum possible value).

These last few parts (where you can't use the tables without a bit of messing around first) are quite awkward, so make sure you get lots of practice.

7) From tables:

a) $P(X \leq 15) = 0.9997$

b) $P(X < 4) = P(X \leq 3) = 0.1302$

c) $P(X > 7) = 1 - P(X \leq 7) = 1 - 0.0639 = 0.9361$

Answers

For parts d)-f) where $X \sim B(n, p)$ with $p > 0.5$, define a new random variable $Y \sim B(n, q)$, where $q = 1 - p$. Then use tables.

d) Define $Y \sim B(50, 0.2)$. Then $P(X \geq 40) = P(Y \leq 10) = 0.5836$

e) Define $Y \sim B(30, 0.3)$. Then $P(X = 20) = P(Y = 10)$
$= P(Y \leq 10) - P(Y \leq 9) = 0.7304 - 0.5888 = 0.1416$

f) Define $Y \sim B(10, 0.25)$. Then $P(X = 7) = P(Y = 3)$
$= P(Y \leq 3) - P(Y \leq 2) = 0.7759 - 0.5256 = 0.2503$

8) a) mean $= 20 \times 0.4 = 8$; variance $= 20 \times 0.4 \times 0.6 = 4.8$

b) mean $= 40 \times 0.15 = 6$; variance $= 40 \times 0.15 \times 0.85 = 5.1$

c) mean $= 25 \times 0.45 = 11.25$;
variance $= 25 \times 0.45 \times 0.55 = 6.1875$

d) mean $= 50 \times 0.8 = 40$; variance $= 50 \times 0.8 \times 0.2 = 8$

e) mean $= 30 \times 0.7 = 21$; variance $= 30 \times 0.7 \times 0.3 = 6.3$

f) mean $= 45 \times 0.012 = 0.54$;
variance $= 45 \times 0.012 \times 0.988 = 0.53352$

Exam Questions

1 a) (i) Define a new random variable $Y \sim B(12, 0.4)$.
Then $P(X < 8) = P(Y > 4) = 1 - P(Y \leq 4)$
$= 1 - 0.4382$ ***[1 mark]*** $= 0.5618$ ***[1 mark]***

(ii) $P(X = 5) = P(Y = 7) = P(Y \leq 7) - P(Y \leq 6)$ ***[1 mark]***
$= 0.9427 - 0.8418 = 0.1009$ ***[1 mark]***

Or you could use the probability function for part (ii):
$P(X = 5) = \binom{12}{5} \times 0.6^5 \times 0.4^7 = 0.1009$

(iii) $P(3 < X \leq 7) = P(X$ is greater than 3 <u>and</u> less than or equal to 7$) = P(Y$ is less than 9 <u>and</u> greater than or equal to 5$) = P(5 \leq Y < 9) = P(5 \leq Y \leq 8)$ ***[1 mark]***
$= P(Y \leq 8) - P(Y \leq 4)$ ***[1 mark]*** $= 0.9847 - 0.4382$
$= 0.5465$ ***[1 mark]***

b) (i) $P(Y = 4) = 0.8^4 \times 0.2^7 \times \frac{11!}{4!7!}$ ***[1 mark]***
$= 0.00173$ (to 3 sig. fig.) ***[1 mark]***

(ii) $E(Y) = 11 \times 0.8 = 8.8$ ***[1 mark]***

(iii) $Var(Y) = 11 \times 0.8 \times 0.2 = 1.76$ ***[1 mark]***

2 a) (i) Let X represent the number of apples that contain a maggot. Then $X \sim B(40, 0.15)$ ***[1 mark]***.
$P(X < 6) = P(X \leq 5) = 0.4325$ ***[1 mark]***

(ii) $P(X > 2) = 1 - P(X \leq 2)$ ***[1 mark]***
$= 1 - 0.0486 = 0.9514$ ***[1 mark]***

(iii) $P(X = 12) = P(X \leq 12) - P(X \leq 11)$ ***[1 mark]***
$= 0.9957 - 0.9880 = 0.0077$ ***[1 mark]***

Or you could use the probability function for part (iii):
$P(X = 12) = \binom{40}{12} \times 0.15^{12} \times 0.85^{28} = 0.0077$

b) The probability that a crate contains more than 2 apples with maggots is 0.9514 (from part a) (ii)).

So define a random variable Y, where Y is the number of crates that contain more than 2 apples with maggots.
Then $Y \sim B(3, 0.9514)$ ***[1 mark]***.
You need to find $P(Y = 2) + P(Y = 3)$. This is:

$0.9514^2 \times (1 - 0.9514) \times \binom{3}{2}$
$+ 0.9514^3 \times (1 - 0.9514)^0 \times \binom{3}{3}$ ***[1 mark]***
$= 0.1320 + 0.8612 = 0.993$ (to 3 d.p.) ***[1 mark]***

3 a) (i) The probability of Simon being able to solve each crossword needs to remain the same ***[1 mark]***, and all the outcomes need to be independent (i.e. Simon solving or not solving a puzzle one day should not affect whether he will be able to solve it on another day) ***[1 mark]***.

(ii) The total number of puzzles he solves (or the number he fails to solve) ***[1 mark]***.

b) $P(X = 4) = p^4 \times (1 - p)^{14} \times \frac{18!}{4!14!}$ ***[1 mark]***

$P(X = 5) = p^5 \times (1 - p)^{13} \times \frac{18!}{5!13!}$ ***[1 mark]***

So $p^4 \times (1 - p)^{14} \times \frac{18!}{4!14!} = p^5 \times (1 - p)^{13} \times \frac{18!}{5!13!}$ ***[1 mark]***

Dividing by things that occur on both sides gives:
$\frac{1 - p}{14} = \frac{p}{5}$ ***[1 mark]***, or $5 = 19p$.

This means $p = \frac{5}{19}$ ***[1 mark]***.

Answers

S2 Section 3 — The Poisson Distribution

Warm-up Questions

1) a) $P(X = 2) = \frac{e^{-3.1} \times 3.1^2}{2!} = 0.2165$ (to 4 d.p.).

b) $P(X = 1) = \frac{e^{-3.1} \times 3.1}{1!} = 0.1397$ (to 4 d.p.).

c) $P(X = 0) = \frac{e^{-3.1} \times 3.1^0}{0!} = 0.0450$ (to 4 d.p.).

d) $P(X < 3) = P(X = 0) + P(X = 1) + P(X = 2)$
$= 0.0450 + 0.1397 + 0.2165 = 0.4012$

e) $P(X \geq 3) = 1 - P(X < 3)$
$= 1 - 0.4012 = 0.5988$ (to 4 d.p.).

2) a) $P(X = 2) = \frac{e^{-8.7} \times 8.7^2}{2!} = 0.0063$ (to 4 d.p.).

b) $P(X = 1) = \frac{e^{-8.7} \times 8.7}{1!} = 0.0014$ (to 4 d.p.).

c) $P(X = 0) = \frac{e^{-8.7} \times 8.7^0}{0!} = 0.0002$ (to 4 d.p.).

d) $P(X < 3) = P(X = 0) + P(X = 1) + P(X = 2)$
$= 0.0002 + 0.0014 + 0.0063 = 0.0079$

e) $P(X \geq 3) = 1 - P(X < 3)$
$= 1 - 0.0079 = 0.9921$ (to 4 d.p.).

3) a) $E(X) = Var(X) = 8$
standard deviation $= \sigma = \sqrt{8} = 2.828$ (to 3 d.p.).

b) $E(X) = Var(X) = 12.11$
standard deviation $= \sigma = \sqrt{12.11} = 3.480$ (to 3 d.p.).

c) $E(X) = Var(X) = 84.2227$
standard deviation $= \sigma = \sqrt{84.2227} = 9.177$ (to 3 d.p.).

4) Using tables:

a) $P(X \leq \mu) = P(X \leq 9) = 0.5874$, $P(X \leq \mu - \sigma) = P(X \leq 6) = 0.2068$

b) $P(X \leq \mu) = P(X \leq 4) = 0.6288$, $P(X \leq \mu - \sigma) = P(X \leq 2) = 0.2381$

5) a) The defective products occur randomly, singly and (on average) at a constant rate, and the random variable represents the number of 'events' (i.e. defective products) within a fixed period, so this would follow a Poisson distribution.

b) There is a fixed number of trials in this situation, and so this situation would be modelled by a binomial distribution. (Or you could say it won't follow a Poisson distribution, as the events don't occur at a constant rate over the 25 trials.)

c) If the random variable represents the number of people joining the queue within a fixed period, and assuming that the people join the queue randomly, singly and (on average) at a constant rate, then this would follow a Poisson distribution.

You do need to make a couple of assumptions here — the Poisson model wouldn't work if you had, say, big groups of factory workers all coming in together a couple of minutes after the lunchtime hooter sounds.

d) The mistakes occur randomly, singly and (on average) at a constant rate, and the random variable represents the number of mistakes within a fixed 'period' (i.e. the number of pages in the document), so this would follow a Poisson distribution.

6) a) The number of atoms decaying in an hour would follow the Poisson distribution Po(2000). So the number decaying in a minute would follow Po(2000 ÷ 60) = Po(33.3).

b) The number of atoms decaying in a day would follow Po(2000 × 24) = Po(48 000).

7) a) If X represents the number of atoms from the first sample decaying per minute, then $X \sim Po(60)$. And if Y represents the number of atoms from the second sample decaying per minute, then $Y \sim Po(90)$. So $X + Y$ (the total number of atoms decaying per minute) $\sim Po(60 + 90) = Po(150)$.

b) The total number of atoms decaying per hour would be distributed as Po(150 × 60) = Po(9000).

8) a) $P(X \leq 2) = 0.0138$

b) $P(X \leq 7) = 0.4530$

c) $P(X \leq 5) = 0.1912$

d) $P(X < 9) = P(X \leq 8) = 0.5925$

e) $P(X \geq 8) = 1 - P(X < 8) = 1 - P(X \leq 7)$
$= 1 - 0.4530 = 0.5470$

f) $P(X > 1) = 1 - P(X \leq 1) = 1 - 0.0030 = 0.9970$

g) $P(X > 7) = 1 - P(X \leq 7) = 1 - 0.4530 = 0.5470$

h) $P(X = 6) = P(X \leq 6) - P(X \leq 5) = 0.3134 - 0.1912 = 0.1222$

i) $P(X = 4) = P(X \leq 4) - P(X \leq 3) = 0.0996 - 0.0424 = 0.0572$

j) $P(X = 3) = P(X \leq 3) - P(X \leq 2) = 0.0424 - 0.0138 = 0.0286$

9) If X represents the number of geese in a random square metre of field, then $X \sim Po(1)$ — since the 'rate' at which geese occur is constant, they're randomly scattered, and geese only occur singly.

a) $P(X = 0) = \frac{e^{-1} \times 1^0}{0!} = 0.3679$

b) $P(X = 1) = \frac{e^{-1} \times 1^1}{1!} = 0.3679$

c) $P(X = 2) = \frac{e^{-1} \times 1^2}{2!} = 0.1839$

d) $P(X > 2) = 1 - P(X \leq 2) = 1 - (0.3679 + 0.3679 + 0.1839)$
$= 1 - 0.9197 = 0.0803$

This is one of those questions where you could use either your Poisson tables or the probability function.

10) a) No — n is not very large, and p is not very small.

b) Yes — n is large, and p is small, so approximate with Po(7).

c) Not really — n is large, but p isn't as small as you'd like.

d) Not really — n is quite small (and so you don't really need to approximate it anyway).

e) This is perfect for a Poisson approximation — n is enormous and p is tiny. It should follow Po(0.1) very closely.

Answers

f) If Y represents the number of 'successes' in 80 trials, then define a new random variable X representing the number of 'failures' in those 80 trials. Then $X \sim B(80, 0.1)$. Since n is quite large, and p is quite small, you could approximate X with $Po(80 \times 0.1) = Po(8)$. Then $Y = 80 - X$.

Exam Questions

1 a) Events need to happen at a constant average rate ***[1 mark]*** and singly ("one at a time") ***[1 mark]***.

You could also have had "events occur randomly" or "independently".

b) (i) If X represents the number of chaffinches visiting the observation spot, then $X \sim Po(7)$ ***[1 mark]***.
Using tables, $P(X < 4) = P(X \leq 3) = 0.0818$ ***[1 mark]***.

(ii) $P(X \geq 7) = 1 - P(X < 7) = 1 - P(X \leq 6)$ ***[1 mark]***
$= 1 - 0.4497 = 0.5503$ ***[1 mark]***

(iii) $P(X = 9) = P(X \leq 9) - P(X \leq 8)$ ***[1 mark]***
$= 0.8305 - 0.7291 = 0.1014$ ***[1 mark]***

Or you could work this last one out using the formula:
$P(X = 9) = \frac{e^{-7}7^9}{9!} = 0.1014$
— you get the same answer either way, obviously.

c) The number of birds of any species visiting per hour would follow the distribution $Po(22 + 7) = Po(29)$ ***[1 mark]***. So the total number of birds visiting in a random 15-minute period will follow $Po(29 \div 4) = Po(7.25)$ ***[1 mark]***.

$$P(X = 3) = \frac{e^{-7.25} \times 7.25^3}{3!} \text{ [1 mark]}$$
$$= 0.045 \text{ (to 3 d.p.) [1 mark]}.$$

2 a) (i) If the mean is 20, then the number of calls per hour follows $Po(20)$. So the number of calls in a random 30-minute period follows $Po(20 \div 2) = Po(10)$ ***[1 mark]***.
Using tables for $\lambda = 10$:
$P(X = 8) = P(X \leq 8) - P(X \leq 7)$ ***[1 mark]***
$= 0.3328 - 0.2202 = 0.1126$ ***[1 mark]***

Or you could work this out using the formula:
$P(X = 8) = \frac{e^{-10}10^8}{8!} = 0.1126.$

(ii) $P(X > 8) = 1 - P(X \leq 8)$ ***[1 mark]***
$= 1 - 0.3328 = 0.6672$ ***[1 mark]***

b) In this context, independently means that receiving a phone call at one particular instant does not affect whether or not a call will be received at a different instant. ***[1 mark]***.

3 a) The number of trials here is fixed (= 400) and the probability of the engineer being unable to fix a fault is constant (= 0.02). This means X (the total number of unsuccessful call-outs) will follow a binomial distribution ***[1 mark]***. In fact, $X \sim B(400, 0.02)$ ***[1 mark]***.

b) (i) To approximate a binomial distribution $B(n, p)$ with a Poisson distribution, n should be large ***[1 mark]*** and p should be small ***[1 mark]***.

(ii) $Po(400 \times 0.02) = Po(8)$ ***[1 mark]***.

(iii) Mean = 8 and variance = 8 ***[1 mark]***.

(iv) P(engineer unable to fix fewer than 10 faults)
$= P(X < 10) = P(X \leq 9)$ ***[1 mark]***.
Using Poisson tables for $\lambda = 8$:
$P(X \leq 9) = 0.7166$ ***[1 mark]***.

S2 Section 4 — Continuous Random Variables

Warm-up Questions

1) a) Sketch the p.d.f.:

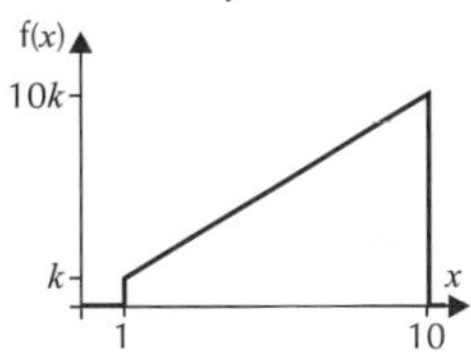

Area under p.d.f. $= \frac{10k + k}{2} \times (10 - 1) = \frac{99k}{2} = 1.$

So $k = \frac{2}{99}$.

b) Sketch the p.d.f.:

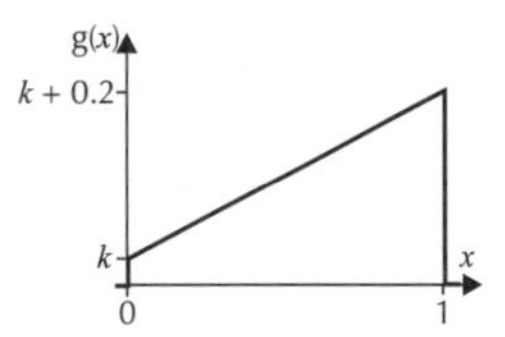

Area under p.d.f. $= \frac{2k + 0.2}{2} \times 1 = k + 0.1 = 1.$

So $k = 0.9$.

2 a) Sketch the p.d.f.:

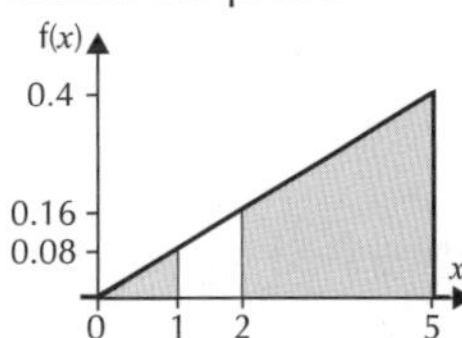

(i) Area under p.d.f. between $x = 0$ and $x = 1$ is:
$1 \times 0.08 \div 2 = 0.04$, so $P(X < 1) = 0.04$.

(ii) Area under p.d.f. between $x = 2$ and $x = 5$ is:
$\frac{0.16 + 0.4}{2} \times 3 = 0.84$, so $P(2 \leq X \leq 5) = 0.84$.

(iii) Area under p.d.f. at the point $x = 4$ is 0.
So $P(X = 4) = 0$.

b) Sketch the p.d.f.:

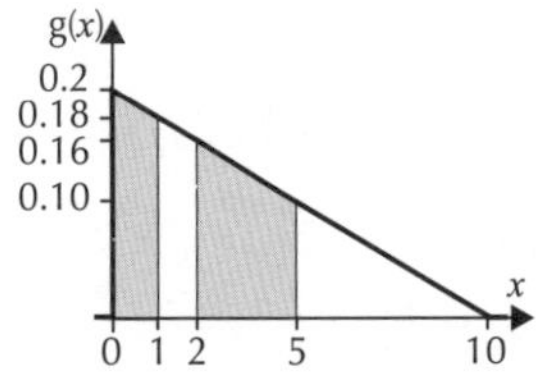

(i) Area under p.d.f. between $x = 0$ and $x = 1$ is:
$\frac{0.2 + 0.18}{2} \times 1 = 0.19$, so $P(X < 1) = 0.19$.

(ii) Area under p.d.f. between $x = 2$ and $x = 5$ is:
$\frac{0.16 + 0.1}{2} \times 3 = 0.39$, so $P(2 \leq X \leq 5) = 0.39$.

(iii) Area under p.d.f. at the point $x = 4$ is 0.
So $P(X = 4) = 0$.

Answers

3 a) $\int_{-\infty}^{\infty} f(x)dx = k\int_0^5 x^2 dx = k\left[\frac{x^3}{3}\right]_0^5 = \frac{125k}{3} = 1$, so $k = \frac{3}{125}$.

$P(X < 1) = \int_0^1 \frac{3}{125}x^2 dx = \frac{3}{125}\left[\frac{x^3}{3}\right]_0^1 = \frac{1}{125}$.

b) $\int_{-\infty}^{\infty} g(x)dx = \int_0^2 (0.1x^2 + kx)dx$

$= \left[\frac{0.1x^3}{3} + \frac{kx^2}{2}\right]_0^2 = \frac{0.8}{3} + 2k = 1.$

So $k = \frac{1}{2} - \frac{0.4}{3} = \frac{15-4}{30} = \frac{11}{30}$

$P(X < 1) = \int_0^1 \left(0.1x^2 + \frac{11}{30}x\right)dx = \left[\frac{0.1x^3}{3} + \frac{11x^2}{60}\right]_0^1$

$= \frac{0.1}{3} + \frac{11}{60} = \frac{13}{60}$

4 a) $\int_{-\infty}^{\infty} f(x)dx = \int_0^2 (0.1x^2 + 0.2)dx$

$= \left[\frac{0.1x^3}{3} + 0.2x\right]_0^2 = \frac{0.8}{3} + 0.4 \neq 1$

So f(x) is not a p.d.f.

b) $g(x) < 0$ for $-1 \leq x < 0$, so g(x) is not a p.d.f.

5 a) $E(X) = \int_{-\infty}^{\infty} xf(x)dx = \int_0^5 0.08x^2 dx = 0.08\left[\frac{x^3}{3}\right]_0^5$

$= \frac{125 \times 0.08}{3} = \frac{10}{3}$

$Var(X) = \int_{-\infty}^{\infty} x^2 f(x)dx - \mu^2 = \int_0^5 0.08x^3 dx - \left(\frac{10}{3}\right)^2$

$= 0.08\left[\frac{x^4}{4}\right]_0^5 - \left(\frac{10}{3}\right)^2 = \frac{625 \times 0.08}{4} - \left(\frac{10}{3}\right)^2$

$= \frac{25}{2} - \left(\frac{10}{3}\right)^2 = \frac{25}{18} = 1.39$ (to 2 d.p.).

$E(Y) = \int_{-\infty}^{\infty} yg(y)dy = \int_0^{10} 0.02y(10-y)dy$

$= 0.02\left[5y^2 - \frac{y^3}{3}\right]_0^{10}$

$= 0.02\left(500 - \frac{1000}{3}\right) = 10 - \frac{20}{3} = \frac{10}{3}$

$Var(Y) = \int_{-\infty}^{\infty} y^2 g(y)dy - \mu^2$

$= \int_0^{10} 0.02y^2(10-y)dy - \left(\frac{10}{3}\right)^2$

$= 0.02\left[\frac{10y^3}{3} - \frac{y^4}{4}\right]_0^{10} - \left(\frac{10}{3}\right)^2$

$= 0.02 \times \left(\frac{10\,000}{3} - \frac{10\,000}{4}\right) - \left(\frac{10}{3}\right)^2$

$= \frac{50}{3} - \left(\frac{10}{3}\right)^2 = \frac{50}{9} = 5.56$ (to 2 d.p.).

b) $E(4X + 2) = 4E(X) + 2 = 4 \times \frac{10}{3} + 2 = \frac{46}{3}$

$E(3Y - 4) = 3E(Y) - 4 = 3 \times \frac{10}{3} - 4 = 6$

$Var(4X + 2) = 16 \times Var(X)$

$= 16 \times \frac{25}{18} = \frac{200}{9} = 22.22$ (to 2 d.p.).

$Var(3Y - 4) = 9 \times Var(Y) = 9 \times \frac{50}{9} = 50$

c) Sketch the p.d.f.:

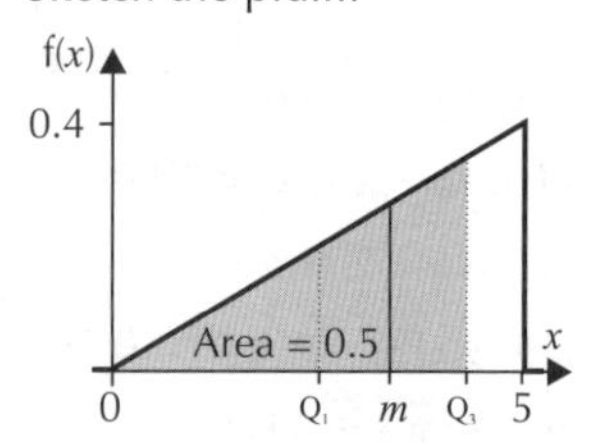

The median is m, where:

$\int_0^m 0.08x dx = 0.08\left[\frac{x^2}{2}\right]_0^m = 0.04m^2 = 0.5$

So the median = $\sqrt{12.5}$ = 3.54 (to 2 d.p.).

The lower quartile is Q_1, where:

$\int_0^{Q_1} 0.08x dx = 0.08\left[\frac{x^2}{2}\right]_0^{Q_1} = 0.04Q_1^2 = 0.25$

So $Q_1 = \sqrt{6.25} = 2.5$.

The upper quartile is Q_3, where:

$\int_0^{Q_3} 0.08x dx = 0.08\left[\frac{x^2}{2}\right]_0^{Q_3} = 0.04Q_3^2 = 0.75$

So $Q_3 = \sqrt{18.75}$.

This means the interquartile range is:
$\sqrt{18.75} - 2.5 = 1.83$ (to 2 d.p.).

Or you could work these out by finding F(x), and then solving $F(m) = 0.5$, $F(Q_1) = 0.25$ and $F(Q_3) = 0.75$.

d) Look at your sketch of the p.d.f. The mode is the value of x where the p.d.f. reaches its maximum, so mode = 5.

6 a) Integrate the pieces of the p.d.f., and then make sure the 'joins' are smooth using a suitable constant of integration (k).

$$F(x) = \begin{cases} 0 \text{ for } x < 0 \\ 0.04x^2 + k \text{ for } 0 \leq x \leq 5 \\ 1 \text{ for } x > 5 \end{cases}$$

Since F(0) = 0 and F(5) = 1, the pieces of this function join smoothly with $\underline{k = 0}$.

b) Integrate the pieces of the p.d.f., and then make sure the 'joins' are smooth using a suitable constant of integration (k).

$$G(x) = \begin{cases} 0 \text{ for } x < 0 \\ 0.2x - 0.01x^2 + k \text{ for } 0 \leq x \leq 10 \\ 1 \text{ for } x > 10 \end{cases}$$

Since G(0) = 0 and G(10) = 1, the pieces of this function join smoothly with $\underline{k = 0}$.

c) Integrate the pieces of the p.d.f., and then make sure the 'joins' are smooth using suitable constants of integration (k_1-k_3).

$$H(x) = \begin{cases} 0 \text{ for } x < 0 \\ x^2 + k_1 \text{ for } 0 \leq x \leq 0.5 \\ x + k_2 \text{ for } 0.5 \leq x \leq 1 \\ 3x - x^2 + k_3 \text{ for } 1 \leq x \leq 1.5 \\ 1 \text{ for } x > 1.5 \end{cases}$$

H(0) = 0 means that $\underline{k_1 = 0}$, which then gives H(0.5) = 0.25.
H(0.5) = 0.25 means that $\underline{k_2 = -0.25}$, giving H(1) = 0.75.
H(1) = 0.75 means that $\underline{k_3 = -1.25}$, giving H(1.5) = 1.
This means all the joins are now 'smooth'.

d) Integrate the pieces of the p.d.f., and then make sure the 'joins' are smooth using suitable constants of integration (k_1 and k_2).

Answers

$$M(x) = \begin{cases} 0 \text{ for } x < 2 \\ 0.5x - 0.05x^2 + k_1 \text{ for } 2 \leq x \leq 4 \\ 0.1x + k_2 \text{ for } 4 \leq x \leq 10 \\ 1 \text{ for } x > 10 \end{cases}$$

M(2) = 0 means that $k_1 = -0.8$, which gives M(4) = 0.4.
M(4) = 0.4 means that $k_2 = 0$, which gives M(10) = 1.
This means all the joins are now 'smooth'.

7 a) Differentiate the different parts of the c.d.f.:

$$f(x) = \begin{cases} 4x^3 \text{ for } 0 \leq x \leq 1 \\ 0 \text{ otherwise} \end{cases}$$

b)

$$g(x) = \begin{cases} \frac{1}{50}(x-1) \text{ for } 1 \leq x < 6 \\ \frac{3}{8} \text{ for } 6 \leq x \leq 8 \\ 0 \text{ otherwise} \end{cases}$$

Exam Questions

1 a) $\int_{-\infty}^{\infty} f(x)dx = \frac{1}{k}\int_0^2 (x+4)dx = \frac{1}{k}\left[\frac{x^2}{2} + 4x\right]_0^2 = \frac{10}{k}$ ***[1 mark]***

This must be equal to 1 ***[1 mark]***.
So $k = 10$ ***[1 mark]***.

There aren't many certainties in life, but "Your S2 exam will test if you know that the total area under a p.d.f. = 1" is one of them.

b) $E(X) = \int_{-\infty}^{\infty} xf(x)dx = \int_0^2 0.1(x^2 + 4x)dx$ ***[1 mark]***

$= 0.1\left[\frac{x^3}{3} + 2x^2\right]_0^2$ ***[1 mark]***

$= 0.1\left(\frac{8}{3} + 8\right) = \frac{32}{30} = \frac{16}{15} = 1.07$ (to 2 d.p.) ***[1 mark]***

c) (i) $\text{Var}(X) = \int_{-\infty}^{\infty} x^2 f(x)dx - \mu^2$

$= 0.1\int_0^2 (x^3 + 4x^2)dx - \left(\frac{16}{15}\right)^2$ ***[1 mark]***

$= 0.1\left[\frac{x^4}{4} + \frac{4x^3}{3}\right]_0^2 - \left(\frac{16}{15}\right)^2$ ***[1 mark]***

$= 0.1\left(4 + \frac{32}{3}\right) - \left(\frac{16}{15}\right)^2$

$= \frac{44}{30} - \left(\frac{16}{15}\right)^2 = \frac{74}{225}$

$= 0.329$ (to 3 d.p.). ***[1 mark]***

If you worked out the integral but forgot to subtract the square of the mean, then you've just thrown a few marks away — at least, you would have done if that had been a real exam.

(ii) $\text{Var}(4X - 2) = 16 \times \text{Var}(X)$ ***[1 mark]***

$= 16 \times \frac{74}{225} = \frac{1184}{225} = 5.262$ (to 3 d.p.). ***[1 mark]***

d) $P(0 \leq X \leq 1.5) = \frac{1}{10}\int_0^{1.5} (x+4)dx$ ***[1 mark]*** $= \frac{1}{10}\left[\frac{x^2}{2} + 4x\right]_0^{1.5}$

$= \frac{1}{10}[(1.125 + 6) - 0] = \frac{7.125}{10} = 0.7125$ ***[1 mark]***

e) Integrate the pieces of the p.d.f., and then make sure the 'joins' are smooth using a constant of integration (k). ***[1 mark]***.

$$F(x) = \begin{cases} 0 \text{ for } x < 0 \\ 0.05x^2 + 0.4x + k \text{ for } 0 \leq x \leq 2 \\ 1 \text{ for } x > 2 \end{cases}$$

[1 mark for each part correctly found]

All the joins are 'smooth' if $k = 0$, so the c.d.f. is:

$$F(x) = \begin{cases} 0 \text{ for } x < 0 \\ 0.05x^2 + 0.4x \text{ for } 0 \leq x \leq 2 \\ 1 \text{ for } x > 2 \end{cases}$$

[1 mark for final answer]

You must define a c.d.f. for all values of x. Don't just do the tricky bits in the middle and assume you're finished.

f) Use the c.d.f. from part e) to find the median.
The median is m, where $0.05m^2 + 0.4m = 0.5$ ***[1 mark]***.
This simplifies to: $m^2 + 8m - 10 = 0$ ***[1 mark]***.
Using the quadratic formula (and choosing the positive answer ***[1 mark]***) gives

$m = \frac{-8 + \sqrt{104}}{2} = 1.099$ (to 3 d.p.) ***[1 mark]***.

g) The mode is at the highest point of the p.d.f. within the range of possible values. Since f(x) has a positive gradient, this must be at the greatest possible value of x, so the mode of X is 2 ***[1 mark]***.

The mode is the easiest of the three 'averages' to work out, but you get fewest marks for doing it. Such is life, I suppose. That's the beauty of S2 really — it's less about statistics and more about the cruelty of life generally.

mean < median < mode, so the skew is negative ***[1 mark]***.

2 a) Using the third part of the c.d.f., F(3) = 0.5 ***[1 mark]***.
So F(3) must also equal 0.5 using the second part of the c.d.f., which means that $2k = 0.5$, or $k = 0.25$ ***[1 mark]***.

Make sure the bits of a cumulative distribution function join together smoothly.

b) Q_1 is given by $F(Q_1) = 0.25$ ***[1 mark]***. Since F(3) = 0.5, the lower quartile must lie in the region described by the second part of the c.d.f., so solve $0.25(Q_1 - 1) = 0.25$, or $Q_1 = 2$ ***[1 mark]***.

Q_3 is given by $F(Q_3) = 0.75$ ***[1 mark]***. Since F(3) = 0.5, the upper quartile must lie in the region described by the third part of the c.d.f., so solve $0.5(Q_3 - 2) = 0.75$, or $Q_3 = 3.5$ ***[1 mark]***.

So the interquartile range is 3.5 – 2 = 1.5 ***[1 mark]***.

c) (i) Differentiate to find the p.d.f.:

$$f(x) = \begin{cases} 0.25 \text{ for } 1 \leq x < 3 \textbf{ [1 mark]} \\ 0.5 \text{ for } 3 \leq x \leq 4 \textbf{ [1 mark]} \\ 0 \text{ otherwise } \textbf{[1 mark]} \end{cases}$$

(ii)

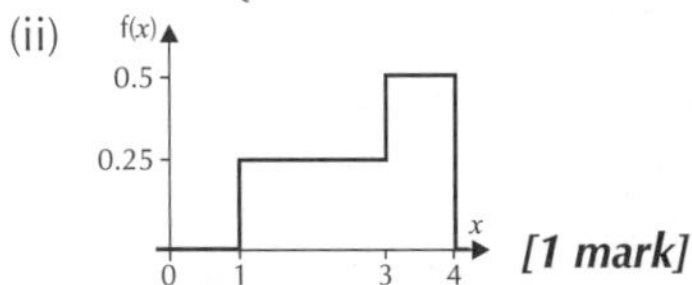

[1 mark]

I always draw a graph of the p.d.f. whether the question asks me to or not. You should too. It not only makes questions easier, but you'll often get marks for doing something that you were going to do anyway. It's like free marks.

Answers

d) (i) $\mu = \int_{-\infty}^{\infty} xf(x)dx$

$= \int_1^3 0.25xdx + \int_3^4 0.5xdx$ ***[1 mark]***

$= [0.125x^2]_1^3 + [0.25x^2]_3^4$ ***[1 mark]***

$= 1 + \frac{7}{4} = \frac{11}{4} = 2.75$ ***[1 mark]***

(ii) $\text{Var}(X) = \sigma^2 = \int_{-\infty}^{\infty} x^2 f(x)dx - \mu^2$

$= \int_1^3 0.25x^2dx + \int_3^4 0.5x^2dx - 2.75^2$ ***[1 mark]***

$= \left[\frac{0.25x^3}{3}\right]_1^3 + \left[\frac{0.5x^3}{3}\right]_3^4 - 2.75^2$ ***[1 mark]***

$= \frac{13}{6} + \frac{37}{6} - \left(\frac{11}{4}\right)^2 = \frac{25}{3} - \frac{121}{16}$

$= \frac{37}{48} = 0.771$ (to 3 d.p.) ***[1 mark]***.

I hope you remembered to subtract the square of the mean.

(iii) $P(X < \mu - \sigma) = P(X < 2.75 - \sqrt{0.771}) = P(X < 1.87)$ ***[1 mark]***. Using the above sketch, the area under the p.d.f. between $x = 1$ and $x = 1.87$ is:
$(1.87 - 1) \times 0.25 = 0.218$ (to 3 d.p.) ***[1 mark]***.

And that, as they say, is that.

S2 Section 5 — Continuous Distributions
Warm-up Questions

1) a)

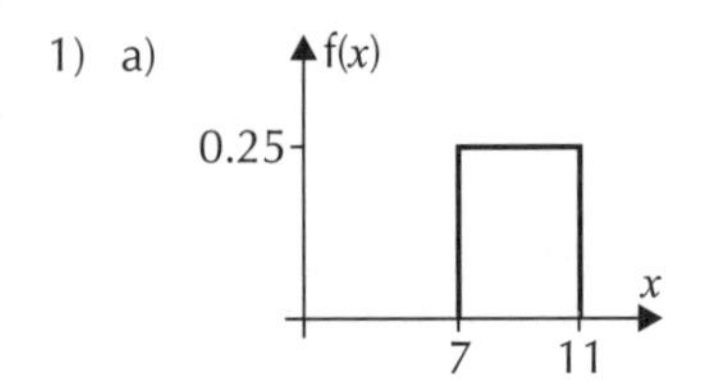

b)

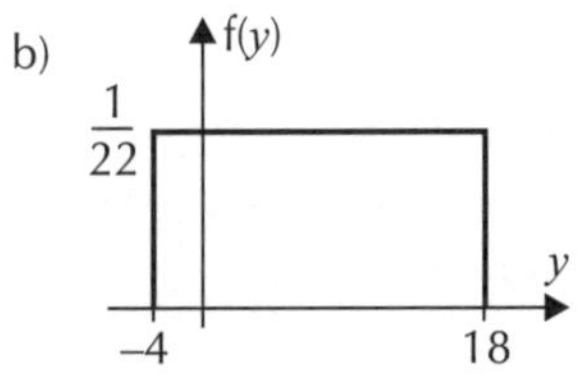

2) a) First sketch the p.d.f.:

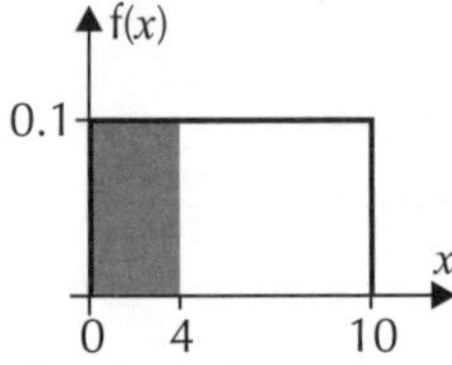

The shaded area represents $P(X < 4)$.
This is $4 \times 0.1 = 0.4$.

b) Similarly, $P(X \geq 8) = 2 \times 0.1 = 0.2$.

c) $P(X = 5) = 0$ [$P(X = k) = 0$ for any k and any continuous random variable X.]

d) $P(3 < X \leq 7) = 4 \times 0.1 = 0.4$.

3) $E(X) = \frac{a+b}{2}$, $\text{Var}(X) = \frac{(b-a)^2}{12}$,

$$F(x) = \begin{cases} 0 \text{ for } x < a \\ \frac{x-a}{b-a} \text{ for } a \leq x \leq b \\ 1 \text{ for } x > b \end{cases}$$

$$\int_{-\infty}^{\infty} x \cdot f(x)dx = \int_a^b x \cdot \frac{1}{b-a}dx = \frac{1}{b-a}\int_a^b x\,dx$$

$$= \frac{1}{b-a}\left[\frac{x^2}{2}\right]_a^b = \frac{1}{b-a}\left[\frac{b^2}{2} - \frac{a^2}{2}\right]$$

$$= \frac{b^2 - a^2}{2(b-a)} = \frac{(b-a)(b+a)}{2(b-a)} = \frac{b+a}{2}$$

This shows $E(X) = \frac{a+b}{2}$.

4) $E(X) = \frac{a+b}{2} = \frac{4+19}{2} = \frac{23}{2}$

$E(Y) = E(6X - 3) = 6E(X) - 3 = 6 \times \frac{23}{2} - 3 = 66$

$\text{Var}(X) = \frac{(b-a)^2}{12} = \frac{(19-4)^2}{12} = \frac{225}{12} = \frac{75}{4}$

$$F(x) = \begin{cases} 0 \text{ for } x < 4 \\ \frac{x-4}{15} \text{ for } 4 \leq x \leq 19 \\ 1 \text{ for } x > 19 \end{cases}$$

5) The error could be anything from –0.5 to 0.5 with equal probability. So $X \sim U[-0.5, 0.5]$.

6) Let X be the number of minutes the train is delayed. Then $X \sim U[0, 12]$, and its p.d.f. would be a rectangle of height $\frac{1}{12}$.

a) P(late for work) $= P(X > 8) = 4 \times \frac{1}{12} = \frac{1}{3}$

b) P(late more than once) = 1 – P(never late) – P(late once).
So P(late more than once)
$= 1 - \left(\frac{2}{3}\right)^5 - 5\left(\frac{2}{3}\right)^4 \times \frac{1}{3} = \frac{243 - 32 - 80}{243} = \frac{131}{243}$

7) Use the Z-tables:

a) $P(Z < 0.84) = 0.7995$

b) $P(Z \geq 1.55) = 1 - P(Z < 1.55)$
$= 1 - P(Z \leq 1.55) = 1 - 0.9394 = 0.0606$

c) $P(Z < -2.10) = P(Z > 2.10) = 1 - P(Z \leq 2.10)$
$= 1 - 0.9821 = 0.0179$

d) $P(0.102 < Z \leq 0.507) = P(Z \leq 0.507) - P(Z \leq 0.102)$
$= 0.6939 - 0.5406 = 0.1533$

I know... these can get a bit fiddly. As always — take it nice and slow, and double-check each step as you do it.

8) a) If $P(Z < z) = 0.9131$, then from the Z-table, $z = 1.360$.

b) If $P(Z > z) = 0.0359$, then $P(Z \leq z) = 0.9641$.
From the Z-table, $z = 1.800$.

9) a) $P(X < 55) = P\left(Z < \frac{55-50}{\sqrt{16}}\right) = P(Z < 1.25) = 0.8944$

b) $P(X < 42) = P\left(Z < \frac{42-50}{\sqrt{16}}\right) = P(Z < -2)$
$= P(Z > 2) = 1 - P(Z \leq 2) = 1 - 0.9772 = 0.0228$

c) $P(X > 56) = P\left(Z > \frac{56-50}{\sqrt{16}}\right) = P(Z > 1.5)$
$= 1 - P(Z \leq 1.5) = 1 - 0.9332 = 0.0668$

d) $P(47 < X < 57) = P(X < 57) - P(X \leq 47)$
$= P(Z < 1.75) - P(Z \leq -0.75)$
$= 0.9599 - P(Z \geq 0.75)$
$= 0.9599 - (1 - P(Z < 0.75))$
$= 0.9599 - (1 - 0.7734) = 0.7333$

Answers

10) $P(X < 15.2) = 0.9783$ means $P\left(Z < \frac{15.2 - \mu}{\sigma}\right) = 0.9783$.

From tables, $\frac{15.2 - \mu}{\sigma} = 2.02$, or $2.02\sigma + \mu = 15.2$.

$P(X > 14.8) = 0.1056$ means $P\left(Z > \frac{14.8 - \mu}{\sigma}\right) = 0.1056$,

or $P\left(Z \leq \frac{14.8 - \mu}{\sigma}\right) = 1 - 0.1056 = 0.8944$.

From tables, $\frac{14.8 - \mu}{\sigma} = 1.25$, or $1.25\sigma + \mu = 14.8$.

Solving these simultaneous equations gives $\sigma = 0.519$ and $\mu = 14.15$.

11) Use the normal approximation $X \sim N(45, 24.75)$.

a) $P(X > 50) \approx P(X > 50.5) = P\left(Z > \frac{50.5 - 45}{\sqrt{24.75}}\right)$
$= P(Z > 1.106)$
$= 1 - P(Z \leq 1.106)$
$= 1 - 0.8655 = 0.1345$

b) $P(X \leq 45) \approx P(X < 45.5) = P\left(Z < \frac{45.5 - 45}{\sqrt{24.75}}\right)$
$= P(Z < 0.101)$
$= 0.5402$

c) $P(40 < X \leq 47) \approx P(X \leq 47.5) - P(X \leq 40.5)$
$= P\left(Z \leq \frac{47.5 - 45}{\sqrt{24.75}}\right) - P\left(Z \leq \frac{40.5 - 45}{\sqrt{24.75}}\right)$
$= P(Z \leq 0.503) - P(Z \leq -0.905)$
$= P(Z \leq 0.503) - (1 - P(Z \leq 0.905))$
$= 0.6925 - 1 + 0.8172 = 0.5097$

12) Use the normal approximation $X \sim N(25, 25)$.

a) $P(X \leq 20) \approx P(X \leq 20.5) = P\left(Z \leq \frac{20.5 - 25}{5}\right)$
$= P(Z \leq -0.9)$
$= 1 - P(Z \leq 0.9)$
$= 1 - 0.8159 = 0.1841$

b) $P(X > 15) \approx P(X > 15.5) = P\left(Z > \frac{15.5 - 25}{5}\right)$
$= P(Z > -1.90)$
$= P(Z < 1.90) = 0.9713$

c) $P(20 \leq X < 30) \approx P(X \leq 29.5) - P(X \leq 19.5)$
$= P\left(Z \leq \frac{29.5 - 25}{5}\right) - P\left(Z \leq \frac{19.5 - 25}{5}\right)$
$= P(Z \leq 0.90) - P(Z \leq -1.10)$
$= P(Z \leq 0.90) - (1 - P(Z \leq 1.10))$
$= 0.8159 - 1 + 0.8643 = 0.6802$

13) People join the queue (on average) at a constant rate. Assuming they join the queue randomly and singly, the total number of people joining the queue in a 15-minute period follows a Poisson distribution, Po(7).

a) If X represents the number of people joining the queue in a 7-hour day, then $X \sim Po(7 \times 28) = Po(196)$. λ is large, so use the normal approximation $X \sim N(196, 196)$.

$P(X > 200) \approx P(X > 200.5) = P\left(Z > \frac{200.5 - 196}{14}\right)$
$= P(Z > 0.321)$
$= 1 - P(Z < 0.321)$
$= 1 - 0.6259 = 0.3741$

b) Let C represent the number of people out of the 200 customers who are seen within 1 minute.
Then $C \sim B(200, 0.7)$.
n is large, and $np = 200 \times 0.7 = 140$ and $nq = 200 \times 0.3 = 60$ are both large, so use the normal approximation, i.e. $C \sim N(140, 42)$.
You need to find P(C < 70% of 200), i.e. $P(C < 140)$.

$P(C < 140) \approx P(C \leq 139.5) = P\left(Z \leq \frac{139.5 - 140}{\sqrt{42}}\right)$
$= P(Z \leq 0.077)$
$= 1 - P(Z \leq 0.077)$
$= 1 - 0.5307 = 0.4693$

Exam Questions

1 a) X is equally likely to take any value between 0 and 20, so $X \sim U[0, 20]$ ***[1 mark for using a continuous uniform distribution, 1 mark for the correct limits]***

b) 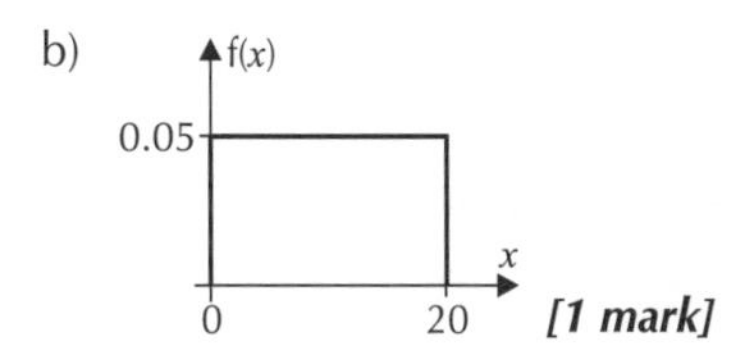

[1 mark]

You probably haven't earned many marks for drawing rectangles since you were about 6 years old and in primary school. So it's nice that the skill you learnt way back then is now helping you pass S2.

c) $E(X) = \frac{0 + 20}{2} = 10$ ***[1 mark]***

$Var(X) = \frac{(20 - 0)^2}{12}$ ***[1 mark]***

$= \frac{400}{12} = 33.3$ (to 3 sig. fig.) ***[1 mark]***.

d) (i) $P(X > 5) = (20 - 5) \times 0.05 = 0.75$ ***[1 mark]***

(ii) $P(X = 2) = 0$ ***[1 mark]***

2 a) (i) Need n to be large ("as large as possible") ***[1 mark]*** and p to be close to 0.5 ***[1 mark]***.

Engrave upon your heart all the conditions required for the various approximations to work. Well... actually, that might be considered to be cheating, so don't do that. But make sure you know them.

(ii) A binomial distribution is discrete ***[1 mark]***, whereas a normal distribution is continuous. The continuity correction means probabilities can be calculated for the continuous normal distribution that correspond approximately to the discrete binomial probabilities ***[1 mark]***.

b) (i) n is large and p is fairly close to 0.5, so use the normal approximation N(60, 24) ***[1 mark]***.

$P(X \geq 65) \approx P(X > 64.5)$
$= P\left(Z > \frac{64.5 - 60}{\sqrt{24}}\right)$ ***[1 mark]***
$= P(Z > 0.919)$
$= 1 - P(Z \leq 0.919)$ ***[1 mark]***
$= 1 - 0.8209 = 0.1791$ ***[1 mark]***

Answers

(ii) $P(50 < X < 62)$

$$\approx P(X < 61.5) - P(X < 50.5) \text{ [1 mark]}$$
$$= P\left(Z < \frac{61.5 - 60}{\sqrt{24}}\right) - P\left(Z < \frac{50.5 - 60}{\sqrt{24}}\right)$$
$$= P(Z < 0.306) - P(Z < -1.939) \text{ [1 mark]}$$
$$= P(Z < 0.306) - (1 - P(Z < 1.939))$$
$$= 0.6201 - (1 - 0.9737) = 0.5938 \text{ [1 mark]}$$

All the usual tricks involved there... normal approximation, continuity correction, subtracting values of $\Phi(z)$ (the function in your big 'normal distribution table'). They'll all be there on exam day too.

3 a) The normal approximation is: $Y \sim N(\mu, \sigma^2)$.

$P(X \leq 151) \approx P(Y \leq 151.5)$ ***[1 mark]***

$= P\left(Z \leq \frac{151.5 - \mu}{\sigma}\right) = 0.8944$ ***[1 mark]***.

From tables, $\frac{151.5 - \mu}{\sigma} = 1.25$.

So $\underline{\mu + 1.25\sigma = 151.5}$ ***[1 mark]***.

$P(X > 127) \approx P(Y > 127.5)$ ***[1 mark]***

$= P\left(Z > \frac{127.5 - \mu}{\sigma}\right) = 0.9970$ ***[1 mark]***.

This means $P\left(Z \leq \frac{127.5 - \mu}{\sigma}\right) = 0.0030$.

But this probability is less than 0.5, so $\frac{127.5 - \mu}{\sigma} < 0$.

So find $1 - 0.0030 = 0.9970$.

$P(Z \leq z) = 0.9970$ means $z = 2.75$.

This tells you that $\frac{127.5 - \mu}{\sigma} = -2.75$,

or $\underline{\mu - 2.75\sigma = 127.5}$ ***[1 mark]***.

Now you can subtract the underlined equations to give $4\sigma = 24$, or $\sigma = 6$ ***[1 mark]***.

This then gives $\mu = 144$ ***[1 mark]***.

I call this question "The Beast" — there's loads to do here. But as always in maths, when something looks hard, the best thing to do is take a deep breath, look at the information you have (here, some probabilities from a normal approximation), and write down some formulas containing that information. Then you can start piecing things together, and try to find out things you don't yet know. The worst thing you can do is panic and start thinking it's too hard. That's what Luke Skywalker did in that film before Yoda told him to chill out a bit. Something like that anyway.

b) You know $\mu = \underline{np = 144}$ ***[1 mark]***

and $\sigma^2 = \underline{np(1 - p) = 36}$ ***[1 mark]***.

Divide the second underlined equation by the first to give $1 - p = 36 \div 144 = 0.25$, or $p = 0.75$ ***[1 mark]***.

Then $n = 144 \div 0.75 = 192$ ***[1 mark]***.

Phew... made it.

4 a) Let X represent the number of items that the new customer could order per week. Then X follows a Poisson distribution with an average of 40, and so $X \sim Po(40)$ ***[1 mark]***.

b) Here, λ is quite large, and so X will approximately follow the normal distribution N(40, 40) ***[1 mark]***.

$$P(X > 50) \approx P(X > 50.5) = P\left(Z > \frac{50.5 - 40}{\sqrt{40}}\right) \text{ [1 mark]}$$
$$= P(Z > 1.660)$$
$$= 1 - P(Z \leq 1.660)$$
$$= 1 - 0.9515 = 0.0485 \text{ [1 mark]}$$

c) The probability that the factory will not be able to meet the new customer's order in two consecutive weeks will be $0.0485^2 = 0.00235...$, which is less than 0.01.

So the manager should sign the contract ***[1 mark for 'yes', with a clear explanation]***.

This question looks quite tough because there are so many words. But it's a pussycat really.

5 Assume that the lives of the batteries are distributed as: $N(\mu, \sigma^2)$. Then $P(X < 20) = 0.25$ and $P(X < 30) = 0.9$. ***[1 mark]***

Transform these 2 equations to get:

$P\left[Z < \frac{20 - \mu}{\sigma}\right] = 0.25$ and $P\left[Z < \frac{30 - \mu}{\sigma}\right] = 0.9$ ***[1 mark]***

Now you need to use your normal table to get:

$\frac{20 - \mu}{\sigma} = -0.674$ ***[1 mark]*** and $\frac{30 - \mu}{\sigma} = 1.282$ ***[1 mark]***

Now rewrite these as:

$20 - \mu = -0.674\sigma$ and $30 - \mu = 1.282\sigma$. ***[1 mark]***

Subtract these two equations to get:

$10 = (1.282 + 0.674)\sigma$

i.e. $\sigma = \frac{10}{1.282 + 0.674} = 5.112$ ***[1 mark]***

Now use this value of σ in one of the equations above:

$\mu = 20 + 0.674 \times 5.112 = 23.45$ ***[1 mark]***

So $X \sim N(23.45, 5.11^2)$ i.e. $X \sim N(23.45, 26.1)$

S2 Section 6 — Hypothesis Tests
Warm-up Questions

1) a) All the members of the tennis club.

b) The individual tennis club members.

c) A full membership list.

It might look like I've written the same answer down three times, but there are important differences. Make sure you know exactly what's meant by the three terms tested here.

2) a) A census would be more sensible. The results will be more accurate and there are only 8 people in the population, so it wouldn't take long to find out the required information from each person.

b) A sample survey should be done. Testing all 500 toys would take too long, but more importantly, it would destroy all the toys.

3) Simple random sampling means the sample will not be affected by sampling bias.

Answers

4) a) Yes

b) No — it contains unknown parameter σ.

c) No — it contains unknown parameter μ.

d) Yes

There's no excuse for getting these ones wrong. You've just got to look for any unknown parameters — if you find one, it's not a statistic.

5) a) $H_0: p = 0.2$, $H_1: p < 0.2$, $\alpha = 0.05$ and $x = 2$:
Under H_0, $X \sim B(20, 0.2)$
$P(X \leq 2) = 0.2061$
$0.2061 > 0.05$, so there is insufficient evidence at the 5% level of significance to reject H_0.

b) $H_0: \lambda = 2.5$, $H_1: \lambda > 2.5$, $\alpha = 0.1$ and $x = 4$:
Under H_0, $X \sim Po(2.5)$
$P(X \geq 4) = 1 - P(X \leq 3) = 1 - 0.7576 = 0.2424$
$0.2424 > 0.1$, so there is insufficient evidence at the 10% level of significance to reject H_0.

6) a) $H_0: p = 0.3$, $H_1: p < 0.3$, $\alpha = 0.05$
Under H_0, $X \sim B(10, 0.3)$
Critical region = biggest possible set of 'low' values of X with a total probability of ≤ 0.05.
$P(X \leq 0) = 0.0282$, $P(X \leq 1) = 0.1493$,
so CR is $X = 0$.

b) $H_0: \lambda = 6$, $H_1: \lambda < 6$, $\alpha = 0.1$
Under H_0, $X \sim Po(6)$
Critical region = biggest possible set of 'low' values of X with a total probability of ≤ 0.1.
$P(X \leq 2) = 0.0620$, $P(X \leq 3) = 0.1512$,
so CR is $X \leq 2$.

These might be getting a bit tedious, but a significant amount of practice is critical when it comes to hypothesis testing.

7) $X \sim N(8, 2) \Rightarrow \overline{X} \sim N(8, \frac{2}{10}) = N(8, 0.2)$

$P(\overline{X} < 7) = P\left(Z < \frac{7-8}{\sqrt{0.2}}\right) = P(Z < -2.236)$
$= P(Z > 2.236) = 1 - P(Z < 2.236)$
$= 1 - 0.9873 = 0.0127$

8) The sample mean is an unbiased estimate of the population mean — this is: $\frac{\sum x}{n} = \frac{80.5}{10} = 8.05$
An unbiased estimate of the population variance is:

$$\frac{n}{n-1}\left[\frac{\sum x^2}{n} - \left(\frac{\sum x}{n}\right)^2\right] = \frac{10}{9}\left[\frac{653.13}{10} - \left(\frac{80.5}{10}\right)^2\right]$$
$$= 0.567 \text{ (to 3 d.p.)}.$$

9) $H_0: \mu = 45$, $H_1: \mu < 45$, $\alpha = 0.05$ and $\sigma^2 = 9$.
Under H_0, $\overline{X} \sim N\left(45, \frac{9}{16}\right)$ and $Z = \frac{42-45}{3/4} = -4$
Critical region = $Z < -1.645$.
$-4 < -1.645$, so there is evidence to reject H_0 at the 5% level.

10) The 1% point of $\chi^2_{(4)}$ = value of x where $P(X \leq x) = 0.99$. So critical value = 13.277. $8.3 < 13.277$, so there is no evidence of an association between the variables.

11) Confidence interval = $\left(\overline{x} - \frac{s}{\sqrt{n}}t_{(n-1)}, \overline{x} + \frac{s}{\sqrt{n}}t_{(n-1)}\right)$.
$n = 12$, so $t_{(11)}$ for $p = 0.975 = 2.201$.
So 95% confidence interval
$= \left(50 - \frac{\sqrt{0.7}}{\sqrt{12}} \times 2.201, 50 + \frac{\sqrt{0.7}}{\sqrt{12}} \times 2.201\right)$
$= (49.5, 50.5)$

12) You'd use a t-test when the population follows a normal distribution, the population variance is unknown and the sample size is small.

Exam Questions

1 The possible samples are: (1, 1, 1), (1, 1, 2), (1, 2, 1), (2, 1, 1), (2, 2, 1), (2, 1, 2), (1, 2, 2) and (2, 2, 2).
[3 marks for showing that there are 8 possible samples, or 2 marks for showing 4 correct samples, or 1 mark for showing at least 1 correct sample.]
So the median could either be 1 or 2. ***[1 mark]***
$P(M = 1) = P(1, 1, 1) + P(1, 1, 2) + P(1, 2, 1) + P(2, 1, 1)$
$= 0.7^3 + (3 \times 0.7^2 \times 0.3) = 0.784$
$P(M = 2) = P(2, 2, 1) + P(2, 1, 2) + P(1, 2, 2) + P(2, 2, 2)$
$= (3 \times 0.3^2 \times 0.7) + 0.3^3 = 0.216$
[1 mark for showing that the probabilities of samples giving the same median value should be added, 1 mark for P(M = 1) = 0.784 and 1 mark for P(M = 2) = 0.216.]

Remember to check that the probabilities you've worked out for the values of the median add up to 1. If not, go back and work out where you've gone wrong.

2 a) First-serve faults must occur randomly (or independently of each other) and at a constant average rate.
[1 mark for saying first-serve faults must occur randomly or independently, and 1 mark for saying first-serve faults must occur at a constant average rate.]

b) $H_0: \lambda = 4$ and $H_1: \lambda < 4$, where λ is the rate of first-serve faults per service game.
X = number of first-serve faults in 5 service games
Under H_0, $X \sim Po(20)$
$\alpha = 0.05$
X is large so you can approximate using $X \sim N(20, 20)$
[2 marks for stating the correct normal approximation, or 1 mark for a normal approximation with only one of the mean or variance correct.]
Applying the continuity correction:
$P(X \leq 12)$ becomes $P(X < 12.5)$ ***[1 mark]***
$= P\left(Z < \frac{12.5 - 20}{\sqrt{20}}\right)$ ***[1 mark]***
$= P(Z < -1.68)$
$= 1 - P(Z < 1.68)$
$= 1 - 0.9535$
$= 0.0465$ ***[1 mark]*** < 0.05, so the result is significant.
There is evidence at the 5% level of significance to reject H_0 and to say that the rate of first-serve faults has decreased. ***[1 mark]***

Answers

3 a) E.g. a random sample will ensure that the observations are independent random variables. The correct population is being sampled from.
[2 marks for two correct statements.]

b) H_0: $p = 0.1$ and H_0: $p \neq 0.1$
X = number of sampled residents against the plan
Under H_0, $X \sim B(50, 0.1)$ ***[1 mark]***
It's a two-tailed test, so the critical region is split into two.
For the lower end: $P(X \leq 2) = 0.1117$, $P(X \leq 1) = 0.0338$ ***[1 mark]***, which is the closest value.
For the upper end: $P(X \geq 10) = 1 - 0.9755 = 0.0245$, $P(X \geq 9) = 1 - 0.9421 = 0.0579$ ***[1 mark]***, which is the closest value.
So CR is $X \leq 1$ ***[1 mark]*** and $X \geq 9$ ***[1 mark]***

Watch out for the wording of these questions. You want the probability in each tail to be as close as possible to 0.05 — which means it can be greater than 0.05.

c) The probability of incorrectly rejecting H_0 is the same as the actual significance level.
So, it's $P(X \leq 1) + P(X \geq 9)$ ***[1 mark]***
$= 0.0338 + 0.0579$
$= 0.0917$ ***[1 mark]***

d) The value 4 doesn't lie in the critical region ***[1 mark]***, so there is insufficient evidence to reject the claim that the proportion of residents against the plan is 10% ***[1 mark]***.
(Allow follow-through for a correct conclusion drawn from an incorrectly calculated critical region in part a).)

4 a) $\bar{x} = \frac{\sum x}{n} = \frac{490}{100} = 4.9\ m$ ***[1 mark]***

$$s^2 = \frac{n}{n-1}\left[\frac{\sum x^2}{n} - \left(\frac{\sum x}{n}\right)^2\right]$$

$$= \frac{100}{99}\left[\frac{2421}{100} - 4.9^2\right] \textbf{\textit{ [1 mark]}}$$

$$= \frac{20}{99} = 0.202 \text{ (to 3 d.p.)} \textbf{\textit{ [1 mark]}}$$

b) Let μ = mean height of trees in 2nd area.
H_0: $\mu = 5.1$ and H_1: $\mu \neq 5.1$ ***[1 mark]***

$$\text{Under } H_0, \bar{X} \sim N\left(5.1, \frac{20/99}{100}\right) \textbf{\textit{ [1 mark]}} = N\left(5.1, \frac{1}{495}\right)$$

$$Z = \frac{4.9 - 5.1}{\sqrt{1/495}} \textbf{\textit{ [1 mark]}} = -4.45 \textbf{\textit{ [1 mark]}}$$

This is a two-tailed test at the 1% level, so the critical values you need are z such that $P(Z < z) = 0.005$ and $P(Z > z) = 0.005$. Looking these up in the normal tables you get critical values of –2.576 and 2.576 ***[1 mark]***. Since $-4.45 < -2.576$, the result is significant. There is evidence to reject H_0 and to suggest that the trees have a different mean height ***[1 mark]***.

5 H_0: no association between age and favourite flavour of ice cream ***[1 mark]***, H_1: there is an association.
Make a table showing the observed frequencies, the expected frequencies and the values of $(O - E)^2 / E$:

Observed frequency (O)	Expected Frequency (E)	$\frac{(O-E)^2}{E}$
10	11.52	0.200...
24	24	0
7	5.28	0.560...
7	7.2	0.005...
14	12.48	0.185...
26	26	0
4	5.72	0.517...
8	7.8	0.005...
100	100	**1.47**

[1 mark for calculating the expected frequencies, 1 mark for all values of E correct, 1 mark for calculating values of (O – E)² / E, 1 mark for all values of (O – E)² / E correct.]
So $\chi^2 = 1.47$ ***[1 mark]***
Under H_0, $X^2 \sim \chi^2_{(\nu)}$ where $\nu = (4-1) \times (2-1) = 3$ ***[1 mark]***.
So the critical value at the 5% level is 7.815 ***[1 mark]***.
$1.47 < 7.815$, so do not reject H_0.
There is no evidence of association between age and favourite flavour of ice cream ***[1 mark]***.

S2 Section 7 — Bivariate Data

Warm-up Questions

1) a)

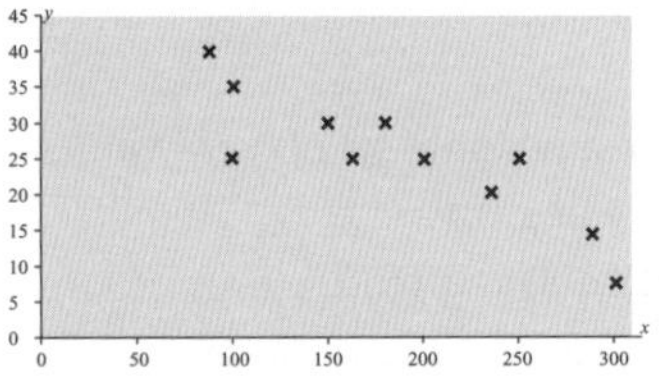

b) First you need to find these values:
$\sum x = 2060$, $\sum y = 277$, $\sum x^2 = 442800$, $\sum y^2 = 7799$ and $\sum xy = 46000$.
Then put these values into the PMCC formula:

$$\frac{46000 - \frac{[2060][277]}{11}}{\sqrt{\left(442800 - \frac{[2060]^2}{11}\right)\left(7799 - \frac{[277]^2}{11}\right)}}$$

$$= \frac{46000 - \frac{570620}{11}}{\sqrt{\left(442800 - \frac{4243600}{11}\right)\left(7799 - \frac{76729}{11}\right)}}$$

$$= \frac{-5874.5455}{\sqrt{57018.1818 \times 823.6364}} = -0.857 \text{ (to 3 sig.fig.)}$$

c) The sample was randomly selected and the bulk of the data-points on the scatter graph look like they would fall inside an ellipse, so the hypothesis test that follows should be valid. This is a 2-tailed test at a 5% significance level with $n = 11$.
H_0: $r = 0$ and H_1: $r \neq 0$.
From the table of critical values for the PMCC, the critical value is 0.6021. Since the PMCC from b) is less than –0.6021, you can reject H_0 and conclude that this data provides evidence at a 5% significance level of a correlation between the volume of alcoholic drinks and their alcohol concentration.

Answers

Don't panic about that nasty ol' PMCC equation. You need to know how to USE it, but they give you the formula in the exam, so you don't need to REMEMBER it. Hurrah.

2)

Physics	54	34	23	57	56	58	13	65	69
English	16	73	89	83	23	81	56	62	61
Physics rank	6	7	8	4	5	3	9	2	1
English rank	9	4	1	2	8	3	7	5	6
d	3	3	7	2	3	0	2	3	5
d^2	9	9	49	4	9	0	4	9	25

So $\sum d^2 = 118$, and

$$r_s = 1 - \frac{6\sum d^2}{n(n^2-1)} = 1 - \frac{6 \times 118}{9(9^2-1)}$$
$$= 1 - \frac{708}{720} = 0.0167 \text{ (to 3 sig.fig.)}$$

You need to carry out a 2-tailed test at a 5% significance level with $n = 9$. H_0: No association between variables, and H_1: Some association.

From the table of critical values for the SRCC, the critical value is 0.7000. Since the SRCC here is not greater than 0.7000, you cannot reject H_0. Therefore this data does not provide evidence of an association between the marks in Physics and English exams.

3) a) **Independent**: the annual number of sunny days
Dependent: the annual number of volleyball-related injuries

b) **Independent**: the annual number of rainy days
Dependent: the annual number of Monopoly-related injuries

c) **Independent**: a person's disposable income
Dependent: a person's spending on luxuries

d) **Independent**: the number of cups of tea drunk per day
Dependent: the number of trips to the loo per day

e) **Independent**: the number of festival tickets sold
Dependent: the number of pairs of Wellington boots bought

4) a) (i) $S_{rr} = 26816.78 - \frac{517.4^2}{10} = 46.504$

(ii) $S_{rw} = 57045.5 - \frac{517.4 \times 1099}{10} = 183.24$

b) $b = \frac{S_{rw}}{S_{rr}} = \frac{183.24}{46.504} = 3.94$

c) $a = \overline{w} - b\overline{r}$, where $\overline{w} = \frac{\sum w}{10} = 109.9$

and $\overline{r} = \frac{\sum r}{10} = 51.74$

So $a = 109.9 - 3.94 \times 51.74 = -94.0$

d) The equation of the regression line is: $w = 3.94r - 94.0$

e) When $r = 60$, the regression line gives an estimate for w of: $w = 3.94 \times 60 - 94.0 = 142.4$ g

f) This estimate might not be very reliable because it uses an r-value from outside the range of the original data. It is extrapolation.

You'll be given the equations for finding a regression line — but you still need to know how to use them, otherwise the formula booklet will just be a blur of incomprehensible squiggles. Oh, and you need to practise USING them of course...

Exam Questions

1) a)

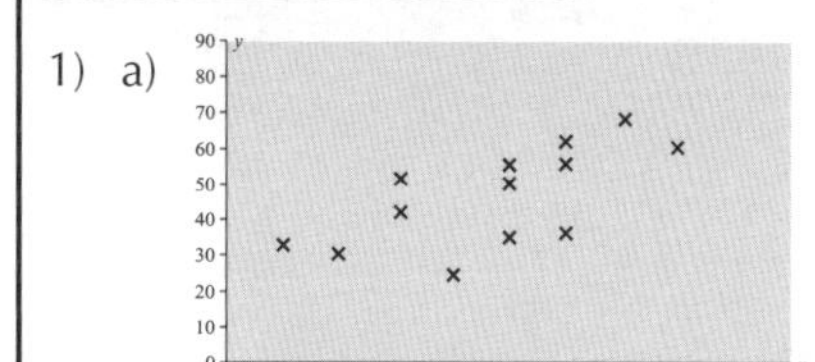

[2 marks for all points plotted correctly, or 1 mark if at least 3 points are plotted correctly.]

Aren't scatter diagrams pretty... Just make sure you're not so distracted by their artistic elegance that you forget to be accurate and lose easy marks.

b) You need to work out these sums:

$\sum x = 61, \sum y = 606,$

$\sum x^2 = 335, \sum y^2 = 30588, \sum xy = 3070$

Then:

$$S_{xx} = \sum x^2 - \frac{(\sum x)^2}{n} = 335 - \frac{61^2}{13} = \frac{634}{13}$$

$$S_{yy} = \sum y^2 - \frac{(\sum y)^2}{n} = 30588 - \frac{606^2}{13} = \frac{30408}{13}$$

$$S_{xy} = \sum xy - \frac{(\sum x)(\sum y)}{n}$$
$$= 3070 - \frac{61 \times 606}{13} = \frac{2944}{13}$$

[3 marks available — 1 for each correct term]

This means:

$$r = \frac{S_{xy}}{\sqrt{S_{xx}S_{yy}}} = \frac{\left(\frac{2944}{13}\right)}{\sqrt{\left(\frac{634}{13}\right)\left(\frac{30408}{13}\right)}}$$
$$= \frac{2944}{\sqrt{634 \times 30408}} = 0.671 \text{ (to 3 d.p.)}.$$

[1 mark]

c) The sample was randomly selected and the bulk of the data-points on the scatter graph look like they would fall inside an ellipse, so the hypothesis test that follows should be valid ***[1 mark]***. This is a 2-tailed test ***[1 mark]*** at a 5% significance level with $n = 13$.
H_0: $r = 0$ and H_1: $r \neq 0$ ***[1 mark]***.
From the table of critical values for the PMCC, the critical value is 0.5529 ***[1 mark]***. Since the PMCC here is greater than the critical value, you can reject H_0 ***[1 mark]*** and conclude that this data provides evidence at a 5% significance level of a correlation between the number of catch-up sessions attended by a Year-7 pupil at that school during the year and their mark on the maths test ***[1 mark]***.

Answers

2 a) 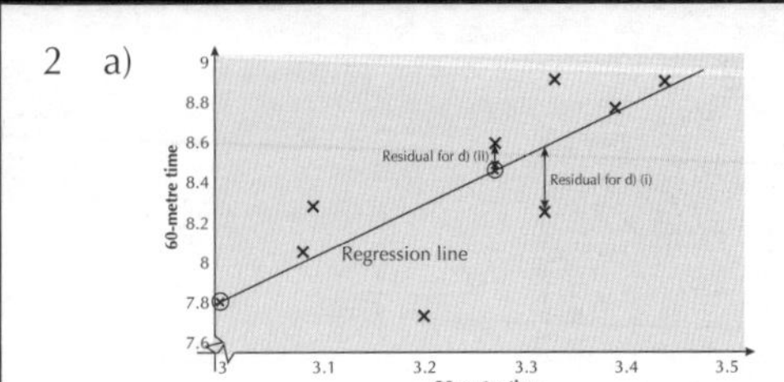

[2 marks for all points plotted correctly, or 1 mark if at least 3 points are plotted correctly.]

b) It's best to make a table like this one, first:

									Totals
20-metre time, x	3.39	3.2	3.09	3.32	3.33	3.27	3.44	3.08	26.12
60-metre time, y	8.78	7.73	8.28	8.25	8.91	8.59	8.9	8.05	67.49
x^2	11.4921	10.24	9.5481	11.0224	11.0889	10.6929	11.8336	9.4864	85.4044
xy	29.7642	24.736	25.5852	27.39	29.6703	28.0893	30.616	24.794	220.645

[2 marks for at least three correct totals, or 1 mark if one total found correctly.]

Then: $S_{xy} = 220.645 - \dfrac{26.12 \times 67.49}{8} = 0.29015$

[1 mark]

$S_{xx} = 85.4044 - \dfrac{26.12^2}{8} = 0.1226$

[1 mark]

Then the gradient b is given by:

$b = \dfrac{S_{xy}}{S_{xx}} = \dfrac{0.29015}{0.1226} = 2.3666$

[1 mark]

And the intercept a is given by:

$$a = \overline{y} - b\overline{x} = \frac{\sum y}{n} - b\frac{\sum x}{n}$$

$$= \frac{67.49}{8} - 2.3666 \times \frac{26.12}{8} = 0.709$$

[1 mark]

So the regression line has equation: $y = 2.367x + 0.709$

[1 mark]

To plot the line, find two points that the line passes through. A regression line always passes through $(\overline{x}, \overline{y})$, which here is (3.27, 8.44). Then put $x = 3$ (say) to find that the line also passes through (3, 7.81).
Now plot these points (in circles) on your scatter diagram, and draw the regression line through them
[1 mark for plotting the line correctly].

Hmm, lots of fiddly things to calculate there. Remember, you get marks for method as well as correct answers, so take it step by step and show all your workings.

c) (i) $y = 2.367 \times 3.15 + 0.709 = 8.17$ (to 3 sig. fig.), (8.16 if $b = 2.3666$ used) ***[1 mark]***

This should be reliable, since we are using interpolation within the range of x for which we have data ***[1 mark].***

(ii) $y = 2.367 \times 3.88 + 0.709 = 9.89$ (to 3 sig. fig.) ***[1 mark]***

This could be unreliable, since we are extrapolating beyond the range of the data ***[1 mark].***

d) (i) residual = 8.25 – (2.367 × 3.32 + 0.709)
= –0.317 (3 sig. fig.), (–0.316 if $b = 2.3666$ used)

[1 mark for calculation, 1 mark for plotting residual correctly]

(ii) residual = 8.59 – (2.367 × 3.27 + 0.709)
= 0.141 (3 sig. fig.), (0.142 if $b = 2.3666$ used)

[1 mark for calculation, 1 mark for plotting residual correctly]

3 a) Put the values into the correct PMCC formula:

$$\text{PMCC} = \frac{S_{xy}}{\sqrt{S_{xx}S_{yy}}} = \frac{6333}{\sqrt{155440 \times 395.5}}$$

$$= \frac{6333}{\sqrt{61476520}} = 0.808 \text{ (to 3 d.p.)}$$

[1 mark for correctly substituting the values into the PMCC formula, and 1 mark for the correct final answer.]

b) This is a 1-tailed test ***[1 mark]*** at a 1% significance level with $n = 10$.
H_0: $r = 0$ and H_1: $r > 0$ ***[1 mark]***.
From the table of critical values for the PMCC, the critical value is 0.7155 ***[1 mark]***. Since the PMCC here is greater than the critical value, you can reject H_0 ***[1 mark]*** and conclude that this data provides evidence at a 1% significance level of a positive correlation between the distance in miles cycled during a training ride and the number of calories eaten at lunch for the members of the cycling club ***[1 mark]***.

c) For this test to be valid, the underlying distribution should be bivariate normal ***[1 mark]***. If this is the case, the bulk of the data points should form a roughly elliptical shape on a scatter diagram ***[1 mark]***.

4 a) At $x = 12.5$, $y = 211.599 + (9.602 \times 12.5) = 331.624$

At $x = 14.7$, $y = 211.599 + (9.602 \times 14.7) = 352.748$

[1 mark for each value of y correctly calculated]

b) Using the equation: 'Residual = Observed y-value – Estimated y-value':
At $x = 12.5$: Residual = 332.5 – 331.624 = 0.876 ***[1 mark]***
At $x = 14.7$: Residual = 352.1 – 352.748 = –0.648 ***[1 mark]***

And that's the end of that — Section 7 done and dusted.

Answers

M2 Section 1 — Kinematics

Warm-up Questions

1)

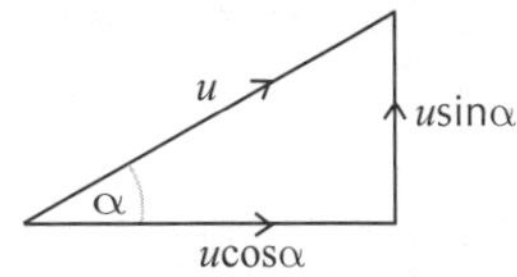

So, parallel to the horizontal, the initial velocity is $u\cos\alpha$.

2) Resolving horizontally (taking right as +ve):
$u = 120;\ s = 60;\ a = 0;\ t = ?$
$s = ut + \frac{1}{2}at^2$
$60 = 120t + \frac{1}{2} \times 0 \times t^2$
$t = 0.5$ s
Resolving vertically (taking down as +ve):
$u = 0;\ s = ?;\ a = 9.8;\ t = 0.5$
$s = ut + \frac{1}{2}at^2$
$= (0 \times 0.5) + (0.5 \times 9.8 \times 0.5^2)$
$= 1.23$ m (to 3 s.f.)

3) Resolving vertically (taking up as +ve):
$u = 22\sin\alpha;\ a = -9.8;\ t = 4;\ s = 0$
s = 0 because the ball lands at the same vertical level it started at.
$s = ut + \frac{1}{2}at^2$
$0 = 22\sin\alpha \times 4 + (0.5 \times -9.8 \times 4^2)$
Rearranging: $\sin\alpha = \frac{78.4}{88}$
$\Rightarrow \alpha = 63.0°$ (3 s.f.)
There are other ways to answer this question — you could use v = u + at and use t = 2, which is the time taken to reach the highest point, when v = 0. I like my way though.

4) a) $a = \frac{dv}{dt} = 16t - 2$
b) $s = \int v\,dt = \frac{8t^3}{3} - t^2 + c$
When $t = 0$, the particle is at the origin, i.e. $s = 0 \Rightarrow c = 0$
So, $s = \frac{8t^3}{3} - t^2$

5) $\dot{\mathbf{r}} = \frac{d\mathbf{r}}{dt}$, which represents the velocity of the particle, and
$\ddot{\mathbf{r}} = \frac{d^2\mathbf{r}}{dt^2}$, which represents the acceleration of the particle.

6) $\mathbf{r} = \int \mathbf{v}\,dt = 2t^2\mathbf{i} + \frac{t^3}{3}\mathbf{j} + \mathbf{C}$
When $t = 0$, the particle is at the origin $\Rightarrow \mathbf{C} = 0\mathbf{i} + 0\mathbf{j}$.
So, $\mathbf{r} = 2t^2\mathbf{i} + \frac{t^3}{3}\mathbf{j}$
$\mathbf{a} = \frac{d\mathbf{v}}{dt} = 4\mathbf{i} + 2t\mathbf{j}$

Exam Questions

1 a) $\tan\alpha = \frac{3}{4} \Rightarrow \sin\alpha = \frac{3}{5}$ **[1 mark]**
Resolving vertically, taking down as +ve:
$u = u_y = 15\sin\alpha = 9;$ **[1 mark]**
$s = 11;\ a = 9.8;\ t = ?$
$s = ut + \frac{1}{2}at^2$ **[1 mark]**
$11 = 9t + 4.9t^2$ **[1 mark]**
Use the quadratic formula to find $t = 0.839$ s (3 s.f.)
[1 mark]. So the stone takes 0.389 s to reach the ground.

b) Resolving horizontally, taking right as +ve:
$u = u_x = 15\cos\alpha = 15 \times \frac{4}{5} = 12;$ **[1 mark]**
$s = ?;\ t = 0.8390$ s
$a = 0$, so $s = ut \Rightarrow OB = 12 \times 0.8390$ **[1 mark]**
So, $OB = 10.07$ m
So stone misses H by $10.07 - 9 = 1.07$ m (3 s.f.) **[1 mark]**

c) Resolving horizontally, taking right as +ve:
$s = 9;\ u_x = u\cos\alpha;\ a = 0;\ t = ?$
$s = u_x t + \frac{1}{2}at^2$ **[1 mark]**
$9 = (u\cos\alpha)t$
$\Rightarrow t = \frac{9}{u\cos\alpha}$ — call this **eqn 1**. **[1 mark]**
Now resolve vertically, taking down as +ve:
$s = 11;\ u_y = u\sin\alpha;\ a = 9.8;\ t = ?$
$s = u_y t + \frac{1}{2}at^2$
$11 = (u\sin\alpha)t + 4.9t^2$ — call this **eqn 2**. **[1 mark]**
t is the same both horizontally and vertically, so substitute **eqn 1** in **eqn 2** to eliminate t:
$11 = 9\left(\frac{u\sin\alpha}{u\cos\alpha}\right) + 4.9\left(\frac{9}{u\cos\alpha}\right)^2$ **[1 mark]**
$11 = 9\tan\alpha + \frac{4.9 \times 81}{u^2\cos^2\alpha}$
$\tan\alpha = \frac{3}{4}$ and $\cos\alpha = \frac{4}{5}$, so substituting and simplifying:
$u^2 = 145.919$
so $u = 12.1$ ms^{-1} (3 s.f.) **[1 mark]**
Wooo. What a beauty part c) is — I'd do that again just for kicks. But then I do love a bit of substituting and eliminating.
If you're confused by this question, then look back over the section — there's an example which is a bit more general, but it uses a lot of the same working.

2 Resolving horizontally, taking right as +ve :
$u = 20\cos30°;\ s = 30;\ a = 0;\ t = ?$

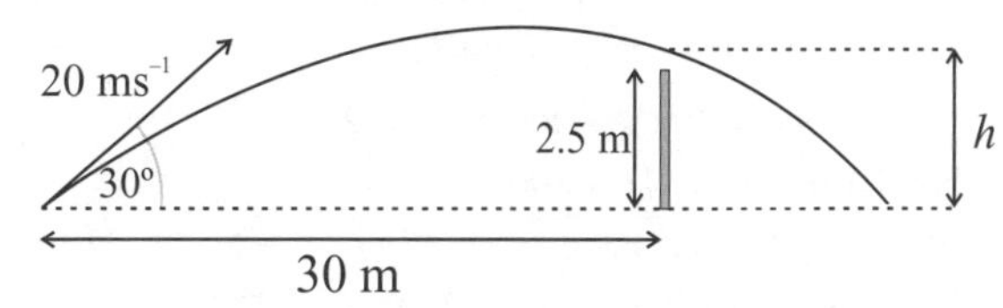

$s = ut + \frac{1}{2}at^2$ **[1 mark]**
$30 = (20\cos30° \times t)$
$t = 1.732$ s **[1 mark]**
Resolving vertically, taking up as +ve:
$s = h;\ u = 20\sin30°;\ t = 1.732;\ a = -9.8$
$s = ut + \frac{1}{2}at^2$ **[1 mark]**
$h = (20\sin30° \times 1.732) + (\frac{1}{2} \times -9.8 \times 1.732^2)$
$= 2.62$ m (to 3 s.f.) **[1 mark]**
2.62 m > 2.5 m, so the ball goes over the crossbar. **[1 mark]**
Assumptions: e.g. ball is a point mass/no air or wind resistance/no spin on the ball **[1 mark]**
That was always my problem when I was taking free kicks — I didn't model the flight of the ball properly before kicking it, so no wonder I never scored.

Answers

3 a) Resolving vertically, taking up as +ve:
$u = U\sin\theta$, $a = -9.8$, $s = 0$, $t = \sin\theta$
$s = ut + \frac{1}{2}at^2 \Rightarrow 0 = U\sin^2\theta - 4.9\sin^2\theta$ ***[1 mark]***
$U = 4.9$ ms^{-1} ***[1 mark]***

b) Resolving horizontally, taking right as +ve:
$u = 4.9\cos\theta$, $a = 0$, $t = \sin\theta$, $s = x$
$s = ut + \frac{1}{2}at^2 \Rightarrow x = 4.9\cos\theta\sin\theta$ ***[1 mark]***.
Use the double angle formula $2\sin\theta\cos\theta = \sin2\theta$ ***[1 mark]***
So $x = 2.45\sin2\theta$ m ***[1 mark]***

Yep, you might be expected to use all those trig identities you thought belonged only in Core. Better make sure you remember them then. Sorry about that.

c) $x = 2 \Rightarrow \sin2\theta = 2 \div 2.45 = 0.8163$ ***[1 mark]***
so, $2\theta = 54.72°$ ***[1 mark]***
or $2\theta = 180° - 54.72° = 125.28°$ ***[1 mark]***.
This is from the symmetry of the graph of sin2θ
$\Rightarrow \theta = 27.36°$ or $\theta = 62.64°$
So $t = \sin\theta = 0.460$ s ***[1 mark]*** or 0.888 s (3 s.f.) ***[1 mark]***.

4 a) $\mathbf{v} = \dot{\mathbf{r}} = (6t^2 - 14t)\mathbf{i} + (6t - 12t^2)\mathbf{j}$
[2 marks in total — 1 mark for attempting to differentiate the position vector, 1 mark for correctly differentiating both components]

b) $\mathbf{v} = \left(\frac{6}{4} - \frac{14}{2}\right)\mathbf{i} + \left(\frac{6}{2} - \frac{12}{4}\right)\mathbf{j}$ ***[1 mark]***
$= -5.5\mathbf{i} + 0\mathbf{j}$
Speed $= \sqrt{(-5.5)^2 + 0^2} = 5.5$ ms^{-1} ***[1 mark]***
The component of velocity in the direction of north is zero, and the component in the direction of east is negative, so the particle is moving due west ***[1 mark]***

c) $\mathbf{a} = \dot{\mathbf{v}}$ ***[1 mark]***
$= (12t - 14)\mathbf{i} + (6 - 24t)\mathbf{j}$ ***[1 mark]***
At $t = 2$, $\mathbf{a} = 10\mathbf{i} - 42\mathbf{j}$ ***[1 mark]***

d) Use $\mathbf{F} = m\mathbf{a}$ to find the force at $t = 2$:
$\mathbf{F} = 10m\mathbf{i} - 42m\mathbf{j}$ ***[1 mark]***
At $t = 2$, $|\mathbf{F}| = \sqrt{(10m)^2 + (-42m)^2} = 43.17m$ ***[1 mark]***
Magnitude of $\mathbf{F}$ at $t = 2$ is 170, so: $43.17m = 170$
$\Rightarrow m = 3.94$ kg (3 s.f.) ***[1 mark]***

e) The vectors **F** and **a** always act in the same direction, so when **F** is acting parallel to **j**, so is **a**. ***[1 mark]***
So, when **F** is acting parallel to **j**, the component of **a** in direction of **i** will be zero ***[1 mark]***, i.e.
$12t - 14 = 0 \Rightarrow t = 1.17$ s (3 s.f.) ***[1 mark]***

5 a) v is at a maximum when $\frac{dv}{dt} = 0$, i.e. when $a = 0$
So, in the interval $0 \le t \le 4$,
$a = \frac{dv}{dt} = 9 - 6t$ ***[1 mark]***
Set $a = 0$:
$0 = 9 - 6t \Rightarrow t = 1.5$ s ***[1 mark]***
So, $v = (9 \times 1.5) - 3(1.5^2)$ ***[1 mark]***
$= 6.75$ ms^{-1} ***[1 mark]***

b) (i) $s = \int v\,dt$
$= \frac{9t^2}{2} - t^3 + c$ for $0 \le t \le 4$. ***[1 mark]***
When $t = 0$, the particle is at the origin, i.e. $s = 0$
$\Rightarrow c = 0$ ***[1 mark]***
So, at $t = 4$:
$s = \frac{9}{2}(16) - 64 = 8$ m ***[1 mark]***

(ii) $s = \int v\,dt$
$= \frac{-192t^{-1}}{-1} + k = \frac{192}{t} + k$ for $t > 4$ ***[1 mark]***
When $t = 4$, $s = 8$, so $8 = \frac{192}{4} + k$ ***[1 mark]***
$\Rightarrow k = -40$ ***[1 mark]***
When $t = 6$,
$s = \frac{192}{6} - 40 = -8$ m ***[1 mark]***

6 a) Resolving horizontally, taking right as +ve:
$u = 14$; $s = x$, $t = ?$, $a = 0$
$s = ut + \frac{1}{2}at^2$ gives:
$x = 14t$, so $t = \frac{x}{14}$ — call this **eqn 1.** ***[1 mark]***
Resolving vertically, taking up as +ve:
$u = 35$; $s = y$; $t = ?$; $a = -9.8$
$s = ut + \frac{1}{2}at^2 \Rightarrow y = 35t - 4.9t^2$ — call this **eqn 2.** ***[1 mark]***
Substitute **eqn 1** into **eqn 2** to eliminate t:
$y = 35\left(\frac{x}{14}\right) - 4.9\left(\frac{x}{14}\right)^2$ ***[1 mark]***
Rearrange:
$y = \frac{5x}{2} - \frac{x^2}{40}$ ***[1 mark]***
If you take down as +ve in the working, then you have to use $a = 9.8$, $s = -y$ (because y is defined as being positive), and $u = -35$. Then all the signs work out right :-)

b) Use formula from part a) with $y = -30$;
$y = -30$, because the ball lands 30 m below the point it's hit from and up was taken as positive when deriving the formula.
$-30 = \frac{5x}{2} - \frac{x^2}{40}$ ***[1 mark]***
Rearrange: $x^2 - 100x - 1200 = 0$
Solve quadratic using quadratic formula ***[1 mark]***
$x = 111$ m (3 s.f.) ***[1 mark]***

c) Distance $AH = 110.8 - 7 = 103.8$ m
Use formula from part a) with $x = 103.8$ m:
$y = \frac{5 \times 103.8}{2} - \frac{(103.8)^2}{40} = -9.861$m ***[1 mark]***
So, when ball is vertically above H, it is 9.86 m below the level of O. Now resolve vertically, taking up as +ve:
$u = u_y = 35$; $s = -9.861$; $a = -g$; $v = v_y$
$v^2 = u^2 + 2as \Rightarrow v_y^2 = 1418$ ***[1 mark]***
In this case, you don't need to take the square root to find the value of v_y as you'd have to square it again to find the speed.
No acceleration horizontally, so $v_x = u_x = 14$ ms^{-1}
Speed, $V = \sqrt{v_x^2 + v_y^2}$ ***[1 mark]***
$= \sqrt{14^2 + 1418} = 40.2$ ms^{-1} (3 s.f.) ***[1 mark]***

Answers

7) a) Using F = ma,

$\frac{1000m}{v} - 0.1mv^2 = m\frac{dv}{dt}$ ***[1 mark]***

Rearranging: $\frac{dv}{dt} = \frac{1000}{v} - 0.1v^2$ ***[1 mark]***

b) $\frac{dv}{dt} = -0.1v^2$ ***[1 mark]***

Once the engines are switched off, the only force acting horizontally on the plane will be the resistive force.

c) Separate the variables and integrate:

$\int \frac{1}{v^2} dv = \int -0.1 dt$ ***[1 mark]***

$\Rightarrow -\frac{1}{v} = -0.1t + c$ ***[1 mark]***

When $t = 0$, $v = 50$, so

$-\frac{1}{50} = -0.1(0) + c$ ***[1 mark]*** $\Rightarrow c = -\frac{1}{50}$ ***[1 mark]***

So, when $v = 25$, $-\frac{1}{25} = -0.1t - \frac{1}{50}$

$\Rightarrow t = 0.2$ seconds ***[1 mark]***

8 a) $\mathbf{v} = \dot{\mathbf{r}} = -3\sin t\mathbf{i} + 3\cos t\mathbf{j}$

Speed is given by the magnitude of the velocity vector:

$|\mathbf{v}| = \sqrt{(-3\sin t)^2 + (3\cos t)^2}$

$= \sqrt{9(\sin^2 t + \cos^2 t)} = 3$

So the speed is a constant 3 ms^{-1}.

[3 marks in total — 1 mark for attempting to differentiate position vector, 1 mark for correct expression for velocity, 1 mark for finding speed correctly]

Remember, $\sin^2\theta + \cos^2\theta = 1$

b) Find the magnitude of the position vector to find the particle's distance from the origin:

$|\mathbf{r}| = \sqrt{(3\cos t)^2 + (3\sin t)^2} = \sqrt{9(\cos^2 t + \sin^2 t)} = 3$, which is constant, so the particle moves in a circle of radius 3 m.

[3 marks in total — 1 mark for attempting to find magnitude of position vector, 1 mark for stating $|\mathbf{r}|$ is constant, 1 mark for finding radius]

The position vector gives the displacement of the particle from the origin at a time t. So the magnitude of the position vector will give the distance from the origin. An object moving in a circle about a point will always be a fixed distance from that point.

9 a) $v = \int a\,dt = \int (8t^2 + 6\sin 2t)\,dt$ ***[1 mark]***

$= \frac{8}{3}t^3 - \frac{6}{2}\cos 2t + c = \frac{8}{3}t^3 - 3\cos 2t + c$ ***[1 mark]***

When $t = 0$, $v = 0$:

$0 = \frac{8}{3}(0)^3 - 3\cos 2(0) + c \Rightarrow 0 = 0 - 3 + c$

$\Rightarrow c = 3$ ***[1 mark]***

So, $v = \frac{8}{3}t^3 - 3\cos 2t + 3$ ***[1 mark]***

b) $v = \frac{8}{3}\left(\frac{\pi}{2}\right)^3 - 3\cos 2\left(\frac{\pi}{2}\right) + 3$ ***[1 mark]***

$v = \frac{\pi^3}{3} - 3(-1) + 3 = \frac{\pi^3}{3} + 6$ ***[1 mark]***

10 a) Using $F = ma$:

$-kv = 2 \times \frac{dv}{dt}$ ***[1 mark]***

So, $\frac{dv}{dt} = \frac{-kv}{2}$ ***[1 mark]***

b) Separate the variables and integrate:

$\frac{dv}{dt} = \frac{-kv}{2} \Rightarrow \frac{1}{v}dv = -\frac{k}{2}dt$

So $\int \frac{1}{v}dv = \int -\frac{k}{2}dt \Rightarrow \ln v = -\frac{kt}{2} + c$ ***[1 mark]***

When $t = 0$, $v = 12$, so:

$\ln 12 = 0 + c \Rightarrow c = \ln 12$ ***[1 mark]***

Take exponentials of both sides to get an equation for v:

$v = e^{-\frac{kt}{2} + \ln 12} = e^{-\frac{kt}{2}} \times e^{\ln 12}$ ***[1 mark]***

So, $v = 12e^{-\frac{kt}{2}} = 12\sqrt{e^{-kt}}$ ***[1 mark]***

M2 Section 2 — Centres of Mass

Warm-up Questions

1) a) Particles in a horizontal line so use $\Sigma mx = \bar{x}\Sigma m$

$m_1x_1 + m_2x_2 + m_3x_3 = \bar{x}(m_1 + m_2 + m_3)$

$\Rightarrow (1 \times 1) + (2 \times 2) + (3 \times 3) = \bar{x}(1 + 2 + 3)$

$\Rightarrow 14 = 6\bar{x} \Rightarrow \bar{x} = 14 \div 6 = 2\frac{1}{3}$.

So coordinates are $(2\frac{1}{3}, 0)$.

b) Particles in a vertical line so use $\Sigma my = \bar{y}\Sigma m$

$m_1y_1 + m_2y_2 + m_3y_3 = \bar{y}(m_1 + m_2 + m_3)$

$\Rightarrow (1 \times 3) + (2 \times 2) + (3 \times 1) = \bar{y}(1 + 2 + 3)$

$\Rightarrow 10 = 6\bar{y} \Rightarrow \bar{y} = 10 \div 6 = 1\frac{2}{3}$.

So coordinates are $(0, 1\frac{2}{3})$.

c) Particles in 2D so use $\Sigma m\mathbf{r} = \bar{\mathbf{r}}\Sigma m$

$m_1\mathbf{r}_1 + m_2\mathbf{r}_2 + m_3\mathbf{r}_3 = \bar{\mathbf{r}}(m_1 + m_2 + m_3)$

$\Rightarrow 1\binom{3}{4} + 2\binom{3}{1} + 3\binom{1}{0} = \bar{\mathbf{r}}(1 + 2 + 3)$

$\Rightarrow \binom{12}{6} = 6\bar{\mathbf{r}} \Rightarrow \bar{\mathbf{r}} = \binom{12}{6} \div 6 = \binom{2}{1}$.

So coordinates are (2, 1).

2) Use $\Sigma m\mathbf{r} = \bar{\mathbf{r}}\Sigma m$

$m\binom{0}{0} + 2m\binom{0}{4} + 3m\binom{5}{4} + 12\binom{5}{0} = (m + 2m + 3m + 12)\binom{3.5}{2}$

$\Rightarrow \binom{15m + 60}{20m} = \binom{21m + 42}{12m + 24}$

$\Rightarrow 20m = 12m + 24 \Rightarrow 8m = 24 \Rightarrow m = 24 \div 8 = 3$ kg.

3) a) Triangle (1) has area $\frac{1}{2} \times 4 \times 3 = 6$, so $m_1 = 6$.

$x_1 = 2$ (symmetry) and $y_1 = 5 - (\frac{2}{3} \times 3) = 3$ ($\frac{2}{3}$ down the median from the top vertex).

Rectangle (2) has area $1 \times 2 = 2$, so $m_2 = 2$.

$x_2 = 2$ and $y_2 = 1.5$ (symmetry).

Combined shape has $\bar{x} = 2$ (symmetry) and:

$m_1y_1 + m_2y_2 = \bar{y}(m_1 + m_2)$

$\Rightarrow (6 \times 3) + (2 \times 1.5) = (6 + 2)\bar{y}$

$\Rightarrow 21 = 8\bar{y} \Rightarrow \bar{y} = 21 \div 8 = 2.625$.

So coordinates are (2, 2.625).

b) Semicircle (1) has area $\frac{1}{2} \times \pi \times 3^2 = 4.5\pi$, so $m_1 = 4.5\pi$.

$x_1 = 8$ (symmetry) and $y_1 = 1 + \frac{2 \times 3 \times \sin\frac{\pi}{2}}{\frac{3\pi}{2}} = \frac{4 + \pi}{\pi}$

(COM is $\frac{2r\sin\alpha}{3\alpha}$ up from the centre of the circle, where $2\alpha = \pi$).

Don't forget — the arc angle is 2α not α...

Triangle (2) has area $\frac{1}{2} \times 2 \times 1 = 1$, so $m_2 = 1$.

Answers

$x_2 = 8$ (symmetry) and $y_2 = \frac{2}{3} \times 1 = \frac{2}{3}$ ($\frac{2}{3}$ up the median from the bottom vertex).

Combined shape has $\bar{x} = 8$ (symmetry) and:

$m_1y_1 + m_2y_2 = \bar{y}(m_1 + m_2)$

$\Rightarrow (4.5\pi \times \frac{4+\pi}{\pi}) + (1 \times \frac{2}{3}) = (4.5\pi + 1)\bar{y}$

$\Rightarrow 18\frac{2}{3} + 4.5\pi = (4.5\pi + 1)\bar{y}$

$\Rightarrow \bar{y} = (18\frac{2}{3} + 4.5\pi) \div (4.5\pi + 1) = 2.167$ (to 3 d.p.).

So coordinates are (8, 2.167).

c) Circle (1) has area $\pi \times 2^2 = 4\pi$, so $m_1 = 4\pi$.

$x_1 = 14$ and $y_1 = 2$ (symmetry).

Square (2) has area $1 \times 1 = 1$, so $m_2 = 1$.

$x_2 = 14.5$ and $y_2 = 2.5$ (symmetry).

Using the removal method:

$m_1\mathbf{r}_1 - m_2\mathbf{r}_2 = \bar{\mathbf{r}}(m_1 - m_2)$

$\Rightarrow 4\pi\binom{14}{2} - \binom{14.5}{2.5} = (4\pi - 1)\bar{\mathbf{r}}$

$\Rightarrow \binom{56\pi - 14.5}{8\pi - 2.5} = (4\pi - 1)\bar{\mathbf{r}}$

$\Rightarrow \bar{\mathbf{r}} = \binom{56\pi - 14.5}{8\pi - 2.5} \div (4\pi - 1) = \binom{13.957}{1.957}$ (to 3 d.p.).

So coordinates are (13.957, 1.957).

4)

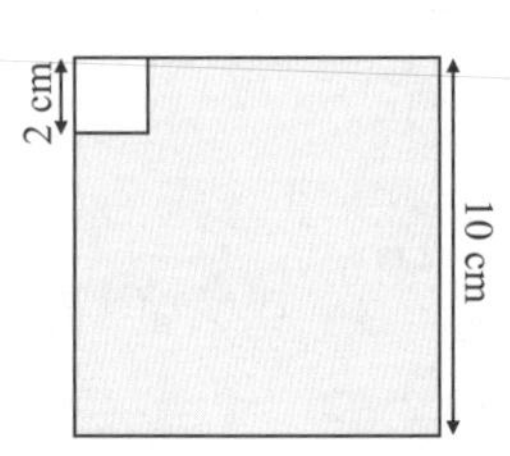

Large square (1) has area $10 \times 10 = 100$, so $m_1 = 100$.

$y_1 = 5$ cm from top edge (symmetry).

Small square (2) has area $2 \times 2 = 4$, so $m_2 = 4$.

$y_2 = 1$ cm from top edge (symmetry).

Using the removal method:

$m_1y_1 - m_2y_2 = \bar{y}(m_1 - m_2)$

$\Rightarrow (100 \times 5) - (4 \times 1) = (100 - 4)\bar{y}$

$\Rightarrow 496 = 96\bar{y} \Rightarrow \bar{y} = 496 \div 96 = 5.167$ cm from the top edge (to 3 d.p.).

You can pick any place to be the origin, but the top edge makes most sense here.

5) a) i) $m_1y_1 + m_2y_2 + m_3y_3 + m_4y_4 = \bar{y}(m_1 + m_2 + m_3 + m_4)$

$(7 \times 0) + (6 \times 0) + (10 \times 5) + (12 \times 5) = \bar{y}(7 + 6 + 10 + 12)$

$110 = 35\bar{y} \Rightarrow \bar{y} = 3.14$ cm.

The framework is 'light' so it has no mass.

ii) $m_1x_1 + m_2x_2 + m_3x_3 + m_4x_4 = \bar{x}(m_1 + m_2 + m_3 + m_4)$

$(7 \times 0) + (6 \times 5) + (10 \times 5) + (12 \times 0) = \bar{x}(7 + 6 + 10 + 12)$

$80 = 35\bar{x} \Rightarrow \bar{x} = 2.29$ cm.

b)

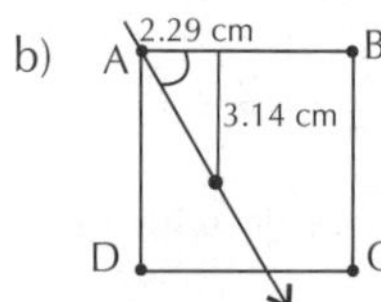

Angle = $\tan^{-1} \frac{3.14}{2.29} = 54°$ to the nearest degree.

Exam Questions

1 (a) Using the formula $\Sigma my = \bar{y}\Sigma m$

$m_1y_1 + m_2y_2 + m_3y_3 = \bar{y}(m_1 + m_2 + m_3)$

$\Rightarrow (4 \times 3) + (3 \times 1) + (2 \times y) = 2 \times (4 + 3 + 2)$ ***[1 mark]***

$\Rightarrow 15 + 2y = 18$ ***[1 mark]***

$\Rightarrow y = (18 - 15) \div 2 = 1.5$ ***[1 mark]***.

(b) Using the formula $\Sigma mx = \bar{x}\Sigma m$

$m_1x_1 + m_2x_2 + m_3x_3 = \bar{x}(m_1 + m_2 + m_3)$

$\Rightarrow (4 \times 1) + (3 \times 5) + (2 \times 4) = \bar{x}(4 + 3 + 2)$ ***[1 mark]***

$\Rightarrow 27 = 9\bar{x}$ ***[1 mark]***

$\Rightarrow \bar{x} = 27 \div 9 = 3$ ***[1 mark]***.

(c) Centre of mass of the lamina is at (3.5, 2.5), due to the symmetry of the shape, and $m_{lamina} = 6$ kg.
Centre of mass of the group of particles is (3, 2) (from (b)) and $m_{particles} = 4 + 3 + 2 = 9$ kg. Using $\Sigma m\mathbf{r} = \bar{\mathbf{r}}\Sigma m$:

$m_{lamina}\mathbf{r}_{lamina} + m_{particles}\mathbf{r}_{particles} = \bar{\mathbf{r}}(m_{lamina} + m_{particles})$

$\Rightarrow 6\binom{3.5}{2.5} + 9\binom{3}{2} = \bar{\mathbf{r}}(6 + 9)$

$\Rightarrow \binom{21 + 27}{15 + 18} = 15\bar{\mathbf{r}}$

$\Rightarrow \bar{\mathbf{r}} = \binom{48}{33} \div 15 = \binom{3.2}{2.2}$,

so the coordinates are (3.2, 2.2).

[6 marks available — 1 mark for the correct x_{lamina}, 1 mark for the correct y_{lamina}, 1 mark for correct entry of horizontal positions in the formula, 1 mark for correct entry of vertical positions in the formula, 1 mark for x coordinate of 3.2, 1 mark for y coordinate of 2.2.]

2 (a) Splitting up the shape into a triangle (1), large square (2) and small square (3), where the mass of each shape is proportional to the area, gives the following masses:

$m_1 = \frac{1}{2} \times 70 \times 30 = 1050.$

$m_2 = 50 \times 50 = 2500.$

$m_3 = 10 \times 10 = 100.$

Taking the point A as the origin, the position vectors of the centres of mass of each shape are as follows:

Triangle:

$x_1 = 25$ (due to the symmetry of the shape) and

$y_1 = 50 + (\frac{1}{3} \times 30) = 60$ (since the COM of a triangle is $\frac{2}{3}$ down the median from the vertex, and so $\frac{1}{3}$ up from the edge). So $\mathbf{r}_1 = \binom{25}{60}$.

Large Square:

$x_2 = 25$ and $y_2 = 25$ (due to the symmetry of the shape) so $\mathbf{r}_2 = \binom{25}{25}$.

Small Square:

$x_3 = 50 + 5 = 55$ and $y_3 = 5$ (due to the symmetry of the shape) so $\mathbf{r}_3 = \binom{55}{5}$.

Using the formula $\Sigma m\mathbf{r} = \bar{\mathbf{r}}\Sigma m$

$m_1\mathbf{r}_1 + m_2\mathbf{r}_2 + m_3\mathbf{r}_3 = \bar{\mathbf{r}}(m_1 + m_2 + m_3) \Rightarrow$

$1050\binom{25}{60} + 2500\binom{25}{25} + 100\binom{55}{5} = \bar{\mathbf{r}}(1050 + 2500 + 100)$

$\Rightarrow \binom{26250 + 62500 + 5500}{63000 + 62500 + 500} = 3650\bar{\mathbf{r}}$

$\Rightarrow \bar{\mathbf{r}} = \binom{94250}{126000} \div 3650 = \binom{25.8219...}{34.5205...}$.

Answers

So, to 3 s.f., the centre of mass of the sign is 25.8 cm from AB and 34.5 cm from AI.

[6 marks available — 1 mark for masses in the correct proportion, 1 mark for each individual centre of mass entered correctly into the formula, 1 mark for correct distance from AB, 1 mark for correct distance from AI.]

(b) For the sign to hang with AI horizontal, the centre of mass of the whole system (sign + particle) must be vertically below D, i.e. $\bar{x}$ must be 25 (taking A as the origin again).

Given that $m_{sign} = 1$ kg and $x_{sign} = 25.82$ (from (a)), and $x_{particle} = 0$ (since it's attached at the origin):

$m_{sign}x_{sign} + m_{particle}x_{particle} = \bar{x}(m_{sign} + m_{particle})$

$(1 \times 25.82) + 0 = 25(1 + m_{particle})$

$\Rightarrow 25.82 \div 25 = 1 + m_{particle}$

$\Rightarrow 1.033 - 1 = m_{particle}$

$\Rightarrow m_{particle} = 0.033$ kg, to 3 s.f.

[3 marks available — 1 mark stating the correct required value of $\bar{x}$, 1 mark for correct entry of values into the formula, 1 mark for correct final answer.]

You should use the slightly less rounded value of 25.82 for x_{sign} in part b). You want your final answer to three significant figures, so your calculations should be with numbers of at least four significant figures. If you took x_{sign} as 25.8, your final answer would be 0.032 kg, and that's wrong dawg.

3 (a) The stencil is a rectangle (1) with a quarter circle (2) of radius (10 – 2) = 8 cm removed. The lamina is uniform so mass is proportional to area, so $m_1 = 12 \times 10 = 120$, and $m_2 = \frac{1}{4} \times \pi \times 8^2 = 16\pi$.

Taking O as the origin, the position of the centre of mass of the rectangle, $\mathbf{r}_1 = \binom{6}{5}$ (from the symmetry of the shape).

The sector angle $2\alpha = \frac{\pi}{2}$, so $\alpha = \frac{\pi}{4}$, and the centre of mass of the sector is $\frac{2r\sin\alpha}{3\alpha}$ from O along the axis of symmetry

$= \dfrac{2 \times 8 \times \sin\frac{\pi}{4}}{\frac{3\pi}{4}} = \dfrac{64}{3\pi\sqrt{2}}$ cm.

This is on the formula sheet — you just have to know how to use it. And you'll have to use trig to find the position vector... In the right-angled triangle below, cos α = x/hyp, and sin α = y/hyp, so with a bit of rearranging you can find x and y for the position vector...

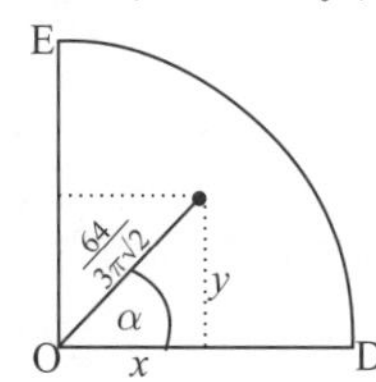

Using trig, the position vector of the centre of mass of the sector, $\mathbf{r}_2 = \begin{pmatrix} \frac{64}{3\pi\sqrt{2}} \times \cos\frac{\pi}{4} \\ \frac{64}{3\pi\sqrt{2}} \times \sin\frac{\pi}{4} \end{pmatrix} = \begin{pmatrix} \frac{32}{3\pi} \\ \frac{32}{3\pi} \end{pmatrix}$.

Using the removal method:

$m_1\mathbf{r}_1 - m_2\mathbf{r}_2 = \bar{\mathbf{r}}(m_1 - m_2)$

$\Rightarrow 120\binom{6}{5} - 16\pi\begin{pmatrix} \frac{32}{3\pi} \\ \frac{32}{3\pi} \end{pmatrix} = \bar{\mathbf{r}}(120 - 16\pi)$

$\Rightarrow \begin{pmatrix} 720 - \frac{512}{3} \\ 600 - \frac{512}{3} \end{pmatrix} = \bar{\mathbf{r}}(120 - 16\pi)$

$\Rightarrow \begin{pmatrix} \frac{1648}{3} \\ \frac{1288}{3} \end{pmatrix} \div (120 - 16\pi) = \bar{\mathbf{r}}$

$\Rightarrow \bar{\mathbf{r}} = \binom{7.8774...}{6.1566...}$.

So, to 3 s.f., the coordinates of the centre of mass of the stencil are (7.88, 6.16).

[7 marks available — 1 mark for the correct total mass, 1 mark for correct r_1, 1 mark for correct r_2, 1 mark for correct entry of horizontal positions in the formula, 1 mark for correct entry of vertical positions in the formula, 1 mark for x coordinate of 7.88, 1 mark for y coordinate of 6.16.]

(b) At the point of toppling, the centre of mass will be vertically above the point D, as shown:

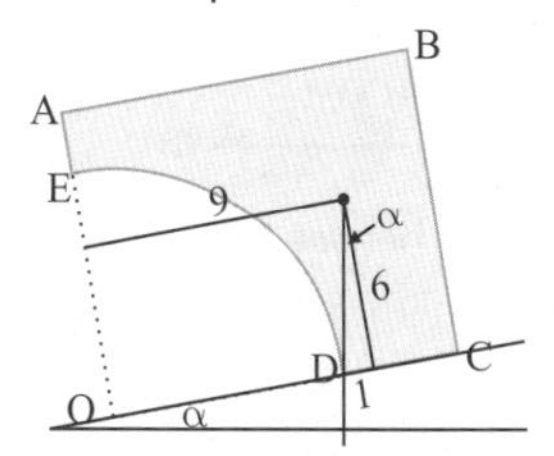

Horizontal distance from D to the centre of mass = 9 – 8 = 1 cm ***[1 mark]***.

Using trig, $\alpha = \tan^{-1}\left(\frac{1}{6}\right)$ ***[1 mark]*** = 0.165 rads to 3 s.f. ***[1 mark]***.

4 (a) (i) Setting M as the origin, the distance of the centre of mass from MP is the horizontal distance $\bar{x}$.

For each element the masses and centres are:

Particle at A: $m_A = 3m$ and $x_A = -10$ (since A is 10 cm to the left of M).

Particle at B: $m_B = 4m$ and $x_B = 10$.

Straight rod: $m_{rod} = 2m$ and $x_{rod} = 0$ (M is the midpoint of the rod, which is the centre of its mass).

Arc: $m_{arc} = \pi m$ and $x_{arc} = 0$ (due to the symmetry of the semicircular arc).

Don't forget to include the masses of all the rods and arcs — unless you're told that they're 'light'.

Combining these elements in the formula $\Sigma mx = \bar{x}\Sigma m$

$m_Ax_A + m_Bx_B + m_{rod}x_{rod} + m_{arc}x_{arc} = \bar{x}(m_A + m_B + m_{rod} + m_{arc})$

$\Rightarrow (3m \times -10) + (4m \times 10) + (2m \times 0) + (\pi m \times 0)$

$= \bar{x}(3m + 4m + 2m + \pi m)$

$\Rightarrow -30m + 40m = (9 + \pi)m\bar{x}$

$\Rightarrow \bar{x} = 10 \div (9 + \pi) = 0.8236$ cm to 4 d.p.

[3 marks available — 1 mark for correct total mass of system, 1 mark for correct entry into formula, 1 mark for correct final answer.]

(ii) With M as the origin still, the distance of the centre of mass from AB is the vertical distance $\bar{y}$.

So: $y_A = y_B = y_{rod} = 0$, since all three lie on the line AB.

For the arc, use the formula $y_{arc} = \frac{r\sin\alpha}{\alpha}$, where $r = 10$, and $\alpha = \frac{\pi}{2}$ (since the angle at the centre, $2\alpha = \pi$), so

Answers

$$y_{\text{arc}} = \frac{10\sin\frac{\pi}{2}}{\frac{\pi}{2}} = \frac{20}{\pi}.$$

This is on the formula sheet if you can't remember in the exam...

Combining these in the formula $\Sigma my = \bar{y}\Sigma m$

$m_A y_A + m_B y_B + m_{rod} y_{rod} + m_{arc} y_{arc} = \bar{y}(m_A + m_B + m_{rod} + m_{arc})$

$\Rightarrow (\pi m \times \frac{20}{\pi}) = (9 + \pi)m\bar{y}$

$\Rightarrow \bar{y} = 20 \div (9 + \pi) = 1.6472$ cm to 4 d.p.

[3 marks available — 1 mark for correct value of y_{arc}, 1 mark for correct entry into formula, 1 mark for correct final answer.]

(b) On a sketch, draw a line from P to the centre of mass to represent the vertical and label relevant lengths and angles:

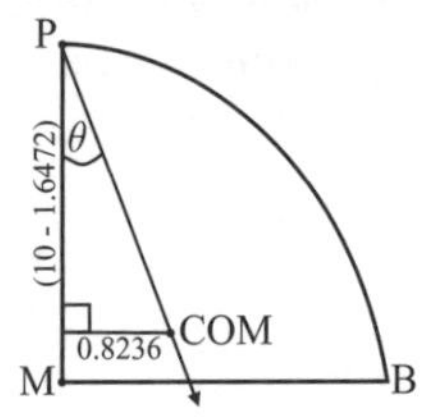

The vertical distance between P and the centre of mass = 10 – 1.6472 (from (a)(ii)) = 8.3528 cm ***[1 mark]***.

Using trig:

$\theta = \tan^{-1}\left(\frac{0.8236}{8.3528}\right)$ ***[1 mark]*** = 0.0983 rads to 3 s.f. ***[1 mark]***.

5 a) Find the centre of mass using the removal method, by subtracting the triangle (2) from the circle (1).

Since both the circle and the triangle that's removed from it are made from the same uniform material, their masses are in proportion to their areas:

Circle $m_1 = \pi r^2 = \pi \times 2^2 = 4\pi$.

Triangle $m_2 = \frac{1}{2} \times 1.5 \times 1.5 = 1.125$.

Taking the point P as the origin, the centre of mass of the circle, $\mathbf{r}_1 = \binom{0}{0}$, since P is the centre of the circle.

The centre of mass of the triangle is at the mean of the coordinates of $P(0,0)$, $Q(0, 1.5)$ and $R(1.5, 0)$, so:

$\mathbf{r}_2 = \begin{pmatrix}(0 + 0 + 1.5) \div 3\\(0 + 1.5 + 0) \div 3\end{pmatrix} = \binom{0.5}{0.5}$.

Using the removal method:

$m_1\mathbf{r}_1 - m_2\mathbf{r}_2 = \bar{\mathbf{r}}(m_1 - m_2)$

$\Rightarrow 4\pi\binom{0}{0} - 1.125\binom{0.5}{0.5} = \bar{\mathbf{r}}(4\pi - 1.125)$

$\Rightarrow \binom{-0.5625}{-0.5625} = 11.4413...\bar{\mathbf{r}}$

$\Rightarrow \bar{\mathbf{r}} = \begin{pmatrix}-0.5625 \div 11.4413...\\-0.5625 \div 11.4413...\end{pmatrix} = \binom{-0.04916...}{-0.04916...}$.

The <u>distance</u> of the COM from P is the magnitude of the position vector:

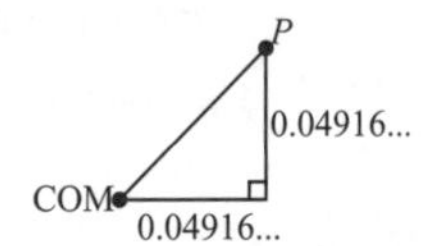

Distance = $\sqrt{0.04916...^2 + 0.04916...^2} = 0.06952...$ = 0.070 cm to 3 d.p.

There are other ways to find the centre of mass of this shape, because you can think of it in a different orientation, or take another point as the origin, but this way's as easy as any.

[5 marks available — 1 mark for correct masses of both shapes, 1 mark for individual centre of mass for both shapes, 1 mark for correct use of removal method formula, 1 mark for correct position vector or coordinates of centre of mass, 1 mark for correct distance from P.]

b) The shape is being hung from Q. Drawing a sketch will make it easier to see what's going on:

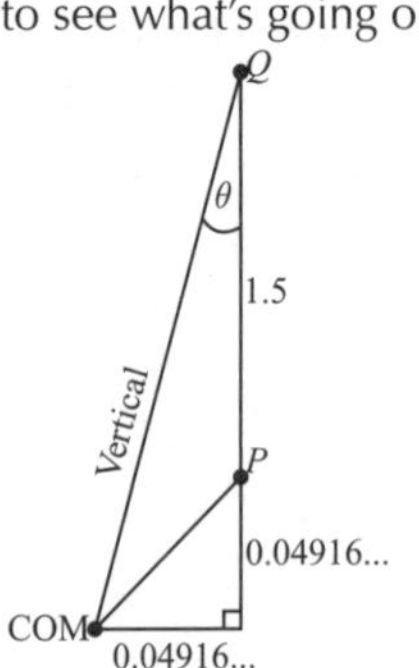

θ is the angle that PQ makes with the vertical.

Using basic trigonometry:

$\theta = \tan^{-1}\left(\frac{0.04916...}{1.5 + 0.04916...}\right) = 1.8177... = 1.8°$ to 1 d.p.

[3 marks available — 1 mark for correct sides of the right-angled triangle, 1 mark for correct working, 1 mark for correct final answer.]

6 Call the 5 × 4 rectangle 'A' and the 5 × 3 rectangle 'B'. The masses of A and B are proportional to their areas, so $m_A = 20$ and $m_B = 15$ ***[1 mark]***. A and B are both uniform, so their centres of mass are at their respective centres: $COM_A = (0, -2, 2.5)$ ***[1 mark]*** and $COM_B = (1.5, 0, 2.5)$ ***[1 mark]***. The shape has a plane of symmetry at $z = 2.5$, so $\bar{z} = 2.5$ ***[1 mark]***. Use $\Sigma m\mathbf{r} = \bar{\mathbf{r}}\Sigma m$ to find the x- and y-coordinates of the COM ***[1 mark]***:

$20\binom{0}{-2} + 15\binom{1.5}{0} = 35\binom{\bar{x}}{\bar{y}}$

$\Rightarrow \binom{\bar{x}}{\bar{y}} = \frac{1}{35}\binom{22.5}{-40} = \binom{0.643}{-1.143}$ ***[1 mark]***

So the coordinates of the centre of mass of the shape are $(\bar{x}, \bar{y}, \bar{z}) = (0.643, -1.14, 2.50)$ (3 s.f.)

7 As the traffic cone is made from uniform material, the masses of the parts are proportional to their areas.

So $m_B = 0.25$. ***[1 mark]***

Find the sloped length of A using Pythagoras:

$l = \sqrt{0.15^2 + 1.2^2} = 1.21$m.

So $m_A = \pi \times 0.15 \times 1.21 = 0.57$. ***[1 mark]***

As B is a uniform square lamina, its centre of mass is at its centre, O, i.e. $y_B = 0$ m from O. ***[1 mark]***

Using the formula for COM of conical shell:

$y_A = \frac{1}{3}(1.2) = 0.4$ m from O. ***[1 mark]***

Now use $\Sigma my = \bar{y}\Sigma m$ to find the COM of C: ***[1 mark]***

$m_B y_B + m_A y_A = m_C y_C$

$\Rightarrow 0.25(0) + 0.57(0.4) = (0.25 + 0.57)y_C$

$\Rightarrow y_C = 0.28$ m (2 d.p.)

So, centre of mass of the traffic cone is 0.28 m vertically above O, on a line through O and the vertex of A. ***[1 mark]***

Answers

M2 Section 3 — Forces and Statics

Warm-up Questions

1) Resolving perpendicular to slope, taking ↖ as +ve:
$F_{net} = ma$
$R - 1.2g\cos25° = 1.2 \times 0$
$R = 1.2g\cos25° = 10.66$ N.
Resolving parallel to slope, taking down slope as +ve:
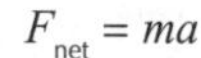
$F_{net} = ma$
$1.2g\sin25° - F = 1.2 \times 0.3$
So, $F = 1.2g\sin25° - 1.2 \times 0.3 = 4.61$ N
Limiting friction, so:
$F = \mu R$
$4.61 = \mu \times 10.66$
$\mu = 0.43$ (to 2 d.p.)

2) Moments about B: $60g \times 3 = T_2 \times 8$
So $T_2 = \frac{180g}{8} = 220.5$ N
Vertically balanced forces, so $T_1 + T_2 = 60g$
$T_1 = 367.5$ N

3) A rod (a long, inextensible particle) where the centre of mass is not at the central point of the rod.
M2 statics is pretty rod-heavy. Think of them as a really simple stick. They don't bend, don't stretch or compress, and have no width.

4)
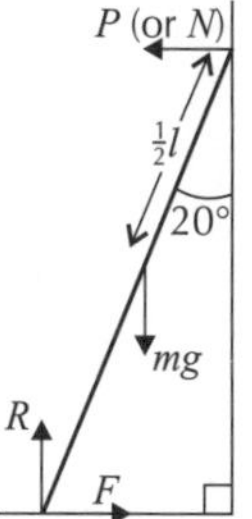

Where:
mg = weight of the ladder
F = friction between ground and ladder
R = normal reaction of the ground
P / N = normal reaction of the wall
As the rod is uniform the weight of the ladder acts at the centre of the rod (i.e. at half of l).
Assumptions: e.g. the ladder can be modelled as a rod, the ladder is rigid, friction is sufficient to keep the ladder in equilibrium, the ladder is perpendicular to the wall when viewed from above.
'Perpendicular' and 'normal' are both used in M2 (as are 'P' and 'N' to label the forces). No need to panic — they mean the same thing in all you'll do here.

5) a)
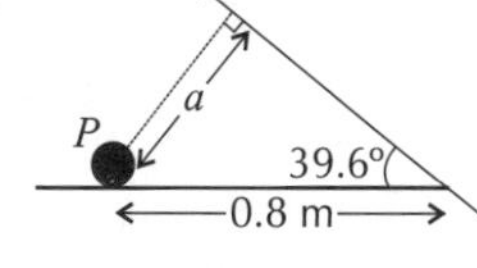

$a = 0.8\sin39.6° = 0.51$ m

b)
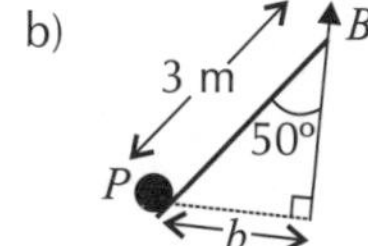

$b = 3\sin50° = 2.30$ m

c)
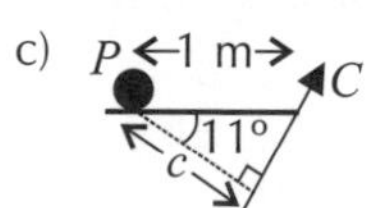

$c = 1\cos11° = 0.98$ m

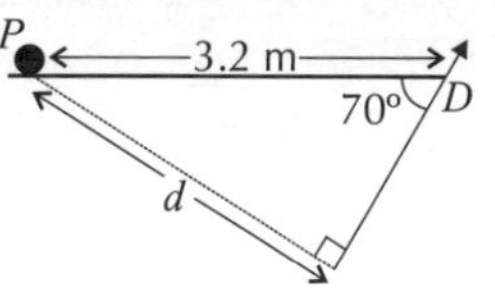

$d = 3.2\sin70° = 3.01$ m

The perpendicular distance is always the shortest distance between a point and a force's line of action. Simple.

6) a) Resolving the forces vertically:
$42 = 28\cos30° + T\cos38.3°$
so $T = \frac{42 - 14\sqrt{3}}{\cos38.3°} = 22.6$ N (3 s.f.)

b) Taking moments about the left-end:
$42xy = 28\cos30(xy + y)$
$42xy - (28\cos30)(xy) = (28\cos30)y$
$17.75xy = 24.25y$
$x = 1.37$

Exam Questions

1) a) Taking moments about A:
$2g \times 0.4 = 30\cos55° \times x$
so $x = \frac{7.84}{17.2} = 0.456$ m

[3 marks available in total]:
- ***1 mark for taking moments about A***
- ***1 mark for correct workings***
- ***1 mark for correct value of x***

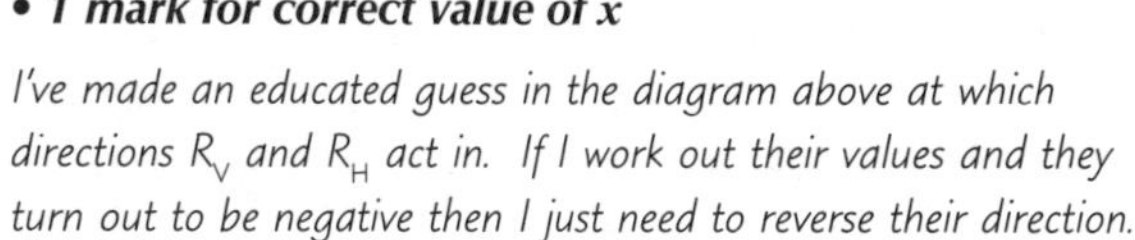
I've made an educated guess in the diagram above at which directions R_V and R_H act in. If I work out their values and they turn out to be negative then I just need to reverse their direction.

b) Resolving vertically:
$R_V = 2g - 30\cos55° = 2.39$ N
Resolving horizontally:
$R_H = 30\sin55° = 24.57$ N

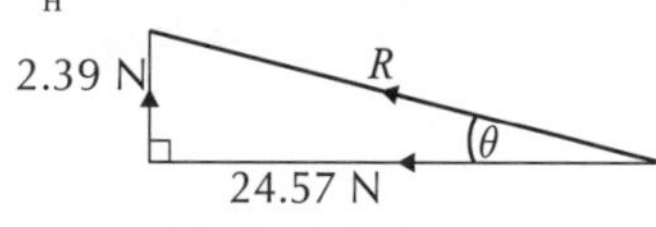

$|R| = \sqrt{2.39^2 + 24.57^2} = 24.7$ N
$\tan\theta = \frac{2.39}{24.57}$
so $\theta = 5.56°$ to the horizontal

[5 marks available in total]:
- ***1 mark for resolving vertically***
- ***1 mark for resolving horizontally***
- ***1 mark for correct workings***
- ***1 mark for correct magnitude of R***
- ***1 mark for correct direction***

2

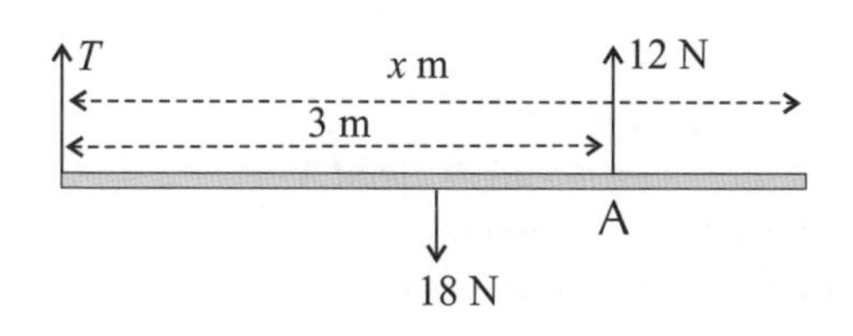

Answers

Taking moments about end string:

$12 \times 3 = 18 \times \frac{x}{2}$ so, $36 = 9x$

$x = 4$ m

[3 marks available in total]:
- ***1 mark for diagram***
- ***1 mark for correct workings***
- ***1 mark for showing that x = 4 m***

3 a)

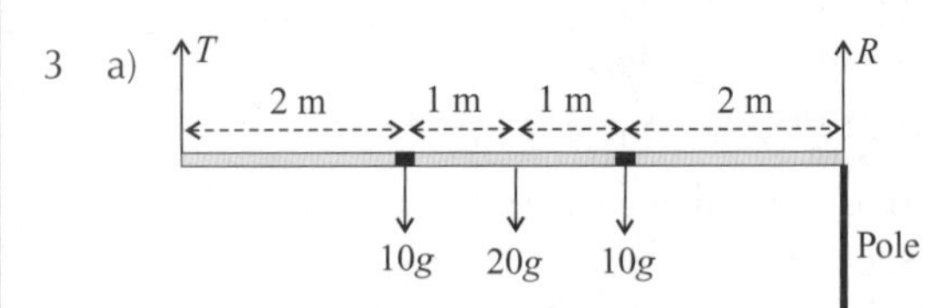

[2 marks available in total]:
- ***1 mark for diagram***
- ***1 mark for correct labelling***

b) Take moments about the pole:

$6T = (2 \times 10g) + (3 \times 20g) + (4 \times 10g)$

$6T = 20g + 60g + 40g = 120g$

So, $T = 20g$

[3 marks available in total]:
- ***1 mark for taking moments about the pole***
- ***1 mark for correct workings***
- ***1 mark for correct value of T***

Why work around the pole? Because we have no idea what the magnitude of R is and working there allows us to ignore it.

c) Resolve vertically:

$T + R = 10g + 20g + 10g$

$20g + R = 40g$

So, $R = 20g$

[2 marks available in total]:
- ***1 mark for resolving vertically***
- ***1 mark for correct value of R***

I suppose we could have found R first and then T, but that's not the order the questions are in, so it's best just to run with it...

4 a) Resolve horizontally:

$T\cos\theta = 72.5$ N

$\tan\theta = \frac{0.8}{1.7}$

$\Rightarrow \theta = 25.2°$

so $T = \frac{72.5}{0.9048} = 80.1$ N

[4 marks available in total]:
- ***1 mark for resolving horizontally***
- ***1 mark for calculating θ or cosθ***
- ***1 mark for correct workings***
- ***1 mark for correct value of T***

b) Taking moments about A:

$(1.2 \times 3g) + (2.4 \times mg) = 1.7 \times 80.13\sin 25.2°$

so $23.52m = 58.00 - 35.28 = 22.72$

and $m = 0.966$ kg (3 s.f)

[3 marks available in total]:
- ***1 mark for taking moments about A***
- ***1 mark for correct workings***
- ***1 mark for correct value of m***

c) Resolving vertically:

$F + 80.13\sin 25.2° = 3g + 0.9660g$

so $F = 38.87 - 34.12 = 4.75$ N (3 s.f)

[3 marks available in total]:
- ***1 mark for resolving vertically***
- ***1 mark for correct workings***
- ***1 mark for correct value of F***

5 a) Taking moments about A:

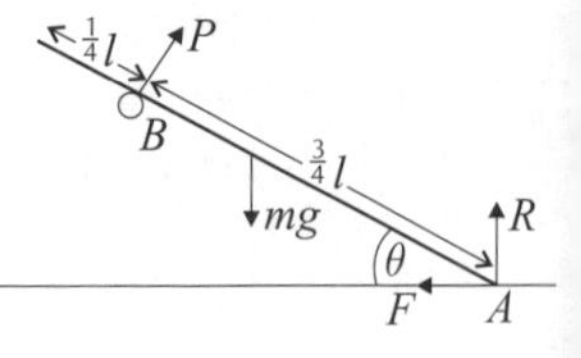

$\frac{1}{2}l \times mg\cos\theta = \frac{3}{4}l \times P$

so $P = \frac{\frac{1}{2}lmg\cos\theta}{\frac{3}{4}l} = \frac{2}{3}mg\cos\theta$

[3 marks available in total]:
- ***1 mark for taking moments about A***
- ***1 mark for correct workings***
- ***1 mark for correct answer***

Mechanics is harder with algebra than with numbers, but once you've mastered it, doing it with numbers will seem trivial. I know it's not nice, but at least it's useful.

b) $\sin\theta = \frac{3}{5}$ so $\cos\theta = \frac{4}{5}$

Using the result of part a) gives $P = \frac{8}{15}mg$

Resolving horizontally:

$F = P\sin\theta = \frac{8}{15}mg \times \frac{3}{5} = \frac{8}{25}mg$

Resolving vertically:

$R + P\cos\theta = mg$

so $R = mg - \frac{4}{5}P = mg - \frac{32}{75}mg = \frac{43}{75}mg$

Non-limiting equilibrium so $F \leq \mu R$

so $\frac{8}{25}mg \leq \mu\frac{43}{75}mg$ and $\mu \geq \frac{24}{43} = 0.56$

[6 marks available in total]:
- ***1 mark for finding P***
- ***1 mark for resolving horizontally***
- ***1 mark for resolving vertically***
- ***1 mark for using F ≤ μR***
- ***1 mark for correct workings***
- ***1 mark for correct value of μ***

That wasn't much fun, but it's good to get some use out of your calculator's fraction button. Unless you did it without a calculator, in which case, congratulations, you're officially 'hardcore'.

6 a) Taking moments about B:

$(mg\cos\theta \times 1.4) + (180\cos\theta \times 2.1)$

$= (490\sin\theta \times 4.2)$

Dividing by $\cos\theta$ gives:

$1.4mg + 378 = 2058\tan\theta = 2058 \times \frac{8}{11}$

so, $1.4mg = 1497 - 378 = 1119$

so $m = \frac{1119}{13.72} = 82$ kg (nearest kg)

[4 marks available in total]:
- ***1 mark for taking moments about B***
- ***1 mark for both sides of moments equation correct***
- ***1 mark for correct workings***
- ***1 mark for the correct value of m***

Answers

b) Resolving horizontally:
$F = 490$ N
Resolving vertically:
$R = 180 + 81.56g = 979.3$ N
As equilibrium is limiting, $F = \mu R$
so $979.3\mu = 490$ N and $\mu = 0.50$

[5 marks available in total]:
- ***1 mark for resolving horizontally***
- ***1 mark for resolving vertically***
- ***1 mark for using F = μR***
- ***1 mark for correct workings***
- ***1 mark for correct value of μ***

7 a)

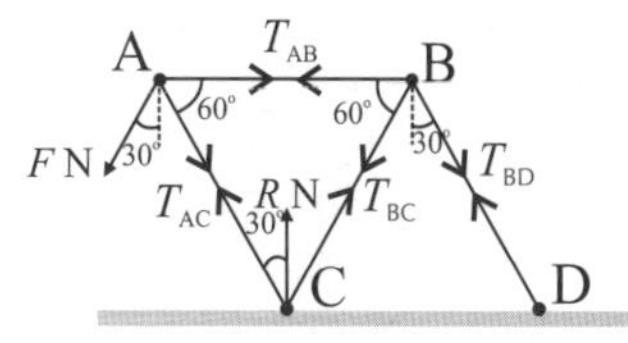

[1 mark]

b) From the diagram, all internal forces assumed to be tensions. Resolve at the joints to find internal forces:

At A: <u>Vertically</u>:
$F\cos30 + T_{AC}\sin60 = 0$
$\Rightarrow T_{AC} = -F\cos30 \div \sin60 = -F$ N.
<u>Horizontally</u>:
$F\sin30 = T_{AC}\cos60 + T_{AB}$
$\Rightarrow T_{AB} = F\sin30 - T_{AC}\cos60$
$= 0.5F - (-0.5F) = F$ N.

At C: <u>Horizontally</u>:
$T_{AC}\sin30 = T_{BC}\sin30$
$\Rightarrow T_{BC} = T_{AC} = -F$ N.
<u>Vertically</u>:
$T_{AC}\cos30 + T_{BC}\cos30 + R = 0$
$\Rightarrow -F\cos30 - F\cos30 + R = 0$
$\Rightarrow R = 2F\cos30 = \sqrt{3}F$ N.

At B: <u>Vertically</u>:
$T_{BC}\cos30 + T_{BD}\cos30 = 0$
$\Rightarrow T_{BD} = -T_{BC} = F$ N.

So T_{AC} and T_{BC} are compressions of F N,
T_{AB} and T_{BD} are tensions of F N and $R = \sqrt{3}F$ N.

[7 marks available in total]:
- ***1 mark for attempting to resolve and apply equilibrium conditions***
- ***1 mark each for T_{AB}, T_{BC}, T_{AC}, T_{BD} correct***
- ***1 mark for thrusts and tensions consistent with diagram***
- ***1 mark for finding R***

8 a) Resolving parallel to slope, taking up the slope as +ve:
$F_{net} = ma$
$8\cos15° + F - 7g\sin15° = 7 \times 0$
$F = 7g\sin15° - 8\cos15°$
$F = 10.03$ N
Resolving perpendicular to slope, taking ↖ as +ve:
$F_{net} = ma$
$R - 8\sin15° - 7g\cos15° = 7 \times 0$
$R = 8\sin15° + 7g\cos15° = 68.33$ N
Limiting friction:

$F = \mu R$, i.e. $10.03 = \mu \times 68.33$,
which gives $\mu = 0.15$ (2 d.p.)

[5 marks available in total]:
- ***1 mark for resolving parallel to slope***
- ***1 mark for correct value of F_{net} in this direction***
- ***1 mark for resolving perpendicular to slope***
- ***1 mark correct value of R***
- ***1 mark for correct value of μ***

b) 8 N removed:
Resolving parallel to the plane, taking down the plane as +ve:
$7g\sin15° - F = 7a$ ①
Resolving perpendicular to slope, taking ↖ as +ve:
$R - 7g\cos15° = 7 \times 0$
$R = 7g\cos15° = 66.26$ N
$F = \mu R$
$F = 0.147 \times 66.26 = 9.74$ N
①: $7g\sin15° - 9.74 = 7a$
$a = \frac{8.01}{7} = 1.14$ ms^{-2} (to 2 d.p.)
$s = 3$; $u = 0$; $a = 1.14$; $t = ?$ $\quad s = ut + \frac{1}{2}at^2$
$3 = 0 + \frac{1}{2} \times 1.14 \times t^2 \quad t = \sqrt{\frac{6}{1.14}} = 2.3$ s (to 2 s.f.)

[7 marks available in total]:
- ***1 mark for resolving parallel to slope***
- ***1 mark for resolving perpendicular to slope***
- ***1 mark for correct value of R***
- ***1 mark for correct value of F***
- ***1 mark for correct value of a***
- ***1 mark for appropriate calculation method for t***
- ***1 mark for correct value of t***

9

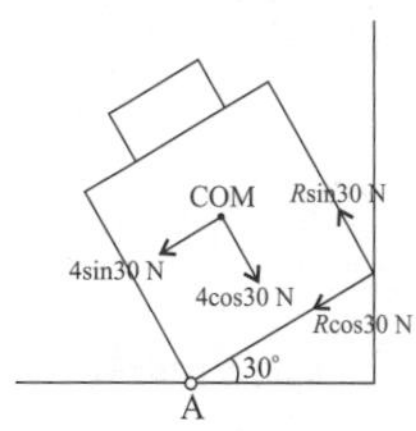

Resolve all forces into components parallel and perpendicular to the base of the shape, as shown above.
Taking moments about A:
$4\sin30 \times 2.28 + R\cos30 \times 0 + R\sin30 \times 4 = 4\cos30 \times 2$
$2 \times 2.28 + 2R = 8\cos30$
$\Rightarrow R = (8\cos30 \div 2) - 2.28 = 1.2$ N (2 s.f.)

[4 marks available in total — 1 mark for resolving, 1 mark for attempting to take moments, 1 mark for using correct values when taking moments, 1 mark for correct answer]

M2 Section 4 — Work and Energy
Warm-up Questions

1) Work done $= F \times s$
$= 250 \times 3 = 750$ J

Answers

2) Work done against gravity = mgh
   $34\,000 = m \times 9.8 \times 12$
   $m = 289$ kg (3 s.f.)

3) Kinetic Energy = $\frac{1}{2}mv^2$
   $= \frac{1}{2} \times 450 \times 13^2$
   $= 38\,025$ J = 38.0 kJ (3 s.f.)

4) Work done = Change in Kinetic Energy
   $800 = \frac{1}{2}m(v^2 - u^2)$
   $u = 0$ and $m = 65$, so:
   $v^2 = \frac{1600}{65} \Rightarrow v = 4.96$ ms^{-1} (3 s.f.)

5) Increase in G.P.E. = mg × increase in height
   $= 0.5 \times 9.8 \times 150$
   $= 735$ J

6) "If there are no external forces doing work on an object, the total mechanical energy of the object will remain constant." You usually need to model the object as a particle, because you have to assume that the object is not acted on by any external forces such as air resistance.

7) When the hat reaches its maximum height, its velocity will be zero. Using conservation of energy:
   Change in G.P.E. = change in K.E.
   $mgh = \frac{1}{2}m(u^2 - v^2)$
   Cancel m from both sides, and substitute $u = 5$, $v = 0$ and $g = 9.8$:
   $9.8h = \frac{1}{2} \times 25 \Rightarrow h = 1.28$ m (3 s.f.)

8) "The work done on an object by external forces is equal to the change in the total mechanical energy of that object." An external force is any force other than an object's weight.

9) Power of engine = driving force × velocity
   $350\,000 = F \times 22 \Rightarrow F = 15\,900$ N (3 s.f.)

Exam Questions

1 a) Increase in Kinetic Energy = $\frac{1}{2}m(v^2 - u^2)$ ***[1 mark]***
   $= \frac{1}{2} \times 90 \times (6^2 - 4^2) = 900$ J ***[1 mark]***
   Increase in Gravitational Potential Energy = mgh ***[1 mark]***
   $= 90 \times 9.8 \times 28\sin30° = 12\,348$ J ***[1 mark]***
   Increase in total Energy = Increase in K.E. + Increase in G.P.E.
   $= 900 + 12\,348 = 13\,248$ J = 13.2 kJ (3 s.f.) ***[1 mark]***

b) Using the work-energy principle:
   Work done on skier = Change in total energy ***[1 mark]***
   $(L - 66) \times 28 = 13\,248$ ***[1 mark]***
   $L = \frac{13\,248 + (66 \times 28)}{28} = 539$ N (3 s.f.) ***[1 mark]***

2 a) K.E. $= \frac{1}{2}mv^2 = \frac{1}{2} \times 0.3 \times 20^2$ ***[1 mark]***
   $= 60$ J ***[1 mark]***

b) Only force acting on the stone is its weight, so use conservation of mechanical energy:
   Change in K.E. = Change in G.P.E. ***[1 mark]***
   $60 - 0 = 0.3 \times 9.8 \times h$ ***[1 mark]***
   $h = 20.4$ m (3 s.f.) ***[1 mark]***

c) Stone's change in K.E. after hitting the water:
   $\frac{1}{2} \times 0.3 \times 1^2 - 60 = -59.85$ J ***[1 mark]***
   Call the depth the stone has sunk x m.
   Change in G.P.E. after hitting the water:
   $-mgx = -2.94x$ ***[1 mark]***
   Work done on the stone by resistive force
   $= Fs = -23x$ ***[1 mark]***
   By the work-energy principle:
   Work done on the stone = Change in total energy ***[1 mark]***
   $-23x = -59.85 - 2.94x$
   Rearrange to find x:
   $x = 2.98$ m (3 s.f) ***[1 mark]***

3 a)

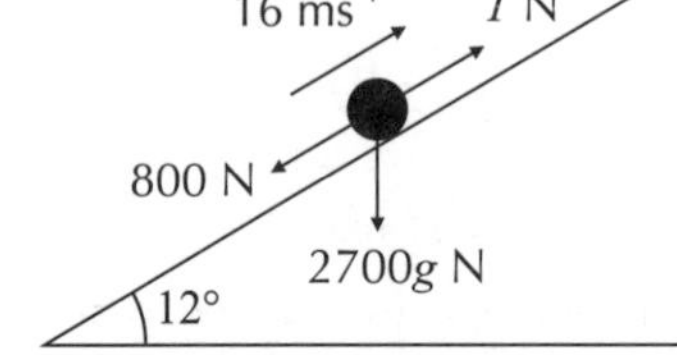

Resolving parallel to the slope using $F = ma$ with $a = 0$:
   $T - 800 - 2700g\sin12° = 0$ ***[1 mark]***
   So, $T = 6301$ N ***[1 mark]***
   Power of engine = Driving Force × Velocity ***[1 mark]***
   $= 6301 \times 16 = 101$ kW (3 s.f.) ***[1 mark]***

b) Work done by resistive force to stop van = $-800x$ ***[1 mark]***
   Change in total energy = Change in G.P.E. + Change in K.E.
   $= (2700 \times g \times x\sin12°) - (\frac{1}{2} \times 2700 \times 16^2)$ ***[1 mark]***
   By work-energy principle,
   $-800x = 2700gx\sin12° - 345\,600$ ***[1 mark]***
   Rearrange to find x:
   $x = \frac{345\,600}{2700g\sin 12° + 800} = 54.8$ m (3 s.f.) ***[1 mark]***

c) Resolve parallel to the slope using $F = ma$ to find a:
   $-800 - 2700g\sin12° = 2700a$ ***[1 mark]***
   $a = -2.334$ ms^{-2} ***[1 mark]***
   Use $v = u + at$ to find the time taken to come to rest:
   $0 = 16 - 2.334t$ ***[1 mark]***
   $t = \frac{16}{2.334} = 6.86$ s (3 s.f.) ***[1 mark]***

4 a) Work done = Force × distance moved
   $= 800\cos40° \times 320 = 196$ kJ (3 s.f.)

   [3 marks available in total]:
   - ***1 mark for using the horizontal component of the force***
   - ***1 mark for correct use of formula for work done***
   - ***1 mark for correct final answer.***

b)

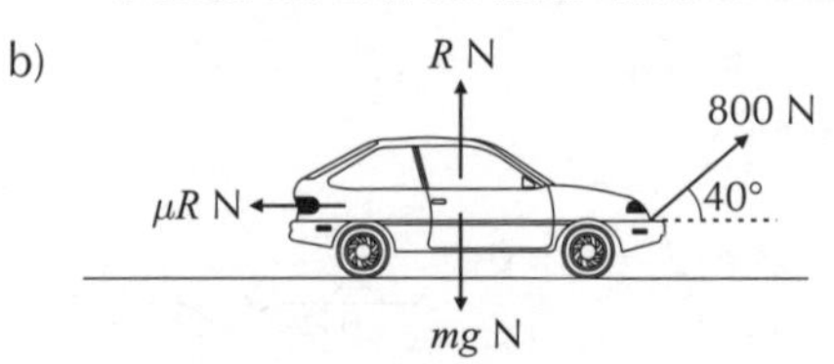

No acceleration vertically, so:
   $R + 800\sin40° = mg$
   $R = 1500g - 800\sin40° = 14\,190$ N ***[1 mark]***
   Car is moving only horizontally, so:

Answers

Work done = change in kinetic energy ***[1 mark]***

$(800\cos40° - \mu R) \times 320 = \frac{1}{2} \times 1500 \times (16^2 - 11^2)$ ***[1 mark]***

Rearrange to find μ:

$\mu = \frac{196\,107 - 101\,250}{4\,541\,000} = 0.0209$ (3 s.f.) ***[1 mark]***

5 a) Use the work rate to find the 'driving' force, F of the cyclist:

$250 = F \times 4$ ***[1 mark]***

$F = 62.5$ N

Resolve parallel to the slope: ***[1 mark]***

$62.5 - 35 - 88g\sin\alpha = 0$ ***[1 mark]***

$\alpha = \sin^{-1}\frac{27.5}{88g} = 1.83°$ (3 s.f.) ***[1 mark]***

b) Use the new work rate to find the new 'driving' force, F':

$370 = F' \times 4$ ***[1 mark]***

$F' = 92.5$ N

Resolve parallel to the slope to find a: ***[1 mark]***

$92.5 - 35 - 88g\sin\alpha = 88a$ ***[1 mark]***

$a = 0.341$ ms^{-2} (3 s.f.) ***[1 mark]***

6 a) The system is in equilibrium, so resolving forces vertically gives $T = mg$, where T is tension in the string. ***[1 mark]***

So, using Hooke's Law:

$T = \frac{\lambda}{l}e = mg$ ***[1 mark]***

$\Rightarrow \lambda = \frac{mgl}{e} = \frac{3 \times 9.8 \times 2}{(5-2)} = 19.6$ N ***[1 mark]***

b) E.P.E. $= \frac{\lambda}{2l}e^2$ ***[1 mark]***

So $\frac{19.6}{2 \times 2} \times 3^2 = 44.1$ J ***[1 mark]***

c) By the Principle of Conservation of Mechanical Energy:

E.P.E. lost = G.P.E. gained + K.E. gained ***[1 mark]***

Assuming that block starts with no K.E. or G.P.E.,

$\frac{\lambda}{2l}(8-2)^2 - \frac{\lambda}{2l}(3-2)^2 = mg(8-3) + \frac{1}{2}mv^2$ ***[1 mark]***

$\frac{19.6}{4}(36) - \frac{19.6}{4}(1) = (3 \times 9.8 \times 5) + \left(\frac{1}{2} \times 3 \times v^2\right)$

[1 mark]

$v^2 = \frac{2}{3}(171.5 - 147) = 16.33$

$v = 4.04$ ms^{-1} (3 s.f.) ***[1 mark]***

7 a) E.P.E. $= \frac{\lambda}{2l}e^2$ ***[1 mark]***

$= \frac{50}{2 \times 5} \times (d-5)^2 = 5(d-5)^2$ ***[1 mark]***

b)

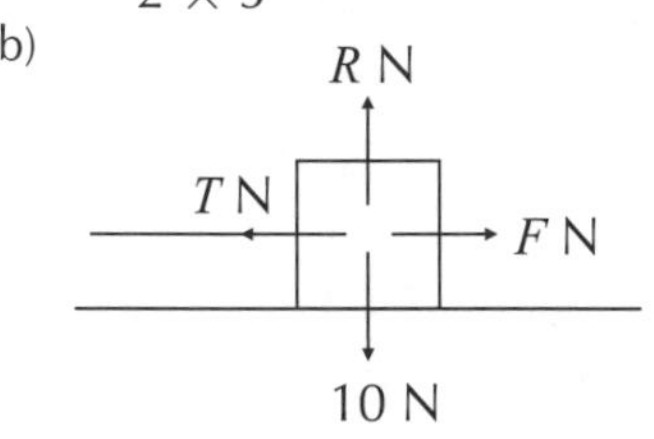

There is no motion vertically, so $R = 10$ N ***[1 mark]***

Block is moving, so friction is limiting $\Rightarrow$

frictional force, $F = \mu R = 0.5 \times 10 = 5$ N ***[1 mark]***

Work done by $F = F \times$ distance moved $= -5d$ ***[1 mark]***.

There is no change in K.E. or G.P.E. between the start and end of motion, so the only change in mechanical energy is the loss of E.P.E. ***[1 mark]***

So, by the work-energy principle:

Change in E.P.E. = Work done on block by friction

$\Rightarrow -5(d-5)^2 = -5d$ ***[1 mark]***

The change in E.P.E. is negative because E.P.E. is lost

Rearranging: $d = d^2 - 10d + 25 \Rightarrow d^2 - 11d + 25 = 0$.

Solve using the quadratic formula: ***[1 mark]***

$d = 7.79$ or $d = 3.21$ (each to 3 s.f.)

Question says that $d > 5$, so take $d = 7.79$ m. ***[1 mark]***

8 a) Find T, the driving force of the car, using Power = Tv:

T = Power $\div v = 20\,000 \div 10 = 2000$ N. ***[1 mark]***

Resolve forces parallel to the slope:

$T - mg\sin\theta - kv = 0$ ***[1 mark]***.

$\Rightarrow 2000 - (1000 \times 9.8 \times 0.1) - 10k = 0$

$\Rightarrow 10k = 1020$

$\Rightarrow k = 102$ ***[1 mark]***

b) (i) Call the new driving force F. Using Power = Fv:

$F = \frac{50\,000}{u}$ ***[1 mark]***

Resolve forces parallel to the slope:

$\frac{50\,000}{u} - mg\sin\theta - 102u = 0$. ***[1 mark]***

Rearranging and substituting known values gives:

$50\,000 - (1000 \times 9.8 \times 0.1)u - 102u^2 = 0$ ***[1 mark]***.

This rearranges to:

$102u^2 + 980u - 50\,000 = 0$ — as required ***[1 mark]***.

(ii) Solve for u using the quadratic formula:

$u = \frac{-980 + \sqrt{980^2 - (4 \times 102 \times -50\,000)}}{2 \times 102}$

$u = 17.9$ ms^{-1} (3 s.f.)

[2 marks available in total]:

- ***1 mark for correctly using quadratic formula***
- ***1 mark for correct value for u***

You don't need to worry about the negative part of ± in the formula, as you're after a speed — which is always positive.

c) Using $F = ma$ gives $T - 102v = ma$. ***[1 mark]***

This time, $T = P \div v = 21000 \div 12$, and so:

$(21000 \div 12) - (102 \times 12) = 1000a$ ***[1 mark]***

$\Rightarrow a = 0.526$ ms^{-2} ***[1 mark]***

M2 Section 5 — Collisions

Warm-up Questions

1 a) $(5 \times 3) + (4 \times 1) = (5 \times 2) + (4 \times v)$

$19 = 10 + 4v$

$v = 2\frac{1}{4}$ ms^{-1} to the right

b) $(5 \times 3) + (4 \times 1) = 9v$

$19 = 9v$

$v = 2\frac{1}{9}$ ms^{-1} to the right

c) $(5 \times 3) + (4 \times -2) = (5 \times -v) + (4 \times 3)$

$7 = -5v + 12$

$5v = 5$

$v = 1$ ms^{-1} to the left

d) $(m \times 6) + (8 \times 2) = (m \times 2) + (8 \times 4)$

$6m + 16 = 2m + 32$

$4m = 16$

$m = 4$ kg

Answers

Collision questions have me bouncing off the ceiling... Be careful with your directions (positive and negative) and it'll all be okay.

2 Impulse acts against motion, so $I = -2$ Ns
$I = mv - mu$
$-2 = 0.3v - (0.3 \times 5)$
$v = -1\frac{2}{3}$ ms^{-1}
Impulse has 2 different equations, I = mv − mu and I = Ft. Remember both.

3 You need to find the particle's velocities just before and just after impact. Falling (down = +ve):

$u = 0$, $s = 2$, $a = 9.8$, $v = ?$
$v^2 = u^2 + 2as$
$v = \sqrt{2 \times 9.8 \times 2}$
$v = 6.261 \text{ms}^{-1}$

Rebound (this time, let up = +ve):

$v = 0$, $u = ?$, $a = -9.8$, $s = 1\frac{1}{3}$
$v^2 = u^2 + 2as$
$0 - u^2 = 2 \times -9.8 \times 1\frac{1}{3}$
$u = 5.112 \text{ms}^{-1}$

Taking up = +ve:
Impulse = $mv - mu$
= $(0.45 \times 5.112) - (0.45 \times -6.261)$
= 5.12 Ns

4 a) $\mathbf{I} = m\mathbf{v} - m\mathbf{u}$, so
$2\mathbf{i} + 5\mathbf{j} = 0.1\mathbf{v} - 0.1(\mathbf{i} + \mathbf{j})$
$2\mathbf{i} + 5\mathbf{j} = 0.1\mathbf{v} - 0.1\mathbf{i} - 0.1\mathbf{j}$
$0.1\mathbf{v} = 2\mathbf{i} + 5\mathbf{j} + 0.1\mathbf{i} + 0.1\mathbf{j} = 2.1\mathbf{i} + 5.1\mathbf{j}$
$\Rightarrow \mathbf{v} = 21\mathbf{i} + 51\mathbf{j}$.

b) $-3\mathbf{i} + \mathbf{j} = 0.1\mathbf{v} - 0.1\mathbf{i} - 0.1\mathbf{j}$
$0.1\mathbf{v} = -3\mathbf{i} + \mathbf{j} + 0.1\mathbf{i} + 0.1\mathbf{j} = -2.9\mathbf{i} + 1.1\mathbf{j}$
$\Rightarrow \mathbf{v} = -29\mathbf{i} + 11\mathbf{j}$.

c) $-\mathbf{i} - 6\mathbf{j} = 0.1\mathbf{v} - 0.1\mathbf{i} - 0.1\mathbf{j}$
$0.1\mathbf{v} = -\mathbf{i} - 6\mathbf{j} + 0.1\mathbf{i} + 0.1\mathbf{j} = -0.9\mathbf{i} - 5.9\mathbf{j}$
$\Rightarrow \mathbf{v} = -9\mathbf{i} - 59\mathbf{j}$.

d) $4\mathbf{i} = 0.1\mathbf{v} - 0.1\mathbf{i} - 0.1\mathbf{j}$
$0.1\mathbf{v} = 4\mathbf{i} + 0.1\mathbf{i} + 0.1\mathbf{j} = 4.1\mathbf{i} + 0.1\mathbf{j}$
$\Rightarrow \mathbf{v} = 41\mathbf{i} + \mathbf{j}$.

5 a) $\mathbf{I} = m\mathbf{v} - m\mathbf{u}$, so
$\mathbf{Q} = 2(-2\mathbf{i} + \mathbf{j}) - 2(4\mathbf{i} - \mathbf{j}) = -4\mathbf{i} + 2\mathbf{j} - 8\mathbf{i} + 2\mathbf{j}$
$\mathbf{Q} = -12\mathbf{i} + 4\mathbf{j}$.

b)

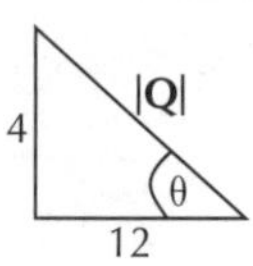

$|\mathbf{Q}| = \sqrt{(-12)^2 + 4^2} = 12.6$ Ns, to 3 s.f.

c) $\theta = \tan^{-1}\left(\frac{4}{12}\right)$
Required angle = $180 - \theta = 162°$ (3 s.f.)

6 a) $m_A\mathbf{u}_A + m_B\mathbf{u}_B = m_A\mathbf{v}_A + m_B\mathbf{v}_B$, so:
$0.5(2\mathbf{i} + \mathbf{j}) + 0.4(-\mathbf{i} - 4\mathbf{j}) = 0.5(-\mathbf{i} - 2\mathbf{j}) + 0.4\mathbf{v}_B$
$\mathbf{i} + 0.5\mathbf{j} - 0.4\mathbf{i} - 1.6\mathbf{j} = -0.5\mathbf{i} - \mathbf{j} + 0.4\mathbf{v}_B$
$0.4\mathbf{v}_B = 1.1\mathbf{i} - 0.1\mathbf{j}$
$\Rightarrow \mathbf{v}_B = 2.75\mathbf{i} - 0.25\mathbf{j}$.
Speed = $|\mathbf{v}_B| = \sqrt{2.75^2 + 0.25^2} = 2.76$ ms^{-1}, to 3 s.f.

b) If they coalesce:
$0.5(2\mathbf{i} + \mathbf{j}) + 0.4(-\mathbf{i} - 4\mathbf{j}) = (0.5 + 0.4)\mathbf{v}$
$\mathbf{i} + 0.5\mathbf{j} - 0.4\mathbf{i} - 1.6\mathbf{j} = 0.9\mathbf{v}$
$0.9\mathbf{v} = 0.6\mathbf{i} - 1.1\mathbf{j} \Rightarrow \mathbf{v} = \frac{2}{3}\mathbf{i} - \frac{11}{9}\mathbf{j}$.
Speed = $|\mathbf{v}| = \sqrt{\left(\frac{2}{3}\right)^2 + \left(-\frac{11}{9}\right)^2} = 1.39$ ms^{-1}, to 3 s.f.

7 $e = \frac{v_2 - v_1}{u_1 - u_2} \Rightarrow 0.3 = \frac{v_2 - v_1}{3 - 0} \Rightarrow v_2 - v_1 = 0.9$ ...*[1]*
And: $m_1u_1 + m_2u_2 = m_1v_1 + m_2v_2$
$\Rightarrow (2 \times 3) + (3 \times 0) = 2v_1 + 3v_2$
$\Rightarrow 6 = 2v_1 + 3v_2 \Rightarrow v_1 + 1.5v_2 = 3$ ...*[2]*
Equation *[1]* + equation *[2]* gives:
$2.5v_2 = 3.9 \Rightarrow v_2 = 1.56$ ms^{-1}.
In equation *[1]*:
$1.56 - v_1 = 0.9 \Rightarrow v_1 = 1.56 - 0.9 = 0.66$ ms^{-1}.

Loss of K.E. = $(\frac{1}{2}m_1u_1^2 + \frac{1}{2}m_2u_2^2) - (\frac{1}{2}m_1v_1^2 + \frac{1}{2}m_2v_2^2)$
$= [(\frac{1}{2} \times 2 \times 3^2) + 0] - [(\frac{1}{2} \times 2 \times 0.66^2) + (\frac{1}{2} \times 3 \times 1.56^2)]$
$= 9 - 4.086 = 4.914$ J.

8 Call the particles A and B. If $u_A = u$ then $u_B = -u$ (as it's going in the opposite direction at the same speed). After the collision, $v_A = 0$ and $v_B = \frac{u}{2}$ (as it's going in the opposite direction to its original motion at half the speed).

$$e = \frac{\text{speed of separation of particles}}{\text{speed of approach of particles}} = \frac{v_B - v_A}{u_A - u_B}$$

$$\Rightarrow e = \frac{\frac{u}{2} - 0}{u - (-u)} = \frac{\frac{u}{2}}{2u} = \frac{u}{4u} = \frac{1}{4}.$$

9 a) For collision with a plane surface, $e = \frac{v}{u}$, so rebound speed $v = eu \Rightarrow v = 0.4 \times 10 = 4$ ms^{-1}.

b) Call the particles A and B, so

$$e = \frac{\text{speed of separation of particles}}{\text{speed of approach of particles}} = \frac{v_B - v_A}{u_A - u_B}$$

$\Rightarrow 0.4 = \frac{v_B - v_A}{10 - (-12)} \Rightarrow v_B - v_A = 0.4 \times 22$
$\Rightarrow v_B - v_A = 8.8$ ...*[1]*

Using the conservation of momentum:
$m_Au_A + m_Bu_B = m_Av_A + m_Bv_B$

$(1 \times 10) + (2 \times -12) = (1 \times v_A) + (2 \times v_B)$
$10 - 24 = v_A + 2v_B \Rightarrow v_A + 2v_B = -14$ ...*[2]*

Equation *[1]* + equation *[2]* gives:
$3v_B = -5.2 \Rightarrow v_B = -1.7333...$ ms^{-1}.

Substituting in equation *[1]* gives:
$-1.7333... - v_A = 8.8$
$\Rightarrow v_A = -1.7333... - 8.8 = -10.5333...$ ms^{-1}.
So, to 3 s.f., the original particle's rebound speed is 10.5 ms^{-1}.

10 Surface is parallel to $\mathbf{i}$, so only the $\mathbf{j}$ component of velocity will change:
$\mathbf{v} = 4\mathbf{i} - e(-1)\mathbf{j} = 4\mathbf{i} + 0.5\mathbf{j}$.

Answers

Kinetic energy lost = $\frac{1}{2}m|\mathbf{v}|^2 - \frac{1}{2}m|\mathbf{u}|^2$

$= \frac{1}{2}m(4^2 + (-1)^2) - \frac{1}{2}m(4^2 + 0.5^2) = \frac{1}{2}(2)(17 - 16.25) = 0.75$ J

11 For the first collision, between A and B:

$e = \frac{v_B - v_A}{u_A - u_B} \Rightarrow \frac{1}{4} = \frac{v_B - v_A}{3u - 2u} \Rightarrow v_B - v_A = \frac{u}{4}$ ...*[1]*

And:

$m_A u_A + m_B u_B = m_A v_A + m_B v_B$

$(1 \times 3u) + (4 \times 2u) = (1 \times v_A) + (4 \times v_B)$

$3u + 8u = v_A + 4v_B \Rightarrow v_A + 4v_B = 11u$ ...*[2]*

Equation *[1]* + equation *[2]* gives:

$5v_B = 11u + \frac{u}{4} \Rightarrow 5v_B = \frac{45u}{4} \Rightarrow v_B = \frac{9u}{4}$.

Substituting in equation *[2]* gives:

$v_A + 9u = 11u \Rightarrow v_A = 11u - 9u = 2u$.

For the second collision, between B and C:

$e = \frac{v_C - v_B}{u_B - u_C} \Rightarrow \frac{1}{3} = \frac{v_C - v_B}{\frac{9u}{4} - u} \Rightarrow v_C - v_B = \frac{5u}{12}$ ...*[3]*

And:

$m_B u_B + m_C u_C = m_B v_B + m_C v_C$

$(4 \times \frac{9u}{4}) + (5 \times u) = (4 \times v_B) + (5 \times v_C)$

$\Rightarrow 4v_B + 5v_C = 14u$ ...*[4]*

4 × Equation *[3]* + equation *[4]* gives:

$9v_C = \frac{5u}{3} + 14u \Rightarrow 9v_C = \frac{47u}{3} \Rightarrow v_C = \frac{47u}{27}$.

Substituting in equation *[3]* gives:

$\frac{47u}{27} - v_B = \frac{5u}{12} \Rightarrow v_B = \frac{47u}{27} - \frac{5u}{12} = \frac{143u}{108}$.

So after both collisions:

A is travelling at $2u = \frac{216u}{108}$,

and B is travelling at $\frac{143u}{108}$,

which means that A is travelling faster than B and so they should collide again.

12 Bounce 1:

Using $v^2 = u^2 + 2as$, where $v = u_1$, $u = 0$, $a = 9.8$ and $s = 1$:

$u_1^2 = 2 \times 9.8 \times 1 = 19.6 \Rightarrow u_1 = \sqrt{19.6} = 4.4271...$

Using $e = \frac{v}{u}$, $v = eu$, where $v = v_1$, $e = 0.5$ and $u = u_1$:

$v_1 = 0.5 \times 4.4271... = 2.2135...$

Then $v^2 = u^2 + 2as$, where $v = 0$, $u = v_1$ and $a = -9.8$:

$0 = 2.2135...^2 + (2 \times -9.8)s_1 \Rightarrow s_1 = \frac{2.2135...^2}{2 \times 9.8} = 0.25$ m.

Bounce 2:

From the symmetry of the vertical motion,

$u_2 = v_1 = 2.2135...$

$v = eu$, where $v = v_2$, $e = 0.5$ and $u = u_2$:

$v_2 = 0.5 \times 2.2135... = 1.1067...$

$v^2 = u^2 + 2as$, where $v = 0$, $u = v_2$ and $a = -9.8$:

$0 = 1.1067...^2 + (2 \times -9.8)s_2 \Rightarrow s_2 = \frac{1.1067...^2}{2 \times 9.8} = 0.0625$ m.

Bounce 3:

From the symmetry of the vertical motion,

$u_3 = v_2 = 1.1067...$

$v = eu$, where $v = v_3$, $e = 0.5$ and $u = u_3$:

$v_2 = 0.5 \times 1.1067... = 0.5533...$

$v^2 = u^2 + 2as$, where $v = 0$, $u = v_3$ and $a = -9.8$:

$0 = 0.5533...^2 + (2 \times -9.8)s_3$

$\Rightarrow s_3 = \frac{0.5533...^2}{2 \times 9.8} = 0.015625$ m.

Exam Questions

1 a) $\mathbf{I} = m\mathbf{v} - m\mathbf{u}$, so

$3\mathbf{i} - 8\mathbf{j} = 0.4\mathbf{v} - 0.4(-6\mathbf{i} + \mathbf{j})$ ***[1 mark]***

$3\mathbf{i} - 8\mathbf{j} = 0.4\mathbf{v} + 2.4\mathbf{i} - 0.4\mathbf{j}$

$0.4\mathbf{v} = 0.6\mathbf{i} - 7.6\mathbf{j}$ ***[1 mark]***

$\Rightarrow \mathbf{v} = 1.5\mathbf{i} - 19\mathbf{j}$ ***[1 mark]***.

Speed is the magnitude of the velocity.

Drawing this as a right-angled triangle:

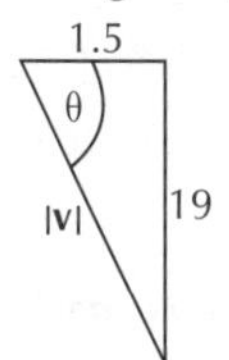

$|\mathbf{v}| = \sqrt{1.5^2 + 19^2}$ ***[1 mark]*** = 19.1 ms⁻¹ to 3 s.f. ***[1 mark]***.

As always, a picture makes everything make a lot more sense.

b) Using the triangle in part a), θ is the angle with the horizontal, so:

$\theta = \tan^{-1}\left(\frac{19}{1.5}\right)$ ***[1 mark]*** = 85.5° to 3 s.f. ***[1 mark]***.

2 Using the principle of conservation of momentum for the collision:

$m_1u_1 + m_2u_2 = m_1v_1 + m_2v_2$

Since marble 2 is stationary before the impact:

$(0.02 \times 2) + (0.06 \times 0) = 0.02v_1 + 0.06v_2$ ***[1 mark]***

$\Rightarrow 0.02v_1 + 0.06v_2 = 0.04$

$\Rightarrow v_1 + 3v_2 = 2$ ...*[1]*

Since the collision is perfectly elastic, and so $e = 1$, the Law of Restitution gives a second equation:

$e = \frac{\text{speed of separation of particles}}{\text{speed of approach of particles}} = \frac{v_2 - v_1}{u_1 - u_2}$

$\Rightarrow 1 = \frac{v_2 - v_1}{2 - 0}$ ***[1 mark]*** $\Rightarrow v_2 - v_1 = 2$ ...*[2]*

Equation *[1]* + equation *[2]* gives:

$4v_2 = 4 \Rightarrow v_2 = 1$ ms⁻¹ ***[1 mark]***.

Substituting in equation *[1]* gives:

$v_1 + (3 \times 1) = 2 \Rightarrow v_1 = -1$ ms⁻¹ ***[1 mark]***.

So after the collision, both particles are travelling at a speed of 1 ms⁻¹ (but the first particle is going in the opposite direction to its initial path).

3 a) Using the Law of Restitution for the collision between P and Q, where P is travelling at u and Q at $-u$ (i.e. in the opposite direction):

Answers

$e = \frac{v_Q - v_P}{u_P - u_Q} \Rightarrow \frac{3}{4} = \frac{v_Q - v_P}{u - (-u)}$ ***[1 mark]*** $\Rightarrow \frac{3}{4} = \frac{v_Q - v_P}{2u}$

$\Rightarrow v_Q - v_P = \frac{3u}{2}$ ...*[1]*

Using conservation of momentum:

$m_P u_P + m_Q u_Q = m_P v_P + m_Q v_Q$

$2mu - mu = 2mv_P + mv_Q$ ***[1 mark]***

$\Rightarrow 2v_P + v_Q = u$ ...*[2]*

Equation *[2]* – equation *[1]* gives:

$3v_P = -\frac{u}{2} \Rightarrow v_P = -\frac{u}{6}$ ***[1 mark]***.

Substituting in equation *[1]* gives:

$v_Q - (-\frac{u}{6}) = \frac{3u}{2} \Rightarrow v_Q = \frac{4u}{3}$ ***[1 mark]***.

Since P's velocity was initially positive, and is now negative, and Q's was initially negative but is now positive, the collision has reversed the directions of both particles ***[1 mark]***.

Sure about that? Yep, positive. I mean negative... erm...

$|v_Q| \div |v_P| = \frac{4u}{3} \div \frac{u}{6} = 8$,

so Q is now going 8 times faster than P ***[1 mark]***.

b) For the collision with the wall, $e_{wall} = \frac{\text{speed of rebound}}{\text{speed of approach}}$.

Q approaches the wall with a speed of $\frac{4u}{3}$ (from a)), so if v_{Qwall} is its rebound speed:

$e_{wall} = \frac{v_{Qwall}}{\frac{4u}{3}} \Rightarrow v_{Qwall} = \frac{4ue_{wall}}{3}$ ***[1 mark]***.

Since Q collides again with P, v_{Qwall} must be greater than v_P which is $\frac{u}{6}$ (from a)), so:

$\frac{4ue_{wall}}{3} > \frac{u}{6}$ ***[1 mark]*** $\Rightarrow e_{wall} > \frac{3u}{6 \times 4u} \Rightarrow e_{wall} > \frac{1}{8}$ ***[1 mark]***.

c) If $e_{wall} = \frac{3}{5}$, then (from b)):

$v_{Qwall} = \frac{4ue_{wall}}{3} = \frac{4u \times 3}{3 \times 5} = \frac{4u}{5}$ ***[1 mark]***.

Q is now travelling in the same direction as P, which is still travelling at a speed of $\frac{u}{6}$ (from a)), and the particles have a coefficient of restitution of $\frac{3}{4}$, so using the Law of Restitution for the second collision between P and Q:

$e = \frac{v_P - v_Q}{u_Q - u_P} \Rightarrow \frac{3}{4} = \frac{v_P - v_Q}{\left(\frac{4u}{5}\right) - \frac{u}{6}}$ ***[1 mark]*** $\Rightarrow \frac{3}{4} = \frac{v_P - v_Q}{\frac{19u}{30}}$

$\Rightarrow v_P - v_Q = \frac{19u}{40}$ ...*[1]*

Using conservation of momentum:

$m_Q u_Q + m_P u_P = m_Q v_Q + m_P v_P$

$\frac{4um}{5} + \frac{2um}{6} = mv_Q + 2mv_P$ ***[1 mark]***

$\Rightarrow v_Q + 2v_P = \frac{17u}{15}$ ...*[2]*

Equation *[1]* + equation *[2]* gives:

$3v_P = \frac{193u}{120} \Rightarrow v_P = \frac{193u}{360}$ ***[1 mark]***.

Substituting in equation *[1]* gives:

$\frac{193u}{360} - v_Q = \frac{19u}{40} \Rightarrow v_Q = \frac{193u}{360} - \frac{19u}{40} = \frac{22u}{360}$ ***[1 mark]***.

Since $v_Q = 0.22$ ms^{-1}:

$\frac{22u}{360} = 0.22$ ***[1 mark]***

$\Rightarrow u = (0.22 \times 360) \div 22 = 3.6$ ms^{-1} ***[1 mark]***.

4 a) Using the Law of Restitution for the collision between particles 1 and 2 gives:

$e = \frac{v_2 - v_1}{u_1 - u_2} \Rightarrow \frac{1}{4} = \frac{v_2 - v_1}{3u - 2u}$ ***[1 mark]***

$\Rightarrow v_2 - v_1 = \frac{u}{4}$ ...*[1]*

Using conservation of momentum:

$m_1 u_1 + m_2 u_2 = m_1 v_1 + m_2 v_2$

$(2m \times 3u) + (3m \times 2u) = 2mv_1 + 3mv_2$ ***[1 mark]***

$\Rightarrow 2v_1 + 3v_2 = 12u$ ...*[2]*

Equation *[1]* × 2 gives:

$2v_2 - 2v_1 = \frac{u}{2}$ ...*[3]*

Equation *[2]* + equation *[3]* gives:

$5v_2 = 12u + \frac{u}{2} \Rightarrow v_2 = \frac{25u}{2 \times 5} = \frac{5u}{2}$ ***[1 mark]***.

Substituting in equation *[1]* gives:

$\frac{5u}{2} - v_1 = \frac{u}{4}$

$\Rightarrow v_1 = \frac{5u}{2} - \frac{u}{4} = \frac{9u}{4}$ ***[1 mark]***.

b) Loss of kinetic energy =

$(\frac{1}{2}m_1 u_1^2 + \frac{1}{2}m_2 u_2^2) - (\frac{1}{2}m_1 v_1^2 + \frac{1}{2}m_2 v_2^2)$

$= [(\frac{1}{2} \times 2m \times (3u)^2) + (\frac{1}{2} \times 3m \times (2u)^2)] -$

$[(\frac{1}{2} \times 2m \times (\frac{9u}{4})^2) + (\frac{1}{2} \times 3m \times (\frac{5u}{2})^2)]$

$= (9mu^2 + 6mu^2) - (\frac{81mu^2}{16} + \frac{150mu^2}{16})$

$= (15 - \frac{231}{16})mu^2 = \frac{9mu^2}{16}$.

[4 marks available — 1 mark for correct values in formula for initial kinetic energy, 1 mark for correct values in formula for final kinetic energy, 1 mark for correct calculation of initial and final energy and 1 mark for correct final answer as the difference between the two.]

5 a) For the collision between A and B, the Law of Restitution gives the following equation:

$e = \frac{v_B - v_A}{u_A - u_B} \Rightarrow e = \frac{v_B - v_A}{4u - 0}$ ***[1 mark]***

$\Rightarrow v_B - v_A = 4ue$ ...*[1]*

Using conservation of momentum:

$m_A u_A + m_B u_B = m_A v_A + m_B v_B$

$4mu + 0 = mv_A + 2mv_B$ ***[1 mark]***

$\Rightarrow v_A + 2v_B = 4u$ ...*[2]*

Equation *[1]* + equation *[2]* gives:

$3v_B = 4u(1 + e) \Rightarrow v_B = \frac{4u}{3}(1 + e)$ ***[1 mark]***.

Substituting in equation *[1]* gives:

$\frac{4u}{3}(1 + e) - v_A = 4ue$

$\Rightarrow v_A = \frac{4u}{3}(1 + e) - 4ue = \frac{4u}{3}(1 - 2e)$ ***[1 mark]***.

Since the coefficient of restitution must be between 0 and 1, and the coefficient of restitution between B and C is $2e$, then $0 \le 2e \le 1 \Rightarrow 1 - 2e \ge 0$ ***[1 mark]***.

i.e. $v_A = \frac{4u}{3}(1 - 2e)$, where $u > 0$ and $1 - 2e \ge 0$. So: v_A cannot be negative ***[1 mark]***, so the collision does not reverse the direction of A's motion ***[1 mark]***.

b) After the collision, A is travelling at $\frac{4u}{3}(1 - 2e)$ and B is travelling at $\frac{4u}{3}(1 + e)$ (from a)). In the time it takes B to travel a distance d, A has travelled $\frac{d}{4}$. So the speed of B must be 4 times the speed of A ***[1 mark]*** i.e.

Answers

$\frac{4u}{3}(1-2e) = \frac{u}{3}(1+e)$ ***[1 mark]***

$\Rightarrow 4 - 8e = 1 + e$

$\Rightarrow 9e = 3 \Rightarrow e = \frac{1}{3}$ ***[1 mark]***

c) Since $e = \frac{1}{3}$ (from b)), the speed of B as it approaches C is: $\frac{4u}{3}(1+\frac{1}{3}) = \frac{16u}{9}$ ***[1 mark]***. The coefficient of restitution between B and C is $2e = \frac{2}{3}$ ***[1 mark]***. So, using the Law of Restitution: $e = \frac{v_C - v_B}{u_B - u_C} \Rightarrow \frac{2}{3} = \frac{v_C - v_B}{\frac{16u}{9} - 0}$ ***[1 mark]***

$\Rightarrow v_C - v_B = \frac{32u}{27}$ ...*[1]*

Using conservation of momentum:

$m_Bu_B + m_Cu_C = m_Bv_B + m_Cv_C$

$(2m \times \frac{16u}{9}) + 0 = 2mv_B + 4mv_C$ ***[1 mark]***

$\Rightarrow v_B + 2v_C = \frac{16u}{9}$ ...*[2]*

Equation *[1]* + equation *[2]* gives:

$3v_C = \frac{80u}{27} \Rightarrow v_C = \frac{80u}{81}$ ***[1 mark]***.

6 Before: (0.8) → 4, (1.2) → 2. After: (0.8) → 2.5, (1.2) → v

$(0.8 \times 4) + (1.2 \times 2) = (0.8 \times 2.5) + 1.2v$

$3.2 + 2.4 = 2.0 + 1.2v$

$v = 3 \text{ ms}^{-1}$

Before: (1.2) → 3, 4 ← (m). After: (1.2+m) 0

$(1.2 \times 3) + (m \times -4) = (1.2 + m) \times 0$

$3.6 = 4m$

$m = 0.9$ kg

[4 marks available in total]:

- ***1 mark for using conservation of momentum***
- ***1 mark for correct value of v***
- ***1 mark for correct workings***
- ***1 mark for correct value of m***

Diagrams are handy for collision questions too, partly because they make the question clearer for you, but they also make it easier for the examiner to see how you're going about answering the question.

7 First, write down everything you're told in the question:

$m_A = 7$, $\mathbf{v}_A = 6\mathbf{i} - 4\mathbf{j}$,

$m_B = 2$, $\mathbf{u}_B = 0$, $\mathbf{v}_B = ?$

$m_C = (7 + 2) = 9$, $\mathbf{v}_C = 3\mathbf{i} + 4\mathbf{j}$

$F = |\mathbf{F}|$, $t = 5$

Now, use conservation of momentum to find $\mathbf{v}_B$:

$m_A\mathbf{v}_A + m_B\mathbf{v}_B = m_C\mathbf{v}_C$ ***[1 mark]***

$7(6\mathbf{i} - 4\mathbf{j}) + 2\mathbf{v}_B = 9(3\mathbf{i} + 4\mathbf{j})$

$42\mathbf{i} - 28\mathbf{j} + 2\mathbf{v}_B = 27\mathbf{i} + 36\mathbf{j}$

$2\mathbf{v}_B = (27 - 42)\mathbf{i} + (36 + 28)\mathbf{j}$

$\mathbf{v}_B = -7.5\mathbf{i} + 32\mathbf{j}$ ***[1 mark]***

Impulse given to B by force $\mathbf{F}$: $\mathbf{I} = \mathbf{F} \times t$ ***[1 mark]***

But impulse is also equal to change in momentum, so:

$\mathbf{F}t = m_B\mathbf{v}_B - m_B\mathbf{u}_B$ ***[1 mark]***

$\mathbf{u}_B = 0$, so $\mathbf{F}t = m_B\mathbf{v}_B = -15\mathbf{i} + 64\mathbf{j}$

$\mathbf{F} = \frac{-15\mathbf{i} + 64\mathbf{j}}{5}$

$= -3\mathbf{i} + 12.8\mathbf{j}$ ***[1 mark]***

$F = |\mathbf{F}| = \sqrt{(-3)^2 + 12.8^2}$ ***[1 mark]***

$F = 13.1$ N (3 s.f.) ***[1 mark]***

When you're given loads of info in a wordy question like this, it really helps to write it all out to start with. That way you can see straight off what you know and what you don't know.

8 a) Split the speed into components parallel and perpendicular to the plate ***[1 mark]***. The plate is vertical, so the vertical components of the speed before and after the collision are equal,

i.e. $3\sqrt{2}\cos 45 = v\cos 30$ ***[1 mark]*** $\Rightarrow$

$v = \frac{3\sqrt{2}\cos 45}{\cos 30} = \frac{3}{(\sqrt{3}/2)} = 2\left(\frac{3}{\sqrt{3}}\right) = 2\sqrt{3}\text{ ms}^{-1}$ ***[1 mark]***.

b) Perpendicular to the plate, the component of the speed after the collision is:

$e \times$ (component of speed before collision) ***[1 mark]***.

Notice there's no minus sign here — we're talking about speed, so only the magnitude matters, not the direction.

So, resolving horizontally, $e3\sqrt{2}\sin 45 = v\sin 30$ ***[1 mark]***

$\Rightarrow e = \frac{v\sin 30}{3\sqrt{2}\sin 45} = \frac{2\sqrt{3}(\frac{1}{2})}{3\sqrt{2}(\frac{1}{\sqrt{2}})}$

So $e = \frac{\sqrt{3}}{3} = 0.577$ (3 s.f.) ***[1 mark]***.

c) Initial kinetic energy $= \frac{1}{2}mu^2 = \frac{1}{2}(1)(3\sqrt{2})^2 = 9$ J.

Final kinetic energy $= \frac{1}{2}mv^2 = \frac{1}{2}(1)(2\sqrt{3})^2 = 6$ J.

Therefore, the kinetic energy lost is $(9 - 6)$ J $= 3$ J.

[2 marks available in total]:

- ***1 mark for both initial and final kinetic energy correct***
- ***1 mark for correct answer***

And that's about your lot for Section 5, a real rollercoaster ride of mathematical wonderment. Hope you had fun...

M2 Section 6 — Uniform Circular Motion

Warm-up Questions

1) a) $\omega = \frac{\theta}{t} = \frac{2\pi}{1.5} = \frac{4\pi}{3}$ radians s^{-1}

$a = r\omega^2 = 3 \times \left(\frac{4\pi}{3}\right)^2 = \frac{16\pi^2}{3}\text{ ms}^{-2}$

Don't forget the units — it's radians per second for angular speed.

b) $\omega = \frac{\theta}{t} = \frac{15 \times 2\pi}{60} = \frac{\pi}{2}$ radians s^{-1}

$a = r\omega^2 = 3 \times \left(\frac{\pi}{2}\right)^2 = \frac{3\pi^2}{4}\text{ ms}^{-2}$

c) $\omega = \frac{\theta}{t} = \frac{\frac{160}{360} \times 2\pi}{1} = \frac{8}{9}\pi$ radians s^{-1}

$a = r\omega^2 = 3 \times \left(\frac{8}{9}\pi\right)^2 = \frac{64\pi^2}{27}\text{ ms}^{-2}$

Answers

d) $\omega = \frac{v}{r} = \frac{10}{3}$ radians s^{-1}

$a = \frac{v^2}{r} = \frac{10^2}{3} = \frac{100}{3}$ ms^{-2}

2 a) $F = mr\omega^2 = 2 \times 0.4 \times (10\pi)^2 = 80\pi^2$ N

b) $F = \frac{mv^2}{r} = \frac{2 \times 4^2}{0.4} = 80$ N

This formula is just the old F = ma in disguise. Just make sure you know the circular motion acceleration formulas.

3 a) Resolving vertically:

$T\cos 45° = 4g \Rightarrow T = 55.4$ N (3 s.f.)

b) Resolving horizontally:

$T\sin 45° = \frac{mv^2}{r} \Rightarrow 55.44\sin 45° = \frac{4v^2}{r}$

$r = 0.102v^2$ m (3 s.f.)

4 a)

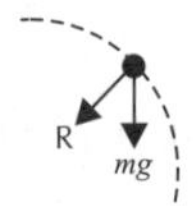

b) Using conservation of mechanical energy (between points A and B):

$\frac{1}{2}m(2.2)^2 + mg(0.2 - 0.2\cos 45°) = \frac{1}{2}mv_B^2$

$4.84 + 2(0.0586g) = v_B^2$

$v_B = \sqrt{4.84 + 0.1172g} = 2.45\,ms^{-1}$ (3 s.f.)

c) Resolving perpendicular to direction of motion:

$R + mg\cos 45° = \frac{mv^2}{r}$

$R + mg\cos 45° = \frac{m(2.447^2)}{0.2}$

$R = \frac{m(2.447^2)}{0.2} - mg\cos 45° = 23.0m$ N (3 s.f.)

5 a) Find the particle's speed, v, at the highest point:

Using conservation of mechanical energy

(between the lowest and highest points on the circle):

$\frac{1}{2}m(9)^2 = \frac{1}{2}mv^2 + 4mg$

$81 = v^2 + 8g$

$v^2 = 2.6 \Rightarrow v = \sqrt{2.6} = 1.61\,ms^{-1}$

The particle's still moving, so the bead on the wire will complete the circle.

Find the tension in the string at the highest point:

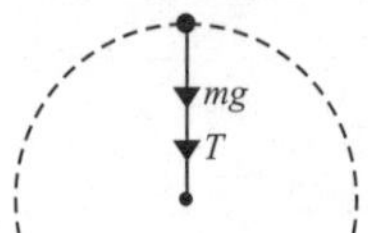

Resolving perpendicular to motion:

$T + mg = \frac{mv^2}{r} \Rightarrow T = \frac{2.6m}{2} - 9.8m$

$T = 1.3m - 9.8m = -8.5m$

The tension is negative, so the particle on the string won't complete the circle.

That's the difference between things that can leave the circle and things that can't. If something's stuck on the circle (like a bead on a wire) it only needs enough energy to get to the top. If something can leave the circle (like a particle on a string), it also needs a positive tension or reaction at the top.

b)

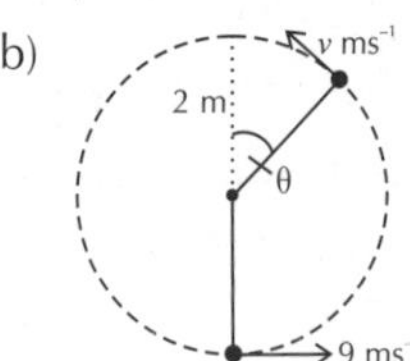

Find the speed, v, when the particle leaves the circle:

Using conservation of mechanical energy

(between the lowest point and the point where particle leaves the circle):

$\frac{1}{2}m(9)^2 = \frac{1}{2}mv^2 + mg(2 + 2\cos\theta)$

$81 = v^2 + 4g(1 + \cos\theta)$

$v^2 = 81 - 4g(1 + \cos\theta)$

When the bead leaves the circular path, $T = 0$, so resolving perpendicular to motion:

$mg\cos\theta = \frac{mv^2}{r} \Rightarrow 2g\cos\theta = 81 - 4g(1 + \cos\theta)$

$2g\cos\theta = 81 - 4g - 4g\cos\theta$

$\cos\theta = \frac{81 - 4g}{6g} = 0.711 \Rightarrow \theta = 44.7°$ (3 s.f.)

Exam Questions

1 a) Using conservation of mechanical energy (between points A and B):

$\frac{1}{2}m(0)^2 + mg(0.5\sin 30°) = \frac{1}{2}mv^2$ **[1 mark]**

$\frac{1}{2}(0)^2 + g(0.5\sin 30°) = \frac{1}{2}v^2 \Rightarrow g(0.5\sin 30°) = \frac{1}{2}v^2$

$2g(0.5\sin 30°) = v^2 \Rightarrow g\sin 30° = v^2$ **[1 mark]**

$v^2 = \frac{g}{2} \Rightarrow v = \sqrt{\frac{g}{2}}$ **[1 mark]**

b) Resolving perpendicular to motion:

$T - mg\sin 30 = \frac{mv^2}{r}$ **[1 mark]**

$\Rightarrow T - g = 2g$ **[1 mark]**

$\Rightarrow T = 3g$ **[1 mark]**

c) Using v from part a):

$v = r\omega \Rightarrow \omega = \frac{v}{r} = \frac{\sqrt{\frac{g}{2}}}{0.5}$ **[1 mark]**

$\omega = \frac{2\sqrt{g}}{\sqrt{2}} = \sqrt{2g}$ radians s^{-1} **[1 mark]**

Don't be put off if you're not shown the whole of the circle. The particle's started moving in a circular path, so you can use the circular motion rules.

Answers

2

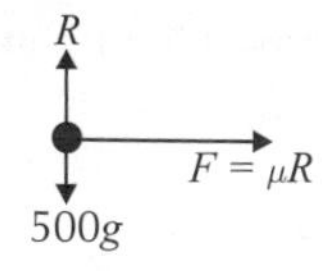

Resolving vertically: $R = 500g$ ***[1 mark]***

Resolving horizontally: $\mu R = \frac{mv^2}{r}$ ***[1 mark]***

$0.5 \times 500g = \frac{500v^2}{30}$ ***[1 mark]***

$0.5g = \frac{v^2}{30}$

$v^2 = 15g$ ***[1 mark]*** $\Rightarrow v = 12.1\text{ms}^{-1}$ (3 s.f.) ***[1 mark]***

3 a)

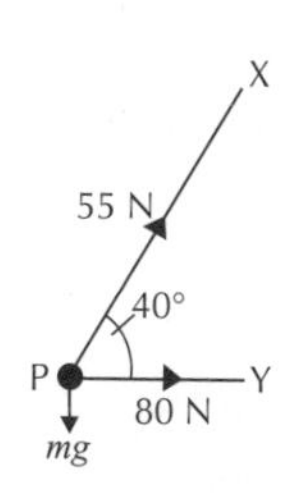

Resolving vertically: $mg = 55\sin 40°$ ***[1 mark]***

$\Rightarrow m = 3.61$ kg (3 s.f.) ***[1 mark]***

b) Resolving horizontally:

$F = \frac{mv^2}{r}$ ***[1 mark]***

$55\cos40° + 80 = \frac{3.607 \times 3^2}{r}$ ***[1 mark]***

$\Rightarrow r = \frac{3.607 \times 3^2}{55\cos40° + 80} = 0.266\text{ m}$ (3 s.f.) ***[1 mark]***

The radius is the same as the length of the horizontal string.

$\omega = \frac{v}{r} = \frac{3}{0.2658} = 11.29\text{ radians s}^{-1}$ ***[1 mark]***

c) 2π radians = 1 revolution

$\frac{11.29}{2\pi} = 1.796$ revolutions per second ***[1 mark]***

$1.796 \times 60 = 108$ revolutions per minute (3 s.f.) ***[1 mark]***

4 a) Using conservation of mechanical energy (between points A and B):

$\frac{1}{2}m(\sqrt{20})^2 = \frac{1}{2}mv^2 + mg$ ***[1 mark]***

$20 = v^2 + 2g$ ***[1 mark]***

$v = \sqrt{0.4}\text{ ms}^{-1}$ ***[1 mark]***

b) Resolving perpendicular to motion at point B:

$mg - R = \frac{mv^2}{r}$ ***[1 mark]***

$mg - R = 0.4m$ ***[1 mark]***

$R = 9.4m$ ***[1 mark]***

c) Using conservation of mechanical energy (between points A and B):

$\frac{1}{2}mv_A^2 = \frac{1}{2}mv_B^2 + mg$ ***[1 mark]***

v_B must be greater than or equal to zero.

If $v_B = 0$: $\frac{1}{2}mv_A^2 = mg$ ***[1 mark]***

$\Rightarrow v_A = \sqrt{2g} = 4.43\text{ ms}^{-1}$ ***[1 mark]***

So the minimum speed at A for the circle to be completed is 4.43 ms^{-1} (3 s.f.).

5 a) Using conservation of mechanical energy (between points J and K):

$\frac{1}{2}m(15)^2 = \frac{1}{2}mv^2 + mg(5 + 5\cos\theta)$

$225 = v^2 + 10g(1 + \cos\theta)$

$v^2 = 225 - 10g(1 + \cos\theta)$

[4 marks available in total]:

- ***1 mark for total mechanical energy at J***
- ***1 mark for total mechanical energy at K***
- ***1 mark for using conservation of mechanical energy***
- ***1 mark for correct expression for v^2***

'Show that' questions are quite nice and friendly. You know when you've got the right answer. Don't make any crazy leaps in logic though — the examiners aren't that daft.

b) Use the formula from part a) with $\theta = 0$:

$v^2 = 225 - 10g(2) = 29$

$\Rightarrow v = \sqrt{29}\text{ ms}^{-1}$ ***[1 mark]***

So, it has enough energy to get to the top, so now find the tension, T, in the string at the top:

$T + mg = \frac{mv^2}{r}$ ***[1 mark]***

$\Rightarrow T + mg = \frac{29m}{5}$ ***[1 mark]***

$\Rightarrow T = \frac{29m}{5} - 9.8m = -4m$ ***[1 mark]***

The tension in the string is negative, so the string is no longer taut and the circle isn't completed ***[1 mark]***.

You have to find the speed first here — you need it to work out the tension. Don't forget, if it was something that couldn't leave the circle — like a bead on a ring — you would only need to show that it had enough energy to reach the top.

c) When the string first becomes slack, the tension in the string is zero.
So resolving perpendicular to the direction of motion:

$mg\cos\theta = \frac{mv^2}{r}$ ***[1 mark]***

$mg\cos\theta = \frac{m(225 - 10g(1 + \cos\theta))}{5}$ ***[1 mark]***

$5g\cos\theta = 225 - 10g(1 + \cos\theta)$

$15g\cos\theta = 225 - 10g$ ***[1 mark]***

$\cos\theta = \frac{127}{147}$ ***[1 mark]*** $\Rightarrow \theta = 30.2°$ (3 s.f.) ***[1 mark]***

Index

Index

Index